Well-Architectured Fluoropolymers: Synthesis, Properties and Applications

Well-Architectured Fluoropolymers: Synthesis, Properties and Applications

Bruno Ameduri and Bernard Boutevin

Laboratory of Macromolecular Chemistry,
Ecole Nationale Superieure de Chimie de Montpellier
8 Rue de l'Ecole Normale,
34296 Montpellier Cedex 5, France

2004

ELSEVIER

Amsterdam – Boston – Heidelberg – London – New York – Oxford – Paris
San Diego – San Francisco – Singapore – Sydney – Tokyo

ELSEVIER B.V.	ELSEVIER Inc.	**ELSEVIER Ltd**	ELSEVIER Ltd
Sara Burgerhartstraat 25	525 B Street, Suite 1900	**The Boulevard, Langford Lane**	84 Theobalds Road
P.O. Box 211, 1000 AE Amsterdam	San Diego, CA 92101-4495	**Kidlington, Oxford OX5 1GB**	London WC1X 8RR
The Netherlands	USA	**UK**	UK

1st edition 2004

Library of Congress Cataloging in Publication Data
A catalog record is available from the Library of Congress.

British Library Cataloguing in Publication Data
A catalogue record is available from the British Library.

ISBN: 0 08 044388 5

♾ The paper used in this publication meets the requirements of ANSI/NISO Z39.48-1992 (Permanence of Paper). Printed in The Netherlands.

Preface

In contrast to organofluorine chemistry, macromolecular fluorine chemistry is rather young. This field was born with the discovery of the first fluoropolymer, polychlorotrifluoroethylene (PCTFE) followed by that of polytetrafluoroethylene (PTFE) in 1934 and 1938, respectively.

Although fluoropolymers are niche market materials, they exhibit remarkable properties and have already found many applications in high-tech fields.

Significant advantages have been made in the synthesis and the applications of these new high-performance materials. The area of these specialty polymers has blossomed and started to flourish just a few decades ago.

Fluorinated polymers are considered as high value-added materials, thanks to their outstanding properties which open up various applications. Such polymers exhibit high thermostability and chemical inertness, low refractive index and coefficient of friction, good water and oil repellency, low surface energy, low relative permittivity, valuable electrical properties, and low friction coefficient. In addition, they are non-sticky and resistant to UV, aging, and to concentrated mineral acids and, for some of them, resistant to alkalies.

Hence, their fields of applications are numerous: paints and coatings (for metals, wood, leather, stone, optical fibers), textile finishings, novel elastomers, high performance resins, membranes-with increasing nowadays uses for fuel cells, functional materials (for photoresists and optical fibers), biomaterials, lubricants, surfactants, wiring insulation, and thermostable polymers for aerospace or microelectronics.

But the processing of these products is still difficult because several of them are not usually soluble, and others are not meltable or exhibit very high melting points. In addition, their prices are still high and even if their properties at high temperatures are still preserved, this is not so at low temperatures because their glass transition temperature is too high or their high crystallinity compared to those of silicones for example.

Hence, these recent years have seen a tremendous interest in fluorine-containing polymers and many teams of research (academic and industrial) which did not have any activities in fluorine chemistry have started to be progressively immersed in macromolecular fluorine chemistry.

Because of the increasing need of better and better performing materials endowed with specific properties, macromolecular engineering has become a useful tool to obtain well-architectured polymers: telomers, telechelics, dendrimers, or alternating, block, star, and graft copolymers. These polymers are nowadays seeing a growing interest.

In the field of fluoropolymers, although excellent books published by Wall (1972), Scheirs (1997), and Hougham et al. (1999) interestingly reported the synthesis and applications of a wide range of fluoropolymers (including thermoplastics, elastomers, and polycondensates), to our knowledge, the coverage of controlled radical reactions in fluoropolymer synthesis has been limited. Exceptions include two chapters regarding ETFE and alternating copolymers of chlorotrifluoroethylene (CTFE) in Scheirs' book, and one chapter in Hougham's book dealing with F-block copolymers.

As a matter of fact, the impressive contribution of the controlled radical polymerisation of hydrogenated monomers, especially enhanced from the mid-90s, has enabled one to achieve not only the synthesis of well-defined polymers with tuned molar masses and narrow polydispersities, but also to obtain well-defined block copolymers. Now, more and more interests are focused on well-controlled fluoropolymers. This is why, we have thought that it was necessary to review most works (from both academics and industries) in this book which completes the three above books.

It is obvious that the literature on fluoropolymers is vast and there is a need for a compact and up-to-date contribution that covers the well-architectured fluoropolymers. This present book represents this effort. It is composed of five chapters (all describing the synthesis, properties and applications of the corresponding macromolecules) and is developped in the following sequence: fluorinated *telomers*, *telechelics*, and *alternating*, *block*, and *graft* copolymers.

Unfortunately, too few articles and patents supplying the synthesis of fluorinated dendrimers, hyperbranched polymers, or star copolymers have been published, hence a chapter encompassing these kinds of fluoropolymers would be too short with respect to the other chapters.

Chapter 1 supplies a quasi-exhaustive review on the synthesis, properties and applications of fluorinated telomers. Indeed, a previous review was published in 1997 and this exciting development has made it necessary to update and to expand the text to reflect the progress made. Although telomers exhibit low molecular weights, they have significant end-groups, are excellent models of polymers, and can be relevant intermediates for further reactions. Interesting fluorotelomers, produced in an industrial scale have already found applications as paints and surfactants.

The second chapter deals with fluorotelechelics (or α,ω-difunctional derivatives) that are useful starting materials involved in polycondensation or polyaddition. A wide range of well-defined fluoropolymers have been hence produced from these precursors. Actually, two distinct kinds of fluorotelechelics have been found. Those with the (per)fluorinated chain located in the polymeric backbone have been developed primarily for their inertness and their thermal stability, while those that exhibit a (per)fluorinated side group find use when enhanced surface properties are desired.

After giving some basic and theoretical concepts of the alternating copolymerisation, Chapter 3 provides a compilation of references on fluorinated alternating copolymers summarised in four main parts: those arising from TFE, from CTFE, from vinylidene fluoride (VDF), and from other fluorinated monomers. Some of them are produced on industrial scales and can be used as original paints or melt-processing fluoropolymers.

The counter part of the fluorinated block copolymers (Chapter 4) consists of three main sections: diblock, triblock, and multiblock copolymers. In both former sections, various pathways of syntheses have been supplied, either from traditional or controlled radical polymerisations, from ionic methods, or from polycondensation. Other routes combining different techniques have also been proposed. The third section reports syntheses and characteristics of original multiblock-containing silicone copolymers.

Finally, the fifth chapter deals with the synthesis, properties, and applications of graft copolymers, and is composed of three main sections : i) that considering the use of fluorinated macromonomers (based mainly on TFE, VDF, and perfluoropolyether units), ii) that arising from the "grafting onto" method, i.e., the addition or the condensation of a fluorinated oligomer, telomer, or polymer onto a polymeric backbone bearing achoring groups; iii) that dealing with the preparation of fluorinated graft copolymers obtained by activation (e.g., radiation, ozone, plasma) of fluoropolymer followed by the grafting of oligomers or polymers on it. Potential applications of these fluorograft copolymers are membranes for fuel cells, compatibilisers, and potential biomaterials.

A concerted effort has been made to include all of the relevant information referenced in the multitude of published sources. However, in this emerging area of well-defined polymers, new processes and products are being invented and discovered almost every week, making it impossible to include every piece of information. Suggestions and inputs from readers are welcome and will be acknowledged gratefully.

We would like to thank Elsevier Ltd for inviting us to write this book, Springer-Verlag for the permission to partly reproduce the chapter on Fluorotelomers published in *Topics Current Chemistry* in 1997, Professors G. Resnati, H. Sawada, B. Pucci, and Dr M. C. Porte-Durrieu for rendering their up-to-date works in this book, and Drs D. Pospiech, M. Apostolo, T. E. Kang, and T. Shimizu for supplying schemes and pictures. We are grateful to Drs D. Jones, Prof. A. Ritcey, and C. Ligon for helpful readings and for Misses C. Euzet and F. Cozon for typing carefully with much patience. We also thank our friend G. Kostov for his valuable comments and constructive criticisims and Dr S. Boileau and Dr F. Guida who agreed to correct several parts of the manuscript.

We have also been fortunate in our PhD students and post-doc fellows, whose enthusiastic participation made significant contributions in the fascinating world of Fluorine Chemistry. Their names appear in the references.

We would like to thank our wives Cathy and Christine for moral support throughout the project. Finally, I (B.A.) would like to acknowledge my sons Damien and Julien for going to sleep on time so I could get to work at night!

BRUNO AMEDURI
BERNARD BOUTEVIN
Montpellier, December 2003

Contents

Nomenclature and Abbreviations

AA	Acrylamide
AAc	Acrylic acid
AEM	Anionic exchange membrane
AFM	Atomic force microscopy
AIBN	Azobisisobutyronitrile
alt	Alternating
ATRP	Atom transfer radical polymerisation
b	Block
BDE	Bond dissociation energy
bipy	Bipyridine
BrTFE	Bromotrifluoroethylene
Bu	Butyl
CL	Caprolactone
CMC	Critical micellar concentration
CRP	Controlled radical polymerisation
CSM	Cure site monomer
C_T	Transfer rate to the telogen (or to the transfer agent)
CTC	Charge transfer complex
CTFE	Chlorotrifluoroethylene
DBP	Dibenzoyl peroxide
DMAEMA	2-(Dimethylamino)ethyl methacrylate
DMFC	Direct methanol fuel cell
d.o.g.	Degree of grafting
DSC	Differential scanning calorimetry
$\overline{DP}_n$	Average degree of polymerisation in number
DTBP	Ditertio-butyl peroxide
E	Ethylene
E-TFE	Poly(ethylene-alt-tetrafluoroethylene) copolymer
EVE	Ethyl vinyl ether
FBSMA	2-(N-methyl perfluorobutyl sulfonamido) ethyl methacrylate
FDA	1,1,2,2-Tetrahydroperfluorodecyl acrylate
FEP	Poly(tetrafluoroethylene-co-hexafluoropropylene)
FOSMA	2-(N-ethyl perfluorooctyl sulfonamido) ethyl methacrylate
g	Graft
HEUR	Hydrophobically modified ethoxylated urethane
HFP	Hexafluoropropylene
HFIB	Hexafluoroisobutylene
HFPO	Hexafluoropropylene oxide
HQ	Hydroquinone
IB	Isobutylene
IEC	Ionic exchange capacity

IPA	Isophthalic acid
iPr	Isopropyl
IR_FI	α,ω-diiodoperfluoroalkane
ITP	Iodine transfer polymerisation
LRP	Living radical polymerisation
LS	Laser scattering
M	Monomer
[M]	Concentration of monomer
MADIX	Macromolecular design through interchange of xanthates
$\overline{M}_n$	Average molar mass in number
$\overline{M}_w$	Average molar mass in weight
MMA	Methyl methacrylate
NMR	Nuclear magnetic resonance
NMP	Nitroxyde mediated polymerisation
NVP	N-vinyl pyrrolidone
P	Propene
PAVE	Perfluoroalkyl vinyl ether
PAAVE	Perfluoroalkoxy alkyl vinyl ether
PCTFE	Polychlorotrifluoroethylene
PDI	Polydispersity index
PDMS	Polydimethylsiloxane
PEO	Poly(ethylene oxide)
PFA	Poly(tetrafluoroethylene-co-perfluoropropylvinyl ether)
PFC	Perfluorocarbon
PFPE	Perfluoropolyether
PMDETA	Pentamethylenetriamine
PMHS	Poly(methylhydrogeno)siloxane
PMVE	Perfluoromethyl vinyl ether
POE	Polyoxyethylene
ppm	Part per million
PPVE	Perfluoropropyl vinyl ether
PS	Polystyrene
PSSA	Polystyrene sulfonic acid
PSU	Polysulfone
PTFE	Polytetrafluoroethylene
PVDF	Polyvinylidene fluoride
PVF	Polyvinyl fluoride
R	Alkyl group
R˙	Alkyl radical
RAFT	Reversible addition fragmentation chain transfer
R_F	Perfluorinated group
$R_{F,Cl}$	Chlorofluorinated group
R_FBr	Perfluoroalkyl bromide
R_FI	Perfluoroalkyl iodide
RT	Room temperature
SAXS	Small angle X-ray scattering
sc	Supercritical

S	Styrene
TAC	Triallyl cyanurate
TAIC	Triallyl isocyanurate
TEM	Transmission electronic microscopy
TEMPO	2,2,6,6-Tetramethyl piperidinyl-1-oxy
TFE	Tetrafluoroethylene
TFS	α,β,β-Trifluorostyrene
T_{dec}	Decomposition temperature
T_m	Melting temperature
T_g	Glass transition temperature
tBu	Tertiobutyl
THF	Tetrahydrofuran
TGA	Thermogravimetric analysis
TPE	Thermoplastic elastomer
TrFE	Trifluoroethylene
UV	Ultraviolet
VBC	Vinylbenzyl chloride
VDF	Vinylidene fluoride (or 1,1-difluoroethylene)
VF	Vinyl fluoride
XPS	X-ray photoelectron spectroscopy
Δ	Heating

Chapter 1

TELOMERISATION REACTIONS OF FLUORINATED ALKENES

1. Introduction

Fluorinated polymers are regarded as high value-added materials due to their outstanding properties which open up various applications [1–5]. Such polymers show low intramolecular and intermolecular interactions, which lead to low cohesive energy and, therefore, to low surface energy. They also exhibit high thermostability and chemical inertness, low refractive index and friction coefficient, good hydrophobicity and lipophobicity, valuable electrical properties and low relative permittivity. In addition, they are non-sticky and resistant to UV, ageing and to concentrated mineral acids and alkalis.

The great value of the unique characteristics of fluorinated polymers in the development of modern industries has ensured an increasing technological interest since the discovery, in 1934, of the first fluoropolymer, poly(chlorotrifluoroethylene).

Hence, their fields of applications are numerous: paints and coatings [6] (for metals [7], wood [8], leather [9], stone [10], optical fibres [11], antifouling [12]), textile finishings [13], novel elastomers [5], high performance resins, membranes [14], surfactants and fire fighting agents [15], functional materials (for photoresists or microlithography [16], conductive polymers [17]), biomaterials [18], thermostable polymers for aerospace, wire and cable industry, chemical process industry, semiconductor industry and for release and anti-stick applications [2].

However, the processing of these products is still difficult because several of them are not usually soluble, and others are not meltable or exhibit very high melting points. In addition, their prices are still high and, even if their properties at high temperatures are still preserved, this is not so at low temperatures because their glass transition temperatures or their crystallinity rates are too high compared to those of silicones for example.

It was thus worth finding a model of the synthesis of fluoropolymers in order to predict their degree of polymerisation, their structure and the mechanism of the reaction.

One of the most interesting strategies is that of telomerisation. Such a reaction, introduced for the first time by Hanford in 1942 [19], in contrast to polymerisation, usually leads to low molecular weight polymers, called telomers, or even to monoadducts with well-defined end-groups. Such products **1** are obtained from the reaction between a telogen or a transfer agent (X–Y), and one or more (n) molecules of a polymerisable compound M (called a taxogen or monomer) having ethylenic unsaturation under radical polymerisation conditions as follows:

$$X-Y + n\,M \xrightarrow{\text{free radicals}} X-(M)_n-Y$$
$$\mathbf{1}$$

Telogen X–Y can be easily cleavable by free radicals (formed according to the conditions of initiation) leading to an X· radical which will be able to react further with monomer. After the propagation of monomer, the final step consists of the transfer of the telogen to the growing telomeric chain. Telomers **1** are intermediate products between organic compounds (e.g., $n = 1$) and macromolecular species ($n = 100$). Hence, in certain cases, end-groups exhibit a chemical importance which can provide useful opportunities for further functionalisations.

The scope of telomerisation was first outlined by Friedlina [20] in 1966, improved upon by Starks in 1974 [21], and further developed by Gordon and

Loftus [22]. We have reviewed such a reaction [23, 24] in which mechanisms and kinetics of radical and redox telomerisations have been described.

All monomers involved in radical polymerisation can be utilised in telomerisation. In this chapter, only monomers containing a fluorine atom or a short perfluorinated group linked to an unsaturated carbon atom will be taken into account.

Below, the initiation step and mechanisms involved in telomerisation will be described, followed by the various processes used, the telogens, the monomers and their respective reactivities. The following part will deal with the pseudo-living radical telomerisation of fluorinated monomers and finally different applications of fluorinated telomers will be given.

2. Initiation and Mechanisms

A telomerisation reaction is the result of four steps: initiation, propagation, termination and transfer. These four steps have been described in previous reviews [21–24] summarised below.

2.1. VARIOUS WAYS OF INITIATION

Telomerisation can be initiated from various processes: photochemical (in the presence of UV), in the presence of radical initiator or redox catalysts, thermally, or initiated by X- or γ-rays. For each case, different mechanisms have been proposed. The initiation processes are explained below and some examples are listed in Table I [25–66].

2.1.1. Photochemical Initiation

An interesting example was proposed by Haszeldine *et al.* [28] in the photochemical-induced addition of alcohol RH to hexafluoropropene (HFP) knowing that no propagation of HFP occurs:

Initiation
$$C_3F_6 \xrightarrow{\text{UV}} (C_3F_6)^*$$

$$(C_3F_6)^* + RH \longrightarrow R^{\cdot} + C_3F_6H^{\cdot}$$

$$R^{\cdot} + C_3F_6 \longrightarrow RC_3F_6^{\cdot}$$

Transfer
$$C_3F_6H^{\cdot} + RH \longrightarrow CF_3CFHCHF_2 + R^{\cdot}$$

$$RC_3F_6^{\cdot} + RH \longrightarrow R(C_3F_6)H + R^{\cdot}$$

TABLE I
DIFFERENT METHODS OF CLEAVAGE IN THE TELOGENS

Bond cleaved	Method	References
C–H	$Cl_3CH + xF_2C=CH_2 \xrightarrow{rad} Cl_3C-(C_2F_2H_2)_xH$	[25]
	$THF + CF_2=CFCF_3 \xrightarrow[peroxide]{UV\ or}$ 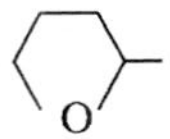 (tetrahydropyranyl ring)—CF_2CFHCF_3	[26, 27]
	$CH_3-OH + nF_2C=CFX \xrightarrow[peroxide]{UV\ or} H-(CFX-CF_2)_n-CH_2OH$ (X = Cl, F, CF_3)	[28–30]
C–F	$Cl_3CF + F_2C=CFCl \xrightarrow{AlCl_3} Cl_3CCF_2CF_2Cl$	[31]
C–Cl	$HCCl_3 + nF_2C=CFCl \xrightarrow{FeCl_3} HCCl_2-(CF_2CFCl)_n-Cl$	[32]
	$CCl_4 + F_2C=CFH \xrightarrow{CuCl} Cl_3CCF_2CFHCl + Cl_3CCFHCF_2Cl$	[33]
	$RCCl_3 + nF_2C=CH_2 \xrightarrow[or\ rad]{redox} RCCl_2(CH_2CF_2)_nCl$ (R: Cl, CH_2OH, CO_2CH_3)	[34]
C–Br	$CFBr_3 + nF_2C=CFH \xrightarrow{UV} Br_2CFCFHCF_2Br$ (major) + Br_2CFCF_2CFHBr (minor)	[35]
	$BrCF_2CFClBr + (m+n-1)F_2C=CFCl \xrightarrow{UV} Br(CF_2CFCl)_n(CFClCF_2)_m-Br$	[36]
	$BrCF_2Br + xF_2C=CFCF_3 \xrightarrow{thermal} BrCF_2(CF_2\overset{\textstyle CF_3}{\underset{\textstyle \vert}{CF}})_x-Br$	[37]
C–I	$iC_3F_7I + nF_2C=CH_2 \xrightarrow{thermal} iC_3F_7(C_2F_2H_2)_nI$ (n = 1–4)	[38–40]
	$C_4F_9I + xHFC=CF_2 \xrightarrow{thermal} C_4F_9(C_2F_3H)_xI$ (x = 1–3)	[41]
	$C_6F_{13}I + xF_2C=CFCF_3 \xrightarrow{thermal} C_6F_{13}(C_3F_6)_xI$ (x = 1–3)	[42]
S–H	$CF_3SH + F_2C=CFH \xrightarrow{UV} CF_3SCFHCF_2H$ (98%) + $CF_3SCF_2CFH_2$ (2%)	[43]
	$HOC_2H_4SH + xF_2C=CH_2 \xrightarrow{rad} HOC_2H_4S(C_2F_2H_2)_xH$ (x = 1–10)	[44, 45]

(continued on next page)

TABLE I (continued)

Bond cleaved	Method	References
S–S	$RS\text{–}SR + nM \xrightarrow{UV} RS(M)_n SR$ (R: CH_3 or CF_3; M: C_2F_4, C_3F_6, C_2F_3Cl, $C_2F_2H_2$)	[46–48]
S–Cl	$FSO_2Cl + xF_2C{=}CH_2 \xrightarrow{peroxide} FSO_2(C_2F_2H_2)_xCl$ (x = 1–3)	[49]
P–H	$PH_3 + xF_2C{=}CH_2 \xrightarrow{UV} H_{3-x}P(C_2F_2H_2)_xH$ (x = 1, 2)	[50]
	$H\overset{O}{\overset{\|}{P}}(OEt)_2 + xCF_2{=}CF_2 \xrightarrow{\gamma\text{-rays}} (EtO)_2\overset{O}{\overset{\|}{P}}{-}(CF_2{-}CF_2)_x{-}H$	[51]
P–Cl	$PCl_5 + nF_2C{=}CFCl \xrightarrow{peroxide} Cl_4P{-}(C_2F_3Cl)_n{-}Cl$	[52]
	$POCl_3 + nF_2C{=}CFCl \xrightarrow{rad} Cl_2P(O)(C_2F_3Cl)_n{-}Cl$	[53]
Si–H	$\underset{Cl}{\overset{CH_3}{Cl{-}Si}}{-}H + nH_2C{=}CHC_6F_5 \xrightarrow{H_2PtCl_6} \underset{Cl}{\overset{CH_3}{Cl{-}Si}}{-}CH_2CH_2C_6F_5$	[54]
	$Cl_3SiH + xCF_3CH{=}CH_2 \xrightarrow{UV} Cl_3Si[CH_2CH(CF_3)]_x{-}H$ (x = 1, 2)	[55]
Br–H	$H{-}Br + F_2C{=}CFCl \xrightarrow{UV} BrCF_2ClCF_2H$	[56]
I–F	$IF_5 \xrightarrow{I_2} [I{-}F] \xrightarrow{xF_2C{=}CFCl} I(C_2F_3Cl)_xF$	[57, 58]
I–Cl	$\begin{cases} I{-}Cl + F_2C{=}CFCl \xrightarrow{thermal} CF_3CFClCF_2Cl(92\%) + CF_3CFClCF_2I(8\%) \\ I{-}Cl + F_2C{=}CFCl \xrightarrow[\text{or redox or rad}]{\Delta \text{ or UV}} ICFClCF_2Cl(95\text{–}100\%) + ICF_2CFCl_2(0\text{–}5\%) \end{cases}$	[59] [60]
I–Br	$I{-}Br + nF_2C{=}CFCl \xrightarrow[\text{or redox}]{\Delta \text{ or UV}} X(C_2F_3Cl)_nY$ X, Y = I, Br	[61, 62]
I–I	$I_2 + nF_2C{=}CF_2 \xrightarrow{\Delta} I(C_2F_4)_nI$	[63]
Cl–Cl	$SO_2Cl_2 + nF_2C{=}CFCl \xrightarrow{rad.} Cl(C_2F_3Cl)_nCl$	[64]
Br–Br	$Br_2 + nF_2C{=}CFCl \xrightarrow{redox} Br(C_2F_3Cl)_n{-}Br$	[65, 66]

$R(C_3F_6)H$ is composed of two isomers: RCF_2CFHCF_3 (major) and $F_3CCFRCHF_2$ (minor). Tedder and Walton [67] suggested a similar mechanism.

2.1.2. Radical Telomerisation

The general process of a telomerisation involving radical initiator A_2 has been suggested as follows (A_2 can be an azo, peroxide or perester initiator which undergoes a homolytic cleavage under heating and generates two radicals [68])

$$\text{Initiation} \qquad A_2 \xrightarrow[\Delta]{k_i} 2A^\cdot$$

$$A^\cdot + XY \longrightarrow X^\cdot + AY$$

$$X^\cdot + F_2C{=}CRR' \longrightarrow X(C_2F_2RR')^\cdot$$

$$\text{Propagation} \quad XC_2F_2RR'' + F_2C{=}CRR' \xrightarrow{k_{p1}} X(C_2F_2RR')_2^\cdot$$

$$X(C_2F_2RR')_n^\cdot + F_2C{=}CRR' \xrightarrow{k_{pn}} X(C_2F_2RR')_{n+1}^\cdot$$

$$\text{Termination} \quad X(C_2F_2RR')_n^\cdot + X(C_2F_2RR')_p^\cdot \xrightarrow{k_{Te}} X(C_2F_2RR')_{n+p}X$$

$$\text{Transfer} \qquad X(C_2F_2RR')_n^\cdot + XY \xrightarrow{k_{tr}} X(C_2F_2RR')_nY + X^\cdot$$

where XY, $F_2C{=}CRR'$, k_i, k_{p1}, k_{Te} and k_{tr} represent the telogen, the monomer, the rate constants of initiation, propagation, termination and transfer, respectively.

The kinetic law, regarding the reverse of average degree of telomerisation, $\overline{DP}_n$, depends upon the transfer constant of the telogen (C_T), the telogen [XY] and monomer [M] molar concentrations, as shown below taking into account both high and low average degrees of telomerisation [21–24]:

(i) For high $\overline{DP}_n$

$$\frac{1}{\overline{DP}_n} = C_T \frac{[XY]}{[M]} \quad \text{with} \quad C_T = \frac{k_{tr}}{k_p}$$

According to this equation, transfer constants of various telogens were determined [69].

Concerning the nature of the initiator, no prediction can be made on the efficiency of such a reactant in the telomerisation of fluoroalkenes. Non-fluorinated initiators led to unsuccessful results whereas investigations on the synthesis of fluorinated peresters, peroxides and azo were performed by

Rice and Sandberg [70], Sawada [71] and Guan *et al.* [72] and in our Laboratory [73, 74], respectively.

(ii) For low $\overline{DP}_n$

The activity of transfer of a telogen in the presence of a monomer is characterised by the transfer constants (or coefficients) C_T^n that may be defined for each growing telomeric radical as the ratio of the transfer rate constant of the telogen k_{tr}^n to the rate constant of propagation k_p^n:

$$C_T^n = \frac{k_{tr}^n}{k_p^n}$$

These rate constants characterise reaction (1) of formation of telomers of n order from n order radicals, and reaction (2) of formation of telomeric radicals of $n + 1$ order from these same radicals, respectively. The telomers produced are unable to participate to a new radical reaction, especially as transfer agents.

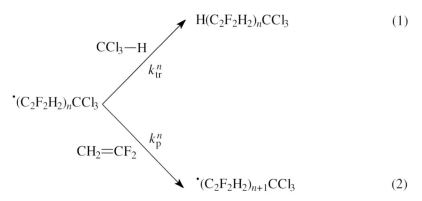

$$H(C_2F_2H_2)_nCCl_3 \qquad\qquad (1)$$

$$\cdot(C_2F_2H_2)_{n+1}CCl_3 \qquad\qquad (2)$$

The telomerisation of VDF with chloroform in solution, from equimolar amounts of these reactants, yielded a mixture composed of the first adducts of the telomerisation, $H(VDF)_nCCl_3$, for which the average cumulated degree of telomerisation in number $\overline{DP}_{n\,cum}$ did not exceed 8 [25, 45]. However, it is necessary to differentiate the rate constants characterising each step of the telomerisation when it primarily produces low $\overline{DP}_{n\,cum}$ (i.e., lower than 10). Thus,

$$k_{tr}^1 \neq k_{tr}^2 \neq k_{tr}^3 \neq \cdots \neq k_{tr}^n, \quad n < 10$$

$$k_p^1 \neq k_p^2 \neq k_p^3 \neq \cdots \neq k_p^n, \quad n < 10$$

Obviously, it is deduced that $C_T^1 \neq C_T^2 \neq C_T^3 \neq \cdots \neq C_T^n$, for $n < 10$. Further, it is considered that the C_T^n values increase with n to reach a limit value C_T^∞ as a reference to characterise the activity of transfer of a telogen in the presence of a given monomer.

David and Gosselain theory [75] enabled us to direct the molar fractions F_n of each telomer to the C_T^n values and to the ratio R of concentrations of telogen to that of the monomer (where R and n represent the [telogen]/[monomer] molar ratio and the degree of telomerisation, respectively):

$$F_n = \frac{C_T^n R}{\prod_{j=1}^{j=n}(1 + C_T^j R)} \tag{1}$$

These authors also deduced the following relationship linking the transfer constants of each order:

$$C_T^n = \frac{F_n}{R \sum_{j=n+1}^{\infty} F_j} \tag{2}$$

Hence, it is possible to assess the values of different constants C_T^n by plotting the ratios

$$\frac{F_n}{\sum_{j=n+1}^{\infty} F_j}$$

vs. the different values of R (Figure 1). The slopes of these straight lines of the experimental data characterising each order n hence give the values C_T^n. For $n = 1, 2, 3, 4, 5$ and 6, the corresponding transfer constants of these orders were 0.056, 0.072, 0.059, 0.071, 0.062 and 0.058, respectively. Consequently, an infinite transfer constant was found of 0.06 at 140 °C [25, 45].

2.1.3. Redox Catalysis

In this case, the reaction requires the use of a catalyst MeL_x (salt of transition metal or metallic complex) which participates in the initiation with an increase in the degree of oxidation of the metal as follows (Y being usually a halogen):

Initiation
$$Me^{n+}L_x + XY \xrightarrow{k_i} Me^{(n+1)}L_xY + X^{\cdot}$$
$$X^{\cdot} + F_2C{=}CRR' \longrightarrow X(C_2F_2RR')^{\cdot}$$

Propagation
$$X(C_2F_2RR')^{\cdot} + F_2C{=}CRR' \xrightarrow{k_{p1}} X(C_2F_2RR')_2^{\cdot}$$
$$X(C_2F_2RR')_n^{\cdot} + F_2C{=}CRR' \xrightarrow{k_{pn}} X(C_2F_2RR')_{n+1}^{\cdot}$$

Transfer $X(C_2F_2RR')_{n+1}^{\cdot} + Me^{(n+1)+}L_xY \xrightarrow{k_{tr}} X(C_2F_2RR')_{n+1}Y + Me^{n+}L_x$

Termination $X(C_2F_2RR')_n^{\cdot} + X(C_2F_2RR')_p^{\cdot} \xrightarrow{k_{Tc}} X(C_2F_2RR')_{n+p}X$

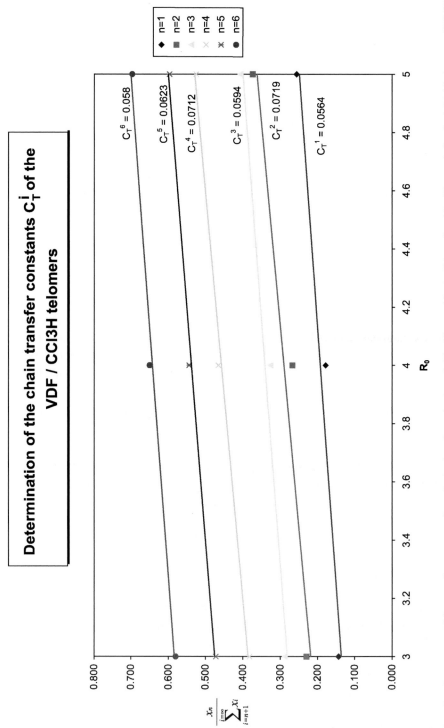

Fig. 1.   Assessment of the transfer constant of *n* order in the telomerisation of chloroform with vinylidene fluoride. (R_0 stands for the initial $[\text{telogen} = CCl_3H]_0/[\text{monomer} = VDF]_0$ molar ratio.)

TABLE II

TRANSFER CONSTANTS OF CATALYSTS C_{Me} AND TELOGEN C_{CCl_4} IN THE
REDOX TELOMERISATION OF CHLOROTRIFLUOROETHYLENE WITH CCl_4 [76]

Nature of the catalyst	C_{Me}	C_{CCl_4}
FeCl$_3$/benzoin	75	0.02
CuCl	700	0.02
RuCl$_2$(PPh$_3$)$_3$	800	0.02

The kinetic law applied to redox telomerisation depends on both the transfer constants of the catalyst and of the telogen but the former one is much greater, as shown in Table II [76]:

$$\frac{1}{\overline{DP}_n} = C_{Me} \frac{[MeL_x]}{[M]} + C_{XY} \frac{[XY]}{[M]}$$

As noted in both radical and redox mechanisms of the telomerisation of fluoroalkenes, termination reactions always occur by recombination and not by disproportionation [70].

In addition, it has been shown that, for the same telogen and fluoroolefin, a radical telomerisation leads to higher molecular weights than those obtained from redox catalysis [21–24].

2.1.4. Thermal Initiation

In certain conditions, sufficient energy supplied by temperature causes fission of the $X-Y$ bond of the telogen.

The mechanism is rather similar to that of the radical telomerisation except for the initiation step in which the radicals are produced from the telogen under temperature as follows:

$$X-Y \xrightarrow{\Delta} X^{\cdot} + Y^{\cdot}$$

$$X^{\cdot} + CF_2{=}CRR' \longrightarrow XC_2F_2RR'^{\cdot}$$

2.2. COTELOMERISATION

In cotelomerisation, the problem is slightly more complex and the kinetics have been little investigated. However, the equation proposed by Tsuchida [77] relating the instantaneous $\overline{DP}_n$ vs. the kinetic constants has been confirmed [78]:

$$(\overline{DP}_n)_i = (r_1[M_1]^2 + 2[M_1][M_2] + r_2[M_2]^2)/(r_1 C_{T1}[M_1][T] + r_2 C_{T2}[M_2][T])$$

where $[M_i]$, $[T]$, r_i and C_{Ti} represent the concentrations of the monomer M_i and of the telogen, the reactivity rate of the monomer M_i and the transfer constant of the telogen to the monomer.

Furthermore, for known r_i, the conversion rates α_1 and α_2 of both monomers vs. time could be determined.

The composition of the cotelomer, i.e., the content of the monomers in the cotelomer, can be predicted for known kinetic constants r_i and C_{Ti} and from $[M_i]_0$ according to the statistical theory of the cotelomerisation. Moreover, the probability to obtain cotelomers of well-defined composition and then the functionality of the cotelomer were calculated [21–24].

2.3. SPECIFIC INITIATIONS FOR FLUOROALKENES

In contrast to classic initiating systems, other less well known processes have been used successfully in telomerisation of fluoroalkenes: these systems involve hypofluorites or hypobromites, or concern ionic (anionic and cationic) or unusual telomerisations described below.

2.3.1. From Hypohalites

It has been known that highly fluorinated alkyl-hypochlorites and hypofluorites R_FOX (with $X = Cl$ or F) are of synthetic value in the preparation of many fluorine-containing products [79–83]. Actually, between -150 and $+50$ °C, such compounds generate $R_FO^\cdot$ radicals able to initiate a telomerisation. However, from higher temperatures (above 70 °C), certain reactions may be explosive.

CF_3OX ($X = F$, Cl) derivatives are the most extensively studied products with respect to their preparation and chemistry. For example, CF_3OF, which is the most studied perfluoroalkyl hypofluorite, was synthesised for the first time by Kellogg and Cady more than 55 years ago [84]. It offers interesting results with vinylidene fluoride [85, 86], chlorotrifluoroethylene (CTFE) [87–89], hexafluoropropene [90], 1,1-dichlorodifluoroethylene [85, 91], trans-1,2-dichlorodifluoroethylene [92] and perfluorobut-2-ene [93] according to the general following reaction:

$$CF_3OF + n\ F_2C{=}CRR' \longrightarrow R_1(C_2F_2RR')_nF + CF_3O(C_2F_2RR')_{2n}OCF_3$$
$$(R_1 = CF_3O \text{ major}; R_1 = F \text{ minor})$$

Johri and DesMarteau [85] have previously shown that such reactions underwent a free radical mechanism. Actually, the radical mechanism was suggested by Di Loreto and Czarnowski [94], the polyfluoroalkyl fluoroxy compound being usually broken homolytically, producing $R_FO^\cdot$ and $^\cdot X$ radicals in the initiation step.

Interestingly, Marraccini et al. [87, 95] reacted CF_3OF, $C_nF_{2n+1}OF$ or $R_{F,Cl}CF_2OF$ to CTFE, yielding the first 10 telomers, whereas they obtained the first three adducts only from the telomerisation of 1,2-dichlorodifluoroethylene with CF_3OF [92]. However, higher molecular weight telomers were produced by modifying the pressures of both gaseous reactants [91].

DesMarteau *et al.* synthesised new perfluoroalkyl hypofluorites [88] and hypobromites [96]. The first ones were produced from perfluorinated acid fluorides or halogenated ketones, as described in the following schemes [88]:

$$R_{F,Cl}C(O)F + XF \xrightarrow{\text{CsF}} R_{F,Cl}CF_2OX \qquad (X = F, Cl)$$

where $R_{F,Cl}$ represents $ClCF_2$, HCF_2, $ClCF_2CFCl$, $BrCF_2CFBr$, FSO_2CF_2 or $FSO_2CF(CF_3)$.

$$R_{F,Cl}C(O)R'_{F,Cl} + XF \xrightarrow{\text{CsF}} R_{F,Cl}R'_{F,Cl}CFOX$$

with $R_{F,Cl} = YCF_2$ (Y = H or Cl) and $R'_{F,Cl} = CF_3$ or $ClCF_2$.

All these polyfluoroalkyl hypohalites successfully reacted to $CF_2=CH_2$ and to $CF_2=CFCl$ [88], in equimolar ratios, from $-145\,°C$ to room temperature for 18–24 h, with better yields in the presence of CTFE. They mainly led to monoadducts composed of both normal and reverse isomers. The authors observed that branched hypohalites were more reactive than linear ones and that, for the same $R_{F,Cl}$ group, $R_{F,Cl}OF$ gave higher yields than those obtained from $R_{F,Cl}OCl$.

In addition, the same team described the synthesis of perfluoroalkyl hypobromites R_FOBr, achieved by reacting $BrOSO_2F$ to R_FONa (where R_F represents $(CF_3)_3C$ or $C_2F_5C(CF_3)_2$ groups) [96]. These compounds reacted with vinylidene fluoride, tetrafluoroethylene, 1,1-dichlorodifluoroethylene and *cis*-1,2-difluoroethylene in the $-93\,°C$–room temperature range for 8–12 h. Monoadducts were produced mainly, the yields ranging from 20 to 37% based on $BrOSO_2F$ used. $(CF_3)_3COBr$ seems more reactive than $C_2F_5C(CF_3)_2OBr$ whatever the fluoroalkene, but less reactive than the hypochlorites or fluoroxy compounds mentioned above. These novel hypobromites readily decompose above $-20\,°C$ by an assumed free radical β-elimination [96]. All the polyfluorinated ethers thus produced are stable, inert and colourless liquids at room temperature.

2.3.2.   Ionic Initiation

Many investigations concerning the ionic or the ring opening polymerisations of numerous monomers have been detailed in many books. However, to our knowledge, very few studies about cationic or anionic polymerisations of fluoroalkenes have been investigated, except the well-known anionic polymerisation of perfluorobutadiene and other specific monomers as studied by Narita [97].

In the examples described below, it can be considered that the anionic telomerisation of hexafluoropropylene oxide (HFPO) and of tetrafluoroethylene, and the cationic telomerisation of fluorinated oxiranes or oxetanes were successful.

2.3.2.1.   Anionic Telomerisation
2.3.2.1.1.   Perfluorinated Oxiranes.   The oldest investigation was conducted by Warnell [98], in 1963, who telomerised HFPO or tetrafluoroethylene epoxide

by ring opening with two ω-iodofluorocarbon ether acid fluorides producing the corresponding adducts, as follows:

$$I(CF_2)_xCOF \underset{T \le -10\ °C}{\overset{CsF}{\rightleftharpoons}} \underset{\mathbf{2}}{I(CF_2)_xCF_2O^{\ominus}Cs^{\oplus}}$$

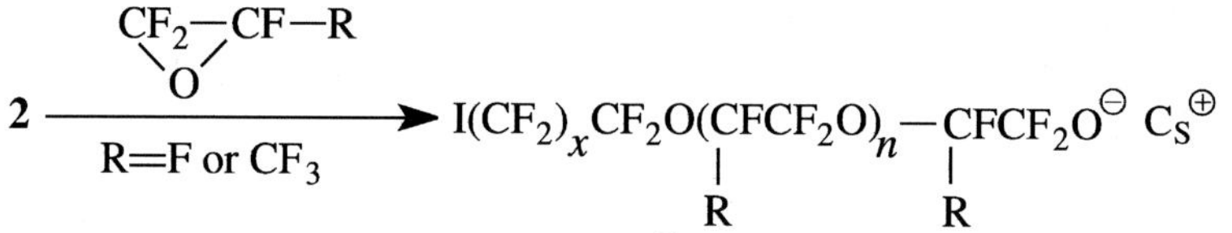

$$\mathbf{2} \xrightarrow[R=F\ or\ CF_3]{\overset{CF_2-CF-R}{\diagdown O \diagup}} \underset{\mathbf{3}}{I(CF_2)_xCF_2O(\underset{R}{CFCF_2O})_n-\underset{R}{CFCF_2O^{\ominus}}\ Cs^{\oplus}}$$

$$\mathbf{3} \rightleftharpoons I(CF_2)_xCF_2O(\underset{R}{CFCF_2O})_n-\underset{R}{CFCOF} + C_SF$$

$$R = CF_3,\ n = 1-4\ \text{and}\ R = F,\ n = 0-2$$

Initiation by ionising radiation or by various catalysts [99], such as monovalent metal fluorides, can be achieved.

Several years later, Ito *et al.* [100] used such a concept to cotelomerise successfully the 4-bromo and 4-chloroheptafluoro-1,2-epoxybutane with HFPO. In the same way, Kvicala *et al.* [101] prepared the first three telomers **4** by telomerising HFPO. Similar telomerisations were also achieved from polyfluoro-ketones [102].

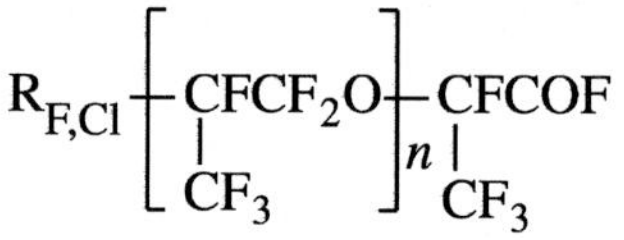

$$R_{F,Cl}\left[\underset{CF_3}{CFCF_2O}\right]_n\underset{CF_3}{CFCOF}$$

$$\mathbf{4}$$

The extrapolation of this work has led to Krytox® products, which are commercially available by the DuPont company [103–105]. In addition, other perfluoropolyethers (PFPEs) containing HFPO units have been marketed by Unimatec under the Aflunox® trade name.

Block copolymers were obtained by a similar process from a fluorinated oxetane instead of HFPO [106], leading to Demnum®, which was marketed by Daikin.

2.3.2.1.2. Tetrafluoroethylene (TFE). Using a similar catalyst as above, Fielding [107] suggested the synthesis of novel branched alkenes from TFE as follows:

$$n\ CF_2{=}CF_2 + CsF \longrightarrow C_nF_{2n}$$

The most abundant oligomer is the pentamer. These perfluorinated olefins are not reactive since the double bond is internal in the chain. However, the ICI company [108] functionalises these alkenes by adding phenol to the perfluorinated olefins:

$$C_{10}F_{20} + RC_6H_4OH \xrightarrow[R'NH_2]{-HF} C_{10}F_{19}OC_6H_4R$$

Another class of telechelic PFPE **5** was achieved from the telomerisation of tetrafluoroethylene, which was photochemically initiated in the presence of oxygen [109]:

$$F_2C{=}CF_2 \xrightarrow[hv]{O_2} \underset{F}{\overset{O}{\diagdown}}C(CF_2O)_x{-}(C_2F_4O)_y{-}CF_2C\overset{O}{\underset{F}{\diagup}}$$

5

with $0.6{-}1.5 \times 10^3 < \overline{M}_n < 4 \times 10^4$.

Such compounds have been marketed by the Montefluos-Ausimont company under the name Fomblin$^\circledR$ [110]. They are composed of two groups: Fomblin Y, produced by a fluorination of **5**, are non-functional polyethers $CF_3(CF_2O)_x$ $(C_2F_4O)_yC_2F_5$, whereas Fomblin Z are obtained by chemical change of **5**, leading to the following telechelics (or α,ω-difunctional oligomers):

$$\text{(G)}{-}(CF_2O)_x(C_2F_4O)_yCF_2\,\text{(G)} \quad \text{with} \quad \text{(G)}:\ OH,\ CO_2R\ or\ NH_2$$

These telechelic oligomers have been described in more detail in Chapter 2, Section 2.1.1.2.1. Similarly, HFP [111] or perfluorobutadiene [112] can also be polymerised.

2.3.2.2. Cationic Telomerisation

To our knowledge, no work dealing with the cationic telomerisation of fluorinated alkenes has been described in the literature in contrast to successful works achieved from fluorinated epoxides [113, 114] or oxetanes [115–117].

2.3.3. Other Initiations

Beyond well-established methods of initiation, other efficient systems have been scarcely used for initiating telomerisation reactions. Pulsed carbon dioxide laser was used to induce telomerisation of tetrafluoroethylene (TFE) with perfluoroalkyl iodides and led to telomeric distributions [118]. This is an exothermic radical chain reaction, producing $R_F(C_2F_4)_nI$ with low n. Dimer proportion dominated in the case of IR multiphoton dissociation. However, at low pressures, n was increased. Similarly, the telomerisation of TFE with BrC_2F_4Br was successful from a KrF laser, a xenon-lamp [119] or simple low-pressure Hg lamps $(5 \times 30\ W)$ [120], and molecular weight distributions depend upon the intensity of the light source [119]. Krespan *et al.* [121] have shown that the initiation by strong fluorooxidisers such as XeF_2, COF_3, AgF_2, MnF_3, CeF_4 or a mixture of them allows telomerisation of various fluoroalkenes. In addition, the same team [122] has recently performed the telomerisation of TFE with perfluorobutyl iodide initiated by fluorographite intercalated with SbF_5 at 80 °C leading to a telomeric distribution showing the first five adducts. Furthermore, a high electrical field during the vacuum-deposition process has yielded the formation of ultrathin ferroelectrical vinylidene fluoride telomeric films [123].

Very interesting surveys were carried out by Paleta's group on the use of aluminium trichloride which favours the cleavage of a C–F bond [124, 125] as follows:

$$Cl_3CF + F_2C{=}CFCl \xrightarrow{AlCl_3} Cl_3CCF_2CF_2Cl + FCCl_2CF_2CFCl_2$$

To our knowledge, no biochemical initiation was efficient, whereas it was successful for the addition of perfluoroalkyl iodides to allyl acetate [126].

Less relevant to the initiating system, but important to process of telomerisation, is the use of supercritical CO_2 [127], which is particularly efficient because it allows a better solubility of fluoroalkenes in organic media, leading to controlled telomerisation with narrow molecular weight distributions.

3. Cleavable Telogens

In this section, the various telogens, or transfer agents, involved in telomerisation of fluoroalkenes will be presented. As mentioned above, a telogen must fulfil three targets:

(i) It requires a weak X–Y linkage able to be cleaved by heating, by γ, UV radiation, X-rays or microwaves or has to react with another species (very often a radical produced by an initiator) to generate a telogen radical.
(ii) This telogen radical must be able to initiate the telomerisation (i.e., such a radical produced has to react with the fluoroolefin).
(iii) The telogen must be efficient enough to transfer to the growing telomeric chain.

More details about the reactivity of telogens are given in Section 5.2, mostly investigated by Tedder and Walton [128, 129].

This is mainly linked to the transfer reaction of the telogen to the growing chain. Theoretical rules have been mentioned in Section 2.1.2. The transfer constant is an intrinsic value of the telogen associated with a monomer and varies between 10^{-4} and 40 (see Section 5.2). The presence of an electron-withdrawing group adjacent to the cleavable bond allows a higher efficiency of the telogen and offers a high selectivity. This is the case of CF_3SSCF_3, which leads to more interesting results than CH_3SSCH_3 [46].

3.1. CLASSICAL CLEAVAGE

Examples of the telomerisation of fluoroalkenes with various transfer agents are given in Table I. All the telogens employed in the telomerisation of hydrogenated olefins [21–24] are also efficient for fluoroalkenes and the use of an initiator also works well. Consequently, cleavage of a range of types of X–Y bonds is possible; these bonds can be located at the chain end or in the centre of the molecule (e.g., disulfides) [46], and access to a range of products is available.

The best-known cleavable bonds are C–H (as in methanol or chloroform), C–X (X = Cl, Br, I, as for polyhalogenomethanes or perfluoroalkyl iodides), X–X, S–H (in mercaptans), S–S, P–H (in dialkyl hydrogenophosphonates) and Si–H (for hydrogenosilanes) (Table I). However, other types have shown to be also been efficient in certain conditions and are described below.

3.2. MORE SPECIFIC CLEAVAGES

Described below are three characteristic examples: C–F, Si–H and O–F cleavages.

3.2.1. C–F Bond Cleavage

The strength of the C–F bond is shown by its high bond dissociation energy (BDE) of 486 kJ mol^{-1}. However, interesting studies were performed by Paleta's group: Cl$_3$CF was used as the telogen in both the telomerisations of CTFE [31] and of 1,2-dichlorodifluoroethylene [124] in the presence of AlCl$_3$, a catalyst efficient for the cleavage of the C–F bond [123, 130]. However, it was not selective when the telogen contains both C–F and C–Cl bonds (see Section 2.3.3).

3.2.2. Si–H Bond Cleavage

To our knowledge, the telomerisation of vinyl acetate with dimethylchlorosilane, initiated by AIBN, is the only example [131] described in the literature which allows the synthesis of real telomers from a Si–H containing telogen. Nevertheless, very low molecular weight TFE telomers could be obtained from activated trichlorosilane [132] or methyldichlorosilane [133] or even, surprisingly, dimethyl silane [134]. It is well known, however, that hydrosilylation reactions, in the presence of hexachloroplatinic acid (i.e., Speier catalyst), lead mainly to monoadducts, even from fluorinated monomers: vinylidene fluoride (VDF) [55], TFE, HFP, chlorotrifluoroethylene (CTFE), or trifluoroethylene (TrFE) [135], C$_6$F$_5$CH=CH$_2$ [54] or C$_6$F$_{13}$CH=CH$_2$ [136]. Peroxides, UV or γ-radiation also allow these hydrosilylations and, interestingly, both monoadduct and diadduct were produced from 3,3,3-trifluoropropene [55] as follows:

$$Cl_3SiH + nCF_3-CH=CH_2 \longrightarrow Cl_3Si(CH_2-\underset{\underset{CF_3}{|}}{CH})_n-H \quad (n=1, 2)$$

3.2.3.   O–F Bond Cleavage

As mentioned in Section 2.3.1, hypohalites are efficient transfer agents for the telomerisation of various fluoromonomers. In contrast to both examples above, such a reaction takes place readily and does not require any catalyst, initiator or heat. Most investigations have shown that various hypofluorites lead to monoadducts but, in certain cases, telomers were obtained. Although the expected $R_FO(M)_nF$ was synthesised, by-products such as $R_FO(M)_nOR_F$ or $F(M)_nF$ were also noted.

4.   Fluoroalkenes Used in (Co)telomerisation

This section contains two main parts: the first one deals with the telomerisation of the most used fluoroolefins, e.g., VDF, TFE and CTFE, for which considerable work has been carried out, TrFE and HFP, and the second summarises investigations performed on less used fluoromonomers, e.g., vinyl fluoride, 1,1,1-trifluoropropene, 1,1-difluoro-2,2-dichloroethylene, and finally on scarcely utilised fluoroalkenes. The syntheses of these fluoroalkenes have been reported in a recent book [4] and in a review [5].

4.1.   Usual or Best-Known Fluoroolefins

4.1.1.   Vinylidene Fluoride

Vinylidene fluoride (or 1,1-difluoroethylene, VDF or VF_2) exhibits many advantages. It is a non-toxic, non-hazardous, non-explosive (in contrast to tetrafluoroethylene or trifluoroethylene) gas and environmentally friendly (it does not contain any chlorine or bromine atoms). Moreover, it can easily polymerise under radical initiation [137] and also under ionic conditions [138]. In addition, it is a non-symmetrical alkene and its propagation may lead to a certain content of defect (i.e., tail to tail or head to head) of VDF chaining. PVDF [137, 139] is an attractive polymer endowed with remarkable properties. It is piezo- and pyroelectric, resistant to acids, to solvents (except DMF, dimethyl acetamide, DMSO and trifluorotoluene), and to nuclear radiation and is a gas barrier polymer. Hence, it has been used in many applications such as loudspeakers, microphones, pianokeys, pyroelectric sensors, IR detectors, paints and coatings and it has been involved in various fields of mines, engineering, biomedical and food industries. However, PVDF's high chemical inertness linked to a high crystallinity rate, its difficult crosslinking and its base sensitivity are limitations. That is why many copolymers of VDF have been synthesised and have been marketed by most international companies producing VDF [2a, 5, 137, 140].

The telomerisation of vinylidene fluoride has been investigated by many authors. Almost all kinds of transfer agents have been used, requiring various means of initiation: thermal, photochemical or from systems involving redox catalysts or radical initiators (Table III [141–172] and Table IV [173–183]), and also hypofluorites, as mentioned in Section 2.3.1.

4.1.1.1. Thermal Initiation

In contrast to many telogens used in the telomerisation of VDF initiated photochemically or in the presence of radical initiators, few transfer agents have been used in the thermal telomerisation (Table III). Only two series have to be considered: those which exhibit C–Br or a C–I cleavable bonds (no work involving a telogen with a C–Cl bond is described in the literature) and hypofluorites (see Section 4.1.1.5).

Hauptschein et al. [141] have shown that VDF reacts with CF_2Br_2 and $CF_3CFBrCF_2Br$ at 190 and 220 °C, respectively, yielding telomeric distributions $R_F(CH_2CF_2)_nBr$, $n = 1-8$ (the $CF(CF_3)Br$ group being more reactive than CF_2Br in the case of the second telogen).

Concerning iodinated transfer agents, almost all perfluoroalkyl iodides and α,ω-diiodoperfluoroalkanes were successfully utilised in thermal telomerisation of VDF. One of the pioneers of such work was Hauptschein et al. [141, 148], who used CF_3I, C_2F_5I, nC_3F_7I, iC_3F_7I, $ClCF_2CFClI$ and $ClCF_2CFICF_3$ at 185–220 °C leading to high telogen conversions. Later, Apsey et al. [38] and Balagué et al. [39, 40] used iC_3F_7I and linear $C_nF_{2n+1}I$ ($n = 4, 6, 8$) telogens, leading to telomeric distributions with a better reactivity of the former branched transfer agent. We have shown that, while the monoadduct exhibits only the structure $R_FCH_2CF_2I$, the diadduct is composed of two isomers [39, 40] (see Section 5.2), while the triadduct had a rather complex structure [139].

An original telogen $(CF_3)_3CI$ was successfully used by a Chinese team [151] as follows:

$$(CF_3)_3CI + n\ H_2C{=}CF_2 \longrightarrow (F_3C)_3C(CH_2CF_2)_nI \quad (n = 1, 2)$$

Another example regarding the thermal initiation of VDF was investigated 3 years ago in the presence of octafluoro[2.2]paracyclophane (bearing trifluoromethyl groups) [171] although this fluorinated derivative can be regarded as initiator. Indeed, from 160 °C such an aromatic fluorinated reactant generates a trifluoromethyl radical able to initiate the telomerisation of VDF. However, the formation of the stable paracyclophane was left inert in the medium while recombination of primary oligo(VDF) radicals occurred.

In conclusion, two main series of transfer agents have been used having specific cleavable bonds (C–X with X = Br and I). They lead to telomeric distributions, in good yields, with more or less high $\overline{DP}_n$ according to the nature of the

TABLE III

THERMAL AND RADICAL TELOMERISATIONS OF VINYLIDENE FLUORIDE

Telogen	Way of initiation	Structure of telomers	References
CF_2Br_2	190 °C	$BrCF_2(C_2H_2F_2)_nBr$ ($n = 1-8$)	[141]
CF_2Br_2	DTBP 140 °C	$BrCF_2(C_2H_2F_2)_nBr$ ($n = 5-25$)	[142]
$CF_3CFBrCF_2Br$	220 °C	$BrCF_2(C_2H_2F_2)_nBr$ ($n = 1-5$)	[141]
$BrCF_2CF_2Br$	Peroxides	$BrCF_2CF_2(VDF)_nBr$	[143–145]
PFPE–Br	Peroxides	PFPE-b-PVDF	[146]
$ClCF_2CFClI$	181 °C, 26 h	$ClCF_2CFCl(VDF)_nI$ ($n = 1-6$)	[141]
CF_3I	Peroxide	$CF_3(VDF)_nI$ ($n = 5-20$)	[147]
$C_nF_{2n+1}I$ ($n = 1-8$)	180–220 °C	$R_F(C_2H_2F_2)_nI$ ($n = 1-7$)	[39, 141, 147, 148]
C_4F_9I	250 °C	$C_4F_9(VDF)_nI$	[149]
iC_3F_7I	185–220 °C	$iC_3F_7(VDF)_nI$ ($n = 1-5$)	[38–40, 150]
$(CF_3)_3CI$	Thermal	$(CF_3)_3C(VDF)_nI$ ($n = 1, 2$)	[151]
$IC_nF_{2n}I$ ($n = 2, 4, 6$)	180–200 °C	$I(VDF)_pC_nF_{2n}(VDF)_qI$ ($p, q = 1-3$)	[152]
Cl_3C-H	DTBP, 140 °C	$Cl_3C(C_2H_2F_2)_nH$ ($n > 3$)	[25, 153]
CCl_4	Acyl peroxide	$CCl_3(VDF)_nCl$	[154]
$RCCl_3$	Peroxide	$RCCl_2(C_2H_2F_2)_nCl$	[25, 34]
$HOCH_3$	DTBP, DBP or AIBN	$HOCH_2(C_2H_2F_2)_nH$ ($n > 10$)	[155, 156]
RBr	Peroxide	Variable n	[25, 157–160]
$BrCF_2CFClBr$	DTBP	$BrCF_2CFCl(C_2H_2F_2)_nBr$	[66]
RI	Peroxide	Variable n	[161–163]
C_4F_9I	AIBN, sc CO_2	$C_4F_9(VDF)_nI$ $n = 1-9$	[127]
PFPE–I	DTBP	$PFPE(VDF)_nI$ $n = 5-50$	[164]
HOC_2H_4SH	AIBN, DBP (or DTBP)	$HOC_2H_4S(VDF)_nH$ $n = 1, 2$ (or 1–6)	[44, 45, 165, 166]
FSO_2Cl	DBP	$FSO_2(C_2H_2F_2)_nCl$ ($n = 1-3$)	[49, 167]
Cl_3CSO_2Br	DBP	$Cl_3C(C_2H_2F_2)_nBr$ ($n = 1, 2$)	[168]
$(EtO)_2P(O)H$	DTBP or perester	$(EtO)_2P(O)(VDF)_nH$ ($n = 1-5$)	[169, 170]
CF_3–paracyclophane	$T > 160$ °C	CF_3–PVDF	[171]
Cyclopentane or cyclohexane	DTBP	c-$C_pH_{2p-1}(VDF)_nH$ ($n > 1$)	[172]
ICH_2I	DTBP, 130 °C	$ICH_2CH_2CF_2I$ (91%)	[162]
THF	DTBP	Monoadduct	[27]

DTBP, di-*tert*-butyl peroxide; DBP, dibenzoyl peroxide; PFPE, perfluoropolyether; sc, supercritical.

telogen (and especially the electrophilicity of the radical generated) and the experimental conditions (initial pressures, [VDF]/[telogen] molar ratios and temperatures). However, our investigations have shown that the monoadduct is composed of $R_FCH_2CF_2X$ as the sole isomer, while the diadduct already consists of two isomers.

4.1.1.2. Radical Initiation

Various transfer agents easily cleavable by radical initiation have been used successfully (Table III). In contrast to numerous works involving telogens which

TABLE IV

PHOTOCHEMICAL, REDOX AND SPECIAL TELOMERISATION OF VINYLIDENE FLUORIDE

Telogen	Way of initiation	Structure of telomers	References
$CFBr_3$	UV	$Br_2CF(C_2H_2F_2)_nBr$ ($n = 1, 2$)	[35]
Cl_3CBr, CF_2Br_2	UV	Low n	[173, 174]
$BrCF_2Br$	Borane/RT	$BrCF_2(VDF)_nBr$ ($n = 5{-}20$)	[142]
CF_2Br_2, CF_3CFBr_2, CF_3CBr_3	γ-ray	$R_F(C_2H_2F_2)_nBr$, variable n	[175]
CF_3I	UV, 28 d/RT	$n = 1$	[176]
CF_3I	UV, 140 °C	$CF_3CH_2CF_2I$ (major), $CF_3CF_2CH_2I$ (minor)	[177]
CF_2HI	UV	$HCF_2CH_2CF_2I$	[161]
H_2S	X-ray	$H_{2-x}S(CH_2CF_2H)_x$ ($x = 1, 2$)	[43]
CF_3SH	X-ray	$CF_3SCH_2CF_2H$	[178]
RSSR (R = CH_3 or CF_3)	UV	$RS(C_2H_2F_2)_nSR$ ($n = 1{-}6$)	[46, 48, 49]
CF_3SO_2SR	UV	$CF_3(VDF)_nSR$ ($n = 1{-}86$)	[179]
PH_3	UV	$H_{3-x}P(CH_2CF_2)_xH$ ($x = 1, 2$)	[50]
ICl	UV or CuCl	ICH_2CF_2Cl	[180]
$RCCl_3$ (R = Cl, CO_2CH_3, CH_2OH)	$FeCl_3$/Ni (or benzoin) or $CuCl_2$	$RCCl_2(VDF)_nCl$ ($n = 1{-}4$)	[34]
C_4F_9I	$FeCl_3$/Ni	$C_4F_9CH_2CF_2I$	[39]
ICF_2I	Lead tetraacetate	$ICF_2CH_2CF_2I$	[181]
R_FOX (X = F or Br)	−145 °C to RT	$R_FOCH_2CF_3$ mainly	[85, 86, 88, 96]
HI (gas)	–	CH_3CF_2I	[182]
R_3SiH	H_2PtCl_6 or UV	$R_3SiCH_2CF_2H$	[55]
H_2O_2	UV	$HO(VDF)_nG$	[183]

exhibit a weak C–H bond, and especially for those which have a cleavable carbon–halogen bond, no investigations have been performed on transfer agents with a C–F bond. Two main telogens having a cleavable C–H group have been utilised in the radical telomerisation of VDF: chloroform and methanol. The former, already mentioned by Toyoda et al. in 1967 [153], requires di-t-butyl peroxide as the best initiator [25]. Besides the formation of several by-products, the $Cl_3C(VDF)_n$–H telomers obtained were used as surfactants after a chemical change of the trichloromethyl into the carboxyl end-group. We carried out similar telomerisations [25], even with a large excess of chloroform, and rather high molecular weight telomers were produced although with poor yields. This was linked to the low transfer constant to chloroform ($C_T = 0.06$ at 140 °C as mentioned in Section 2.1.2).

The radical telomerisation of VDF with methanol was investigated in 1986 by Oku et al. [155] who synthesised novel $H_2C{=}CHCO_2CH_2(VDF)_nH$ macro-monomers (yield = 40%).

A more extensive approach of this telomerisation was investigated in our laboratory a few years ago [156], and high molecular weight telomers were obtained in medium yields whatever the nature of the initiator (azo, peroxides, peresters, percarbonates), or even the method of initiation (photochemical, thermal or in the presence of redox catalysts, all of these three systems being unsuccessful). Starting from a fivefold excess of methanol, telomers of DP_{10-12} were obtained.

Indeed, there were concomitant reactions of both expected telomerisation (way 2, ca. 80%) and unexpected direct addition of the radical initiator onto VDF (way 1, ca. 20%), clearly seen by ^{13}C and ^{1}H NMR spectroscopy [156] and leading to both telomers and oligomers, as follows:

$$\text{Way 1}: \quad (t\text{BuO})_2 \xrightarrow[\Delta < 140\ °C]{} (\text{CH}_3)_3\text{CO}^{\cdot} \xrightarrow[\Delta > 140\ °C]{} {}^{\cdot}\text{CH}_3 \xrightarrow{\text{H}_2\text{C}=\text{CF}_2} \text{CH}_3(\text{C}_2\text{H}_2\text{F}_2)_n^{\cdot} \xrightarrow[\text{CH}_3\text{OH}]{\text{transfer}} \text{CH}_3(\text{VDF})_n\text{H}$$

$$\text{Way 2}: \text{HOCH}_3 \xrightarrow{(\text{CH}_3)_3\text{CO}^{\cdot} \text{ or } {}^{\cdot}\text{CH}_3} \text{HOCH}_2^{\cdot} \xrightarrow{\text{H}_2\text{C}=\text{CF}_2} \text{HOCH}_2\text{CH}_2{}^{\cdot}\text{CF}_2 \quad + \quad \text{HOCH}_2\text{CF}_2\text{CH}_2^{\cdot}$$

$$\downarrow x\,\text{VDF} \qquad\qquad \downarrow y\,\text{VDF}$$

$$\text{HOCH}_2\text{CH}_2\text{CF}_2(\text{VDF})_x^{\cdot} \qquad \text{HOCH}_2\text{CF}_2\text{CH}_2(\text{VDF})_y^{\cdot}$$

$$\text{transfer} \downarrow \text{CH}_3\text{OH} \qquad\qquad \text{transfer} \downarrow \text{CH}_3\text{OH}$$

$$\text{HOCH}_2\text{CH}_2\text{CF}_2(\text{VDF})_x\text{H} \quad \text{HOCH}_2\text{CF}_2\text{CH}_2(\text{VDF})_y\text{H}$$

Interestingly, recent results on the homopolymerisation of VDF initiated by di-*tert*-butyl peroxide, *t*-butyl peroxypivalate, and azo di-*tert*-butyl [139] have confirmed the presence of the same methyl end-group in the oligomers series. The presence of these radicals was shown by the rearrangement of *tert*-butoxy radicals generated by the di-*tert*-butyl peroxide after homolytic scission [66, 139]. The surprising formation of non-hydroxylated oligomers could be explained by the very low transfer constant to methanol: $C_{\text{MeOH}} = 0.008$ at 140 °C [184].

Such a case is rather unusual and demonstrates how the assessment of the transfer constant of the telogen is crucial to be sure that real telomers have been obtained.

We also attempted reacting fluorinated alcohols with VDF in the same conditions as above [166]; indeed, CF_3CH_2OH was poorly reactive while hexafluoroisopropanol did not react at all. These *quasi* unsuccessful results were rather surprising since it was expected that the electron-withdrawing groups bearing the fluorine atoms enabled the fluorinated transfer agents to be more reactive.

Whereas only CCl_4, $Cl_3CCO_2CH_3$ and Cl_3CCH_2OH [34] have been used as telogens involving cleavage of the C−Cl bond [25, 34, 153], all kinds of brominated telogens were successfully used in the presence of peroxides: CCl_3Br [25], $ClCF_2Br$ [157], $BrCF_2Br$ [25, 142, 158], CF_3CF_2Br [159],

CF_3CFBr_2 [159], $BrCF_2CF_2Br$ [143–145], $BrCF_2CFClBr$ [66] for which CFClBr is quite reactive unlike the $BrCF_2$ end-group, $CHBr_3$ [160], CBr_4 [160] or ω-bromoperfluoropolyethers $CF_3(OC_2F_4)_n(OCF_2)_pBr$ yielding original fluorinated block cotelomers [146].

Recently, the transfer constant to $BrCF_2CFClBr$ was assessed when the telomerisation of VDF with this transfer agent was investigated at 75 °C, and initiated by t-butyl peroxypivalate. It was shown $C_T = 1.3$ (at 75 °C) which indicates the high efficiency of this telogen [66].

Various iodinated telogens have been used in radical initiation reaction: $ClCH_2I$, CH_2I_2 [162], CH_3I [163], CF_3I [147], $ClCF_2CFClI$ [141], C_4F_9I [127], and $I(C_2F_4)_nI$ (with $n = 1$, 2 and 3, Table V). C_4F_9I was used in the telomerisation of VDF in the presence of AIBN in supercritical carbon dioxide as the solvent [127].

TABLE V

TELOMERISATION OF VINYLIDENE FLUORIDE WITH IODINATED TRANSFER AGENTS

Telogen	Method of initiation	Structure of telomers	References
ICl	Various initiations	$ClCF_2CH_2I$	[180]
HI	Thermal	CH_3CF_2I	[182]
CF_3I	UV, 28 d/RT	$CF_3(VDF)_1I$	[176]
CF_3I	UV, 0–100 °C	$CF_3(C_2F_2H_2)I$	[188]
CF_3I	UV, 140 °C	$CF_3CH_2CF_2I$ (major), $CF_3CF_2CH_2I$ (minor)	[177]
CF_3I	TBPPI	$CF_3(VDF)_nI$ ($n = 10–30$)	[147]
CF_2HI	UV	$HCF_2CH_2CF_2I$	[161]
iC_3F_7I	185–220 °C	$iC_3F_7(VDF)_nI$ ($n = 1–5$)	[38–40]
nC_3F_7I	20 h, UV, 140–210 °C	$nC_3F_7(C_2F_2H_2)I$	[177]
$(CF_3)_3CI$	Thermal	$(CF_3)_3C(VDF)_nI$ ($n = 1, 2$)	[151]
C_4F_9I	230 °C, 15 h	$C_4F_9(VDF)_nI$	[149]
C_4F_9I	AIBN, sc CO_2	$C_4F_9(VDF)_nI$ ($n = 1–9$)	[127]
C_4F_9I	$FeCl_3$/Ni	$C_4F_9CH_2CF_2I$	[39, 40]
$C_pF_{2p+1}I$ ($p = 1–8$)	180–220 °C	$C_nF_{2p+1}(VDF)_nI$, low n	[38–40]
$C_5F_{11}CFICF_3$	180–190 °C	$C_5F_{11}CF(CF_3)(CH_2CF_2)_nI$	[40]
$ClCF_2CFClI$	181 °C, 26 h	$ClCF_2CFCl(VDF)_nI$	[141]
ICH_2I	DTBP, 130 °C	$ICH_2–CH_2CF_2I$ (91%)	[162]
ICF_2I	LTA, 70 °C	$ICF_2CH_2CF_2I$	[181]
$I(C_2F_4)_nI$ ($n = 1–3$)	180 °C or radical	$I(VDF)_p(C_2F_4)_n(VDF)_qI$, variable $p + q$	[152, 189]
PFPE–I	DTBP, 140 °C	Diblock $PFPE(VDF)_nI$ ($n = 5–50$)	[164]

d, day; RT, room temperature; sc, supercritical; DTBP, di-*tert*-butyl peroxide; LTA, lead tetraacetate; PFPE, perfluoropolyether group; TBPPI, *tert*-butyl peroxypivalate.

More recently, VDF telomers obtained from CF_3I were characterised by ^{19}F NMR spectroscopy and by matrix analysed light desorption ionisation time of flight MS (MALDI-TOF MS) as the first example [147] on a perfluoroaromatic matrix and they were silver cationised. Indeed, this technique is complementary to NMR and could satisfy the analysis of various fluoropolymers [185]. Figure 2 represents the MALDI-TOF mass spectra of $CF_3(VDF)_{11}IAg^+$ and $CF_3(VDF)_{20}$ IAg^+ that nicely show gaussian distributions of the different telomers ranging from the 5th to 21st telomers (for average $n = 11$) and 9th to 27th telomers (for average $n = 20$). These spectra also show the difference of mass between peaks of $64 \, g \, mol^{-1}$ corresponding to exactly one VDF unit and the end-groups deduced from monoisotopic mass correspond to the expected I and CF_3 extremities.

Interestingly, the assessment of the average molecular weights in number and in weight ($\overline{M}_n = 1240$ and $\overline{M}_w = 1250$, respectively, also confirmed those obtained by ^{19}F NMR spectroscopy; Figure 3 from the integral of the signal centred at -61 ppm assigned to characteristic CF_3 end-groups as a label) enabled us to demonstrate that narrow polydispersity indices (PDI $= 1.09$) were obtained.

Similarly, electron spray ionisation was realised on VDF [186] and CTFE [187] telomers.

A few sulfurated transfer agents have been studied. In our laboratory, 2-hydroxyethyl mercaptan reacted with VDF, leading mainly to the first two telomers in the presence of dibenzoyl peroxide or to the first six telomers with di-*tert*-butyl peroxide, used with an excess of VDF [44, 165]. More recently, we revisited this reaction and investigated the kinetics of telomerisation of VDF with the same transfer agent, leading to high values of the first two order transfer constants: $C_T^1 = 19.9$ and $C_T^2 = 25.3$ at $140 \, °C$, showing a high selectivity of the telomerisation and a high efficiency of this thiol. However, it was observed that, if the fluoroalkene was introduced in large excess about the mercaptan [45, 166], this latter was consumed much quicker than the olefin (because of its high reactivity) and, consequently, a polymerisation of VDF was noted after total consumption of the thiol.

Grigor'ev *et al.* [49] and Tiers [167] used FSO_2Cl producing FSO_2 $(CH_2CF_2)_nCl$ ($n = 1–3$). Zhu *et al.* [168] tried Cl_3CSO_2Br initiated by dibenzoyl peroxide and obtained the first two adducts in low yields. Neither disulfides nor thiocarbamates were tested in the radical telomerisation of VDF.

Among the phosphorated telogens, diethyl hydrogenophosphonate (or diethyl phosphite) (DEHP) behaves effectively for the telomerisation of VDF initiated by di-*t*-butyl peroxide [169, 170], perester [170] or azo [45, 170]. As an example, the gas chromatogram of the total product mixture of the telomerisation of VDF with DEHP (Figure 4) exhibits the first nine telomers. The first four telomeric adducts were isolated and characterised by ^1H, ^{19}F, ^{13}C and ^{31}P NMR, and the infinite transfer constant to DEPH was assessed to be 0.34 at $140 \, °C$ [170].

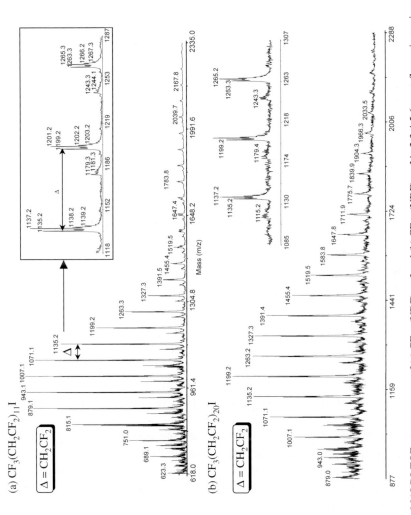

FIG. 2. Positive-ion MALDI-TOF mass spectrum of (a) $CF_3-(VDF)_{11}-I$ and (b) $CF_3-(VDF)_{20}-I$ in 2,3,4,5,6-pentafluorocinnamic acid (PFCA). The telomers are Ag^+ cationised (addition of AgTFA in telomer solution). The inset shows expanded zones of spectra. The data were acquired in a linear mode.

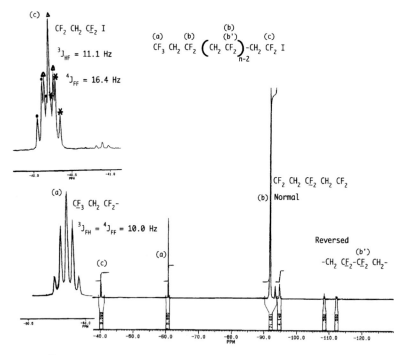

FIG. 3. ^{19}F NMR spectrum of CF$_3$(VDF)$_{20}$I (recorded in deuterated acetone/DMF).

4.1.1.3. Photochemical Initiation

The photoinduced telomerisation of VDF has been investigated by many authors (Table IV). The first work was started in 1954 by Haszeldine [176], who used CF$_3$I as the transfer agent leading to the monoadduct after 28 days of irradiation (using a wavelength $\lambda > 300$ nm) at room temperature. Cape *et al.* [177] studied the same reaction at 140 °C for 12 h and determined the Arrhenius parameters for the addition of CF$_3$ radicals to both sites of VDF; that of the isomer CF$_3$CH$_2$CF$_2$I was about three times higher than that of CF$_3$CF$_2$CH$_2$I.

Tedder and Walton also used brominated methane-derived telogens: CFBr$_3$ [35], even in excess, led to the expected normal and reverse monoadducts, five fluorobrominated by-products (mainly formed by recombination of radicals) and to a diadduct. As described above, the kinetics of the reaction were investigated as for CCl$_3$Br [173] and CF$_2$Br$_2$ [174].

In addition, various chain length telomers were produced by γ-ray-induced telomerisation of VDF with CF$_2$Br$_2$, CF$_3$CFBr$_2$ or CF$_3$CBr$_3$ [175] by careful control of the molar ratio of the initial reactants.

Telogens containing sulfur atom(s) were also utilised and, as for the above halogenated transfer agents, led mainly to monoadducts. For instance, Harris and Stacey [43] chose a large excess of H$_2$S leading to HSCH$_2$CF$_2$H in 69% yield

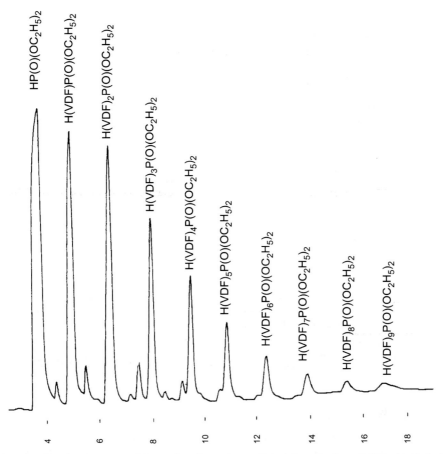

FIG. 4. Gas chromatogram of the total product mixture of the telomerisation of VDF with diethyl-phosphite (reproduced with permission from M. Due *et al.*, *J. Fluorine Chem.*, **112**, 2–12 (2001)).

whereas, in the presence of a lower amount of H_2S, the monoadduct and diadduct were produced in 62 and 19% yield, respectively. The former product is a fluorinated mercaptan able to further react with other fluoromonomers [43]. The same authors [178] also used CF_3SH as the telogen with X-ray initiation yielding the monoadduct selectively.

UV initiation of reaction using CH_3SSCH_3 [46] and CF_3SSCF_3 [47] was also investigated. The hydrogenated transfer agent led to the monoadduct but by-products were also formed, whereas the fluorinated one allowed a total conversion of VDF and yielded a telomeric distribution $CF_3S(C_2H_2F_2)_n$ SCF_3 ($n = 1-6$) with reverse and normal adducts [46–48]. The diadduct has the structure $CF_3SCH_2CF_2CF_2CH_2SCF_3$ and shows that the telomers are formed by recombination of primary radicals.

In addition, PH_3 has successfully reacted with VDF under photochemical initiation leading to mono- and diadduct from an equimolar starting materials ratio [50].

VDF and I–Cl were photolysed producing ICH_2CF_2Cl as the sole product in fair yield [180]. This monoadduct was characterised from ^1H and ^{19}F NMR spectra of CH_3CF_2Cl, obtained after addition of $SnBu_3H$.

More recently, alkyl or aryl trifluoromethanethiosulfonates were involved in the telomerisation of VDF under photochemical initiation [179]. Indeed, in that photochemically induced condition, this transfer agent underwent a sulfur–sulfur cleavage that generates a $CF_3SO_2^\cdot$ radical. That radical loses SO_2 to produce a trifluoromethyl radical able to initiate the telomerisation of VDF as follows:

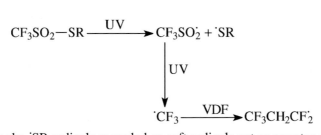

Interestingly, $^\cdot SR$ radicals regarded as soft radicals act as counter radicals and enable the formation of $CF_3(VDF)_nSR$ telomers. Telomers of structure $CF_3(VDF)_nSR$ were obtained with average n VDF units up to 86 [179].

In conclusion, photochemical telomerisation of VDF with iodoalkanes involving the cleavage of the C–I bond mainly produces the monoadduct, and reaction kinetics have been investigated. Addition onto the CH_2 side of VDF was favoured but non-negligible formation of various by-products was also observed. From brominated telogens, the presence of several by-products, besides the expected telomers, was also noted. The literature does not mention any work involving a telogen with a C–Cl cleavable bond, probably due to its high BDE. Sulfurated and phosphorated telogens were successfully utilised, with a special mention for CF_3SO_2SR which tends to control the telomerisation.

4.1.1.4. Redox Telomerisation

Few investigations have been performed on the redox telomerisation of VDF (Table IV) showing a good selectivity in terms of low VDF content in the telomers. Several studies carried out in our laboratory [34] have used CCl_4 in the presence of $FeCl_3/Ni$, $FeCl_3$/benzoin mixtures or $CuCl_2$ as catalysts, with a twofold excess of VDF over CCl_4. The first four telomers were produced and the diadduct was composed of two isomers: $Cl_3C(C_2H_2F_2)_2Cl$ and $ClCF_2CH_2CCl_2$ CH_2CF_2Cl. In similar conditions, $CCl_3CO_2CH_3$ led to the first three adducts (monoadduct 60%, diadduct 32% and triadduct 8%), whereas Cl_3CCH_2OH, catalysed by $FeCl_3$/benzoin in the same conditions, yielded 81% of monoadduct and 19% of diadduct. Usually, the yield of these reactions was poor to medium.

From C_4F_9I and $FeCl_3/Ni$, only $C_4F_9CH_2CF_2I$ was obtained in a 55% yield [39, 40]. The structure of this monoadduct is the same as that produced by thermal initiation. Such a selectivity was confirmed by Chen and Li [181], who obtained $ICF_2CH_2CF_2I$ from ICF_2I in the presence of lead tetraacetate.

4.1.1.5. *Other Ways of Initiation*

Belfield *et al.* [142] have recently shown that boranes behaved efficiently as catalysts/initiators in the telomerisation of VDF with $BrCF_2Br$. These authors obtained telomers in good yields with average degrees of telomerisation ranging between 10 and 20.

Hypohalites and silanes also enable the telomerisation of VDF. DesMarteau *et al.* have performed the most interesting research using fluorinated hypohalites (also mentioned in Section 2.3.1). Johri and DesMarteau [85] achieved the addition of trifluoromethylhypofluorite to $CH_2{=}CF_2$, leading to $CF_3OCH_2CF_3$, $CF_3OCF_2CH_2F$ and CF_3CH_2F in proportion 97.5, 2.0 and 0.5%, respectively. A similar investigation conducted by Sekiya and Ueda [86] selectively produced the first isomer.

Linear CF_3CFZCF_2OX ($Z = Cl$, Br and $X = F$, Cl) and branched $R_{F,Cl}R'_{F,Cl}$ CFOX ($R_{F,Cl} = R'_{F,Cl} = CF_2Y$ with $Y = H$, Cl or $R_{F,Cl} = CF_2Y$; $R'_{F,Cl} = CF_3$ polyhalogenoalkyl hypochlorite and hypofluorite were reacted with VDF in equimolar ratio from $-145\,°C$ to room temperature for 18–24 h giving the monoadduct in 20–70% yield with a major amount of normal $R_{F,Cl}OCH_2CF_3$ isomer [88]. The authors noted that, for the same $R_{F,Cl}$ group, $R_{F,Cl}OF$ led to better yields than those obtained from $R_{F,Cl}OCl$, that $CF_3CFBrCF_2OF$ was more reactive than $CF_3CFClCF_2OF$, and that branched hypohalites reacted more easily than linear ones.

The same group [96] also synthesised novel branched perfluorohypobromites, which reacted with VDF from $-93\,°C$ to room temperature for 8–12 h as follows:

$$R_FOBr + H_2C{=}CF_2 \longrightarrow R_F\,OCF_2CH_2Br\ (20–31\%) + \text{by products } (R_FOH)$$
$$R_F = (CF_3)_3C \text{ or } C_2F_5C(CF_3)_2$$

Surprisingly, both hypobromites produced monoadducts composed exclusively of the reverse $R_FCF_2CH_2Br$ isomer. Hydrosilylation mainly led to monoadducts [55].

4.1.1.6. *Conclusion*

Many investigations have been performed on the telomerisation of VDF with a wide range of transfer agents using various means of initiation: thermal, photochemical (mainly UV and few studies with γ- or X-rays), redox catalysts or radical initiators.

Thermal initiation has the advantage of leading to the formation of well-defined telomers with either selective production of the monoadduct or with higher $\overline{DP}_n$ than those obtained from photochemical initiation. It appears easy and attractive since, except for an autoclave, no special equipment or solvent or expensive reagents (e.g., initiators) are required. In addition, this route provides "clean" reactions with no formation of by-products.

Work on photoinduced telomerisation has led to most impressive results, with interesting kinetics studies being performed by Tedder and Walton [35, 173, 174, 177]. VDF was used to obtain lower telomeric distributions, mainly for monoadducts except from CF_3SO_2SR (from which DP_n up to 86 were achieved [179]), but also yielded several by-products from brominated telogens.

Radical initiation is an "intermediate" means involving many kinds of telogens but formation of by-products from the initiator end-group sometimes occurred (e.g., the telomerisation with methanol [156]).

Few surveys have been developed in redox-type telomerisation. Most involve CCl_3R, others C_4F_9I, or ICF_2I and produced mainly the monoadduct [34, 39, 40, 181] in the presence of ferric or lead catalysts.

Many investigations were realised from (per)fluorinated iodides or diiodides in different conditions of initiations and solvents, summarised in Table V. Highly fluorinated telomers of various molecular weights were thus obtained.

Concerning the structure of the telomers, chaining defects have been observed, since already a head-to-head addition occurred in the diadduct. New processes or the use of several ligands should be worth investigating and might afford regioselective telomers.

Most industrial production of VDF is generated from F141b (Cl_2CFCH_3) or F142b ($ClCF_2CH_3$) and, since this process is unlikely to be stopped, it can be imagined that (co)telomerisation and (co)polymerisation of such olefins still have a prosperous future.

4.1.2. Tetrafluoroethylene

Tetrafluoroethylene is obviously the most used monomer in telomerisation because it gives the best compromise between good ability to polymerise (in comparison to other monomers) and the properties of the perfluorinated chains obtained. Furthermore, because of its symmetry, there are no problems of regioselectivity or defects in chaining. However, it is the most difficult to handle owing to its hazardous behaviour towards oxygen and under pressure.

Whatever the method of initiation, successful results have been obtained involving all kinds of telogens. A summary of the literature is listed in Tables VI–X corresponding to photochemical, thermal, radical, redox and miscellaneous initiations, respectively [190–240].

On the industrial scale, thermal telomerisation appears the most widely used, probably because of the thermal stability of the fluorinated telomers produced. Moreover, it is well recognised that the poor reactivity with redox catalysts leads to low molecular weight telomers, as for thermal initiation at high temperature. On the other hand, thermal initiation at low temperatures and chemically induced radical telomerisation provide telomers with high $\overline{DP}_n$ and wider polydispersities. The syntheses of $C_nF_{2n+1}I$ and $C_nF_{2n+1}Br$ and their functionalisations have been reviewed by Wakselman and Lantz [241].

TABLE VI

PHOTOCHEMICAL TELOMERISATION OF TETRAFLUOROETHYLENE

Telogen	Initiation way	$\overline{DP}_n$	References
$R_1R_2C(OH)-H$ (MeOH, EtOH, iPrOH)	UV (RT)	1, 2	[29]
$X_3CS-SCX_3$ (X = H, F)	UV (RT)	1–5	[47, 48]
Cl_3Si-H	UV (RT)	1	[132]
$(CH_3)_3Si-H$	UV (RT)	1	[134]
CF_3I	UV (RT)	1–10	[193, 194]
R_FI	UV (185–220 °C)	1.3	[150]
$BrCF_2CF_2Br$	UV (RT)	Variable	[119, 120, 196]

RT, room temperature.

4.1.3. Chlorotrifluoroethylene

4.1.3.1. Introduction

Chlorotrifluoroethylene (CTFE) has been one of the most used monomers after tetrafluoroethylene and vinylidene fluoride, although its availability is endangered because the production of its precursor $ClCF_2CFCl_2$ was stopped a few years ago. All kinds of transfer agents have been successfully utilised.

The first and longest section here reviews the telogens having a cleavable C–X bond. The exception arises from Paleta's work [31], in which a C–F linkage is cleaved in the presence of $AlCl_3$ as the catalyst. The results concerning the

TABLE VII

THERMAL TELOMERISATION OF TFE

Telogen	Catalyst	T (°C)	$\overline{DP}_n$	References
$C_6H_5SSC_6H_5$	I_2	175	1, 2	[197]
ICF_2COF	–	220	≤ 5	[98]
PCl_5	–	178	> 4	[52, 198]
$POCl_3$	–	150	1–10	[53]
SF_5Cl	–	150	> 2	[199]
CCl_4	PCl_5	180	2–5	[52]
$BrCF_2Cl$	Ni	360	1.4	[200]
SF_5Br	–	90	1–16	[201]
$ClCF_2CF_2I$	–	180	2–5	[202]
$ClCF_2CFICF_3$	–	170	≤ 6	[203]
CF_3CCl_2I	–	142	1.3	[204]
C_3F_7I	–	190–240	1–3	[150]
$C_nF_{2n+1}I$	–	> 300	1–8	[205]
$I(C_2F_4)_nI$ ($n = 0, 1$)	–	230	2–4	[63]
IC_2F_4I/F_5SSF_5	–	150	4–9	[206]

TABLE VIII

RADICAL TELOMERISATION OF TFE

Telogen	Initiator	Solvent	T (°C)	$\overline{DP}_n$	References
Alkane	DBP	F 113	125	1–3	[207]
Methylcyclohexane	DTBP	–	150	32	[208]
CH_3OH, C_2H_5OH,	DBP	–	100	3	[209]
$(CH_3)_2CHOH$	DIPC	F 113	45	–	[210]
CH_3OH	AIBN	CH_3CN	70	1–4	[30]
CH_3OH	DTBP	–	125–150	≤4	[211]
CH_3OH	H_2O_2	–	70	3–5	[212]
$(CH_3)_2CHOH$	BEPH	–	80	1.2	[213]
$CH_3OC_2H_4OH$	DTBP	–	115–130	1, 2	[214]
CH_3CO_2H	DBP	–	100	≤3	[121]
Cl_3CBr, CH_3OH	AIBN/DBP	–	94–100	1–5	[215]
CH_3Cl	DBP	–	125	>10	[216]
CF_3CCl_3	DTBP	–	150	20	[217]
CBr_4, $CHBr_3$	DTBP	–	150	1	[160]
Cl_3CSO_2Br	DBP	–	90–150	1–12	[168]
$ClSO_2F$	DTBP	–	135	1–10	[167]
$(EtO)_2P(O)H$	DBP	–	130	≤4	[51, 169, 218]

R_F–H	DTBP	–	150	260–500	[208]
CH_3CF_2I	Acyl peroxide	–	–	1–7	[182]
C_2F_5I	CH_3CO_3H	F 113	78–125/P	Variable	[219]
C_2F_5I	DIPC or AIBN or DBP	–	40–60	1–3	[220]
C_2F_5I	$(Cl_2C=CClCO_2)_2$	–	70–75	1.9	[221]
C_2F_5I	R_FCO_3H	–	60–150	3	[222]
C_2F_5I	$(C_mF_{2m+1}COO)_2$ ($m = 3, 7, 8, 11$)	–	25–55	1.8–2.5	[223]
C_2F_5I, iC_3F_7I	$(RC_6H_{10}OCO_2)_2$	–	60–90	1–6	[224]
iC_3F_7I, R_FI	DTBP	–	130	1–10	[225a, b]
R_FI	AIBN or DTBP	–	90–116	2.2	[225c]
C_4F_9I	AIBN	sc CO_2	50–80	1–15	[226]
IC_2F_4I	DBP	–	–	1–3	[227]
R_FI/R_FI_2	Ammonium persulfate	–	80	High n	[228]

DBP, dibenzoyl peroxide; DIPC, diisopropyl peroxydicarbonate; BEPH, butyl ethyl peroxyhexanoate; DTBP, di-*tert*-butyl peroxide; F 113, $ClCF_2CFCl_2$.

TABLE IX

REDOX TELOMERISATION OF TFE

Telogen	Catalyst	T (°C)	$\overline{DP}_n$	References
CCl_4	Cu, $CuCl_2$ or $Fe(CO)_5$	100	3	[229]
CCl_4	$FeCl_3$/benzoin	140	1.7–5.0	[230]
C_2F_5I	Cu/Ni	80–100	1–10	[231a, b]
C_2F_5I	Zn, Mn, V, Re, Rh, Ru, Ag	70–120	1–4	[231c]
$C_nF_{2n+1}I$ ($n = 2, 4, 6$)	IF_5/SbF_5	60	1.6–4.0	[232]
CF_3I	$CuCl/H_2NC_2H_4OH$	200	1–10	[233]
R_FI	$ZrCl_4/H_2NC_2H_4OH$	150	1–6	[234]
C_2F_5I	Metal salt/amine $AlCl_3$	140	3	[235]

TABLE X

TELOMERISATION OF TFE (MISCELLANEOUS)

Telogen	Catalyst or initiation way	T (°C)	$\overline{DP}_n$	References
CH_3COCH_3	γ-rays	RT	5, 6	[190]
$Cl_2C=CCl_2$	γ-rays	RT	5–15	[191]
CCl_4, CCl_3H	Et_4NF/pinene	150	10–24	[236]
iPrOH	–	–	≤4	[237]
I_2	$As(C_2F_5)_2$	–	1–6	[238]
BrC_2F_4Br	KrF laser, Xe lamp or low P Hg lamp	RT	Variable according to light source	[119, 120]
CF_3Br	CO_2 TEA laser	RT	1–5 and dibromide	[118a]
CF_3I	Pulsed CO_2 laser	RT	–	[118b]
$ClCF_2CFClI$	γ-rays	RT	<3	[192]
CF_3I, C_2F_5I, C_3F_7I	γ-rays	RT	2–3	[195]
C_4F_9I	$MetF_n$ (Met = Xe, Co, Ag, Mn, Ce)	RT to 80	1–9	[121]
C_4F_9I	SbF_5/fluorographite	80	1–5	[122]
$FSO_2C_2F_4I$	–	–	–	[239a]
$FSO_2C_2F_4OC_2F_4I$	–	–	–	[239b]
CCl_3SO_2Br	No initiator	–	Variable	[240]
$(CH_3)_3Si–H$	γ-rays	RT	$n = 1$ (66%), $n = 2$ (34%)	[135]
$(CF_3)_3COBr$	–	−85	1	[96]
$C_2F_5C(CF_3)_2OBr$	–	−93	1	[96]

RT, room temperature.

cleavage of C–Cl, C–Br and C–I bonds are reviewed in Tables XI–XIV, respectively [242–282]. The second part of this section describes other minor kinds of telogens, such as hypohalites [87, 89], disulfides [47, 48], alcohols [278–281], boranes [282] and SO_2Cl_2 [276, 277] (Table XV).

TABLE XI

TELOMERISATION OF CTFE WITH TELOGENS CONTAINING CLEAVABLE C–Cl BOND

Telogen	Catalyst	Solvent	T (°C)	$\overline{DP}_n$	References
CCl_4	CuCl or $CuCl_2$	CH_3CN	130	1	[242–245]
CCl_4	$FeCl_3$/benzoin	CH_3CN	110	1–20	[76, 246]
CCl_4	$FeCl_3$/benzoin	MeOH	110	1–20	[247]
CCl_4	$FeCl_3$/Ni	CH_3CN	110	1–10	[248]
CF_3CCl_3	$FeCl_3$/benzoin or CuCl	CH_3CN	130–150	1–5	[249]
CCl_4	$AlCl_3$	CCl_4	50	1	[250]
$RCCl_3$	$FeCl_3$/benzoin or $CuCl_2$	CH_3CN	110–150	1–10	[251]
$CFCl_2$–CF_2Cl	$FeCl_3$/Ni	CH_3CN	150–200	1–5	[252]
RO_2C–CCl_3	$FeCl_3$/benzoin or $CuCl_2$	CH_3CN	110–150	1–10	[253]
Cl–$(CFCl$–$CF_2)_n$–CCl_3	$FeCl_3$/benzoin or $CuCl_2$	CH_3CN	110–150	1–3	[243]

4.1.3.2. Telomerisation of CTFE with Transfer Agents Containing C–Cl Bonds

Redox telomerisation of CTFE with CCl_4 has been extensively studied either in the presence of copper salts (copper of various oxidation states) or from iron salts (mainly Fe^{2+}). In this latter case, a reducing agent is required: organic

TABLE XII

TELOMERISATION OF CTFE WITH TELOGENS HAVING C–Br BONDS

Telogen	Catalyst or initiating system	Solvent	T (°C)	$\overline{DP}_n$	References
HBr	UV	–	RT	1	[56, 256]
Br_2	–	CF_2Cl–$CFCl_2$	RT	1	[257, 258]
$BrCF_2$–$CFClBr$	Thermal $CuBr_2$/Cu	MeCN	120	1–10; $n = 1$ (50%)	[65]
$BrCF_2$–$CFClBr$	UV	–	RT	1	[36, 259]
CF_3–$CClBr_2$	$FeBr_3$/Ni	CH_3CN	110	1–10; $n = 1$ (52%)	[260]
CF_2Br_2, CF_2ClBr	DBP	–	100	6–50	[261]
CF_2Br_2	Boranes	–	RT	5–20	[142]
CCl_3Br	$AlCl_3$	–	RT	1	[262]
CCl_3Br	UV	–	RT	1	[56]
CCl_3Br	UV	–	RT	1, 2	[259, 263]
CCl_3Br	$FeCl_3$/benzoin, $CuCl_2$, DBP	CH_3CN	110–130	1	[264]
CCl_2Br_2	UV	–	RT	1	[259]

DBP, dibenzoyl peroxide; RT, room temperature; n, CTFE base units in the telomers.

TABLE XIII

TELOMERISATION OF CTFE WITH TELOGENS BEARING –CFClI AND –CCl$_2$I END-GROUPS

Telogen	Catalyst	Conditions	$\overline{DP}_n$	References
ClCF$_2$CCl$_2$I	None	18 h, 140 °C	10–15	[59]
CF$_3$CCl$_2$I	None	17 h, 145 °C	4–6	[204, 265]
CF$_3$CFClI	t-Butylperoxypivalate	16 h, 70 °C	1–10	[266]
CF$_3$CFClI	None	18 h, 180 °C	1–3	[266]
ClCF$_2$CFClI	None	5 h, 180 °C	1–20	[267]
ClCF$_2$CFClI	UV	RT	1–11	[267]
FCl$_2$CCFClI	DBP	140–150 °C	1	[268]
ICF$_2$CFClI	Sun light, UV or γ-ray	RT	2–10	[269]

species such as benzoin have been traditionally employed in our Laboratory whereas later the Occidental Chemical company preferred using nickel metal.

Other telogens have also been efficient such as RCCl$_3$ (where R represents CCl$_3$, CHCl$_2$, CH$_2$Cl, CH$_3$ groups), CCl$_3$CO$_2$CH$_3$, ClCF$_2$CFCl$_2$ and CF$_3$CCl$_3$ (Table XI). The presence of copper salts usually favours the formation of the monoadduct, in contrast to iron salts which lead to polydispersed telomeric distributions.

Interestingly, when the telogen was totally consumed in the telomerisation of CTFE with CCl$_4$, it was observed that the telomers produced containing the CCl$_3$ end-group are able to reinitiate telomerisation yielding products with the following formulae: Cl(CFClCF$_2$)$_n$CCl$_2$(CF$_2$CFCl)$_p$Cl [243].

In addition, the chlorofluorinated Cl(CFClCF$_2$)$_n$ chain activates the CCl$_3$ end-group. Thus, the telomers produced can be used as original telogens for various telomerisations of non-fluorinated monomers leading, for example, to the diblock cotelomer 6 [254, 255]:

$$Cl(CFClCF_2)_n CCl_2(CH_2\underset{\underset{CO_2R}{|}}{CH})_m Cl$$

6

The main results are listed in Table XI.

TABLE XIV

TELOMERISATION OF CTFE WITH TELOGENS EXHIBITING A CLEAVABLE C–I BOND

Telogen	Catalyst	Conditions	$\overline{DP}_n$	References
CHI$_3$	(C$_2$F$_5$CO)$_2$O$_2$	12 h, 50–55 °C	13	[265]
CF$_3$I	UV	RT	1 (85%)	[262, 270]
CF$_3$I	t-Butylperoxypivalate	8 h, 100 °C	1–4	[271, 272]
C$_2$F$_5$I	–	–	1–6	[272, 273]
C$_n$F$_{2n+1}$I	UV	RT		[274]
iC$_3$F$_7$I or C$_2$F$_5$I	TiCl$_4$, ZnCl$_2$, YrCl$_3$ with H$_2$N–C$_2$H$_4$OH	150 °C	1–5, 2 (yield = 25%)	[234]
C$_n$F$_{2n+1}$I (n = 4, 6)	CuCl/CuCl$_2$	150 °C	n = 1–10	[266, 275]

TABLE XV

OTHER TELOGENS USED IN THE TELOMERISATION OF CTFE

Telogen	Catalyst	Conditions	$\overline{DP}_n$	References
SO$_2$Cl$_2$	DBP	CCl$_4$, 95 °C	2–5	[276, 277]
CCl$_3$–F	AlCl$_3$	CCl$_4$, 20 °C	1	[31]
SO$_2$ClF	DTBP		1–10	[167]
Cl$_3$CSO$_2$Br	No initiator	Various temperatures	1–10	[240]
CF$_3$S–SCF$_3$	UV	–	1–3	[47, 48]
HSiCl$_2$CH$_3$	H$_2$PtCl$_6$	160 °C	>1	[132]
H–PO(OEt)$_2$	DTBP	6 h, 130 °C	1 (mainly)	[169]
H–PO(OEt)$_2$	DTBP	140 °C	1–5	[187]
RR'CH–OH	UV	–	≤3	[278]
CH$_3$OH	UV	–	1–20	[278, 279]
CH$_3$OH	Peroxides	7 h, 134 °C	1–3	[280]
iPrOH	γ-rays	RT	1	[281]
Boranes	DTBP	4 h, 135 °C	Variable	[282]
CF$_3$OF	–	− 78 °C	F(C$_2$F$_3$Cl)$_n$F (7%) F$_3$CO(C$_2$F$_3$Cl)$_n$F (43%) 1–4	[87]
C$_2$F$_5$OF	–	− 72 °C	1–6	[89]
IBr	UV	CH$_2$Cl$_2$, RT	Monoadduct + by-products	[62]
ICl	Δ, $h\nu$, redox catalysts, radical initiator		ClCF$_2$CFClI (90–100) + Cl$_2$CF–CF$_2$I (0–10)	[60]
I$_2$	γ-rays	RT	1–6	[269]
Cycloalkanes	DTBP	140 °C, 24 h	Monoadduct (33%) Telomers (52%)	[172]

DBP, dibenzoyl peroxide; DTBP, di-*tert*-butyl peroxide; n, CTFE units.

4.1.3.3. Telomerisation of CTFE with Telogens Containing C–Br Bonds

Brominated telogens have also been widely used and Table XII lists various brominated telogens [256–264]. Using Br$_2$, the BrCFClCF$_2$Br adduct was easily obtained. It led to compounds containing an active CFClBr end-group, offering a broader telomeric distribution, as in the case of the telomerisation of CTFE [267] or, more recently, of VDF that produced the first seven telomers [66]. Further, the synthesis of the telogen was recently optimised [283].

The most studied telogen is certainly CCl$_3$Br, initiated either by UV, peroxide or by redox systems. However, we have demonstrated that, in this last case, the redox catalyst, especially ferric chloride, induces a disproportionation, which leads to a mixture of new telogens as follows [264]:

$$BrCCl_3 \rightleftharpoons Br_2CCl_2 + CCl_4$$

In addition to these polybrominated telogens, CF_3CCIBr_2 has also been shown to be efficient. It was also observed that the higher the number of bromine atoms in the telogen, the lower the $\overline{DP}_n$ and the polydispersity.

However, all these products are very interesting because they allow easy $C-Br$ chemical interconversions giving access to functional compounds (e.g., RCO_2H, $R-OH$, etc.) [284, 285].

4.1.3.4. Telomerisation of CTFE with Telogens Containing CFClI and CCl₂I End-Groups

In order to increase the reactivity of the telogens having $C-I$ groups (see Section 4.1.3.5), the authors have introduced one or two chlorine atoms into the iodinated extremity. The synthesis of these new products is by addition of IX (X being Cl or F) to chlorofluoroalkenes as follows:

$$ICl + Cl_2C{=}CF_2 \longrightarrow ICCl_2CF_2Cl + \text{traces of } Cl_3CCF_2I \qquad [59]$$

$$IF + Cl_2C{=}CF_2 \longrightarrow ICCl_2CF_3 + \text{traces of } Cl_2CFCF_2I \qquad [59, 286]$$

$$IF + ClCF{=}CF_2 \longrightarrow ICFClCF_3 + \text{traces of } ICF_2CFCl_2 \qquad [57, 58, 286]$$

Special mention should be made of the addition of IBr onto CTFE that led to the formation of four products: expected $ICFClCF_2Br$, $BrCFClCF_2I$, and unexpected ICF_2CFClI and $BrCF_2CFClBr$ [62].

Thus, versatile telogens have been produced with an increased activation of the cleavable bond. In fact, in 1955, Haszeldine [267] already observed a kind of living character of the telomerisation of CTFE according to the following reaction:

$$ClCF_2CFClI + F_2C{=}CFCl \longrightarrow Cl(CF_2CFCl)_nI$$

Non-exhaustive studies dealing with the telomerisation of CTFE with these above telogens are summarised in Table XIII.

4.1.3.5. Telomerisation of CTFE with Perfluoroalkyl Iodides

In contrast to the telomerisation of tetrafluoroethylene with perfluoroalkyl iodides (R_FI), the addition of these telogens to CTFE has not been fully investigated. This may be explained by the fact that the $R_F(CF_2CFCl)_nI$ telomers produced are more efficient telogens than the starting R_FI transfer agents, as can be seen in the previous section. Table XIV sums up the results in the literature.

When CF_3I is used as the telogen, the question arises of the position of addition of the ˙CF_3 radical to CTFE [270, 271] leading to CF_3CF_2CFClI (mainly, 90%) and $CF_3CFClCF_2I$.

Interestingly, Gumbrech and Dettre [265] used a mixture of $(C_2F_5CO_2)_2$ and CHI_3 (as a iodine donor) to obtain $C_2F_5(C_2F_3Cl)_nI$ telomers. However, a rather high $\overline{DP}_n$ (ca. 13) was observed.

More recent work [266] has indicated that redox catalysis provides a better selectivity in polydispersity and $\overline{DP}_n$. However, some transfers from the catalyst ($CuCl$, $CuCl_2$ or $FeCl_3$) occurred and led to by-products [275].

It can be concluded from all these results that the reaction between R_FI and CTFE is very complex and difficult to perform successfully.

4.1.3.6. Telomerisation of CTFE with Other Telogens

CTFE is very efficient with hypohalites [87, 89], in contrast to other fluoroalkenes which have mainly led to the monoadduct. The reaction is as follows:

$$R_FOX + x\ CF_2{=}CFCl \xrightarrow{-75\ ^{\circ}C} R_FO(C_2F_3Cl)_nX + X(C_2F_3Cl)_mX$$
$$+ R_FO(C_2F_3Cl)_pOR_F$$

$$(X = F, Cl\ or\ Br\ \ and\ \ R_F = CF_3, C_2F_5)$$

Various other telogens undergoing S–Cl [167, 276, 277], S–S [47, 48], Si–H [132], P–H [169, 187], C–H [172, 278–281] cleavages or from boranes [282] have also been used successfully (Table XV).

4.1.4. Trifluoroethylene

One of the most widely known developments of this monomer concerns its interesting copolymers with VDF for piezoelectrical properties. Unlike vinylidene fluoride, tetrafluoroethylene and chlorotrifluoroethylene, which have been extensively used in polymerisation and telomerisation, trifluoroethylene (TrFE) has been investigated in telomerisation by only a few authors. Non-exhaustive results are listed in Table XVI. This olefin exhibits a non-symmetric aspect which is particularly interesting.

Powell and Chambers [287] have prepared telomers by radiochemical initiation in the presence of γ-rays. This reaction was successful with transfer agents activated by oxygen, leading to variable $\overline{DP}_n$ (2.75–5.75) depending on the experimental conditions.

Sloan et al. [35, 160] reported the formation of reverse adducts when trifluoroethylene was telomerised either photochemically with fluorotribromomethane [35] or in the presence of a radical initiator with CBr_4 or CBr_3H [160]. Haszeldine et al. [288–291] investigated the telomerisation of TrFE with different perfluoroalkyl iodides (CF_3I, iC_3F_7I) and they showed that the thermal initiation led to a higher amount of reverse adduct than with the photochemically induced reaction. We have also shown that true telomerisation occurs when the reaction is initiated thermally, since the first five adducts were formed [41].

Bissel [292] investigated the redox telomerisation of trifluoroethylene with iodine monochloride and observed the selective formation of $ClCF_2CFHI$ as the sole product. When thermal or photochemical initiations were used, this isomer was the major product (95–97%) but the reverse $ClCFHCF_2I$ was also observed [293]. Kotora and Hajek [33] showed that redox telomerisation from carbon tetrachloride also led to two isomers. Anhudinov et al. [294] and Cosca [296] used polyhalogenated or brominated telogens, respectively, whereas Harris and Stacey [43, 178] and Haran and Sharp [48] investigated reaction with

TABLE XVI

TELOMERISATION OF TRIFLUOROETHYLENE WITH DIFFERENT TELOGENS

Telogens	Initiation	Structures of telomers	References
MeOH, Et$_2$O, MeCHO, THF	γ-rays	ROCHR'(C$_2$F$_3$H)$_n$H (n = 1–10)	[287]
THF	DTBP, 140 °C	Monoadduct isolated	[27]
c-C$_6$H$_{12}$, dioxane	γ-rays	Monoadduct mainly	[287]
NEt$_3$	γ-rays	(bis)monoadduct and telomers	[287]
H$_2$S	X-rays	HCF$_2$CFHSH (85%)	[43]
		H$_2$CFCF$_2$SH (15%)	
CF$_3$SH	hν	CF$_3$SCFHCF$_2$H (98%)	[178]
		CF$_3$SCF$_2$CFH$_2$ (2%)	
CF$_3$SSCF$_3$	UV	CF$_3$S(C$_2$F$_3$H)$_n$SCF$_3$	[48]
Cl$_3$SiH	γ-rays	Cl$_3$SiCHFCF$_2$H (100%)	[135]
(CH$_3$)$_3$SiH	γ-rays	(CH$_3$)$_3$SiCF$_2$CH$_2$F (50%)	[135]
		(CH$_3$)$_3$SiCFHCF$_2$H (50%)	
CF$_3$I, (CF$_3$)$_2$CFI	hν	R$_F$CHFCF$_2$I (96%)	[288–291]
		R$_F$CF$_2$CFHI (4%)	
(CF$_3$)$_2$CFI	γ-rays	iC$_3$F$_7$(C$_2$F$_3$H)$_n$I (n = 1–3)	[287]
R$_F$I	Thermal (190 °C)	R$_F$CHFCF$_2$I (85%)	[41, 288–291]
		R$_F$CF$_2$CHFI (15%)	

CCl_4	$CuCl_2$ (80 °C)	CCl_3CHFCF_2Cl $CHFClCF_2CCl_3$	[33]
ICl	UV	$ClCF_2CFHI$	[292]
ICl	Thermal	$ClCF_2CFHI$ (95%) $ClCFHCF_2I$ (5%)	[293]
$ClCF_2CFClC_2F_4I$	DBP (3 h, 110 °C)	$ClCF_2CFClC_2F_4(CFHCF_2)_nI$ ($n = 1-3$)	[294]
$Cl_2CFCFClI$	DBP	$Cl_2CFCFCl(C_2F_3H)_nI$	[295]
CF_2Br_2	DBP	$BrCF_2CF_2CFHBr$ (minor) $BrCF_2CFHCF_2Br$ (major)	[296]
CF_2Br_2	Borane	$BrCF_2(C_2F_3H)_nBr$	[142]
$CFBr_3$	hv	$CFBr_2CFHCF_2Br$ (major) $CFBr_2CF_2CFHBr$ (minor)	[35]
CBr_4, $CHBr_3$	DTBP	CBr_3CFHCF_2Br (major) CBr_3CF_2CFHBr (minor)	[160]

DBP, dibenzoyl peroxide; DTBP, di-*tert*-butyl peroxide.

sulfur-containing transfer agents. In all the above four cases, reverse isomers were also produced.

It appears from the literature survey that most investigations have been performed photochemically and, in most cases, the monoadduct is composed of two isomers, the ratio of which depends upon the electrophilicity of the telogen radical. TrFE exhibits the same reactivity towards fluoroalkyl iodides, whatever their structure. In addition, trifluoroethylene seems less reactive than vinylidene fluoride but more reactive than hexafluoropropene.

4.1.5.   Hexafluoropropene

4.1.5.1.   Introduction

Hexafluoropropene (or hexafluoropropylene) (HFP) is one of those fluorinated monomers which are the most difficult to telomerise. This is in contrast to those mentioned above. HFP is regarded as one of the most electrophilic of fluorinated olefins and may be considered to be polarised as follows [178, 297]:

$$\overset{\delta+}{F_2C}=\overset{\delta-}{CF}-CF_3$$

Although several articles and patents deal with the cotelomerisation or copolymerisation of HFP with other fluorinated olefins (e.g., with vinylidene fluoride leading to well-known Dai-el®, Fluorel®, Tecnoflon®, Viton® elastomers [5], Cefral®, Foraflon®, KF Polymer®, Kynar® or Solef® thermoplastics [140] or Cefral® and Tecnoflon® thermoplastic elastomers), its homotelomerisation does not occur readily. This may come from the fact that hexafluoropropene homopolymerises only with great difficulty in the presence of free radicals [267], as shown by various studies (actually the reactivity ratio of HFP in copolymerisation is close to zero [5, 298]). Only one investigation has successfully led to high molecular weight PHFP when the pressures used were greater than 10^8 Pa [299]. Non-exhaustive results illustrating telomerisations of HFP are listed in Table XVII [300–318].

4.1.5.2.   Redox Telomerisation of HFP

The only result obtained by redox catalysis with yttrium salts was carried out by Jaeger [300]. The expected telomers were not observed but, instead, a secondary fluorinated amine containing HFP base units was formed. However, the microstructure of HFP chaining was not described in detail.

4.1.5.3.   Photochemical Telomerisation of HFP

In contrast, the addition of various telogens containing heteroatoms was successfully initiated photochemically. The addition of various mercaptans (CF_3SH, CF_3CH_2SH and CH_3SH) to HFP led to relative amounts of normal and reverse isomers of 45 : 55, 70 : 30 and 91 : 9 [178]. These results can be correlated to the relative electrophilicity of RS˙ radicals, the decreasing order

TABLE XVII

TELOMERISATION OF HEXAFLUOROPROPENE WITH DIFFERENT TELOGENS

Telogens	Initiation	Structure of telomers	References
$(CF_3)_2CFI$	YCl_3, $NH_2C_2H_4OH$	$(CF_3)_2CF(C_3F_6)_nNHC_2H_4I$ ($n = 1-3$)	[300]
RSH (R = CH_3, CF_3CH_2, CF_3)	UV	$RSCF_2CFHCF_3/RSCF(CF_3)CF_2H$ (variable ratio)	[178]
CF_3SSCF_3	UV	$CF_3SCF_2CF(CF_3)SCF_3$	[47]
$HSiCl_3$	γ-ray	$Cl_3SiCF_2CFHCF_3$ (35%) $Cl_3SiCF(CF_3)CF_2H$ (65%)	[135]
PH_3	UV	$PH_2CF_2CF(CF_3)H$ (66%) $PH_2CF(CF_3)CF_2H$ (34%)	[50]
$(EtO)_2P(O)H$	Peroxides	$(EtO)_2P(O)CF_2CFHCF_3$ (80%) $(EtO)_2P(O)CF(CF_3)CF_2H$ (20%)	[169]
THF	DTBP	Monoadduct	[27]
THF	γ-rays	Monoadduct	[308]
ROH	UV (96 h)	$HOCR_1R_2CF_2CFHCF_3$	[28, 301]
R = Me, Et, iPr, Bu	Peroxides (150 °C)	$HCF_2CF(CF_3)CR_1R_2OH$	[28, 302]
$(CF_3)_2CH$, CF_3CH_2	Thermal (280 °C)	Monoadduct	[28, 301]
Diol	UV	Mono- and diadduct	[303]
1,3-Dioxolane	UV	Mono- and diadduct	[304]
$R_{F,Cl}H$	275 °C (3–6 d)	$R_{F,Cl}C_3F_6H$	[301]
CF_3I	UV	$λ < 3000$ Å ($n = 1$) $λ > 3000$ Å ($n = 1, 2$)	[305, 306]
C_2F_5I	$C_nX_{2n+1}CO_3H$ (X = H, F, Cl) γ-rays or peroxides	$C_2F_5(C_3F_6)_pI$, variable p	[222] [26, 307]

(continued on next page)

TABLE XVII (continued)

Telogens	Initiation	Structure of telomers	References
ICl	Thermal (98–200 °C)	$CF_3CFClCF_2Cl$ (92%)	[59, 310]
		$CF_3CFClCF_2I$ (8%)	
R_FI ($R_F = CF_3$, C_3F_7)	Thermal (200 °C)	$R_F(CF_2CFCF_3)_nI$ ($n = 1$–10)	[37, 305, 309]
$CF_2ClCFClI$		$ClCF_2CFClCF_2CFICF_3$	[311, 313]
CF_3CFClI	Thermal (200–230 °C)	$CF_3CFCl(C_3F_6)I$	[266]
$C_nF_{2n+1}I$ ($n = 4, 6, 8$)	Thermal (250 °C)	$C_nF_{2n+1}(C_3F_6)_pI$ ($p = 1$–3)	[42]
iC_3F_7I		$(CF_3)_2CF(C_3F_6)I$ (30% yield)	[42]
$IC_2F_4CH_2CF_2I$	210 °C	$IC_2F_4CH_2CF_2CF_2CFICF_3$ selectively	[316]
$C_4F_9(C_3F_6)I$	250 °C, 48 h	$C_4F_9(C_3F_6)_2I$ (8%)	[42]
$I(C_2F_4)_nI$ ($n = 2$–4)	Thermal (210–260 °C)	$I(C_3F_6)_m(C_2F_4)_n(C_3F_6)_pI$	[63, 312, 315]
		($m = 0$–2; $n = 2$–4; $p = 1, 2$)	
CF_3Br, CF_2Br_2, $CF_3CF_2CF_2Br$	Thermal (250 °C)	$R_F(C_3F_6)_nBr$ ($n = 1$–6)	[37]
H_2CO	Thermal (130–150 °C)	$HOCH_2CF(CF_3)CO_2H$ (41%)	[317]
CF_3OF	20–75 °C	$CF_3O(C_3F_6)F$	[90]
Cyclopentadiene	180 °C, 72 h	$endo : exo = 32 : 68$	[318]

being: $CF_3S^.$ > $CF_3CH_2S^.$ > $CH_3S^.$. Dear and Guilbert [47] only obtained the monoadduct $CF_3SCF_2CF(CF_3)SCF_3$ (in contrast to the first six VDF telomers) just like Burch et al. [50] and Haszeldine et al. [169], who synthesised two phosphorated isomers from PH_3 and $HP(O)(OEt)_2$, respectively. The electrophilic $PH_2^.$ or $(EtO)_2P(O)^.$ radicals mainly react with the CF_2 side of HFP, in particular in the case of the bulkier latter radical.

A series of various alcohols investigated in the photochemical-induced telomerisation of HFP at room temperature led mainly to monoadducts composed of both isomers. This can be explained according to the mechanism proposed by Haszeldine et al. [28] presented in Section 2.1.

Haszeldine et al. [301] also performed the photochemical addition of various R_1R_2CHCl (with R_1, R_2 = CH_3, F; C_2H_5, H; CH_3, CH_3; nC_3H_7, H and C_2H_5, CH_3) telogens to HFP at about 40 °C. Besides the formation of monoadducts produced mainly from the cleavage of C–H bond, they observed small amounts of compounds obtained by the cleavage of the C–Cl bond and cyclic or rearranged products.

No reaction occurred from hexafluoroisopropanol, whereas trifluoroethanol was efficient with HFP [28]. Nevertheless, up to 45% of reverse adduct $HOCR_1R_2[CF(CF_3)CF_2]H$ was produced.

Interestingly, for a series of hydrogenated telogens, Lin et al. [302] suggested a similar decreasing order: CH_3OH > C_2H_5OH > $(CH_3)_2CHOH$. Furthermore, initiation by peroxides led to the same results.

Under photochemical initiation, Paleta's team [303] succeeded in introducing two HFP units onto the methylene groups adjacent to hydroxy end-groups in 1,4-butanediol. Good results were also obtained by these Czech authors from dioxolanes [304].

Amongst transfer agents involving a C–I bond, for trifluoroiodomethane the wavelength λ plays an important role on the telomeric distribution since, at $\lambda < 3000$ Å, only $CF_3C_3F_6I$ was formed whereas, at higher wavelengths, the first two telomers were produced [305]. However, the photochemically induced addition of perfluoro-t-butyl iodide to HFP, even at 163 °C, was unsuccessful [306], probably because of the steric hindrance that prevents the bulky perfluorinated radical to attack HFP.

4.1.5.4. Radical Initiation

Besides the telomerisation of HFP initiated by peroxides as presented above, interesting investigations were performed by Rudolph and Massonne [221], who used a fluorinated carboxylic peracid leading to the first three telomers. No reaction occurred, in contrast with hydrogenated peroxides (t-butyl peroxypivalate), perester (dibenzoyl peroxide) or azo (AIBN) initiators [307]. However, the reaction initiated by γ-rays led to successful results as in the following examples. First, the hydrosilylation reaction of trichlorosilane with HFP yielded both expected isomers with a rather high amount of reverse adduct [135]. Second, γ-rays (or di-$tert$-butyl peroxides [27]) also allowed the addition of one HFP unit to THF

[28b] or one or two units to cyclic ethers, as achieved by Chambers *et al.* [308]:

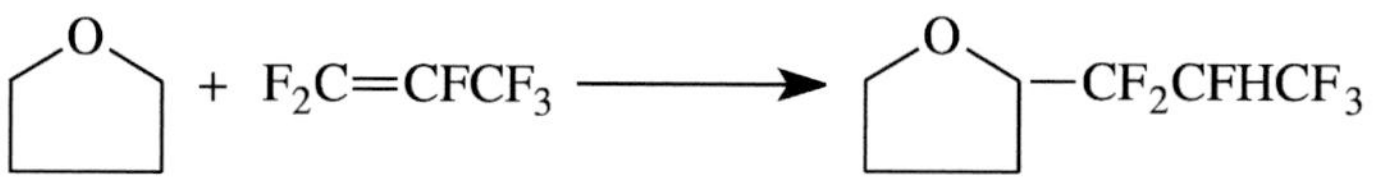

$$\text{(cyclic ether)} + F_2C{=}CFCF_3 \longrightarrow \text{(cyclic ether)}{-}CF_2CFHCF_3$$

The former procedure was successfully applied to poly(THF), leading to novel fluorinated telechelic diol precursors of α,ω-diacrylates that are useful for coating optical fibres [11a]:

$$PTHF + F_2C{=}CFCF_3 \longrightarrow HO{-}(C_4H_8O)_x{-}(C_4H_7O)_y{-}H$$
$$|$$
$$C_3F_6H$$

with $x = y = 4.5$.

An interesting comparison between photochemical and radical initiations was made by Low *et al.* [309], who studied the competitive addition of CF_3I to HFP and ethylene at different temperatures (up to 170 and 189 °C from di-*t*-butyl peroxide and UV-induced reactions, respectively). These authors noted that, in all cases, $CF_3C_2H_4I$, $CF_3(C_3F_6)I$ and $CF_3(C_3F_6)C_2H_4I$ were formed. The latter two products are composed of two isomers, the normal one ($CF_3CF_2CFICF_3$) being produced in higher amount than the reverse one (($CF_3)_2CFCF_2I$). They also observed that the higher the temperature, the higher the yield of these three products and that, below 140 °C, photochemical initiation was more efficient. Above this temperature, radical initiation led to better results. In addition, from UV-induced reaction, $CF_3(C_3F_6)I$ was always produced in the higher amount whereas the opposite was noted in the other process.

4.1.5.5. *Thermal Initiation*

In most studies concerning the telomerisation of HFP, the reactions were carried out thermally.

At rather low temperatures (-50 °C), Dos Santos Afonso and Schumacher [90] succeeded in adding trifluoromethyl hypofluorite to HFP according to the following scheme:

$$CF_3OF + CF_2{=}CFCF_3 \xrightarrow{20-75\ °C} CF_3OCF_2CF_2CF_3\ (68\%)$$
$$+\ CF_3OCF(CF_3)_2\ (32\%)$$

However, except for ICl which led to both normal CF_3CFICF_2Cl and reverse $CF_3CFClCF_2I$ isomers at 98 °C [59] sometimes with up to 40% of reversed adduct [310], most other transfer agents required higher temperatures, especially higher than 200 °C.

Amongst telogens involving a cleavable C–H bond, two series can be considered. First, chlorinated and fluorinated methanes or ethanes R–H [301] were shown to react with HFP above 275 °C for 3–6 d with 14–97% HFP conversion. This reaction mainly led to the monoadduct, usually composed of two isomers (normal RCF_2CFHCF_3 and reverse $RCF(CF_3)CF_2H$) except for chloroform and fluorine-containing alkanes which yielded the normal adduct selectively. However, the reaction between HFP and chlorinated propanes or butanes gave by-products (e.g., cyclic or thermally rearranged derivatives or compounds produced from hydrogen abstraction or from C–Cl cleavage) [301]. These telogens gave better yields from photochemical reactions. Second, a series of alcohols was shown to be efficient above 280 °C [28] due to the adjacent oxygen electron-withdrawing atom.

Allyl chloride or polychlorinated ethanes with a cleavable C–Cl bond were also investigated thermally (250–290 °C) by Haszeldine *et al.* [301]. Besides normal and reverse monoadducts produced in poor yields, cyclic or thermally rearranged compounds were obtained. Perfluoroalkyl bromide and CF_2Br_2 led successfully to telomeric distributions above 250 °C [37].

The most interesting results, however, come from perfluoroalkyl iodides or α,ω-diiodoperfluoroalkanes for which the terminal CF_2–I group has a lower BDE than that of CF_2–Br. The pioneer of such investigations from 1958 was Hauptschein [37], who used CF_3I, C_3F_7I and $ClCF_2CFClI$ as the transfer agents. The originality of such thermal telomerisation is that the mechanism of this reaction does not proceed by propagation. The authors suggested that the mechanism is composed of a succession of addition steps according to the following scheme:

$$n\text{-}C_3F_7I + F_2C{=}CFCF_3 \longrightarrow n\text{-}C_3F_7CF_2CFICF_3$$

$$n\text{-}C_3F_7CF_2CFICF_3 + F_2C{=}CFCF_3 \longrightarrow n\text{-}C_3F_7[CF_2CF(CF_3)]_2I \text{ and so on}$$

Kirschenbaum *et al.* [311] proposed that each step involves an addition complex:

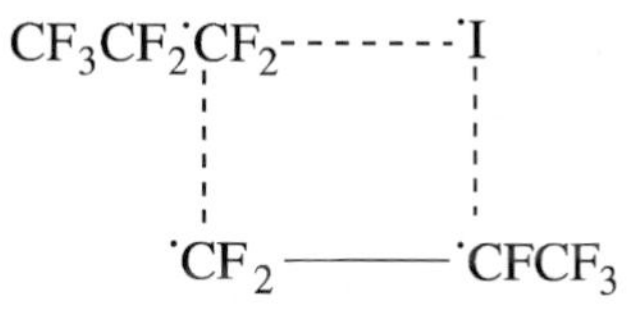

$$
\begin{array}{ccc}
CF_3CF_2^{\cdot}CF_2 & \text{------} & {}^{\cdot}I \\
\vert & & \vert \\
\vert & & \vert \\
{}^{\cdot}CF_2 & \text{------} & {}^{\cdot}CFCF_3
\end{array}
$$

Amongst other fluorinated transfer agents, R_FCFClI telogens are more reactive than R_FCF_2I, as shown by Hauptschein [37] and by Amiry *et al.* [266]. Tortelli and Tonelli [63] and Baum and Malik [312] have confirmed the stepwise mechanism suggested by the pioneered work of Hauptschein *et al.* [313]. However, we could not confirm the work of Hauptschein, who showed that the thermal addition of $i\text{-}C_3F_7I$ to HFP was as successful as that of linear transfer agents [37]. In our work, this telogen reacted with HFP in a low conversion, up to ca. 30% at 250 °C [42]. In order to confirm this result, the telomerisation of HFP with the sterically hindered $C_4F_9(C_3F_6)I$ was carried out at 250 °C for 48 h, and led to 8% of $C_4F_9(C_3F_6)_2I$, although this was not in good agreement with Hauptschein's results [37] in which better yields were obtained.

The nature of the autoclave greatly influences the formation of by-products. A Russian team [314] has shown that, for a vessel made of nickel, the addition of CF_3I to HFP led mainly to $(CF_3)_3CI$ whereas, for similar reaction in a Hastelloy vessel, no by-product was observed.

Interestingly, Tortelli and Tonelli [63], Baum and Malik [312] and, more recently, Manseri et al. [315] succeeded in telomerising HFP with α,ω-diiodoperfluoroalkanes, leading to normal $I(C_3F_6)_x(CF_2)_nI$ and "false" $I(C_3F_6)_y (CF_2)_n(C_3F_6)_zI$ adducts (where x, y or $z = 1$ or 2).

Another unexpected example concerns the radical telomerisation of HFP with $ICF_2CF_2CH_2CF_2I$ (i.e., the monoadduct from IC_2F_4I with VDF), which selectively led to $ICF_2CF_2CH_2CF_2CF_2CFICF_3$ [316].

In addition, a particular case deals with the condensation of paraformaldehyde with HFP in the presence of chlorosulfonic acid at $130-150\,°C$, leading to α-fluoro-α-trifluoromethylhydracrylic acid 7 in 41% yield as follows [317]:

$$F_2C{=}CFCF_3 + H_2CO + H_2O \rightarrow [HOCF_2CF(CF_3)CH_2OH]$$
$$\rightarrow HO_2CCF(CF_3)CH_2OH$$
$$\mathbf{7}$$

Finally, the Diels–Alder reaction of HFP with cyclopentadiene was achieved by Feast et al. [318] and led to a 32 : 68 mixture of the racemic endo and exo isomers, respectively.

4.1.5.6. Conclusion

Apart from redox catalysis, for which only one investigation has been reported, several types of initiation are possible when reacting HFP with various transfer agents. Usually, however, the reaction is selective since low molecular weight telomers are produced. Thermal initiation appears the most efficient, especially for R_FX (X = Br or I) and IR_FI. The telomeric distribution is limited to the first two or three telomers, while the monoadduct is composed of two isomers, the amount of which depends upon the temperature and the electrophilic character of the telogenic radical. On the other hand, linear R_FI are much more reactive than branched ones, and this is probably due to the high electrophilicity of the telogen radical and to the steric hindrance of the CF_3 side group.

4.2. LESS USED MONOMERS

Besides these fluoroolefins, numerous other fluorinated alkenes have been used in telomerisation. Most of them are asymmetric and thus lead to adducts composed of at least two isomers. Among them, vinyl fluoride, 3,3,3-trifluoropropene and 1,2-difluoro-1,2-dichloroethylene have been involved with various telogens.

4.2.1. Vinyl Fluoride

Vinyl fluoride is an interesting monomer, a precursor of PVF or Tedlar[®] [319, 320] (produced by the DuPont company) and known for its good resistance to UV radiation. In telomerisation, the most intensive studies have been performed by Tedder and Walton, who assessed the relative rates of addition of several telogens exhibiting cleavable C–Br or C–I bonds, under UV at various temperatures (Table XVIII [321–326]). They suggested correlations between the orientation ratios for addition of various halogenated radicals, taking into account their size and their polarity. These authors also reported interesting discussions on the estimation of the Arrhenius parameters. However, their work was mostly devoted to the preparation of monoadduct and to the study of their kinetics.

However, telomers were obtained by Rondestvedt [324], who added $C_4F_9IF_2$ (prepared by reacting C_4F_9I with Cl_3F) as the catalyst at 120 °C.

An investigation with trifluoromethyl hypofluorite led mainly to the monoadduct but, to the best of our knowledge, no work was performed using alcohol, mercaptan, disulfide and phosphorous or silicon-containing telogens.

4.2.2. 3,3,3-Trifluoropropene

This fluoroalkene, prepared for the first time by Haszeldine [327] by dehydroiodination of $F_3CCH_2CH_2I$, is the precursor of a commercially available

TABLE XVIII

TELOMERISATION OF VINYL FLUORIDE WITH VARIOUS TRANSFER AGENTS

Telogen	Initiation	Structure of telomers (%)	References
$Cl_3C–Br$	UV, 150 °C	Cl_3CCH_2CHFBr (95)	[321]
		$Cl_3CCHFCH_2Br$ (5)	
$BrCF_2Br$	UV	$BrCF_2CH_2CHFBr$ (96)	[174, 322]
		$BrCF_2CHFCH_2Br$ (4)	
$Br_2CF–Br$	UV, 100–200 °C	(a)	[35]
$CHBr_3$, CBr_4	150 °C	Monoadducts mainly	[160]
$F_3C–I$	UV, 14 d/RT	F_3CCH_2CFHI (84)	[323]
		$F_3C(CH_2CFH)_2I$ (7)	
$F_3C–I$	UV, various T	(a)	[177]
$F_3C–I$	$C_4F_9IF_2$ catalyst/120 °C	$F_3C(C_2H_3F)_nI$	[324]
$nC_pF_{2p+1}–I/iC_3F_7I$ ($p = 2, 3, 4, 7, 8$)	UV, various T	$R_F(C_2H_3F)I$ (a)	[325]
$(F_3C)_3C–I$	UV, 72–163 °C	(b)	[305, 326]
$F_3CO–F$	− 50 °C	$F_3COCH_2CHF_2$ (87–90)	[85, 86]
		$F_3COCHFCH_2F$ (13–10)	
THF	DTBP	Monoadduct	[27]

(a) denotes that both monoadduct isomers are produced with a high amount of normal adduct; (b) denotes that a slight increase of reverse adduct with temperature is observed.

fluorinated silicone from Dow Corning called Silastic® (see Chapter 2, Section 3.4 and Chapter 4, Section 2.3). This polymer preserves interesting properties (inertness, fair surface tension, chain mobility) at low and quite high temperatures. Interestingly, this alkene is quite reactive, in contrast to longer perfluorinated chain vinyl-type monomers (e.g., $C_nF_{2n+1}CH{=}CH_2$ with $n > 1$). Nowadays, this fluoroolefin is marketed by the Great Lakes Chemical Corporation.

Several authors have preferred using this monomer in telomerisation involving a quite wide variety of telogens (Table XIX). It has already given satisfactory results using different types of initiation: redox, UV or γ-radiations, or in the presence of peroxides. In this field, the most pertinent studies were performed by Russian teams (those of Terent'ev and Vasil'eva in the 1980s and Zamyslov in the 1990s) which chose esters [328], alcohols [329] or halogenated compounds [330–335]. Zamyslov *et al.* [329] studied the telomerisation of 3,3,3-trifluoropropene with different aliphatic alcohols (C_1–C_4) under γ-irradiation

TABLE XIX

TELOMERISATION OF 3,3,3-TRIFLUOROPROPENE WITH VARIOUS TRANSFER AGENTS

Telogen	Initiation	Structure of telomers	References
$H_3COCOC(CH_3)_2H$	Peroxides	$H_3COCOC(CH_3)_2$ $[CH_2CH(CF_3)]_nH$	[328]
$(CH_3)_2C(OH)H$	γ-rays ($T < 90\,°C$)	$(CH_3)_2C(OH)$ $[CH_2CH(CF_3)]_nH$ ($n = 1, 2$)	[329]
$C_6H_5CH_2{-}Cl$	$Fe(CO)_5$	$C_6H_5CHX\,[CH_2CH(CF_3)]_nY$ ($n = 1, 2; X = Cl, H; Y = H, Cl$)	[330]
CCl_4	$Fe(CO)_5$	$Cl_3C[CH_2CH(CF_3)]_nCl$ ($n = 1{-}3$)	[331]
CCl_4	$CuCl_2/LiCl$	$Cl_3C[CH_2CH(CF_3)]_nCl$ ($n = 1{-}7$)	[332]
$C_6H_5CH_2{-}Br$	$Fe(CO)_5$	As above with $X = Br, H$ and $Y = H, Br$	[333]
$Br_2CH{-}Br$	$Fe(CO)_5$	$Br_2CH[CH_2CH(CF_3)]_nBr$ ($n = 1{-}3$) $CF_3CBrHCH_2CBr_2[CH_2CH(CF_3)]_2H$	[334]
$Br_2CH{-}Br$	Peroxides	$XCBr_2[CH_2CH(CF_3)]_nY$ ($Y = X = H$ or Br; $n = 1, 2$)	[334]
$BrCH_2{-}Br$	$Fe(CO)_5$	$BrCH_2[CH_2CH(CF_3)]_nBr$ ($n = 1, 2$)	[334]
CBr_4	DBP	$Br_3C[CH_2CH(CF_3)]_nBr$ ($n = 1{-}3$)	[335]
CF_3I	UV, 5 d	$CF_3[CH_2CH(CF_3)]_nI$ ($n = 1, 2$)	[336]
CF_3I	$225\,°C$, 36 h	$CF_3[CH_2CH(CF_3)]_nI$ ($n = 1{-}3$)	[336]
CF_3I	UV, various T	Normal and reverse monoadducts, few amounts of $n = 2$	[327]
$Cl_3Si{-}H$	UV	$Cl_3Si[CH_2CH(CF_3)]_nH$ ($n = 1, 2$)	[55]
$(C_2H_5O)_2P(O)H$	$(t\text{-}BuO)_2$, $130\,°C$	$(C_2H_5O)_2P(O)CH_2CH_2CF_3$ (39%)	[169, 337]
THF	DTBP, $140\,°C$	Monoadduct (48%)	[27]
2-Me-1,3-dioxolane	$Fe(CO)_5$	2,4-bis(TFP)-2-Me-dioxane	[338]
CH_3SSCH_3	UV	$CH_3SCH_2CH(CF_3)SCH_3$	[48]
HBr	UV	$CF_3CH_2CH_2Br$	[336]
Cyclopentadiene	$180\,°C$, 72 h	*exo* : *endo* = 62 : 38	[318]

DBP, dibenzoyl peroxide; d, day; DTBP, di-*tert*-butyl peroxide.

to obtain fluoroalcohols. The kinetics of telomerisation and the reactivity of these alcohols were also discussed. In addition, Vasil'eva's team even proposed mechanisms of formation for unexpected telomers [334] and determined various chain transfer constants of telogens showing that, in certain cases, the transfer to the telogen is more efficient than the radical chain growth [333, 335]. They have also observed that $C_6H_5CH_2Cl$ [330] and $C_6H_5CH_2Br$ [333] undergo both C–H and C–halogen cleavages since their reaction products are $C_6H_5CH_2(M)_n$–X and $C_6H_5CHX(M)_nH$, where X = Cl, Br and M represents the 1,1,1-trifluoropropene. In the same way, bromoform, in the presence of iron complex or peroxide (such as dibenzoyl peroxide or di-t-butyl peroxide), also leads to two series of telomers arising from its C–H and C–Br cleavages. Consequently, the formation of "false adduct" $CF_3CHBrCH_2CBr_2$ $(CH_2CH(CF_3))_2H$ is not unexpected.

Haszeldine [336] noted that CF_3I is an effective telogen both thermally or photochemically initiated, the latter being more selective than the former. Low et al. [309] investigated this reaction under UV radiation at various temperatures and obtained a mixture of normal and reverse isomers for the monoadduct. They observed that a higher proportion of reverse derivative was formed at higher temperatures. In addition, they noted the formation of higher adducts above 175 °C.

Amongst other transfer agents, silanes, phosphonates, disulfides and hydrogen bromide are able to react with 3,3,3-trifluoropropene, mainly yielding the monoadduct (Table XIX) [328–338].

4.2.3. 1,2-Difluoro-1,2-dichloroethylene

1,2-Difluoro-1,2-dichloroethylene was telomerised by different transfer agents (Table XX) [339–348], containing halogenated, hydroxylated or phosphonate end-groups and hypofluorites. Various initiating processes were used. This monomer requires specific hard conditions (^{60}Co irradiations or peroxides under high temperatures) to enable its reactivity with telogens. In most cases, the results have shown that only the monoadduct was produced. Unfortunately, in the presence of $AlCl_3$, fluorotrichloromethane also led to several by-products [125, 348], whereas several ethers, alcohols (the radical addition of methanol initiated by peroxides offered the monoadduct only [339] in different yields according to the nature of the peroxide used) or esters are not sufficiently efficient (only 8–30% monomer conversion) [339]. Perhaloalkyl and trifluoromethyl hypofluorites gave the most encouraging results since telomers were produced [92].

4.2.4. Other Fluoroolefins

Table XXI lists non-exhaustive results of the telomerisation of various fluoroalkenes (classified by an increasing number of fluorine atom(s) or perfluorinated

TABLE XX

TELOMERISATION OF 1, 2-DICHLORO-1, 2-DIFLUOROETHYLENE WITH VARIOUS TRANSFER AGENTS

Transfer agent	Method of initiation	Overall yield (%)	References
CH_3OH	γ-rays, ^{60}Co, 40 °C, 700 h, $R_0 = 3$	16	[339, 340]
CH_3OH	UV, 40 °C, $R_0 = 5$	11	[341]
CH_3OH	DTBP, 150 °C, 5 h, $R_0 = 9$	10	[342]
CH_3OH	DHBP, 134 °C, 7 h, $R_0 = 40$	65	[343]
C_2H_5OH	γ-rays	–	[339]
C_3H_7OH	γ-rays	–	[339]
CH_3OAc	Peroxides	–	[342]
i-PrOR (R = H, Ac)	Peroxides	<30	[342]
CF_3CH_2OH	Peroxides	–	[344]
THF	γ-rays	–	[345]
THF	Peroxides	<30	[342]
Dioxane	γ-rays	–	[345]
$(C_2H_5)_2O$	γ-rays	–	[345]
$(RO)_2P(O)H$ (R: CH_3, C_2H_5, n-Pr)	γ-rays, 230 h	–	[346]
RSH	γ-rays	–	[346]
$ClCH_2F$ or Cl_2CHF	$AlCl_3$	–	[347]
Cl_3CF	$AlCl_3$	–	[124, 348]
CF_3OF	Low T	–	[92]
$R_fO–F$	Low T	–	[87]

group(s) directly linked to ethylenic carbons) with different telogens, most of them being halogenated compounds.

First, it is noted that several kinds of initiation are possible. Second, although many olefins have been investigated for the production of the monoadduct only (especially for kinetics or studies of regioselectivity), the formation of higher boiling telomers is not excluded since the authors used an excess of telogens. This is mentioned in most cases presented in Table XXI [349–356].

The more interesting features or reactions leading to telomeric distributions are described below. For hydrosilylation reactions, many fluorinated monomers have been used [357, 358] and the results have been very well reviewed by Marciniec [359]. Such a reaction leads mainly to monoadducts, except for 3,3,3-trifluoro-propene [55], TFE [132–134] and CTFE [132].

The reactivity of monomers is described in Section 5, but those of $CF_3CH=CF_2$ and $CF_3–CF=CH_2$ with CF_3I are compared here. Only the latter olefin leads to telomers useful for hydraulic fluids, lubricants and for heat transfer media [360]. Similarly, $C_2F_5CF=CF_2$ has been telomerised thermally by Paciorek et al. [361], who suggested a probable mechanism. Longer perfluoroolefins, or perfluoroalkyl vinyl ethers, investigated by Paleta's group mainly [303, 304] led to monoadducts, as for the radical addition of THF onto $C_5F_{11}CF=CF_2$ [27].

TABLE XXI

TELOMERISATION OF VARIOUS MINOR FLUOROALKENES WITH DIFFERENT TRANSFER AGENTS

Fluoroolefin	Telogen	Way of initiation	Structure of telomers (%)	References
$F_2C{=}CCl_2$	Cl_3CBr	UV	$Cl_3CCF_2CCl_2Br$ (28)	[321]
	ICl	$-10\ ^{\circ}C$	$ClCF_2CCl_2I$ only	[59]
	ICl	Fe, $20\ ^{\circ}C$	$\left\{\begin{array}{l} ClCF_2CCl_2I\ (70) \\ ICF_2CCl_3\ (16) \\ ClCF_2CCl_3\ (14) \end{array}\right.$	[59]
	CF_3OF	$-7\ ^{\circ}C$ to RT	$\left\{\begin{array}{l} CF_3OCF_2CCl_2F\ (major) \\ CF_3O(C_2F_2Cl_2)_tOCF_3 \\ CF_3CCl_2F \end{array}\right.$	[85, 91]
	R_FOBr	$-83\ ^{\circ}C$ to RT	$\left\{\begin{array}{l} R_FOCF_2CCl_2Br\ (75{-}90) \\ R_FOCCl_2CF_2Br\ (25{-}10) \end{array}\right.$	[96]
	Cycloalkanes	γ-rays or DTBP	Monoadduct + diadduct	[172]
	CH_3OH	Peroxide	$HOCH_2CF_2CCl_2H$	[343]
	RSH	Peroxide	$RSCF_2CCl_2H$	[349]
	$(RO)_2P(O)H$	γ-rays	$(RO)_2P(O)CF_2CCl_2H$	[346]
	CF_3I	UV, $30{-}100\ ^{\circ}C$	$\left\{\begin{array}{l} (CF_3)_2CHCF_2I\ (55{-}63) \\ CF_3CHIC_2F_5\ (45{-}37) \end{array}\right.$	[350]
$F_2C{=}CHCF_3$	CF_3I	$212\ ^{\circ}C$	$\left\{\begin{array}{l} (CF_3)_2CHCF_2I\ (17) \\ CF_3CHIC_2F_5\ (83) \end{array}\right.$	[350]
	HBr	UV	$\left\{\begin{array}{l} CF_3CH_2CF_2Br\ (28) \\ CF_3CHBrCF_2H\ (12) \\ BrCF_2CHBrCHF_2\ (60) \end{array}\right.$	[350]

(continued on next page)

TABLE XXI (*continued*)

Fluoroolefin	Telogen	Way of initiation	Structure of telomers (%)	References
$F_2C=CFBr$	Cl_3CBr	UV	$Cl_3CCF_2CFBr_2$ (97)	[321]
			$Cl_3CCFBrCF_2Br$ (3)	
	Cl_3CF	UV	$Cl_2CF(CF_2CFBr)_nCl$ ($n = 1–13$)	[351]
	Br_3CX (X = F, Br)	UV	$XCBr_2(CF_2CFBr)_nBr$ ($n = 1, 2$)	[352]
	F_3CSSCF_3	UV	$F_3CS(CF_2CFBr)_nSCF_3$ ($n = 1–3$)	[47]
	CH_3OH	Radical	$HOCH_2(C_2F_3Br)_n$–H	[353]
	Cycloalkanes	γ-rays	Monoadduct	[172]
$F_2C=CFOCH_3$	F_3CSH	UV	$F_3CSCF_2CFHOCH_3$ (71)	[43]
$F_2C=CFR_F$ (R_F: C_6F_5, C_5F_{11})	CCl_4	Radical	$Cl_3CCF_2CFClR_F$	[354]
	THF	DTBP	Monoadduct (55)	[27]
$F_3CCF=CFCF_3$	F_3CSH	UV	$CF_3SCF(CF_3)CFHCF_3$ (low yield)	[43]
	F_3COF	$-50\,°C$	$CF_3OCF(CF_3)CF_2CF_3$	[93]
	CH_3OH	γ-rays	$HOCH_2CF(CF_3)CFHCF_3$	[355]
$F_2C=CHCl$	CF_3I	UV or 225 °C	$CF_3(CF_2CHCl)_nI$ ($n = 1, 2$)	[356]
	Cl_3CF	$AlCl_3$	Cl_3CCF_2CHFCl	[347]
	CH_3CH_2OH	γ-rays	Mixture	[287]
	Cycloalkanes	DTBP	Mixture	[172]

The photochemically induced reaction of $F_2C=CHCl$ with CF_3I, offering 92% of monoadduct but much lower yields, was observed by Haszeldine's team [356] for thermal initiation above 225 °C, however, no reaction occurred after 100 h at 190 °C. In successful reactions, both isomers were obtained (Table XXI), whereas only one was produced from the photochemical addition of HBr to $F_2C=CHCl$.

In contrast, similar reaction of $F_2C=CHCF_3$ formed the following three isomers [350]:

$$H-Br + F_2C=CHCF_3 \longrightarrow \begin{cases} CF_3CH_2CF_2Br \ (24\%) \ \textbf{8} \\ CF_3CHBrCF_2H \ (9\%) \ \textbf{9} \\ BrCF_2CHBrCHF_2 \ (50\%) \ \textbf{10} \end{cases}$$

The authors assumed that dibromide **10** formed as a major product may arise from the dehydrofluorination of monoadduct **9** followed by a radical addition of HBr to the intermediate olefin according to the following scheme:

$$\textbf{9} \xrightarrow{-HF} F_2C=CBrCHF_2 \xrightarrow{HBr} BrCF_2CHBrCHF_2$$
$$\textbf{10}$$

This would explain the low amount of **9** isomer and the high proportion of **10**. Investigations performed on the same fluoroalkene with regards to the addition of trifluoroiodomethane have shown that the mode of initiation has a drastic effect on the amount of both isomers [350]. For both photochemical and thermal initiations, the proportion of isomers is inverse.

Similarly, 1,1-dichlorodifluoroethylene was also telomerised with various telogens using different types of initiation [59, 85, 91, 96, 321, 349]. The kinetics were investigated using Cl_3CBr and perfluoroalkyl hypohalites. An interesting study concerns the addition of iodine monochloride to this olefin, producing exclusively $ClCF_2CCl_2I$ isomer at -10 °C. However, at higher temperatures and in the presence of iron as the catalyst, the Cl_3CCF_2I isomer was also produced, and the formation of chlorinated product $ClCF_2CCl_3$ was also observed under these conditions [59].

When methanol was used as the telogen [343] under peroxidic initiation, we noted a quantitative conversion of this fluorochloroolefin producing $HOCH_2CF_2CCl_2H$ selectively.

Bromotrifluoroethylene (BrTFE) telomerises well photochemically and no other way of initiation seems to have been mentioned in the literature. Even disulfides have provided telomers [47], and BrTFE telomers have found applications as gyroscope flotation oils [352].

Finally, perfluorinated alkenes, prepared according to various routes (decarboxylation of perfluorinated acids or anhydrides, action of organolithium, magnesium reagents or metallic zinc–copper couple to perfluoroalkyl iodides, oligomerisation of TFE and HFP, addition of perfluoroalkyl iodides to perfluoroallyl chloride [362] and addition of KF to perhalogenated esters or alkanes [363, 364]), are shown to be unreactive in telomerisation except for radical addition of CCl_4 [354], photochemical addition of CF_3SH [178] or reaction of CF_3OF [93]. 1,3-Perfluorobutadiene (e.g., synthesised either by Narita [97], Bargigia *et al.* [365], or Dedek and Chvatal [36]) cannot be telomerised in a radical process, but was efficiently polymerised anionically [97]. Similarly, perfluoro-1,5-hexadiene did not lead to any telomers [366].

Among all remaining monomers, those containing (per)fluorinated side chains such as fluorinated acrylates, vinyl ethers or vinyl esters, maleimides and styrenic monomers are also very interesting and have been studied in (co)telomerisation. Most of them have been previously reviewed [13a] and are not mentioned in this chapter, although more information is reported in Chapter 5, Section 2.1.

4.3. Cotelomerisation of Fluoroalkenes

Interesting commercially available fluorinated copolymers are known for their excellent properties and most of them have been synthesised for use as elastomers [3, 5] or thermoplastics [140]. The literature is abundant on the copolymerisation of two fluorinated olefins and the copolymerisation of fluoroalkenes with non-halogenated monomers (e.g., vinyl ethers [367]), regarded as electron withdrawing and electron donating, respectively (see Chapter 3). However, few articles or patents describe the cotelomerisation of fluoromonomers (Section 4.3.1). Brace [368, 369] provided interesting reviews on the radical addition of mercaptans [368] and perfluoroalkyl iodides [369] to different kinds of monomers, and yet other transfer agents (e.g., brominated telogens, methanol) have already shown efficiency in cotelomerisation. This part deals with non-exhaustive examples and only two or three fluoroalkenes are considered.

Two alternatives are possible: either direct cotelomerisation according to a kind of batch procedure or the step-by-step or sequential addition (i.e., the telomerisation of a second coalkene with a telomer produced from a first monomer via a two-step process). Hence, fluorinated diblock cotelomers have been produced and we invite the readers to find more information in Chapter 4, Section 2.1.2.4.1.2.

4.3.1. Cotelomerisation of Two Fluoroolefins

4.3.1.1. CTFE/VDF Cotelomers

Cotelomers produced by cotelomerisation of CTFE and VDF seem to have attracted much interest because they constitute an interesting model for Cefral®, Foraflon®, Kel F® or Solef® copolymers [5]. Barnhart [64] used thionyl chloride as the telogen to produce $Cl(VDF-CTFE)_nCl$ ($n < 10$) while Hauptschein and Braid thermally cotelomerised these olefins with CF_3I [370] or $ClCF_2CFClI$ [371]. The cotelomers obtained exhibit the following random structure $ClCF_2$ $CFCl[(CTFE)_n(VDF)_p]I$ where $n + p$ varies in the range of 3–20 depending on experimental conditions.

Furukawa [228] used perfluoroalkyl iodides and α,ω-diiodoperfluoroalkanes as telogens for the synthesis of high molecular weight cotelomers.

More recently, we used 2-hydroxyethyl mercaptan to cotelomerise both these co-olefins by radical initiation [44, 165], leading to $HOC_2H_4S(CTFE)_x(VDF)_yH$ with x and y close to 4. Interestingly, such cotelomers exhibit an almost alternating structure.

4.3.1.2. VDF/TFE Cotelomers

Of the few investigations performed, the oldest was conducted by Wolff who used isobutane [372] or diisopropyl ether [373] under radical initiation.

Ono and Ukihashi [195] successfully carried out the radical cotelomerisation of VDF and TFE with perfluoroisopropyl iodide but, in similar conditions, that of TFE and ethylene failed.

In addition, Banhardt [64] prepared α,ω-dichloropoly(VDF-co-TrFE) random cotelomers from SO_2Cl_2. We also reported the synthesis of TrFE telomers containing TFE, VDF and HFP [41].

4.3.1.3. VDF/HFP Cotelomers

Copolymers of VDF with HFP (and TFE) are well-known KF Polymer®, Kynar®, Solef® thermoplastics [140] or Dai-el®, Kynarflex®, Fluorel®, Tecnoflon® or Viton® elastomers [5] as a non-exhaustive list. Several investigations of cotelomerisation have been described in the literature and a few of them are presented below.

Rice and Sandberg [70] first synthesised α,ω-dihydroxyl terminated VDF/HFP cotelomers from perfluoroadipic, perfluorosuccinic or perfluoroglutaric anhydrides under hydrogen peroxide initiation. Later, researchers from the Ausimont company studied the peroxide-induced cotelomerisation of VDF and HFP with 1,2-dibromoperfluoroethane leading to fluorinated products having an average number molecular weight of 660 and a T_g of $-94\,°C$ [374].

More recently, we synthesised telechelic dihydroxy poly(VDF-co-HFP) oligomers ($M_n = 600-1200$) by dead end radical copolymerisation of VDF and HFP initiated by hydrogen peroxide [375].

In addition, Chambers et al. [175] have used brominated telogens under γ-rays or thermal initiations leading to various $\overline{DP}_n$ by controlling the initial molar ratio of reactants. More recent investigations were carried in our laboratory for the preparation of $CF_3[(VDF)_xHFP]_ySR$ (where R represents C_6H_{11} or C_6H_5) [179] random cotelomers and PFPE-b-poly(VDF-co-HFP) diblock cotelomers [164].

Fluoroiodinated transfer agents were also used in iodine transfer polymerisation (ITP) in a controlled way (see Chapter 4, Section 2.1.2.4.1) and enables Ausimont [376], Daikin [377] and DuPont [378, 379] to produce novel fluorinated copolymers with targeted molar masses and narrow polydispersities.

Tatemoto and Nakagawa [377] started from $i\text{-}C_3F_7I$ under peroxide initiation to cotelomerise VDF and HFP, useful for the preparation of block and graft copolymers. Several examples of the tertelomerisation of VDF with HFP and TFE are given in Section 4.3.2.

4.3.1.4. HFP/TFE Cotelomers

Haupstchein and Braid [317, 380] studied the cotelomerisation of HFP and TFE with IC_3F_6Cl at 190–220 °C and obtained various molecular weight cotelomers depending on initial molar ratios of the reactants.

4.3.1.5. Stepwise Cotelomerisation

The stepwise cotelomerisation leading to original block cotelomers is reported in Chapter 4, Section 2.1.2.4.1.2 and is not reported in this chapter. However, we suggest below one typical example where the telomer formed acts as an interesting transfer agent for a further telomerisation [40, 316]:

$$I(VDF)_p(TFE)_n(VDF)_qI + HFP \rightarrow I(HFP)_x(VDF)_p(TFE)_n(VDF)_q(HFP)_yI$$

$$I(HFP)(TFE)_n(HFP)_tI + VDF \rightarrow I(VDF)_z(HFP)(TFE)_n(HFP)_t(VEF)_\omega I$$

These oligo- or poly(TFE-co-VDF-co-HFP) are interesting models for Dai-el®, Fluorel®, Tecnoflon® and Viton® elastomers.

4.3.2. Telomerisation of Three Fluoroalkenes

An interesting tertelomerisation of VDF with TFE and CTFE, or VDF with TFE and HFP, was performed by Bannai et al. [381], who used methanol as the transfer agent under peroxidic initiation. Excellent yields were reached and average number molecular weights were ca. 1000.

These fluorinated cotelomers are novel precursors of the following original (meth)acrylic macromonomers (R = H or CH_3):

$$H_2C=CRCO_2CH_2[(VDF)_{0.5}(TFE-CTFE)_{0.5}]_n-H$$

In similar radical conditions, perfluoro-α,ω-dibrominated telogens successfully cotelomerise VDF with HFP and TFE (and CTFE) [374].

Bromofluoroalkyl iodide [382] or α,ω-diiodoperfluoroalkanes [383] also allows the cotelomerisation of TFE with VDF and HFP.

Interestingly, the even less activated CH_2I_2 telogen was efficient in similar cotelomerisations, leading to fluorinated cotelomers which were peroxide vulcanised and moulded into a sheet showing good adhesion to metal plates through curable adhesives [384].

4.4. CONCLUSION

A large variety of telogens have been used in telomerisation of traditional fluoroolefins and numerous studies have been performed on the telomerisation of

VDF, TFE and CTFE. Fewer investigations have been carried out on trifluoroethylene and hexafluoropropene. The appropriate means of initiation depends upon the desired telomer. For instance, peroxide-induced telomerisation of VDF with CCl_4 leads to higher molecular weight telomers than when catalysed by copper salts, which appear more selective [34].

Besides the most important and commercially available fluoroalkenes, a large variety of different fluoroolefins have been successfully telomerised. Although most of them led mainly to monoadducts (mainly involved in basic research), a couple of them produced higher molecular weight telomers, and several of them have found interesting applications.

Vinyl fluoride, 3,3,3-trifluoropropene and 1,2-dichlorodifluoroethylene have shown great efficiency in telomerisation with a range of telogens, but less interest has been focused on other fluoroalkenes. This may be due to the lack of availability of monomers, and consequently studies on the monoadduct are more frequent. In spite of its toxicity, 1,1-dichlorodifluoroethylene is a special monomer which has been studied mainly for its monoadducts but it can be imagined that, because of steric hindrance of CCl_2–CCl_2 chaining, propagation may occur by tail-to-head addition, which could be of much interest.

Bromotrifluoroethylene, although less available than CTFE, can be easily telomerised, and the bulkiness of the bromine atom does not seem to hinder this reaction.

Hence, the synthesis of the telomers of these fluoroalkenes and, above all, their properties and their applications have not attracted research groups and it can be hoped that future investigations will be performed.

Cotelomerisation is an increasing means of modelling copolymerisation and of explaining the structure of complex copolymers. It can be expected that future studies will be carried out in order to design novel well-architectured polymers. In addition, further stepwise cotelomerisation that led to block cotelomers is reported in Chapter 4, Section 2.1.2.4.1.2.

5. Reactivity of Monomers and Telogens

5.1. INTRODUCTION AND THEORETICAL CONCEPTS

In the literature, examples of the reactivity of radicals with olefins are abundant. In Section 4, non-exhaustive examples dealing with the reactivity of radicals with various olefins have already been mentioned.

According to Tedder and Walton [128, 129, 173, 174, 306, 322], "no simple property can be used to determine the orientation of additions of free radicals" which depends upon a complex mixture of polar and steric parameters and of bond strengths.

As seen in Section 3, a telogen requires cleavage of a weak X–Y to be efficient in telomerisation. Hence, the BDE of the telogen X–Y (yielding X˙ and Y˙ radicals)

TABLE XXII

MEAN BOND DISSOCIATION ENERGY (BDE) OF VARIOUS CLASSIC CLEAVABLE BONDS
OF TELOGENS [21, 385, 386]

X	BDE (kJ mol^{-1} at 25 °C)					
	(X–H)	(X–F)	(X–Cl)	(X–Br)	(X–I)	(X–C)
H	432	561	427	362	295	416
C	416	486	326	285	213	350
N	391	272	193	–	–	303
O	463	193	205	201	201	368
F	561	155	–	–	191	452
Cl	427	–	239	218	208	330
Br	362	–	218	193	–	285
I	295	191	208	–	151	213
Si	318	565	381	310	234	–
P	322	490	326	264	184	–
S	326	285	255	218	–	255

has to be taken into account as an indicator of intrinsic reactivity. The BDE can be calculated according to the following equation, where $\Delta H_f^0 i$ designates the enthalpy of formation of species:

$$BDE_{XY} = \Delta H_f^0 X^{\cdot} + \Delta H_f^0 Y^{\cdot} - \Delta H_f^0 XY$$

Tables XXII and XXIII list the BDEs of various bonds that the transfer agent may contain [385, 386], enabling the efficiency of different telogens to be compared. For instance, as can be expected, the cleavage of a C–F bond is the most difficult of all to perform, whereas CF_2–I has a better chance to be cleaved (Table XXIII). The table shows that the bond strength cannot predict the reactivity, as seen by numerous examples.

Concerning the ability of the telogen to initiate the telomerisation, the reactivity and the obtained regioselectivity of the telogen radical $X^{\cdot}$ to the olefin can be driven from several rules. Actually, the radical $X^{\cdot}$ produced can be either nucleophilic (e.g., $RS^{\cdot}$ or CH_3 [387]) or electrophilic (e.g., C_nF_{2n+1}).

TABLE XXIII

RELATIVE RATE CONSTANT k OF THE PHOTODISSOCIATION OF R_FI
(ABOUT THAT OF CF_3I) AND CF_2–I BOND DISSOCIATION ENERGY

R_FI	k [404]	BDE (kJ mol^{-1}) [405]
CF_3I	1	220
C_2F_5I	8	212
n-C_3F_7I	15	206
i-C_3F_7I	49	206
n-C_4F_9I	–	202

Usually, fluoroalkenes are regarded as electron-poor olefins, and thus will react more easily with nucleophilic radicals.

However, in contrast to ionic addition to alkenes for which the regioselectivity is achieved according to a unique rule (Markovnikov's law), radical addition generally occurs at the less hindered carbon atom of the double bond. Tedder and Walton [128, 129, 306] suggested five rules which govern the radical additions related mainly to polar and steric effects—these latter being more important— and subsequently confirmed by Geise [388] in 1983 and then by Rüchardt [389] and Beckwith [390]. Other interesting theoretical surveys and concepts were provided by Koutecky *et al.* [391] and Canadell *et al.* [392], who suggested that the preferential attack of the radical occurs on the carbon which exhibits the highest linear combination of atomic orbitals coefficient in the highest occupied molecular orbital. In addition to this rule, Delbecq *et al.* [393] have considered a law of steric control (preferential attack occurs on the less substituted carbon) and a thermodynamic control rule (the most exothermic reaction is the easiest). More recently, Wong *et al.* [394] have shown that the exothermicity of the reaction is the main parameter governing the reactivity of the radical about the olefin.

In contrast to these concepts obtained from numerous ab initio molecular orbital calculations (usually from small radicals, e.g., $H^{.}$, $F^{.}$, $H_3C^{.}$ or $HOCH_2^{.}$ to fluoroalkenes), some semi-empirical calculation on longer halogenated radicals was performed by Roshkov *et al.* [395, 396] and Xu *et al.* [397]. The former team determined equilibrium geometries and electronic properties of perfluoroalkyl halogenides and showed that fluorine atoms in vinyl position strongly stabilise all the sigma molecular orbitals.

The Chinese group [397] proposed the electronic structures of 17 perchloro-fluoroolefins and perfluoroolefins via MNDO calculation. They have shown that the direction of the nucleophilic attack is governed not only by the perturbation energy of the ground state, but also by the stability of the anionic intermediate and the activation energy of the reaction. They suggested the following decreasing reactivity series:

$$F_2C{=}CFR_F > TFE > R_FCF{=}CFCl > R_FCF{=}CClR_F' > R_FCF{=}CFR_F'$$
$$> R_FCF{=}CCl_2$$

In the case of the addition of halogenated free radicals to fluoroalkenes, investigations were performed by Tedder and Walton. For example, this team has extensively studied the kinetics of addition of $H_3C^{.}$ [163, 398], $Cl_3C^{.}$ [399], $Br_3C^{.}$ [35, 160], $Br_2CF^{.}$ [35], $CF_2Br^{.}$ [322, 400], $CH_2I^{.}$ [162], $CFHI^{.}$ [161], $F_3C^{.}$ [177, 325], $C_2F_5^{.}$ [325, 401], $nC_3F_7^{.}$ [325], $iC_3F_7^{.}$ [325, 402], $CF_3(CF_2)_n^{.}$ ($n = 6, 7$) [325], $(CF_3)_3C^{.}$ [305] to specific sites of unsymmetrical fluoroalkenes. They determined the rate constants of these radicals to both sites of these olefins and the Arrhenius parameters. The results were mainly correlated to the electrophilicities of radicals which add more easily to nucleophilic (or less electrophilic) alkenes. For example, Table XXIV illustrates the probability of addition of various halogenated radicals to VDF. It is noted that the more electrophilic the radicals

TABLE XXIV

PROBABILITIES OF ADDITION OF CHLORO- AND
FLUORORADICALS TO BOTH SITES OF VDF

Radicals	$H_2C=CF_2$	
$\dot{C}H_2F$	1	0.440
$\dot{C}HF_2$	1	0.150
$\dot{C}Cl_3$	1	0.012
$\dot{C}F_3$	1	0.032
$\dot{C}F_2CF_3$	1	0.011
$\dot{C}F_2CF_2CF_3$	1	0.009
$\dot{C}F(CF_3)_2$	1	0.001

(i.e., the more branched the perfluoroalkyl radical), the more selective the addition to the CH_2 site.

As a matter of fact, Dolbier [403] comprehensively reviewed the syntheses, structures, reactivities and chemistry of fluorine-containing free radicals in solution. He also reported the influence of the fluorinated substituents on the structure stability, the reactivities of (per)fluoro-n-alkyl and branched chain (per)fluoroalkyl radicals, and supplied an interesting strategic overview of all aspects of organofluorine radical chemistry. In addition, this review also gives a summary of relative rates of addition of various radicals onto some fluoroethylenes, and especially the following increasing series of electrophilicity first suggested by El Soueni *et al.* [326]:

$$CF_3 < CF_3\dot{C}F_2 < C_2F_5\dot{C}F_2 < nC_3F_7\dot{C}F_2 < nC_5F_{11}\dot{C}F_2 < (CF_3)_2\dot{C}F < (CF_3)_3\dot{C}$$

This is in good agreement with the relative rate constants of the photodissociation of perfluoroalkyl iodides into $\dot{R}_F$ and $\dot{I}$ [404] linked also to the BDE [405] as shown in Table XXIII.

5.2. REACTIVITY OF TELOGENS

The following examples may illustrate the theories above.

First, if one considers such a concept applied to VDF, according to Haupstchein *et al.* [141, 148], the monoadduct coming from the radical addition of $\dot{R}_F$ to VDF is composed of two isomers $R_FCH_2CF_2I$ (95%) and $R_FCF_2CH_2I$ (5%). This result was confirmed by Chambers [38, 150] from i-C_3F_7I. However, from recent investigations, we have demonstrated by ^1H and ^{19}F NMR of the $R_F(C_2H_2F_2)I$ monoadducts and of the $R_F(C_2H_2F_2)H$ (produced by reduction of these monoadducts), that $C_4F_9CH_2CF_2I$, $C_6F_{13}CH_2CF_2I$, $C_8F_{17}CH_2CF_2I$ [39] and $IC_nF_{2n}CH_2CF_2I$ ($n = 2, 4, 6$) [152] were the only isomers produced by thermal telomerisation of VDF from the corresponding transfer agents. This was mainly explained by the selective addition of the electrophilic $\dot{R}_F$ radical to CH_2 group regarded as the less electrophilic site of VDF [161].

TABLE XXV

RADICAL TELOMERISATION OF TRIFLUOROETHYLENE WITH FLUOROALKYL
IODIDES; INFLUENCE OF THE STRUCTURE OF THE TELOGEN ONTO THE RATIO
OF NORMAL/REVERSE MONOADDUCT ISOMERS

Telogen	Normal adduct (%)	Reversed adduct (%)
$C_4F_9CH_2CF_2I$	60	40
C_4F_9I	75	25
$(CF_3)_2CFI$	90	10
$C_5F_{11}CF(CF_3)I$	93	7

However, for each case the diadduct was composed of two isomers $R_FCH_2CF_2CH_2CF_2I$ (92%) and $R_FCH_2CF_2CF_2CH_2I$ (8%) [39] because R_FCH_2-CF_2 appears less electrophilic than $R_F^{\cdot}$ radical.

Furthermore, the higher the temperature, the higher the telogen conversion, and the higher the $[R_FI]/[VDF]$ initial molar ratio, the more selective the telomerisation [39, 141, 148, 152]. Chambers $et\ al.$ [150] suggested the following reactivity series about VDF: $F_3C-I < C_2F_5-I < n\text{-}C_3F_7-I < i\text{-}C_3F_7-I$ mainly linked to the decrease of the strength of the C–I bond.

In the case of diiodides, a higher amount of α,ω-diadduct (50% in the case of $ICF_2CH_2(C_2F_4)CH_2CF_2I$ and 95% for $ICF_2CH_2(C_6F_{12})CH_2CF_2I$ [152]) is formed at greater lengths of $I(C_2F_4)_nI$ telogen.

With regards to the reactivity of free radicals produced from various telogens (especially (per)fluoroalkyl iodides) to trifluoroethylene, the electrophilic character of the (per)fluoroalkyl radical generated has a great influence on the proportion of reverse adduct (Table XXV). It was noted that the higher its electrophilicity, the higher the amount of "normal" isomer. For instance, from $C_4F_9CH_2CF_2I$ the normal/reverse ratio is 1.5, whereas it is 13.3 from $C_5F_{11}CFICF_3$ [41].

It is also worth comparing both the transfer constants of the telogens and the dissociation energy of the cleavable bond. Table XXVI lists these values

TABLE XXVI

TRANSFER CONSTANT VALUES OF TELOGENS VS. THE DISSOCIATION ENERGY OF THE CLEAVABLE BOND
(BDE) OF THE TELOGEN, FOR THE TELOMERISATION OF VINYLIDENE FLUORIDE

Transfer agent	C_T	T (°C)	BDE (kJ mol^{-1})	References
$H-CH_2OH$	0.008	140	411	[156, 184]
$H-CCl_3$	0.060	140	393	[25]
$Cl-CCl_3$	0.250	140	306	[25]
$H-P(O)(OEt)_2$	0.340	140	322	[170]
$Br-CFClCF_2Br$	1.30	75		[66]
$Br-CCl_3$	>30	140	234	[25]
$H-SC_2H_4OH$	>40	140	<340	[166]
$I-Cl$	Very high	25–100	208	[180]

compared to the nature of the transfer agent in the telomerisation of VDF. Values of C_T were assessed from the David and Gosselain law [75] (see Section 2.1.2) at low VDF conversion (for polyhalogenomethane [25], mercaptan [166], diethyl hydrogenophosphonate [170] or 1,2-dibromo-1-chlorotrifluoroethane [66]) or determined at high VDF conversions (for methanol [156]). It is clear that the higher the BDE, the lower the transfer constant and the higher the molecular weights of the telomer produced.

5.3. REACTIVITY OF FLUOROALKENES

Under the same conditions, two fluoroolefins are not supposed to react similarly with a given telogen. For instance, HFP requires much energy to react with a telogen compared to TFE or VDF and it homopolymerises with difficulty, in contrast to other fluoroalkenes. Several examples of compared reactivities of fluoroolefins in telomerisation involving the same telogen are given below.

5.3.1. Telomerisation of Fluoroalkenes with Methanol

Concerning methanol, in contrast to VDF which produces rather high molecular weights [155, 156] (the telomerisation of VDF involving a fivefold excess of methanol led to telomers of DP_{10-12} [156]) or trifluoroethylene [251] which gives telomers containing 1–10 base units, TFE [26, 29, 211], HFP [28], CTFE [278–280, 406] and difluorodichloroethylene ($CFCl=CFCl$ and $F_2C=CCl_2$) [343] have led mainly to monoadducts. We have tried to explain such results by showing, under semi-empirical calculations, that in the case of VDF, the propagation rate is higher than the transfer rate [134] (Table XXVII). Table XXVII lists the experimental conditions and the results obtained.

As a matter of fact, under radical initiation, methanol undergoes the cleavage of the C–H bond in the methyl end-group while an ionic initiation breaks the H–O bond [353, 407]. Kostov et al. [40] noted that the same mole number of methanol and tetrafluoroethylene led to the first four telomers and by using the David Gosselain law [75], they could assess the transfer constant to the methanol of n order (AIBN was chosen as the initiator at 70 °C and $C_T = 0.150$).

Recent investigations have dealt with the telomerisation of VDF [155, 156], of CTFE [278–280, 407] and of difluorodichloroethylene ($CFCl=CFCl$ and $F_2C=CCl_2$) [343] with methanol under peroxidic initiation as follows:

$$5\ CH_3OH + H_2C=CF_2 \longrightarrow HOCH_2(VDF)_nH \quad (n = 8-12)$$

$$40\ CH_3OH + F_2C=CFCl \longrightarrow HOCH_2CF_2CClFH\ (95\%)$$
$$+ HOCH_2CFClCF_2H\ (5\%)$$

$$40\ CH_3OH + FCCl=CFCl \longrightarrow HOCH_2CFClCFClH$$

$$60\ CH_3OH + F_2C=CCl_2 \longrightarrow HOCH_2CF_2CCl_2H$$

TABLE XXVII

RADICAL TELOMERISATION OF VARIOUS FLUOROALKENES WITH METHANOL. (EXPERIMENTAL CONDITIONS AND CHARACTERISTICS)

Fluoroalkene	Initiation conditions	R_0^a	Average units in telomers	$C_T^{\infty b}$	References
$F_2C{=}CF_2$	Peroxides	Variable	Variable	n.d.	[26, 211]
	UV	Variable	Variable	n.d.	[29]
	AIBN, 70 °C	1	1–4	0.150	[30]
$F_2C{=}CFCF_3$	Thermal, UV or peroxides	1	1	n.d.	[28]
$F_2C{=}CFCl$	DTBP, UV or γ-rays, peroxides	1–40	1	n.d.	[278, 279, 406]
$F_2C{=}CFCl$	DHBP, 134 °C, 7 h	40	1	n.d.	[280]
$F_2C{=}CFBr$	Radical			n.d.	[353]
$F_2C{=}CFH$	γ-rays	1	3–8	n.d.	[287]
$F_2C{=}CCl_2$	γ-rays	1	1	n.d.	[339]
$F_2C{=}CCl_2$	Peroxides	60	1	n.d.	[343]
$F_2C{=}CHCl$	Peroxides	40	1	n.d.	[343]
$F_2C{=}CH_2$	Peroxides	3	10–15	n.d.	[155]
$F_2C{=}CH_2$	Peroxides	5	8–12	0.008	[156, 184]
$F_3C{-}CH{=}CH_2$	γ-rays	>50		n.d.	[329b, d]
$FCCl{=}CFCl$	^{60}Co irradiations, 700 h	3	1	n.d.	[339, 340]
$FCCl{=}CFCl$	UV	5	1	n.d.	[341]
$FCCl{=}CFCl$	DTBP, 5 h	9	1	n.d.	[342]
$FCCl{=}CFCl$	DHBP, 134 °C, 7 h	40	1	n.d.	[343]
$FCCl{=}CCl_2$	γ-rays	1	1	n.d.	[339]
$F_3C{-}CF{=}CF{-}CF_3$	γ-rays	1	1	n.d.	[355]

n.d., not determined.

[a] R_0 denotes the $[\text{transfer agent}]_0/[\text{fluoroalkene}]_0$ initial molar ratio ($R_0 = [\text{MeOH}]_0/[\text{fluoroolefin}]_0$).

[b] Infinite transfer constant to methanol.

Although the initiation conditions and the stoichiometries varied from one study to another, it can be concluded that the higher the number of chlorine atoms borne by the ethylenic carbon atoms, the lower the average degree of telomerisation ($\overline{DP}_n$), and the higher the number of hydrogen atoms, the higher the $\overline{DP}_n$.

In contrast, TFE and VDF behave differently when they react with Cl_3CSO_2Br from the peroxide initiation, yielding $Cl_3C(C_2F_4)_nBr$ ($n = 1-10$) and Cl_3C $(C_2H_2F_2)_pBr$ ($p = 1, 2$), respectively [168].

Trifluoromethyl hypofluorite, known for achieving higher yields than those obtained from the corresponding hypochlorites, also behaves differently with respect to various fluoroalkenes. HFP leads to the monoadduct as expected [90], just like VDF [85, 86], 1,1-dichlorodifluoroethylene [91], 2-perfluorobutene [93], in contrast to the first three adducts or the first 10 telomers obtained from 1,2-dichlorodifluoroethylene [92] and CTFE [87–89, 92], respectively.

5.3.2. Radical Addition of ICl onto Fluoroalkenes

A typical example concerns the telomerisation of various fluoroalkenes with iodine monochloride. Indeed, the radical addition of ICl onto different F-olefins was achieved by many authors [59, 60, 180, 292, 293, 310]. However, in the case of asymmetric alkenes, these authors paid little attention to the presence of non-expected by-products arising from the reverse addition. Taking into account that the polarity of iodine monochloride is $^{\delta+}I \cdots Cl^{\delta-}$ and regarding polar, ionic and steric effects, some reversed isomers cannot be neglected.

That is why the comparison of the reactivity of various fluoroalkenes, such as chlorotrifluoroethylene [60], trifluoroethylene [292, 293], vinylidene fluoride [180], hexafluoropropylene [59, 310] and 1,1-difluorodichloroethylene [59], was achieved in the presence of iodine monochloride. In most cases, the different radical initiations (photochemical, thermal, or the use of redox catalysts or radical initiators) were investigated and the following results were obtained:

$$I-Cl + F_2C=\underset{\underset{CF_3}{|}}{C}F \longrightarrow ICF_2\underset{\underset{CF_3}{|}}{C}FCl \ (40\%) + ICFCF_2Cl \ (60\%)$$

$$I-Cl + F_2C=CFCl \rightarrow ICF_2CFCl_2 \ (8-11\%) + ICFClCF_2Cl \ (89-92\%)$$

$$I-Cl + F_2C=CFH \rightarrow ICF_2CFHCl \ (5\%) + ICFHCF_2Cl \ (95\%)$$

$$I-Cl + F_2C=CH_2 \rightarrow ICF_2CH_2Cl \ (0-1\%) + ICH_2CF_2Cl \ (99-100\%)$$

$$I-Cl + F_2C=CCl_2 \rightarrow ICF_2CCl_3 \ (100\%) + ICCl_2CF_2Cl \ (0\%)$$

Interestingly, it is observed that the lower the fluorine content, the higher the selectivity.

The monoadduct was obtained in all cases, leading to original activated telogens containing CF_2I or, better, $CFClI$ end-groups [59, 60, 180]. The addition of ICl to 1,1-dichlorodifluoroethylene was found to be bidirectional, but to a lesser extent

[59] than in the case of CTFE [60]; the use of a catalyst (e.g., iron) affects the yield and the amount of by-product. HFP requires heat, unlike other monomers [59]. Hence, the following decreasing reactivity series can be suggested:

$$CTFE > VDF > CF_2=CCl_2 > HFP$$

Perfluoroalkyl iodides have been used as telogens with most fluoroalkenes. An extensive kinetic research on the synthesis of monoadducts was performed by Tedder and Walton. From various results in the literature [39–42, 205, 286], the following decreasing reactivity scale may be proposed:

$$VDF > TFE > F_2C=CFH > HFP > CTFE$$

and for less reactive monomers, the more electrophilic the telogen radical, the less easier the reaction.

5.3.3. Telomerisation of fluoroalkenes with dialkyl hydrogen phosphonates (or dialkyl phosphites)

Various telomerisations of fluoroolefins with dialkyl hydrogenophosphonates were investigated (Table XXVIII). Different experimental conditions (sometimes specific, such as those using γ irradiations) were used, several authors looking for the monoadduct only. Hence, it is difficult to give a real comparison on the reactivity of these fluoroalkenes. However, few kinetics were studied, only those concerning the telomerisation of chlorotrifluoroethylene [193] and vinylidene

TABLE XXVIII

RADICAL TELOMERISATION OF VARIOUS FLUOROALKENES WITH DIALKYL HYDROGEN PHOSPHONATE (OR DIALKYL PHOSPHITE) AND CHARACTERISTICS

Fluoroalkene	Initiation conditions	R_0^a	$\overline{DP}_n$	$C_T^{\infty b}$	References
$F_2C=CF_2$	AIBN	0.1–2	$n = 1$–10	n.d.	[51]
	Peroxide	–	$n = 1$–6	n.d.	[218]
$F_2C=CFCF_3$	Peroxide, thermal	–	$n = 1$	n.d.	[169]
$F_2C=CFCl$	Peroxides, 140 °C	0.3–2.0	$n = 1$–10	0.34	[187]
$F_2C=CH_2$	DTBP, 140 °C	0.1–2.0	$n = 1$–15	0.34	[170]
	DBP, 92 °C				
	AIBN, 80 °C				
FClC=CFCl	γ-rays, 9.3 d	3.0	–	n.d.	[346]
$F_2C=CCl_2$	γ-rays, 16 d	2.1	–	n.d.	[346]
$CFCl=CCl_2$	γ-rays, 15 d	2.0	–	n.d.	[346]
$H_2C=CHCF_3$	DTBP, 130 °C	1.0	Monoadduct (39%)	n.d.	[169]
$H_2C=CHR_F$	Peroxide, AIBN	1.0	$n = 1$–3	n.d.	[337]

n.d., not determined.
$^a R_0 = [HP(O)(OR)_2]_0/[\text{fluoroalkene}]_0$.
$^b C_T^{\infty}$ means the infinite transfer constant to $HP(O)(OR)_2$.

fluoride [208] initiated by peroxides. Interestingly, the assessments of the infinite transfer constants to diethyl hydrogenophosphonate were 0.34 in both cases (at 140 °C).

5.4. CONCLUSION

Dissociation energy of cleavable bonds of telogens appears to be a key parameter to take into account as the starting point of the reactivity. Then, various factors controlling the reactivity (energetic, electronic, polar, steric, conformational effects, role of orbitalar interactions, etc.) may affect the reactivity of a radical and its sense of addition.

However, three targets deserve to be investigated in telomerisation: the regioselectivity, the tacticity and the molecular weight of telomers. For the first one, Tedder and Walton have extensively explored numerous additions of various radicals to different olefins, especially to determine the kinetics of the formation of monoadducts. To achieve desired molecular weights, even if some trends have been given from the different results of telomerisation of fluoroalkenes described in the literature, it is still difficult to predict which olefin could be more easily telomerised than another one, even by a chosen initiation method. As a matter of fact, in the case of halogenated transfer agents, an increasing reactivity series can be proposed:

$$CFH-Br < CFCl-Br < CF_2-I < CFCl-I$$

6. Radical Living or Controlled Telomerisation and Polymerisation of Fluoromonomers

6.1. INTRODUCTION

Since the mid-1990s, controlled radical polymerisation (CRP) has drawn much interest from both academic and industrial researchers. To date, new techniques have been proposed and developed to control the reactivity of free radicals. Such a control may lead to giving a "living" character to the radical polymerisation. However, despite considerable progress, the truly living character is far from being attained and it seems preferable to use the term "controlled" process, rather than "living" process. Typically, such polymerisations can be controlled by the use of specific counter radicals [408], usually organic ($^{\bullet}$CR) issued from nitroxyl, alkoxyamines, arylazooxyls compounds or triphenyl methyl (called trityl) groups, CPh_2-G where G represents $OSi(CH_3)_3$, OPh or CN, or sulfurated radical $^{\bullet}SR$ with R = alkyl, aryl, $C(O)R'$ or $C(S)R''$, where R'' represents R', OR', NR'_2 (R' being an alkyl group) and organometallic complexes (C) able to generate stable free radicals.

Equilibrium may be written as follows:

$$I-M_n^{\cdot} + {}^{\cdot}CR \underset{\text{or UV}}{\overset{\Delta}{\rightleftarrows}} I-M_n-CR$$

$$I-M_n-M^{\cdot} + C \underset{\text{or UV}}{\overset{\Delta}{\rightleftarrows}} I-M_n-(MC)^{\cdot}$$

where I and M represent the initiator radical and the monomer, respectively.

The organometallic complexes are quite various: for example, cobaltocene/bis(ethyl acetoacetato) copper (II), CuCl or CuBr/bipyridyl, cobaltoxime complex, reduced nickel/halide system, organoborane and ruthenium complex/trialkoxyaluminium system [408].

Such chemical equations as these show the essential difference between the traditional radical polymerisation and the controlled radical polymerisation in the presence of counter radicals or metallic complexes.

Difficulties arise from the various possible interactions between the counter radical, the initiator and the monomer. The truly living character is only demonstrated when some requirements are fulfilled. Molecular weight must increase in a linear fashion with the monomer conversion. Polydispersity must be narrow and lower than that in the classical process (theoretical value, $\overline{M}_\omega/\overline{M}_n = 1.1-1.5$), which supposes, in particular, a rapid initiation step. The obtaining of high and strictly controlled molecular weight must be possible. The controlled process can also be evidenced by the preparation of block copolymers. Various methods to synthesise fluorinated block copolymers by radical polymerisation or telomerisation are reviewed in Chapter 4, Section 2.1.2, which also extensively reports the different possibilities for using CRP.

According to the literature, the only examples of controlled radical telomerisation of fluoroolefins require adequate fluorinated telogens. The oldest method is based on the cleavage of the C–I bond that has already led to industrial applications.

The cleavage of the C–I bond can be achieved by various methods. However, according to well-chosen monomers, two principal ways have been developed in order to control telomerisation from alkyl iodides: the ITP or degenerative transfer [376–379] and sequential telomerisation as mentioned in Section 4.3. Interestingly, ITP can be easily applied to fluorinated monomers.

6.2. Iodine Transfer Polymerisation

ITP was one of the radical living processes developed in the late 1970s by Tatemoto at the Daikin company [377, 409]. Actually, it is required to use (per)fluoroalkyl iodides because their highly electron-withdrawing (per)fluorinated group R_F allows the lowest level of the CF_2-I BDE. Such a C–I cleavage is also possible in $R_F CH_2 CH_2 I$ [379a]. Various fluorinated monomers have been successfully used in ITP [376–379, 409]. Basic similarities in these living polymerisation systems are found in the stepwise growth of polymeric chains at each active species [408]. The active living centre, generally located at the end-groups of the growing polymer, has the same reactivity at any time during polymerisation, even when the reaction is stopped. In the case of ITP of fluoroolefins, the terminal active bond is always the $C-I$ bond originated from the initial iodine-containing chain transfer agent and monomer, as follows:

$$C_nF_{2n+1}-I + (p+1)H_2C=CF_2 \xrightarrow{R^\cdot \text{ or } \Delta} C_nF_{2n+1}-(C_2H_2F_2)_p-CH_2CF_2-I$$

Nowadays, this concept is still applied at the Dai-Act company (for preparing Dai-el® thermoplastic elastomers) and has also been used by the Solvay Solexis (formerly Ausimont S.p.A.) and DuPont companies for the production of Tecnoflon® [376] and Viton® [379], respectively (see Chapter 4, Section 2.1.2.4.1).

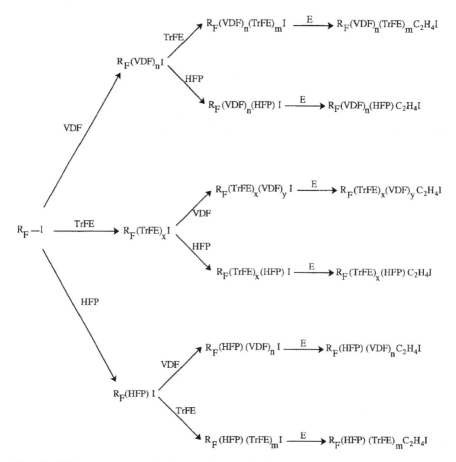

Fig. 5. Various routes to obtain fluorinated telomers containing vinylidene fluoride (VDF), hexafluoropropylene (HFP), trifluoroethylene (TrFE) and ethylene (E) by stepwise cotelomerisation of VDF, HFP and TrFE (reproduced with permission from J. Balagué *et al.*, *J. Fluorine Chem.*, **102**, 253–268 (2000)).

This concept was also exploited in order to prepare "living" and well-defined tetrafluoroethylene telomers in which the telomer produced acts as a further telogen, in a sequential telomerisation, as follows:

$$C_2F_5I \xrightarrow{C_2F_4} C_4F_9I \xrightarrow{C_2F_4} C_6F_{13}I \xrightarrow{C_2F_4} C_8F_{17}I \longrightarrow C_nF_{2n+1}I$$

Such a living telomerisation can be initiated either thermally or in the presence of radical initiators or redox catalysts [410]. Similarly, by stepwise cotelomerisations of VDF, HFP and TrFE, a wide variety of highly fluorinated telomers of various structures were obtained (Figure 5). Interestingly, the position of a branched CF_3 group arising from an HFP unit drastically affects the crystallinity of the telomers (e.g., a telomer that exhibits a C_7F_{15} chain is crystalline while a $C_5F_{11}CF(CF_3)$ is totally amorphous [40]).

6.3. CONCLUSION

Even if, in 1955, Haszeldine [267] or, in 1958, Hauptschein et al. [59, 311] started to show a certain livingness of the radical telomerisation of CTFE with $ClCF_2CFClI$, the up-to-date research has attracted many academic or industrial chemists towards such a fascinating area. In addition, Dear and Guilbert [47] and Sharp et al. [46, 48] performed the telomerisation of various fluoroolefins with disulfides but they did not examine the living character of this reaction.

However, the stepwise cotelomerisation of fluoroalkenes has already shown original livingness [40, 410], and the polymerisation via organometallic systems, which is quite successful for hydrogenated monomers, could be an encouraging route for these halogenated olefins.

7. Applications of Fluorinated Telomers

The applications of fluorinated telomers or fluorooligomers are linked to their exceptional properties [1–18]: excellent thermostability, remarkable surface properties, low refractive indexes, low dielectric constants and very high chemical stabilities.

However, the only application of the oligomers, obtained directly, is lubricant oils, for example Kel F® oils but, generally, the oligomers need some chemical transformations in order to lead to applications. In this latter case, the number of functional groups (generally one or two) and the nature of the linkage between the functional group and the fluorinated chain play an important role in these kinds of applications. Several examples of non-functional, monofunctional and telechelic telomers are given below.

7.1. From Non-Functional Telomers Mainly Used as Lubricants

Three series of products are used as lubricants: silicone-oils or poly(trifluoro-propyl methyl siloxane) (Silastic® produced by the Dow Corning company [411]), perfluorinated polyethers (Demnum® [106], Krytox® [103–105], Fomblin® [110] and Aflunox® marketed by Daikin, DuPont, Ausimont and nowadays by Solvay Solexis and Unimatech, formerly Nippon Mekktron, respectively), and Kel F® oils (3M Co.). Only the last series of products are obtained directly by telomerisation.

Initially, the telomerisation of CTFE with SO_2Cl_2 was used by the Kellogg company to prepare $Cl-(CF_2-CFCl)_n-Cl$ thermostable oligomers [277]. A few years later, inspired by the redox telomerisation of CTFE with CCl_4 [32], the Occidental Chemical company prepared $CCl_3-(CF_2-CFCl)_n-Cl$ telomers subsequently fluorinated by ClF_3 in order to increase their thermal stability [412].

Recently, other new methods have been investigated by Marraccini [87, 95] using CF_3OF as the telogen for the production of original $CF_3O(C_2F_3Cl)_nF$ telomers (see Section 4.1.3).

However, research programs in this area are difficult to develop because of several chemical problems: e.g., inversion of addition in CTFE oligomerisation which leads to a decrease of thermal stability brought about by $-CFCl-CFCl-$ linkages and, above all nowadays, the availability of CTFE because the production of chlorofluorinated ethane $CF_2Cl-CFCl_2$ stopped a few years ago. In addition, perfluorooctyl bromide exhibits interesting biomedical applications and is likely to be used as a blood substitute [241].

7.2. From Monofunctional Telomers

Monofunctional oligomers can be synthesised either by chemical change of the end-groups [153] or by direct telomerisation with a functional telogen (e.g., $Cl_3CCO_2CH_3$ [34], HOC_2H_4SH [44], CH_3OH [155, 156], $HP(O)(OEt)_2$ [45, 169, 170]). The other method is the electrochemical fluorination of carboxylic or sulfonic hydrogenated acid [413].

Extensive chemical reactions, starting from TFE telomers, have been proposed in Chapter 5, Section 2.1 for the synthesis of fluorinated alcohols (and consequently (meth)acrylates), mercaptans, amines, vinyl ethers, vinyl esters, maleimides, styrenics, oxetanes and other fluoromonomers.

In any case, the field of applications is mainly devoted to surfactants [15a, 203c] and numerous compounds are commercially available (anionic, cationic, zwitterionic, non-ionic but also mono- or polydispersed compounds).

One of the well-known applications concerns the fire-fighting foam agents (A3F) [15b], but there is also the use of fluorosurfactants in emulsions (for fluoromonomers) and microemulsions, or in polymerisations of hydrogenated monomers (e.g., styrene and MMA) in supercritical CO_2.

In addition, these fluorinated monofunctional telomers can be interesting precursors of original macromonomers [155, 381], leading to novel polymeric optical fibres [414] or grafted copolymers (Chapter 5).

Monofunctional oligomers are also used as precursors of numerous monomers. One of the best-known ones is the series of fluoroacrylates for which the synthesis, the properties and the applications have been described in detail [13a]. Their use for the protection of textiles has challenged several companies: Asahiguard® (Asahi Chem. Corp.), Foraperle® (Atofina up to 2002 and nowadays DuPont that also produces Zonyl®) and Lezanova® (Daikin), while Scotchguard® was withdrawn a few years ago. These products are also efficient for the protection of leathers [9], stones [10] and in coatings with high weatherability. The second important application of fluoroacrylates concerns materials for optics [11]: optical fibres (core and cladding), optical components (guides) and optical glasses. The third is the coating of amorphous silicon in order to prepare positive photoresists for electronics. Other minor applications concern fluoroelastomers [5], solid propellants and dental materials [381].

Other relevant strategies deal with the preparation of original crosslinkable or curable fluoroelastomers [5]. For example, $C_6F_{13}CH=CF_2$ (obtained from the dehydroiodination of $C_6F_{13}CH_2CF_2I$ arising from the telomerisation of VDF with $C_6F_{13}I$ [149]) and $F_2C=CF(CH_2)_xR$ (with $x = 1-3$ and R: H, OAc or SAc, and synthesised from a straightforward way via the radical addition of 1-chloro-2-iodotrifluoroethane onto allyl alcohol [415]) were copolymerised with commercially available fluoroalkenes (such as VDF, TFE, HFP or PMVE). The resulting copolymers were vulcanised either in the presence of bisphenol AF [149] or with non-conjugated dienes [415b], leading to novel fluoroelastomers with excellent processability [149].

Two other families of polymers with fluorinated side groups have been investigated in the field of elastomers: polysiloxanes and polyphosphazenes [416, 417]. These latter macromolecules, called NPF, have been obtained by chemical modification of poly(dichlorophosphazene) as for classic phosphazenes:

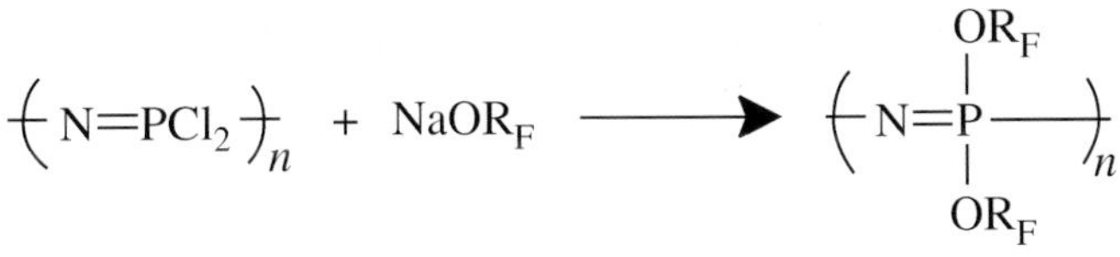

$$\left(N{=}PCl_2\right)_n + NaOR_F \longrightarrow \left(N{=}\overset{\displaystyle OR_F}{\underset{\displaystyle OR_F}{P}}\right)_n$$

The most well known, and marketed by Firestone, is the following:

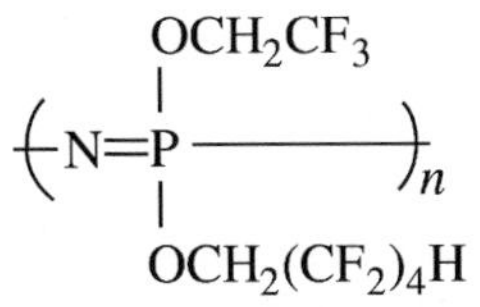

$$\left(N{=}\overset{\displaystyle OCH_2CF_3}{\underset{\displaystyle OCH_2(CF_2)_4H}{P}}\right)_n$$

It is obtained with a higher molecular weight ($\overline{M}_n > 10^6$) and its glass transition temperature is close to $-68\,°C$ [417]. Interestingly, this fluoropolyphosphazene preserves its good properties up to 175 °C and exhibits excellent chemical inertness. However, the synthesis of such a polymer is not easy (mainly because of corrosion) and the purification of the trimer precursor is difficult.

Concerning fluorinated polysiloxanes, the most common and commercially available (by Dow Corning) has the following formula:

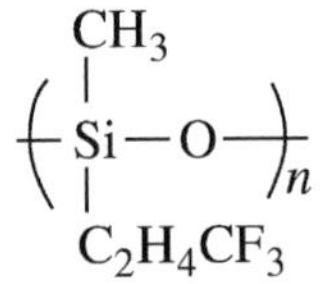

$$\left(\begin{array}{c} CH_3 \\ | \\ Si-O \\ | \\ C_2H_4CF_3 \end{array} \right)_n$$

It is synthesised from the corresponding D_3 cyclosiloxane and the corresponding polymers exhibit T_gs of -68 °C. Such a polymer can easily be crosslinked by various processes of silicone chemistry (e.g., peroxides, SiH or Si–vinyl derivatives). Fluorosilicones can be used for their resistance to swelling in non-polar medium (e.g., Sifel® fluids, see Chapter 4, Section 2.3) event at 100 °C; but their main application is to replace elastomers of fluorinated monomers at low temperatures.

Furthermore, important and interesting investigations have been performed by Kobayashi and Owen [418] who prepared fluorosilicones with perfluorinated lateral group:

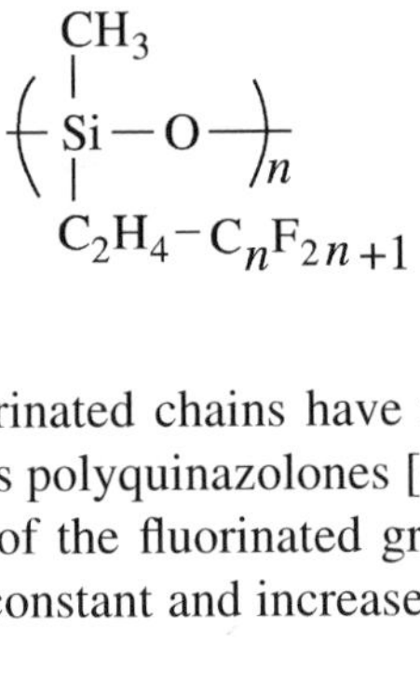

$$\left(\begin{array}{c} CH_3 \\ | \\ Si-O \\ | \\ C_2H_4-C_nF_{2n+1} \end{array} \right)_n$$

with $n = 1-8$.

At the same time, the fluorinated chains have found applications in various thermostable polymers such as polyquinazolones [419] and polyimides [420]. In this latter case, the presence of the fluorinated group decreases both the water absorption and the dielectric constant and increases the solubility in solvents for coatings.

7.3. FROM TELECHELIC OLIGOMERS

Telechelic (or α,ω-difunctional) oligomers exhibit functional groups at both ends of the oligomeric backbone. These compounds are quite useful precursors for well-defined architectured polymers (e.g., polycondensates, polyadducts and other fluoropolymers previously reviewed [421]). As mentioned above, a large effort has been made over the last 10 years in order to prepare new telechelic fluorinated telomers. These investigations have provided compounds other than perfluoro polyethers (Krytox®, Fomblin®, Aflunox® and Demnum®).

Thus, according to appropriate strategies, from fluorinated α,ω-diiodinated telomers or their derivatives, various α,ω-dienes, diols, diacids, diesters, diacrylates, dicyanates and diamines have been prepared. Interestingly, fluorinated polyurethanes, polyesters, polyamides, polyimides, silicones and other fluorinated networks from dicyanates and diacrylates were synthesised. Indeed, their properties depend upon the position of the fluorinated chain: when it is located in the polymeric backbone, thermal stability and chemical inertness are obtained, while surface properties result from polymers containing fluorinated side chains. Extensive information can be found in Chapter 2.

In addition, α,ω-diiodofluoroalkanes are relevant precursors of triblock copolymers [376–379] as original thermoplastic elastomers (comprehensively described in Chapter 4, Section 2.2.2.2).

7.4. FROM MULTIFUNCTIONAL TELOMERS

One example is reported in a Japanese patent [422] which claims the tetratelomerisation of VDF, CTFE, HFP and vinyl acetate with methanol. The hydrolysis of the acetoxy side groups in the resulting tetratelomers led to fluorinated polyalcohols. These latter were methacrylated and then polymerised in the presence of organic peroxide to produce cured elastomeric fluoropolymers.

8. Conclusion

This chapter summarises the important number of investigations performed on the telomerisation of well-known fluorinated alkenes. These fluorotelomers allow synthesis of novel fluoropolymers regarded as high value-added materials usually with exceptional properties.

The telomerisation reaction appears a powerful tool to prepare well-defined molecules and their well-characterised end-groups allow further key reactions. However, an adequate method of initiation has to be chosen and, even if traditional initiations are still up to date, recent processes in clean (or green) media appear promising (e.g., in supercritical fluids).

In addition, it is an elegant model of (co)polymerisation and the resulting telomers can also be used for the characterisation of higher molecular weight polymers: they can act as original standards for GPC or can be relevant in the assignments of chemical shifts in NMR spectroscopy (e.g., to determine the defects of chaining and the end-groups in fluoropolymers [139]).

However, it is known that some defects in chaining may enhance dramatic issues (e.g., in polychlorofluorinated oils, two vicinal chlorine atoms induce weak points which affect the thermostability).

Fortunately, much work is still being developed, especially to provide a better knowledge of the synthesis of tailor-made polymers with well-defined architecture in order to improve the properties by better control of the regioselectivity and of the tacticity. Even if much work on the reactivity has been carried out by Tedder and Walton or, more recently, by Dolbier, methods are sought for a better orientation of the sense of addition, since most fluoroalkenes are unsymmetrical.

In our opinion, two main processes should bring original solutions. The first one concerns the step-by-step or sequential addition of fluoroalkenes (i.e., homostepwise telomerisation as described by Tatemoto and by researchers at Solvay Solexis and at DuPont) in order to favour the generated radical to react on the same site of the olefin.

The second opportunity deals with the use of organometallic systems which may allow a very good control of both the regioselectivity and the tacticity. This kind of polymerisation has been used successfully on hydrogenated monomers but has not yet been investigated on fluoroalkenes.

Hence, such systems should induce interesting consequences on the structures and the properties of materials thus obtained. For example, in the case of fluoroacrylic monomers, it is well known that the crystallisation of perfluorinated chains ensures improvement of surface properties. In addition, by introducing well-defined fluorotelomers in the backbone or in the side chains, building block silicones can be synthesised with monitored thermal properties, especially at low temperatures.

These telomers are also interesting precursors of various fluoropolymers (such as telechelics and alternating, block or graft copolymers). These special well-architectured fluoropolymers are described in Chapters 2–5.

References

1. (a) Wall, L. (1972) *Fluoropolymers*, Wiley, New York; (b) Fiering, A. E. (1994) Fluoroplastics. In *Organofluorine Chemistry: Principles and Applications*, Vol. 15 (Eds, Banks, R. E., Smart, B. and Tatlow, J. C.) Plenum Press, New York, pp. 339–362; (c) Johns, K. and Stead, G. (2000) Fluoroproducts—the extremophiles. *J. Fluorine Chem.*, **104**, 5–18.

2. (a) Scheirs, J. (1997) *Modern Fluoropolymers*, Wiley, New York; (b) Ajroldi, G. (1997) Structural and physical properties of some fluoropolymers. *Chim. l'Industria*, **79**, 483–487.

3. Hougham, G., Cassidy, P., Jonhs, K. and Davidson, T. (1999) *Fluoropolymers*, Kluvert, New York.

4. (a) Ebnesajjad, S. (2000) *Fluoroplastics, Melt Processible Fluoroplastics*, Plastic Design Library. Handbook Series, Vol. 1, Norwich, New York; (b) Ebnesajjad, S. (2003) *Fluoroplastics, Non-Melt Processible Fluoroplastics*, Plastic Design Library. Handbook Series, Vol. 2, Norwich, New York; (c) Imae, T. (2003) Fluorinated polymers. *Curr. Opin. Colloids Interface Sci.*, **8**, 308–325.

5. Ameduri, B., Boutevin, B. and Kostov, G. (2001) Fluoroelastomers: synthesis, properties and applications. *Prog. Polym. Sci.*, **26**, 105–187.

6. (a) Vecellio, M. (2000) Opportunities and developments in fluoropolymeric coatings. *Prog. Org. Coat.*, **40**, 225–242; (b) Wood, K. (2002) The effect of the fluoropolymer architecture on the exterior weathering of coatings. *Macromol. Symp.*, **187**, 469–479.

7. Castelvetro, V., Aglietto, M., Ciardelli, F., Chiantore, O., Lazzari, M. and Toniolo, L. (2002) Structure control, coating properties, and durability of fluorinated acrylic-based polymers. *J. Coat. Technol.*, **74**, 57–66.

8. (a) Bongiovanni, R., Montefusco, F., Priola, A., Macchioni, N., Lazzeri, S., Sozzi, L. and Ameduri, B. (2002) High performance UV-cured coatings for wood protection. *Prog. Org. Coat.*, **45**, 359–363; (b) Bongiovanni, R., Montefusco, F., Priola, A., Macchioni, N., Lazzeri, S., Sozzi, L. and Ameduri, B. (2003) Photocurable wood coatings containing fluorinated monomers. *Eur. Coat., Pitture Vernici*, **11**, 25–30.

9. Jariwala, C. P., Eggleston, J. D., Yandrasit, S. M. A. and Dams, R. J. (2003) Alkylated fluorochemical oligomers and use thereof, US Patent 6,288,157 (assigned to 3M).

10. (a) Ciardelli, F., Aglietto, M., Montagnini di Mirabello, L., Passaglia, E., Gianscristoforo, S., Castelvetro, V. and Ruggeri, G. (1997) New fluorinated acrylic polymers for improving weatherability of building stone materials. *Prog. Org. Coat.*, **32**, 43–50; (b) Castelvetro, V., Aglietto, M., Ciardelli, F., Chiantore, O. and Lazzari, M. (2001) Design for fluorinated acrylic-based polymers as water repellent intrinsically photostable coating materials for stone. In *Proc. Am. Chem. Soc. Symp. Ser. 787* (Eds, Castner, D. G. and Grainger, D. W.) Fluorinated Surfaces, Coatings, and Films, Washington, DC. pp. 129–142, Chap. 10.

11. (a) Head, R. A. and Johnson, S. (1988) Coating compositions for optical fibers. Eur. Patent Appl. 260,842 (assigned to Imperial Chem. Industries PLC, UK); (b) Barraud, J., Gervat, S., Ratovelomanana, V., Boutevin, B., Parisi, J. P., Cahuzac, A. and Octeur, R. J. (1992) French Patent 92 04,222 (assigned to Alcatel); (c) Schuman, P. D. (1996) Curable, inter-polymer optical fiber cladding compositions. PCT Int. Appl. WO 96 03,609 Al (assigned to Optical Polymer Research, Inc.).

12. Youngblood, J. P., Andruzzi, L., Ober, C. K., Hexemer, A., Kramer, E. J., Callow, J. A., Finlay, J. A. and Callow, M. E. (2003) Coatings based on side-chain ether-linked poly(ethylene glycol) and fluorocarbon polymers for the control of marine biofouling. *Biofouling*, **19**, 91–98.

13. (a) Boutevin, B. and Pietrasanta, Y. (1988) Les Acrylates et Polyacrylates Fluorés: Dérivés et Applications. Puteaux (Fr): Erec; (b) Timperley, C. M., Arbon, R. E., Bird, M., Brewer, S. A., Parry, M. W., Sellers, D. J. and Willis, C. R. (2003) Bis(fluoroalkyl)acrylic and methacrylic phosphate monomers, their polymers and some of their properties. *J. Fluorine Chem.*, **121**, 23–31.

14. (a) Doyle, M. and Rajendran, G. (2003) Perfluorinated Membranes. In *Handbook of Fuel Cells— Fundamentals, Technology and Applications; Fuel Cell Technology and Applications*, Vol. 3 (Eds, Vielstich, W., Hubert, A., Gasteiger, M. and Lamm, A.) pp. 351–395, Chap. 30
(b) Arcella, V., Ghielmi, A. and Tommasi, G. (2003) High performance perfluoropolymer films for membranes. *Ann. NY Acad. Sci.*, **984**, 226–244.

15. (a) Kissa, E. (1994) *Fluorinated Surfactants: Synthesis, Preparations, Applications*, Marcel Dekker, New York; (b) Pabon, M. and Corpart, J. M. (2002) Fluorinated surfactants: synthesis, properties, effluent treatment. *J. Fluorine Chem.*, **114**, 149–156.

16. (a) Ober, C. K., Douki, K., Vohra, V. R., Kwark, Y.-J., Liu, X.-Q., Conley, W., Miller, D. and Zimmerman, P. (2002) New strategies for high resolution photoresists. *J. Photopolym. Sci. Technol.*, **15**, 603–611; (b) Kishimura, S., Endo, M. and Sasago, M. (2002) Dissolution characteristics of acidic groups for 157-nm resists. *J. Photopolym. Sci. Technol.*, **15**, 625–628; (c) Feiring, A. E., Wonchoba, E. R., Fischel, B. E., Thieu, T. V. and Nassirpour, M. R. (2002) Amorphous fluoropolymers from tetrafluoroethylene and bulky vinyl esters or vinyl ethers. *J. Fluorine Chem.*, **118**, 95–98; (d) Vohra, V. R., Liu, X. Q., Douki, K., Ober, C. K., Conley, W., Zimmerman, P. and Miller, D. (2003) Fluoropolymer resists for 157 nm lithography. *Proc. SPIE Int. Soc. Opt. Eng.*, **5039**, 539–547; (e) Feiring, A. E., Crawford, M. K., Farnham, W. B., Feldman, J., French, R. H., Leffew, K. W., Petrov, V. A., Schadt, F. L. III, Wheland, R. C. and Zumsteg, F. C. (2003) Design of very transparent fluoropolymer resists for semiconductor manufacture at 157 nm. *J. Fluorine Chem.*, **122**, 11–16.

17. Krebs, F. C. and Jensen, T. (2003) Fluorinated molecules relevant to conducting polymer research. *J. Fluorine Chem.*, **120**, 77–84.

18. (a) Krafft, M. P., Riess, J. G. and Weers, J. G. (1998) The design and engineering of oxygen-delivering fluorocarbon emulsions. In *Submicron Emulsion in Drug Targeting and Delivery* (Ed, Benita, S.) Harwood Academic Publishers, Toronto, pp. 235–333, Chap. 10; (b) Krafft, M. P. and Riess, J. G. (1998) Highly fluorinated amphiphiles and colloidal system, and their applications in the biomedical field. A contribution. *Biochimie*, **80**, 489–514; (c) Krafft, M. P. and Riess, J. G. (2003) Fluorinated colloids and interfaces. *Curr. Opin. Colloids Interface Sci.*, **8**, 213–214; (d) Krafft, M. P., Chittofrati, A. and Riess, J. G. (2003) Emulsions and microemulsions with a fluorocarbon phase. *Curr. Opin. Colloids Interface Sci.*, **8**, 251–258.

19. Hanford, W. E. and Joyce, R. M., Jr. (1948) Halogenated hydrocarbons. US Patent 2,440,800 (assigned to E. I. du Pont de Nemours & Co.).

20. Freidlina, R. K., Terent'ev, A. B., Khorlina, M. Y. and Aminov, S. N. (1966) Rearrangement of radicals in the process of telomerization of ethylene by oxygen-containing telogens. *Zh. Vses. Khim.*, **11**, 211–215.

21. Starks, C. (1974) *Free Radical Telomerization*, Academic Press, New York.

22. Gordon, R. and Loftus, R. (1989) Telomerization. In *Ency. Polym. Sci. Technol.* (Eds, Kirk, R. O. and Othmer, D. F.) Wiley, New York, pp. 533–572.

23. (a) Boutevin, B. and Pietrasanta, Y. (1989) Telomerization. In *Comprehensive Polymer Science* (Eds, Allen, G. B., Bevington, J. C., Eastmond, A. L. and Russo, A.) Pergamon Press, Oxford, pp. 185–198; (b) Améduri, B. and Boutevin, B. (1994) Telomerization. In *Encyclopedia Advanced Materials* (Eds, Bloor, D. B., Brook, R. J., Flemings, R. D. and Mahajan, S.) Pergamon Press, Oxford, pp. 2767–2777.

24. Améduri, B. and Boutevin, B. (1997) Telomerisation of fluoroalkenes. *Top. Curr. Chem.*, **192**, 165–233.

25. Duc, M., Améduri, B., Kharroubi, M. and Boutevin, B. (1998) Radical telomerization of 1,1-difluoroethylene with trichloromethyled telogens. *Polym. Prepr. (Am. Chem. Soc., Polym. Div.)*, **39**(2), 845–846.

26. Joyce, R. M., Jr. (1951) Polyfluoroalkanols. US 2,559,628 (assigned to E.I. du Pont de Nemours & Co.).

27. Chen, J., Zhang, Y.-F., Zheng, X., Vij, A., Wingate, D., Meng, D., White, K., Kirchmeier, R. L. and Shreeve, J. M. (1996) Synthesis of cyclic ethers with fluorinated side chains. *Inorg. Chem.*, **35**, 1590–1601.

28. (a) Haszeldine, R. N., Rowland, R., Sheppard, R. P. and Tipping, A. E. (1985) Fluoroolefin chemistry. Part 20. Reaction of hexafluoropropene with alcohols. *J. Fluorine Chem.*, **28**, 291–302; (b) Chambers, R. D., Grievson, B., Drakesmith, F. G. and Powell, R. L. (1985) Free radical chemistry. Part 5. A new approach to the synthesis of perfluorinated ethers. *J. Fluorine Chem.*, **29**, 323–339.

29. Paleta, O., Dedek, V., Reutschek, H. and Timpe, H. J. (1989) Photochemical synthesis of fluoroalkanols based on tetrafluoroethylene. *J. Fluorine Chem.*, **42**, 345–353.

30. Kostov, G., Kotov, S. and Balbolov, E. (1998) Some kinetic aspects regarding the preparation of telomeric ω-hydroperfluoroalkanols. *React. Kinet. Catal. Lett.*, **63**, 107–115.

31. Posta, A. and Paleta, O. (1966) The addition reaction of carbon tetrachloride to trifluorochloroethylene. *Collect. Czech. Chem. Commun.*, **31**, 2389–2398.

32. Boutevin, B. and Pietrasanta, Y. (1976) Telomerization by redox catalysis. 6. Chemical transformation of chlorotrifluoroethylene. *Eur. Polym. J.*, **12**, 231–238.

33. Kotora, M. and Hajek, M. (1993) Ligand effect on regioselectivity in the addition reaction of tetrachloromethane with trifluoroethene catalyzed by copper(I) complexes. *J. Fluorine Chem.*, **64**, 101–105.

34. Boutevin, B., Furet, Y., Lemanach, L. and Vial-Reveillon, F. (1990) Telomerization of vinylidene fluoride with carbon tetrachloride and RCCl$_3$. *J. Fluorine Chem.*, **47**, 95–109.

35. Sloan, J. P., Tedder, J. M. and Walton, J. C. (1973) Free radical addition to olefins. 10. Addition of dibromofluoromethyl radicals to fluoroethylenes. *J. Chem. Soc., Faraday Trans. 1*, **69**, 1143–1152.

36. Dedek, V. Z. (1986) Addition of 1,2-dibromo-1-chlorotrifluoroethane to chlorotrifluoroethylene induced by UV-radiation. Synthesis of perfluoro-1,3-butadiene and perfluoro-1,3,5-hexatriene. *J. Fluorine Chem.*, **31**, 363–379.

37. Hauptschein, M., Braid, M. and Fainberg, A. H. (1958) Thermal syntheses of telomers of fluorinated olefins. III. Perfluoropropene telomer bromides. *J. Am. Chem. Soc.*, **80**, 851–853.

38. Apsey, G. C., Chambers, R. D., Salisbury, M. J. and Moggi, G. (1988) Polymer chemistry. Part 1. Model compounds related to hexafluoropropene and vinylidene fluoride elastomer. *J. Fluorine Chem.*, **40**, 261–282.

39. Balague, J., Améduri, B., Boutevin, B. and Caporiccio, G. (1995) Synthesis of fluorinated telomers. Part 1. Telomerization of vinylidene fluoride with perfluoroalkyl iodides. *J. Fluorine Chem.*, **70**, 215–223.

40. Balague, J., Améduri, B., Boutevin, B. and Caporiccio, G. (2000) Controlled stepwise telomerization of vinylidene fluoride, hexafluoropropene and trifluoroethylene with iodofluorinated transfer agents. *J. Fluorine Chem.*, **102**, 253–268.

41. Balague, J., Améduri, B., Boutevin, B. and Caporiccio, G. (1995) Synthese de telomeres fluores. Partie II. Telomerisation du trifluoroethylene avec des iodures de perfluoroalkyle. *J. Fluorine Chem.*, **73**, 237–246.

42. Balague, J., Ameduri, B., Boutevin, B. and Caporiccio, G. (1995) Synthesis of fluorinated telomers. Part III. Telomerization of hexafluoropropene with perfluoroalkyl iodides. *J. Fluorine Chem.*, **74**, 49–58.

43. Harris, J. and Stacey, F. (1963) The free radical addition of hydrogen sulfide to fluoroethylenes. *J. Am. Chem. Soc.*, **85**, 749–754.

44. Boutevin, B., Furet, Y., Hervaud, Y. and Rigal, G. (1994) Study of the telomerization and cotelomerization of vinylidene fluoride (VF$_2$). Part I. Cotelomerization of VF$_2$ with vinyl acetate and 2-hydroxyethyl mercaptan. *J. Fluorine Chem.*, **69**, 11–18.

45. Duc, M. (1997) Télomérisation du fluorure de vinylidène. Application à la synthèse d'oligomères fluorés fonctionnels. PhD Dissertation, University of Montpellier.

46. Sharp, D. W. A. and Miguel, H. T. (1978) Fluorine-containing dithioethers. *Isr. J. Chem.*, **17**, 144–147.

47. Dear, R. and Gilbert, E. (1974) Telomerization of bis(trifluoromethyl)disulfide with polyfluoroolefins. *J. Fluorine Chem.*, **4**, 107–110.

48. Haran, G. and Sharp, D. W. A. (1972) Photochemically initiated reactions of bis(trifluoromethyl) disulfide with olefins. *J. Chem. Soc., Perkin Trans. 1*, 34–38.

49. (a) Grigor'ev, N. A., German, L. S. and Freidlina, R. K. (1979) Telomerization of vinylidene fluoride by sulfuryl chloride fluoride. *Izv. Akad. Nauk SSSR, Ser. Khim.*, 918–921; (b) Grigor'ev, N., German, L. and Freidlina, R. (1979) Telomerization of vinylidene fluoride with sulfuryl chloride fluoride. *Bull. Acad. Sci. USSR*, **4**, 863–865.

50. Burch, G. M., Goldwhite, H. and Haszeldine, R. N. (1963) Organophosphorus chemistry. I. The reaction of phosphine with fluoroolefins. *J. Chem. Soc.*, 1083–1091.

51. Brace, N. O. (1961) ω-Hydroperfluoroalkylphosphonic acids, esters, and dichlorides. *J. Org. Chem.*, **26**, 3197–3201.

52. Roberts, H. L. (1967) Tetrafluoroethylene telomers of the structure Cl(CF$_2$CF$_2$)$_n$Cl. German Patent (assigned to Imperial Chem. Industries Ltd) (*Chem. Abstr.*, **66**, 85452).

53. Barnhart, W. S. (1957) The preparation of phosphorus oxychloride-halogen-containing telomers. US Patent 2,786,827 (assigned to M. W. Kellogg Co.).

54. (a) Boutevin, B., Pietrasanta, Y. and Youssef, B. (1986) Synthesis of fluorinated polysiloxanes. I. Hydrosilylation of fluorinated allylic olefins and fluorinated styrenes. *J. Fluorine Chem.*, **31**, 57–73; (b) Boutevin, B., Caporiccio, G., Guida-Pietrasanta, F. and Ratsimihéty, A. (1993) Study of alkylation of chlorosilanes. Part IV. Influence of the introduction of branched chains on the synthesis and properties of tetra(fluoroalkyl)silanes and α,ω-fluoroalkylene disilanes. *J. Fluorine Chem.*, **60**, 211–218.

55. Haszeldine, R. N., Newlands, M. J. and Plumb, J. B. (1965) Polyfluoroalkyl compounds of silicon. VI. Reaction of 3,3,3-trifluoropropene with silane, and the conversion of the products into silicones and polysiloxanes. *J. Chem. Soc.*, 2101–2107.

56. Haszeldine, R. N. and Steele, B. R. (1954) Addition of free radicals to unsaturated systems. X. The reaction of HBr with $F_2C : CF_2$ and $F_2C : CFCl$. *J. Chem. Soc.*, 3747–3751.

57. Bissell, E. R. and Shaw, G. C. (1962) Addition of iodine monochloride to chlorotrifluoro-ethylene. *J. Org. Chem.*, **27**, 1482–1483.

58. Boutevin, B., Gornowicz, G. and Caporiccio, G. (1992) Process for preparing an active telogen. Eur. Patent Appl. 3,045,015 (assigned to Dow Corning).

59. Haupstchein, M., Braid, M. and Fainberg, A. (1961) Addition of iodine halides to fluorinated olefins. I. The direction of addition of iodine monochloride to perhaloölefins and some related reactions. *J. Am. Chem. Soc.*, **83**, 2495–2500.

60. Améduri, B., Boutevin, B., Kostov, G. K. and Petrova, P. (1995) Novel fluorinated monomers bearing reactive side groups. Part 1. Preparation and use of $ClCF_2CFClI$ as the telogen. *J. Fluorine Chem.*, **74**, 261–267.

61. Haszeldine, R. N. (1952) Fluoroolefins. I. The synthesis of hexafluoro-1,3-butadiene. *J. Chem. Soc.*, 4423–4431.

62. Améduri, B., Boutevin, B., Kostov, G. K. and Petrova, P. (1999) Synthesis and polymerization of fluorinated monomers bearing reactive side groups. Part 5. Radical addition of iodine monobromide to chlorotrifluoroethylene to form a useful intermediate in the synthesis of 4,5,5-trifluoro-4-ene-pentanol. *J. Fluorine Chem.*, **93**, 117–125.

63. Tortelli, V. and Tonelli, C. (1990) Telomerization of tetrafluoroethylene and hexafluoropropene: synthesis of diiodoperfluoroalkanes. *J. Fluorine Chem.*, **47**, 199–217.

64. Barnhart, W. S. (1959) Halogenated cotelomers. US Patent 2,898,382 (assigned to Minnesota Mining and Manufacturing Co.).

65. Dannels, B. F., Fifolt, M. J. and Tang, D. Y. (1989) Telomer preparation from chlorotrifluoroethylene and 1,2-dibromo-2-chlorotrifluoroethane (Occidental Chem. Corp., USA). US Patent 4,808,760 (assigned to Occidental Chem. Corp.).

66. Ameduri, B., Boutevin, B. and Kostov, G. (2002) Synthesis and copolymerisation of vinylidene fluoride (VDF) with trifluorovinyl monomers. 11. Telomers of VDF with 1-chloro-1,2-dibromotrifluorethane as precursors of bromine-containing copolymers. *Macromol. Chem. Phys.*, **203**, 1763–1771.

67. Tedder, J. M. and Walton, J. C. (1967) Reactions of halomethyl radicals. *Prog. React. Kinet.*, **4**, 37–61.

68. Masson, J. (1989) Decomposition rates of organic free radical initiators. In *Polymer Handbook* (Eds, Brandrup, J. and Immergut, E. H.) Wiley, New York, pp. 1–76.

69. Bauduin, G., Boutevin, B., Mistral, J. P. and Sarraf, L. (1985) Kinetics of telomerization. 4. New method for determination of transfer constants in radical catalysis. *Makromol. Chem.*, **186**, 1445–1455.

70. Rice, D. E. and Sandberg, C. L. (1971) Functionally-terminated copolymers of vinylidene fluoride and hexafluoropropene. *Polym. Prepr. (Am. Chem. Soc., Div. Polym. Chem.)*, **12**, 396–402.

71. (a) Sawada, H. (1996) Fluorinated peroxides. *Chem. Rev.*, **96**, 1779–1808; (b) Sawada, H. (2000) Chemistry of fluoroalkanoyl peroxide, 1980–1998. *J. Fluorine Chem.*, **105**, 219–220; (c) Sawada, H., Ikeno, K. and Kawase, T. (2002) Synthesis of amphiphilic fluoroalkoxy end-capped cooligomers containing oxime-blocked isocyanato segments: architecture and applications of new self-assembled fluorinated molecular aggregates. *Macromolecules*, **35**, 4306–4313.

72. Guan, C. L., Chen, L., Deng, C. H. and Zhao, C. X. (2003) Synthesis and characterization of perfluoro[1-(2-fluorosulfonyl)ethoxy] ethyl end-capped styrene oligomers. *J. Fluorine Chem.*, **119**, 97–100.

73. Bessière, J. M., Boutevin, B. and Loubet, O. (1993) Synthesis of a perfluorinated azo initiator. Application to the determination of the recombination/disproportionation rate during the polymerisation of styrene. *Eur. Polym. J.*, **31**, 673–677.

74. Lebreton, P. and Boutevin, B. (2000) Primary radical termination and unimolecular termination in the heterogeneous polymerization of acrylamide initiated by a fluorinated azo-derivative initiator: kinetic study. *J. Polym. Sci., Part A: Polym. Chem.*, **38**, 1834–1843.

75. David, C. and Gosselain, P. (1962) Etude de la réaction de télomérisation de l'éthylène et du tétrachlorure de carbone initiée par rayonnement gamma. *Tetrahedron*, **18**, 639–651.

76. Boutevin, B. and Pietrasanta, Y. (1976) Telomerization by redox catalysis. 4. Chlorotrifluoro-ethylene telomerization with carbon tetrachloride. *Eur. Polym. J.*, **12**, 219–223.

77. Tsuchida, E., Kitamura, K. and Shinohara, I. (1972) Simplified treatment of cooligomerization kinetics. *J. Polym. Sci., Polym. Chem. Ed.*, **10**, 3639–3650.

78. (a) Boutevin, B., Parisi, J. P. and Vaneeckhoutte, P. (1991) Kinetics of cotelomerization. II. Evolution of degree of polymerization in cotelomerization: use of the Lewis and Mayo relation. *Eur. Polym. J.*, **27**, 159–163; (b) Boutevin, B., Parisi, J. P. and Vaneeckhoutte, P. (1991) Kinetics of cotelomerization. III. Statistical study of cotelomerization. *Eur. Polym. J.*, **27**, 1029–1034.

79. Lustig, M. and Shreeve, J. M. (1973) Fluoroxyfluoroalkanes and perfluoroacyl and inorganic hypofluorites. *Adv. Fluorine Chem.*, **7**, 175–198.

80. Shreeve, J. (1983) Fluorinated hypofluorites and hypochlorites. *Adv. Inorg. Radiochem.*, **26**, 119–168.

81. Mukhametshin, F. (1980) Advances in the chemistry of fluoroorganic hypohalites and related compounds. *Russ. Chem. Rev.*, **49**, 1260.

82. Schack, C. J. and Christe, K. O. (1978) Reactions of electropositive chlorine compounds with fluorocarbons. *Isr. J. Chem.*, **17**, 20–30.

83. Storzer, W. and DesMarteau, D. (1991) Two fluorinated, fluorosulfonyl-containing hypochlorites and their alkali–metal precursors. *Inorg. Chem.*, **30**, 4821–4826.

84. Kellog, K. and Cady, G. (1948) Trifluoromethyl hypofluorite. *J. Am. Chem. Soc.*, **70**, 3986–3990.

85. Johri, K. and DesMarteau, D. (1983) Comparison of the reactivity of CF_3OX (X = Cl, F) with some simple alkenes. *J. Org. Chem.*, **48**, 242–250.

86. Sekiya, A. and Ueda, K. (1990) Synthesis of perfluoroethylmethylether by direct fluorination. *Chem. Lett.*, 609–612.

87. Marraccini, A., Pasquale, A. and Vincenti, M. (1990) Process for preparing chlorotrifluoro-ethylene telomers and telomers obtained thereby. Eur. Patent 348,980 (assigned to Ausimont S.p.A.).

88. Randolph, B. and DesMarteau, D. (1993) Synthesis of functionalized polyfluoroalkyl hypochlorites and fluoroxy compounds and their reactions with some fluoroalkenes. *J. Fluorine Chem.*, **64**, 129–149.

89. Campbell, D. H., Fifolt, M. J. and Saran, M. S. (1985) Chlorotrifluoroethylene telomers. German Patent 3,438,934 (assigned to Occidental Chem. Corp.) (*Chem. Abstr.*, **104**, 5565).

90. Dos Santos Afonso, M. and Schumacher, H. J. (1984) Kinetics and mechanism of the thermal reaction between trifluoromethyl hypofluorite and hexafluoropropene. *Int. J. Chem. Kinet.*, **16**, 103–115.

91. Dos Santos Afonso, M. and Czarnowski, J. (1988) Kinetics and mechanism of the thermal reaction between trifluoromethyl hypofluorite and 1,1-dichlorodifluoroethylene. *Z. Phys. Chem.*, **158**, 25–34.

92. Marraccini, A., Pasquale, A., Fiurani, T. and Navarrini, W. (1990) Preparation of perhalogenated ethers by perfluoroalkoxylation of perhaloalkyenes or perhaloalkyl vinyl ethers with perhaloalkoxy fluorides. Eur. Patent 404,070 (assigned to Ausimont S.r.l., Italy).

93. Dos Santos Afonso, M. and Schumacher, H. J. (1988) Kinetics and mechanism of the thermal reaction between perfluoro-2-butene and trifluoromethyl hypofluorite. *Z. Phys. Chem.*, **158**, 15–23.

94. Di Loreto, H. and Czarnowski, J. (1994) The thermal addition of trifluoromethylhypofluorite, CF_3OF, to trichloroethene. *J. Fluorine Chem.*, **66**, 1–4.

95. Marraccini, A., Perego, G. and Guastalla, G. (1989) New chlorotrifluoroethylene telomers with inert end groups and process for preparing them. Eur. Patent 321,990 (assigned to Ausimont S.r.l., Italy).

96. Anderson, J. and DesMarteau, D. (1996) Perfluoroalkyl hypobromites: synthesis and reactivity with some fluoroalkenes. *J. Fluorine Chem.*, **77**, 147–152.

97. (a) Narita, T. (1994) Anionic polymerization of hexafluoro-1,3-butadiene and characterization of the polymer. *Macromol. Symp.*, **82**, 185–199; (b) Narita, T. (1999) Anionic polymerization of vinyl monomers. *Prog. Polym. Sci.*, **24**, 1095–1148.

98. Warnell, J. L. (1967) Fluorocarbon ethers containing iodine. US Patent 3,311,658 (assigned to du Pont de Nemours, E. I., and Co.).

99. Moore, E. P., Milian, A. S., Jr. and Eleuterio, H. S. (1966) Fluorocarbon ethers from hexafluoropropylene oxide. US Patent 3,250,808 (assigned to E. I. du Pont de Nemours & Co.).

100. Ito, T., Kaufman, J., Kratzer, R., Nakahara, J. and Paciorek, K. (1979) Synthesis and co-telomerization of 4-bromo- and 4-chloroheptafluoro-1,2-epoxybutanes. *J. Fluorine Chem.*, **14**, 93–103.

101. Kvicala, J., Paleta, O. and Dedek, V. (1990) Ionic telomerization of chlorofluoropropionyl fluorides with hexafluoropropene oxide. *J. Fluorine Chem.*, **47**, 441–457.

102. Flynn, R. (1988) Preparation of (fluoroalkoxy)carbonyl fluorides. US Patent 4,749,526 (assigned to 3M).

103. Hill, J. T. (1974) Polymers from hexafluoropropylene oxide (HFPO). *J. Macromol. Sci., Chem.*, **8**, 499–520.

104. Eleuterio, H. S. (1972) Polymerization of perfluoro epoxides. *J. Macromol. Sci., Chem.*, **6**, 1027–1052.

105. Sokolov, S. (1975) Polymerisation of fluorooxiranes. *Zh. Win. Org.*, **11**, 303–319.

106. Ohsaka, Y., Tohzuka, T. and Takaki, S. (1985) Halogen-containing polyether. Eur. Patent 148, 482 (assigned to Daikin Kogyo Co., Ltd).

107. Fielding, H. C. (1968) Fluorinated compounds containing functional groups. British Patent 1,130,822 (assigned to Imperial Chemical Industries Ltd).

108. Fielding, H. C., Deem, W. R., Houghton, L. E. and Hutchinson, J. (1973) Tetrafluoroethylene oligomer derivatives. German Patent 2,215,385 (assigned to Imperial Chemical Industries Ltd).

109. Caporiccio, G., Viola, G. T. and Corti, C. (1983) Perfluoropolyethers. Eur. Patent 89,820 (assigned to Montedison S.p.A., Italy).

110. Scheirs, J. (1997) Perfluoropolyethers (synthesis, characterization, and applications). In *Modern Fluoropolymers* (Ed, Scheirs, J.) Wiley, New York, pp. 435–486, Chap. 24.

111. Tanesi, D. P. A. and Corti, C. (1969) Fluorinated polyethanes. US Patent 3,442,942 (assigned to Montefluos).

112. Sianesi, D., Pasetti, A. and Belardinelli, G. (1984) High-molecular-weight polymeric perfluorinated copolyethers. US Patent 4,451,646 (assigned to Montedison S.p.A., Italy).

113. Trischler, F. D. and Hollander, J. (1967) Preparation of fluorine-containing polyethers. *J. Polym. Sci., Polym. Chem. Ed.*, **5**, 2343–2349.

114. Collet, A., Commeyras, A., Hirn, B. C. and Viguier, M. (1994) Perfluorinated polyethers polyols for use in polyurethanes. PCT Int. Demand WO 94, 10222.

115. Améduri, B., Boutevin, B. and Karam, L. (1993) Synthesis and properties of poly[3-chloromethyl-3-[(1,1,2,2-tetrahydroperfluorooctyloxy)methyl]oxetane]. *J. Fluorine Chem.*, **65**, 43–47.

116. Malik, A. A., Manser, G. E., Archibald, T. G., Duffy-Matzner, J. L., Harvey, W. L., Grech, G. J. and Carlson, R. P. (1996) Mono-substituted fluorinated oxetane monomers. Int. Demand WO 96/2 1657 (assigned to Aerojet General Corp.).

117. (a) Thomas, R., Kausch, C. M., Medsker, R. E., Russell, V. M., Malik, A. A. and Kim, Y. (2002) Surface activity and interfacial rheology of a series of novel architecture, water dispersible

poly(fluorooxetane)s. *Polym. Prepr. (Am. Chem. Soc., Div. Polym. Chem.)*, **43**(2), 890–891; (b) Fluorine in Coatings V Conference, Orlando, January 21–22, 2003, Paper 3; (c) Weinert, R. J., Robbins, J. E., Medsker, R. and Woodland, D. (2003) Surface improvements in coatings using polyfluorooxetane modified polyesters. Fluorine in Coatings V Conference, Orlando, January 21–22, 2003, Paper 13.

118. (a) Monteiro dos Santos, A. (1992) Carbon dioxide TEA laser-induced reaction of bromotri-fluoromethane with tetrafluoroethene. MPQ (MPQ 168) (*Chem. Abstr.*, **118**, 38275); (b) Gong, M., Fuss, W., Kompa, K., 1990. Carbon dioxide laser induced chain reaction of tetrafluo-roethene + trifluoroiodomethane. *J. Phys. Chem.* **94**, 6332–6337. *Chem. Abstr.*, **123**, 8893.

119. Zhang, L. and Zhang, J. (1995) Effect of light intensity on the photoinduced telomerization reaction of BrC_2F_4Br and C_2F_4. *Wuli Huaxue Xuebao*, **11**, 308–314 (*Chem. Abstr.*, **123**, 8893).

120. Zhang, L., Zhang, J., Yang, W., Wang, Y., Fuss, W. and Weizbauer, S. (1998) Highly selective photochemical synthesis of perfluoroalkyl bromides and iodides. *J. Fluorine Chem.*, **88**, 153–168.

121. Krespan, C. G., Petrov, V. A. and Smart, B. E. (1995) Initiators for telomerization of polyfluoroalkyl iodides with fluoroolefins. Int. Demand WO 9,532,936 (assigned to du Pont de Nemours, E. I., and Co., USA).

122. Krespan, C. G. and Petrov, V. A. (1995) Fluorinated carbon-based initiators for polymerization and telomerization of vinyl monomers. US Patent 5,459,212 (assigned to du Pont de Nemours, E. I., and Co., USA).

123. (a) Yoshida, Y., Ishizaki, K., Horiuchi, T. and Matsushige, K. (1993) Study on molecular orientation and phase transition behavior in vinylidene fluoride telomer evaporated films. *Kogaku Shuho—Kyushu Daigaku*, **66**, 199–204 (*Chem. Abstr.*, **120**, 271981); (b) Yoshida, Y., Kenji, I., Ishizaki, K., Horiuchi, T. and Matsushige, K. (1997) Effect of substrate on molecular orientation in evaporated thin films of vinylidene fluoride oligomers. *Jpn. Appl. Phys. Part 1*, **36**, 7389–7394 (*Chem. Abstr.*, **128**, 161641).

124. Paleta, O. (1977) Ionic addition reactions of halomethanes with fluoroolefins. *Fluorine Chem. Rev.*, **8**, 39–71.

125. Paleta, O., Posta, A. and Liska, F. (1978) Haloacrylic acids. IX. Laboratory scale production of some trifluoroacrylic acid derivatives. *Sbornik Vys. Chem. Technol., C*, **C25**, 105–121 (*Chem. Abstr.*, **95**, 6026).

126. Ikeya, T. and Kitazume, T. (1987) 12th Symp. Fluorine Chem. Hakuta, Japan, Abst. 3R17.

127. Combes, J., Guan, Z. and De Simone, J. M. (1994) Homogeneous free-radical polymerizations in carbon dioxide. 3. Telomerization of 1,1-difluoroethylene in supercritical carbon dioxide. *Macromolecules*, **27**, 865–867.

128. Tedder, J. M. and Walton, J. C. (1978) Directive effects in gas-phase radical addition reactions. *Adv. Phys. Org. Chem.*, **16**, 51–86.

129. Tedder, J. M. and Walton, J. C. (1980) The importance of polarity and steric effects in determining the rate and orientation of free radical addition to olefins. Rules for determining the rate and preferred orientation. *Tetrahedron*, **36**, 701–707.

130. Boutevin, B., Pietrasanta, Y., Rousseau, A. and Bosc, D. (1987) Synthesis of monomers for transparent polymers. Part 1. Synthesis of trideuteriomethyl trifluoroacrylate. *J. Fluorine Chem.*, **37**, 151–169.

131. Tezuka, Y. and Imai, K. (1984) Synthesis of poly(vinyl alcohol)–poly(dimethylsiloxane) block copolymer. *Makromol. Chem., Rapid Commun.*, **5**, 559–565.

132. Ponomarenko, V. A., Cherkaev, V. G., Petrov, A. D. and Zadorozhnyi, N. A. (1958) Chloroplatinic acid as catalyst in the reaction of addition of silane hydrides to unsaturated compounds. *Izv. Akad. Nauk SSSR, Ser. Khim.*, 247–249.

133. Geyer, A. M. and Haszeldine, R. N. (1957) Perfluoroalkyl compounds of silicon. III. A polyfluoroalkyl silicone. *J. Chem. Soc.*, 3925–3927.

134. Geyer, A. M. and Haszeldine, R. N. (1957) Polyfluoroalkyl compounds of silicon. II. Free radical reaction of dialkylsilanes with fluoroolefins. *J. Chem. Soc.*, 1038–1043.

135. Zimin, A. V., Matyuk, V. M., Yankelevich, A. Z. and Shapet'ko, N. N. (1976) Study of the characteristics of the radiation-induced-Chem hydrosilylation of halogenated and nonhalogenated olefins. *Dokl. Akad. Nauk SSSR, Ser. Khim.*, **231**, 870–873.

136. Boutevin, B., Guida-Pietrasanta, F., Ratsimihety, A., Caporiccio, G. and Gornowicz, G. (1993) Study of the alkylation of chlorosilanes. Part I. Synthesis of tetra(1H,1H,2H,2H-polyfluoroalkyl)silanes. *J. Fluorine Chem.*, **60**, 211–223.

137. Seiler, D. A. (1997) PVDF in the chemical industry. In *Modern Fluoropolymers* (Ed, Scheirs, J.) Wiley, New York, pp. 487–506, Chap. 25.

138. Liepins, R., Surles, J. R., Morosoff, N., Stannett, V. T., Timmons, M. L. and Wortman, J. J. (1978) Poly(vinylidene fluoride) with low content of head-to-head chain imperfections. *J. Polym. Sci., Polym. Chem. Ed.*, **16**, 3039–3044.

139. Guiot, J., Ameduri, B. and Boutevin, B. (2002) Radical homopolymerization of vinylidene fluoride initiated by *tert*-butyl peroxypivalate. Investigation in the microstructure by ^{19}F and ^{1}H NMR spectroscopies and mechanisms. *Macromolecules*, **35**, 8694–8707.

140. (a) Tournut, C. (1994) New copolymers of vinylidene fluoride. *Macromol. Symp.*, **82**, 99–110; (b) Tournut, C. (1997) Thermoplastic copolymers of vinylidene fluoride. In *Modern Fluoropolymers* (Eds., Scheirs, J.) Wiley, New York, pp. 577–596, Chap. 31.

141. Hauptschein, M., Braid, M. and Lawlor, F. E. (1958) Thermal syntheses of telomers of fluorinated olefins. II. 1,1-Difluoroethylene. *J. Am. Chem. Soc.*, **80**, 846–851.

142. Belfield, K. D., Albel-Sadek, G. G., Huang, J. and Ting, R. Y. (2002) Radical telomerization of vinylidene fluoride in the presence of dibromodifluoromethane as telogen, presented at the 223rd ACS National Meeting, Orlando, POLY-192. *Polym. Prepr. (Am. Chem. Soc., Polym. Div.)*, **43**, 644–645.

143. Kim, Y. K. (1974) Low temperature fluorosilicone compositions. US Patent 3,818,064 (assigned to Dow Corning).

144. Modena, S., Pianca, M., Tato, M., Moggi, G. and Russo, S. (1989) Radical telomerization of vinylidene fluoride in the presence of 1,2-dibromotetrafluoroethane. *J. Fluorine Chem.*, **43**, 15–25.

145. (a) Zhang, Z., Ying, S. and Shi, Z. (1998) Synthesis of ABA triblock copolymers containing both fluorocarbon and hydrocarbon blocks for compatibilizers. *Polym. Prepr. (Am. Chem. Soc., Div. Polym. Chem.)*, **39**(2), 843–844; (b) Zhang, Z., Ying, S. and Shi, Z. (1999) Synthesis of fluorine-containing block copolymers via ATRP. 1. Synthesis and characterization of PSt-PVDF-PSt triblock copolymers. *Polymer*, **40**, 1341–1344.

146. Moggi, G., Modena, S. and Marchionni, G. (1990) A vinylidene fluoride block copolymer containing perfluoropolyether groups. *J. Fluorine Chem.*, **49**, 141–146.

147. Ameduri, B., Ladavière, C., Boutevin, B. and Delolme, F., MALDI TOF spectroscopy of fluorinated telomers, *Macromolecules*, submitted for publication.

148. Hauptschein, M. and Oesterling, R. E. (1960) Novel elimination reactions of telomer iodides of 1,1-difluoroethylene. *J. Am. Chem. Soc.*, **82**, 2868–2871.

149. Hung, M. (2002) Fluoro elastomer compositions having excellent processability. US Patent 002, 248 (assigned to DuPont).

150. Chambers, R. D., Hutchinson, J., Mobbs, R. H. and Musgrave, W. K. R. (1964) Telomerization reactions in the synthesis of models for some fluorocarbon polymers. *Tetrahedron*, **20**, 497–506.

151. Chen, Q.-Y., Ma, Z.-Z., Jiang, X.-K., Zhang, Y.-F. and Jia, S.-M. (1980) Telomerizations of vinylidene fluoride with perhaloalkanes and the study of their products. *Huaxue Xuebao*, **38**, 175–184 (*Chem. Abstr.*, **94**, 174184).

152. Manseri, A., Améduri, B., Boutevin, B., Chambers, R. D., Caporiccio, G. and Wright, A. P. (1995) Synthesis of fluorinated telomers. Part 4. Telomerization of vinylidene fluoride with commercially available α,ω-diiodoperfluoroalkanes. *J. Fluorine Chem.*, **74**, 59–67.

153. Toyoda, Y. S. and Nobuo, C. (1967) Telomerization of fluoromonomers. Japanese Patent 77,319 (assigned to Kureha Chem. Ind. Co. Ltd).

154. (a) Herman, T. U., Kubono, A., Umemoto, S., Kibutani, T. and Okui, N. (1997) Effect of molecular weight and chain end groups on crystal forms of poly(vinylidene fluoride) oligomers. *Polymer*, **38**, 1677–1683; (b) Herman, T. U., Umemoto, S., Kikutani, T. and Okui, N. (1998) Chain length effects on crystal formation in vinylidene fluoride oligomers. *Polym. J.*, **30**, 659–663.

155. Oku, J., Chan, R. J. H., Hall, H. K. Jr. and Hughes, O. R. (1986) Synthesis of a poly(vinylidene fluoride) macromonomer. *Polym. Bull.*, **16**, 481–485.

156. Duc, M., Améduri, B., Boutevin, B., Kharroubi, M. and Sage, J. M. (1998) Telomerization of vinylidene fluoride with methanol. Elucidation of the reaction process and mechanism by a structural analysis of the telomers. *Macromol. Chem. Phys.*, **199**, 1271–1288.

157. Tarrant, P. and Lovelace, A. M. (1955) Free radical addition involving fluorine compounds. III. The addition of bromochlorodifluoromethane to olefins. *J. Am. Chem. Soc.*, **77**, 768–772.

158. Tarrant, P., Lovelace, A. M. and Lilyquist, M. R. (1955) Free-radical additions involving fluorine compounds. IV. The addition of dibromodifluoromethane to some fluoroolefins. *J. Am. Chem. Soc.*, **77**, 2783–2787.

159. Modena, S., Pianca, M., Tato, M., Moggi, G. and Russo, S. (1985) Radical reactions of 1,1-difluoroethene. *J. Fluorine Chem.*, **29**, 154–162.

160. Ashton, D. S., Shand, D. J., Tedder, J. M. and Walton, J. C. (1975) Free radical addition to olefins. XV. Addition of bromoform and carbon tetrabromide to fluoroethylenes. *J. Chem. Soc., Perkin Trans. 2*, 320–325.

161. Sloan, J. P., Tedder, J. M. and Walton, J. C. (1975) Free radical addition to olefins. XVII. Addition of fluoroiodomethane to fluoroethylenes. *J. Chem. Soc., Perkin Trans. 2*, 1846–1850.

162. McMurray, N., Tedder, J. M., Vertommen, L. L. T. and Walton, J. C. (1976) Free radical addition to olefins. XVII. Addition of chloroiodo- and diiodomethane to fluoro alkenes. *J. Chem. Soc., Perkin Trans. 2*, 63–67.

163. Low, H. C., Tedder, J. M. and Walton, J. C. (1976) Free radical addition to olefins. Part 20. A reinvestigation of the addition of methyl radicals to fluoroethylenes. *J. Chem. Soc., Faraday Trans. 1*, **72**, 1707–1714.

164. Gelin, M. P. and Ameduri, B. (2003) Fluorinated block copolymers containing poly(vinylidene fluoride) or poly(vinylidene fluoride-*co*-hexafluoropropylene) blocks from perfluoropolyethers: synthesis and thermal properties. *J. Polym. Sci., Part A: Polym. Chem.*, **41**, 160–171.

165. Boutevin, B., Furet, Y., Hervaud, Y. and Rigal, G. (1995) Study of vinylidene fluoride (VF$_2$) telomerization and cotelomerization. Part II. VF$_2$ and chlorotrifluoroethylene (CTFE) cotelomerization with 2-hydroxyethylmercaptan by radical initiation. *J. Fluorine Chem.*, **74**, 37–42.

166. Duc, M., Ameduri, B. and Boutevin, B. Radical telomerisation of vinylidene fluoride with mercaptans, in preparation.

167. Tiers, G. V. D. (1958) Organic sulfonyl fluorides. US Patent 2,846,472 (assigned to Minnesota Mining and Manufacturing Co.).

168. Zhu, Y., Sun, S. X., Zhang, Y. H. and Jiang, X. (1993) Can trichloromethanesulfonyl bromide be used as an addendum as well as a telogen in its addition reaction to vinylidene fluoride? *Chinese Chem. Lett.*, **4**, 583–586.

169. Haszeldine, R. N., Hobson, D. L. and Taylor, D. R. (1976) Organophosphorus chemistry. Part 19. Free-radical addition of dialkyl phosphites to polyfluoroolefins. *J. Fluorine Chem.*, **8**, 115–124.

170. Duc, M., Améduri, B. and Boutevin, B. (2001) Radical telomerization of vinylidene fluoride with diethyl hydrogenophosphonate. Characterization of the first telomeric adducts and assessment of the transfer constants. *J. Fluorine Chem.*, **112**, 3–12.

171. Dolbier, W. R. Jr., Duan, J. X., Abboud, K. and Ameduri, B. (2000) Synthesis and reactivity of a novel, dimeric derivative of octafluoro[2,2] paracyclophane. A new source of trifluoromethyl radicals. *J. Am. Chem. Soc.*, **122**, 12083–12086.

172. Cooper, J. A., Copin, E. and Sandford, G. (2002) Free radical addition of cyclopentane and cyclohexane to halogeno derivatives of 1,2-difluoroethene. *J. Fluorine Chem.*, **115**, 83–90.

173. Tedder, J. M. and Walton, J. C. (1966) Free radical addition to olefins. II. Addition of trichloromethyl radicals to fluoroethylenes. *Trans. Faraday Soc.*, **62**, 1859–1865.

174. Tedder, J. M. and Walton, J. C. (1974) Free radical addition to olefins. 11. Addition of bromodifluoromethyl radicals to fluoroethylenes. *J. Chem. Soc., Faraday Trans. 1*, **70**, 308–319.

175. Chambers, R. D., Proctor, L. D. and Caporiccio, G. (1995) Polymer chemistry. Part V. Gamma-ray induced telomerization reactions involving 1,1-difluoroethene and hexafluoropropene. *J. Fluorine Chem.*, **70**, 241–247.

176. Haszeldine, R. N. and Steele, B. R. (1954) Addition of free radicals to unsaturated systems. VII. 1,1-Difluoroethylene. *J. Chem. Soc.*, 923–925.

177. Cape, J. N., Greig, A. C., Tedder, J. M. and Walton, J. C. (1975) Free radical addition to olefins. 14. Addition of trifluoromethyl radicals to fluoroethylenes. *J. Chem. Soc., Faraday Trans. 1*, **71**, 592–601.

178. Harris, J. and Stacey, F. (1961) The free radical addition of trifluoromethanethiol to fluoroolefins. *J. Am. Chem. Soc.*, **83**, 840–845.

179. Ameduri, B., Billard, T. and Langlois, B. (2002) Telomerization of vinylidene fluoride with alkyl (or aryl) trifluoromethanethiosulfonates. *J. Polym. Sci., Part A: Polym. Chem.*, **40**, 4538–4549.

180. Kharroubi, M., Manseri, A., Ameduri, B. and Boutevin, B. (2000) Radical addition of iodine monochloride to vinylidene fluoride. *J. Fluorine Chem.*, **103**, 102–110.

181. (a) Li, A. R. and Chen, Q. Y. (1997) Lead tetraacetate induced addition reaction of difluorodiiodomethane to alkenes and alkynes. Synthesis of fluorinated telechelic compounds. *Synthesis*, 1481–1488; (b) Li, A. R. and Chen, Q. Y. (1997) One pot synthesis of dialkyldifluoromethane and dialkyldifluoromethyl iodides from the reaction of difluorodiiodomethane with alkenes in sulphinatodehalogenation systems. *J. Fluorine Chem.*, **82**, 151–155.

182. Rondestvedt, C. S. Jr. (1977) Methyl-terminated perfluoroalkyl iodides and related compounds. *J. Org. Chem.*, **42**, 1985–1990.

183. Saintloup, R. and Ameduri, B. (2002) Photochemical induced polymerisation of vinylidene fluoride (VDF) with hydrogen peroxide to obtain original telechelic PVDF. *J. Fluorine Chem.*, **116**, 27–37.

184. Duc, M., Ameduri, B., Boutevin, B. and Karroubi, M. (1998) Radical telomerization of 1,1-difluoroethylene with methanol. *Polym. Prepr. (Am. Chem. Soc., Polym. Div.)*, **39**(2), 816–817.

185. (a) Latourte, L., Blais, J.-C., Tabet, J.-C. and Cole, R. B. (1997) Desorption behavior and distributions of fluorinated polymers in MALDI and electrospray ionization mass spectrometry. *Anal. Chem.*, **69**, 2742–2750; (b) Gooden, J. K., Gross, M. L., Mueller, A., Stefanescu, A. D. and Wooley, K. L. (1998) Cyclization in hyperbranched polymer syntheses: characterization by MALDI-TOF mass spectrometry. *J. Am. Chem. Soc.*, **120**, 10180–10186; (c) Marie, A., Fournier, F. and Tabet, J. C. (2000) Characterization of synthetic polymers by MALDI-TOF/MS: investigation into new methods of sample target preparation and consequence on mass spectrum finger print. *Anal. Chem.*, **72**, 5106–5114; (d) Ming, W., Lou, X., Van de Grampel, R. D., Van Dongen, J. L. J. and Van der Linde, R. (2001). *Macromolecules*, **34**, 2389–2393; (e) Grimm, B., Krüger, R.-P., Schrader, S. and Prescher, D. (2002) Molecular structure investigations on fluorine containing polyazomethines by means of the MALDI-TOF-MS technique. *J. Fluorine Chem.*, **113**, 85–91; (f) Chen, G., Cooks, R. G., Jha, S. K., Oupicky, D. and Green, M. M. (1997) Block microstructural characterization of copolymers formed from fluorinated and non-fluorinated alkyl polyisocyanates using desorption chemical ionization mass spectrometry. *Int. J. Mass Spectrom. Ion Proc.*, **165/166**, 391–404.

186. Marie, A., Fournier, F., Tabet, J. C., Ameduri, B. and Walker, J. (2002) Collision-induced dissociation studies of polyvinylidene fluoride telomers in an electrospray-ion trap mass spectrometer. *Anal. Chem.*, **74**, 3213–3220.

187. Gaboyard, M., Boutevin, B. and Hervaud, Y. (2001) Synthesis of new phosphonates bearing fluorinated chains. *J. Fluorine Chem.*, **107**, 5–12.
188. Braslavsky, S. E., Casas, F. and Cifuntes, O. (1970) Photolysis of trifluoromethyl iodide in the presence of ethylene and 1,1-difluoroethylene. *J. Chem. Soc. (B)*, 1059–1060.
189. Jo, S. M., Lee, W. S., Ahn, B. S., Park, K. Y., Kim, K. A. and Paeng, I. S. R. (2000) New AB or ABA type block copolymers: atom transfer radical polymerization (ATRP) of methyl methacrylate using iodine-terminated PVDFs as macroinitiators. *Polym. Bull., Berlin*, **44**, 1–9.
190. Kiryukhin, D. P., Nevel'skaya, T. I., Kim, I. P. and Barkalov, I. M. (1982) Telomerization of tetrafluoroethylene in acetone initiated by cobalt-60 gamma-rays and by radical initiators. Soluble telomers. *Vys. Soed. Ser. A*, **24**, 307–311.
191. Jeffrey, G. C. (1966) Tetrafluoroethylene and tetrachloroethylene telomers. US Patent 3,235,611 (assigned to Dow Chem. Co.).
192. Fearn, J. E. (1971) Synthesis of fluorodienes. *J. Res. Natl. Bur. Stand.*, **75**, 41–56.
193. Haszeldine, R. N. (1949) Reactions of fluorocarbon radicals. I. The reaction of iodotrifluoro-methane with ethylene and tetrafluoroethylene. *J. Chem. Soc.*, 2856–2861.
194. Ashton, D. S., Tedder, J. M. and Walton, J. C. (1974) Free radical addition to olefins. 12. Telomerization of tetrafluoroethylene with dibromodifluoromethane and trifluoroiodomethane. *J. Chem. Soc., Faraday Trans. 1*, **70**, 299–307.
195. Ono, Y. and Ukihashi, H. (1970) Telomerization of fluorinated olefins with perfluoroalkyl iodides. *Asahi Garasu Kenkyu Hokoku*, **20**, 55–65 (*Chem. Abstr.*, **74**, 99234).
196. Gao, Y. and Arakawa, H. (2001) Method for preparation of medium-chain length halogenated hydrocarbons or carbons by photochemical telomerization. Japanese Patent 348,348 A2 (assigned to Kanto Denka Kogyo, Ltd) (*Chem. Abstr.*, **136**, 37320).
197. Zaitseva, E. L., Rozantseva, T. V., Chicherina, I. I. and Yakubovich, A. Y. (1971) Telomerization of tetrafluoroethylene by diphenyl disulfide. *Zh. Org. Khim. (Engl. Transl.)*, **7**, 2548–2551 (*Chem. Abstr.*, **76**, 72156).
198. Roberts, H. (1965) Tetrafluoroethylene telomers. Netherlands Patent 6,402,992 (assigned to ICI) (*Chem. Abstr.*, **62**, 13043).
199. (a) Roberts, H. L. (1961) Polymers and telomers of tetrafluoroethylene. British Patent 877,961 (assigned to imperial Chem. Industries Ltd); (b) Roberts, H. L. (1962) Tetrafluoroethylene telomers. British Patent 898,309 (assigned to Imperial Chem. Industries Ltd).
200. Darragh, J. I. (1974) Tetrafluoroethylene telomers. German Patent 2,416,261 (assigned to Imperial Chemical Industries Ltd).
201. Terjeson, R. and Gard, G. (1987) New pentafluorothio(SF5)fluoropolymers. *J. Fluorine Chem.*, **35**, 653–662.
202. Zhang, Y., Guo, C., Zhou, X. and Chen, Q. (1982) Perfluoro- and polyfluorosulfonic acids. III. Thermal telomerization of 1-chloro-2-iodotetrafluoroethane with tetrafluoroethylene and chemical transformation of the products. *Huaxue Xuebao*, **40**, 331–336 (*Chem. Abstr.*, **97**, 126982).
203. (a) Hauptschein, M. and Braid, M. (1962) Tetrafluoroethylene telomers. Belgium Patent 612,520 (assigned to Pennsalt Chemicals Corp.); (b) Water- and oil-repellent polymer coatings (1967) Netherlands Patent 6,609,483 (assigned to Pennsalt Chemicals Corp.) (*Chem. Abstr.*, **67**, 12621); (c) Meissner, E., Myszkowski, J. and Szymanowski, J. (1995). Synthesis and surface activity of surfactants containing fluorocarbon hydrophobes. *Tenside Surf. Det.* **32**, 261-2271.
204. Hauptschein, M. and Braid, M. (1965) Telomer iodides. US Patent 3,219,712 (assigned to Pennsalt Chemicals Corp.).
205. (a) Bertocchio, R., Lacote, G. and Verge, C. (1993) Preparation of perfluoroalkyliodides. Eur. Patent 552,076 (assigned to Elf Atochem S.A., Fr.); (b) Bauduin, G., Boutevin, B., Bertocchio, R., Lantz, A., Verge, C. (1998). Etude de la télomérisation de C_2F_4 avec des iodures de perfluoroalkyle. Partie 1. Télomérisation radicalaire. *J. Fluorine Chem.* **90**, 107–115.

206. Hutchinson, J. (1973) The preparation of telomers of "pentafluorosulphur iodide" and related reactions. *J. Fluorine Chem.*, **3**, 429–432.

207. Moore, L. O. (1983) Radical reactions of highly polar molecules. Hydrocarbons as chain-transfer agents in fluoroolefin telomerizations. *Macromolecules*, **16**, 357–359.

208. (a) Peavy, R. E. (1994) Fluoroalkene/hydrofluorocarbon telomers and their synthesis. US Patent 5,310,870 (assigned to du Pont de Nemours, E. I., and Co., USA); (b) Peavy, R. E. (1995) Preparation of fluoroalkene/chlorofluorocarbon telomers. US Patent 5,552,500 (assigned to du Pont de Nemours, E. I., and Co.).

209. Afanas'ev, I. B., Safronenko, E. D. and Beer, A. A. (1967) Kinetics of radical telomerization of tetrafluoroethylene with alcohols. *Vys. Soed. Ser. B*, **9**, 802–804 (*Chem. Abstr.*, **68**, 30099).

210. Asahi Glass Co. (1975) Tetrafluoroethylene telomers. Japanese Patent 47,908.

211. (a) Satokawa, T., Fujii, T., Ohmori, A. and Fujita, Y. (1979) Telomerization of tetrafluoro-ethylene. Japan Patent 154,707 (assigned to Daikin Kogyo Co., Ltd); (b) Satokawa, T., Fujii, T., Ohmori, A. and Fujita, Y. (1982) Telomerization of tetrafluoroethylene. US Patent 4,346,250 (assigned to Daikin Kogyo Co., Ltd, Japan).

212. No, V. B., Bakhmutov, Y. L. and Pospelova, N. B. (1976) Effect of hydrogen peroxide on the telomerization of tetrafluoroethylene with methanol. *Zh. Org. Khim.*, **12**, 1825–1830 (*Chem. Abstr.*, **85**, 159134).

213. Daikin Kogyo Co. (1981) Telomerization of tetrafluoroethylene. Japanese Patent 43,225 (*Chem. Abstr.*, **95**, 186628).

214. Huang, W.-Y., Liang, W.-X., Lin, W.-D., Chen, J.-H. and Zhou, X.-Y. (1980) Structures of the telomerization products of tetrafluoroethene with ethylene glycol mono- and dimethyl ether. *Huaxue Xuebao*, **38**, 283–291 (*Chem. Abstr.*, **94**, 30115).

215. (a) Chkhubianishvili, N. G., Baer, A. A. and Dzheiranishvili, M. S. (1969) Kinetics of some telomerization reactions. *Soobshch. Akad. Nauk Gruz. SSR*, **56**, 329–332 (*Chem. Abstr.*, **72**, 79598); (b) Chkhubianishvili, N. G., Baer, A. A. and Khananashvili, L. M. (2000) Investigation of tetrafluoroethylene and bromotrichloromethane telomerization. *Russ. Polym. News*, **5**, 14–20 (*Chem. Abstr.*, **133**, 282125).

216. Moore, L. O. (1971) Radical reactions of highly polar molecules. Reactivities in atom abstractions from chloroalkanes by fluoroalkyl radicals. *J. Phys. Chem.*, **75**, 2075–2099.

217. EI Du Pont and de Nemours and Co. (1967) Tetrafluoroethylene and perfluorinated polyether lubricants. French Patent 1,471,692.

218. Bittles, J. A. and Joyce, J. R. (1951) Fluoroalkanephosphonic compounds. US Patent 2,559,754 (assigned to DuPont).

219. Rudolph, W. and Massonne, J. (1975) Perfluoroethyl iodide telomers. German Patent 2,162,368 (assigned to Kali-Chemie A.-G.) (*Chem. Abstr.*, **79**, 79519).

220. Ono, Y. (1973) Perfluoroalkyl iodides. Japan Patent 4,842,852 (assigned to Asahi Glass Co., Ltd).

221. Rebsdat, S., Schuierer, E. and Hahn, H. (1970) Linear polyfluoroalkyl iodide compounds. German Patent 1,915,395 (assigned to Farbwerke Hoechst A.-G.) (*Chem. Abstr.*, **73**, 76644).

222. Rudolph, W. and Massonne, J. (1973) High-molecular perfluoroalkyl iodide telomers weight. French Patent 2,163,444 (assigned to Kali-Chemie A.-G.) (*Chem. Abstr.*, **80**, 3066).

223. Felix, B. (1973) Perfluoroalkyl iodides. German Patent 2,164,567 (assigned to Ferbuerke Hoechst AG) (*Chem. Abstr.*, **79**, 78077).

224. Felix, B. and Weiss, E. (1977) Perfluoroalkyl iodide telomers. German Patent 2,542,496 (assigned to Ferbuerke Hoechst A.-G.).

225. (a) Blanchard, W. A. and Rhode, J. C. (1965) Perfluoroalkyl iodides. US Patent 3,226,449 (assigned to E. I. du Pont de Nemours & Co.); (b) Nagai, M., Shinkai, H., Kato, T., Asaoka, M., Nakatsu, T. and Fujii, Y. (1974) Tetrafluoroethylene telomers. Japanese Patent 61,103 (assigned to Daikin Kogyo Co., Ltd); (c) Nagai, M., Shinkai, H., Kato, T., Asaoka, M., Nakatsu, T. and

Fujii, Y. (1974) Tetrafluoroethylene telomers. Japanese Patent 61,104 (assigned to Daikin Kogyo Co., Ltd).

226. Romack, T. J., Combes, J. and De Simone, J. (1995) Free radical telomerization of tetrafluoroethylene in supercritical carbon dioxide. *Macromolecules*, **28**, 1724–1726.

227. Iserson, H., Magazzu, J. J. and Osborn, S. W. (1972) α,ω-Diiodoperfluoroalkanes. German Patent 2,130,378 (assigned to Thiokol Chem. Corp.) (*Chem. Abstr.*, **76**, 99090).

228. Furukawa, Y. (1982) Fluorine-containing rubber. Eur. Patent Appl. 0,045,070 (assigned to Daikin Kogyo Co., Ltd).

229. Tittle, B. and Platt, A. E. (1965) Tetrafluoroethylene telomers. British Patent 1,007,542 (assigned to ICI) (*Chem. Abstr.*, **64**, 1956).

230. Battais, A., Boutevin, B., Bertocchio, R. and Lantz, A. (1989) Télomérisation du tetrafluoro-éthylène avec le tétrachlorure de carbone par catalyse redox. *J. Fluorine Chem.*, **42**, 215–224.

231. (a) Chen, Q., Su, D. and Zhu, R. (1987) Copper-induced telomerization of tetrafluoroethylene with fluoroalkyl iodides. *J. Fluorine Chem.*, **36**, 483–489; (b) Von Werner, K. (1996) Preparation of perfluoroalkyl iodides telomers. Eur. Patent 718,262 (assigned to Hoechst); (c) Von Werner, K. (1996) Metal-catalyzed production of perfluoroalkyl iodides telomers. Eur. Patent 718,263 (assigned to Hoechst).

232. Parsons, R. E. (1965) Perfluoroalkyl iodides. French Patent 1,385,682 (assigned to E. I. du Pont de Nemours & Co.).

233. Jaeger, H. (1969) Perfluoroalkyl iodide–perfluoroolefin telomers. South African Patent 2,001, 140 (assigned to CIBA Ltd).

234. Jaeger, H. (1970) High molecular weight perfluoroalkyl iodides. German Patent 2,001,140 (assigned to CIBA Ltd) (*Chem. Abstr.*, **73**, 109259).

235. Rudolph, W., Massonne, J., Jaeger, H. H. and Gress, H. (1972) Metal salt–amine complex catalysts for the telomerization of pentafluoroiodoethane with tetrafluoroethylene. German Patent 2,034,472 (assigned to Kali-Chemie A.-G.) (*Chem. Abstr.*, **77**, 6148).

236. Fielding, H. C. (1968) Tetrafluoroethylene–chloromethane adducts. British Patent 1,127,045 (assigned to Imperial Chem. Industries Ltd).

237. Hong, L., Hu, C., Ying, S. and Mao, S. (1987) Telomerization of tetrafluoroethylene with isopropanol. *Gaofenzi Xuebao*, 30–34 (*Chem. Abstr.*, **107**, 116038).

238. Tittle, B. (1971) Perfluoroalkyl iodides. British Patent 1,242,712 (assigned to Imperial Chem. Industries Ltd) (*Chem. Abstr.*, **75**, 129321).

239. (a) Caporiccio, G., Bargigia, G. and Guidetti, G. (1978) Fluorooxaalkanesulfonic acids and their derivatives. German Patent 2,735,210 (assigned to Montedison S.p.A., Italy) (*Chem. Abstr.*, **88**, 153252); (b) Shanghai Guangming Electroplating Factory(1977). Preparation and application of potassium oxaperfluoroalkane sulfonates. *Huaxue Xuebao*, **35**, 209–220 (*Chem. Abstr.*, **89**, 196–929).

240. Fu, W. M., Zhu, Y., Sun, S. X., Zhang, Y. N. and Ji, G. Z. (1996) Telomerization of tetrafluoroethylene and chlorotrifluoroethylene with trichloromethanesulfonyl bromide as a telogen. *Chinese Chem. Lett.*, **7**, 717–720 (*Chem. Abstr.*, **125**, 276730).

241. Wakselman, C. and Lantz, A. (1994) Perfluoralkyl bromides and iodides. In *Organofluorine Chemistry: Principles and Applications*, Vol. 15 (Eds, Banks, R. E., Smart, B. and Tatlow, J. C.) Plenum Press, New York, pp. 177–194.

242. Boutevin, B., Doheim, M., Pietrasanta, Y. and Rigal, G. (1979) Influence des alcools dans les réactions de télomérisation du chlorotrifluoroéthylène par catalyse redox. *J. Fluorine Chem.*, **14**, 29–43.

243. Boutevin, B. and Pietrasanta, Y. (1988) Hydrocarbures chlorofluorés et leur procédé de préparation. French Patent 01,882 (assigned to Atochem).

244. Boutevin, B., Hervaud, Y. and Rolland, L. (1984) Applications de mélanges stables de composés chlorofluorés à base de fluides calogènes pour pompe à chaleur à absorption. French Patent 01, 128 (assigned to Gaz de France).

245. Dannels, B. F. (1984) Chlorotrifluoroethylene telomerization. Eur. Patent Appl. 140,385 (assigned to Occidental Chem. Corp., USA).

246. (a) Boutevin, B., Maubert, C., Mebkhout, A. and Pietrasanta, Y. (1981) Telomerization kinetics by redox catalysis. *J. Polym. Sci., Polym. Chem. Ed.*, **19**, 499–509; (b) Boutevin, B., Maubert, C., Pietrasanta, Y. and Sierra, P. (1981) Telomerization kinetics by redox catalysis: a study of reactions leading to telomers with low degrees of polymerization. *J. Polym. Sci., Polym. Chem. Ed.*, **19**, 511–522.

247. Boutevin, B., Doheim, M., Pietrasanta, Y. and Rigal, G. (1979). *J. Polym. Sci., Polym. Chem. Ed.*, **14**, 29–36.

248. Saran, M. S. (1983) Chlorotrifluoroethylene telomerization. Eur. Patent Appl. 93,580 (assigned to Occidental Chem., USA).

249. Boutevin, B., Pietrasanta, Y. and Sideris, A. (1975) Synthesis and chemical transformation of chlorotrifluoroethylene and 1,1,1-trichlorotrifluoroethane telomers. *CR Acced. Sci., Ser. C*, **281**, 405–408.

250. Henne, A. K. and Kraus, D. W. (1951) The direction of addition of carbon tetrachloride to $CFCl=CF_2$ under polar conditions. *J. Am. Chem. Soc.*, **73**, 5303–5304.

251. Pietrasanta, Y., Rabat, J. P. and Vernet, J. L. (1974) Redox-catalyzed telomerization. II. Telomerization of styrene, acrylonitrile, and chlorotrifluoroethylene by chlorinated telogens. *Eur. Polym. J.*, **10**, 633–637.

252. Dannels, B. F. and Olsen, D. J. (1989) Preparation of telomers of chlorotrifluoroethylene and trichlorotrifluoroethane. German Patent 3,837,394 (assigned to Occidental Chem. Corp., USA) (*Chem. Abstr.*, **112**, 21425).

253. Battais, A., Boutevin, B. and Pietrasanta, Y. (1979) Study of the reactivity of polyhalogenated esters with potassium fluoride. *J. Fluorine Chem.*, **14**, 467–484.

254. Boutevin, B. and Maliszewicz, M. (1983) Synthesis of photocrosslinkable cotelomers. 6. Utilization of cotelomers with two sequences. *Makromol. Chem.*, **184**, 977–989.

255. Boutevin, B., Hervaud, Y. and Pietrasanta, Y. (1984) Synthesis of phosphonic compounds with a chlorofluorinated chain. Part II. Telomerization of diethyl allyl- and vinylphosphonates. *Phosphorus Sulfur Related Elem.*, **20**, 189–196.

256. Park, J. and Lacher, J. P. (1973) Addition of hydrogen bromide to chlorotrifluoroethylene. Identification of the minor products. *Daehan Hwahak Hwoejee*, **17**, 70 (*Chem. Abstr.*, **73**, 158818).

257. Paleta, O., Liska, F. and Posta, A. (1970) Preparation of fluoroacetic acids based on trifluorochloroethylene. *Collect. Czech. Chem. Commun.*, **35**, 1302–1306.

258. Rogozinski, M., Shorr, L., Hasman, U. and Ader-Barcas, D. (1968) Bromine telomerization. *J. Org. Chem.*, **33**, 3859–3864.

259. Ehrenfeld, R. L. (1957) Oxygen-free halopolyfluoro compounds. US Patent 2,788,375 (*Chem. Abstr.*, **51**, 11759).

260. Dannels, B. F. and Tang, D. Y. (1989) Preparation of telomers of chlorotrifluoroethylene and 1,1-dibromo-1-chlorotrifluoroethane. German Patent 3,820,934 (assigned to Occidental Chem. Corp., USA) (*Chem. Abstr.*, **111**, 177752).

261. Barnhart, W. S. (1959) Telomerization with fluorobromoalkanes. US Patent 2,875,253 (assigned to Minnesota Mining and Manufacturing Co.).

262. Henne, A. L. and Kraus, D. W. (1954) The direction of free radical addition to chlorotrifluoroethylene. *J. Am. Chem. Soc.*, **76**, 1175–1176.

263. Kim, Y. K. (1968) ω-Bromoperfluoroalkanoic acids and derivatives. French Patent 1,533,794 (assigned to Dow Corning Corp.).

264. Boutevin, B., Cals, J. and Pietrasanta, Y. (1976) Telomerization by redox catalysis. 5. Telomerization of chlorotrifluoroethylene with bromotrichloromethane by redox catalysis. *Eur. Polym. J.*, **12**, 225–230.

265. Gumprecht, W. and Dettre, R. H. (1975) A novel telomerization procedure for controlled introduction of perfluoro-*n*-alkyl end groups. *J. Fluorine Chem.*, **5**, 245–263.

266. Amiry, M. P., Chambers, R. D., Greenhall, M. P., Améduri, B., Boutevin, B., Caporiccio, G., Gornowicz, G. A. and Wright, A. P. (1993) The peroxide-initiated telomerization of chlorotrifluoroethylene with perfluorochloroalkyl iodides. *Polym. Prepr. (Am. Chem. Soc., Div. Polym. Chem.)*, **34**, 411–412.

267. Haszeldine, R. N. (1955) Fluoroolefins. IV. Synthesis of polyfluoroalkanes containing functional groups from chlorotrifluoroethylene, and the short-chain polymerization of olefins. *J. Chem. Soc.*, 4291–4302.

268. Kremlev, M. M., Cherednichenko, P. G., Moklyachuk, L. I. and Yagupol'skii, L. M. (1989) Addition of 1-iodo-1,2-difluoro-1,2,2-trichloroethane and 1-iodo-1,2-dichloro-1,2,2-trifluoro-ethane with trifluorochloroethene and 1,2-difluorochloroethene. *Zh. Org. Khim. (Engl. Transl.)*, **25**, 2582–2588 (*Chem. Abstr.*, **113**, 39872).

269. Chambers, R. D., Greenhall, M. P., Wright, A. P. and Caporiccio, G. (1995) A new telogen for telechelic oligomers of chlorotrifluoroethylene. *J. Fluorine Chem.*, **73**, 87–94.

270. Haszeldine, R. N. and Steele, B. R. (1953) The addition of free radicals to unsaturated systems. III. Chlorotrifluoroethylene. *J. Chem. Soc.*, 1592–1600.

271. Henne, A. L. and Kraus, D. W. (1951) The direction of addition to chlorotrifluoroethylene under free radical conditions. *J. Am. Chem. Soc.*, **73**, 1791–1792.

272. Caporiccio, G. (1991) Chemically reactive fluorinated organosilicon compounds and their polymers. US Patent 5,041,588 (assigned to Dow Corning Corp., USA) (*Chem. Abstr.*, **115**, 281856).

273. Matsuo, H., Kawakami, S. and Ito, K. (1985) Chlorotrifluoroethylene oligomers. Japanese Patent 60,184,032 (assigned to Asahi Glass Co., Ltd, Japan) (*Chem. Abstr.*, **104**, 109007).

274. Kiseleva, L. N., Cherstkov, V. F., Sterlin, S. R., Mysov, E. I., Velichko, F. K. and German, L. S. (1989) Radical reactions of D2-perfluoroisohexenyl iodide with fluorine-containing ethylenes. *Izv. Akad. Nauk SSSR, Ser. Khim.*, 1130–1133 (*Chem. Abstr.*, **112**, 76318).

275. Balagué, J. (1994) Synthèse, caractérisations et applications de nouveaux télomères et silanes fluorés. PhD dissertation, Université de Montpellier.

276. Barnhart, W. S. (1969) Trifluorochloroethylene telomeric plasticizer. German Patent 1,296,383 (assigned to Minnesota Mining and Manufacturing Co.) (*Chem. Abstr.*, **71**, 39817).

277. Barnhart, W. S. (1958) Plasticized halogenated olefin polymers. US Patent 2,820,772 (assigned to Minnesota Mining & Manufacturing Co.) (*Chem. Abstr.*, **52**, 6845).

278. Liska, F., Nemec, M. and Dedek, V. (1974) Radical addition of secondary alcohols to trifluorochloroethylene and cyclization of fluorochloroalkanols under the formation of fluorinated oxetanes and tetrahydropyrans. *Collect. Czech. Chem. Commun.*, **39**, 580–586.

279. Chutny, B. (1980) Fluorochloroalkylmethanols and di(fluorochloroalkyl)methanols. Czech. Patent 182,999 (*Chem. Abstr.*, **94**, 140412).

280. Guiot, J., Alric, J., Ameduri, B., Boutevin, B. and Rousseau, A. (2001) Telomerization of chlorotrifluoroethylene with methanol. *New J. Chem.*, **25**, 1185–1190.

281. Chutny, B., Liska, F. and Dedek, V. (1969) Gamma-induced telomerization of trifluorochloro-ethylene in 2-propanol solution. *Large Radiat. Sources Ind. Processes, Proc. Symp. Util. Large Radiat. Sources Accel. Ind. Process*, 31–50.

282. Emrick, D. D. (1964) Telomerization of fluorinated hydrocarbons with alkylene glycol boranes. US Patent 3,136,808 (assigned to Standard Oil Co.).

283. Neouze, M. A., Guiot, J., Sauguet, L., Ameduri, B., Boutevin, B. and Lannuzel, T. Synthesis and copolymerization of fluoromonomer bearing a lateral functional group. Part 18. Copolymerization of vinylidene fluoride with 1,1,2-trifluoro-4-bromo-1-butene. Submitted to *J. Polym. Sci. Part A, Polym. Chem.*

284. Boutevin, B., Cals, J. and Pietrasanta, Y. (1975) Télomérisation par catalyse rédox. VI. Transformation chimique des télomères du chlorotrifluoroéthylène. *Eur. Polym. J.*, **12**, 231–240.

285. Kim, Y. K. (1967) Synthesis of monobromoperfluoroalkanecarboxylic acids and derivatives. *J. Org. Chem.*, **32**, 3673–3675.

286. (a) Chambers, R. D., Musgrave, W. K. R. and Savory, J. (1961) The addition of "halogen monofluorides" to fluoroolefins. *Proc. Chem. Soc.*, 113–118; (b) Chambers, R. D., Musgrave, W. K. R. and Savory, J. (1961) Mixtures of halogens and halogen polyfluorides as effective sources of the halogen monofluorides in reactions with fluoroolefins. *J. Chem. Soc.*, 3779–3786.

287. (a) Powell, R. L. and Chambers, R. D. (1996) Gamma-ray-induced preparation of trifluoroethylene telomers in the presence of organic chain transfer agents. UK Patent 2,292, 151 (assigned to Imperial Chem. Industries Plc, UK); (b) Chambers, R. D., Gilani, A. H. S., Gilbert, A. F., Hutchinson, J., Powell, R. L. (2000). Free radical chemistry. Part XII: Radical reactions of trifluorethene. *J. Fluorine Chem.* **106**, 53–67.

288. Fleming, G. L., Haszeldine, R. N. and Tipping, A. E. (1973) Addition of free radicals to unsaturated systems. XX. Direction of radical addition of heptafluoro-2-iodopropane to vinyl fluoride, trifluoroethylene, and hexafluoropropene. *J. Chem. Soc., Perkin Trans. 1*, 574–577.

289. Haszeldine, R. N., Mir, I. D., Tipping, A. E. and Wilson, A. G. (1976) Addition of free radicals to unsaturated systems. Part XXII. Photochemical addition of trifluoroiodomethane and iodine to perfluoro-(3-methylbut-1-ene) and photochemical dimerization of the olefin. *J. Chem. Soc., Perkin Trans. 1*, 1170–1173.

290. Haszeldine, R. N. and Steele, B. R. (1957) Addition of free radicals to unsaturated systems. XIV. The direction of radical addition to trifluoroethylene. *J. Chem. Soc.*, 2800–2806.

291. Haszeldine, R. N., Keen, D. W. and Tipping, A. E. (1970) Free-radical additions to unsaturated systems. XVII. Reaction of trifluoroiodomethane with mixtures of ethylene and vinyl fluoride and of ethylene and propene. *J. Chem. Soc. C: Org.*, 414–421.

292. Bissel, E. R. (1964) Some isomers of chloroiodotrifluoroethane. *J. Org. Chem.*, **29**, 252–254.

293. Améduri, B., Boutevin, B., Kharroubi, M., Kostov, G. and Petrova, P. (1998) Radical addition of iodine monochloride to trifluoroethylene. *J. Fluorine Chem.*, **91**, 41–48.

294. Ankudinov, A. K., Ryazanova, R. M. and Sokolov, S. V. (1974) Reactions of telomers of 1,2-dichloro-4-iodoperfluorobutane with trifluoroethylene. *Zh. Org. Khim.*, **10**, 2503–2507.

295. Kremlev, M. M., Moklyachuk, L. I., Fialkov, Y. A. and Yagupol'skii, L. M. (1984) Synthesis and properties of 1-chloroperfluoro-1,3-butadiene. *Zh. Org. Khim.*, **20**, 1162–1167.

296. Coscia, A. (1995) Free radical addition to trifluoroethylene. *J. Org. Chem.*, **26**, 1961–1963.

297. Haszeldine, R. N. and Osborne, J. E. (1956) Addition of free radicals to unsaturated systems. XII. Free-radical and electrophilic attack on fluoroolefins. *J. Chem. Soc.*, 61–71.

298. Putnam, R. (1989) Polymerization of fluoro monomers. In *Comprehensive Polymer Science* (Eds, Allen, G., Bevington, J. C., Eastmond, A. L. and Russo, A.) Pergamon Press, Oxford, pp. 321–326.

299. (a) Eleuterio, H. S. (1960) Perfluoropropylene polymers. US Patent 2,958,685 (assigned to E. I. du Pont de Nemours & Co.); (b) Eleuterio, H. S. and Pensak, D. (2001) Polymerization of hexafluoropropylene. Presented at the 13th European Symposium in Fluorine Chemistry, Bordeaux, July 2001, Presentation C-2.

300. Jaeger, H. (1978) Perfluoralkyl iodides. US Patent 4,067,916 (assigned to Ciba-Geigy A.-G.).

301. Haszeldine, R. N., Rowland, R., Tipping, A. E. and Tyrrell, G. (1982) Reaction of hexafluoropropene with haloalkanes. *J. Fluorine Chem.*, **21**, 253–259.

302. Lin, W. C., Chen, C.-H., Hung, H. C., Chou, H. Y. and Hsu, T. F. (1964) Telomerization of fluoroolefins. I. Telomerization of tetrafluoroethylene with aliphatic alcohols. *Ko Fen Tzu T'ung Hsun*, **6**, 363–367 (*Chem. Abstr.*, **63**, 16147).

303. (a) Paleta, O., Cirkva, V. and Kvicala, J. (1994) Photoaddition reactions of fluoroolefins with diols and cyclic ethers. *Macromol. Symp.*, **82**, 111–114; (b) Cirkva, V., Polak, R. and Paleta, O. (1996) Radical addition to fluoroolefins. Photochemical fluoroalkylation of alkanols and alkane diols with perfluoro vinyl ethers; photo-supported *O*-alkylation of butane-1,4-diol with hexafluoropropene. *J. Fluorine Chem.*, **80**, 135–144.

304. Cirkva, V. and Paleta, O. (1999) Radical addition reactions of fluorinated species. Part 7. Highly selective two-step synthesis of 1-(polyfluoroalkyl)ethane-1,2-diols; regioselectivity

of the additions of methylated 1,3-dioxolanes to perfluoroolefins. *J. Fluorine Chem.*, **94**, 141–156.

305. Haszeldine, R. N. (1953) Addition of free radicals to unsaturated systems. IV. The direction of radical addition to hexafluoropropene. *J. Chem. Soc.*, 3559–3564.

306. Tedder, J. M., Walton, J. C. and Vertommen, L. L. T. (1979) Free radical addition to olefins. Part 25. Addition of perfluoro-*tert*-butyl radicals to fluoroolefins. *J. Chem. Soc., Faraday Trans. 1*, **75**, 1040–1049.

307. Pollet, T. (1991) Etude de la télomérisation radicalaire d'oléfines fluorées. CNAM, Engineer Dissertation, University of Montpellier.

308. (a) Chambers, R. D., Fuss, R. W., Jones, M., Sartori, P., Swales, A. P. and Herkelmann, R. (1990) Free-radical chemistry. Part 8. Electrochemical fluorination of partly fluorinated ethers. *J. Fluorine Chem.*, **49**, 409–419; (b) Chambers, R. D., Diter, P., Dunn, S. N., Farren, C., Sandford, G., Batsanov, A. S. and Howard, J. A. K. (2000) Free radical chemistry. Part 11. Additions of cyclic and acyclic alcohols and diols to hexafluoropropene. *J. Chem. Soc., Perkin Trans. 1*, 1639–1649.

309. Low, H. C., Tedder, J. M. and Walton, J. C. (1976) Free radical addition to olefins. Part 19. Trifluoromethyl radical addition to fluoro-substituted propenes and its absolute rate of addition to ethylene. *J. Chem. Soc., Faraday Trans. 1*, **72**, 1300–1309.

310. Manseri, A. and Ameduri, B., unpublished results.

311. Kirshenbaum, A. D., Streng, A. G. and Hauptschein, M. (1953) The thermal decomposition of silver salts of perfluoro carboxylic acids. *J. Am. Chem. Soc.*, **75**, 3141–3145.

312. Baum, K. and Malik, A. (1994) Difunctional monomers based on perfluoropropylene telomers. *J. Org. Chem.*, **59**, 6804–6807.

313. Hauptschein, M., Braid, M. and Lawlor, F. E. (1957) Thermal syntheses of telomers of fluorinated olefins. I. Perfluoropropene. *J. Am. Chem. Soc.*, **79**, 2549–2553.

314. Deev, L. E., Nazarenko, T. I. and Pashkevich, K. I. (1988) New routes to perfluoro-*tert*-butyl iodide. *Izv. Akad. Nauk SSSR, Ser. Khim.*, 402–404 (*Chem. Abstr.*, **109**, 169809).

315. Manseri, A., Boulahia, D., Ameduri, B., Boutevin, B. and Caporiccio, G. (1999) Synthesis of fluorinated telomers. Part 6. Telomerization of hexafluoropropene with α,ω-diiodoperfluoro-alkanes. *J. Fluorine Chem.*, **94**, 175–182.

316. Manseri, A., Ameduri, B., Boutevin, B. and Caporiccio, G. (1996) Unexpected telomerization of hexafluoropropene with dissymmetrical halogenated telechelic telogens. *J. Fluorine Chem.*, **78**, 145–153.

317. Dyatkin, B. L., Mochalina, E. P. and Knunyants, I. L. (1961) Condensation of formaldehyde with perfluoroolefins–tetrafluoroethylene, hexafluoropropylene, and trifluorochloroethylene. *Dokl. Akad. Nauk SSSR*, **139**, 106–109.

318. Feast, W. J., Gimeno, M. and Koshravi, E. (2003) Approaches to highly polar polymers with low glass transition temperatures. 1. Fluorinated polymers via a combination of ring-opening metathesis polymerisation and hydrogenation. *Polymer*, **44**, 6111–6121.

319. Kalb, G. H., Coffman, D. D., Ford, T. A. and Jonhston, F. L. (1960) Polymers from vinyl fluoride. *J. Appl. Polym. Sci.*, **4**, 55–61.

320. Iracki, E. S. (2001) Fluorosurfacing with polyvinyl fluoride film: performance in selected applications. Fluorine in Coatings IV Conference, Brussels, Paper No. 31.

321. Walton, J. C., Johari, D. P., Sidebottom, H. W. and Tedder, J. M. (1971) Free-radical addition to olefins. VII. Addition of trichloromethyl radicals to chloro olefins. *J. Chem. Soc. B: Phys. Org.*, 95–99.

322. Walton, J. C. and Tedder, J. M. (1978) Kinetics and mechanisms of the addition of fluorine-containing radicals to olefins. *ACS Symp. Ser.*, **66**, 107–127 (Fluorine containing free radicals).

323. Haszeldine, R. N. and Steele, B. R. (1953) The addition of free radicals to unsaturated systems. II. Radical addition of olefins of the type RCH : CH$_2$. *J. Chem. Soc., Abstracts*, 1199–1206.

324. Rondestvedt, C. S., Jr. (1968) Perfluoroalkyl fluoriodides and their use for telomerization of perfluoroalkyl iodides with olefins. French Patent 1,521,775 (assigned to du Pont de Nemours, E.I., and Co.) (*Chem. Abstr.*, **71**, 2955).

325. Ashton, D. S., Mackay, A. F., Tedder, J. M., Tipney, D. C. and Walton, J. C. (1973) Effect of radical size and structure on the orientation of addition of perfluoroalkyl radicals to fluoroalkenes. *J. Chem. Soc., Chem. Commun.*, **14**, 496–497.

326. El-Soueni, A., Tedder, J. M., Vertommen, L. L. T. and Walton, J. C. (1978) Reactivity and selectivity in fluoroalkyl radical addition reactions. *Coll. Int. Centre Natl Recherche Scientifique*, **278**, 411–413.

327. Haszeldine, R. N. (1951) Reactions of fluorocarbon radicals. V. Alternative syntheses for (trifluoromethyl)acetylene (3,3,3-trifluoropropyne) and the influence of polyfluoro groups on adjacent hydrogen and halogen atoms. *J. Chem. Soc. Abstracts*, 2495–2504.

328. Ikonnikov, N. S., Lamova, N. I. and Terent'ev, A. B. (1988) Telomerization of 3,3,3-trifluoropropene with methyl isobutanoate. *Izv. Akad. Nauk SSSR, Ser. Khim.*, 117–121 (*Chem. Abstr.*, **109**, 169831).

329. (a) Zamyslov, R. A., Shostenko, A. G., Dobrov, I. V. and Myshkin, V. E. (1980). Radical telomerization of 3,3,3-trifluoro-1-propene with 2-propanol. *Zh. Org. Khim.*, **16**, 897–901 (*Chem. Abstr.*, **93**, 167229); (b) Shostenko, A. G., Dobrov, I. V. and Chertorizhskii, A. V. (1983) Preparation of telomeric fluorine-containing alcohols during gamma-radiation. *Khim. Prom.*, 339–340 (*Chem. Abstr.*, **99**, 123027); (c) Zamyslov, R. A. (1986) Reactivity of C1–C4 aliphatic alcohols in radical telomerization with 3,3,3-trifluoropropylene. *Zh. Vses. Khim. Obsh.*, **31**, 589–591 (*Chem. Abstr.*, **106**, 175716); (d) Zamyslov, R. A., Shostenko, A. G., Dobrov, I. V. and Tarasova, N. P. (1987) Kinetics of the reaction of 2-propanol with tri- and hexafluoropropylene initiated by γ-radiation. *Kinet. Katal.*, **28**, 977–979 (*Chem. Abstr.*, **108**, 74741).

330. Terent'ev, A. B. and Vasil'eva, T. T. (1995) Organobromine compounds in reactions of homolytic addition and telomerization. *Ind. Chem. Library*, **7**, 180–201.

331. Vasil'eva, T. T., Fokina, I. A., Vitt, S. V. and Dostovalova, V. I. (1990) Radical telomerization of 3,3,3-trifluoropropene-1 with CCl$_4$. *Izv. Akad. Nauk SSSR, Ser. Khim.*, **8**, 1807–1811 (*Chem. Abstr.*, **114**, 41974).

332. Keim, W., Raffeis, G. H. and Kurth, D. (1990) Transition metal catalysed carbon–carbon coupling reactions of 3,3,3-trifluoropropene. *J. Fluorine Chem.*, **48**, 229–237.

333. (a) Gasanov, R. G., Vasil'eva, T. T. and Gapusenko, S. I. (1991) Mechanism and rate constant of rearrangement of radicals C$_6$H$_5$(CH$_2$)$_2$CH(CF$_3$)CH$_2$CHCF$_3$ into C$_6$H$_5$CHCH$_2$CH(CF$_3$)CH$_2$CH$_2$CF$_3$ with 1,5-migration of hydrogen. *Kinet. Katal.*, **32**, 1466–1470 (*Chem. Abstr.*, **116**, 105413); (b) Vasil'eva, T. T., Fokina, I. A. and Vitt, S. V. (1991) Reaction of benzyl chloride with propylene and trifluoropropene under metal complex initiation conditions. *Izv. Akad. Nauk SSSR, Ser. Khim.*, 1384–1388 (*Chem. Abstr.*, **116**, 40981).

334. Vasil'eva, T. T., Kochetkova, V. A., Dostovalova, V. I., Nelyubin, B. V. and Freidlina, R. K. (1989) Reaction of 3,3,3-trifluoro-1-propene with bromoform and methylene bromide. *Izv. Akad. Nauk SSSR, Ser. Khim.*, 2558–2562 (*Chem. Abstr.*, **113**, 5666).

335. Vasil'eva, T. T., Kochetkova, V. A., Nelyubin, B. V., Dostovalova, V. I. and Freidlina, R. K. (1987) Radical telomerization of 3,3,3-trifluoro-1-propene with tetrabromomethane. Individual chain transfer constants. *Izv. Akad. Nauk SSSR, Ser. Khim.*, 808–811.

336. Haszeldine, R. N. (1952) Addition of free radicals to unsaturated systems. I. The direction of radical addition to 3,3,3-trifluoropropene. *J. Chem. Soc., Abstracts*, 2504–2513.

337. Block, H. D. (1976) Verfahren zur herstelling von 2-perfluoroalkyl phosphon-saeureestern. German Patent DE 2,514,640 (assigned to Bayer).

338. Terent'ev, A. B., Pastushenko, E. V., Kruglov, D. E. and Rybininia, R. E. (1992) Radical telomerization of 3,3,3,-trifluoropropene with 2-methyl-1,3-dioxolane. *Izv. Akad. Nauk SSSR, Ser. Khim.*, 2768–2772 (*Chem. Abstr.*, **120**, 106821).

339. Muramatsu, H. (1962) The radiation-induced addition reaction of alcohols to perhalogeno olefins. *J. Org. Chem.*, **27**, 2325–2328.

340. Kondo, K. and Yanagihara, T. (1980) Fluorine-containing dienes. Japanese Patent 55,000,305 (assigned to Toyo Soda Mfg. Co., Ltd).

341. Kondo, K. and Yanagihara, T. (1987) Fluorine-containing dienes. Japanese Patent 62,053,941 (assigned to Toyo Soda Mfg. Co., Ltd).

342. Modena, S., Fontana, A. and Moggi, G. (1985) Radical additions to 1,2-dichlorodifluoroethene. *J. Fluorine Chem.*, **30**, 109–121.

343. Guiot, J., Ameduri, B., Boutevin, B. and Fruchier, A. (2002) Synthesis of fluorinated telomers. Part 7. Telomerization of 1,1-difluoro-2-chloroethylene and 1,2-difluoro-1,2-dichloroethylene with methanol. *New J. Chem.*, **26**, 1768–1773.

344. Huang, Q. and DesMarteau, D. (2001) Synthesis and reaction chemistry of fluoroxydifluoromethyl fluoroformyl peroxide. *J. Fluorine Chem.*, **112**, 363–368.

345. Muramatsu, H., Inukai, K. and Ueda, T. (1964) The radiation-induced addition reaction of ethers to chloroolefins. *J. Org. Chem.*, **29**, 2220–2223.

346. (a) Inukai, K., Ueda, T. and Muramatsu, H. (1967) The syntheses of some phosphonates and sulfides containing a fluorovinyl group. *Bull. Chem. Soc. Jpn*, **40**, 1288–1290; (b) Inukai, K., Ueda, T. and Maramatsu, H. (1964) The radiation-induced addition reaction of dialkyl phosphonates to chlorofluoroolefins. *J. Org. Chem.*, **29**, 2224–2226.

347. Paleta, O. (1971) Addition reactions of haloolefins. X. The reactivity of monofluorochloromethanes and influence of solvents on the addition rate of halogenomethanes. *Collect. Czech. Chem. Commun.*, 2062–2066.

348. Paleta, O., Kvicala, J., Gunter, J. and Dedek, V. (1986) Synthesis of perfluoroallylchloride and some chlorofluoropropenes. *Bull. Soc. Chim. France*, **6**, 920–924.

349. Beaune, O., Bessière, J. M., Boutevin, B. and El Bachiri, A. (1995) Study of the alternating radical copolymerization of dichlorodifluoroethylene with ethyl vinyl ether. *J. Fluorine Chem.*, **73**, 27–32.

350. Gregory, R., Haszeldine, R. N. and Tipping, A. E. (1969) Addition of free radicals to unsaturated systems. XVI. A reinvestigation of radical addition to 1,1,3,3,3-pentafluoropropene. *J. Chem. Soc., C: Org.*, 991–995.

351. Nema, S. K., Francis, A. U., Narendranath, P. K. and Rao, K. V. C. (1979) Synthesis of telomers of bromotrifluoroethylene by UV radiation. *Ind. Polym. Radiat., Proc. Symp.*, 89–102 (*Chem. Abstr.*, **94**, 122008).

352. Dittman, A. L. (1972) High-density fluorobromoalkanes for gyroscope flotation. US Patent 3,668,262 (assigned to Halocarbon Products Corp.).

353. Demiel, A. (1960) Addition of alcohols to fluorinated ethylenes. *J. Org. Chem.*, **25**, 993–996.

354. Chen, L. (1990) Free-radical initiated addition of carbon tetrachloride to fluoroolefins. *J. Fluorine Chem.*, **47**, 261–272.

355. Kurykin, M. A., German, L. S., Kartasheva, L. I. and Pikaev, A. K. (1996) Radiochemical addition of alcohols to perfluoro-2-alkenes. *J. Fluorine Chem.*, **77**, 193–194.

356. Gregory, R., Haszeldine, R. N. and Tipping, A. E. (1968) Addition of free radicals to unsaturated systems. XV. Investigation of the direction of radical addition to chloro-1,1-difluoroethylene. *J. Chem. Soc. C: Org.*, 3020–3025.

357. Boutevin, B. and Pietrasanta, Y. (1985) The synthesis and applications of fluorinated silicones, notably in high-performance coatings. *Prog. Org. Coat.*, **13**, 297–331.

358. Boutevin, B., Guida-Pietrasanta, F. and Ratsimihety, A. (2000) In *Silicon-Containing Polymers* (Eds, Jones, R. G., Ando, W. and Chojnowski, J.) Kluwer Academic Publishers, Dordrecht, pp. 79–112.

359. Marciniec, B. (1992) *Comprehensive Handbook on Hydrosilylation*, Pergamon Press, Oxford.

360. Hauptschein, M., Braid, M. and Lawlor, F. E. (1966) US Patent 3,240,825 (assigned to Pennsalt Chem. Co.) (*Chem. Abstr.*, **64**, 17422).

361. Paciorek, K. L., Merkl, B. A. and Lenk, C. T. (1962) Reaction of 4,4-dihydro-3-iodoperfluoroheptane and 2,2-dihydro-3-iodoperfluoropentane. *J. Org. Chem.*, **27**, 1015–1018.

362. Cirkva, V., Paleta, O., Améduri, B. and Boutevin, B. (1995) Radical additions to fluoroolefins. Thermal reaction of perfluoroallyl chloride with perfluoroalkyl iodides as a selective synthesis of terminal perfluoroolefins. *J. Fluorine Chem.*, **75**, 87–92.

363. Battais, A., Boutevin, B., Cot, L., Granier, W. and Pietrasanta, Y. (1979) Fluorination of polyhalo telomers by potassium fluoride: synthesis of perfluoroolefins. *J. Fluorine Chem.*, **13**, 531–550.

364. Battais, A., Boutevin, B., Cot, L., Granier, W. and Pietrasanta, Y. (1979) Etude de la réactivité du fluorure de potassium sur des esters polyhalogénés. *J. Fluorine Chem.*, **14**, 467–484.

365. Bargigia, G., Tortelli, V., Tonelli, C. and Modena, S. (1988) Process for the synthesis of perfluoroalkadienes. Eur. Patent Appl. 270,956 (assigned to Ausimont S.p.A., Italy).

366. Fearn, J. E., Brown, D. W. and Wall, L. A. (1966) Polymers and telomers of perfluoro-1,4-pentadiene. *J. Polym. Sci., Polym. Chem. Ed.*, **4**, 131–140.

367. (a) Boutevin, B. and Améduri, B. (1994) Copolymerization of fluorinated monomers with nonfluorinated monomers. Reactivity and mechanisms. *Macromol. Symp.*, **82**, 1–17; (b) Ameduri, B. and Boutevin, B. (2000) Copolymerisation of fluorinated monomers: recent developments and future trends. *J. Fluorine Chem.*, **104**, 53–62.

368. Brace, N. O. (1993) Free-radical addition of 2-(perfluoroalkyl)ethanethiols to alkenes, alkadienes, cycloalkenes, alkynes and vinyl monomers. *J. Fluorine Chem.*, **62**, 217–241.

369. (a) Brace, N. O. (1999) Syntheses with perfluoroalkyl radicals from perfluoroalkyl iodides. A rapid survey of synthetic possibilities with emphasis on practical applications. Part one: alkenes, alkynes and allylic compounds. *J. Fluorine Chem.*, **93**, 1–25; (b) Brace, N. O. (1999) Syntheses with perfluoroalkyl iodides. Part II. Addition to non-conjugated alkadiene; cyclization of 1,6-heptadiene, and of 4-substituted 1,6-heptadienoic compounds: *bis*-allyl ether, ethyl diallylmalonate, *N,N′*-diallylamine and *N*-substituted diallylamines; and additions to homologous *exo*- and *endo*-cyclic alkenes, and to bicyclic alkenes. *J. Fluorine Chem.*, **96**, 101–127 (and references herein).

370. Hauptschein, M. and Braid, M. (1963) Halogenated organic compounds. US Patent 3,089,991 (assigned to Pennsalt Chemicals Corp.).

371. Hauptschein, M. and Braid, M. (1963) Cotelomer oils from vinylidene fluorides. US Patent 3,091,648 (assigned to Pennsalt Chemicals Corp.).

372. Wolff, N. E. (1959) Fluorine-containing co-telomers. US Patent 2,907,795 (assigned to E. I. du Pont de Nemours & Co.).

373. Wolff, N. E. (1958) Fluorine containing cotelomers. US Patent 2,856,440 (assigned to E. I. du Pont de Nemours & Co.).

374. Modena, S., Fontana, A., Moggi, G. and Bargigia, G. (1988) Cotelomers of vinylidene fluoride with fluorinated olefins and process for their preparation. Eur. Patent Appl. 251,284 (assigned to Ausimont S.p.A., Italy).

375. Saint-Loup, R., Manseri, A., Ameduri, B., Lebret, B. and Vignane, P. (2002) Copolymerization of vinylidene fluoride with hexafluoropropene initiated by hydrogen peroxide. *Macromolecules*, **35**, 1524–1536.

376. (a) Arcella, V., Brinati, G., Albano, M. and Tortelli, V. (1995) New fluorinated thermoplastic elastomers having superior mechanical and elastic properties, and preparation process thereof. Eur. Patent Appl. 0,683,186 A1 (assigned to Ausimont S.p.A.); (b) Arcella, V., Brinati, G., Albano, M. and Tortelli, V. (1995) Fluorinated thermoplastic elastomers having superior mechanical and elastic properties. Eur. Patent Appl. 0,661,312 A1 (assigned to Ausimont S.p.A.); (c) Arcella, V., Brinati, G. and Apostolo, M. (1997). New high performance fluoroelastomers. *Chim. Oggi.* **79**, 345–351; (d) Brinati, G. and Arcella, V. (2001). Branching and pseudo-living technology in the synthesis of high performance fluoroelastomers. *Rubber World.* **224**, 27–32; (e) Arrigoni, S. and Apostolo, M. (2003). Fluoropolymers architecture design: the branching and pseudo-living technology. *Trends Polym. Sci.* **7**, 23–40.

377. Tatemoto, M. and Nakagawa, T. (1978) Segmented polymers. German Patent 2,729,671 (assigned to Daikin Kogyo Co., Ltd, Japan) (*Chem. Abstr.*, **88**, 137374).

378. Oka, M. and Tatemoto, M. (1984) Vinylidene fluoride–hexafluoropropylene copolymer having terminal iodides. *Contemp. Top. Polym. Sci.*, **4**, 763–777.

379. (a) Hung, M. H. (1993) Iodine-containing chain transfer agents for fluoroolefin polymerization. US Patent 5,231,154 (assigned to DuPont); (b) Carlson, D. P., Fluorinated thermoplastic elastomers with improved base stability. Eur. Patent Appl. 0,444,700 A2 (1991) and US Patent 5, 284,920 (1994) (assigned to E.I. DuPont de Nemours and Co.).

380. Hauptschein, M. and Braid, M. (1961) Halogenated organic compounds. US Patent 3,002,031 (assigned to Pennsalt Chemicals Corp.).

381. Bannai, N., Yasumi, H. and Hirayama, S. (1986) Fluorine-containing monomer for dental plate. US Patent 4,580,981 (assigned to Kureha Chem. Industry Co., Ltd, Japan).

382. Tatsu, H., Okabe, J., Naraki, A., Abe, M. and Ebina, Y. (1987) Peroxide-vulcanizable fluororubbers. German Patent 3,710,818 (assigned to Nippon Mecktron Co., Ltd) (*Chem. Abstr.*, **108**, 76908).

383. Caporiccio, G. (1994) Fluorinated compounds containing hetero atoms and polymers thereof. US Patent 5,350,878 (assigned to Dow Corning Corp., USA).

384. Kasahara, M. and Kai, M. (1994) Fluoro rubber oil seals. Japan Patent 633,040 (Asahi Chem. Ind.) (*Chem. Abstr.*, **122**, 190156).

385. Fossey, J., Lefort, D. and Sorba, J. (1995) *Free Radicals in Organic Chemistry*, Wiley, New York.

386. March, J. (1992) *Advanced Organic Chemistry*, Wiley, New York.

387. (a) Arnaud, R., Subra, R., Barone, V., Lelj, F., Olivella, S., Sole, A. and Russo, N. (1986) Ab-initio mechanistic studies of radical reactions. Directive effects in the addition of methyl radical to unsymmetrical fluoroethenes. *J. Chem. Soc., Perkin Trans. 2*, 1517–1524; (b) Arnaud, R. and Vidal, S. (1992) Energetics of CH_3, CF_3 and OCH_3 addition reactions to substituted olefins. *New J. Chem.*, **16**, 471–474; (c) Arnaud, R. and Vidal, S. (1986). *J. Chem. Soc., Perkin Trans. II*, 1517.

388. (a) Geise, B. (1983) Formation of carbon–carbon bonds by addition of radicals to alkenes. *Angew. Chem.*, **95**, 771–782; (b) Geise, B. (1997) Structure–reactivity correlation of radical reactions with alkenes. *Polym. Prepr. (Am. Chem. Soc.)*, **38**(1), 639–640; (c) Ghosey-Geise, A. and Geise, B. (1998) Factors influencing the addition of radicals to alkenes. *ACS Symp. Ser.*, **685**, 50–61 (Controlled radical polymerization).

389. Ruechardt, C. (1980) Steric effects in free radical chemistry. *Top. Curr. Chem.*, **88**, 1–32.

390. Beckwith, A. (1981) Regio-selectivity and stereo-selectivity in radical reactions. *Tetrahedron*, **37**, 3073–3100.

391. Koutecky, V. B., Koutecky, J. and Salem, L. (1977) A theory of free radical reactions. *J. Am. Chem. Soc.*, **99**, 842–850.

392. (a) Canadell, E., Eisenstein, O., Ohanessian, G. and Poblet, J. M. (1985) Theoretical analysis of radical reactions: on the anomalous behavior of methyl toward fluoro-substituted olefins. *J. Phys. Chem.*, **89**, 4856–4861; (b) Poblet, J. M., Canadell, E. and Sordo, T. (1983) Concerning the orientation in free radical additions to olefins. *Can. J. Chem.*, **61**, 2068–2069.

393. Delbecq, F., Ilavsky, D., Anh, N. and Lefour, J. (1985) Theoretical study of regioselectivity in radical additions to substituted alkenes. 1. Hydrogen addition to ethylene, vinyl amine, and vinyl borane. *J. Am. Chem. Soc.*, **107**, 1623–1631.

394. Wong, M., Pross, A. and Radom, L. (1994) Addition of *tert*-butyl radical to substituted alkenes: a theoretical study of the reaction mechanism. *J. Am. Chem. Soc.*, 11938–11943.

395. Rozhkov, I. N. and Borisov, Y. A. (1989) Electronic structure and reactivity of organofluorine compounds. 1. Electronic structure of fluoroethylenes calculated by MNDO and STO-3GF methods. *Izv. Akad. Nauk SSSR, Ser. Khim.*, 1801–1805.

396. Rempel, G. D., Borisov, Y. A., Raevskii, N. I., Igumnov, S. M. and Rozhkov, I. N. (1990) Electronic structure and reactivity of organofluorine compounds. 3. MNDO and AM1 calculations of the ground state of perfluoroalkyl chlorides, bromides, and iodides. *Izv. Akad. Nauk SSSR, Ser. Khim.*, **5**, 1059–1063.

397. Xu, Z., Li, S. S. and Hu, C. M. (1991) Reactions of perchlorofluoro compounds. VIII. A structure–reactivity relationship of perchlorofluoroolefins based on MNDO calculations. *Chinese J. Chem.*, **9**, 335–342.

398. Tedder, J. M., Walton, J. C. and Winton, K. D. R. (1971) Orientation of methyl radical addition to fluoroethylenes. *J. Chem. Soc. D: Chem. Commun.*, 1046–1047.

399. Sidebottom, H. W., Tedder, J. M. and Walton, J. C. (1972) Arrhenius parameters for the reactions of trichloromethyl radicals. *Int. J. Chem. Kinet.*, **4**, 249–252.

400. Tedder, J. M. and Walton, J. C. (1980) The halogenation of the cycloalkanes and their derivatives. *Adv. Free-Radical Chem. (London)*, **6**, 155–184.

401. El Soueni, A., Tedder, J. M. and Walton, J. C. (1978) Free radical addition to olefins. Part 23. The addition of pentafluoroethyl radicals to fluoroolefins. *J. Fluorine Chem.*, **11**, 407–417.

402. Vertommen, L. L. T., Tedder, J. M. and Walton, J. C. (1977) Free radical addition to olefins. Part 21. The orientation and kinetics of the addition of perfluoroisopropyl radicals to fluoroolefins. *J. Chem. Res., Part B: Synop.*, 18–19.

403. Dolbier, W. R. (1997) Fluorinated free radicals. *Top. Curr. Chem.*, **192**, 97–164.

404. Klabunde, K. J. (1970) Study of the photolytic stabilities of fluoroalkyl iodides by electron spin resonance trapping techniques and the temperature dependence of the nitroxide splitting constants. *J. Am. Chem. Soc.*, **92**, 2427–2432.

405. Okafo, E. N. and Whittle, E. (1975) Competitive study of the reactions Br + RfI.far. IBr + Rf and determination of bond dissociation energies D(Rf–I) where Rf = trifluoromethyl, pentafluoroethyl, heptafluoropropyl, iso-heptafluorpropyl, and nonafluorobutyl. *Int. J. Chem. Kinet.*, **7**, 287–300.

406. Liska, F. and Simek, S. (1970) Radical addition of methanol to trifluorochloroethylene. *Collect. Czech. Chem. Commun.*, **35**, 1752–1759.

407. (a) Estrine, B., Soler, R., Damez, C., Bouquillon, S., Henin, F. and Muzart, J. (2003) Recycling in telomerization of butadiene with methanol and phenol: Pd-KF/Al$_2$O$_3$ as an active heterogenous catalyst system. *Green Chem.*, **5**, 686–689; (b) Behr, A. and Urschey, M. (2003) Highly selective biphasic telomerization of butadiene with glycols: scope and limitations. *Adv. Synth. Catal.*, **345**, 1242–1246.

408. Boutevin, B. (2000) From telomerization to living radical polymerization. *J. Polym. Sci., Part A: Polym. Chem.*, **38**, 3235–3243.

409. Tatemoto, M. (1985) Iodine transfer polymerization of fluoroalkenes. *Int. Polym. Sci. Technol.*, **12**(4), 85–97 (translated in English from *Nippon Gomu Kyokaishi*, **57**, 761–770 (1984)).

410. Vergé, C. (1991) Télomérisation des iodures de perfluoroalkyle avec le tétrafluoréthylène, Université de Montpellier II, France.

411. (a) Owen, M. J. and Kobayashi, H. (1994) Surface active fluorosilicone polymers. *Macromol. Symp.*, **82**, 115–123; (b) Ameduri, B., Boutevin, B., Caporiccio, G., Guida-Pietrasanta, F. and Ratsimihéty, A. (1999) Use of fluorinated telomers in the preparation of silicones. In *Fluoropolymers Synthesis*, Vol. 1 (Eds, Hougham, G., Davidson, T., Jonhs, K. and Cassidy, R.) Plenum Press, New York, pp. 65–74, Chap. 5.

412. Saran, M. S. (1983) Fluorinating chlorofluoro telomers. Eur. Patent Appl. 93,579 A2 (assigned to Occidental Chem. Corp., USA).

413. Abe, T. and Nagase, S. (1982) Electrochemical fluorination as a route to perfluorinated organic compounds of industrial interest. In *Preparation, Properties and Industrial Applications of Organofluorine Compounds* (Ed, Banks, R.) Ellis Horwood, Chichester, pp. 19–34.

414. Optical fibers (1983). Japan Patent 58,007,602 (assigned to Mitsubishi Rayon Co., Ltd, Japan) (*Chem. Abstr.*, **99**, 72093).

415. (a) Ameduri, B., Boutevin, B., Kostov, G. and Petrova, P. (1999) Synthesis and polymerization of fluorinated monomers bearing a reactive side group. Part 9. Bulk copolymerization of vinylidene fluoride with 4,5,5-trifluoro-4-ene-pentyl acetate. *Macromolecules*, **32**, 4544–4550;

(b) Petrova, P., Ameduri, B., Boutevin, B. and Kostov, G. (2000) Functional trifluorvinyl monomers and their copolymerization with fluorinated olefins. Int. Demand WO 00/31009 (assigned to Solvay).

416. Valaitis, J. K. and Kyker, G. S. (1979) Thermal stability of phosphonitrilic fluoroelastomers. *J. Appl. Polym. Sci.*, **23**, 765–775.

417. (a) Tate, D. P. (1974) Polyphosphazene elastomers. *J. Polym. Sci., Polym. Symp.*, **48**, 33–45; (b) Kyker, G. S. and Antkowiak, T. A. (1974) Phosphonitrilic fluoroelastomers. New class of solvent resistant polymers with exceptional flexibility at low temperature. *Rubber Chem. Technol.*, **47**, 32–39; (c) Tate, D. P. and Antkowiak, T. A. (1980). In *Elastomers*, Vol. 10 (Eds, Kirk, R. E. and Othmer, D. F.) Ency. Polym. Sci. Technol., pp. 935–954; (d) Allcock, H. R., Rutt, J. S. and Fitzpatrick, R. J. (1991) Surface reaction of poly[bis(trifluoroethoxy)phosphazene] films by basic hydrolysis. *Chem. Mater.*, **3**, 442–453; (e) Allcock, H. R. (2002). *Chemistry and Applications of Polyphosphazenes*, Wiley, New York.

418. Kobayashi, H. and Owen, M. J. (1995) Surface properties of fluorosilicones. *Trends Polym. Sci.*, **3**, 330–335.

419. Boutevin, B., Ranjalahy-Rasoloarijao, L., Rousseau, A., Garapon, J. and Sillion, B. (1992) Synthesis of polyquinazolones with fluorinated side chains. *Makromol. Chem.*, **193**, 1995–2006.

420. (a) Takekoshi, T. (1990) Polyimides. *Adv. Polym. Sci.*, **94**, 1–25; (b) Hougham, G., *Fluorine-Containing Polyimides*. In Ref. [3], pp. 233–276, Chap. 13; (c) Ando, S., Matsuura, T. and Sasaki, S., *Synthesis and Properties of Perfluorinated Polyimides*. In Ref. [3], pp. 277–304, Chap. 14; (d) Matsuura, T., Ando, S. and Sasaki, S., *Synthesis and Properties of Partially Fluorinated Polyimides for Optical Applications*. In Ref. [3], pp. 305–350, Chap. 15; (e) Harris, F. W., Li, F. and Cheng, S. Z. D. *Novel Organo-soluble Fluorinated Polyimides for Optical, Microelectronic, and Fiber Applications*. In Ref. [3], pp. 351–370, Chap. 16.

421. (a) Boutevin, B. and Robin, J. J. (1992) Synthesis and properties of fluorinated diols. *Adv. Polym. Sci.*, **102**, 105–131; (b) Améduri, B. and Boutevin, B. (1992) Synthesis and applications of fluorinated telechelic monodispersed compounds. *Adv. Polym. Sci.*, **102**, 133–169.

422. Japan Patent 59,117,504 (1984) Polymerizable fluorine-containing compounds (assigned to Kureha Chem. Ind. Co. Ltd) (*Chem. Abstr.*, **102**, 26116).

Chapter 2

SYNTHESIS OF FLUORINATED TELECHELICS AS PRECURSORS OF WELL-DEFINED FLUOROPOLYMERS

1. Introduction

Fluorinated polymers exhibit a unique combination of high thermal stability, chemical inertness (to acids, bases, solvents and petroleum), low dielectric constants and dissipation factors, low water absorptivities and refractive index, excellent weatherability and a good resistance to oxidation and ageing, low flammabilities and very interesting surface properties [1–3]. Therefore, such products are involved in many applications (aerospace, aeronautics, engineering, optics, textile finishing and microelectronics). In spite of their high price and difficult processing due to their high melting points, fluoropolymers have an increasing market.

Well-architectured polymers can be prepared either from controlled radical polymerisation [4–6] or from polycondensation or polyaddition. In these last two processes, telechelic [7] or α,ω-difunctional (i.e., the functional groups are located at both extremities of the chain) precursors are required to obtain high-molecular-weight materials with satisfactory properties [8]. In the case of fluorinated telechelics, the literature describing their synthesis is abundant and it was reviewed several years ago [9–11].

Two kinds of fluorinated telechelics exhibiting different properties are suggested and, hence, the polymers prepared from them can be involved in different applications. On the one hand, those possessing the fluorinated group in a lateral position about the polymeric backbone are exploited for their enhanced surface properties, but cannot be used as thermostable and chemical resistant materials. On the other hand, those containing the fluorinated chain in the backbone exhibit much better thermal properties and are also resistant to aggressive media, UV radiation and ageing.

The syntheses of both types of telechelics are described hereafter. These products range from molecular to polymeric compounds and from mono-dispersed to polydispersed.

2. Synthesis of Fluorinated Telechelics

This section is split into two parts corresponding to both series of fluorotelechelics mentioned above, the difference arising from the location of the fluorinated group. Various strategies have been proposed to prepare fluorinated telechelics.

2.1. Telechelics Having a Fluorinated Chain in the Backbone

First, the industrial derivatives will be presented, starting with molecular telechelics.

2.1.1. Industrial Products

2.1.1.1. Short Fluorotelechelics

To our knowledge, Exfluor Research Corp. is the only company marketing telechelic diols $HOCH_2(CF_2)_nCH_2OH$ with $n = 2$, 4, 6, 8, 10 and diacids prepared by direct fluorination [12] of the hydrogenated precursors [13].

2.1.1.2. Longer Fluorotelechelics

A non-exhaustive list of two classes of fluorinated telechelic polymers consists of the perfluoropolyethers and the fluorosilicones.

2.1.1.2.1. *Perfluoropolyethers.* Nowadays, among fluorinated telechelics prepared on an industrial scale, only four commercially available polyethers have been obtained. Their synthesis is described as follows:

(i) the anionic polymerisation of perfluoroepoxides (especially hexafluoropropylene oxide, HFPO) led to Krytox® produced by the DuPont de Nemours company [14, 15].

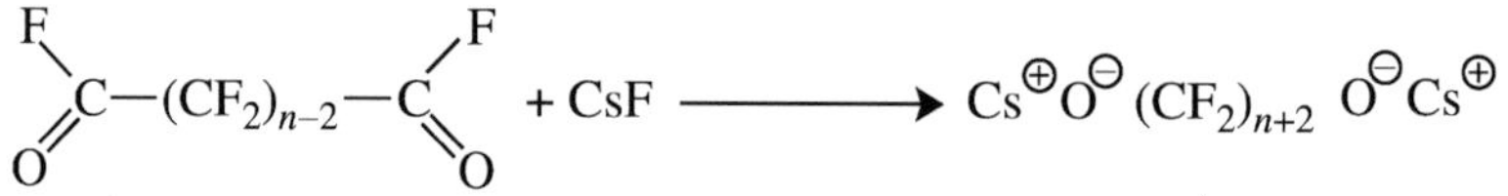

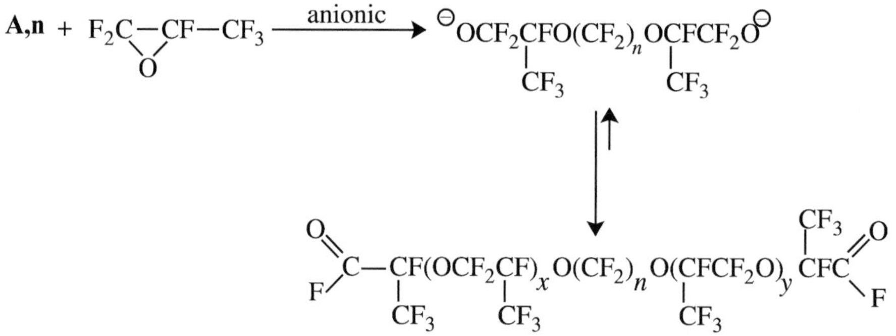

(ii) the photo-oxidation of perfluoroolefins (e.g., tetrafluoroethylene [16], hexafluoropropene (HFP) [17] and perfluorobutadiene [18]) was performed by the Ausimont company, yielding functional or non-functional (or neutral) Fomblin® oligomers ranging from 1000 to 4000 in molecular weights [19, 20];

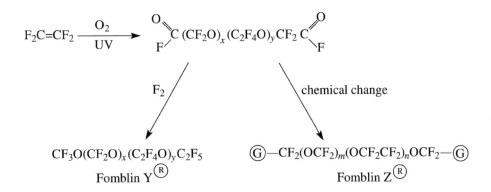

(iii) the Daikin Company markets Demnum® oligomers, which can be inert or functional, from the ring opening polymerisation of fluorinated oxetanes followed by fluorination to obtain thermostable oils [21].

$$\begin{matrix} CF_2{-}CF_2 \\ | \qquad | \\ CH_2{-}O \end{matrix} \xrightarrow{F^{\ominus}} F(CH_2CF_2CF_2O)_nCH_2CF_2COF \xrightarrow{F_2} \text{stable oil}$$

(iv) The ring opening polymerisation of HFPO also leads to PFPE marketed by Unimatec (and formerly by NOK Corp. or Nippon Mektron Ltd) under the Aflunox® trade name.

The syntheses, properties and applications of perfluoropolyethers have been extensively reviewed by several authors [22–27].

Perfluoropolyethers (PFPE), which also constitute a special class of fluoropolymers, have found various applications such as lubricants [22], elastomers [24], pump fluids and heat transfer fluids under demanding conditions. These liquid perfluoropolymers have very high gas solubilities which make them useful. Advantage is taken of gas permeable thin films in the development of cosmetics and barrier creams based on PFPE that offer a very high degree of skin protection and moisture retention but still allow the skin to breathe. Further, they can offer protection to ancient buildings where application of a monomolecular film of PFPE to the stone surface can provide protection against the aggressive acid rain but can still allow internal absorbed water in the stone to escape as vapour. These PFPEs are also useful for lubrication of thin film magnetic media [25, 26].

Four main families of fluorinated polyethers are produced on an industrial scale. These are Fomblin®, Krytox®, Demnum® and Aflunox® produced by Ausimont, DuPont, Daikin and Unimatec companies, respectively [24].

Lagow et al. [27] and Chambers et al. [28] achieved their syntheses by direct fluorination of hydrogenated polyethers; this process requires much care to avoid side-reactions such as cleavages of chains. Nowadays, the production of PFPE (by direct fluorination) is developed by Exfluor Research Corp. [29].

Some interesting characteristics are their low T_gs (lower than $-50\,°C$) and their totally amorphous state thanks to their aliphatic structure and to their ether links [24, 27]. In addition, these products, endowed with high thermostability, excellent chemical inertness and low surface tension, find relevant applications as lubricants and heat carrier-fluids in hostile conditions.

PFPE can be homopolymers or copolymers and can exhibit linear or branched structures, as follows:

$CF_3(OCF_2CF_2)_p(OCF_2)_qOCF_3$ $(1000 < \bar{M}_n < 4000)$ Fomblin® Z

$CF_3CF_2CF_2O(CF_2CF_2CF_2O)_nCF_2CF_3$ Demnum®
   $(1600 < \bar{M}_n < 7000)$

$CF_3CF_2[OCF_2CF(CF_3)]_nF$ $(1300 < \bar{M}_n < 10{,}300)$ Krytox®

$CF_3O[CF_2CF(CF_3)O]_p(CF_2O)_nCF_3$ Fomblin® Y
   $(600 < \bar{M}_n < 7000)$

$CF_3(OCF_2CF_2)_p[OCF_2CF(CF_3)]_q(OCF_2)_rOCF_3$ Fomblin® K
   $(900 < \bar{M}_n < 10{,}000)$

$-[O-CF(CF_3)CF_2]_n-$ $(1350 < \bar{M}_n < 16{,}600)$ Aflunox® or Aflunox® V

On a smaller scale, the Hoescht company developed HFPO polymers leading to fluorinated polyether acryl fluoride which, after hydrolysis and decarboxylation, becomes Hostinert® PFPE [30].

Functionalised PFPEs are well reported [20, 23–25] with most investigations having been performed by Ausimont. They are:

$X-CF_2(OCF_2)_m(OCF_2CF_2)_nOCF_2-X$ with X: COOH   Fomblin® Z DIAC

                                              X: COOCH$_3$ Fomblin® Z-DEAL

                                              X: CH$_2$OH   Fomblin® Z-DOL

                                   X: CH$_2$(OCH$_2$CH$_2$)$_p$OH   Fomblin® Z DOL TX

                         X: CH$_2$OCH$_2$CH(OH)CH$_2$OH   Fomblin® Z TETRAOL

These polymers are usually insoluble in common organic solvents and are in some cases prone to reaction.

2.1.1.2.2. *Fluorosilicones.* A few chemical industries produce fluorinated silicones, the main leader being the Dow Corning company. Poly(trifluoropropyl methyl siloxanes) represent more than 95% of commercially available fluorosilicones. They are prepared from the ring opening polymerisation of this monomer being prepared by hydrosilylation of

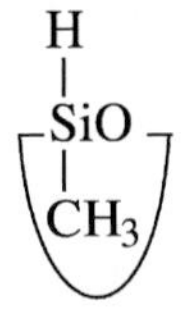

$$
\begin{array}{c}
H \\
| \\
-SiO- \\
| \\
CH_3
\end{array}
$$

with $CF_3CH{=}CH_2$. In convenient conditions,

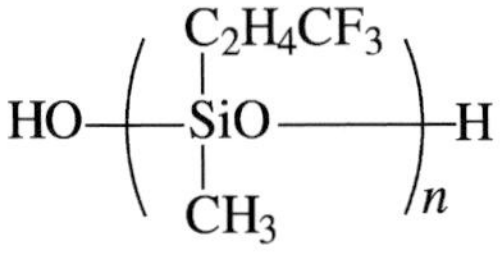

$$
HO\!\!\left(\!\!\begin{array}{c}
C_2H_4CF_3 \\
| \\
SiO \\
| \\
CH_3
\end{array}\!\!\right)_{\!\!n}\!\!H
$$

can be produced [31].

Additional information on fluorosilicones, with emphasis on hybrid systems, is reported in Section 3.5 and also in Chapter 4, Section 2.3.2.

In spite of the lack of industrial fluorotelechelics (different from polyethers) to give fluorinated non-functional homo- and copolymers, the current trend shows a growing interest as evidenced by numerous investigations [9–11]. It seems that many reactions involve α,ω-diiodoperfluoroalkanes, while fewer require α,ω-bis(trichloromethyl) chlorofluorinated telomers as reactants to prepare telechelic intermediates preserving the fluorinated chain in the backbone. Thus, in the following section, the syntheses of these diiodides are first reported, followed by their functionalisation into fluorotelechelics. Later, the synthesis and use of chlorofluorotelomers are reported.

2.1.2. Non-industrial Products

2.1.2.1. Short Fluorotelechelics
2.1.2.1.1. By Direct Fluorination. Classical fluorinations by electrochemical fluorination (known as the Simons process) or those realised in the presence of HF or F_2 have already shown their efficiency and are reported in many books [1–3, 32]:

$$HO_2C-(CH_2)_x-CO_2H \xrightarrow{HF} HO_2C-(CF_2)_x-CO_2H$$

2.1.2.1.2. From Telechelic Fluorinated Diiodides
2.1.2.1.2.1. Synthesis of α,ω-Diiodoperfluoroalkanes. In contrast to perfluoroalkyl iodides, which are widely used as reactants for the synthesis of fluorinated derivatives [1, 33, 34], α,ω-diiodo(per)fluoroalkanes have been involved in fewer experiments because of the difficulty of their synthesis and their consequent high price.

However, a large variety of preparations of α,ω-diiodo(per)fluoroalkanes have been investigated and can be gathered in two different ways: from organic synthesis (mainly by chemical change of fluorotelechelics) and from telomerisation reactions of fluoroolefins with either iodine or α,ω-diiodoperfluoroalkanes.

2.1.2.1.2.1.1. From Organic Synthesis. The literature reports various methods of syntheses leading to different chain length diiodides exhibiting specific groups or functions in the fluorinated chain (e.g., ether linkage or perfluorinated group).

2.1.2.1.2.1.1.1. Synthesis of ICF₂I. Difluorodiiodomethane is the shortest fluorinated diiodide, synthesised for the first time by Mahler [35] in 1963. His low-yield procedure involved the addition of difluorocarbene to molecular iodine. Later, Mitsch [36], followed by Burton *et al.* [37, 38] and then by Chen and Zhu [39], slightly improved the yield.

In 1984, Elsheimer *et al.* [40] reported the synthesis of ICF_2I by fluorination of readily available tetraiodomethane [41, 42] with mercury (II) fluoride as follows:

$$CI_4 + HgF_2 \rightarrow ICF_2I + HgI_2$$

However, the preparation described by Su *et al.* [43] substantially improved the yield of ICF_2I up to 60%. This Chinese team reacted chloro (or bromo) difluoromethylacetate with potassium iodide and molecular iodine in the presence of a catalytic amount of cuprous iodide with a quantitative conversion of the polyhalogeno acetate. Additional data on the synthesis of ICF_2I can be found in a recent review [44].

Interestingly, difluorodiiodomethane is a source of difluoroiodomethyl or, better, of difluorocarbene radicals [45]. However, its use as a polymerisation initiator was unsuccessful [46].

2.1.2.1.2.1.1.2. Iodination of Carboxylic Diacids or Acid Dihalides.

(a) Hunsdiecker reaction

The Hunsdiecker reaction is one of the most utilised methods to prepare α,ω-diiodides in moderate yields [47] from the pyrolytic reaction of silver salts of perfluorocarboxylic acids in the presence of iodine (Scheme 1). This rather old

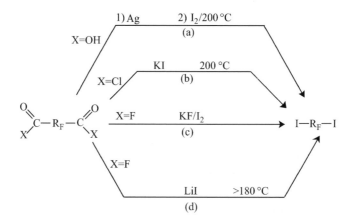

references: a [33, 34] ; b [38a, 38b] ; c [39] ; d [40]

references: a [47, 48] ; b [53, 54] ; c [55] ; d [56]

SCHEME 1.   Telechelic fluorinated diiodides from α,ω-dicarbonylated derivatives.

reaction has been often applied to monoadduct compounds and Rice successfully used it from $O(C_2F_4CO_2H)_2$ [48] or other diacids [49] to prepare an original diiodide containing an ether linkage. It is interesting to note that although the synthesis of $HO_2C(CF_2CFCl)_nCF_2CO_2H$ was achieved in two steps [10, 50] from the telomerisation of chlorotrifluoroethylene (CTFE) with carbon tetrachloride [10, 51] (more details are given in Section 2.1.2.1.3.1) or with bromotrichloromethane [10, 52], no attempt to obtain $I(CF_2CFCl)_nCF_2I$ has been performed so far.

(b) *Other methods*

As mentioned above, fluorinated telechelic esters, acids, acid chlorides or acid fluorides are useful precursors of the corresponding diiodides (Scheme 1 [53–55]).

Obviously, an excess of iodinated reactant is required to achieve the synthesis of the telechelic compound, otherwise only one end-group is converted. For example, this is the case with the iodination involving lithium iodide [56]. The authors reported that an equimolar amount of LiI gave only the α-iodo-ω-acyl fluoride in 50% yield while an excess of LiI produced $I(CF_2)_nI$ ($n = 3$) in 72% yield.

Like Rice [48, 49], Riley *et al.* [55] obtained original telechelic diiodides containing ether bridges useful for preserving the softness of the fluorosilicones generated from them.

On the other hand, Mc Loughlin prepared aliphatic perfluoroalkanes from preheated crystallised potassium iodide in anhydrous medium [53] as useful intermediates of telechelic aromatic diiodides according to an Ullman coupling, as follows [57]:

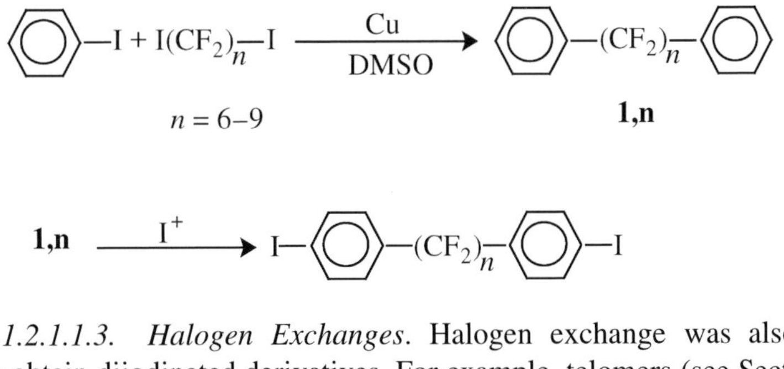

$$\langle\bigcirc\rangle\!\!-\!\!I + I(CF_2)_n\!\!-\!\!I \xrightarrow[\text{DMSO}]{\text{Cu}} \langle\bigcirc\rangle\!\!-\!\!(CF_2)_n\!\!-\!\!\langle\bigcirc\rangle$$

$$n = 6\text{--}9 \qquad\qquad \mathbf{1,n}$$

$$\mathbf{1,n} \xrightarrow{\text{I}^+} I\!\!-\!\!\langle\bigcirc\rangle\!\!-\!\!(CF_2)_n\!\!-\!\!\langle\bigcirc\rangle\!\!-\!\!I$$

2.1.2.1.2.1.1.3. Halogen Exchanges. Halogen exchange was also found useful to obtain diiodinated derivatives. For example, telomers (see Section 2.2) of tetrafluoroethylene (TFE) with iodine monochloride [58, 59] or iodine monobromide [59, 60] afforded the corresponding diiodides by further reaction with molecular iodine.

2.1.2.1.2.1.2. Radical Reactions Involving Iodine or α,ω-Diiodofluoroalkanes
2.1.2.1.2.1.2.1. From Iodine.

(a) *Reaction with tetrafluoroethylene (TFE)*

The addition of molecular iodine to fluorinated alkenes is one of the simplest methods to prepare telechelic diiodides since it does not require other reactants (e.g., catalyst or radical initiators). In 1949, 1,2-diiodotetrafluoro-ethane was synthesised for the first time by Coffman et al. [61]. After 15 h at 60 °C in a silver-lined tube under pressure, the addition of iodine to TFE led to 74% of IC_2F_4I (based on I_2). Four years later, Haszeldine and Leedham [62] reported a slightly higher yield when the reaction was carried out in a stainless steel autoclave. Then, Brace [44] slightly improved the process while Knunyants et al. [63] reached 83% yield of 1,2-diiodotetrafluoroethane in a steel ampoule. Although these reactions were initiated thermally, the photochemically induced addition of iodine to TFE by means of a halogen lamp (wavelength of 250–800 nm) successfully led to IC_2F_4I with very high yields (94%) [64].

As a matter of fact, prior to this recent work, most additions of iodine to TFE were performed under pressure at rather high temperatures. Moreover, as expected, an excess of TFE, under thermal telomerisation [65, 66], led to higher $I(C_2F_4)_nI$ adducts ($n = 1-5$) showing the good propagation of TFE in such a radical process [67, 68].

$$I_2 + n\,F_2C{=}CF_2 \xrightarrow[\Delta]{} I(C_2F_4)_nI$$

Indeed, this procedure, called telomerisation [65, 66] (see Section 2.1.2.1.2.1.2.3 and Chapter 1), was pioneered by Bedford and Baum [67] while Tortelli and Tonelli [68] interestingly optimised the reaction (the best conditions were in an autoclave at 180–200 °C for 8 h with a 2–3-fold excess of TFE). Similarly, Kotov et al. [69] modified this procedure with TFE being produced from PTFE wastes. Nowadays, the thermal process for obtaining these α,ω-diiodoperfluoroalkanes is industrialised by the Asahi [70] and Daikin companies.

Like perfluoroalkyl iodides, these diiodides exhibit weak carbon–iodine bonds [34, 66] which allowed 1,2-diiodoperfluoroethane to produce higher telomers under heating.

This can be explained by the well-known β scission (or reversion) [71, 72] of that reactant leading back to TFE and iodine as follows:

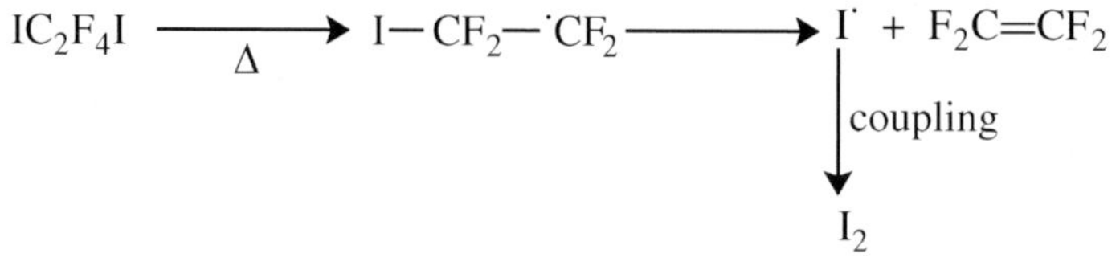

$$IC_2F_4I \xrightarrow{\Delta} I{-}CF_2{-}{^\cdot}CF_2 \longrightarrow I^{\cdot} + F_2C{=}CF_2$$

$$\downarrow \text{coupling}$$

$$I_2$$

Furthermore, Knunyants et al. [63] noted that the higher the temperature, the more homologues formed.

Surprisingly, however, when $As(C_2F_5)_2$ was used as the catalyst for the addition of molecular iodine to TFE, the diiodides were not produced [73].

(b) *Reaction with chlorotrifluoroethylene (CTFE)*

In contrast to the addition of molecular iodine to TFE discussed above, its addition to CTFE has been sparsely described in the literature. Nevertheless, the most interesting study was reported by Chambers et al. [74, 75], who showed that iodine is slightly soluble in liquid CTFE under autogenous pressure at room temperature. Indeed, the radical initiations were performed under sunlight, UV or gamma-radiations, yielding a complete conversion of all solid iodine to ICF_2CFClI (**2**) which, in the presence of light or heat, led back to I_2 and CTFE.

$$I_2 + F_2C{=}CFCl \underset{}{\overset{h\nu}{\rightleftharpoons}} \underset{\mathbf{2}}{ICF_2CFClI} \xrightarrow{\text{CTFE}} I(CF_2CFCl)_nI$$

We were able to confirm such a statement [76] and showed that from 40 °C, **2** decomposed. However, additional irradiation of **2** with an excess of liquid CTFE yielded irreversible formation of telomers $I(CF_2CFCl)_nI$, $n = 2-10$ [74, 75], much more stable than **2**.

Furthermore, even in the course of the addition of iodine monobromide to CTFE, traces of **2** were noticed in addition to the main product, $I(C_2F_3Cl)Br$, mainly produced [77].

(c) Reaction with hexafluoropropylene oxide (HFPO)

Few studies have been carried out but Yang [78] investigated the nickel catalysed addition of I_2 to HFPO at 185 °C, producing $I(CF_2)_nI$, $n = 1$ (mainly), 2, 3, in 78–90% yield. Further, similar reactions starting from ICl or IBr led to ICF_2I, ICF_2X, CF_2X_2 (X = Cl or Br) and CF_3COF.

2.1.2.1.2.1.2.2.   From Iodobromodifluoromethane. Ashton et al. [79] noted that the UV pyrolysis of $BrCF_2I$ with TFE led to $I(CF_2)_nI$, with $n = 2-5$, after 45 min at 100 °C under telomerisation conditions.

2.1.2.1.2.1.2.3.   Telomerisation of Commercially Available Fluoroolefins with α,ω-Diiodoperfluoroalkanes. Telomerisation reactions have been extensively reported in Chapter 1 but we have found it interesting to recall here several specific syntheses not detailed in that chapter. Such telomers (**3**) are obtained from the reaction between a telogen or a transfer agent (X–Y, assuming that in this case iodine or diiodides, $I–R_F–I$, will be taken into account) and one or more (*n*) molecules of a polymerisable compound M (called monomer or taxogen) having ethylenic unsaturation, under radical polymerisation conditions, as follows:

$$X{-}Y + n\,M \xrightarrow{\text{free radical}} \underset{\mathbf{3}}{X{-}(M)_n{-}Y}$$

Various telomerisations of fluoroalkenes (especially tetrafluoroethylene, vinylidene fluoride and hexafluoropropene) with different diiodides of increasing fluorinated chain lengths are described below.

Considering the thermal behaviour of these fluorinated diiodides, IC_4F_8I or higher TFE adducts still preserve their good thermostability at least up to 260 °C for 3 days [63, 80]. This property is enhanced for telomers containing an increasing TFE base unit number [81]. In contrast, IC_2F_4I starts to degrade at 180 °C [63, 80], whereas ICF_2I decomposes below 90 °C [44, 45] and ICF_2CFClI undergoes a reversion from 40 °C [76] or at room temperature under sunlight [74, 75]. As a matter of fact, the cleavage of the $CF_2–I$ bond is rather easy, in contrast to that of $CF_2–Br$ and $CF_2–Cl$.

(a) *From diiododifluoromethane*

A Chinese team [45] detailed the synthesis of original end-cappings of diiododifluoromethane by TFE, VDF and HFP in the presence of lead tetraacetate (LTA) as the catalyst. Such a redox initiation, surprisingly performed at mild temperature (70 °C), was never achieved before in the presence of HFP. The only previous redox catalysis was realised in the presence of yttrium salts leading to by-products [82].

In addition, the use of LTA seemed to produce a regioselective reaction:

$$ICF_2I + F_2C=CFCF_3 \xrightarrow[\text{70 °C, 30 h}]{\text{LTA}} ICF_2CF_2CFICF_3 \ (85\%)$$

Interestingly, this lead complex was very selective in monoaddition, since with a 3-fold excess of TFE, which telomerises easily in the presence of metal salts or complexes [66], the authors only obtained traces of $ICF_2(C_2F_4)_2I$ diadduct. The yields of monoadduct were very high (85, 90 and 92% in the case of end-cappings with HFP, TFE and VDF, respectively). However, no stepwise monoaddition was described.

(b) *From α,ω-diiodoperfluoroalkanes containing TFE base units*

(b1) *Telomerisation of TFE*

Various ways of initiating the telomerisation of TFE with $I(C_2F_4)_nI$ ($n = 1$ or 2) are possible thermally (at 230 °C) [68], in the presence of dibenzoyl peroxide [83] or by gamma rays from a ^{60}Co source [84]. These initiators lead to molecular distributions of $I(C_2F_4)_pI$ with $p = 2\text{–}4$, $1\text{–}3$ and $1\text{–}5$, respectively.

(b2) *Telomerisation of VDF*

Although the redox catalysis in the presence of $FeCl_3/Ni$ at 160 °C was selective [85], the thermal-induced telomerisation of VDF led to a telomeric distribution (up to the first six adducts) [80]. The monoadduct contained the only $I(C_2F_4)_nCH_2CF_2I$ isomer while the diadduct was composed of two isomers: $ICF_2CH_2(C_2F_4)_nCH_2CF_2I$ (A) and $I(C_2F_4)_n(VDF)_2I$ (B), the ratio of which interestingly varied according to the number of TFEs: for $n = 1$, A and B were equimolar, whereas for $n = 2$, the A/B molar ratio was 9 [80].

Further, these TFE/VDF cotelomers are also original transfer agents to be end-capped by HFP. An elegant example demonstrates the regioselectivity of addition of the $IC_2F_4CH_2CF_2$ radical toward the less hindered side of HFP [86], as follows:

$$IC_2F_4CH_2CF_2I + F_2C=CFCF_3 \xrightarrow{\text{230 °C/65 h}} IC_2F_4CH_2CF_2CF_2CFICF_3 \ (55\%)$$

(b3) *Telomerisation of hexafluoropropene*

HFP is known to homopolymerise in drastic conditions (at high temperatures and at ca. 1000 bar [87, 88]); consequently, its telomerisation is also limited. However, under thermal conditions, the first three adducts were produced [89] with more than 85% of the monoadduct.

Tortelli and Tonnelli [68], followed by Baum and Malik [90], pioneered the thermal telomerisation of HFP with 1,4-diiodoperfluorobutane, leading to the following adducts:

$$IC_4F_8I + nF_2C{=}CFCF_3 \xrightarrow{\Delta} IC_4F_8(C_3F_6)I + I(C_3F_6)C_4F_8(C_3F_6)I$$

$$+ I(C_3F_6)_2C_4F_8I + I(C_3F_6)C_4F_8(C_3F_6)_2I$$

Recently, the optimisation of such a reaction toward the formation of the diadduct was achieved (after 48 h reaction at 260 °C with an initial $[HFP]_0/[IC_4F_8I]_0$ molar ratio of 3.2, $I(HFP)C_4F_8(HFP)I$ was mainly produced and in good yield) [91].

As above, such highly fluorinated diiodides were successful candidates to allow the telomerisation of VDF [92, 93], as shown in the monoiodide series [94].

$$I(HFP)_xC_4F_8(HFP)_yI + pVDF \rightarrow I(VDF)_a(HFP)_xC_4F_8(HFP)_y(VDF)_bI$$

$$(x + y = 1{-}4, \quad a + b = 4{-}15)$$

Such original cotelomers are interesting models of fluoroelastomers, as mentioned in Table I [95], which gives a non-exhaustive list of commercially available fluoropolymers based on most commercial fluoroolefins produced by various companies. Indeed, 2 two-step routes can be used to achieve the synthesis of functionalisable diiodide cotelomers based on TFE, VDF and HFP base units (Scheme 2).

Indeed, as the "telomerisation" of HFP occurs at quite high temperatures (> 200 °C), it is required to perform this reaction as a first step prior to the step involving the telomerisation of VDF. Because this latter reaction cannot produce $ICF_2CH_2(C_2F_4)_nCH_2CF_2I$ selectively, two consecutive VDF units induce thermal weak points which make VDF telomers unstable at high temperatures.

In addition to the thermal properties, such fluorinated cotelomers exhibit glass transition temperatures below -40 °C, obviously explained by their molecular weights being lower than those of commercially available fluoroelastomers.

Interestingly, these highly fluorinated diiodides have been functionalised into original, non-conjugated dienes as precursors of hybrid fluorosilicones (see Section 3.4).

(c) *From $I(CF_2CFCl)_nI$*

Chambers *et al.* [74, 75] reported the thermal addition of CTFE telomers to TFE leading to monoadduct **4** without any trace of higher TFE adducts, as follows:

$$I(C_2F_3Cl)_6I + F_2C{=}CF_2 \xrightarrow[\text{HFP (trace)}]{16\ h/165\ °C} \underset{\mathbf{4}}{I(C_2F_3Cl)_6CF_2CF_2I}$$

2.1.2.1.2.1.3.   Use of IR_FI in Iodine Transfer Polymerisation. Iodine transfer polymerisation (ITP) is one of the radical-controlled or pseudo-living processes developed in the late 1970s by Tatemoto *et al.* [5, 96–99]. This truly living character is usually demonstrated when the molecular weight increases with

TABLE I

MAIN COMMERCIALLY AVAILABLE FLUOROELASTOMERS

	HFP	PMVE	CTFE	P	HPFP
VDF	Daiel 801 (Daikin) Fluorel (3M/Dyneon) Tecnoflon (Solvay Sol) SKF-26 (Russia) Viton A (DuPont)	–	Kel F (Dyneon) SKF-32 (Russia) Voltalef (Elf Atochem)	–	Tecnoflon SL (Montecatini)
TFE	–	Kalrez (DuPont)		Aflas (Asahi Glass)	–
VDF + TFE	Daiel 901 (Daikin) Fluorel (Dyneon) Tecnoflon (Solvay Sol) Viton B (DuPont) +Ethylene: Tecnoflon (Solvay Sol) +X: Viton GH (DuPont)	Viton GLT (DuPont)	–	–	Tecnoflon T (Montecatini)

CTFE, chlorotrifluoroethylene ($F_2C=CFCl$); HFP, hexafluoropropene ($F_2C=CFCF_3$); HPFP, 1-hydro-pentafluoropropene ($FHC=CF-CF_3$); P, propene ($H_2C=CHCH_3$); PMVE, perfluoromethyl vinyl ether ($F_2C=CFOCF_3$); TFE, tetrafluoroethylene ($F_2C=CF_2$); VDF, vinylidene fluoride or 1,1-difluoroethylene ($F_2C=CH_2$); X, cure site monomer ($XCY=CZ-R-G$; R: spacer; G: crosslinkable function).

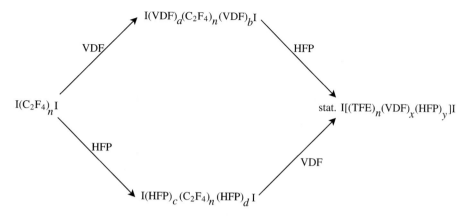

SCHEME 2. Synthesis of highly fluorinated α,ω-diiodoalkanes containing TFE, VDF and HFP base units as original models of fluoroelastomers.

conversion in a linear fashion. Polydispersity must be narrower and lower than that observed in classical processes (theoretical value, $\bar{M}_{\omega}/\bar{M}_n = 1.1-1.5$), which suppose, in particular, a rapid initiation step. Obtaining high and strictly controlled molecular weights has thus been possible. Actually, the ITP process requires (per)fluoroalkyl iodides and also diiodides because their highly electron-withdrawing (per)fluorinated group R_F allows the lowest level of the CF_2-I bond dissociation energy. Various fluorinated monomers (e.g., VDF) or non-fluorinated ones (styrene, acrylates or methacrylates) have been successfully used in ITP. Basic similarities in these living polymerisation systems are found in the stepwise growth of polymeric chains at each active site. The active living centre, generally located at the end-groups of the growing macroradical, has the same reactivity at any time during polymerisation even when the reaction is stopped [96, 97]. In the case of ITP of fluoroalkenes, the terminal active bond is always the $C-I$ bond originated from the initial iodine-containing chain transfer agent and monomer, as follows:

$$I-C_nF_{2n}-I + n\,H_2C{=}CF_2 \xrightarrow{R'\ or\ \Delta} I-CF_2CH_2(VDF)_qC_nF_{2n}(VDF)_pCH_2CF_2-I$$

Usually, molecular weights are not higher than 30,000 and yet polydispersity is narrow (ca. 1.5) [5, 98, 99]. Tatemoto et al. used peroxides as initiators of polymerisation. Improvement is also possible by using polyiodide compounds [5, 96].

Several investigations have shown that ITP can occur by radical initiation in emulsion, microemulsion or solution. When emulsion is chosen, a perfluoroalkyl iodide or a diiodide is involved and limits the molecular weights [96–99].

The Daikin company (recently called Dai-Act) has thus been able to produce a wide range of thermoplastic elastomers called Dai-El® [96, 99] (Table I).

These commercially available Dai-El Thermoplastics exhibit very interesting properties such as a high specific volume (1.90), a high melting point (160–220 °C), a high thermostability up to 380–400 °C, a refractive index of 1.357 and good surface properties ($\gamma_c \sim$ 19.6–20.5 dyne cm^{-1}). These characteristics offer excellent resistance against aggressive chemicals and strong acids, fuels and oils. In addition, tensile modulus is close to that of cured fluoroelastomers.

Interestingly, the DuPont [100] and Ausimont (now Solvay Solexis) [101] companies have also been attracted by this concept (using IC_4F_8I or IC_2H_4 $(TFE)_nC_2H_4I$), in the controlled radical polymerisations of VDF, TFE, trifluoroethylene, and in the copolymerisation of VDF and HFP. In this last case, polydispersity indexes (PDIs) as low as 1.87 could be obtained [100]. Extensive information is supplied in Chapter 4, Sections 2.1.2.4.1 and 2.2.2.2.

2.1.2.1.2.1.4. Coupling Reactions of Diiodoperfluoroalkanes. Although the Wurtz coupling can easily be realised in aromatic series [102] (see Section 2.1.2.1.2.2.4), few studies were applied to aliphatic ones. Ford *et al.* [103] studied such a reaction involving α,ω-diiodoperfluoroalkanes in the presence of cadmium at mild temperatures. ^{19}F NMR at 80 °C revealed an increase in the molecular weights, and the fluorodiiodides produced exhibited surprisingly low PDIs:

$$I(CF_2)_nI \xrightarrow[80\ °C]{Cd} I(CF_2)_{16n}I, \qquad PDI = 1.00$$

2.1.2.1.2.2. Chemical Change of α,ω-Diiodo(per)fluoroalkanes into Fluorotelechelics. Several ways are possible to enable the diiodides to be functionalised. Basically, these routes can be gathered into four families:

(i) Direct transformation of both iodine atoms into functions, as follows (Met designates a metal):

$$I-R_F-I + 2\ Met-\text{\textcircled{G}} \longrightarrow \text{\textcircled{G}}-R_F-\text{\textcircled{G}} + 2\ Met\ I$$

(ii) Functionalisation from the bis(monoaddition) of the diiodides to ω-functional-α-ethylenic derivatives (\text{\textcircled{G}} and R designate a functional and a spacer group, respectively):

$$I-R_F-I + 2H_2C{=}CH-R-\text{\textcircled{G}} \longrightarrow \text{\textcircled{G}}-R-CHICH_2-R_F-CH_2CHI-R-\text{\textcircled{G}} \qquad (1)$$
$$\mathbf{5}$$
$$\mathbf{5} + H_2\ \text{or}\ H^{\ominus} \longrightarrow \text{\textcircled{G}}-R-C_2H_4-R_F-C_2H_4-R-\text{\textcircled{G}} \qquad (2)$$

The first step is possible thanks to the easily cleaved CF_2-I bond already mentioned above.

(iii) As a spacer is required between the function and the (per)fluorinated chain, the bis(ethylenation) of the diiodide (Eq. (3)) followed by a disubstitution (Eq. (4)) or a bis(dehydroiodination) (Eq. (5)) produce fluorotelechelics or α,ω-divinyl (per)fluorinated alkanes, respectively.

$$IR_FI + 2H_2C{=}CH_2 \rightarrow \underset{\mathbf{6}}{I{-}C_2H_4{-}R_F{-}C_2H_4I} \tag{3}$$

$$\mathbf{6} \begin{cases} \xrightarrow{2X{-}\text{\textcircled{G}}} \text{\textcircled{G}}{-}C_2H_4{-}R_F{-}C_2H_4{-}\text{\textcircled{G}} & (4) \\[2ex] \xrightarrow{-2HI} H_2C{=}CH{-}R_F{-}CH{=}CH_2 & (5) \end{cases}$$

(iv) Coupling reactions between α,ω-diiodides and ω-functional-α-iodinated compounds:

$$I{-}R_F{-}I + 2I{-}\text{\textcircled{G}} \longrightarrow \text{\textcircled{G}}{-}R_F{-}\text{\textcircled{G}} + 2I_2$$

These four classes of reactions are summarised in Scheme 3 [104–148].

2.1.2.1.2.2.1. Direct Chemical Change. α,ω-Diiodo(per)fluoroalkanes do not undergo normal S_N1 or S_N2 substitutions but react with selected nucleophiles by an $S_{RN}1$ process [121], as follows:

$$I(C_2F_4)_nI \xrightarrow[\text{(2) } H_2]{\text{(1) } (CH_3)_2CNO_2Li} H_2N{-}\overset{\overset{\displaystyle CH_3}{|}}{\underset{\underset{\displaystyle CH_3}{|}}{C}}(C_2F_4)_n\overset{\overset{\displaystyle CH_3}{|}}{\underset{\underset{\displaystyle CH_3}{|}}{C}}{-}NH_2$$

In this manner, a Chinese team [149] performed the sulfination of fluorinated diiodides with $Na_2S_2O_4$ in aqueous $NaHCO_3$ in moderate to high yields to produce $NaO_2S{-}R_F{-}SO_2Na$, converted into fluorinated sulfonyl chlorides.

2.1.2.1.2.2.2. Fluorinated Telechelics from the bis(monoaddition) of Diiodides to Unsaturated ω-Functional Reactants. As in Eqs. (1) and (2), both steps are described below.

2.1.2.1.2.2.2.1. Radical Addition of Diiodides to Unsaturated Functional Reagents (Eq. (1)). Brace [150, 151] has recently reviewed the addition of perfluoroalkyl iodides to unsaturated derivatives, showing that many investigations have been performed. Indeed, the use of these R_FI is a versatile means to enable the fluoroalkylation of compounds.

In the case of the diiodide series, fewer studies have been attempted and sometimes several authors did not pursue their investigations to the point where they obtained fluorotelechelics.

As reported [152], the weak $CF_2{-}I$ bond can easily be cleaved in the presence of various systems already described [150, 151, 153]: thermally, photochemically, biochemically, electrochemically, from various initiators or catalysts (organic peroxides and diazo compounds, hydrogen peroxide, sodium dithionite

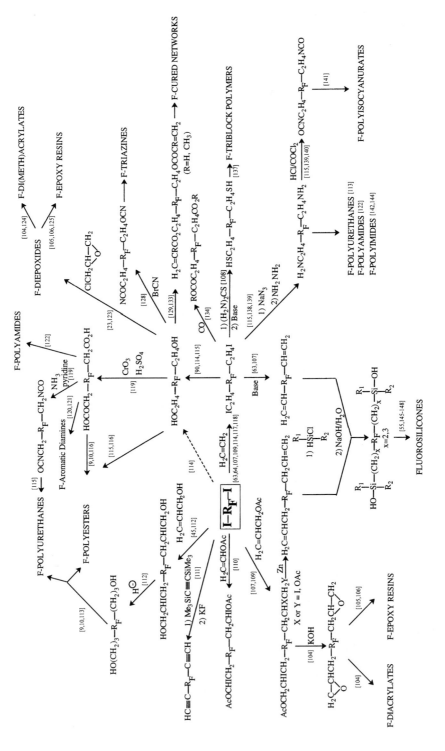

SCHEME 3.   α,ω-Diiodofluoroalkanes as useful precursors of fluorinated telechelics and polymers.

and related agents, triethyl borane, alkyl phosphines, various metals, metallic salts and metallic complexes).

Among the unsaturated ω-functional derivatives, allyl alcohol, vinyl and allyl acetate and acrylates have been primarily studied, whereas Metzger's group [154, 155] tried fatty ethylenic compounds successfully.

Usually, the $ICF_2R_FCF_2$ (or $\cdot CF_2R_FCF_2$) radical reacts toward the less hindered side of the unsaturated compound leading to the expected $-CF_2R_FCF_2CH_2CHI-R-$Ⓖ

Brace [102] pioneered such research by reacting $IC_6F_{12}I$ with a 2–3-fold excess of vinyl acetate to give $(AcOCHICH_2)_2C_6F_{12}$, which was then reduced to $AcO(CH_2)_2(CF_2)_6(CH_2)_2OAc$.

Interestingly, Khrlakyan et al. [156] compared the reactivity of allyl alcohol and allyl acetate toward $I(C_2F_4)_nI$ ($n = 1, 2$), in the presence of AIBN (at 80–90 °C) or dicyclohexyl peroxydicarbonate at 58 °C. They noted the higher reactivity of the unsaturated acetate with a quasi-selective formation of telechelic (82%) in the presence of the dicarbonate. Slight exotherms occurred when these initiators were utilised with allyl acetate.

However, from the same monomer, initiated by dibenzoyl peroxide at 90 °C, a sharp exotherm occurred at 170–180 °C followed by a thermal rearrangement of the intermediate radical leading unexpectedly to $IR_FCH_2CH(OAc)CH_2I$, as the mechanism depicted in Scheme 4 suggests [153].

The fluorinated diiododiacetates produced were chemically changed into non-conjugated dienes [107, 108] or diepoxides [104], whatever the amount of rearranged diacetates since they underwent the bis(deiodoacetalisation) or bis(cyclisation), respectively.

For bulkier unsaturated reactants, two examples are given below.

(1) 1,4-Diiodoperfluorobutane reacted cleanly with a 5-fold excess of norbornene leading to the 1 : 2 adduct in high yield, in the presence of sodium p-toluene sulfinate [157].

(2) Huang et al. [158] reacted a 2-fold excess of ω-ethylenic sterol in the presence of sodium dithionite giving a fluorinated bisterol in 76% yield.

Another interesting study was achieved by Wilson and Griffin [159] and more recently revisited by Percec et al. [160], who synthesised fluorinated copolymers containing fluorinated and hydrogenated microblocks, by radical polyadditions of diiodoperfluoroalkanes to non-conjugated dienes as follows:

$$I(CF_2)_nI + H_2C=CH(CH_2)_pCH=CH_2 \xrightarrow[\text{(2) SnBu}_3\text{H, AIBN, 90 °C}]{\text{(1) AIBN (85–145 °C)}} \left[(CF_2)_n (CH_2)_m \right]_q$$

$$n = 4, 6 \qquad p = 2, 4, 6 \qquad m = p + 4$$

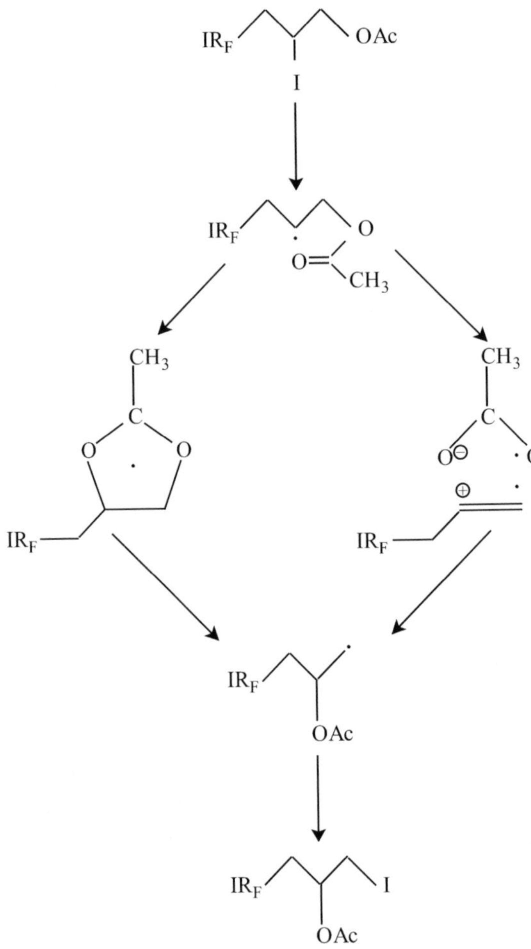

SCHEME 4. Mechanism of the formation of rearranged diiodoacetate.

These polymers, having molecular weights of about 20,000, showed thermotropic mesophases upon heating. This new class of liquid crystals was also investigated by Russell *et al.* [161].

Finally, although several authors have focused on the addition of perfluoro-alkyl iodides to acetylenic compounds, to our knowledge, only two surveys of diiodides were performed:

(i) by Baum *et al.* [111], leading to $HC{\equiv}C{-}R_F{-}C{\equiv}CH$ in three steps from $HC{\equiv}C{-}Si(CH_3)_3$;

(ii) by Amato *et al.* [162], producing $I(CF_2)_4CH{=}CIC(CH_3)_2OH$ mono-alcohol only from the addition of IC_4F_8I onto an excess of 2-methyl-3-butyn-2-ol.

Allyl amine does not react with perfluoroalkyl iodides [163]; hence, it is necessary to protect the amino group. Thus, the bismonoaddition of 1,6-diiodoperfluorohexane to allyl phthalimide **7** enabled us to prepare telechelic diamines [164] as follows:

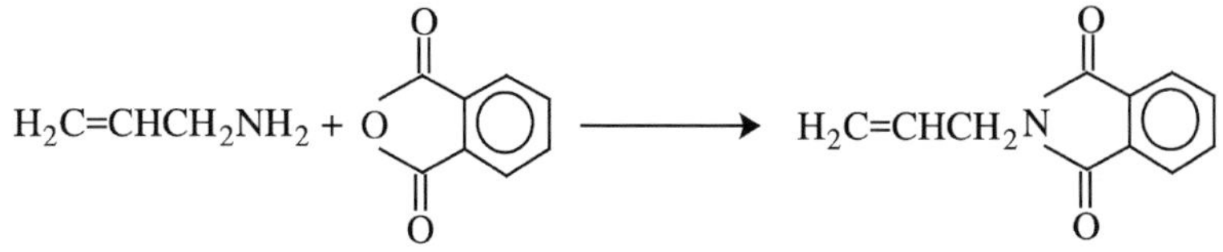

7

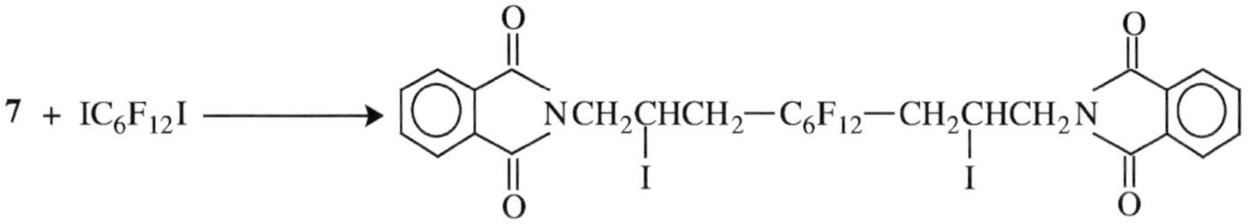

8

$$8 \xrightarrow[\text{(2) H}_2\text{N--NH}_2,\ \text{EtOH}]{\text{(1) SnBu}_3\text{H}} \text{H}_2\text{NC}_3\text{H}_6 - \text{C}_6\text{F}_{12} - \text{C}_3\text{H}_6\text{NH}_2$$

More recently, in a *quasi* similar approach, Amato and Calas [165] could prepare $\text{H}_2\text{NC}_3\text{H}_6(\text{CF}_2)_n(\text{CH}_2)_{10}\text{CH}_2\text{SH}$ ($n = 6, 8$) from $\text{I(CF}_2)_n\text{I}$.

2.1.2.1.2.2.2.2. Reduction of Iodine Atoms (Eq. (2)). Various ways to selectively reduce both iodine atoms into hydrogens have been successfully reported.

(*a*) *In the presence of tributylstannane (see also above)*

The simplest, but more expensive, method requires the use of tributylstannane, known as an efficient reducer [166]. Such a reactant has been employed frequently in the case of monoadduct $\text{R}_\text{F}\text{CH}_2\text{CHI}-\text{R}-Ⓖ$ [118, 150, 151], but few papers have described its use from telechelic diiodides.

Besides the example above, we have recently synthesised diol **9** according to the following scheme [112]:

$$\text{IC}_6\text{F}_{12}\text{I} + n\,\text{H}_2\text{C}=\text{CHCH}_2\text{OH} \xrightarrow{\text{AIBN}} \text{HOCH}_2\text{CHICH}_2\text{C}_6\text{F}_{12}\text{CH}_2\text{CHICH}_2\text{OH}$$

$$\xrightarrow{\text{SnBu}_3\text{H}} \underset{\textbf{9}}{\text{HOC}_3\text{H}_6\text{C}_6\text{F}_{12}\text{C}_3\text{H}_6\text{OH}}$$

Such a concept was successfully applied onto $\text{I(VDF)}_x\text{C}_3\text{F}_6(\text{C}_2\text{F}_4)_2\text{C}_3\text{F}_6$ $\text{(VDF)}_y\text{I}$ (with $x + y = 2-15$) to generate original telechelic diols containing VDF, HFP and TFE base units [93].

(*b*) *In the presence of hydrogen*

Alternatively, the reduction may be performed with hydrogen in the presence of a palladium/charcoal complex as carried out by Metzger *et al.* [154, 155], Qui and Burton [167] or Gresham [168].

(c) *In the presence of zinc or zincic catalysts*

A reduction realised in the presence of zinc/HCl [110, 169] or a Zn/NiCl$_2$ complex [170] was also successful.

Interestingly, Qui and Burton [167] compared various conditions of reduction of fluorinated diiododiol and showed that the most efficient one, which avoided the dehydroiodination, was in the presence of the following system: H$_2$/Pd–C (5%)/NaHCO$_3$/MeOH.

(d) *From the EtOH/K$_2$CO$_3$ system*

Gresham [171] reported the selective reduction of an iodine atom in R$_F$CH$_2$CHICH$_2$OCH$_2$CH(OH)CH$_2$OH in the presence of the ethanol/potassium carbonate system.

Regarding the perfluoroalkyl iodide series, Brace [150, 151] mentioned that several authors performed the synthesis of R$_F$–C$_2$H$_4$–R–Ⓖ in one step. However, to our knowledge, no similar reference was reported in the literature from diiodides.

2.1.2.1.2.2.3.   Functionalisation by Substitution After Ethylenation. This alternative is a two-step process, as depicted by Eqs. (3) and (4) and in Scheme 3.

2.1.2.1.2.2.3.1.   Ethylenation (Eq. (3)). Various authors have studied the ethylenation of (per)fluoroalkyl iodides or diiodides and several initiations were efficient. Four routes are given below.

(a) *Thermal initiation*

This process was pioneered by Brace at 180–220 °C [118]. However, Knunyants *et al.* [63] used it at 220 °C to allow the bis(monoethylenation) of 1,2-diiodotetrafluoroethane with at least a 3.5-fold excess of ethylene, as follows:

$$IC_2F_4I + n\,H_2C=CH_2 \rightarrow IC_2H_4C_2F_4C_2H_4I + IC_2H_4C_4F_8C_2H_4I$$

1	3.5	90%	–
1	>10	80%	8%

The formation of the polyhalogenooctane was explained by the ethylenation of 1,4-diiodooctafluorobutane obtained from the addition of IC$_2$F$_4$I to TFE, this monomer being produced by reversion of IC$_2$F$_4$I at these high temperatures (see Section 2.2.1.1).

Attempts at higher temperatures (350 °C) were performed by Bloechl [172]. More recently, Yang [173] proposed the ethylenation of ICF$_2$I at 120–240 °C.

(b) *Photochemical initiation*

Renn *et al.* [64] achieved the bis(monoethylenation) of 1,2-diiodotetrafluoroethane from UV radiation by means of a halogen lamp (250–800 nm wavelength).

(c) *From redox catalysis*

Several teams have used various metallic catalysts that were efficient in ethylenation of (per)fluoroalkyl iodides or diiodides. Interesting surveys were performed by Von Werner [117], who used different triphenyl phosphine complexes, while Li and Chen [44, 45] obtained the expected bis(ethylenated) derivative in good yield from diiododifluoromethane, in the presence of lead tetraacetate (LTA):

$$ICF_2I + n\,H_2C{=}CH_2 \xrightarrow[\text{diglyme}]{\text{LTA 70 °C}} IC_2H_4CF_2C_2H_4I \ (86\%)$$

Researchers at the Ausimont Company [114] employed CuI as the catalyst and obtained the expected $IC_2H_4(C_2F_4)_nC_2H_4I$ ($n = 2, 3$).

In the presence of the same catalyst, we also noted satisfactory results (with minimum yield of 70%) from commercially available $I(C_2F_4)_nI$ ($n = 1, 2, 3$) [107] or from cotelomers of VDF, TFE and HFP [108].

$$IR_FI + n\,H_2C{=}CH_2 \xrightarrow[\text{160 °C}]{\text{CuCl/H}_2\text{NC}_2\text{H}_4\text{OH}} IC_2H_4R_FC_2H_4I$$

with R_F: $(VDF)_a(C_2F_4)_n(VDF)_b$ (with $n = 1, 2$ and $a = 0$, $b = 1$; $a = b = 1$ and $a = 0$, $b = 2$); $(HFP)_xC_4F_8(HFP)_y$ ($x = y = 1$; $x = 0$ and $y = 1$); $C_2F_4(VDF)(HFP)$.

Urata et al. [134, 135] used $Fe(CO)_{12}$, $Co_2(CO)_8$ and $Ru_3(CO)_{12}$, to effectively catalyse this reaction, leading to $IC_2H_4(CF_2)_nC_2H_4I$ ($n = 4, 6$) in 91% yield.

Finally, Chambers et al. [74, 75] successfully achieved the ethylenation of CTFE telomers in the presence of platinum at 70 °C in 32% yield:

$$I(CF_2CFCl)_6I + n\,H_2C{=}CH_2 \xrightarrow[\text{70 °C}]{\text{10\% Pt/C}} IC_2H_4(CF_2CFCl)_6C_2H_4I$$

(d) From radical initiation

Percarbonates have also shown efficiency in the bis(ethylenation) of α,ω-diiodo(per)fluoroalkanes containing VDF and/or HFP base units [108]. Although it was necessary to fill the autoclave three times with the initiator and ethylene, the reactions have the advantage of being carried out at only 60 °C. The yields ranged between 60 and 80%.

Peroxide initiators were also used: Baum et al. [111] performed the bis(monoethylenation) of $I(CF_2)_nI$ ($n = 4, 6$) in the presence of di-*tert*-butyl peroxide at 130 °C.

Hence, various kinds of initiations are possible to allow bis(ethylenation) reactions. Interestingly, these methods are very selective and also point out that two consecutive ethylene units were never observed.

2.1.2.1.2.2.3.2. Substitution of Both Iodine Atoms by Two Functions (Eq. (4)). These above α,ω-bis(monoethylenated) diiodides are versatile intermediates for obtaining a large variety of fluorinated telechelics.

Scheme 3 shows non-exhaustive possibilities to carry out the chemical transformation:

$$IC_2H_4{-}R_F{-}C_2H_4I \longrightarrow \text{Ⓖ}\,C_2H_4{-}R_F-C_2H_4\,\text{Ⓖ}$$

Some of them are given as examples below.

(a) Synthesis of telechelic diols

Diols are very interesting intermediates for the synthesis of various fluoropolymers [9–11, 113, 174] (see Section 3.1). Telechelic diols from the bis(monoethylenated) forerunners were prepared by the Daikin [175], Ausimont [114] or Fluorochem Inc. [115] companies, starting from $NaNO_2$/betaine as surfactant, $CuSO_4 \cdot 5H_2O$ or AgOAc, and H_2SO_4/SO_3, respectively, in yields higher than 92%. Such methods are more convenient than those requiring a water/DMF mixture which led to olefins or dienes.

These fluorodiols are particularly interesting precursors of a wide variety of fluorotelechelics, as shown in Scheme 3.

(b) Synthesis of fluorinated dienes (Eq. (5))

The bis(ethylenated) diiodides were easily changed into fluorinated non-conjugated dienes in alkali medium [63, 107, 108, 111]. They are interesting precursors of hybrid fluorosilicones [55, 145–148]. Interestingly, 3,3,4,4,5,5,6,6-octafluoro-1,7-octadiene underwent hydrosilylation with trichlorosilane as precursors of fluorinated silsesquioxanes [176a] as hybrid gels with a much better thermostability than the gels prepared from the corresponding hydrogenated dienes:

$$H_2C{=}CHC_4F_8CH{=}CH_2 + 2.5HSiCl_3 \xrightarrow[\text{(2) MeOH}]{\text{(1) } H_2PtCl_6}$$

$$(CH_3O)_3SiC_2H_4C_4F_8C_2H_4Si(OCH_3)_3 \xrightarrow{F^{\ominus}} gel$$

The same strategy was used at the DuPont de Nemours company [176b] for the hydrosilylation of $H_2C{=}CH(CH_2)_4C_{10}F_{20}(CH_2)_4CH{=}CH_2$ with triethoxysilanes in the preparation of hydrophobic coatings onto glass endowed with a surface energy as low as $15.2\,mN\,cm^{-1}$.

(c) Synthesis of telechelic dinitriles

Telechelic dinitriles, precursors of diamines, diacids and then fluoropoly-amides, were prepared by Anton and MacKinney [177] by reacting HCN with divinyl dienes in the presence of tritolyl phosphite and a nickel complex. The synthesis of fluorotelechelic dinitriles, diamines and diisocyanates was reviewed by Baum [111b].

(d) Synthesis of fluorinated thiols

Although numerous studies report the synthesis of fluorinated mercaptans from $R_FC_2H_4I$, no fluorodithiol has been previously described in the literature. Wathier [136] has recently succeeded in obtaining $HSC_2H_4C_6F_{12}C_2H_4SH$ from the corresponding 3,3,4,4,5,5,6,6,7,7,8,8-dodecafluoro-1,10-diiodide in the presence of thiourea. Such a dithiol is a potential precursor of triblock copolymers as Fokina *et al.* [137] produced previously from α,ω-dimercapto polybutadienes.

Conversely, original fluorinated mercaptan **10**, prepared from the dodeca-fluoro-1,10-diiodide, acts as an efficient transfer agent for the telomerisation of tri(hydroxymethyl)acrylamido methane (THAM) as follows [178, 179]:

$$IC_2H_4C_6F_{12}C_2H_4I \xrightarrow[\text{(2) }(NH_2)_2CS/NH_4OH]{\text{(1) }SnBu_3H} CH_3CH_2C_6F_{12}C_2H_4SH$$
$$\underset{\textbf{10}}{}$$

$$\textbf{10} \xrightarrow{\text{THAM}} C_2H_5C_6F_{12}C_2H_4S(THAM)_n\text{-}H$$

These telomers were original precursors of non-ionic surfactants to stabilise proteins in solution.

(e) *Miscellaneous*

The literature also reports useful fluorinated α,ω-dinitro derivatives [180] and telechelic bis(quaternary ammonium salts) [122, 139, 181]; especially dihydro-bromide salts which, after neutralisation, undergo interfacial polymerisation with diacid chlorides to give polyamides [139].

2.1.2.1.2.2.4. Coupling Reactions from α,ω-Diiodo(per)fluoroalkanes.
Mc Loughlin *et al.* [182] and, several years later, Chen *et al.* [183] carried out extensive research on the synthesis of diaromatic difunctional compounds linked to fluorinated chains according to the following Ullman coupling reaction:

where Ⓖ represents a functional group such as hydroxy (i.e., bisphenol), carboxylate, isocyanate [140] or nitro (precursor of amine) in *para* or *meta* positions about the fluorinated chain. The first team also branched several functional groups onto each aromatic ring [184], in 85% yield, such as the following tetracarboxylates:

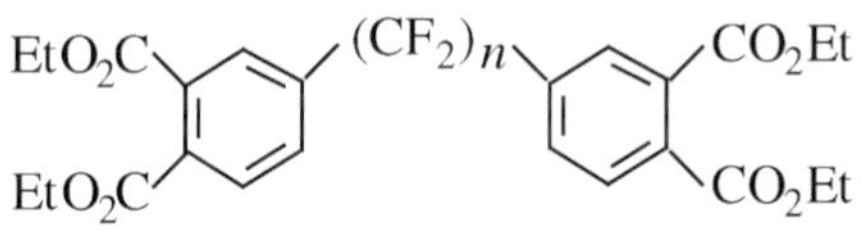

From these compounds, novel polymers were prepared such as polyesters [185, 186], silicones [57] and polyimides [57].

2.1.2.1.3. From α,ω-bis(trichloromethyl)chlorofluoroalkanes
2.1.2.1.3.1. Synthesis of α,ω-bis(trichloromethyl)chlorofluoroalkanes.
Fluorinated $Cl_3C-R_F-CCl_3$ were synthesised in a two-step procedure starting from the telomerisation of chlorotrifluoroethylene (CTFE) [10, 50, 51, 187–196] or tetrafluoroethylene (TFE) [197] with carbon tetrachloride, as follows:

$$CCl_4 + F_2C{=}CFX \rightarrow Cl_3C-(CF_2CFX)_x-CF_2CFClX$$
$$(X = F \text{ or } Cl)$$

$$Cl_3C-(CF_2CFX)_x-CF_2CFClX \xrightarrow[CCl_4]{AlCl_3} Cl_3C-(CF_2CFX)_x-CF_2CCl_3$$

Redox [187–190, 192–195] and radical [191] telomerisations were successfully achieved, the former one being more selective (i.e., leading to low-molecular-weight telomers), especially in the presence of copper salts (whatever its degree of oxidation) or from iron salts (mainly ferric) for which a reducing agent (such as benzoin [188, 192] or nickel metal [194]) is required.

These telomers are very stable to acid and alkali media and thermal stable since they have been used as Kel F® oils.

2.1.2.1.3.2. Synthesis of Chlorofluoro Telechelics. These chlorofluorinated α,ω-bis(trichloromethyl) telomers are key intermediates for obtaining telechelic diacids [198, 199] and diols [200, 201] either by direct chemical change in the presence of oleum or by bismonoaddition of these telomers into functional monomers (e.g., allyl alcohol [200] or allyl acetate [201, 202]).

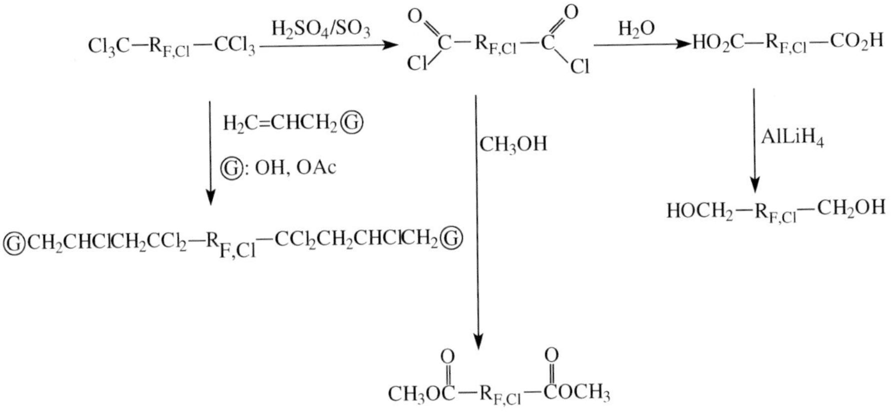

This latter reaction involves either redox catalysts (e.g., CuCl or FeCl₃/ benzoin) or dichlorotris(triphenyl phosphine) ruthenium as an efficient complex for 1 : 1 monoaddition. Further complexation of CuCl by isopropylamine [203] also led to a high yield of telechelic.

2.1.2.2. Long Fluorinated Telechelics

Interestingly, fluorinated telechelic peresters [204–206], synthesised by oxidation of fluorinated acid chlorides [204, 205] or by the direct addition of acid fluorides [206] to hydrogen peroxide, were successfully used by Rice and Sandberg in the copolymerisation of VDF and HFP leading to original fluoroelastomers [204–207] as follows:

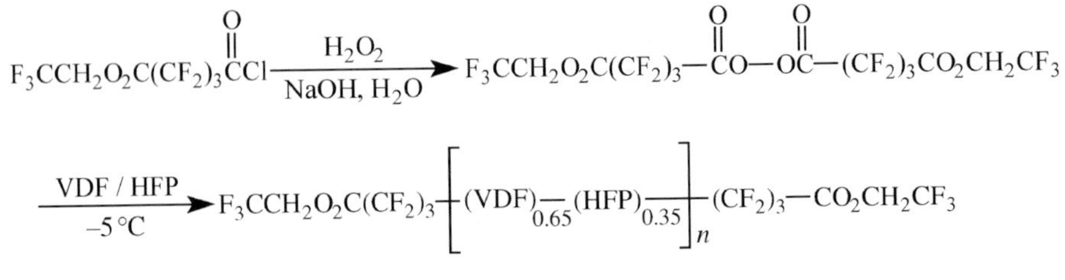

The average molecular weight as determined by vapour pressure osmometry (VPO) was ca. 4000. Similarly, the same authors prepared new fluorinated initiators useful for the preparation of fluoroelastomers:

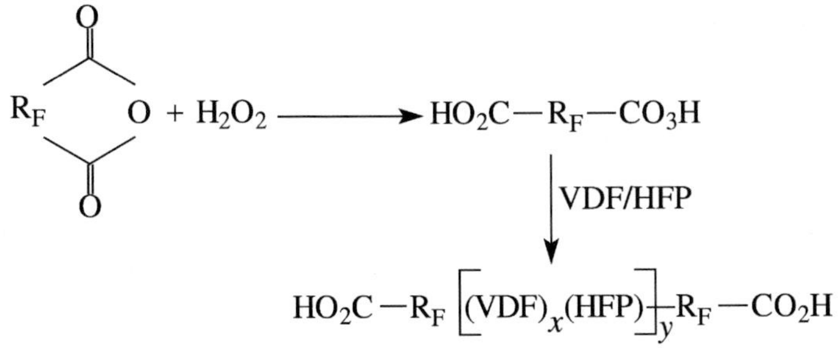

$$R_F \underset{O}{\overset{O}{\diagdown}} O + H_2O_2 \longrightarrow HO_2C-R_F-CO_3H$$

$$\downarrow VDF/HFP$$

$$HO_2C-R_F\left[(VDF)_x(HFP)\right]_y R_F-CO_2H$$

Interestingly, a Russian team reported the synthesis of other functional fluorinated peresters and their kinetics of thermal decomposition [208].

Moreover, Oka and Morita [209] also prepared hydroxytelechelic poly(VDF-co-HFP) copolymers from $(t\,BuOCH_2CF_2CO_2)_2$, after a subsequent hydrolysis step.

More recently, hydrogen peroxide was used to prepare novel α,ω-dihydroxylated poly(VDF-co-HFP) ($\bar{M}_n = 300-2000$) in a two-step procedure [210, 211] (G designates CH_2OH or CO_2H):

$$H_2O_2 + n\,VDF + p\,HFP \longrightarrow HO\left[(VDF)_x(HFP)\right]_z G$$

$$A$$

$$A \xrightarrow{AlLiH_4} HO\left[(VDF)_x(HFP)\right]_z CH_2OH$$

A similar investigation was recently realised by photo-induced dead-end polymerisation of VDF leading to original telechelic PVDFs [210b].

While propagation of monomers occurs in a classical way (although the propagation rates of all monomers have not yet been assessed even if various reactivity ratios have already been determined [95, 212]), termination proceeds by recombination only.

Original fluorotelechelic oligodiols as random copolymers containing VDF and HFP base units (e.g., $HOCH_2CF_2(CH_2CF_2)_p[CF(CF_3)CF_2]_qCF_2CH_2OH$) have been synthesised by Chan et al. [213]. However, in this patent, the description of the preparation is not reported.

2.1.3. Conclusion

Although they have not been as thoroughly reviewed as the perfluoroalkyl monoiodides, α,ω-diiodo(per)fluoroalkanes have been investigated by various authors. Basically, two main groups of reactions have been developed: a quite wide range of organic syntheses have led to short diiodides while telomerisations of fluoroolefins with molecular iodine or with diiodoperfluoroalkanes have successfully supplied higher molecular weight diiodides. Interestingly, this process has been scaled up and nowadays α,ω-diiodoperfluoroalkanes are commercially available. In contrast to TFE mainly and also to CTFE, no addition of iodine to VDF, HFP or trifluoroethylene has previously been reported in the literature.

Further, telomerisation of various fluoroolefins with I–R_F–I has provided a wide range of highly fluorinated original diiodides. Mostly thermal but also photochemical initiation, from ^{60}Co gamma source or peroxidic initiations were attempted successfully.

For much higher molecular weight diiodides, the Ullman coupling reaction of I–R_F–I and the ITP of fluoroolefins with I–R_F–I led to original polymeric diiodo materials with narrow polydispersities. All these α,ω-diiodo(per)fluoroalkanes are suitable precursors for obtaining telechelic derivatives.

Basically, among the four main reactions enabling the α,ω-diiodo(per)fluoroalkanes to produce fluorotelechelics, two of them have attracted much interest: (i) that involving the bismonoaddition of I–R_F–I to various α-functional unsaturated derivatives as a direct functionalisation and followed by selective reductions of both iodine atoms; (ii) that requiring first a bis(monoethylenation) of I–R_F–I (for which many investigations have been developed showing that most initiation processes were successful) followed by a substitution.

Less extensively studied was the Ullman reaction between I–R_F–I and α-functional-ω-iodo derivatives. More investigations should be expected. In addition, α,ω-bis(trichloromethyl) telomers are also useful intermediates to prepare a wide range of various chlorofluoro telechelics.

Using functional telechelic peresters or, better, cheap hydrogen peroxide enables one to obtain original longer fluorinated telechelic fluoroelastomers and deserves more investigations, even if a careful synthesis of the perester is required.

2.2.   TELECHELICS BEARING A FLUORINATED SIDE GROUP

Very few industrial products exist in contrast to the large variety of fluorotelechelics prepared in a synthetic way. As in the previous section, it is necessary to distinguish short and longer fluorinated telechelics.

2.2.1.   *Short Fluorotelechelics*

In this part, a more exhaustive bibliographical approach is given on the aliphatic telechelics (mainly on the α,ω-diols) rather than the aromatic ones.

2.2.1.1. Aliphatic Fluorotelechelics

Several ways to synthesise telechelics containing a fluorinated side chain are presented. The most popular methods are (i) the condensation of a fluorinated compound terminated by a halogen atom onto a malonate; (ii) the radical addition of a fluorinated mercaptan or a perfluoroalkyl iodide onto the double bond of a telechelic derivative; (iii) from the ring opening polymerisation of cyclic ethers such as epoxides, acetals or dioxolanes. Each method is reported below.

2.2.1.1.1. By Condensation. Smeltz [214] carried out the condensation of a 1H,1H,2H,2H-1-iodoperfluoroalkane onto a malonate to obtain in two steps a fluorotelechelic diol bearing a perfluorinated lateral group, as follows:

$$C_nF_{2n+1}C_2H_4I + H_2C(CO_2Et)_2 \xrightarrow{Na} C_nF_{2n+1}C_2H_4-CH(CO_2Et)_2$$
$$\xrightarrow{\text{reduction}} R_FC_2H_4CH(CH_2OH)_2 \qquad (n = 6, 8)$$

This synthesis was improved by Geribaldi's group [215].
A similar method can also be applied onto diethanol amine as follows [216]:

$$R_FC_2H_4I + HN(C_2H_4OH)_2 \longrightarrow HOC_2H_4-N-C_2H_4OH$$
$$\overset{|}{\underset{}{C_2H_4-R_F}}$$

This strategy was previously used in monofunctional series when the Ugine Kuhlman Company [217] investigated the addition of $R_FC_2H_4I$ onto functional amines such as ethanolamine. Besides obtaining the corresponding $R_FC_2H_4NHC_2H_4OH$, $(R_FC_2H_4)_2NC_2H_4OH$ was noted as a by-product.

In the above reaction, if a diamine is used, the resulting fluorinated diamino product, $R_FC_2H_4-NH-(CH_2)_x-NHR$, can react with propiolactone [217] or with propanesultone [218] to produce:

$$R_F-C_2H_4-N-(CH_2)_{\overline{x}}-N-R \qquad \text{and} \qquad R_F-C_2H_4-N-(CH_2)_{\overline{x}}-N-R \quad \text{respectively}$$
$$\overset{|}{C=O} \quad \overset{|}{C=O} \qquad\qquad\qquad \overset{|}{SO_2} \quad \overset{|}{SO_2}$$
$$\overset{|}{C_2H_4OH} \ \overset{|}{C_2H_4OH} \qquad\qquad\qquad \overset{|}{C_3H_6OH} \ \overset{|}{C_3H_6OH}$$

Telechelic diols containing a sulfonate group can also be obtained from sulfuryl halogenides [219] and from esters [220]:

$$R_F(CH_2)_xSO_2Cl + HN(C_2H_4OH)_2 \xrightarrow{-HCl} R_F(CH_2)_xSO_2-N(C_2H_4OH)_2$$

$$(x = 2, 4)$$

Further syntheses of primary–primary fluorodiols [221, 222] such as:

$$R_F-SO_2-N-(C_2H_4O)_x-H$$
$$(C_2H_4O)_y-H \qquad \text{with } x+y=3-30$$

or

$$R_F-CO-N-C_2H_4OH$$
$$C_2H_4OH$$

were successfully achieved.

Similarly, primary–secondary diols [223]:

$$R_F-SO_2-N-CH_2CH(OH)CH_2OH$$
$$CH_3$$

or secondary–secondary diols [221]:

$$R_F-SO_2-N(CH_2CHCH_3)_2$$
$$OH$$

were also obtained.

Another pathway involves the condensation of 1H,1H,2H,2H-1-iodoperfluoroalkane onto hydrogenated telechelic diols bearing a mercapto function [224]:

$$R_FC_2H_4I + HSCH_2CH(OH)CH_2OH \xrightarrow{\text{MeOH}} R_FC_2H_4SCH_2CH(OH)CH_2OH$$

Further, Bloechl [218] realised the condensation of a fluorinated thiol with glycidol and obtained, after basic hydrolysis, the corresponding fluorinated diol:

$$R_FC_2H_4S-CH_2CH(OH)CH_2OH$$

2.2.1.1.2. By Radical Way. The radical addition of fluorinated thiols or perfluoroalkyl iodides onto unsaturated hydrogenated telechelic diols [171, 225], $H_2C{=}CHCH_2CH(CO_2Et)_2$ [171], or itaconic acid [226] led to original fluorotelechelics bearing a fluorinated side chain. Initiators used were AIBN or peroxides.

$$C_8F_{17}C_2H_4SH + H_2C{=}CHCH_2OCH_2CH(OH)CH_2OH \longrightarrow C_8F_{17}C_2H_4SC_3H_6OCH_2CHCH_2OH$$
$$OH$$

Gresham [171, 225] obtained fluorotelechelic diols by radical addition of fluorinated thiols onto 2-butene-1,4-diol, as follows:

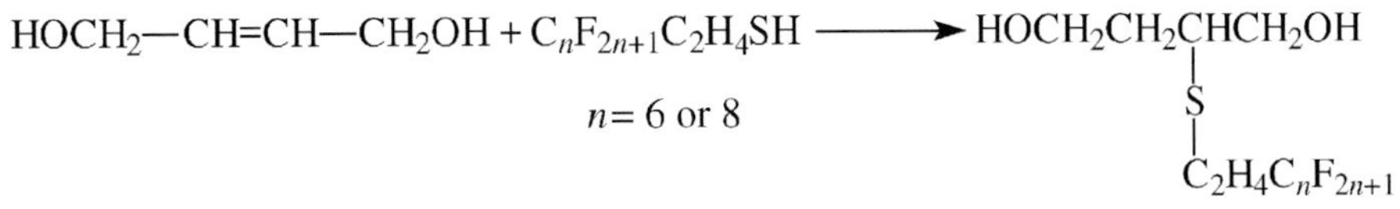

$$HOCH_2-CH=CH-CH_2OH + C_nF_{2n+1}C_2H_4SH \longrightarrow HOCH_2CH_2CHCH_2OH$$

$$n = 6 \text{ or } 8$$

$$\underset{\underset{C_2H_4C_nF_{2n+1}}{|}}{\overset{|}{S}}$$

Hager [227] performed the same reaction with maleic acid using piperidine as catalyst.

Dear *et al.* [228] and later Hamiff and Deisenroth [229] successfully prepared the following telechelic diols by a similar reaction from a telechelic alkyne:

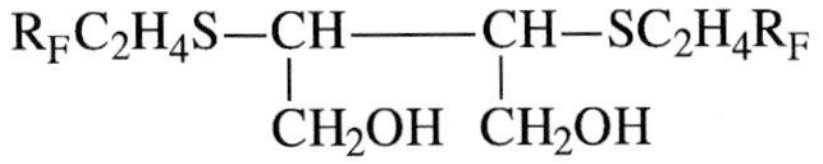

$$R_FC_2H_4S-\underset{\underset{CH_2OH}{|}}{CH}\text{————}\underset{\underset{CH_2OH}{|}}{CH}-SC_2H_4R_F$$

This last team also reported the synthesis of $(HOCH_2)_2C(CH_2SC_2H_4R_F)_2$ from a perfluoroalkyl ethyl mercaptan [229].

Further, zinc also enables the radical addition of perfluoroalkyl iodides onto unsaturated functional monomers [230] to produce, by the major monoadduct, a telechelic derivative:

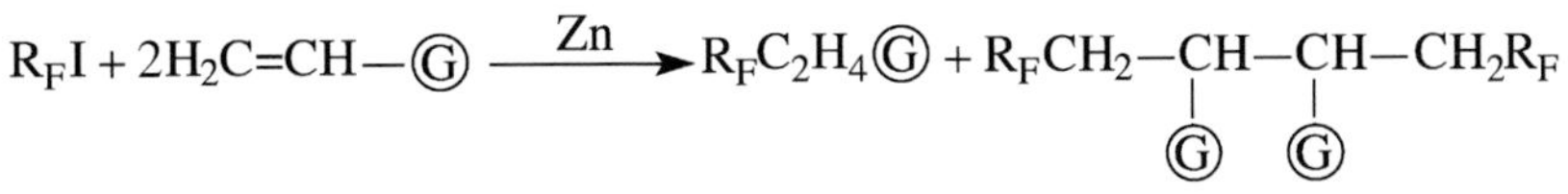

$$R_FI + 2H_2C=CH-\textcircled{G} \xrightarrow{\text{Zn}} R_FC_2H_4\textcircled{G} + R_FCH_2-\underset{\underset{\textcircled{G}}{|}}{CH}-\underset{\underset{\textcircled{G}}{|}}{CH}-CH_2R_F$$

$\textcircled{G}$: CN, OAc, CO_2Et, $CONH_2$

The same procedure was successfully achieved on itaconic acid [231].

2.2.1.1.3. By Ring Opening of Fluorinated Cyclic Ethers

2.2.1.1.3.1. From Fluorinated Epoxides. Oxiranes or epoxides are very important synthetic and industrial intermediates. Fluorinated oxiranes have also been a subject of much interest because such compounds are versatile and valuable intermediates for special products and various polymeric materials. For example, fluorinated oxiranes can be transformed to new alkenes [232, 233], polyhalogenated fluorohydrins [234] and to new surfactants [235–237]. They are used in the preparation of polymeric materials such as telechelic diols or polyethers with fluorinated side chains for water uptake [238, 239], novel acrylates for adhesive compositions [240], coatings [241] including ones of low surface tension [242] and telechelic epoxides [243] with particular naval applications [113, 243].

A large variety of syntheses of fluorinated epoxides has been proposed in the literature as previously reviewed [242, 244]. Among them, Cambon *et al.* [244, 245] prepared:

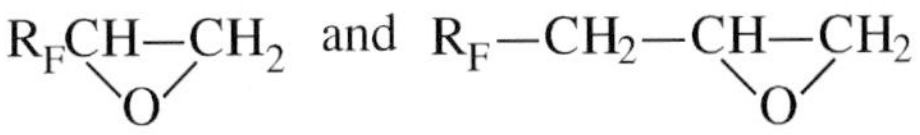

$$R_F\underset{\diagdown O\diagup}{CH-CH_2} \text{ and } R_F-CH_2-\underset{\diagdown O\diagup}{CH-CH_2}$$

The most classical method is the radical addition of perfluoroalkyl iodides onto either allyl alcohol or allyl acetate, and the corresponding fluorinated iodohydrine or iodoacetate are chemically changed into fluoroepoxides:

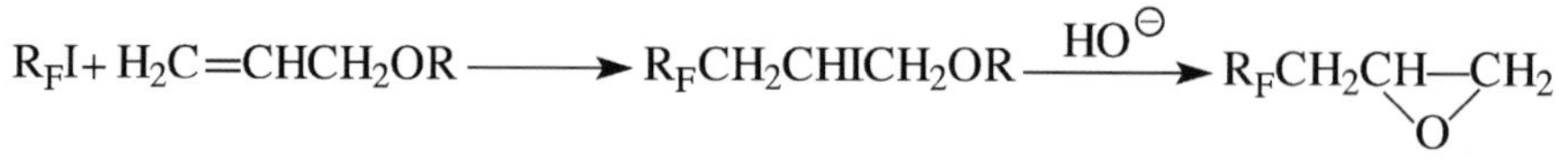

$$R_FI + H_2C=CHCH_2OR \longrightarrow R_FCH_2CHICH_2OR \xrightarrow{HO^{\ominus}} R_FCH_2CH\!-\!\!CH_2$$

R: H or COCH$_3$

Various types of fluorinated groups have been used:

$(CF_3)_2CFOC_2F_4$ [246], C_nF_{2n+1} [247, 248], $(CF_3)_2CF(CF_2)_4$ [249, 250]

Cirkva *et al.* [251, 252] optimised the reaction involving allyl acetate and reached an overall yield of F-epoxides of 85–87% from R_FI.

Several authors described the syntheses of perfluoroalkyloxiranes starting from perfluoroalkyl ethylenes [244, 245, 253, 254] or from branched 3-(perfluoroalkyl)propenes [255]. There was also developed a direct epoxidation of 3-(perfluoroalkyl)propenes with the fluorine–water agent [256, 257]. In addition, the reaction of fluorinated hydroxyderivatives with epichlorohydrin gave epoxy compounds with an ether bridge [258].

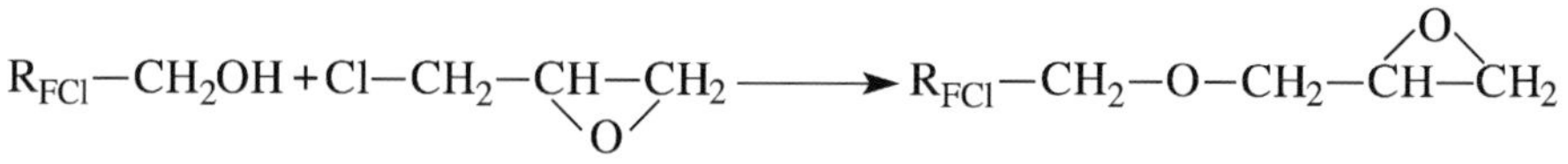

$$R_{FCl}\!-\!CH_2OH + Cl\!-\!CH_2\!-\!CH\!-\!CH_2 \longrightarrow R_{FCl}\!-\!CH_2\!-\!O\!-\!CH_2\!-\!CH\!-\!CH_2$$

$R_{FCl} = CF_3$, $Cl(CFCl\!-\!CF_2)_n\!-\!$

Most simple telechelics from these fluoroepoxides were 2-fluoroalkyl ethane-1,2-diols $R_FCH(OH)CH_2OH$ where R_F designates CF_3 [259], C_3F_7 [260] or more complex-branched perfluoroalkyl groups [261]. These fluorodiols exhibit good hydrophobic and oleophobic properties and have been used for heating porous (paper, wood, leather, fibres, textiles) or non-porous materials (surfaces of glass, metals or synthetic products).

These fluorinated oxiranes can be changed into 1,2-diols as reported by the following example [258]:

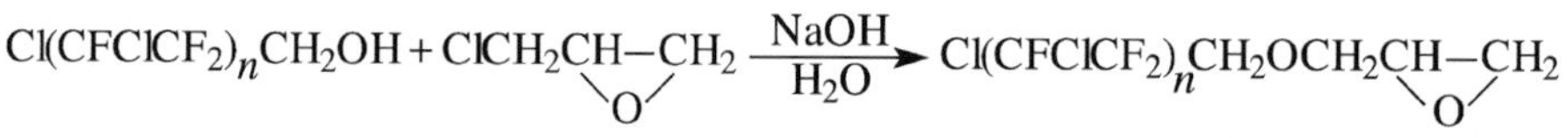

$$Cl(CFClCF_2)_nCH_2OH + ClCH_2CH\!-\!CH_2 \xrightarrow[H_2O]{NaOH} Cl(CFClCF_2)_nCH_2OCH_2CH\!-\!CH_2$$

$$\textbf{11}$$

$$\textbf{11} \xrightarrow{H^+/CH_3OH} Cl(CFClCF_2)_nCH_2OCH_2CHCH_2OH$$
$$\qquad\qquad\qquad\qquad\qquad\qquad\qquad OH$$

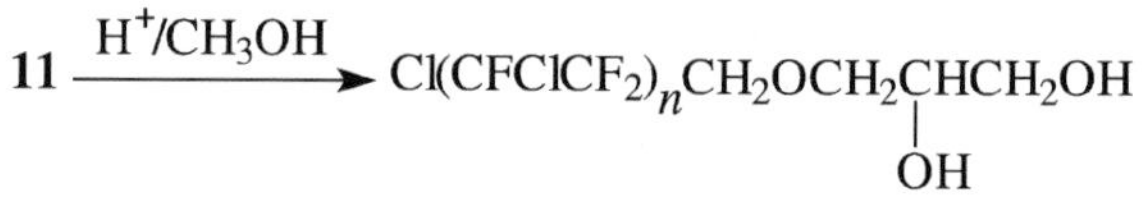

Ayari *et al.* [254] have shown that the longer the length of the perfluorinated chain, the greater the yields of the diols: 60, 64 and 75% for C_4F_9, C_6F_{13} and C_8F_{17}, respectively.

However, such an acidic hydrolysis required rather high temperatures (100–140 °C) for R_F: CF_3, C_2F_5 and C_3F_7 [262].

In addition, they can be converted into longer diols, as follows [252]:

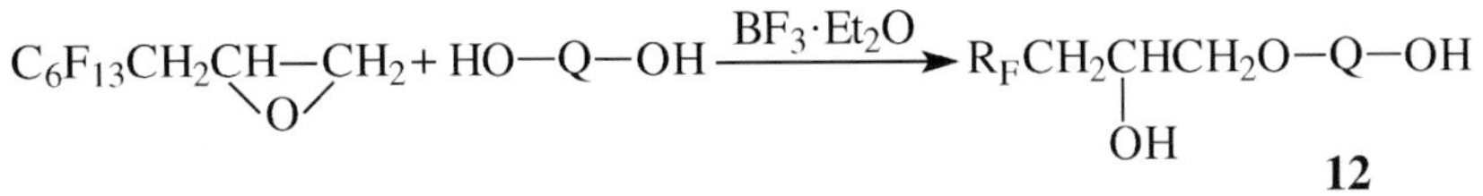

$$C_6F_{13}CH_2CH\overset{\diagdown}{\underset{O}{\diagup}}CH_2 + HO-Q-OH \xrightarrow{BF_3\cdot Et_2O} R_FCH_2\underset{\underset{OH}{|}}{C}HCH_2O-Q-OH$$

12

$$\mathbf{12} + C_6F_{13}CH_2CH\overset{\diagdown}{\underset{O}{\diagup}}CH_2 \xrightarrow{BF_3\cdot Et_2O} HO-\underset{\underset{CH_2R_F}{|}}{C}HCH_2O-Q-OCH_2\underset{\underset{CH_2R_F}{|}}{C}H-OH$$

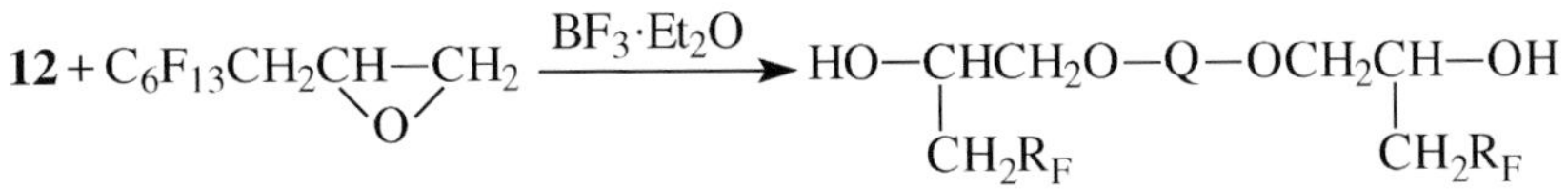

with Q: $(CH_2)_n$, $n = 2, 3, 4$ or $(CH_2CH_2)_2O$.

Starting from these F-epoxides, Mastouri *et al.* [263] prepared five hydrazinoalcohols by ring opening reaction, as follows:

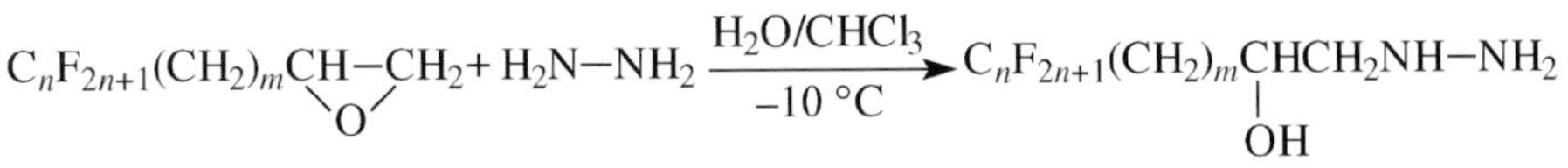

$$C_nF_{2n+1}(CH_2)_mCH\overset{\diagdown}{\underset{O}{\diagup}}CH_2 + H_2N-NH_2 \xrightarrow[-10\,°C]{H_2O/CHCl_3} C_nF_{2n+1}(CH_2)_m\underset{\underset{OH}{|}}{C}HCH_2NH-NH_2$$

$n = 6, 8$; $m = 0, 1$

Temperature seems to play an important role since, by an increase to 0 °C, the formation of hydrazino telechelic diol was produced in 8–10% yield. These authors [264] also reported the preparation of bis(F-alkylated) aminodiols via ring opening reaction of these F-oxiranes with ammonia at room temperature.

Furthermore, the direct addition of fluorinated amines onto ethylene oxide has led to original primary–primary fluorodiols [265]:

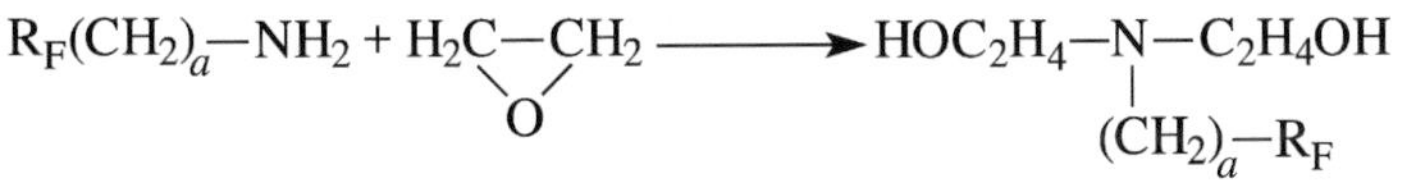

$$R_F(CH_2)_a-NH_2 + H_2C\overset{\diagdown}{\underset{O}{\diagup}}CH_2 \longrightarrow HOC_2H_4-\underset{\underset{(CH_2)_a-R_F}{|}}{N}-C_2H_4OH$$

R_F: CF_3 to $C_{20}F_{41}$

$a = 2$ or 4

Using the similar strategy with different fluorinated amines and hydrogenated epoxides, Foulletier *et al.* [216, 266] synthesised additional fluorinated diols:

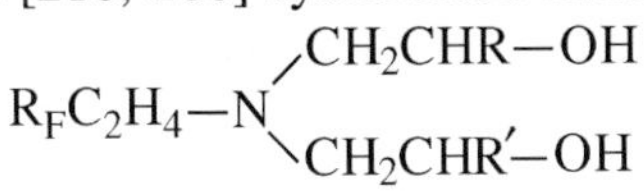

$$R_FC_2H_4-N\overset{\diagup CH_2CHR-OH}{\diagdown CH_2CHR'-OH}$$

with R: CH_3 and R': H or CH_3.

Further, the condensation of a fluorinated thiol with glycidol was successfully carried out by Bloech [218], who synthesised $R_FC_2H_4SCH_2CH(OH)CH_2OH$ as follows:

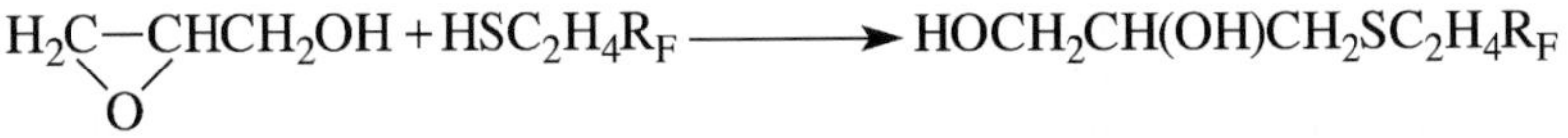

$$H_2C\overset{\diagdown}{\underset{O}{\diagup}}CHCH_2OH + HSC_2H_4R_F \longrightarrow HOCH_2CH(OH)CH_2SC_2H_4R_F$$

Nowadays, such a condensation is developed by several companies, e.g., by Atofina [267] using both glycidol and epichlorohydrin [268].

2.2.1.1.3.2. Acetals. An original fluorinated C_3 diol was also suggested [269, 270], from 2-ethyl-2-hydroxymethyl-1,3-propane diol, as follows:

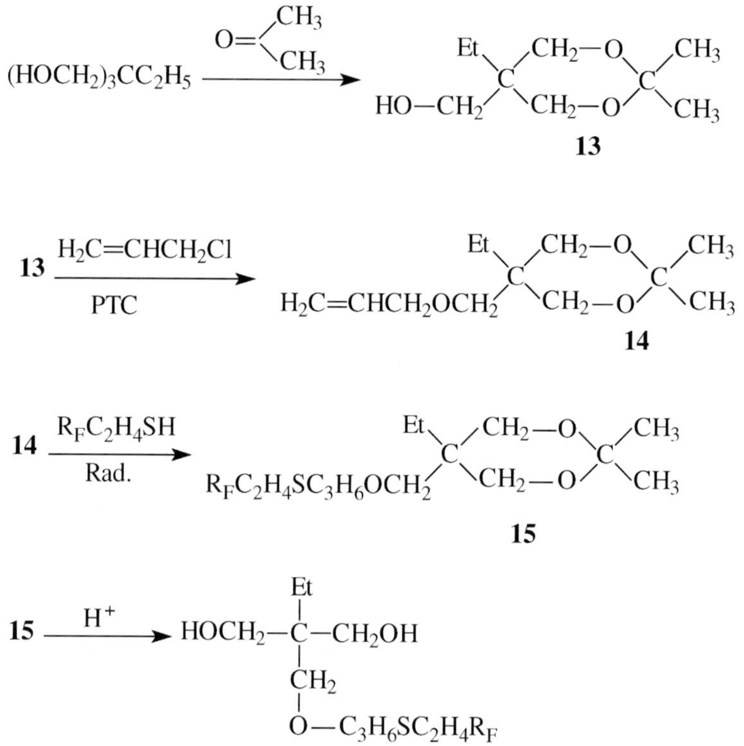

$$(HOCH_2)_3CC_2H_5 \xrightarrow{\quad O=C(CH_3)_2 \quad}$$

13

$$\textbf{13} \xrightarrow[\text{PTC}]{\quad H_2C=CHCH_2Cl \quad}$$

14

$$\textbf{14} \xrightarrow[\text{Rad.}]{\quad R_FC_2H_4SH \quad}$$

15

$$\textbf{15} \xrightarrow{\quad H^+ \quad} HOCH_2-\underset{\underset{O-C_3H_6SC_2H_4R_F}{\overset{|}{CH_2}}}{\overset{\overset{Et}{|}}{C}}-CH_2OH$$

More recently, another diol has been produced, starting from the radical addition of R_FI onto **14** followed by the reduction of iodine atom and deprotection of the acetal.

2.2.1.1.3.3. Dioxolanes. 1,2-Diols were prepared by the hydrolysis of 4-(2-chloro-1,1,2-trifluoroethyl)-2,2-dimethyl-1,3-dioxolane [271], as follows:

$$(CH_3)_2C\underset{O-CH_2}{\overset{O-CH(CF_2CFCl)_nH}{<}} \longrightarrow H(CFCl-CF_2)_n-\underset{\underset{OH}{|}}{CH}-CH_2OH$$

2.2.1.1.4. Miscellaneous Methods. The reaction between 1,3-butadiene and hexafluoroacetone at 170 °C for 48 h led to:

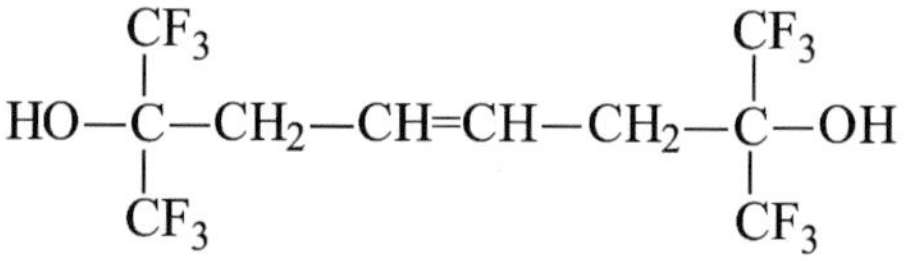

$$HO-\underset{\underset{CF_3}{|}}{\overset{\overset{CF_3}{|}}{C}}-CH_2-CH=CH-CH_2-\underset{\underset{CF_3}{|}}{\overset{\overset{CF_3}{|}}{C}}-OH$$

The *trans* isomer was produced in 82% while the *cis* was obtained in 4% [272].

2.2.1.2. Aromatic Fluorotelechelics

Most simple fluorinated aromatic telechelics were obtained by reacting benzene with 2 mol of hexafluoroacetone [273], leading to 90% of

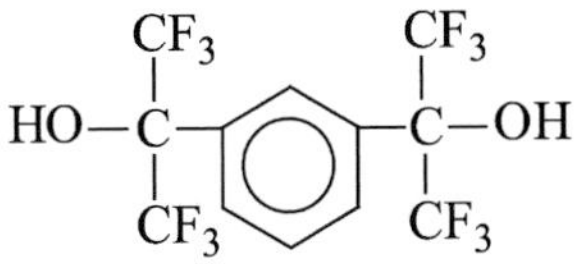

$$\text{HO}-\underset{\text{CF}_3}{\overset{\text{CF}_3}{C}}-\text{(benzene ring)}-\underset{\text{CF}_3}{\overset{\text{CF}_3}{C}}-\text{OH}$$

and 10% of

$$\text{HO}-\underset{\text{CF}_3}{\overset{\text{CF}_3}{C}}-\text{(benzene ring)}-\underset{\text{CF}_3}{\overset{\text{CF}_3}{C}}-\text{OH.}$$

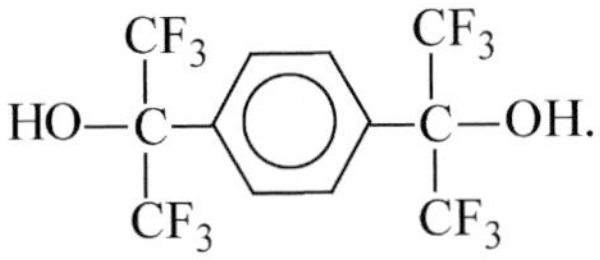

Another example concerns the preparation of an aromatic nitrogenous diol by fluorination of a carboxylic acid function in the presence of sulfur tetrafluoride [274]:

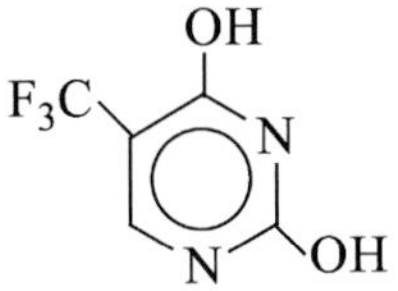

$$F_3C\text{—(pyrimidine ring with OH, N, N, OH substituents)}$$

Interesting investigations on the synthesis of fluorinated polyurethanes (FPUs) were achieved by a Russian team involving original fluorinated diisocyanates [275, 276]:

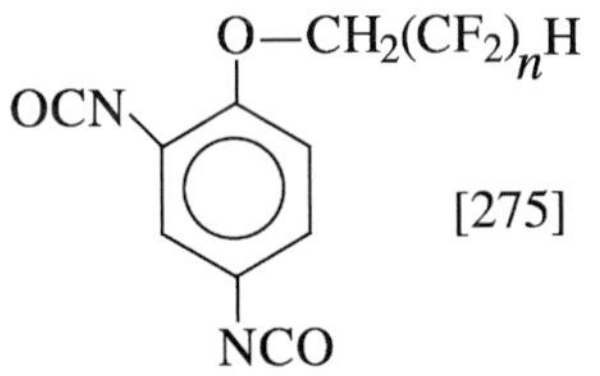

$$\text{OCN}\text{—(ring)}\text{—O—CH}_2(\text{CF}_2)_n\text{H, NCO} \quad [275] \qquad \text{OCN—(ring)—CF}_3,\ \text{NCO} \quad [276]$$

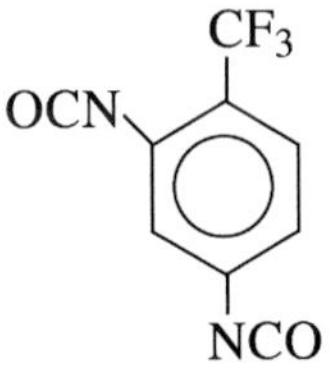

2.2.2. Long Fluorotelechelics

2.2.2.1. Grafting of Fluorinated Derivatives onto Hydroxy Telechelic Polybutadienes

2.2.2.1.1. From the Addition of Fluorinated Iodides or Thiols. New strategies were successfully applied for the synthesis of higher molecular weight macrodiols bearing perfluorinated side groups, starting from the hydroxytelechelic polybutadiene (HTPBD). First, the Thiokol Company [277] achieved the radical or photochemical addition of C_{6-20} perfluoroalkyl iodides onto HTPBD while, several years later, the grafting of fluorinated thiols led to sulfur-containing fluorinated macrodiols, as follows [278]:

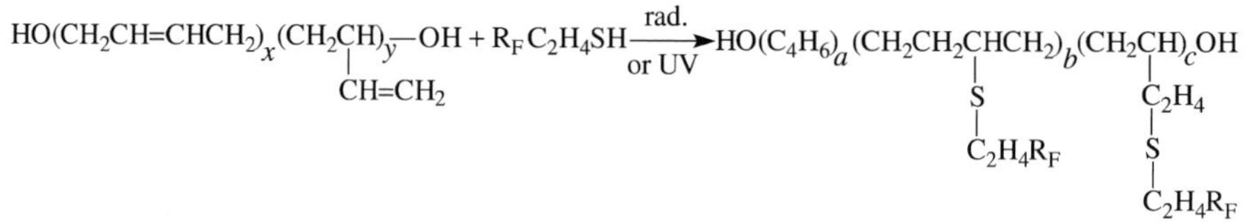

$$HO(CH_2CH=CHCH_2)_x(CH_2CH)_y—OH + R_FC_2H_4SH \xrightarrow[\text{or UV}]{\text{rad.}} HO(C_4H_6)_a(CH_2CH_2CHCH_2)_b(CH_2CH)_cOH$$

with pendant groups CH=CH$_2$ on the left, and on the right C$_2$H$_4$R$_F$ (via S) and S—C$_2$H$_4$R$_F$ with C$_2$H$_4$.

The initiation was achieved either photochemically [278a] or in the presence of a radical initiator [278b].

Interestingly, a wide variety of different fluorotelechelic standards were synthesised with fluorine content ranging from 3% (i.e., 5% of grafting) to 60% (i.e., 100% grafting). It was noted that the 1,2-double bonds were more reactive than the 1,4-ones and that, in the course of the grafting, the saturation of the former double bonds was achieved first, followed by the reaction of the fluorothiol onto the 1,4-double bonds. These compounds show interesting rheological (especially viscometric) and thermal behaviour [278]. Further information on the modification of poly(diene) (e.g., polybutadiene and polyisoprene) with fluorinated derivatives, leading to original grafted copolymers, is more fully developed in Chapter 5, Section 2.2.2.

2.2.2.1.2.   From the Hydrosilylation of Fluorosilanes. A similar investigation was studied by the same authors from the hydrosilylation of HTPBD with fluorinated silanes [279]. These silanes were prepared in a two-step reaction, as follows [280]:

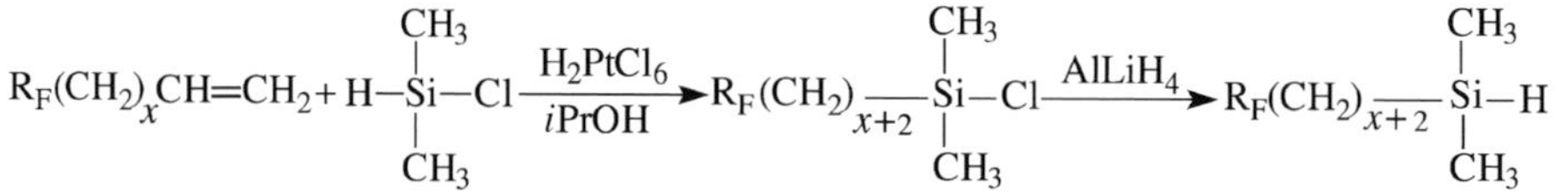

$$R_F(CH_2)_xCH=CH_2 + H—\underset{CH_3}{\overset{CH_3}{Si}}—Cl \xrightarrow[\text{iPrOH}]{H_2PtCl_6} R_F(CH_2)_{x+2}—\underset{CH_3}{\overset{CH_3}{Si}}—Cl \xrightarrow{AlLiH_4} R_F(CH_2)_{x+2}—\underset{CH_3}{\overset{CH_3}{Si}}—H$$

R$_F$: C$_6$F$_{13}$, C$_8$F$_{17}$

$x = 0$ or 1

Such a hydrosilylation of a fluorinated alkene is well known and applied to copoly(dimethyl) (methyl-hydrogeno) siloxane [281, 282]. The resulting telechelics can exhibit liquid-crystal properties [283] or may undergo crosslinking to give composite membranes with a high volatile organic compound permeability as well as good mechanical and solvent stability [284].

2.2.2.2.   From the Ring Opening Polymerisation of Cyclic Ethers
2.2.2.2.1.   From the Ring Opening Polymerisation of Fluoroepoxides. Fluorinated epoxides have been reported above (see Section 2.2.1.1.3.1) and their acidic ring opening polymerisations were investigated by numerous authors.

First, oxiranes containing a short fluorinated groups were polymerised by Trischler and Hollander [285]:

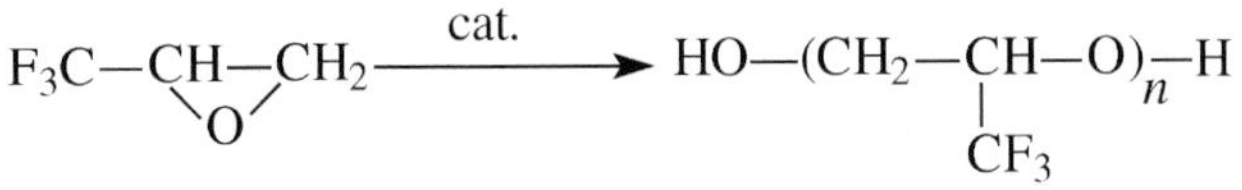

$$F_3C-CH-CH_2 \xrightarrow{\text{cat.}} HO-(CH_2-CH-O)_n-H$$
$$\underset{\displaystyle O}{\diagdown\diagup} \qquad\qquad\qquad \underset{\displaystyle CF_3}{|}$$

or by Gosnell and Hollander [238]:

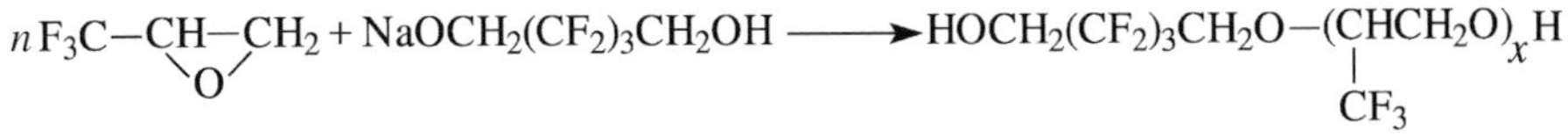

$$n\,F_3C-CH-CH_2 + NaOCH_2(CF_2)_3CH_2OH \longrightarrow HOCH_2(CF_2)_3CH_2O-(CHCH_2O)_x H$$
$$\underset{\displaystyle O}{\diagdown\diagup} \qquad\qquad\qquad\qquad\qquad\qquad\qquad\quad \underset{\displaystyle CF_3}{|}$$

The authors utilised two kinds of catalysts to initiate the ring opening polymerisation: either $BF_3 \cdot OEt_2$ that enabled them to achieve the synthesis of polymers with double bonds as end-groups, or $AlCl_3$ leading to polymers containing a hydroxylated end-group.

Knunyants [286] and O'Rear and Griffith [287] achieved the ring opening polymerisation of the following original fluoroepoxides in the presence of benzylamine or dibutyl amine.

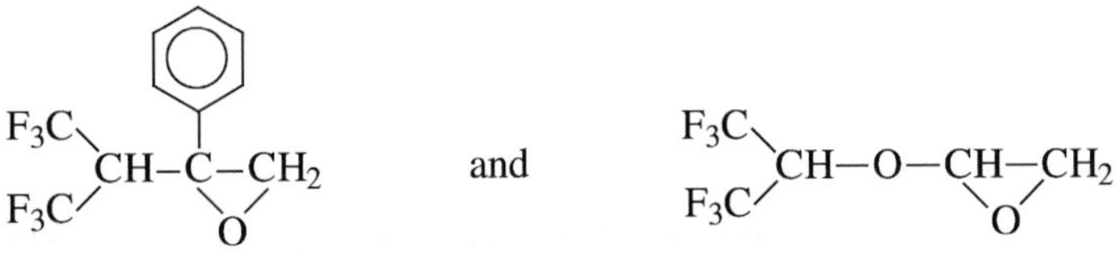

$$\underset{F_3C}{\overset{F_3C}{\diagdown}}CH-\underset{\underset{\displaystyle O}{\diagdown\diagup}}{\overset{\displaystyle \text{⬡}}{C}}-CH_2 \qquad \text{and} \qquad \underset{F_3C}{\overset{F_3C}{\diagdown}}CH-O-CH-CH_2$$
$$\underset{\displaystyle O}{\diagdown\diagup}$$

An English team [288] cationically polymerised different fluoroepoxides leading to fluoropolyethers with interesting surface properties.

Oxiranes containing longer fluorinated chains have been polymerised by Evans [289], Brace [110] and Pittman *et al.* [290]:

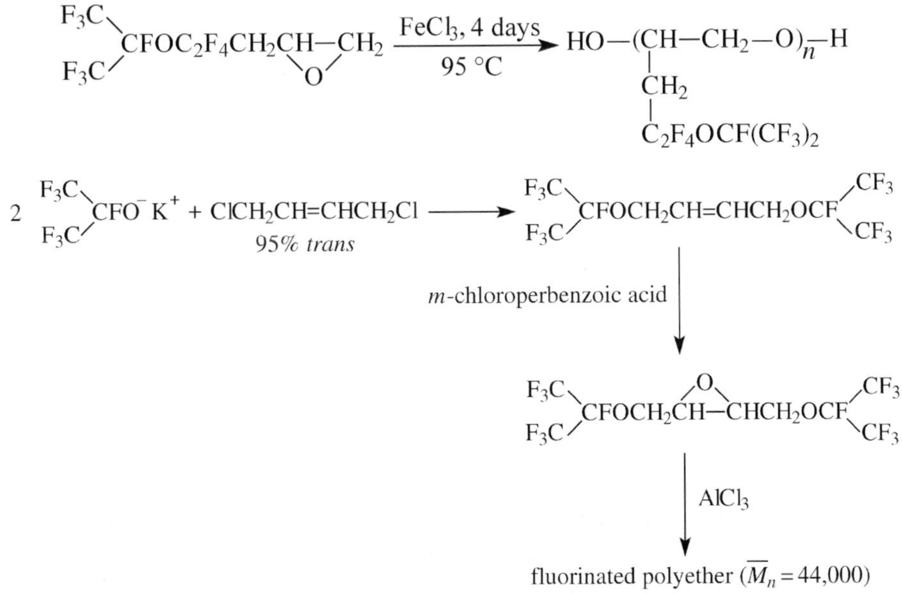

$$\underset{F_3C}{\overset{F_3C}{\diagdown}}CFOC_2F_4CH_2CH-CH_2 \xrightarrow[95\,°C]{FeCl_3,\,4\;days} HO-(CH-CH_2-O)_n-H$$
$$\underset{\displaystyle O}{\diagdown\diagup} \qquad\qquad\qquad \underset{\displaystyle CH_2}{|}$$
$$\underset{\displaystyle C_2F_4OCF(CF_3)_2}{|}$$

$$2\;\underset{F_3C}{\overset{F_3C}{\diagdown}}CFO^-\,K^+ + ClCH_2CH{=}CHCH_2Cl \longrightarrow \underset{F_3C}{\overset{F_3C}{\diagdown}}CFOCH_2CH{=}CHCH_2OCF\underset{\diagdown CF_3}{\overset{\diagup CF_3}{}}$$
$$95\%\;trans$$

m-chloroperbenzoic acid ↓

$$\underset{F_3C}{\overset{F_3C}{\diagdown}}CFOCH_2\underset{\underset{\displaystyle}{\diagdown\diagup}}{\overset{\displaystyle O}{CH}}-CHCH_2OCF\underset{\diagdown CF_3}{\overset{\diagup CF_3}{}}$$

↓ $AlCl_3$

fluorinated polyether ($\overline{M}_n = 44{,}000$)

2.2.2.2.2.   From the Ring Opening Polymerisation of Fluorooxetanes. Oxetanes are known to be ring opening polymerised under acidic cationic conditions leading to polyethers. Original fluorooxetanes can be synthesised by the Williamson reaction of the bischloromethyl oxetane (BCMO) with fluorinated alcohols according to the following scheme [291]:

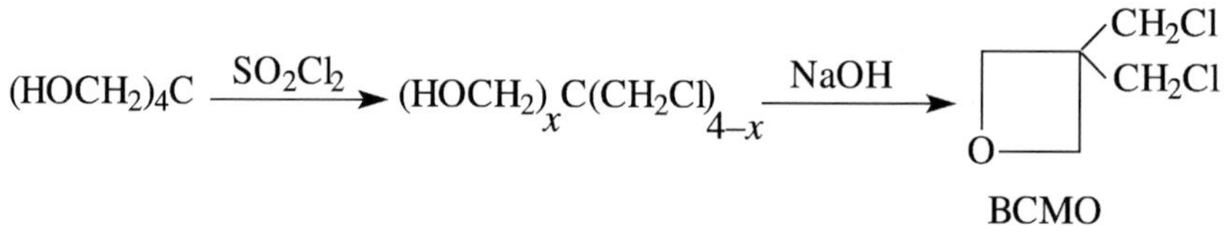

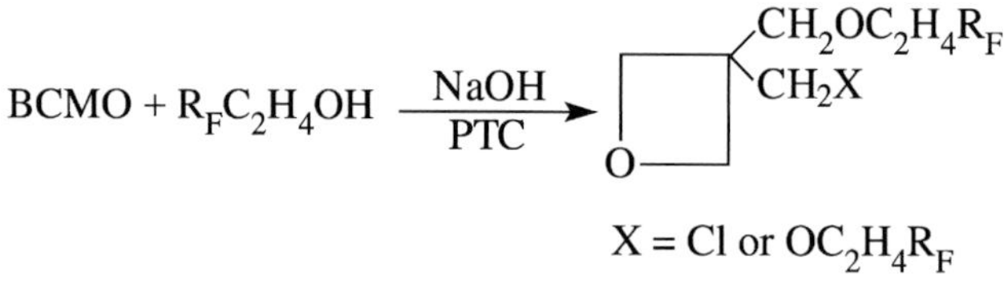

$$(HOCH_2)_4C \xrightarrow{SO_2Cl_2} (HOCH_2)_x C(CH_2Cl)_{4-x} \xrightarrow{NaOH}$$

BCMO

$$BCMO + R_F C_2 H_4 OH \xrightarrow[PTC]{NaOH}$$

$$X = Cl \text{ or } OC_2 H_4 R_F$$

Telechelic macrodiols with molecular weights ranging between 3000 and 5000 were obtained by cationic ring opening polymerisation.

$$\xrightarrow[HOC_2H_4OH]{BF_3-OEt_2} HO \left[CH_2 - \underset{CH_2OC_2H_4R_F}{\overset{CH_2X}{\underset{|}{\overset{|}{C}}}} - CH_2 - O \right]_n C_2H_4OH$$

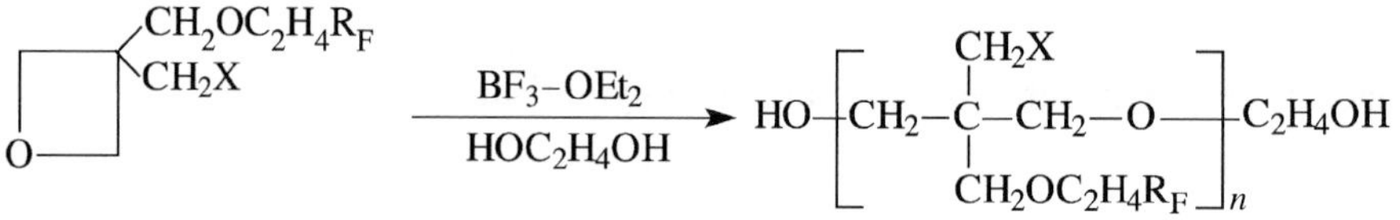

Improved surface and thermal properties were noted compared to similar fluorinated polyethers bearing $CH_2OCH_2(C_2F_4)_nH$ ($n = 2, 3, 4$) side group [292]. Oxetanes with shorter fluorinated groups were prepared by various authors [293].

2.2.2.2.3.   From the Ring Opening Polymerisation of Fluorofuranes. The Chambers' team [294] investigated the electron beam grafting of HFP onto THF leading to:

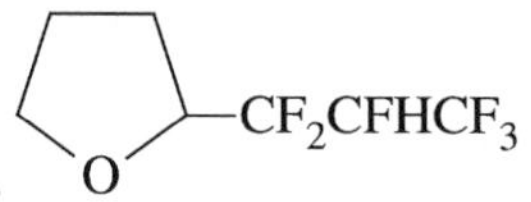

$$-CF_2CFHCF_3$$

This fluorinated furan was not ring opening polymerised but the authors investigated the e-beam grafting of HFP onto poly(THF), leading to:

$$HO - (C_4H_8O)_x - (C_4H_7O)_y - H$$
$$\underset{C_3F_6H}{|}$$

with $x = y = 4.5$.

The acrylation of the hydroxy end-groups of these fluorinated oligomers enabled the ICI Company to supply a fluoroacrylated polyether resin (named Neorad®) as precursor of original photopolymers for optical fibres [295].

2.2.2.3. By Radical Polymerisation of Fluoro(meth)acrylates with Telechelic Initiators

Three examples have been chosen to show that the choice of the telechelic initiator is crucial for obtaining interesting well-defined oligomers or polymers. The first two examples concern the use of initers or iniferters in the polymerisation of fluorinated (meth)acrylate, while the third one requires a reactivable C–Br bond to allow the bromine atom radical polymerisation of fluorinated acrylates (FA).

(a) The radical polymerisation of 2,2,2-trifluoroethyl methacrylate (TFEMA), initiated by a telechelic tetraphenyl ethane-type initer and leading to telechelic poly(TFEMA) oligomers, was successfully achieved by Roussel [296], as follows:

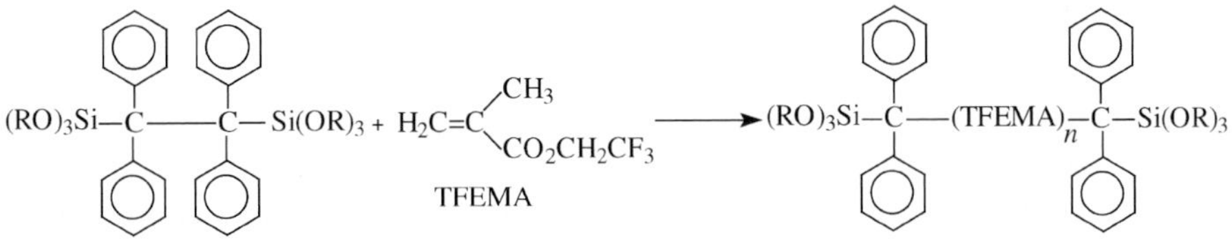

By a thermal cleavage at the bonds between the last TFEMA unit and the end-group from the initiator, these telechelic oligomers behaved as efficient macroiniters for further free-radical polymerisation of styrene(s) yielding telechelic poly(TFEMA)-*b*-PS block copolymers with a macroinitiator efficiency close to 85%.

(b) Telechelic poly(1,1-dihydroperfluorooctyl acrylate) was prepared by radical polymerisation in the presence of bis(*N,N*-diethyldithiocarbamate) as iniferter [297a]. This polymer could reinitiate further polymerisation of styrene and acrylic acid to enable the formation of triblock copolymers:

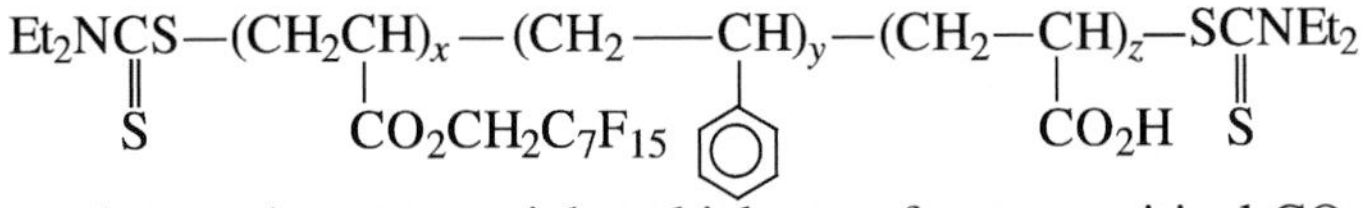

$$Et_2NCS-(CH_2CH)_x-(CH_2---CH)_y-(CH_2-CH)_z-SCNEt_2$$

with substituents: $\overset{\|}{S}$, $CO_2CH_2C_7F_{15}$, phenyl, CO_2H, $\overset{\|}{S}$

These copolymers have potential as thickeners for supercritical CO_2 used as a displacing agent in enhanced oil recovery.

A similar strategy was also efficient: telechelic polystyrenes capped with *N,N*-diethyldithiocarbamate groups as macroinitiators enabled the polymerisation of the FA leading to a PFA-*b*-PS-*b*-PFA triblock copolymer [297b]. These copolymers were used as lipophilic surfactants in supercritical CO_2 applications.

(c) Original semifluorinated ABA triblock copolymers with different segment lengths were synthesised in a two-step procedure [298]. The first one concerns the preparation of telechelic dibromo polystyrene by bromine transfer radical polymerisation (BrTRP) of styrene with α,ω-dibromopolystyrene (achieved by BrTRP of styrene with α,ω-dibromo-*p*-xylene in the presence of cuprous bromide/bipyridyl complex as the catalyst). In the second step, this telechelic polystyrene enabled the BrTRP of 2-[(perfluorononenyl)oxy] ethyl methacrylate and ethylene glycol monomethacrylate monoperfluorooctanoate (FMA) to lead to PFMA-*b*-PS-*b*-PFMA triblock copolymers. As expected, the longer the poly(fluoromethacrylate) block, the higher the contact angle value of the block copolymer.

More detailed information on the synthesis, properties and applications of fluorinated block copolymers is supplied in Chapter 4.

2.2.3. *Conclusion*

Telechelic products bearing a fluorinated side chain have been synthesised by means of a wide variety of different methods, although more commercially available derivatives have been prepared on a large scale.

Short fluorotelechelics can be obtained via condensation, radical addition of fluorinated reactants to hydrogenated telechelics or from the ring opening of fluorinated cyclic ethers. Besides the availability of all these reactants above, these methods enable one to prepare telechelics of various chain lengths (from C2 diol up to C5, from the ring opening of fluoroepoxide to available hydrogenated telechelics containing five carbon atoms). Less exhaustive methods to produce fluoroaromatic telechelics have been suggested.

Macrodiols ($\bar{M}_n$ ranging between 1500 and 5000) can be synthesised from the radical grafting of fluorinated thiols and from the hydrosilylation of hydroxy-telechelic polybutadienes with hydrogenofluorosilanes, or from the ring opening polymerisation of cyclic ethers.

Hence, well-tailored molecular and oligomeric molecules (ranging from secondary–secondary or primary–primary diols) have been prepared with specific reactivities. They constitute relevant monomers for polycondensations or polyadditions leading to fluoropolymers, as reported below.

3. Synthesis of Well-Architectured Fluoropolymers from Fluorinated Telechelics

Telechelic diols, diacids, diisocyanates, diamines and dienes are among the most promising starting materials for the synthesis of well-architectured fluoropolymers via condensation, polymerisations or polyaddition polymerisations. Several examples of preparation of these fluoropolymers are presented in the following sections.

3.1. FROM FLUORINATED DIOLS

Diols are interesting forerunners for the synthesis of various polymers (e.g., polyurethanes, polyesters, polyethers) and are also useful intermediates to be chemically changed into diepoxides, di(meth)acrylates or dicyanates (Scheme 3). Some of them are depicted below.

3.1.1. Fluorinated Polyurethanes

The first FPU was patented in 1958 [299]. Twelve years later, a relevant survey [300] detailed the comparison of the reactivity of fluorinated diols with that of non-halogenated ones for the preparation of such polymers. A Russian team performed interesting work by reacting 2,2,3,3,4,4-hexafluoro-1,5-pentanediol and hexamethylene diisocyanate [301], as follows:

$$OCN-(CH_2)_6-NCO + HOCH_2-(CF_2)_3-CH_2OH \longrightarrow$$

$$\left[\overset{\overset{\displaystyle O}{\displaystyle \|}}{C}-NH-(CH_2)_6-NH-\overset{\overset{\displaystyle O}{\displaystyle \|}}{C}-OCH_2-(CF_2)_3-CH_2O \right]_n$$

These authors have shown that such FPUs exhibit better thermostability than hydrogenated analogues because of the fluorinated groups that prevent thermal decomposition of the urethane group.

Then, Keller [302] developed interesting polyurethanes by polycondensation of various fluorodiols with hexamethylene diisocyanate. When prepared at temperatures above 75 °C, they were amorphous, insoluble and transparent from brittle to tough as the reaction progressed. In contrast, they became brittle and soluble in common organic solvents when the reaction was performed at a temperature below 75 °C.

Additional investigations for the synthesis of FPUs have been realised by using either fluoro-containing diisocyanates [303], fluorinated chain extenders [303–310] or fluorinated soft segments (some of them are based on telechelic perfluoropolyethers [303, 311–318]).

Kajiyama et al. [307] introduced a fluorocarbon-containing diol as chain extender into a polyurethane and investigated its effect on the surface properties of the PU. Similarly, Ho et al. [313, 314] obtained FPUs based on a series of aliphatic fluorinated diols that contained TFE and HFP to improve their surface properties with minimum adhesion.

$$HOCH_2CH_2(C_3F_6)_x(C_2F_4)_2(C_3F_6)_yCH_2CH_2OH \xrightarrow[\text{bulk or THF}]{OCN(CH_2)_6NCO} FPUs$$

$(x + y = 1.00$ or 1.67 or $2.47)$

Physical properties were also assessed: the T_gs ranged from 18 to 27 °C and their viscosities from 0.22 to 0.46 dl g^{-1} while their surface tensions ranged between 25.5 and 31.5 erg cm^{-2}.

Tonelli *et al.* [311, 312] used perfluoropolyethers as soft segments in the synthesis of FPUs that exhibit low-temperature elastomeric behaviour, thermal stability and superior chemical resistance.

In addition, these FPUs have been investigated for use in biomedical applications. For example, Kachiwagi *et al.* [318] synthesised original polyurethanes bearing a perfluorinated side chain and showed that the *in vitro* thrombus formation was reduced as the fluoroalkyl content was increased.

Takakura *et al.* [303] investigated the structure and mechanical properties of fluorinated segmented poly(urethane urea)s, concluding that their structures were highly ordered and that such polymers behaved as elastomers. In addition, such FPU exhibited antithrombogenic properties [119] and durability for use in an artificial heart pump [303].

Further, Sbarbati Del Guerra *et al.* [315] showed that FPUs with different fluorine contents are not thrombogenic and not cytotoxic.

As a matter of fact, Tang *et al.* [319, 320] mixed fluorine-containing polyurethanes, used as surface-modifying macromolecules, with base PUs to alter the surface chemistry of the base PU, thereby altering their biostability and/or biocompatibility.

Interestingly, Chen and Kuo [310] synthesised aliphatic FPUs chain extended with 1,4-butanediol or fluorinated diols such as 2,2,3,3,4,4,5,5-octafluoro-1,6-hexanediol or 2,2,3,3-tetrafluoro-1,4-butanediol. The influence of the type of chain extender and the average length of hard segments on the physical properties (microstructure and thermal transitions), surface properties and blood compatibility were investigated. The fluorinated PUs exhibit weaker hydrogen bonding, a smaller fraction of hydrogen-bonded carbonyls, less phase-separated soft segment domains, a lower melting temperature and less heat of fusion of crystalline hard segment domains than the PUs chain extended with 1,4-butanediol.

Indeed, it is noteworthy to compare the nature of the diisocyanate: the bulky fluorocarbon chains seem to have a larger effect on the hard segment packing of hexamethylene diisocyanate-based PUs than 4,4'-methylene diphenyl diisocyanate [305, 307]. This has a larger influence on the macroscopic physical properties of the former. Moreover, the longer CF_2 groups of the chain extender enhance the effect.

The *in vitro* platelet adhesion experiments (which may lead to thrombus formation) show that the interaction between platelets and the material surface is affected by the incorporation of fluorocarbon chains in PUs [310]. The authors have noted that a higher fluorocarbon chain content significantly reduces the degree of platelet adhesion and platelet activation on the PU surface.

In addition, polyurethanes based on aromatic diisocyanates were successfully prepared where ring rigidity led to increased T_gs of the FPUs. The same fluorinated diol as above was involved in the synthesis of fluorinated polymers with 3,5-toluene diisocyanate and with a mixture of 2,4-toluene diisocyanate and 3,3'-bistolylene-4,4'-diisocyanate [321]:

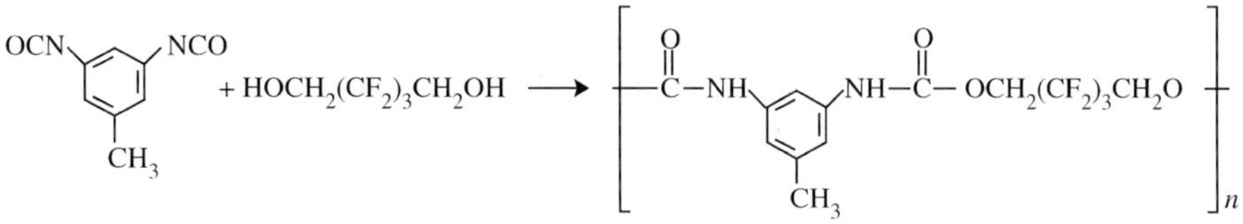

$$OCN-\underset{\underset{CH_3}{|}}{\text{⟨⟩}}-NCO \;+\; HOCH_2(CF_2)_3CH_2OH \;\longrightarrow\; \left[\!\!-\!\overset{O}{\overset{\|}{C}}\!-\!NH\!-\!\underset{\underset{CH_3}{|}}{\text{⟨⟩}}\!-\!NH\!-\!\overset{O}{\overset{\|}{C}}\!-\!OCH_2(CF_2)_3CH_2O\!-\!\!\right]_{\!n}$$

Interestingly, Yoon and Ratner [304–306, 309] synthesised additional aromatic FPUs by using various perfluorinated chain extenders and, by X-ray photoelectron spectroscopy, they studied the surface structure of segmented poly(ether urethane)s and poly(ether urethane urea)s. They noticed that the surface topography of such polymers depended strongly on the extent of the phase separation.

Interestingly, Wall [322], Ho *et al.* [174] and, more recently, Brady [113] reported relevant strategies of syntheses, properties and applications of fluorinated aromatic and aliphatic polyurethanes, even using a perfluorinated diisocyanate and a fluorinated diol, as follows:

$$OCN-(CF_2)_3-NCO + HOCH_2-(CF_2)_3-CH_2OH \longrightarrow \left[\!\!-\!\overset{O}{\overset{\|}{C}}\!-\!NH\!-\!(CF_2)_3\!-\!NH\!-\!\overset{O}{\overset{\|}{C}}\!-\!OCH_2\!-\!(CF_2)_3\!-\!CH_2O\!-\!\!\right]_{\!n}$$

However, such a polyurethane was shown to be hydrolytically unstable [323]. In addition, these authors reviewed the FPUs prepared from polyesters of fluorinated diols, and polyethers and detailed their properties [323].

Malichenko *et al.* [301] also used fluoroaromatic diisocyanates for better materials useful as containers of liquid oxygen and novel adhesives. In addition, other FPUs prepared from fluorodiols containing HFP base units were also reported [90, 314].

Polyurethanes bearing fluorinated side chains [221–223, 308, 309] exhibit water- and oil-repellant properties, have good abrasion resistance and have very low surface energy.

An example is the original block 3-[3-(perfluorooctyl) propoxy]-1,2-propanediol-isophorone diisocyanate-polyethylene glycol polymer [171] and FPUs prepared from telechelic diols bearing a perfluorinated side chain [225] used in soil-release finishes.

Original polyurethanes bearing fluorinated side groups have been synthesised from fluorinated polyethers diols having fluorolateral groups prepared from fluoroepoxides [324]:

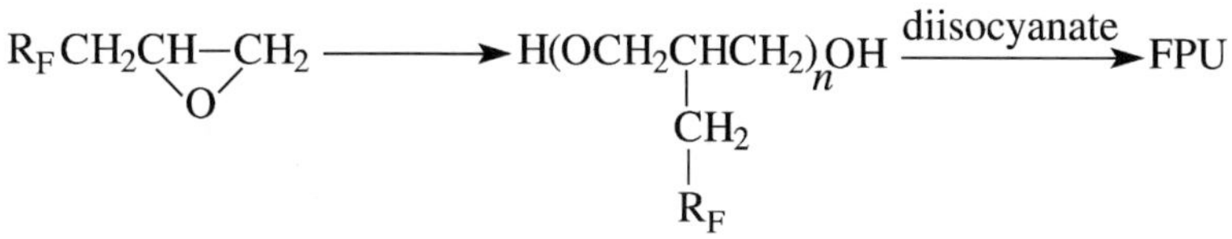

$$R_FCH_2\underset{\underset{O}{\diagdown\diagup}}{CH-CH_2} \longrightarrow H(OCH_2\underset{\underset{\underset{R_F}{|}}{CH_2}}{CHCH_2})_nOH \xrightarrow{\text{diisocyanate}} FPU$$

or from fluorinated oxetanes [325].

Yu *et al.* [316] investigated on a series of FPUs from the following polyether glycols bearing various fluoroalkyl chain length:

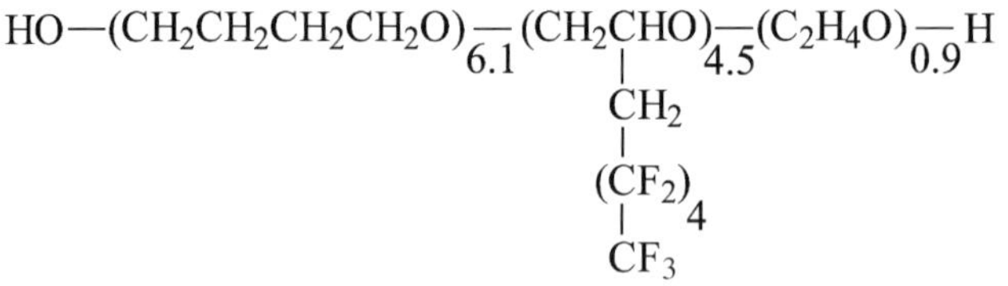

$$HO-(CH_2CH_2CH_2CH_2O)\underset{6.1}{}-(CH_2CHO)\underset{4.5}{}-(C_2H_4O)\underset{0.9}{}-H$$
$$\overset{|}{CH_2}$$
$$\overset{|}{(CF_2)_4}$$
$$\overset{|}{CF_3}$$

Similarly, Collet *et al.* [326] and then Cathebras [327] prepared various FPUs that exhibit interesting surface properties using a wide range of fluorinated oligomers as follows:

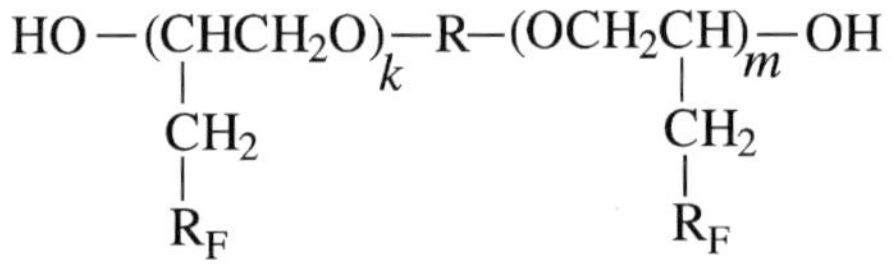

$$HO-(CHCH_2O)\underset{k}{}-R-(OCH_2CH)\underset{m}{}-OH$$
$$\overset{|}{CH_2} \qquad\qquad\qquad \overset{|}{CH_2}$$
$$\overset{|}{R_F} \qquad\qquad\qquad \overset{|}{R_F}$$

$$R_F: C_4F_9, C_6F_{13}, C_8F_{17}$$

R: diethylene glycol or ethoxylated bisphenol A

Original telechelic fluorodiols containing shorter fluorinated side groups brought by hexafluoropropylene units have also been successfully used for obtaining FPUs:

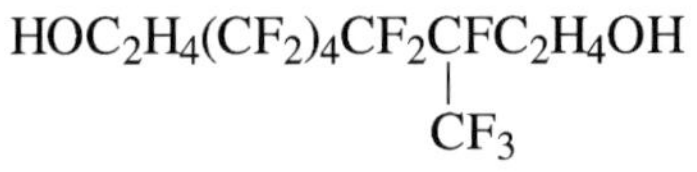

$$HOC_2H_4(CF_2)_4CF_2CFC_2H_4OH \qquad [292]$$
$$\overset{|}{CF_3}$$

$$HOC_2H_4CFCF_2-(CF_2)_4-CF_2CFC_2H_4OH \qquad [293]$$
$$\overset{|}{CF_3} \qquad\qquad\qquad\qquad \overset{|}{CF_3}$$

3.1.2. Fluorinated Polyesters

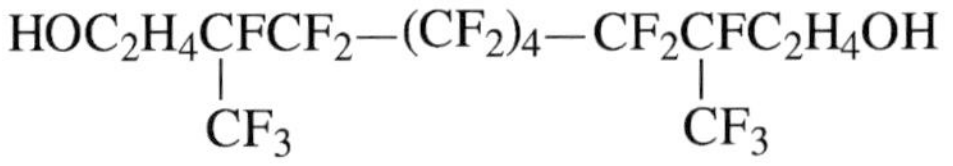

From 1952, Filler *et al.* [328] studied the polycondensation of fluorinated acid dichlorides and diols according to the following reaction:

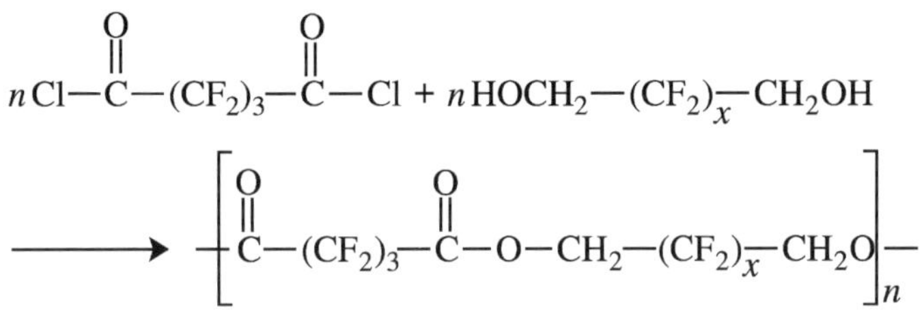

$$n\,Cl-\overset{\overset{O}{\|}}{C}-(CF_2)_3-\overset{\overset{O}{\|}}{C}-Cl + n\,HOCH_2-(CF_2)\overline{x}-CH_2OH$$

$$\longrightarrow \left[\overset{\overset{O}{\|}}{C}-(CF_2)_3-\overset{\overset{O}{\|}}{C}-O-CH_2-(CF_2)\overline{x}-CH_2O\right]_n-$$

$$x = 3, 4$$

In addition, Gosnell and Hollander [238] carried out the following synthesis:

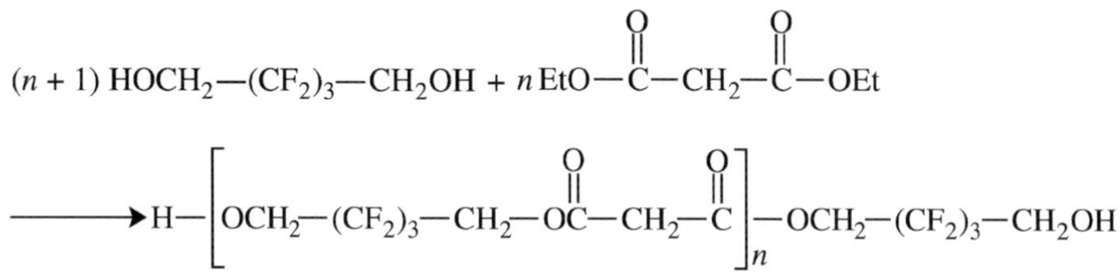

$$(n + 1)\ HOCH_2-(CF_2)_3-CH_2OH + n\ EtO-\overset{\overset{O}{\|}}{C}-CH_2-\overset{\overset{O}{\|}}{C}-OEt$$

$$\longrightarrow H-\left[OCH_2-(CF_2)_3-CH_2-O\overset{\overset{O}{\|}}{C}-CH_2-\overset{\overset{O}{\|}}{C}\right]_n-OCH_2-(CF_2)_3-CH_2OH$$

leading to a polyester having a molecular weight of 1800.

These authors also used fluoroaromatic diesters to prepare original hydrophobic fluorinated polyesters, which still preserve their properties at low temperatures ($T_g = -65\ °C$) and are resistant to oxidation, aggressive chemicals, petroleum and fire, with good mechanical properties. They found relevant applications as containers for liquid oxygen.

In our laboratory, the following polytransesterification was used to obtain polyester diol 16 containing CTFE base units [329]:

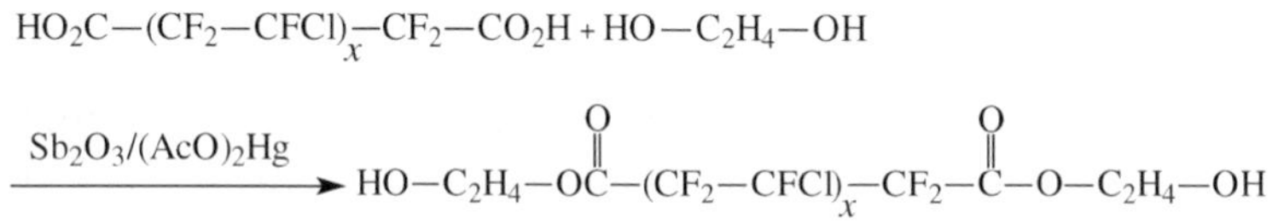

$$HO_2C-(CF_2-CFCl)_x-CF_2-CO_2H + HO-C_2H_4-OH$$

$$\xrightarrow{Sb_2O_3/(AcO)_2Hg} HO-C_2H_4-O\overset{\overset{O}{\|}}{C}-(CF_2-CFCl)_x-CF_2-\overset{\overset{O}{\|}}{C}-O-C_2H_4-OH$$

16

$$16 \xrightarrow{Ti(OBu)_4/200\ °C} HO-CH_2-CH_2-O\left[\overset{\overset{O}{\|}}{C}-(CF_2-CFCl)_x-CF_2-\overset{\overset{O}{\|}}{C}-O-CH_2-CH_2-O\right]_y H$$

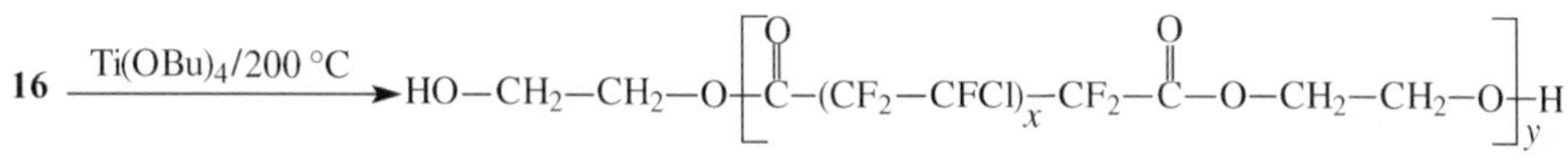

Another approach was used by Schweiker and Robitschek [330] in the development of other fluorinated polyesters.

In addition, the polymerisation of fluorinated epoxides leading to new polyether macrodiol precursors of original fluoropolyesters was performed by Evans and Morton [246] and by Gosnell and Hollander [238]:

$$n\ F_3C-HC\overset{\diagdown\diagup}{\underset{O}{}}CH_2 + NaO-CH_2-(CF_2)_3-CH_2OH$$

$$\longrightarrow HOCH_2-(CF_2)_3-CH_2-O-(\overset{\overset{CF_3}{|}}{CH}-CH_2-O)_x H$$

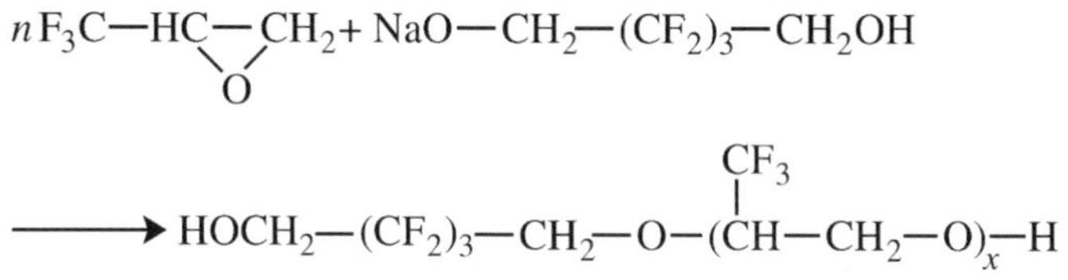

Pilati et al. [331] prepared polyesters from dimethylterephthalate, ethylene glycol and the diester $CH_3O_2C-R_F-O-R_F-CO_2CH_3$. The copolycondensation by transesterification was studied and the authors investigated the morphological, thermal and surface properties of these obtained polyesters. It was noted that the PET/copolymer (PET–PFPE) blend thus obtained contained up to 35 wt% of PFPE but only 7% were bonded with the prepared PET.

Further investigations by the same group involving commercially available Fomblin® Z (see Section 2.1.1.2.1) telechelic perfluoropolyethers successfully led to original fluorinated polyesters [332].

Interestingly, Pospeich et al. [333] investigated the synthesis of original aromatic polyesters [333a] and polysulfone-b-polyester segmented block copolymers [333b] bearing a perfluorinated side chain. These former fluorinated polyesters (which $\bar{M}_n = 3000-9500$) were prepared by direct polycondensation of fluorinated isophthalic or terephthalic methyl esters with hydroquinone and exhibit the following formulae of the main base unit:

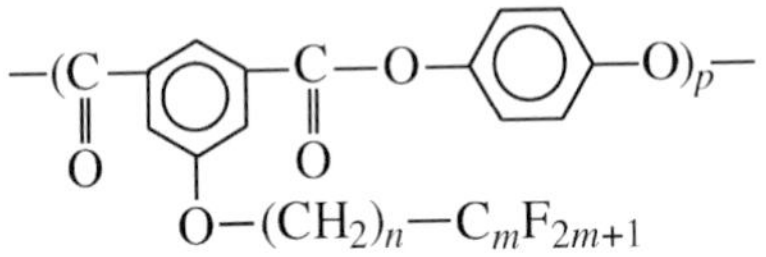

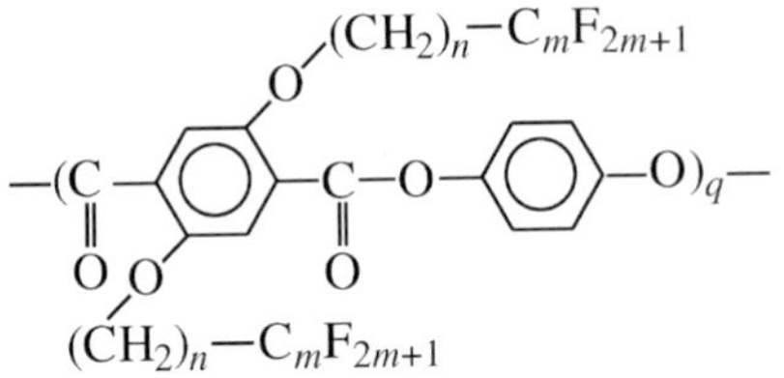

Both fluorinated telechelic diesters were obtained from the Mitsunobu [334] reaction of semifluorinated alcohols $HO(CH_2)_n-C_mF_{2m+1}$ (with $n = 6$ or 10 and $m = 6$, 8 or 10) with 5-hydroxy-dimethylisophthalate or 2,5-dihydroxy-dimethylterephthalate.

The presence of the fluorinated side group drastically changed the wetting behaviour of these polymers. Indeed, self-organising structures were observed, caused by microphase separation between the polymer backbones and the fluorinated substituents.

3.1.3. Fluorinated Polyethers

Besides Krytox® [14, 15], Fomblin® [16–20, 23–25], Demnum® [21] and Aflunox® marketed by DuPont, Ausimont, Daikin and Unimatec (formerly Nippon Mektron), respectively, the DuPont Company [121] also suggested the synthesis of the following fluoropolyether from the ionic polymerisation of hydroxy trifluorovinyl ethers with fluorodiol in the presence of CsF:

$$H \left[(OCH_2C_3F_6CH_2O)_{\overline{n}}(CF_2CHFOCF_2CF(CF_3)OCF_2CF_2CH_2O)_m \right] H$$

Interestingly, fluoropolyethers can also be prepared from the reaction between a dialcoholate with a dihalogenide, as studied by Johncock et al. [335]:

$$HOCH_2(CF_2)_nCH_2OH + X-R-X \xrightarrow[T < 50\,°C]{NaH} -(R-O-CH_2-R_F-CH_2O)_{\overline{n}}$$

R: CH_2, CH_2OCH_2, $CH_2-C_6H_4-Z-C_6H_4-CH_2$ (with Z: O, S, CO, CH_2, $C(CH_3)_2$)

Fluoroelastomers, thermostable up to about 300 °C, were obtained with T_gs ranging between -57 and -5 °C.

Similarly, Cook [336] synthesised original polyethers from the following highly fluorinated diene: $F_2C=CFCF_2CFClCF_2CF=CF_2$ in alkali medium.

3.1.4. Other Fluoropolymers Produced from F-Diols

3.1.4.1. Fluorinated Diglycidyl Polyethers

The synthesis of fluorinated epoxides was reported in Section 2.2.1.1.3.1 and an elegant pathway to prepare telechelic glycidyl polyethers consists of the addition of a epichlorohydrin in excess onto a diol [21, 123, 337–339]:

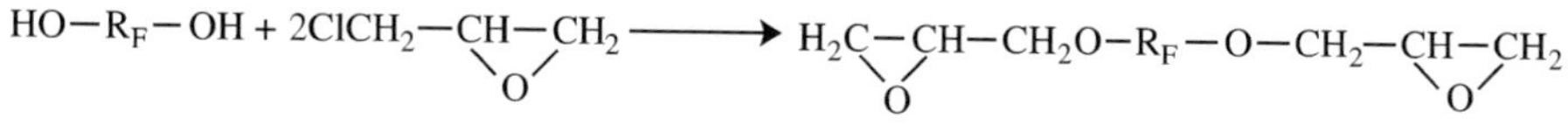

$$HO-R_F-OH + 2ClCH_2-\underset{\diagdown O \diagup}{CH}-CH_2 \longrightarrow H_2C-\underset{\diagdown O \diagup}{CH}-CH_2O-R_F-O-CH_2-\underset{\diagdown O \diagup}{CH}-CH_2$$

$$R_F = C_nF_{2n}$$

According to the amount of epichlorohydrin, diglycidic ethers of various chain lengths were obtained:

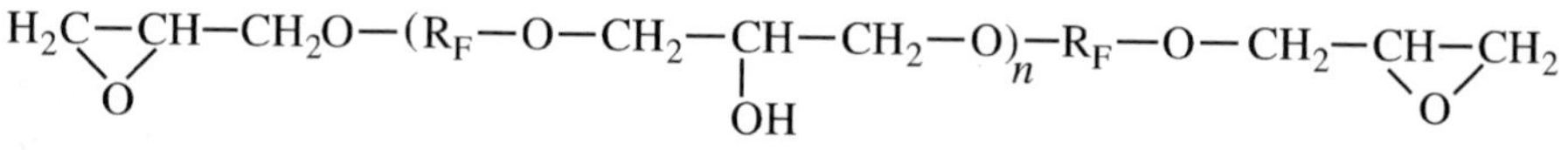

$$H_2C-\underset{\diagdown O \diagup}{CH}-CH_2O-(R_F-O-CH_2-\underset{|\ \ OH}{CH}-CH_2-O)_n-R_F-O-CH_2-\underset{\diagdown O \diagup}{CH}-CH_2$$

Numerous authors have studied this reaction according to the nature of the R radical, mainly fluoroaromatic, and found that these fluoropolyethers exhibit low T_gs.

In the same way, O'Rear et al. [337] obtained prepolymer diols from a stoichiometric amount of diol and diglycidyl ether. Such products also led to FPUs.

Griffith's investigations on fluorinated epoxy resins are numerous [116, 139] and it is difficult to summarise them. An interesting patent [243] concerns a large variety of aliphatic and aromatic telechelic diepoxides, cured by various dianhydrides and even fluorinated as the following one:

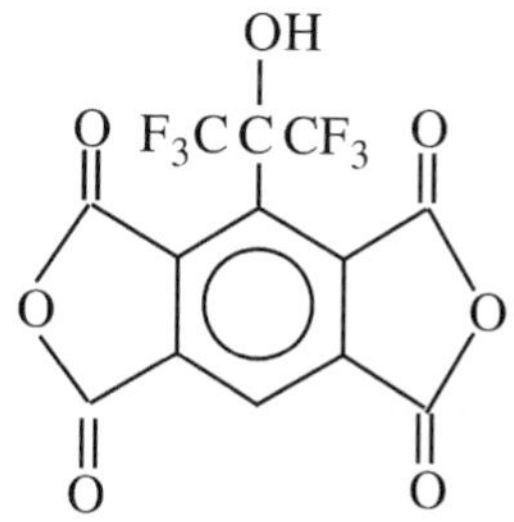

$$\begin{array}{c} OH \\ | \\ O \quad F_3C\,C\,CF_3 \quad O \\ \| \qquad\qquad \| \end{array}$$

Thanks to their good resistance to heat, light and to oxidation, such fluorinated epoxy resins have found interesting applications as hydrophobic and oleophobic coatings [243, 337, 339].

3.1.4.2. *Fluorinated Di(meth)acrylates*

As a matter of fact, fluorodiols are also interesting precursors for obtaining telechelic (meth)acrylates [104, 124]. Guyman and Bowman [340–342] investigated the photopolymerisation of the latter and found them to behave as stabilised ferroelectric liquid crystals. Such fluorotelechelics have also led to original fluorinated cured networks [129, 132, 243].

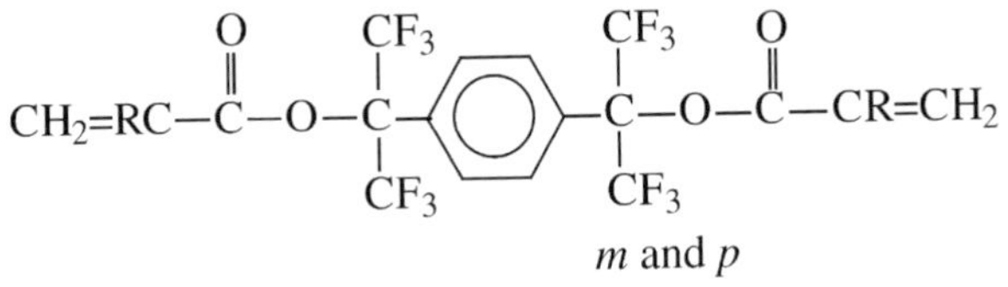

$$CH_2=RC-\overset{\overset{\displaystyle O}{\|}}{C}-O-\overset{\overset{\displaystyle CF_3}{|}}{\underset{\underset{\displaystyle CF_3}{|}}{C}}-\underset{}{\bigcirc}-\overset{\overset{\displaystyle CF_3}{|}}{\underset{\underset{\displaystyle CF_3}{|}}{C}}-O-\overset{\overset{\displaystyle O}{\|}}{C}-CR=CH_2$$

m and *p*

3.1.4.3. *Fluorinated Dicyanates*

The condensation of an excess of cyanogen bromide to fluorodiols led to fluorinated α,ω-dicyanates (Scheme 3) that Snow and Buckley [126–128] investigated extensively. When heated, such products underwent a cyclotrimerisation reaction into triazine as thermostable derivatives exhibiting low electric field permittivity and hydrophobicity.

3.1.4.4. *Use of Telechelic Fluorodiol as Precursor of Original Initiator in Free Radical Controlled Polymerisation*

An original diol bearing a perfluorinated end-group was successfully condensed with 2-bromo-2-methylpropionyl bromide to generate an α,ω-dibrominated initiator used in atom transfer radical polymerisation (ATRP) of methyl methacrylate (MMA) and styrene [270]. Hence, a series of narrow molecular weight distribution fluorinated polymers (with molar masses increasing linearly with monomer conversion) were prepared via copper-mediated living radical polymerisation. These original well-defined PMMA and PS were endowed with improved surface properties.

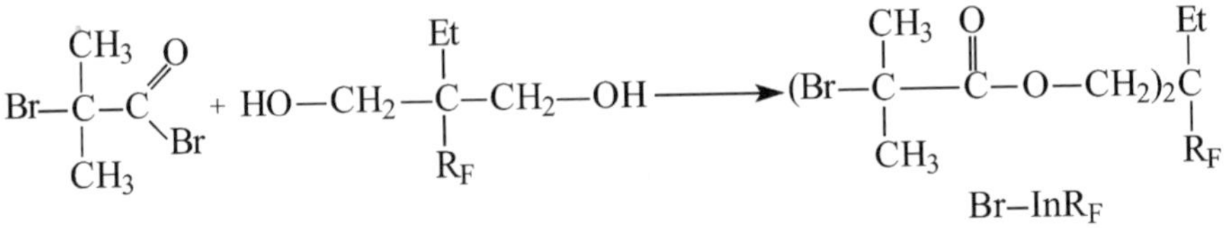

$$Br-\overset{\overset{\displaystyle CH_3}{|}}{\underset{\underset{\displaystyle CH_3}{|}}{C}}-\overset{\overset{\displaystyle O}{\diagup}}{\underset{\diagdown Br}{C}} \; + \; HO-CH_2-\overset{\overset{\displaystyle Et}{|}}{\underset{\underset{\displaystyle R_F}{|}}{C}}-CH_2-OH \longrightarrow (Br-\overset{\overset{\displaystyle CH_3}{|}}{\underset{\underset{\displaystyle CH_3}{|}}{C}}-\overset{\overset{\displaystyle O}{\|}}{C}-O-CH_2)_2\overset{\overset{\displaystyle Et}{|}}{\underset{\underset{\displaystyle R_F}{|}}{C}}$$

$$Br-InR_F$$

where R_F: $CH_2O(CH_2)_3S(CH_2)_2(CF_2)_7-CF_3$.

3.2. FROM FLUORINATED DIACIDS

These telechelics are also precursors of a wide variety of compounds (polyesters, polyamides). In aliphatic series, Anton and Mc Kinney [177] synthesised fluorinated telechelic diacids from the oxidation of α,ω-dinitriles prepared by addition of HCN to fluorinated non-conjugated dienes. These diacids polycondensed with 1,6-diaminohexane are interesting precursors of fluorinated hydrophobic polyamides with a high melting point (268 °C).

Later, Steinhauser and Mülhaupt [343] proposed the synthesis of two families of fluorinated copolyamides (6,6 F and 4,8 F) containing various amounts of bis- and tris(tetrafluoroethylene) segments by polycondensing polymethylene diamines with α,ω-dicarboxyloligo(tetrafluoroethylene)s. In the presence of various stoichiometric quantities, the fluorine content varied from 0 to 39 wt% and inherent viscosities ranged from 0.27 to 1.02 dl g^{-1}. The authors also noted that the longer the fluorinated sequences, the higher the solubilities in organic solvents and the contact angles to water, and the lower the melting temperatures.

In aromatic series, from acid dichlorides, Webster et al. [344] obtained original telechelic molecules as follows (Ⓖ being a functional group):

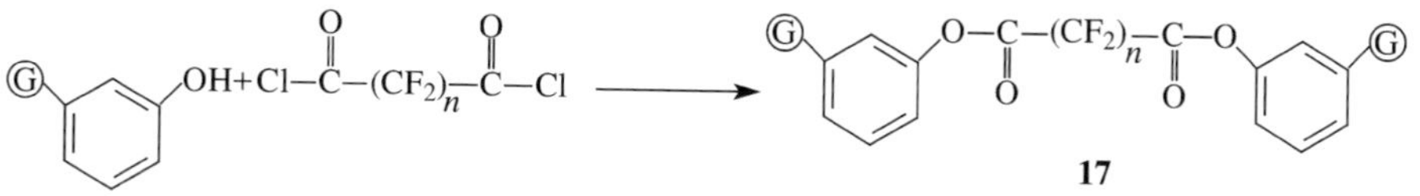

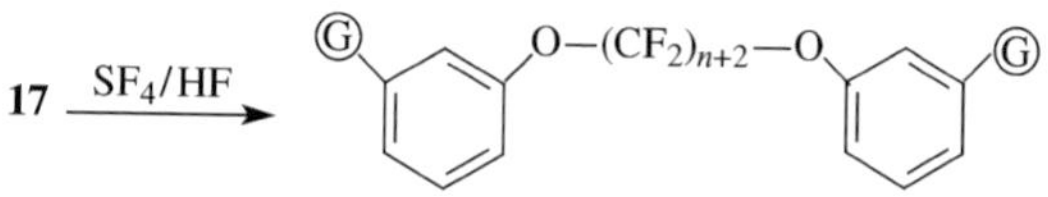

Obviously, this latter fluorotelechelic was more thermostable than **17** and a decreasing scale of thermostability was suggested:

$$\text{ArCF}_2 \sim \text{ArOCF}_2 \gg \text{ArCH}_2\text{OCF}_2 \sim \text{ArOCH}_2\text{CF}_2$$

From this method, this team [141] performed the synthesis of diisocyanates to yield:

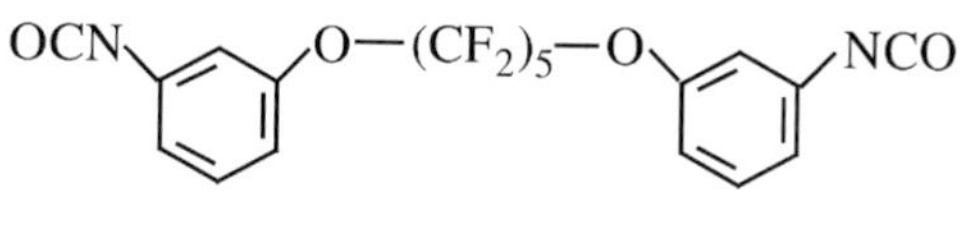

which produced polyisocyanurates. These polymers were compared to those prepared from diisocyanates which did not possess the ether bridges and they exhibited identical thermal properties. However, these authors showed that the hydrolytic stability of the polymers with the $C_6H_4-OCF_2$ group was better than the other one.

3.3.   FROM FLUORINATED DIAMINES

Telechelic diamines can be precursors of polyurethanes, polyamides, polyimides and polyureas.

3.3.1. Fluorinated Polyamides

Polyamides are prepared from the polycondensation between α,ω-diacids and α,ω-diamines. Twenty-five years ago, fluorinated polyamides containing CTFE units were obtained in our laboratory [345] from fluorotelechelic diesters and aliphatic diamines.

More recently, fluorine-containing aromatic polyamides were produced by Yamazaki *et al.* [346] by the direct polycondensation of aromatic diamines with fluoroaliphatic telechelic dicarboxylic acids in *N*-methyl pyrrolidone in the presence of triphenyl phosphite and pyridine as the condensing agent:

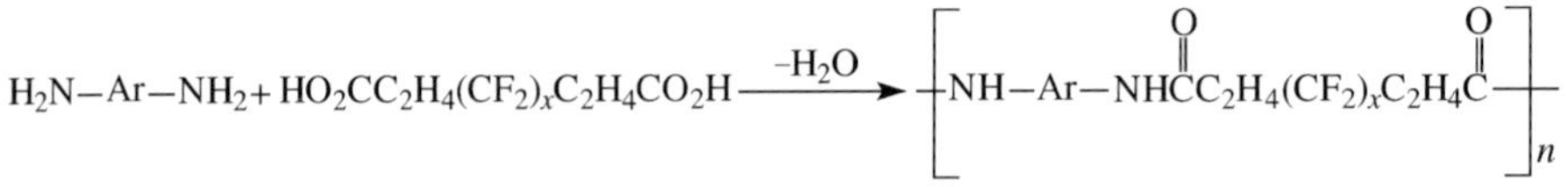

$$H_2N-Ar-NH_2 + HO_2CC_2H_4(CF_2)_xC_2H_4CO_2H \xrightarrow{-H_2O} \left[NH-Ar-NH\overset{O}{\overset{\|}{C}}C_2H_4(CF_2)_xC_2H_4\overset{O}{\overset{\|}{C}} \right]_n$$

with $x = 4$ or 6 and Ar:

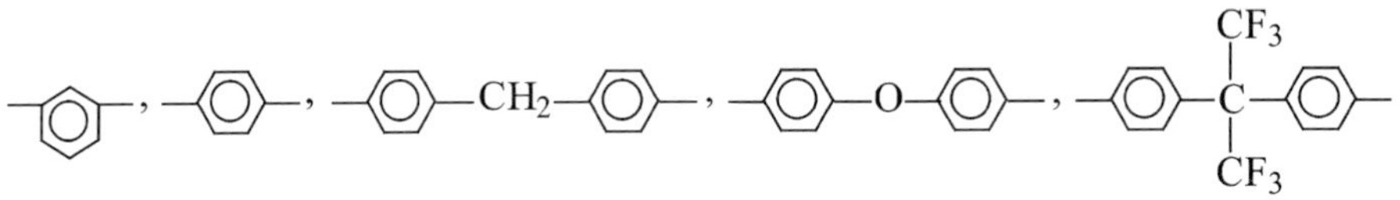

Other original fluorinated polyamides containing Krytox® units, which supply high chemical resistance, were synthesised by Webster *et al.* [141]:

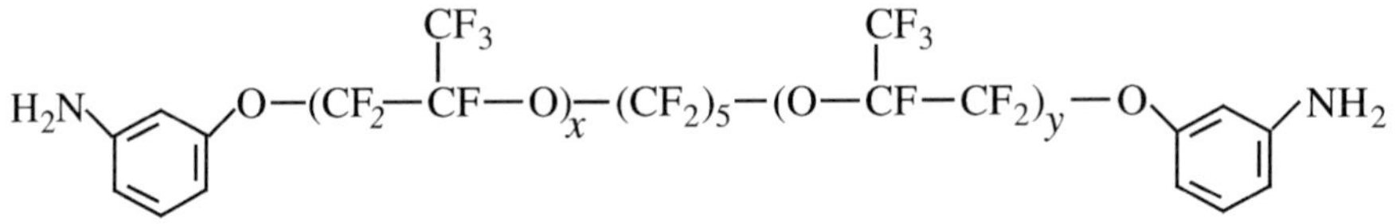

(where $x + y = 0$, 1, 2 and 3) also useful for the synthesis of novel fluoropolyamides and polyimides exhibiting high thermostabilities.

In addition, fluorinated polyamide/polyurethanes were synthesised by Chapman *et al.* [347, 348].

3.3.2. Fluorinated Polyimides

Telechelic diamines are also relevant intermediates for the synthesis of fluorinated polyimides (FPIs) [349, 350]. Such fluoropolymers exhibit very interesting properties such as high thermal, chemical and dimensional stability, low moisture uptake, dielectric constants and coefficients of thermal expansion and good mechanical properties which are of special importance in electronic, space and automotive applications. Many investigations have been performed on the preparation of FPI which possesses hexafluoropropylidene groups. In contrast, few results have been obtained on FPI containing an aliphatic fluorinated group. Critchley *et al.* [351], followed by Webster [352], used phthalamide containing $(CF_2)_n$ with $n = 3-8$ or $O(CF_2)_5O$ groups, respectively, while Strepparola *et al.* [353] utilised perfluoropolyether (such as Fomblin®), which led to the following elastomeric FPI:

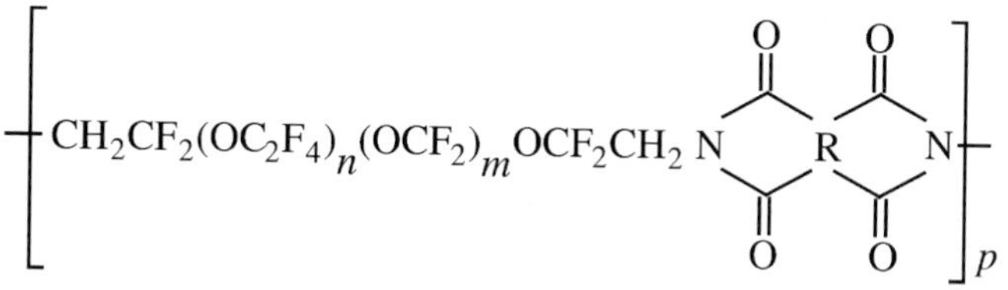

In addition, Labadie and Hedrick [142, 143] synthesised original aromatic diamines:

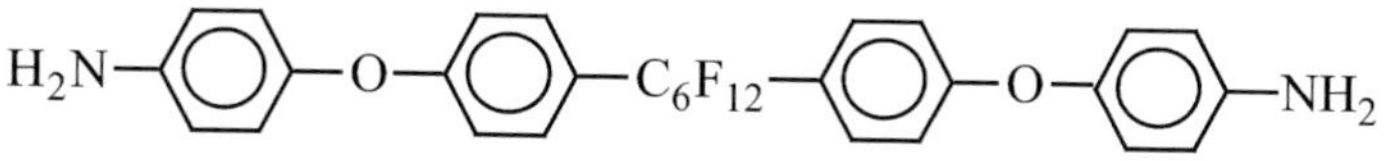

(from 1,6-diiodoperfluorohexane) which displayed good reactivity with pyromellitic dianhydride and yielded random FPI under conventional poly(amic acid) polycondensation conditions. These fluoropolymers exhibit low dielectric constants.

On the other hand, Eastmond and Paprotny [354] have developed an original method to prepare telechelic fluoroaromatic diamines in high yield (>95%), as follows:

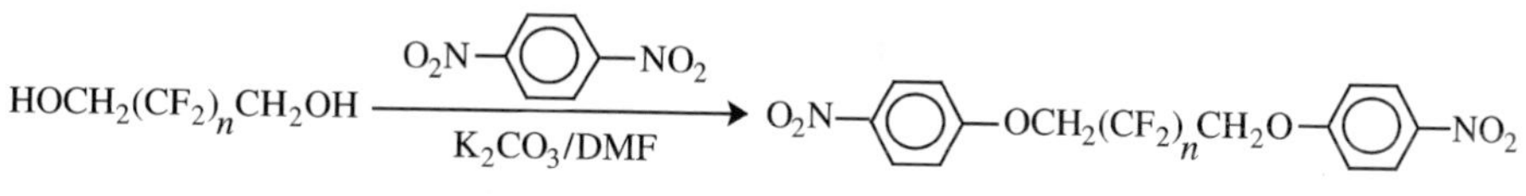

$n = 3, 4, 8, 10$

They condensed the corresponding diamines with 6 FDA and produced FPIs with high T_g and thermostability and low dielectric constants:

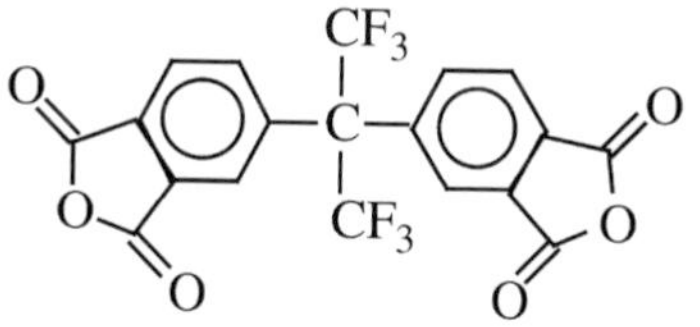

6 FDA

The same team also polycondensed these aromatic diamines with the following aromatic dianhydride to generate a wide range of FPIs where the fluorinated segments distributed between the dianhydride and diamine moieties led to polymers with a wide range of morphologies (T_m reaching 330 °C), some even exhibiting liquid crystallinity [354].

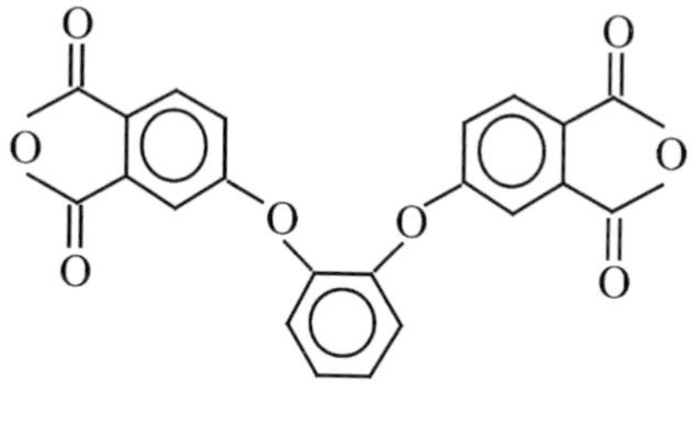

More recently, fluorinated telechelic oligoimides bearing trialkoxysilane end-groups were synthesised in two ways, starting from the condensation of 6 FDA with an aromatic fluorinated (or not [355]) telechelic diamine (BTDB) [356]. According to the stoichiometry, α,ω-diamino or α,ω-dianhydride oligoimides of different molecular weights were obtained. In the first method, the produced diamino PIs were condensed with *cis*-5-norbornene-endo-2,3-dicarboxylic anhydride (or nadic anhydride) leading to original α,ω-dinadic oligoimides (FOI).

The solution was heated at different temperatures to achieve a complete cyclisation as follows:

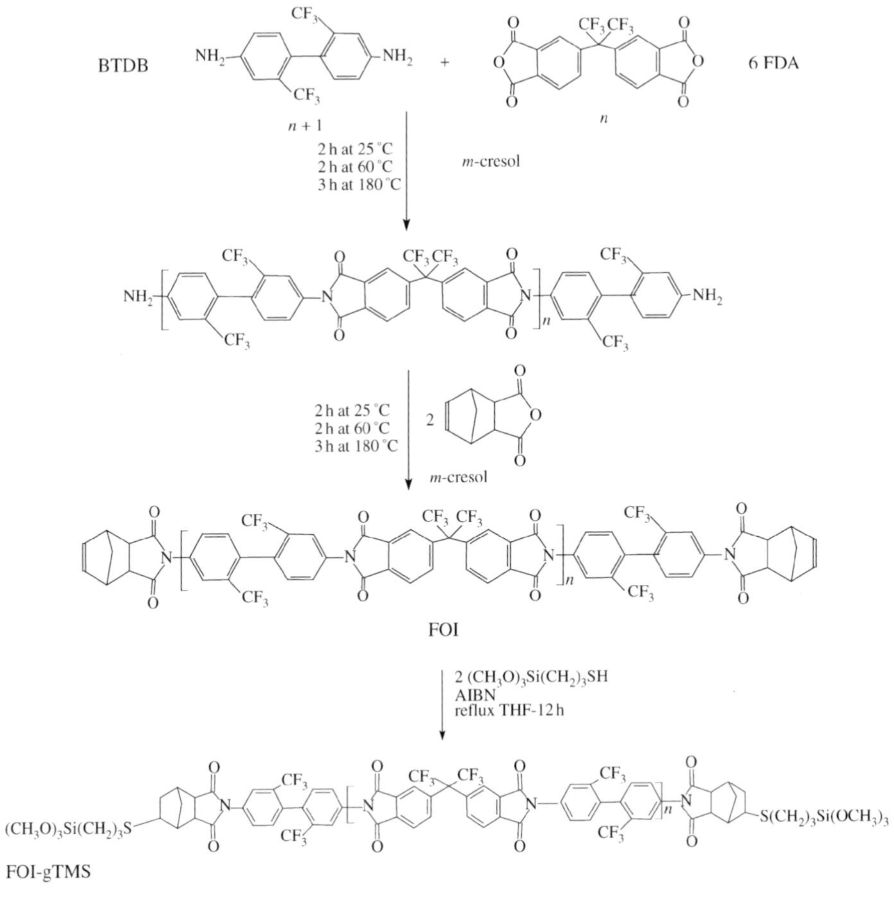

FOI-gTMS

In the second pathway, a reverse stoichiometry led to novel α,ω-dianhydride oligoimide that could react onto allyl amine to generate α,ω-diallyl oligoimides.

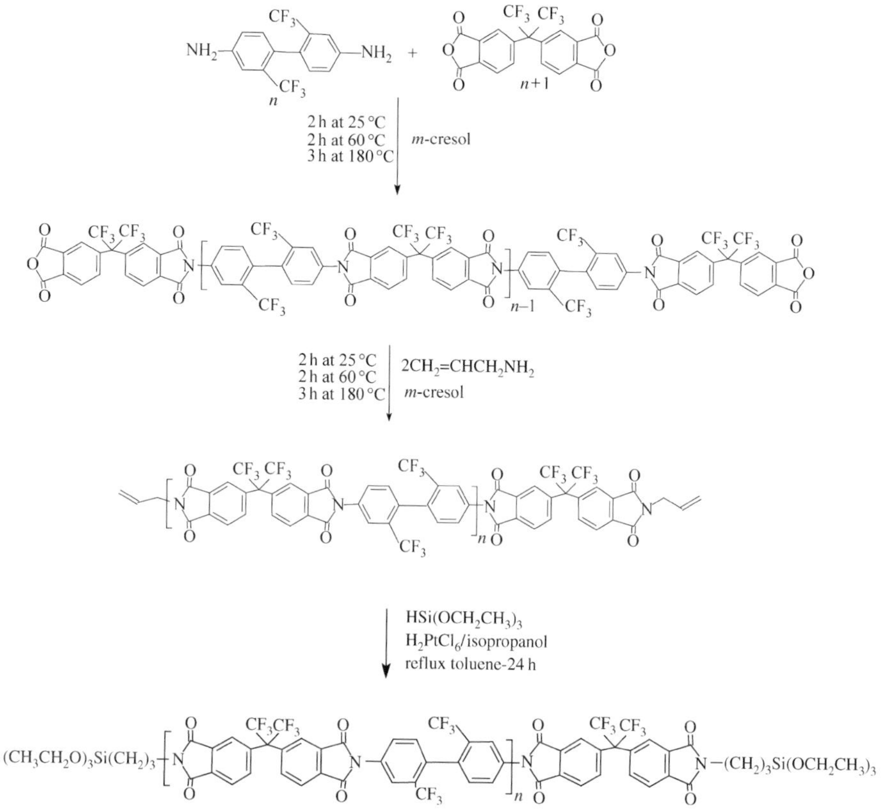

These fluorotelechelic nadic and allylic double-bond terminated PI oligomers were interesting precursors of original α,ω-bis(trialkoxysilane)s fluorinated oligoimides via two methods (model reactions were achieved onto non-fluorinated reactants [357]): either from the radical addition of (3-mercapto propyl) trimethoxysilane onto the nadic double bonds initiated by AIBN [356b] or by hydrosilylation of triethoxysilane with the allylic double bond in the presence of platinic Speier's catalyst [356b].

Interestingly, these telechelic bis(trialkoxysilyl) fluorooligoimides were self-crosslinked either thermally (from $T > 250\,°C$) or in the presence of 1,1,1-tris (4-hydroxyphenyl)ethane (TRIOH) [356b] (Scheme 5).

Besides the synthesis of these linear oligoimides, original polyimides bearing a fluorinated chain as a side group were also prepared from fluorinated telechelic diamine.

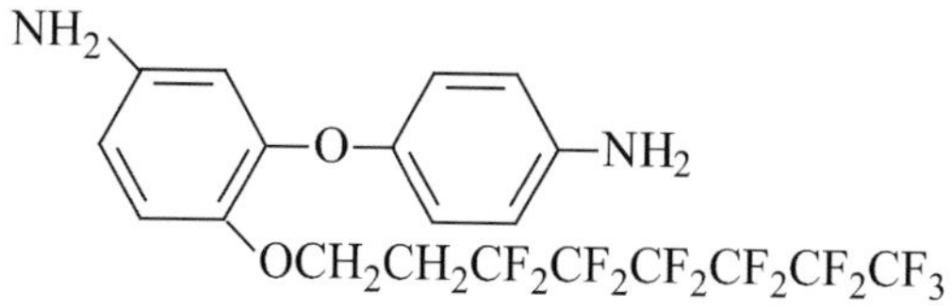

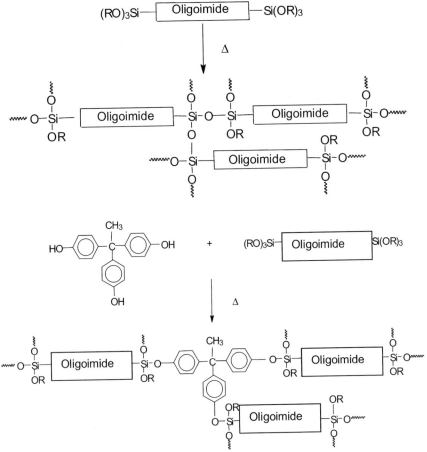

SCHEME 5.   General scheme of the thermal self-crosslinking reaction of the trialkoxysilyl groups (a) and with TRIOH as crosslinker agent (b).

This α,ω-diamino compound was synthesised in a five-step procedure from 2-methoxy-5-nitrophenol [357]. The polycondensation of this diamine with 6 FDA dianhydride led to original PTs bearing a fluorinated lateral group. Interestingly, these PIs show a high T_g (195 °C) and thermal stability (T_{dec} = 380 °C for 5 wt% loss) and the presence of this fluorinated group enabled these PTs to exhibit good surface properties (γ_s = 25.3 mN m^{-1}) for 37 wt% of fluorine.

Instead of adding a fluorinated side group linked to aromatic rings, a French group [358] introduced sulfonic acid functions, carried out by sulfonation. Interestingly, the sulfonated PIs act as novel membranes for fuel cells with high conductivities.

3.4. FROM FLUORINATED NON-CONJUGATED DIENES

The synthesis of fluorinated dienes (Scheme 3) has been achieved:

(a) from the bis(dehydroiodination) of $IC_2H_4-R_F-C_2H_4I$ [55, 63, 90, 107, 108, 359, 360],
(b) from the bis(deiodoacetalisation) of $R_F(CH_2CHICH_2OAc)_2$ [107, 108],
(c) from coupling of 1,3-diioperfluoropropane with an excess of 1-iodo-2-chloroethylene [361] or
(d) from other routes [54, 362].

Fluorinated dienes can be the precursor of original telechelic dinitriles [177], involved in cycloadditions with TFE or CTFE [360] or, from the thermal addition of C_2F_5I [363], used in novel fluoroelastomers [101] or α,ω-disilanols. The latter is prepared by hydrosilylation reactions of hydrogenochlorosilanes with these dienes (Scheme 3).

Although the hydrosilylation is quite efficient from α,ω-diallyl compounds, even containing a bulky CF_3 side group from the hexafluoropropene unit adjacent to these allyl end-groups [148], a similar reaction from α,ω-divinyl dienes led to lower yields. This was explained by the formation of side-product **18** coming from the unexpected $ClR_1R_2SiC_2H_4R_FCF_2CH(CH_3)$ SiR_1R_2Cl isomer:

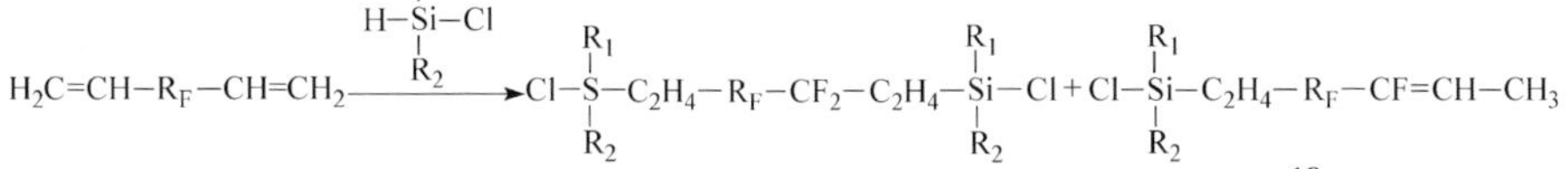

$$H_2C=CH-R_F-CH=CH_2 \xrightarrow[R_2]{\overset{R_1}{\underset{\,}{H-Si-Cl}}} Cl-\overset{R_1}{\underset{R_2}{S}}-C_2H_4-R_F-CF_2-C_2H_4-\overset{R_1}{\underset{R_2}{Si}}-Cl + Cl-\overset{R_1}{\underset{R_2}{Si}}-C_2H_4-R_F-CF=CH-CH_3$$

18

The hydrolysis of these dichlorosilanes yielded quantitatively the α,ω-disilanols. Interestingly, these disilanols can be either homopolycondensed or copolycondensed with α,ω-dichlorodisilanes to produce original hybrid silicones [145–148], as follows:

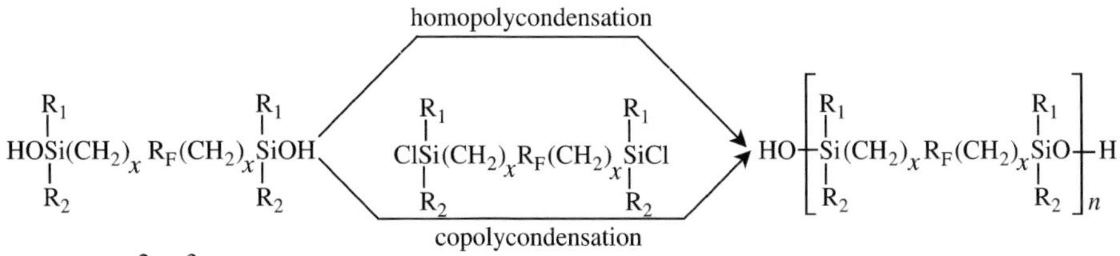

homopolycondensation

$$HO\overset{R_1}{\underset{R_2}{Si}}(CH_2)_x R_F(CH_2)_x \overset{R_1}{\underset{R_2}{Si}}OH \quad Cl\overset{R_1}{\underset{R_2}{Si}}(CH_2)_x R_F(CH_2)_x \overset{R_1}{\underset{R_2}{Si}}Cl \longrightarrow HO\left[\overset{R_1}{\underset{R_2}{Si}}(CH_2)_x R_F(CH_2)_x \overset{R_1}{\underset{R_2}{Si}}O\right]_n H$$

copolycondensation

$x = 2$ or 3

Such fluorosilicones exhibit very good properties at low and high temperatures. According to the nature of the fluorinated group and to their molecular weights, their T_gs ranged between -26 and $-52\,°C$. They are also quite thermostable and start to decompose at $380\,°C$, which is $180\,°C$ higher than the commercially available Silastic® (poly(methyltrifluoropropyl siloxane)) produced by the Dow Corning Company. They have been used as gaskets or O-rings in the aerospace industry. More examples devoted to the synthesis, properties and applications of fluorinated multiblock copolymers based on siloxanes are reported in Chapter 4 (Section 2.3).

In addition, telechelic dienes were shown to be efficient crosslinking agents. For example, in the technology of thiol-ene systems, Ameduri *et al.* [364] could achieve the radical crosslinking of PVDFs bearing thiol side-functions in the presence of non-conjugated dienes (see Chapter 5, Section 2.3).

Interestingly, the Ausimont Company [365] used $H_2C=CH(CF_2)_6CH=CH_2$ in its copolymerisation with VDF or its terpolymerisation with VDF and HFP and hence enabled a potential grafting of PVDF and poly(VDF-*co*-HFP) using a "branching and pseudo-living technology" [366]. The PDIs of the resulting co- and terpolymers were lower than those synthesised without any diene; more details have been added in Chapter 4, Section 2.2.2.2.

A similar process which was not found *pseudo*-living was discovered ca. 10 years before by Tatemoto *et al.* [367] in the polymerisation of tetrafluoroethylene with $0.1-1.0\,mol\%$ of the above fluorodiene.

3.5. Miscellaneous

The monomers for bisbenzoxazoles can also be mentioned. Evers [368] had been investigating this topic for several years. The reaction is as follows:

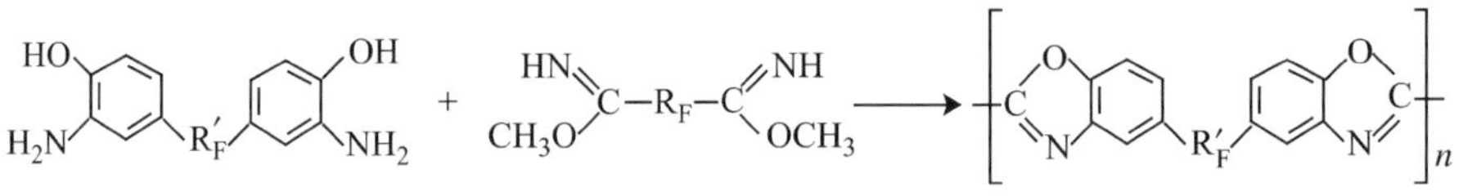

where R_F' represents $(CF_2)_n$, with $n = 3$, 8 or $(CF_2)_2-O-(CF_2)_5-O-(CF_2)_2$.

The dimethyl perfluorodiimidate was obtained by addition of methanol onto the corresponding dinitrile, under nitrogen, with a tertiary amine as catalyst.

Evers [368] published numerous investigations on the outstanding thermal properties of these polybenzoxazoles, regarded as one of the most thermostable polymers.

Conversely, in the cases of FPUs or polyesters, the urethane or ester links did not exhibit chemical or thermal resistances comparable to those of the fluorinated polyethers. Hence, original research began on the formation of novel stable bridges. Several examples are listed below:

– Chain extension thanks to nitrile bonds:

3.5.1. From Fluorinated Dinitriles [369]

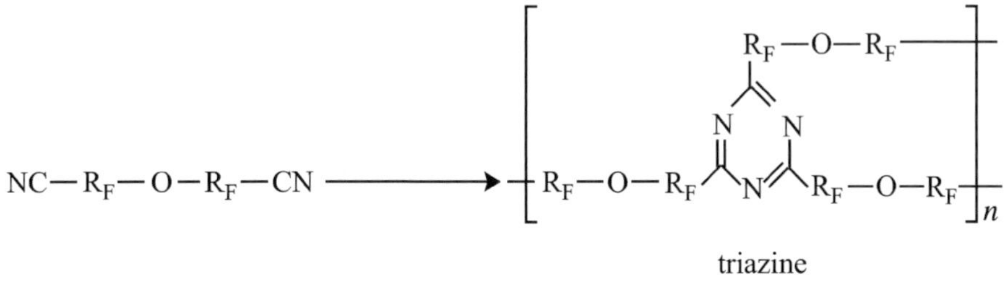

$$NC-R_F-O-R_F-CN \longrightarrow \left[R_F-O-R_F \underset{\substack{N \quad N \\ }}{\overset{R_F-O-R_F}{\bigtriangleup}} R_F-O-R_F \right]_n$$

triazine

3.5.2. From Fluorinated Dialcoholates [370]

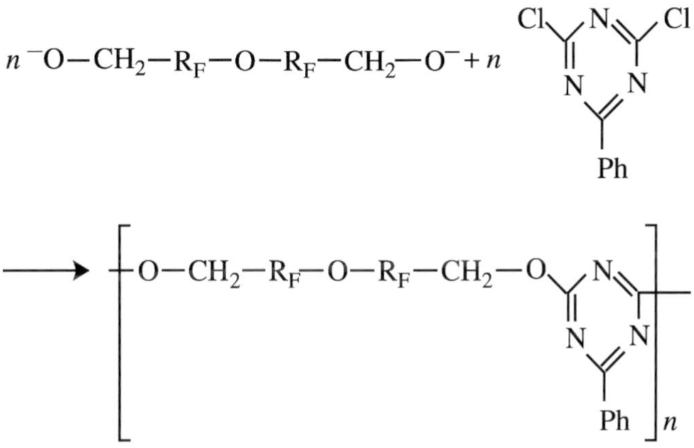

$$n\ ^-O-CH_2-R_F-O-R_F-CH_2-O^- + n \quad \underset{Ph}{Cl\diagdown\underset{N}{\bigtriangleup}Cl}$$

$$\longrightarrow \left[O-CH_2-R_F-O-R_F-CH_2-O \diagdown \underset{\substack{N \quad N}}{\bigtriangleup} \underset{Ph}{} \right]_n$$

3.5.3. From a Dinitrile/Nitrile Blend [371]

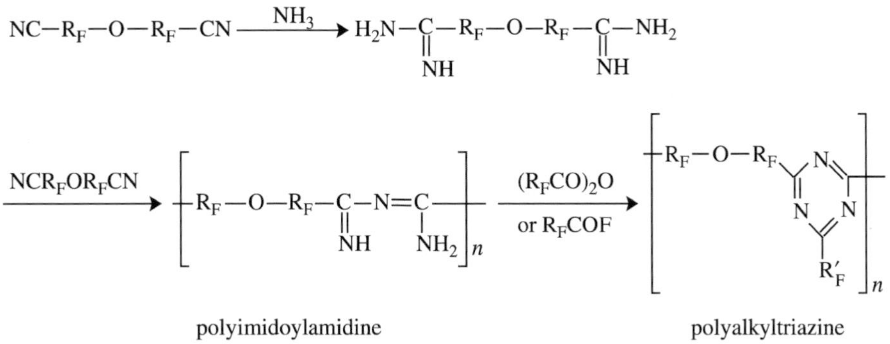

$$NC-R_F-O-R_F-CN \xrightarrow{NH_3} H_2N-\underset{NH}{\overset{\parallel}{C}}-R_F-O-R_F-\underset{NH}{\overset{\parallel}{C}}-NH_2$$

$$NCR_FOR_FCN \longrightarrow \left[R_F-O-R_F-\underset{\substack{\parallel \\ NH}}{C}-N=\underset{\substack{| \\ NH_2}}{C} \right]_n \xrightarrow[\text{or } R_FCOF]{(R_FCO)_2O} \left[R_F-O-R_F \diagdown \underset{\substack{N \quad N \\ R'_F}}{\bigtriangleup} \right]_n$$

polyimidoylamidine   polyalkyltriazine

This latter method is interesting since a linear polymer was obtained rather than the network formed from the direct trimerisation of dicyanated compounds.

3.5.4.   Chain Extension Thanks to Isocyanurate Bonds

Webster *et al.* [141] prepared diisocyanated ethers with catalysts such as bis(N,N'-dimethyl)-1,4-diaminobutane leading to polyisocyanurates with the following structure:

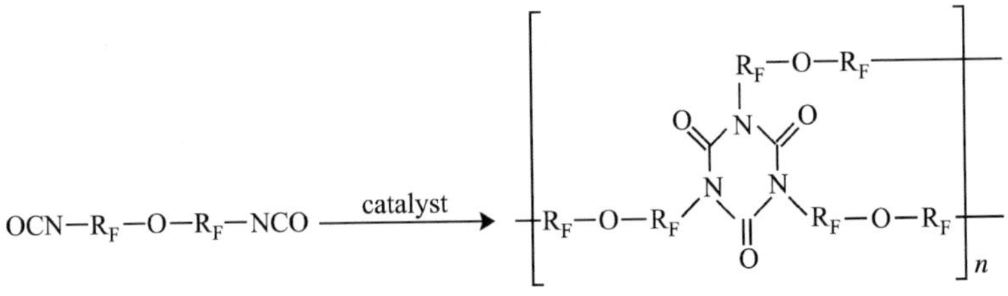

$$OCN-R_F-O-R_F-NCO \xrightarrow{\text{catalyst}}$$

where: $-R_F-O-R_F-$ is

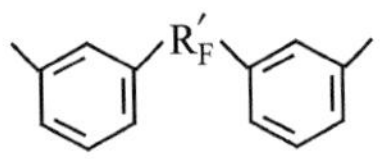

with $R'_F = -(CF_2)_n- ; -O-(CF_2)_n-O- ; -O-(CF_2-\underset{\underset{CF_3}{|}}{CF}-O)_m-(CF_2)_5-(O-\underset{\underset{CF_3}{|}}{CF}-CF_2)_n-O-$

3.5.5.   Chain Extension with Benzenic Bonds

Such a reaction was carried out by De Pasquale *et al.* [372] according to the following scheme:

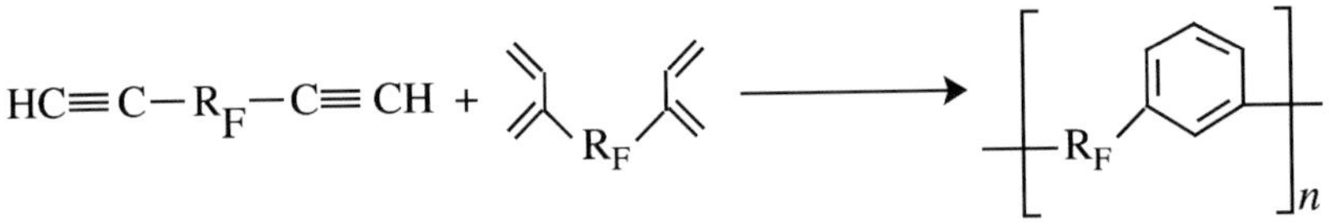

$$HC\equiv C-R_F-C\equiv CH +$$

The trimerisation of these α,ω-diacetylenic compounds at 300 °C usually leads to thermostable polymers with high T_g [373], high dimensional stability, satisfactory adhesion to metals, low moisture uptake and easy processability. Such polymers have found useful applications such as prepregs for composite matrices in electronics and space industries.

In addition, stable glassy resins have been produced by thermal polymerisation of fluorinated α,ω-diacetylenes $HC\equiv C(CF_2)_nC\equiv CH$ ($n = 6, 8, 10$) [111, 374].

3.5.6.   From Fluorinated α,ω-Diiodides

3.5.6.1.   Involved in Supramolecular Chemistry

α,ω-Diiodoperfluorocarbons are not only useful synthons in the covalent synthesis of many difunctional compounds but have also been employed as supramolecular synthons in the non-covalent synthesis of self-assembled architectures recently developed by Resnati's group [375–381]. In iodoper-fluoroalkanes and -arenes, the high electronegativity of fluorine atoms and perfluorinated residues enhances the ability of iodine atoms to work as electron acceptors. A wide diversity of lone pair possessing atoms, either neutral or anionic, behaves as effective electron donors and the resulting attractive interaction has been named "halogen bonding" to stress the similarities with the hydrogen bonding [376]. When telechelic bidentate electron donors are used (e.g., N,N,N',N'-tetramethylethylenediamine, 4,4'-dipyridyl, 1,4,10,13-tetraoxa-7,16-diazacyclooctadecane (K.2.2) or 4,7,13,16,21,24-hexaoxa-1,10-diazacy-clo[8,8,8]hexacosane (K.2.2.2)), the interaction is reiterated at either ends of both modules and a range of one-dimensional non-covalent co-polymers have been obtained as crystalline solids which are stable in the air at room temperature. Trimeric systems, obtained from monoiodoperfluorocarbons (electron acceptor species) and bidentate electron donors (and vice versa from diiodoperfluorocarbons and monodentate electron donors), have also been reported [376, 377].

The self-assembly of perfluoroalkyl iodides and bromides has afforded more interesting structures than those obtained from perfluoroaryl analogues. Layered supramolecular architectures have been invariably obtained starting from both mono- and bidentate electron acceptors of different lengths and from a range of electron-donating motifs. The hydrophobic effect is responsible for the layered (or bilayered) disposition of long alkyl-chain protic acids in polar media. Similarly, the fluorophobic effect is responsible for the layered disposition of long perfluoroalkyl-chain Lewis acids (haloperfluoroalkanes) in hydrocarbon media. Indeed, the fluorophobic effect forces micro-segregation that efficiently controls the packing of halogen-bonded infinite chains, or discrete aggregates, into layered supramolecular architectures. The unique properties that perfluoro-carbon derivatives have with respect to hydrocarbon compounds result in the low affinity existing between perfluorocarbon and hydrocarbon derivatives. This low affinity survives in the hybrid materials assembled through the halogen bonding and causes the remarkable segregation of the two species into layered systems.

Experimental results and theoretical calculations show that the halogen bonding between nitrogen donors and primary perfluoroalkyl iodides is approximately $28 \, \text{kJ mol}^{-1}$. Bromoperfluorocarbons give rise to weaker interactions ($15 \, \text{kJ mol}^{-1}$ when nitrogen atoms work as donor sites) [378]. The halogen bonding strength increases when moving from primary to secondary to tertiary iodoperfluoroalkanes and iodoperfluoroarenes show a behaviour similar to that of primary alkyl iodides [379]. As to the donor site, nitrogen [375, 380], oxygen [381] and sulfur [382] atoms have been used, nitrogen usually working as the strongest donor. The higher the electron density on the donor site, the stronger the halogen bonding it gives rise to. It is thus not surprising that fluoride, bromide and iodided anions work as effective electron donors towards iodo- and bromoperfluorocarbons [383].

The halogen bonding driven self-assembly of perfluorocarbons and hydrocarbons occurs also in solution, as revealed by the non-minor low field shifts observed in signals of ^{19}F NMR spectra. The stronger the interaction, the larger the observed effect, and shifts as large as 18 ppm [379] have been measured. Indeed, the technique is a sensitive, simple and reliable tool to detect the occurrence of the interaction and to rank the relative strength of different halogen bondings [384].

Interesting applications of this new and non-covalent approach to the compatibilisation of hydrocarbon and perfluorocarbon derivatives have been in the obtainment of enantiopure perfluorocarbons through diastereisomeric adducts formation starting from racemic perfluorocarbons [383a], in the solid phase synthesis [385], in the fluorous coating of hydrocarbon polymers [386] and in the solubilisation of organic and inorganic salts in fluorous media [383d].

Another interesting, general and useful application is related to the fact that the obtained hybrid materials can be routinely studied through single crystal X-ray analyses. Due to their unique combination of chemical and physical properties (e.g., low frictional character, high compressibilities, high surface activity, low cohesive pressures), it is very difficult to grow suitable crystals for the solid state analysis of perfluorocarbon compounds through X-ray diffraction. Consistent with the great tendency of perfluorocarbon derivatives to form waxes, the number of long perfluoroalkyl chains described in the Cambridge Crystallographic Data Centre is quite limited. Iodo- and bromoperfluorocarbons involvement in the formation of halogen-bonded networks appears to be a reliable strategy to obtain their X-ray structures.

3.5.6.2.   Involved in Pseudo-living Radical Polymerisation: Synthesis of Block Copolymers

α,ω-Diiodoperfluoroalkanes enable control of the radical (co)polymerisation of fluoroalkenes via "ITP", "degenerative transfer" or "*pseudo*-living and branching technology". The diiodides frequently transfer from chain to chain. A comprehensive study was recently realised by Apostolo et al. [387] on the free radical controlled polymerisation of VDF and HFP in microemulsion. This study provides a kinetic model able to identify the most appropriate operating conditions to produce VDF/HFP copolymers with desired characteristics in terms of microstructure, especially branched macromolecules with uniform chain lengths and molecular distributions. Actually, a significant reduction of the polydispersity was noted by this Italian group (with a final value of 2.3 in contrast to 4–5 without any diiodide). For more information, we refer the reader to Chapter 4, Section 2.2.2.2, where quasi-exhaustive results are supplied.

Regarding original architectured molecules, interesting studies involved telechelic dihalogenated (diiodo- and dibromo-) PVDFs as precursors of block copolymers.

For example, α,ω-diiodinated PVDF telomers were synthesised to initiate the ATRP of methyl methacrylate (MMA) leading to a triblock PMMA-*b*-PVDF-*b*-PMMA copolymer [388], as follows:

$$I-(CF_2)_4-I + nH_2C=CF_2 \rightarrow I-PVDF-I \xrightarrow[\text{CuI/bipy}]{\text{MMA}} \text{PMMA-}b\text{-PVDF-}b\text{-PMMA}$$

The kinetic plots ($\ln([M]_0/[M])$ vs. time) showed nearly a linear behaviour, the $\bar{M}_n$ (assessed by NMR) of block copolymers increasing linearly with the monomer conversion as evidence of a controlled radical polymerisation of MMA.

On the other hand, α,ω-dibrominated PVDF may also be mentioned, although the C–Br bond is stronger than the C–I bond but weaker than the C–Cl bond. As above, the ATRP of styrene in the presence of α,ω-dibrominated intermediates **19**, regarded as "macroinitiators", successfully led to triblock copolymers [389]. These dihalogenated compounds were prepared by telomerisation of vinylidene fluoride with 1,2-dibromotetrafluoroethane [390].

$$BrCF_2CF_2Br + n\,VDF \xrightarrow{\text{peroxide}} \underset{\mathbf{19}}{Br-(VDF)_m-R_F-(VDF)_n-Br}$$

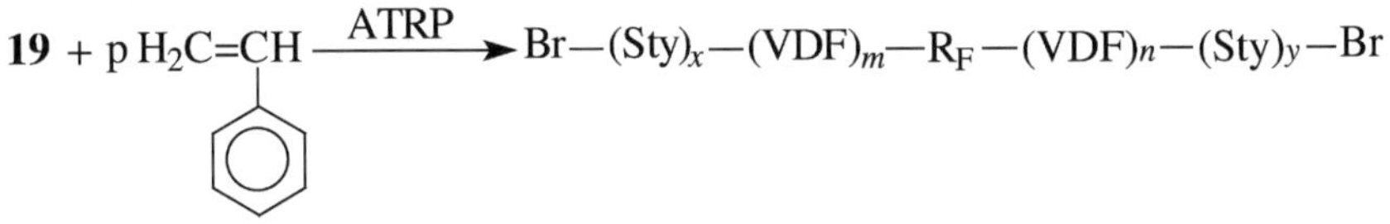

$$\mathbf{19} + p\,H_2C=CH \xrightarrow{\text{ATRP}} Br-(Sty)_x-(VDF)_m-R_F-(VDF)_n-(Sty)_y-Br$$

3.6. CONCLUSION

This section has mentioned the large variety of well-architectured fluoropolymers that can be prepared from telechelic molecular derivatives: polyesters, polyurethanes, polyamides, polyimides, polyureas, silicones, or original networks from diacrylates, dicyanates or diacetylenics.

In addition, certain precursors are quite useful to control radical polymerisation (from non-conjugated dienes or from α,ω-diiodides) for obtaining thermoplastic elastomers, and are developed further in Chapter 4.

According to the desired property of the final fluoropolymer, the fluorinated chain must be introduced in the telechelic precursor in the appropriate way: either in the oligomeric backbone to bring chemical inertness and thermal stability, or in the side chain to improve the surface properties of the resulting well-defined fluoropolymer. However, efforts must still be made to provide a good anchoring of this fluorinated lateral group.

4. Conclusion and Perspectives

The synthesis of fluorinated molecular or macromolecular telechelic compounds, whatever the position of the fluorinated chain, has been developed over many years and continues to draw numerous investigations. This is due to properties of fluorinated derivatives, although several ice, insolubility, sometimes high melting points and onal groups in positions about perfluorinated groups, t the compromise between the expected properties and has not been reached yet.

stimulating and valuable are fluorinated telechelics diates to obtain well-defined fluoropolymers. Various plied to successfully achieve the preparation of such telechelics. The driving force of the fluoropolymers lies on the properties searched: if thermostable or inert polymers are required, it is more convenient to build a telechelic in which the backbone is fluorinated while, for expected good surface properties, it is wiser to synthesise telechelics bearing a perfluorinated side chain. Hence, in the former series, typical examples deal with the introduction of aromatic rings between the fluorinated chain and the chemical function to improve the thermal properties. The same behaviour occurs for the fluorinated chains/aromatic nucleus links ($C_6H_5-CF_2$, $C_6H_5-O-CF_2$, $C_6H_5-CH_2-OCF_2$, $C_6H_5-CH_2-OCH_2CF_2$ groups). In the latter series, for which little attention is paid to the thermal stability, the fluorinated chain/functional group junction is not as critical as above; the authors thus look for the simplest method that possesses the most available or accessible products.

Although perfluoroalkyl iodides are quite versatile compounds involved in many investigations to introduce a perfluorinated group on a molecule thanks to its low CF_2-I bond dissociation energy, this is not true of α,ω-diiodofluoroalkanes, regarded as key starting reactants for the former series, since fewer researches have been performed from them. This is primarily due to their difficult syntheses and their high price. However, it may be assumed that these drawbacks can be overcome by a more efficient synthesis to make them more readily available. Nevertheless, these iodinated compounds act as efficient transfer agents in stepwise cotelomerisation of fluoroalkenes and also efficiently enable the free living radical (co)polymerisation of fluorinated olefins. In addition, they are relevant precursors of supramolecular systems or original block copolymers. Such derivatives are quite interesting and represent versatile precursors of a wide variety of fluorinated telechelics, containing a (per)fluorinated chain on their backbone to enhance thermal properties, chemical inertness and resistance to oxidation and ageing.

Although only 3–4 fluorotelechelics are commercially available from the rather wide variety of fluorinated homopolymers and copolymers, they are still very useful intermediates and represent a window for the fluoropolymers regarded as speciality polymers. Hopefully, the market and the production are in the process of expansion; thermoplastic elastomers such as Dai-El® offer real evidence since they are already produced on an industrial scale.

The syntheses of α,ω-diiodofluoroalkanes have been performed by various authors and obviously from companies producing TFE. For CTFE, the future is not optimistic because of environmental problems caused by this chlorofluoro-alkene (e.g., ozone depletion). No investigation has yet been attempted to use VDF, HFP, or a mixture of both, with molecular iodine. Nevertheless, stepwise cotelomerisation of different monomers with $I(C_2F_4)_nI$ appears as an attractive way.

More interesting is the preparation of model fluoroelastomers from these telechelic diiodides by means of telomerisation. This reaction enables one to introduce key fluorinated base units in a controlled way, leading to easily functionalised, well-architectured telomers. So far, few works have been performed, but it can be expected that future surveys will especially be focused on the end-capping of HFP or the (co)telomerisations of VDF, tetrafluoroethylene or trifluoroethylene to prepare new blocks able to introduce soft segments in thermostable polymers.

Attempts to functionalise these diiodides have been done via 2 two-step processes. The first one requires a bis(monoethylenation) for which many efforts have been developed already, leading to interesting results. However, investigations still deserve to be continued to allow the bis(ethylenated) diiodide to be chemically changed into a telechelic intermediate in higher yields. As for the second step, it can be expected that the direct functionalisation of IR_FI with ethylenic ω-functional derivatives should find the same success as that involving perfluoroalkyl iodides. However, since vinyl and allyl acetate and allyl alcohol have mainly been used, a more diverse choice of monomers should be expected in future investigations.

Although α,ω-bis(trichloromethyl) chlorofluoroalkanes have been known for more than 25 years, they might be given up because of environmental conse-quences. However, making these telechelics can be achieved by various routes.

In addition, this review proposes various ways to synthesise telechelics bearing a fluorinated side group (by condensation, radical additions, from the ring opening of fluorinated cyclic ethers or by grafting fluorinated mercaptans or silanes onto telechelic polybutadienes). However, no outstanding applications have been pointed out although further research is in progress.

Such intermediates allow the synthesis of a wide range of well-defined polymers as high value added materials (fluorinated silicones, polyethers, polyurethanes, polyesters, polyamides, polyimides, epoxy resins, polybenzo-xazoles, fluoropolymers containing triazine or isocyanurate rings and crosslinked networks from fluorinated dicyanates). The fluorinated block in the backbone enables them to achieve good thermal properties. For example, recent investigations have shown that thermostable telechelic dicyanates, fluorosili-cones and fluorinated polyamides and polyimides have found interesting applications in aerospace, microelectronics and other high-tech fields (e.g., membranes for fuel cells). Although their price is the main drawback, the properties and applications of these materials are still promising, and several industrial applications are already relevant: optical fibres prepared from a primary and, more recently, from a secondary cladding that contains a fluorinated group, allowing one to monitor the refractive index and also providing good hydrophobicity. Another example deals with fluoropolyurethanes for which the fluorinated chain seems to reduce the hydrolytic degradation of urethane bridges.

As for fluoropolyimides, if many investigations concern the use of gem-difluoromethyl groups for improving the solubility of the polymers, only a few involve reactants containing a linear perfluorinated group. One could expect more surveys in this area. In the same way, recent investigations on hybrid silicones have shown that those with a fluorinated chain in the backbone have a higher thermostability (even above 200 °C) than that of commercially available fluorosilicones.

Few fluoroaromatic telechelics have been described in this review. However, one interesting approach was pioneered by Mc Loughlin and closely studied by Eapen in the coupling of paraiodo functionalised benzene with fluorodiiodides. It was noted that the electron-withdrawing nature of the functional group is a problem and more work is required for further progress.

The corresponding fluoroaromatic polymers exhibit glass transition temperatures higher than 200 °C, decomposition temperatures that reach 400 °C and a satisfactory softness which represents the weak point of the non-fluorinated thermostable products.

Although few companies have so far managed to produce diiodides, it can be expected that a further step will be taken in the industrial production of new fluorinated telechelic products that would open up new areas.

It must be taken into account that much research is still required since the fluorinated groups modify the reactivity of a function negating its use in certain syntheses. However, introducing the adequate spacer or protective groups gives hope that such preparations may be successfully realised.

References

1. Smart, B. E. (1995) Properties of fluorinated compounds, physical and physicochemical properties. In *Chemistry of Organic Fluorine Compounds II*, ACS Monograph 187 (Eds, Hudlicky, M. and Pavlath, S. E.) American Chemical Society, Washington, DC, pp. 979.
2. Scheirs, J. (1997) *Modern Fluoropolymers*, Wiley & Sons, New York.
3. Hougham, G., Johns, K., Cassidy, P. E. and Davidson, T. (1999) *Fluoropolymers: Synthesis and Properties*, Plenum Press, New York.
4. Matyjaszewski, K. (2000) *Controlled/Living Radical Polymerisation, Progress in ATRP, NMP, and RAFT*, American Chemical Society Symposium Series 768, Washington, DC, pp. 1–440.
5. Oka, M. and Tatemoto, M. (1984) Vinylidene fluoride–hexafluoropropylene copolymer having terminal iodines. *Contemp. Top. Polym. Sci.*, **4**, 763–774.
6. Lacroix-Desmazes, P., Ameduri, B. and Boutevin, B. (2002) Use of fluorinated organic compounds in living radical polymerisation. Use of fluorinated organic compounds in living radical polymerizations. *Collect. Czech. Chem. Commun.*, **67**, 1383–1415.
7. Telechelics: from the Greek words "telos" and "khêlê" that stand for length and clamp, respectively.
8. Goetals, E. J. (1989) *Telechelic Polymers and their Applications*, CRC Press, Boca Raton, FL.

9. Boutevin, B. and Robin, J. J. (1992) Synthesis and properties of fluorinated diols. *Adv. Polym. Sci.*, **102**, 105–132.

10. Améduri, B. and Boutevin, B. (1992) Synthesis and applications of fluorinated telechelic monodispersed compounds. *Adv. Polym. Sci.*, **102**, 133–169.

11. Améduri, B. and Boutevin, B. (1999) Use of telechelic fluorinated diiodides to obtain well-defined fluoropolymers. *J. Fluorine Chem.*, **100**, 97–116.

12. Lagow, R. J. and Wei, H. C., *Direct Fluorination of Polymers*. In Ref. [3], Vol. 2, Chap. 14, pp. 209–222.

13. Kawa, H. (1999) In *Synthesis of Unique Fluorinated Diols*, Proceedings of Fluorine in Coatings III, Conference, Orlando, Paint Research Association ed., January 25–27, 1999, Paper No. 6.

14. Moore, E. P. (1967) Polymerisation of hexafluoropropylene epoxide. US Patent 3,322,826 (assigned to DuPont de Nemours).

15. Eleuterio, H. S. (1972) Polymerisation of perfluoro epoxides. *J. Macromol. Sci. Chem.*, **6**, 1027–1052.

16. Caporiccio, G., Viola, G. and Corti, C. (1983) Perfluoropolyethers, and a process for the preparation of the same. Eur. Patent 89,820 (assigned to Montedison).

17. Tanesi, D., Pasetti, A. and Corti, C. (1969) Fluorinated oxygen containing acyl fluorides. US Patent 3,442,942 (assigned to Montefluos).

18. Sianesi, D., Pasetti, A. and Belardinelu, G. (1984) High molecular weight polymeric perfluorinated copolyethers and process for their preparation from tetrafluoroethylene. US Patent 4,451,646 (assigned to Montefluos).

19. Sianesi, D., Marchionni, G. and De Pasquale, R. J. (1994) In *Organofluorine Chemistry: Principles and Commercial Applications* (Ed, Banks, R. E.) Plenum Press, New York, pp. 431–460.

20. Tonelli, C., Gavezotti, P. and Strepparala, E. (1999) Linear perfluoropolyether difunctional oligomers: chemistry, properties and applications. *J. Fluorine Chem.*, **95**, 51–70.

21. Yohnosuke, O., Takashi, T. and Shogi, T. (1983) Halogen-containing polyether. Eur. Patent Appl. 0,148,482 (assigned to Daikin) (*Chem. Abstr.*, **104**, 69315 (1986)).

22. Slinn, D. S. and Green, S. W. (1982) Fluorocarbon fluids for the use in the electronic industry in preparation, properties and industrial applications of organofluoride compounds, Vol. 2 (Ed, Banks, R. E.) Ellis Horwood, Chichester, pp. 45–82.

23. Turri, S., Scicchitano, M. and Tonelli, C. (1996) End-group chemistry of fluorooligomers: highly selective syntheses of diepoxy, diallyl and tetraol derivatives. *J. Polym. Sci., Part A: Polym. Chem.*, **34**, 3263–3275.

24. Scheirs, J. (1997) Perfluoropolyethers. Modern Fluoropolymers, Chapter 24 (Ed, Scheirs, J.) Wiley & Sons, New York, pp. 435–485.

25. Scicchitano, M. and Turri, S. (1999) Cyclic acetals of fluorinated polyether macrodiols. *J. Fluorine Chem.*, **95**, 97–103.

26. Gilson, R. and Grundy, P. J. (1994) Fluorine compound lubrication of thin film magnetic media. In *Fluorine in Coatings I, Conference*, Paint Research Association, Salford, UK, September 28–30, 1994.

27. Lagow, R. J., Bierschenk, T. R., Juhlke, T. J. and Kawa, H. (1992) A New Synthetic procedure for the preparation and manufacture of perfluoropolyethers. In *Synthetic Fluorine Chemistry* (Eds, Olah, G. A., Chambers, R. D. and Surya Prakash, G. K.) Wiley & Sons, New York, pp. 97–126.

28. Chambers, R. D., Hutchinson, J., Kelly, N. M., Joel, A. K., Rees, A. J. and Telford, P. T. (1999) Free radical chemistry. Part 9. Approaches to fluorinated polyethers-model compounds studies. *Isr. J. Chem.*, **39**, 133–140.

29. (a) Bierschenk, T. R., Kawa, H., Juhlke, T. J. and Lagow, R. J. (1988) Perfluorination of polyethers prepared by polymerisation of acetals, ketals, and ortho esters. NASA Contract Report-182155; (b) Jones, W. R., Bierschenk, T. R., Juhlke, T. J., Kawa, H. and Lagow, R. J. (1988) Preparation of new perfluoro ether fluids exhibiting excellent thermal-oxidative. *Ind. Eng. Chem. Prod. Res. Rev.*, **27**(8), 1497–1502.

30. (a) Meyer, M. (1993) Process for the preparation of perfluorinated ethers. Eur. Patent EP 543,288 (assigned to Hoechst AG) (*Chem. Abstr.*, **119**, 161082 (1993)); (b) Meyer, M., Stapel, R., Kottmann, H. and Gries, T. (1991) Preparation of perfluorinated polyethers. Eur. Patent EP 444, 554 (assigned to Hoechst) (*Chem. Abstr.*, **115**, 257001k (1991)).

31. Maxson, M. T., Norris, A. W. and Owen, M. J., *Fluorosilicones*. In Ref. [2], Chap. 20, pp. 359–372.

32. (a) Simons, J. H. (1950) Fluorine Chemistry, Vol. 1, Academic Press, New York, p. 40; (b) Lovelace, A. M., Rausch, D. A. and Postelnek, W. (1958) *J. Am. Chem. Soc. Monograph Series*, Reinhold, New York.

33. Tarrant, P. (1984) Fluorocarbon iodides—versatile reagents. *J. Fluorine Chem.*, **25**, 69–74.

34. Haszeldine, R. N. (1986) Perfluoroalkyl iodides. *J. Fluorine Chem.*, **33**, 307.

35. Mahler, W. (1963) Inorganic chemistry of carbon difluoride. *Inorg. Chem.*, **2**, 230.

36. Mitsch, R. A. (1964) Difluorodiazirine. II. Difluoromethane derivatives. *J. Heterocycl. Chem.*, **1**, 233–234.

37. Kesling, H. S. and Burton, D. J. (1975) Halodifluoromethylphosphonium salts—halomethyl-transfer agents. *Tetrahedron Lett.*, 3355–3358.

38. Wheaton, G. A. and Burton, D. J. (1978) Generation and reactions of halodifluoromethide ions. *J. Org. Chem.*, **43**, 2643–2651.

39. Wu, G., Zheng, P., Chen, Q. Y. and Zhu, S. Z. (1987) Structures of an ylide 4,4,5,5,6,6-hexafluoro-2-(phenyliodonio)dithianide 1,1,3,3-tetroxide (I) and its complex with dimethyl sulfoxide (II). *Sci. Sinica, Ser. B*, **30**, 561.

40. Elsheimer, S., Dolbier, W. R. Jr., Murla, M., Sepplet, K. and Paprott, G. (1984) Difluorodiiodomethane: its preparation, properties, and free-radical reactions. *J. Org. Chem.*, **49**, 205–207.

41. Mc Arthur, R. E. and Simons, J. H. (1950) Inorganic Synthesis, Vol. 3 (Ed, Audrieth, L. F.) McGraw-Hill, New York, pp. 37.

42. Soroos, H. and Hinkamp, J. B. (1945) The redistribution reaction. XI. Application to the preparation of carbon tetra-iodide and related halides. *J. Am. Chem. Soc.*, **67**, 1642–1643.

43. Su, D. B., Duan, J. X. and Chen, Q. Y. (1992) A simple, novel method for the preparation of trifluoromethyl iodide and diiododifluoromethane. *J. Chem. Soc., Chem. Commun.*, 807–815.

44. Brace, N. O. (2001) Synthesis with perfluoroalkyl iodides. A review—Part III. Addition of R_FI to norbornene esters, acids and anhydrides, alkenoic acids and esters, alkenylsuccinic anhydrides or diesters, and to vinyl monomers; lactonization and other reactions of adducts; hydroper-fluoroalkylation by R_FI; synthesis and reactions of $I(CF_2)_nI$ homologues ($n = 1-3$); perfluoroalkylation of arenes by R_FI or $[R_FCO_2]_2$; R_FI in the synthesis of R_FSR and segmented $R_F(CH_2)_nSH$; and, useful derivatives therefrom. *J. Fluorine Chem.*, **108**, 147–175.

45. (a) Li, A. R. and Chen, Q. Y. (1997) Lead tetraacetate induced addition reaction of difluorodiiodomethane to alkenes and alkynes. Synthesis of fluorinated telechelic compounds. *Synthesis*, 1481–1488; (b) Li, A. R. and Chen, Q. Y. (1997) One-pot synthesis of dialkyldifluoromethane and alkyldifluoromethyl iodides from the reaction of difluorodiiodo-methane with alkenes in sulphinatodehalogenation systems. *J. Fluorine Chem.*, **82**, 151–155.

46. Smart, B. E., Private Communication.

47. Hauptschein, M. and Grosse, A. V. (1951) Perfluoroalkyl halides from silver perfluoro-fatty acid salts. I. Perfluoroalkyl iodides. *J. Am. Chem. Soc.*, **73**, 2461–2463.

48. Rice, D. E. (1968) Original functional initiators for fluoroalkenes. *Polym. Lett.*, **6**, 335–342.

49. Rice, D. E. (1970) Perfluoroalkylene oxide polymers. US Patent 3,505,411 (assigned to 3M) (*Chem. Abstr.*, **72**, 122178w (1970)).

50. Boutevin, B., Ranjalahy, L. and Rousseau, A. (1992) Synthèse des α,ω-dichlorures d'acides à chaînes chlorofluorées par hydrolyse des α,ω-bis(trichlorométhylés) correspondants. *J. Fluorine Chem.*, **56**, 153–162 (and references herein).

51. Boutevin, B. and Pietrasanta, Y. (1973) Télomérisation du monochlorotrifluoroéthylène avec le tétrachlorure de carbone. *Tetrahedron Lett.*, **12**, 887–894.

52. Boutevin, B., Cals, J. and Pietrasanta, Y. (1975) Télomérisation par catalyse rédox. V. Télomérisation du chlorotrifluoroéthylène avec le bromotrichlorométhane par catalyse rédox. *Eur. Polym. J.*, **12**, 225–231.

53. Mc Loughlin, V. C. R. (1968) Some novel perfluoroalkanedioic acid derivatives and α,ω-iodoperfluoroalkanes. *Tetrahedron Lett.*, **46**, 4761–4762.

54. Dolbier, W. R., Rong, X. X., Bartberger, M. D., Koroniak, H., Smart, B. E. and Yang, Z. Y. (1998) Cyclisation reactivities of fluorinated hex-5-enyl radicals. *J. Chem. Soc., Perkin Trans.*, **2**, 219–230.

55. Riley, M. O., Kim, Y. K. and Pierce, O. R. (1977) The synthesis of fluoroether–fluorosilicone hybrid polymers. *J. Fluorine Chem.*, **10**, 85–110.

56. Fukaya, H., Hayashi, E., Hayakawa, Y. and Abe, T. (1997) Synthesis of new nitrogen-containing perfluoroalkyl iodides. *J. Fluorine Chem.*, **83**, 117–123.

57. Critchley, J. P., Mc Loughlin, V. C. R., Thrower, J. and White, I. M. (1970) Perfluoroalkylene-linked aromatic polymers. *Br. Polym. J.*, **2**, 288–294.

58. Guo, C. and Zhang, Y. (1978) Perfluoro- and polyfluorosulfonic acids. II. Preparation of some oxapolyfluoroalkane sulfonic acids. *Q. Chem. Hua. Hsuch Pao*, **37**, 315 (*Chem. Abstr.*, **93**, 113886g (1980)).

59. Caporiccio, G., Bargigia, G., Tonelli, C. and Tortelli, V. (1988) Process for preparing α,ω-haloperfluoroalkanes. US Patent 4,731,170 (assigned to Ausimont).

60. Caporiccio, G., Bargigia, G., Tonelli, C. and Tortelli, V. (1986) α,ω-Dihaloperfluoroalkanes. Eur. Patent Appl. 194,781 (assigned to Ausimont) (*Chem. Abstr.*, **106**, 69110 (1987)).

61. Coffman, D. D., Raasch, M. S., Rigby, G. W., Barrick, P. L. and Hanford, W. E. (1949) Addition reactions of tetrafluoroethylene. *J. Org. Chem.*, **14**, 747–753.

62. Haszeldine, R. N. and Leedham, K. J. (1953) The reaction of fluorocarbon radicals. Part IX. Synthesis and reactions of pentafluoropropionic acid. *J. Chem. Soc.*, 1548–1552.

63. Knunyants, I. L., Khrakyan, S. P., Zeifman, Yu. Y. and Shokina, V. V. (1964) Fluorinated diiodoalkanes and diolefins. *Bull. Acad. Sci. USSR, Div. Chem. Sci.*, 359–361.

64. Renn, J. A., Toney, A. D., Terjeson, R. J. and Gard, G. L. (1997) A facile preparation of ICF₂CF₂I and its reaction with ethylene. *J. Fluorine Chem.*, **86**, 113–114.

65. Ameduri, B. and Boutevin, B. (1994) Telomerisation. In *Encyclopedia of Advanced Materials* (Eds, Bloor, D., Brook, R. J., Flemings, M. C. and Mahajan, S.) Pergamon Press, Oxford, pp. 2777–2784, and see Chap. I.

66. Ameduri, B. and Boutevin, B. (1997) Telomerisation reactions of fluoroalkenes. In *Topics in Current Chemistry*, Vol. 192 (Ed, Chambers, R. D.) Springer, Heidelberg, pp. 165–233.

67. Bedford, C. D. and Baum, K. (1980) Preparation of α,ω-diiodoperfluoroalkanes. *J. Org. Chem.*, **45**, 347–348.

68. Tortelli, V. and Tonelli, C. (1990) Telomerisation of tetrafluoroethylene and hexafluoropropene: synthesis of diiodoperfluoroalkanes. *J. Fluorine Chem.*, **47**, 199–217.

69. Kotov, S. V., Ivanov, G. P. and Kostov, G. K. (1988) A convenient one-stage synthesis of some diiodoperfluoroalkanes by using tetrafluoroethylene derived from poly(tetrafluoroethylene) waste. *J. Fluorine Chem.*, **41**, 293–295.

70. Yamabe, M., Arai, K. and Kaneko, J. (1976) α,ω-Diiodopolyfluoroalkanes. Japanese Patent 76, 133,206 (assigned Asahi Glass Co., Ltd) (*Chem. Abstr.*, **87**, 5365d (1977)).

71. Knunyants, I. L., Li, C. Y. and Shokina, V. V. (1961) α,ω-Perfluorodiolefins and some of their transformations. *Izv. An. SSSR, Otd. Khim. M.*, 1462–1468 (*Chem. Abstr.*, **56**, 1899 (1962)).

72. Brace, N. O. (1962) Radical addition of iodoperfluoroalkanes to vinyl and allyl monomers. *J. Org. Chem.*, **27**, 3033–3038.

73. Tittle, B. (1971) Perfluoroalkyl iodides. British Patent 1,242,712 (assigned to ICI) (*Chem. Abstr.*, **75**, 129321 (1971)).

74. Chambers, R. D., Greenhall, M. P., Wright, A. P. and Caporiccio, G. (1991) A new telogen for telechelic oligomers of chlorortrifluoroethylene. *J. Chem. Soc., Chem. Commun.*, 1323–1326.

75. Chambers, R. D., Greenhall, M. P., Wright, A. P. and Caporiccio, G. (1995) A new telogen for telechelic oligomers of chlorotrifluoroethylene. *J. Fluorine Chem.*, **73**, 87–94.
76. Ameduri, B., Boutevin, B., Kostov, G. and Petrova, P. (1995) Novel fluorinated monomers bearing reactive side groups. Part 1. Preparation and use of $ClCF_2CFClI$ as the telogen. *J. Fluorine Chem.*, **74**, 261–267.
77. Ameduri, B., Boutevin, B., Kostov, G. and Petrova, P. (1999) Synthesis and polymerisation of fluorinated monomers bearing a reactive lateral group. Part 5. Radical addition of iodine monobromide to chlorotrifluoroethylene to form a useful intermediate in the synthesis of 4,5,5-trifluoro-4-ene-pentanol. *J. Fluorine Chem.*, **93**, 117–127.
78. (a) Yang, Z. Y. (1996) Nickel-catalysed reaction of highly fluorinated epoxides with halogens. *J. Am. Chem. Soc.*, **118**, 8140–8141; (b) Yang, Z. Y. (2000) Diiododifluoromethane: an excellent telogen for the preparation of 1,3-diiodofluoropropane derivatives. *J. Fluorine Chem.*, **102**, 239–241.
79. Ashton, D. S., Tedder, J. M. and Walton, J. C. (1974) Gas chromatographic properties of tetrafluoroethylene telomers. *J. Chromatogr.*, **90**, 315–324.
80. Manseri, A., Améduri, B., Boutevin, B., Caporiccio, G., Chambers, R. D. and Wright, A. P. (1995) Synthesis of fluorinated telomers. Part IV. Telomerisation of vinylidene fluoride with commercially available α,ω-diiodoperfluoroalkanes. *J. Fluorine Chem.*, **74**, 59–67.
81. Knunyants, I. L., Khrlakyan, S. P., Zeifman, Yu. Y. and Shokina, V. V. (1964) Fluorinated diiodoalkanes and diolefins. *Izv. Akad. Nauk USSR, Ser. Khim.*, **2**, 384–386 (*Chem. Abstr.*, **60**, 67762 (1964)).
82. Jaeger, H. (1978) Photoabsorption and photoionization measurements on constituent atmospheric gases in the extreme VUV. US Patent 4,067,916 (assigned to Ciba Ltd) (*Chem. Abstr.*, **89**, 120598 (1978)).
83. Iserson, H., Magazzu, J. J. and Osborn, S. W. (1972) α,ω-Diiodoperfluoroalkanes. German Patent 2,130,378 (assigned to Thiokol Chem. Co.) (*Chem. Abstr.*, **76**, 99090h (1972)).
84. (a) Fearn, J. E. (1971) Synthesis of fluorodienes. *J. Res. Natl. Bur. Stand., Sect. A*, **75**, 41–56; (b) Fearn, J. E. (1972) Polymerisation of fluoroolefins. In *Fluoropolymers* (Ed, Wall, L. A.) Intersc. Wiley, New York, pp. 23–24.
85. Balagué, J., Améduri, B., Boutevin, B. and Caporiccio, G. (1995) Synthesis of fluorinated telomers. Part 1. Telomerisation of vinylidene fluoride with perfluoroalkyl iodides. *J. Fluorine Chem.*, **70**, 215–223.
86. Manseri, A., Ameduri, B., Boutevin, B., Caporiccio, G., Chambers, R. D. and Wright, A. P. (1996) Unexpected telomerisation of hexafluoropropene with dissymmetrical halogenated telechelic telogens. *J. Fluorine Chem.*, **78**, 145–150.
87. Eleuterio, H. S. and Herbert, S., Perfluoropropylene polymers. US Patent 2,958,685 (assigned to DuPont de Nemours) (*Chem. Abstr.*, **60**, 20875 (1960)).
88. Eleuterio, H. S. and Pensak, D. (2001) Polyhexafluoropropylene: an epic polymer. Oral Communication *C2*, 13th European Symposium on Fluorine Chemistry, Bordeaux (F), July 15–20, 2001.
89. Balagué, J., Améduri, B., Boutevin, B. and Caporiccio, G. (1995) Synthèse de télomères fluorés. Partie III. Télomérisation de l'hexafluoropropène avec des iodures de perfluoroalkyle. *J. Fluorine Chem.*, **74**, 49–58.
90. Baum, K. and Malik, A. (1994) Difunctional monomers based on perfluoropropylene telomers. *J. Org. Chem.*, **59**, 6804–6807.
91. Boulahia, D., Manseri, A., Ameduri, B., Boutevin, B. and Caporiccio, G. (1999) Synthesis of fluorinated telomers. Part 6. Telomerisation of hexafluoropropene with α,ω-diiodoperfluoroalkanes. *J. Fluorine Chem.*, **94**, 175–182.
92. Manseri, A. (1994) Synthèse de nouveaux télomères téléchéliques diiodés précurseurs de silicones fluorés. PhD Dissertation, University of Montpellier.
93. Lebret, B., Vignane, P. and Ameduri, B., Telechelic diols containing vinylidene fluoride, hexafluoropropylene and tetrafluoroethylene, in preparation.

94. Balagué, J., Améduri, B., Boutevin, B. and Caporiccio, G. (2000) Controlled step-wise telomerisation of vinylidene fluoride, hexafluoropropene and trifluoroethylene with iodofluorinated transfer agents. *J. Fluorine Chem.*, **102**, 253–268.

95. Ameduri, B., Boutevin, B. and Kostov, G. (2001) Fluoroelastomers: synthesis, properties and applications. *Prog. Polym. Sci.*, **26**, 105–187.

96. Tatemoto, M. and Nakagawa, T. (1978) Segmented polymers. German Patent 2,729,671 (assigned to Daikin Kogyo Co. Ltd) (*Chem. Abstr.*, **88**, 137374 (1978)).

97. Tatemoto, M., Tomoda, M. and Ueta, Y. (1980) Co-crosslinkable mixture. German Patent DE 2, 940,135 (assigned to Daikin Kogyo Co. Ltd, Japan) (*Chem. Abstr.*, **93**, 27580).

98. Tatemoto, M. (1985) Fluorinated thermoplastic elastomers. *Int. Polym. Sci. Technol.*, **12**, 85–96 (translated in English from *Nippon Gomu Kyokaishi*, **57**, 761 (1984)).

99. Tatemoto, M. and Morita, S. (1982) Liquid fluoropolymer rubber. Eur. Patent Appl. EP 27,721 (assigned to Daikin Kogyo Co. Ltd) (*Chem. Abstr.*, **95**, 170754 (1981)).

100. (a) Hung, M. H. (1993) Iodine-containing chain-transfer agents for fluoroolefin polymerisation. US Patent 5,231,154 (assigned to DuPont de Nemours) (*Chem. Abstr.*, **120**, 136811 (1994)); (b) Carlson, D. P. Fluorinated thermoplastic elastomers with improved base stability. Eur. Pat. Appl. 0,444,700 A2; US Patent 5,284,920 (assigned to E.I. DuPont de Nemours and Co.) 01-03-1991 and 08-02-1994.

101. Arcella, V., Brinati, G. and Apostolo, M. (1997) Microstructure control in the emulsion polymerisation of fluorinated monomers. *La Chim. l'Industria*, **79**, 345–351.

102. Hassan, J., Sevignon, M., Gozzi, C., Schulz, E. and Lemaire, M. (2002) Aryl–aryl bond formation one century after the discovery of the Ullmann reaction. *Chem. Rev.*, **102**, 1359–1469.

103. Ford, L. A., Aubele, D. L. and Des Marteau, D. D. (1999) The coupling of perfluoroalkyl iodides to form novel difunctional compounds. 14th Winter Fluorine Conference Proceedings, Saint Petersburg Beach, January 17–22, 1999, Poster No. 27.

104. Cirkva, V., Gaboyard, M. and Paleta, O. (2000) Fluorinated epoxides. 5. Highly selective synthesis of diepoxides from α,ω-diiodoperfluoroalkanes. Regioselectivity of nucleophilic epoxide-ring opening and new amphiphilic compounds and monomers. *J. Fluorine Chem.*, **102**, 349–361.

105. Mascia, L., Zitouni, F. and Tonelli, C. (1994) Miscibilization of telechelic fluoroalkeneoxide oligomers in epoxy resins and effects on morphology and physical properties. *J. Appl. Polym. Sci.*, **51**, 905–923.

106. Re, A., Tonelli, C. and Tortelli, V. (1988) Fluorinated epoxy resins and process for preparing them. Eur. Patent Sp. 0,293,889 B1 (assigned to Ausimont).

107. Manséri, A., Améduri, B., Boutevin, B., Kotora, M., Hajek, M. and Caporiccio, G. (1995) Synthesis of telechelic dienes from fluorinated α,ω-diiodoperfluoroalkanes. Part I. Divinyl and diallyl derivatives from model $I(C_2F_4)_nI$ compounds. *J. Fluorine Chem.*, **73**, 151–158.

108. Manséri, A., Boulahia, D., Améduri, B., Boutevin, B. and Caporiccio, G. (1997) Synthèse de diènes téléchéliques à partir d'α,ω-diiodoalcanes fluorés—Partie II—divinyls et diallyls présentant des motifs constitutifs tetrafluoroéthylène, fluorure de vinylidène et hexafluoropropène. *J. Fluorine Chem.*, **81**, 103–113.

109. Cirkva, V., Améduri, B., Boutevin, B. and Paleta, O. (1995) Radical induced reaction of monoiodo and diiodoperfluoroalkanes with allyl acetate: telomer and rearranged products, mass-spectral distinguishing of regioisomers. *J. Fluorine Chem.*, **74**, 97–105.

110. Brace, N. O. (1964) Addition of polyfluoroalkyl iodides to unsaturated compounds. US Patent 3, 145,222 (assigned to DuPont de Nemours) (*Chem. Abstr.*, **61**, 10589g (1964)).

111. (a) Baum, K., Bedford, C. D. and Hunadi, R. J. (1982) Synthesis of fluorinated acetylenes. *J. Org. Chem.*, **47**, 2251–2257; (b) Baum, K. (1992) Fluorinated condensation monomers. In *Synthetic Fluorine Chemistry* (Eds, Olah, G., Chambers, R. D. and Surya Prakash, G. K.), Wiley & Sons, New York, Chap. 17, pp. 381–393.

112. Lahiouel, D., Ameduri, B. and Boutevin, B. (2001) A telechelic fluorinated diol from 1,6-diiodoperfluorohexane. *J. Fluorine Chem.*, **107**, 81–88.

113. Brady, R. F. Jr. (1997) In *Fluorinated Polyurethanes in Modern Fluoropolymers* (Ed, Scheirs, J.) Wiley & Sons, New York, Chap. 6, pp. 127–163.

114. Bargigia, G., Tonelli, C. and Tortelli, V. (1987) Process for the synthesis of mono- or di-hydroxyfluoroalkanes as intermediates for textile treatment agents. Eur. Patent Appl. 0,247,614 (assigned to Ausimont).

115. Baum, K., Archibald, T. G. and Malik, A. A. (1993) Polyfluorinated, branched-chain diols and diisocyanates and fluorinated polyurethanes prepared therefrom. US Patent 5,204,441 (assigned to Fluorochem Inc.).

116. Mera, A. E. and Griffith, R. (1994) Melt condensation and solution polymerisation of highly fluorinated aliphatic polyesters. *J. Fluorine Chem.*, **69**, 151–155.

117. (a) Von Werner, K. (1985) Fluoro-substituted alkanes or alkenes. German Patent DE 3,338,299; (b) Von Werner, K. (1985) Fluoroalkyl-substituted iodoalkanes. DE 3,338,300 (assigned to Hoescht) (*Chem. Abstr.*, **103**, 214858–214859 (1985)).

118. Brace, N. O. (1962) Ethylene-1,2-diiodotetrafluoroethane telomers. US Patent 3,016,407 (assigned to DuPont).

119. Takakura, T., Yamabe, M. and Kato, M. (1988) Synthesis of fluorinated difunctional monomers. *J. Fluorine Chem.*, **41**, 173–183.

120. Nadji, S., Tesoro, G. C. and Pendharkar, S. (1991) Aromatic diamines of high fluorine content. *J. Fluorine Chem.*, **53**, 327–338.

121. Feiring, A. E. (1994) Synthesis of new fluoropolymers: tailoring macromolecular properties with fluorinated substituents. *J. Macromol. Sci., Pure Appl. Chem.*, **A31**(11), 1657–1673.

122. Mera, A. E., Griffith, J. R. and Baum, K. (1988) Peptide blocking group and coupling techniques as applied to the synthesis of fluoroaliphatic polyamides. *Polym. Prepr. (Am. Chem. Soc., Div. Polym. Chem.)*, **29**, 398–399.

123. Bassilana Serrurier, C. and Cambon, A. (1998) Synthèse de nouveaux diépoxides hautement fluorés, précurseurs de tensioactifs bolaphiles à espaceur fluoré. *J. Fluorine Chem.*, **87**, 37–40.

124. Stansbury, J. W. and Choi, K. M. (1998) Homopolymerisation studies of new fluorinated dimethacrylates monomers. *Polym. Prepr. (Am. Chem. Soc., Div. Polym. Chem.)*, **39**, 878–879.

125. Lee, S. Y. (1988) The use of fluoroepoxy compounds as adhesives to bond fluoroplastics without any surface treatment. *SAMPE Quart.*, **19**, 44–48 (*Chem. Abstr.*, **110**, 174550 (1989)).

126. Snow, A. W. and Buckley, L. J. (1997) Fluoromethylene cyanate ester resins. Synthesis, characterization and fluoromethylene chain length effects. *Macromolecules*, **30**, 394–405.

127. Snow, A. W., Buckley, L. J. and Armistead, J. P. (1999) NCOCH$_2$(CF$_2$)$_6$CH$_2$OCN cyanate ester resin. A detailed study. *J. Polym. Sci., Part A: Polym. Chem.*, **37**, 135–150.

128. Snow, A. W. and Buckley, L. J., *Fluorinated Cyanate Ester Resins*. In Ref. [3], Vol. 2, Chap. 2, pp. 11–24.

129. Bongiovanni, R., Malucelli, G., Priola, A. and Tonelli, C. (1997) UV-curable systems containing perfluoropolyether structures: synthesis and characterization. *Macromol. Chem. Phys.*, **198**, 1893–1907.

130. Bongiovanni, R., Malucelli, G., Pollicino, A., Tonelli, C., Simeone, G. and Priola, A. (1998) Perfluoropolyether structures as surface modifying agents of UV-curable systems. *Macromol. Chem. Phys.*, **199**, 1099–1105.

131. Ameduri, B., Bongiovanni, R., Malucelli, G., Pollicino, A. and Priola, A. (1999) New fluorinated acrylic monomers for the surface modification of UV-curable systems. *J. Polym. Sci., Part A: Polym. Chem.*, **37**, 77–86.

132. Bongiovanni, R., Malucelli, G., Lombardi, V., Priola, A., Siracusa, V., Tonelli, C. and Di Meo, A. (2001) *Polymer*, **42**, 2299–2305.

133. Barraud, J. Y., Gervat, S., Ratovelomanana, V., Boutevin, B., Parisi, J. P., Cahuzac, A. and Octeur, R. J. (1993) Matériau polymère de type polyuréthane acrylate pour revêtement de fibre optique ou pour ruban à fibres optiques. Eur. Patent Appl. 930,400,880 (assigned to Alcatel).

134. Fuchikami, T. and Urata, H. (1990) Fluorine-containing α,ω-bifunctional compounds and process for their production. Eur. Patent Appl. 0,383,141 (assigned to Sagami Chemical Research).

135. Urata, H., Kinoshita, Y., Asanuma, T., Kosukegawa, O. and Fuchikami, T. (1991) A facile synthesis of α,ω-dicarboxylic acids containing perfluoroalkylene groups. *J. Org. Chem.*, **56**, 4996–4999.

136. Chaudier, Y., Zito, F., Barthélémy, P., Stroebel, D., Améduri, B., Popot, J. L. and Pucci, B. (2002) Synthesis and preliminary biochemical assessment of ethyl-terminated perfluoroalkylamine oxide surfactants. *Bioorg. Med. Chem. Lett.*, **12**, 1587–1590.

137. Fokina, T. A., Apukhtina, N. P., Klebanskii, A. L. K., Nelson, K. N. and Solodovnikova, G. S. (1996) Telomers with terminal functional groups and elastomers based on them. *Vysokomol. Soyed*, **8**, 2197–2198.

138. Malik, A. A., Tzeng, D., Cheng, P. and Baum, K. (1991) Synthesis of fluorinated diisocyanates. *J. Org. Chem.*, **56**, 3043–3044.

139. Mera, A. E., Griffith, J. R. and Baum, K. (1990) Synthesis of reactive fluoroaliphatic diamines. *J. Fluorine Chem.*, **49**, 313–320.

140. Baum, K. (1992) Fluorinated condensation monomers. In *Synthetic Fluorine Chemistry* (Eds, Olah, G. A., Chambers, R. D. and Surya Prakash, G. K.) Wiley, New York, Chap. 17, pp. 381.

141. Webster, J. A., Butler, J. M. and Morrow, T. J. (1972) Synthesis and properties of particular fluorocarbon polymers. *Nuovo Chim.*, **48**, 51–54.

142. Labadie, J. W. and Hedrick, J. L. (1990) New low dielectric constant polyimide block and random copolymers. Proc. Int. SAMPE Electron. Conf. (Electron. Mater.—Our Future), **4**, 495–504 (*Chem. Abstr.*, **114**, 229914 (1991)).

143. Labadie, J. W. and Hedrick, J. L. (1990) New low dielectric constant polyimide block and random copolymers. *SAMPE J.*, **26**, 19–26 (*Chem. Abstr.*, **114**, 165458 (1991)).

144. Ando, S., Matsuura, T. and Sasaki, S., *Synthesis and Properties of Perfluorinated Polyimides*. In Ref. [3], Vol. 1, Chap. 14, pp. 277–304.

145. (a) Boutevin, B. and Pietrasanta, Y. (1985) The synthesis and applications of fluorinated silicones, notably in high-performance coatings. *Prog. Org. Coat.*, **13**, 297–331; (b) Améduri, B., Boutevin, B., Caporiccio, G., Guida-Pietrasanta, F., Manséri, A. and Ratsimihéty, A. (1996) Synthesis of hybrid fluorinated silicones. Part I. Influence of the spacer between the silicon atom and the fluorinated chain in the preparation and the thermal properties of hybrid homopolymers. *J. Polym. Sci., Part A: Polym. Chem.*, **34**, 3077–3086.

146. Boutevin, B., Guida-Pietrasanta, F., Ratsimihety, A. and Caporiccio, G. (1997) Synthesis, characterization and thermal properties of a new fluorinated silalkylene–polysiloxane. *Main Group Metal Chem.*, **20**, 133–138.

147. (a) Boutevin, B., Caporiccio, G., Guida-Pietrasanta, F. and Ratsimihéty, A. (1998) Hybrid fluorinated silicones. II. Synthesis of homopolymers and copolymers. Comparison of their thermal properties. *Macromol. Chem. Phys.*, **199**, 61–70; (b) Boutevin, B., Caporiccio, G., Guida-Pietrasanta, F. and Ratsimihéty, A. (2003) Polysilafluoroalkyleneoligosiloxanes: a class of fluoroelastomers with low glass transition temperature. *J. Fluorine Chem.*, **124**, 131–138.

148. Améduri, B., Boutevin, B., Caporiccio, G., Manséri, A., Guida-Pietrasanta, F. and Ratsimihéty, A. (1999) Use of fluorotelomers in the synthesis of hybrid silicones. In *Fluoropolymers: Synthesis and Properties*, Vol. 1 (Eds, Hougham, G., Johns, K., Cassidy, P. E. and Davidson, T.) Plenum Press, New York, Chap. 5, pp. 67–79.

149. Huang, W., Huang, B. and Wang, W. (1985) Sulfinatodehalogenation. III. Sulfinatodeiodination of primary perfluoroalkyl iodides and α,ω-perfluoroalkylene diiodides by sodium dithionite. *Huaxue Xuebao*, **43**, 663 (*Chem. Abstr.*, **104**, 206683 (1986)).

150. Brace, N. O. (1999) Syntheses with perfluoroalkyl radicals from perfluoroalkyl iodides. A rapid survey of synthetic possibilities with emphasis on practical applications. Part one: alkenes, alkynes and allylic compounds. *J. Fluorine Chem.*, **93**, 1–25.

151. Brace, N. O. (1999) Syntheses with perfluoroalkyl iodides. Part II. Addition to non-conjugated alkadienes; cyclisation of 1,6-heptadiene, and of 4-substituted 1,6-heptadienoic compounds:

bis-allyl ether, ethyl diallylmalonate, N,N'-diallylamine and N-substituted diallylamines; and additions to homologous exo- and endocyclic alkenes, and to bicyclic alkenes. *J. Fluorine Chem.*, **96**, 101–127.

152. Thomas, R., *Material Properties of Fluoropolymers and Perfluoroalkyl-Based Polymers.* In Ref. [3], Vol. 2, Chap. 4, pp. 47–67.

153. Kotora, M., Kvicala, J., Améduri, B., Hajek, M. and Boutevin, B. (1993) Rearrangement of 2-iodo-3-perfluoroalkyl-1-propyl acetates to 1-iodo-3-perfluoroalkyl-2-propyl acetates. *J. Fluorine Chem.*, **64**, 259–267 (and references herein).

154. Metzger, J. O., Linker, U. and Mahler, R. (1995) Electron transfer initiated free radical additions to unsaturated fatty compounds, Proceedings World Congress on Soc. Fat. Res., Vol. 3, (Eds, Barnes, R. J. and Association) pp. 477–480 (*Chem. Abstr.*, **128**, 129408 (1998)).

155. Metzger, J. O., Mahler, R. and Schmidt, A. (1996) Electron transfer initiated free radical additions of perfluoroalkyl iodides and diiodides to alkenes. *Liebigs Ann.*, **5**, 693–696.

156. Khrlakyan, S. P., Shokina, V. V. and Knunyants, I. L. (1965) Fluorinated mono- and diepoxy compounds. *Izv. Akad. Nauk, USSR Ser. Khim. (Engl. Transl.)*, 72–75 (*Chem. Abstr.*, **62**, 66356 (1965)).

157. Feiring, A. E. (1985) Reaction of perfluoroalkyl iodides with electron donor nucleophiles. Addition of perfluoroalkyl iodides to olefins initiated by electron transfer. *J. Org. Chem.*, **50**, 3269–3274.

158. Huang, W. Y., Hu, L. Q. and Ge, W. Z. (1989) Synthesis of cholesterol and its analogs with fluorinated side-chains. *J. Fluorine Chem.*, **43**, 305–318.

159. Wilson, L. M. and Griffin, A. C. (1993) Liquid-crystalline fluorocarbon–hydrocarbon microblock polymers. *Macromolecules*, **26**, 6312–6314.

160. Percec, V., Schlueter, D. and Ungar, G. (1997) Rational design of a hexagonal columnar mesophase in telechelic alternating multicomponent semifluorinated polyethylene oligomers. *Macromolecules*, **30**, 645–648.

161. (a) Twieg, R., Russell, T. P., Siemens, R. L. and Rabolt, J. F. (1985) Observations of a gel phase in binary mixtures of semifluorinated *n*-alkanes with hydrocarbon liquids. *Macromolecules*, **18**, 1361–1362; (b) Russell, T. P., Rabolt, J. F., Twieg, R. J., Siemens, R. L. and Farmer, B. L. (1986) Structural characterization of semifluorinated *n*-alkanes. 2. Solid–solid transition behavior. *Macromolecules*, **19**, 1135–1143.

162. Amato, C., Naud, C., Calas, P. and Commeyras, A. (2002) Unexpected selective monoadduct formation from 2-methyl-3-butyn-2-ol and α,ω-diiodoperfluorobutane. *J. Fluorine Chem.*, **113**, 55–63.

163. (a) Laurent, P., Blancou, H. and Commeyras, A. (1993) Synthèse de diamines à chaînes perfluoroalkylées. *J. Fluorine Chem.*, **62**, 173–182; (b) Mas, A., Laurent, P., Blancou, H. and Schue, F. (2002) Synthesis of aliphatic diamine and polyether imide with long perfluoroalkyl side group. *J. Fluorine Chem.*, **117**, 27–33.

164. Bradaï, M., Ameduri, B., Boutevin, B. and Belbachir, M., Synthesis of telechelic diamines by radical way, in preparation.

165. Amato, C. and Calas, P. (2003) Synthesis of amino terminated semifluorinated long-chain alkanethiols. *J. Fluorine Chem.*, **124**, 169–176.

166. Neumann, W. J. (1987) Tri-*n*-butyl hydride as reagent in organic synthesis. *Synthesis*, **8**, 665–683.

167. Qui, W. and Burton, D. J. (1993) Preparation of fluorinated 1,2- and α,ω-diols. *J. Org. Chem.*, **58**, 419–423.

168. Gresham, J. T. (1970) *N*-(heptadecafluoro-2,5-dioxahexadecyl)pyridinium chloride for oil- and water-repellant textiles. German Patent 2,016,019 (assigned to FMC Corp.) (*Chem. Abstr.*, **74**, 23610 (1971)).

169. Brace, N. O. and Mac Kenzie, A. K., Perfluoroalkyl phosphates. US Patent 3,083,224 (assigned to DuPont) (*Chem. Abstr.*, **59**, 5023f (1963)).

170. Yang, Z. Y. and Burton, D. J. (1992) A novel and practical preparation of α,α-difluoro functionalised phosphonates from iododifluoromethylphosphonate. *J. Org. Chem.*, **57**, 4676–4683.

171. Gresham, J. T. (1973) Fluorinated polyurethanes as soil release agents. US Patent 3,759,874 (assigned to FMC Co.).

172. Bloechl, W. (1965) The strength and impact resistance of medium density ferrous sinterings. Dutch Patent 6,506,069 (*Chem. Abstr.*, **64**, 17421 (1966)).

173. Yu Yang, Z. (1997) Process using CF_2I_2 and olefins for producing diiodofluorocompounds, and products thereof. PCT Int. Appl. WO 9,744,300 (assigned to DuPont) (*Chem. Abstr.*, **128**, 34511 (1998)).

174. Ho, T., Malik, A. A., Wynne, K. J., Mc Carthy, T. J., Zhuang, K. H. Z., Baum, K. and Honeychuck, R. V. (1996) *Polyurethanes Based on Fluorinated Diols*, ACS Symposium Series, Vol. 624, Chap. 23, pp. 362–376.

175. Daikin, Production of fluoroalcohol. Japanese Patent 02,157,238.

176. (a) Ameduri, B., Boutevin, B., Moreau, J. J. E. and Wonchiman, M. (2000) Hybrid organic gels containing perfluoro-alkyl moieties. *J. Fluorine Chem.*, **104**, 185–194; (b) Michalczyk, M. J. and Sharp, K. G. (1994) Single component inorganic/organic network materials and precursors thereof. PCT Int. Appl. WO 94 06,807 (assigned to DuPont de Nemours).

177. Anton, D. R. and Mac Kinney, R. J. (1990) Partially fluorinated compounds and polymers. Eur. Patent Appl. EP 352,718 (assigned to DuPont de Nemours) (*Chem. Abstr.*, **113**, 6989 (1990)).

178. Barthelemy, P., Ameduri, B., Chabaud, E., Popot, J. L. and Pucci, B. (1999). *Org. Lett.*, **1**, 1689–1692 (and Fourth French Symposium in Organofluorine Chemistry, Chatenay-Malabry (France) 9–12 May, 1999 (Communication No. O-16)).

179. Chaudier, Y., Zito, F., Barthelemy, P., Stroebel, D., Ameduri, B., Popot, J. L. and Pucci, B. (2002) Synthesis and preliminary physico-chemical and biological assessments of ethyl terminated perfluoroalkylamine oxide surfactants. *Bioorg. Med. Chem. Lett.*, **12**, 1587–1590.

180. Malik, A. A., Archibald, T. G., Tzeng, D., Garver, L. C. and Baum, K. (1989) Synthesis of fluorine-containing nitro compounds. *J. Fluorine Chem.*, **43**, 291–300.

181. Yagupol'skii, L. M., Kondratenko, N. V., Timofeeva, G. N., Dronkina, M. I. and Yagupol'skii, Yu. L. (1980) Quaternary ammonium salts with perfluoroalkyl groups at the nitrogen atom. *Zh. Org. Kh. (Engl. Transl.)*, **16**, 2139–2143.

182. Mc Loughlin, V. C. R. and Thrower, J. (1969) Route to fluoroalkyl substituted aromatic compounds involving fluoroalkylcopper intermediates. *Tetrahedron*, **25**, 5921–5940.

183. Chen, G. C., Chen, L. S. and Eapen, K. C. (1993) Perfluoroalkylations and perfluorooxaalkylations. Part 2. Copper-mediated cross-coupling of secondary perfluorooxaalkyl iodides and aryl halides. *J. Fluorine Chem.*, **65**, 59–65.

184. Mc Loughlin, V. C. R. and Thrower, J. (1970) Fluorinated alkylaryl compounds. British Patent 1,156,912 (assigned to the Minister of Technology, UK).

185. Mc Loughlin, V. C. R., Thrower, J., Hewins, M. A. H., Pipett, J. S. and White, M. A. (1996) A new route to fluoroalkyl-substituted aromatic compounds (Royal Aircraft Establish), Technical Report 66341, UDC 678,674.

186. Mc Loughlin, V. C. R. and Thrower, J. (1968) Fluoroalkyl aromatics. US Patent 3,408,411 (assigned to the Minister of Technology, UK).

187. Henne, A. L. and Krauss, D. W. (1951) The Direction of addition of carbon tetrachloride to $CFCl=CF_2$ under polar conditions. *J. Am. Chem. Soc.*, **73**, 5303–5304.

188. Boutevin, B., Pietrasanta, Y. and Sideris, A. (1976) Télomérisation par catalyse rédox. VII. Synthèse et transformation chimique des télomères du chlorotrifluoroéthylène et du 1,1,1-trichloroéthane. *Eur. Polym. J.*, **12**, 283–289.

189. Boutevin, B., Doheim, M., Pietrasanta, Y. and Rigal, G. (1979) Influence des alcools dans les réactions de télomérisation du chlorotrifluoroéthylène par catalyse rédox. *J. Fluorine Chem.*, **13**, 29–37.

190. Boutevin, B., Doheim, M., Pietrasanta, Y. and Sideris, A. (1978) Modification chimique des résines novalaques et alcool polyvinylique par addition de composés fluorés. *Double Liaison*, **XXV**, 273–280.

191. Battais, A., Boutevin, B., Cals, J., Hervaud, Y., Hugon, J. P., Pietrasanta, Y. and Sideris, A. (1979) Synthèse de nombreux monomères, télomères et polymères à partir du chlorotrifluoroéthylène. *Inform. Chim.*, 209–220.

192. Boutevin, B., Maubert, C., Mebkhout, A. and Pietrasanta, Y. (1981) Telomerisation kinetics by redox catalysis. *J. Polym. Sci., Part A: Polym. Chem.*, **19**, 499–511.

193. Boutevin, B., Hervaud, Y. and Rolland, L. (1983) Applications de mélanges stables de composés chlorofluorés à titre de fluides calogènes pour pompe à chaleur à absorption. French Patent FR 8, 301,128 (assigned to Gaz de France).

194. Mohan, S. (1983) Chlorotrifluoroethylene telomerisation process. Eur. Patent Appl. 93,580 (assigned to Occidental Chem.).

195. Dannels, B. (1985) Chlorotrifluoroethylene telomerisation process. Eur. Patent Appl. 140,385 (assigned to Occidental Chem.).

196. Boutevin, B. and Pietrasanta, Y. (1988) Hydrocarbures chlorofluorés et leur procédé de préparation. French Patent FR 8,801,882 (assigned to Atochem).

197. Battais, A., Boutevin, B., Pietrasanta, Y., Bertocchio, R. and Lanz, A. (1989) Télomérisation du tétrafluoroéthylène avec le tétrachlorure de carbone par catalyse rédox. *J. Fluorine Chem.*, **42**, 215–224.

198. Boutevin, B. and Pietrasanta, Y. (1975) Télomérisation par catalyse rédox. VI. Transformation chimique des télomères du chlorotrifluoroéthylène. *Eur. Polym. J.*, **12**, 231–238.

199. Battais, A., Boutevin, B., Hugon, J. P. and Pietrasanta, Y. (1980) Synthèse de diols fluorés à partir de dérivés des télomères du chlorotrifluoroéthylène et du tétrachlorure de carbone. *J. Fluorine Chem.*, **16**, 397–405.

200. (a) Battais, A., Boutevin, B., Pietrasanta, Y. and Sarraf, T. (1982) Synthèse de diols polyhalogénés par télomérisation de l'undécylénate de méthyle 2. *Makromol. Chem.*, **183**, 2359–2366; (b) Boutevin, B., Cals, J. and Pietrasanta, Y. (1974) Transformation chimique des télomères du chlorotrifluoroéthylène et du tétrachlorure de carbone. *Tetrahedron Lett.*, **12**, 939–945.

201. Ameduri, B., Boutevin, B., Lecrom, C., Pietrasanta, Y. and Parsy, R. (1988) Etude de la diaddition de l'acétate d'allyle sur les télogènes chlorofluorés possédant deux groupements CCl₃ en extrémité de chaîne. *Makromol. Chem.*, **189**, 2545–2558.

202. Ameduri, B., Boutevin, B., Lecrom, C. and Garnier, L. (1992) Synthesis of halogenated monodispersed telechelic oligomers. III. Bistelomerisation of allyl acetate with telogens which exhibit α,ω-di(trichloromethyled) end groups. *J. Polym. Sci., Part A: Polym. Chem.*, **30**, 49–62.

203. Kotora, M., Ameduri, B., Boutevin, B. and Hajek, M., unpublished results.

204. Rice, D. E. and Sandberg, C. L. (1969) Carboxy- and ester-terminated copolymers of vinylidene fluoride and hexafluoropropene. US Patent 3,438,953 (assigned to 3M) (*Chem. Abstr.*, **70**, 115875b (1969)); Carboxy-terminated vinylidene fluoride–perfluoropropene copolymers. US Patent 3,457,245 (assigned to 3M) (*Chem. Abstr.*, **71**, 71183p (1969)).

205. Rice, D. E. (1969) Bis (α-substituted perfluoroacyl) peroxides. US Patent 3,461,155 (assigned to 3M) (*Chem. Abstr.*, **71**, 90811p (1969)).

206. Zhao, C. Y., Zhou, R., Pan, H., Jin, X., Qu, Y., Wu, C. and Jiang, X. (1982) Thermal decomposition of some perfluoro and polyfluorodiacyl peroxides. *J. Org. Chem.*, **47**, 2009–2013.

207. Rice, D. E. and Sanberg, C. L. (1971) Functionally-terminated copolymers of vinylidene fluoride and hexafluoropropene. *Polym. Prepr. (Am. Chem. Soc., Div. Polym. Sci.)*, **12**(1), 396–402.

208. Novikov, V. A., Sass, V. P., Ivanova, L. S., Sokolov, L. F. and Sokolov, S. V. (1975) Kinetics of thermal decomposition of perfluorodiacyl peroxides. *Polym. Sci., USSR Ser. A*, **9**, 1414–1417.

209. Oka, M. and Morita, S. (1985) Fluorine-containing diacyl peroxides and use thereof. Eur. Patent 0,186,215 (assigned to Daikin).

210. (a) Saint-Loup, R., Manseri, A., Ameduri, B., Lebret, B. and Vignane, P. (2001) Procédé de préparation de polymères fluorés téléchéliques et polymères ainsi préparés. French Patent FR 2, 810,668 (assigned to CEA); (b) Saint-Loup, R. and Ameduri, B. (2002) Photochemical induced polymerisation of vinylidene fluoride (VDF) with hydrogen peroxide to obtain original telechelic PVDF. *J. Fluorine Chem.*, **116**, 27–34.

211. Saint-Loup, R., Manseri, A., Ameduri, B., Lebret, B. and Vignane, P. (2002) Synthesis and properties of novel fluorotelechelic macrodiols containing vinylidene fluoride, hexafluoropropene and chlorotrifluoroethylene. *Macromolecules*, **35**, 1524–1536.

212. Kostov, G., Ameduri, B. and Boutevin, B. (2002) New approaches to the synthesis of functionalised fluorine-containing polymers. *J. Fluorine Chem.*, **114**, 171–176.

213. Chan, M. L., Reed, R. Jr., Gollmar, H. G., Gotzmer, C. and Gill, R. C. (1991) Plastic bonded explosives using fluorocarbon binders. US Patent 5,049,213 (assigned to US Navy).

214. (a) Smeltz, K. (1969) Fluorinated alcohols. US Patent 3,478,116; (b) Smeltz, K. (1969) Fluorinated polyesters. US Patent 3,504,016 (assigned to DuPont de Nemours).

215. Trabelsi, H., Szönyi, F. and Geribaldi, S. (2001) Nucleophilic displacements of 2-perfluoroalkyl-1-iodoethanes: improved synthesis of fluorine-containing malonic esters. *J. Fluorine Chem.*, **107**, 177–181.

216. (a) Foulletier, L. and Lalu, J. P. (1970) Wahlweise mit zwei oder drei walzen betreibbares vorgespanntes walzgerüst. German Patent DE 2,004,1956 (assigned to Atochem) (*Chem. Abstr.*, **71**, 110486c (1970)); (b) Foulletier, L. and Lalu, J. P. (1970) Steuereinrichtung für brenner. DE 2,004,196 (assigned to Atochem).

217. Ugine Khulmann (1967) Nouveaux composés organiques fluorés. French Patent 1,532,284.

218. (a) Bloechl, W. (1969) Ester and ether derivatives of fluoroalcohols containing polar bonding groups. German Patent DE 1,925,555 (*Chem. Abstr.*, **74**, 87347t (1969)); (b) Bloechl, W. (1972) Ester and ether derivatives of fluoroalcohols containing polar bonding groups. German Patent DE 2,052,579 (*Chem. Abstr.*, **77**, 100756j (1972)).

219. Bouvet, P. and Lalu, J. P. (1969) Nouveaux sulfamido alcools polyfluorés. French Patent FR 2, 034,142 (assigned to Ugine Kuhlman).

220. Anello, L. (1971) Telomers for use as surfactants. French Patent FR 1,959,703 (assigned to Montedison) (*Chem. Abstr.*, **74**, 99476 (1971)).

221. Koemm, U. and Geisler, K. (1984) Perfluoroalkylgruppen-enthaltende polyurethane and verfahren zu ihrer herstellung. German Patent DE 3,319,368 (assigned to Bayer) (*Chem. Abstr.*, **102**, 1985 (1984)).

222. Polyurethane resin (1986) Japanese Patent 61,252,220.

223. Lazerte, J. D. and Guenthner, R. A. (1973) Polyesters, alkyd resins, polyurethanes and epoxy resins. German Patent 1,620,965 (assigned to 3M) (*Chem. Abstr.*, **79**, 67291v (1973)).

224. (a) Toukan, S. S. and Hauptschein, M. (1973) Fluorierte alkylsulfide und verfahren zu ihrer herstellung. German Patent DE 2,239,709 (assigned to DuPont de Nemours); (b) Toukan, S. S. and Hauptschein, M., Fluorine and sulfur-containing compositions. US Patent 3,948,887 (1976) and US Patent 3,906,049 (1977) (assigned to DuPont de Nemours).

225. Gresham, J. T. and Skillman, N. J. (1973) Fluorinated polyurethanes as soil release agents. US Patent 3,759,874 (assigned to FMC Co.).

226. Boutevin, B., unpublished results.

227. Hager, R. B. (1969) Fluoroalkyl dicarboxylic acids and derivatives. US Patent 3,471,518 (assigned to Pennsalt Chem. Corp.).

228. (a) Dear, R. E. A. and Falk, R. A. (1975) Nouveaux glycols contenant deux groupes perfluoroalkylthio et leur procédé de fabrication. Brevet Français FR 2,259,819 (assigned to Ciba-Geigy); (b) Dear, R. E. A., Falk, R. A. and Mueller, K. L. (1985) Polyurethanes containing perfluoroalkylthio groups. German Patent DE 2,503,872 (assigned to Ciba-Geigy) (*Chem. Abstr.*, **84**, 45224b (1976)).

229. Hamiff, M. and Deisenroth, T. (1998) Synthesis of bisperfluoroalkyldiols. poster PII-29 presented at the 12th European Symposium in Fluorine Chem., Berlin.

230. (a) Laurent, Ph., Blancou, H. and Commeyras, A. (1991) Synthesis of fluorinated diols and diamines. *J. Fluorine Chem.*, **54**(P41), 291; (b) Requirand, N., Blancou, H. and Commeyras, A. (1993) F-alkylation de l'acide undécylenique. *Bull. Soc. Chim. Fr.*, **130**, 798–806.

231. Laurent, Ph., Blancou, H. and Commeyras, A. (1993) Synthesis of polyfluoroalkylsuccinic acids from perfluoroalkyl iodides and methylenebutanedioic acid (itaconic acid). *Synthesis*, **1**, 77–83.

232. Coe, P. L., Sellars, A., Tatlow, J. C., Fielding, H. C. and Whittaker, G. (1985) Polyfluoro-1,2-epoxyalkanes and cycloalkanes. Part IV. Thermal reactions of the epoxides of the pentamer and hexamer oligomers of tetrafluoroethylene. *J. Fluorine Chem.*, **27**, 71–84.

233. Clive, D. L. J. and Denyer, C. V. (1973) New method for converting epoxides into olefins: use of triphenylphosphine selenide and trifluoroacetic acid. *J. Chem. Soc., Chem. Commun.*, **7**, 253–256.

234. Chaabouni, M. M. and Baklouti, A. (1989) Ouverture des F-alkyl et Cl-alkyl époxydes par les fluorhydrates d'amines. *Bull. Soc. Chem. Fr.*, **4**, 549–556.

235. Szönyi, S. and Cambon, A. (1984) New perfluoroalkylated organic surfactants synthon intermediates and method of obtaining them. Fr. Demand 2,530,623.

236. Szönyi, S., Vandamme, R. and Cambon, A. (1985) Synthèse de nouveaux tensio-actifs F-alkylés bifonctionnels et application à la préparation de mousses extinctrices polyvalentes. *J. Fluorine Chem.*, **30**, 37–57.

237. Paleta, O., Cirkva, V., Polak, R., Kefurt, K., Moravcova, J., Kodieek, M., Forman, S. and Bares, M. (1996) New perfluoroalkylated polyhydroxy compounds as potential biosurfactants. Some of their physical and biological properties. 3ème Colloque Francophone sur la Chimie Organique du Fluor, Reims, Poster No. 10.

238. Gosnell, R. B. and Hollander, J. (1967) Synthesis of fluorinated polyurethanes. *J. Macromol. Sci. Phys.*, **1**, 831–850.

239. Hollander, J., Trischler, F. D. and Harrison, E. S. (1967) Highly fluorinated polyurethanes. *Polym. Prepr. (Am. Chem. Soc., Div. Polym. Chem.)*, **8**, 1149–1152 (*Chem. Abstr.*, **70**, 97293 (1969)).

240. Maruno, T., Nakamura, K., Ohmori, A., Shimizu, Y., Kubo, M. and Kobayashi, H. (1988) Fluorine-containing alicyclic and aromatic cyclic compounds, process, and adhesive composition containing the compounds. Eur. Patent Appl. 0,295,639 (assigned to Daikin Ind. Ltd) (*Chem. Abstr.*, **111**, 79484 (1989)).

241. Ohmori, A. (1980) Epoxy resin composition. German Patent DE 2,939,550 (assigned to Daikin Kogyo Co.) (*Chem. Abstr.*, **93**, 47875 (1980)).

242. Guery, C., Hugues, T., Kotea, N., Viguier, M. and Commeyras, A. (1989) Fluoro acrylic polymers and copolymers: surface properties and thermal behaviour. *J. Fluorine Chem.*, **45**(P26), 206.

243. Griffith, J. R. and O'Rear, J. G. (1977) Fluoro-anhydride curing agents and precursors for fluorinated epoxy resins. US Patent 4,045,408 (assigned to US Navy) (*Chem. Abstr.*, **87**, 168854 (1997)).

244. Coudures, C., Pastor, R. and Cambon, A. (1984) Synthèse de F-alkyl oxirannes. *J. Fluorine Chem.*, **24**, 93–104.

245. Chaabouni, B. M., Baklouti, A., Szonyi, S. and Cambon, A. (1990) Nouvelles méthodes de préparation des F-alkyl oxiranes à partir des acétates de bromo-2 F-alkyl-2-éthyle et des bromo-2 F-alkyl-2-éthanol. *J. Fluorine Chem.*, **46**, 307–315.

246. Evans, F. W. and Litt, M. H. (1968) Fluorinated epoxy compounds and polymers. US Patent 3, 388,078 (assigned to Allied Chem. Co.) (*Chem. Abstr.*, **69**, 28654q (1968)).

247. Calas, P., Moreau, P. and Commeyras, A. (1980) 3-(perfluoroalkyl)propargyl alcohols. British Patent GB 2,486,526 (assigned to PCUK).

248. Commeyras, A., Blancou, H. and Lantz, A. (1978) Perfluoroalkanecarboxylic and perfluoroalkanesulfinic acid derivatives. German Patent DE 2,756,169 (assigned to PCUK).

249. Fluorinated esters and their polymers (1967) French Patent 1,535,485 (assigned to Daikin Kogyo Co.) (*Chem. Abstr.*, **71**, 12585y (1969)).

250. Daikin Kogyo Kabushiki Kaishu (1966) Diols fluorés, notamment pour le traitement de produits fibreux. French Patent 1,475,237.

251. Cirkva, V., Ameduri, B., Boutevin, B. and Paleta, O. (1997) Highly selective synthesis of [(perfluoroalkyl)methyl]oxiranes (by the addition of iodoperfluoroalkanes to allyl acetate). *J. Fluorine Chem.*, **83**, 151–158.

252. Cirkva, V., Ameduri, B., Boutevin, B. and Paleta, O. (1997) Chemistry of [(perfluoroalkyl) methyl] oxiranes. Regioselectivity of ring opening with O-nucleophiles and the preparation of amphiphilic monomers. *J. Fluorine Chem.*, **84**, 53–61.

253. Abenin, P. S., Szönyi, F. and Cambon, A. (1991) Synthèse de 3-[2F-alkyléthylamino]-1,2-époxypropanes et obtention de nouveaux tensioactifs mono ou bicaténaires à tête β-hydroxylée. *J. Fluorine Chem.*, **55**, 1–11.

254. Ayari, A., Szönyi, S., Rouvier, E. and Cambon, A. (1990) Méthodes simples de *cis*-hydroxylation des F-alkyl éthylènes. *J. Fluorine Chem.*, **50**, 67–75.

255. Dmowski, W., Plenkiewicz, H. and Porwisiak, J. (1988) Synthetic utility of 3-(perfluoro-1,1-dimethylbutyl)-1-propene. Part I. Conversion to the epoxide and to alcohols. *J. Fluorine Chem.*, **41**, 191–212.

256. Hung, M. H., Smart, B. E., Feiring, A. E. and Rozen, S. (1991) Direct epoxidation of fluorinated olefins using the F_2–H_2O–CH_3CN system. *J. Org. Chem.*, **56**, 3187–3189.

257. Rozen, S. M. and Smart, B. E. (1991) Epoxidation of fluorine-containing olefins. PCT Int. Appl. WO 91 05778 (assigned to DuPont de Nemours) (*Chem. Abstr.*, **115**, 71372 (1991)).

258. Boutevin, B., Hugon, J. P. and Pietrasanta, Y. (1981) Synthèse d'époxydes et d'éthers glycidiques à chaîne chlorofluorée. *J. Fluorine Chem.*, **17**, 357–362.

259. Mc Bee, E. T. and Burton, T. M. (1952) The preparation and properties of 3,3,3-trifluoro-1,2-epoxypropane. *J. Am. Chem. Soc.*, **74**, 3022–3023.

260. Rausch, D. A., Lovelace, A. M. Jr. and Coleman, L. E. (1956) The preparation and properties of some fluorine-containing epoxides. *J. Org. Chem.*, **21**, 1328–1329.

261. Ciba Geigy (1974) Perfluoroalkyléthanediols, leur procédé de préparation et leur utilisation. French Patent FR 2,235,105.

262. Park, J. D., Rogers, F. E. and Lacher, J. R. (1961) Free-radical catalysed addition of unsaturated alcohols to perhaloalkanes. *J. Org. Chem.*, **26**, 2089–2095.

263. Mastouri, I., Hedhli, A. and Baklouti, A. (2001) Synthesis of F-alkylated hydrazinoalcohols. *J. Fluorine Chem.*, **108**, 121–124.

264. Charrada, B., Hedhli, A. and Baklouti, A. (2000) Aqueous ammonia: a versatile reagent for symmetrical di(fluoroalkylated) amino diols synthesis. *Tetrahedron Lett.*, **41**, 7347–7349.

265. Foulletier, L. and Lalu, J. P. (1970) Nouvelle méthode de préparation d'aminoalcools polyfluorés. French Patent FR 1,588,865 (assigned to Ugine Khulman).

266. Foulletier, L. and Lalu, J. P. (1969) Nouveaux aminodiols polyfluorés. French Patent FR 2,031, 650 (assigned to Ugine Khulman).

267. Hager, R. B., Toukan, S. S. and Walter, G. J. (1972) Sulfur-containing fluorocarbons. German Patent DE 2,342,888 (assigned to Pennwalt) (*Chem. Abstr.*, **81**, 153076b (1974)).

268. Hager, R. B., Toukan, S. S. and Walter, G. J. (1975) Oil water repellent fluorine and sulfur-containing polymers. US Patent 3,893,984 (assigned to Pennwalt) (*Chem. Abstr.*, **84**, 5826t (1976)).

269. Coussot, A. (1994) Etude de nouveaux monomères et oligomères photoréticulables: Applications aux revêtements étanches à l'eau pour fibres optiques télécommunications. PhD Dissertation, University of Montpellier.

270. (a) Perrier, S., Jackson, S. G., Haddleton, D. M., Ameduri, B. and Boutevin, B. (2002) Preparation of fluorinated methacrylic copolymers by copper mediated living radical polymerisation. *Tetrahedron*, **58**, 4053–4059; (b) Perrier, S., Jackson, S. G., Haddleton, D. M., Ameduri, B. and Boutevin, B. (2003) Preparation of fluorinated copolymers from bromine transfer radical polymerisation. *Macromolecules*, **36**, 9042–9049.

271. Dedek, V. and Hemer, I. (1985) Chemistry of organic fluorine compounds. Part XXIX. Photochemical chlorotrifluoroethylation of 1,2-, 1,3-, and 1,4-diols. *Collect. Czech. Chem. Commun.*, **50**, 2743–2752.

272. Urry, W. H., Niu, J. H. Y. and Lundsted, L. G. (1968) Multiple multicenter reactions of perfluoro ketones with olefins. *J. Org. Chem.*, **33**, 2302–2310.

273. Farah, B. S., Gilbert, E. E. and Sibilia, J. P. (1965) Perhalo ketones. V. The reaction of perhalocetones with aromatic hydrocarbons. *J. Org. Chem.*, **30**, 998–1001.

274. Hudlicky, M. (1976) *The Chemistry of Organic Fluorine Compounds*, Wiley & Sons, New York.

275. Yazlovitskii, A. V. and Malichenko, B. F. (1991) Effect of fluorine atoms on the properties of linear polyurethanes. *Vysokomol. Svedin Ser. A*, **13**, 734 (*Chem. Abstr.*, **74**, 142711 (1971)).

276. Malichenko, B. F., Tsypina, O. N. and Nestorov, A. E. (1969) Polyurethanes from isomeric trifluoromethyl phenylene diisocyanates and 1,4-butanediol. *Vysokomol. Svedin Ser. B*, **11**, 67 (*Chem. Abstr.*, **70**, 97296 (1969)).

277. Villa, J. L. and Iserson, H. (1974) Modified polybutadiene. German Patent DE 2,325,561 (assigned to Thiokol Chem. Corp.) (*Chem. Abstr.*, **81**, 38326w (1974)).

278. (a) Boutevin, B., Hervaud, Y. and Nouiri, M. (1990) Addition de thiols fluorés sur les polybutadiènes hydroxy téléchéliques par voie photochimique. *Eur. Polym. J.*, **26**, 877–886; (b) Ameduri, B., Boutevin, B. and Nouiri, M. (1993) Synthesis and properties of fluorinated telechelic macromolecular diols prepared by radical grafting of fluorinated thiols onto hydroxyl-terminated poly(butadienes). *J. Polym. Sci., Part A: Polym. Chem.*, **31**, 2069–2082.

279. Nouiri, M. (1995) Synthèse et caractérisation de nouveaux polybutadiènes hydroxytéléchéliques fluorosoufrés et fluorosiliciés. Moroccan State PhD, University of Casablanca, Morocco.

280. Ameduri, B., Boutevin, B., Nouiri, M. and Talbi, M. (1995) Synthesis and properties of fluorosilicon-containing polybutadienes by hydrosilylation of fluorinated hydrogenosilanes. Part 1. Preparation of the silylation agents. *J. Fluorine Chem.*, **74**, 191–198.

281. Furukawa, Y. and Kotera, M. (1998) Process for the preparation of fluorosilicone compounds having hydrolyzable groups. Eur. Patent 0,978,526A1 (assigned to Asahi Glass Comp. Ltd).

282. Boutevin, B., Guida-Pietrasanta, F. and Ratsimihety, A. (2000) Synthesis of photocrosslinkable fluorinated polydimethylsiloxanes: direct introduction of acrylic pendant groups via hydro-silylation. *J. Polym. Sci., Part A: Polym. Chem.*, **38**, 3722–3728.

283. (a) Bracon, F., Guittard, F., Taffin de Givenchy, E., Cambon, A. and Geribaldi, S. (1999) Highly fluorinated monomers precursors of side chain liquid crystalline polysiloxannes. *J. Polym. Sci., Part A: Polym. Chem.*, **37**, 4487–4496; (b) Bracon, F., Guittard, F., Taffin de Givenchy, E., Cambon, A. and Geribaldi, S. (2000) New fluorinated monomers containing an ester function in the spacers precursors of side chain liquid crystalline polysiloxannes. *Polymer*, **41**, 7905–7913.

284. Guizard, C., Boutevin, B., Guida, F., Ratsimihety, A., Amblard, P., Lassere, J. C. and Naiglin, S. (2001) VOC vapour transport properties of new membranes based on cross-linked fluorinated elastomers. *Sep. Purif. Technol.*, **22–23**, 23–30.

285. Trischler, F. D. and Hollander, J. (1967) Preparation of fluorine-containing polyethers. *Polym. Prepr. (Am. Chem. Soc., Div. Polym. Chem.)*, **8**, 491–496.

286. (a) Bekker, R. A., Dyatkin, B. L., Asratyan, G. V. and Knunyants, I. L. (1972) Fluorinated epoxides. Russian Patent SU352,890; (b) Knunyants, I. L., Chih-Yuan, L. and Shokina, V. V. (1963) Polyfluorinated linear bifunctional compounds (containing like functions) as potential monomers. *Russ. Chem. Rev.*, **32**, 461–476.

287. Griffith, J. R., O'Rear, J. R. and Reines, S. A. (1971) Fluoroepoxies. Candidate matrix type for future composites of ultra-high performance. *J. Am. Chem. Soc. Div. Org. Coat. Plast. Chem.*, **31**, 546–551.

288. Matuszczak, S. and Feast, W. J. (2000) An approach to fluorinated surface coatings via photoinitiated cationic cross-linking of mixed epoxy and fluoroepoxy systems. *J. Fluorine Chem.*, **102**, 269–277.

289. Evans, F. W. and Litt, M. H. (1968) Fluorinated epoxy compounds and polymers. US Patent 3, 388,078 (assigned to DuPont de Nemours).

290. Pittman, A. G., Allen, G., Wasley, W. L. and Roitman, J. (1970) Polymers derived from fluoroketones. IV. Poly[1,4-bis(heptafluoroisopropioxy)-2-butene oxide]. *J. Polym. Sci., Part B*, **8**, 873–877.

291. Ameduri, B., Boutevin, B. and Karam, L. (1993) Synthesis and properties of poly-[3-chloromethyl-3-(1,1,2,2-tetrahydroperfluorooctyloxy) methyl oxetane]. *J. Fluorine Chem.*, **65**, 43–52.

292. Vakhlamova, L. S. H., Kashkin, A. V., Krylov, A. I. and Kashkina, G. S. (1978) Synthesis of fluorine-containing derivatives of oxacyclobutane. *V.S. Sukhinin, Zh. Vses. Khim.*, **23**, 357 (*Chem. Abstr.*, **89**, 110440 (1978)).

293. (a) Cook, E. W. and Landrum, B. F. (1965) Synthesis of partially fluorinated oxetanes. *J. Heterocycl. Chem.*, **2**, 327–328; (b) Malik, A. A., Manser, G. E., Archibald, T. G., Duffy-Matzner, J. L., Harvey, W. L., Grech, G. L. and Carlson, R. P. (1996) Mono-substituted fluorinated oxetane monomers from 3-haloalkyl-3-alkyloxetanes, copolymers and prepolymers, and elastomers, WO 96/21657 (assigned to Aerojet—General Corp.); (c) Fujiwara, T., Makal, U. and Wynne, K. J. (2003) Synthesis and characterization of novel amphiphilic telechelic polyoxetanes. *Macromolecules*, **36**, 9383–9389.

294. (a) Chambers, R. D., Grievson, B., Drakesmith, F. G. and Powell, R. L. (1985) Free radical chemistry. Part 5. A new approach to the synthesis of perfluorinated ethers. *J. Fluorine Chem.*, **29**, 323–339; (b) Chambers, R. D., Joel, A. K. and Rees, A. J. (2000) Elemental fluorine. Part 11. Fluorination of modified ethers and polyethers. *J. Fluorine Chem.*, **101**, 97–105.

295. Head, R. A. and Johnson, S. (1987) Coating compositions. Eur. Patent Appl. 0,260,842 A2 (assigned to Imperial Chemical Industries).

296. Roussel, J. and Boutevin, B. (2001) Control of free-radical polymerisation of 2,2,2-trifluoroethyl methacrylate (TEMA) by a substituted fluorinated tetraphenylethane-type initer. *J. Fluorine Chem.*, **108**, 37–46.

297. (a) Guan, Z., Mc Clain, J. B., Samulski, E. T. and DeSimone, J. M. (1994) Fluorocarbon/hydrocarbon block copolymers for CO$_2$ viscosity enhancement. *Polym. Prepr. (Am. Chem. Soc., Div. Polym. Chem.)*, **35**, 725–726; (b) Guan, Z. and DeSimone, J. M. (1994) Fluorocarbon-based heterophase polymeric materials. 1. Block copolymer surfactants for carbon dioxide applications. *Macromolecules*, **27**, 5527–5532.

298. Zhang, Z. B., Ying, S. K., Hu, Q. H. and Xu, X. D. (2002) Semifluorinated ABA triblock copolymers: synthesis, characterization, and amphiphilic properties. *J. Appl. Polym. Sci.*, **83**, 2625–2633.

299. Fluorinated polyurethane resins (1958) British Patent 797,795.

300. Kresta, J. E. and Eldred, E. W. (1998) *60 years of Polyurethanes*, Technomic Publishers, Basel, Switzerland.

301. Malichenko, B. F., Yazloviskii, A. V. and Nesteron, A. E. (1970) Polyurethanes based on phenylene diisocyanates containing fluorinated alkoxy group. *Vysokomol. Svedin Ser. A*, **12**, 1700 (*Chem. Abstr.*, **73**, 120958x (1973)).

302. Keller, T. M. (1985) Polyurethanes from a fluorinated tertiary diol. *J. Polym. Sci., Polym. Chem. Ed.*, **23**, 2557–2563.

303. (a) Takakura, T., Kato, M. and Yamabe, M. (1990) Fluorinated polyurethanes. 1. Synthesis and characterization of fluorine-containing segmented polyurethane ureas. *Makromol. Chem.*, **191**, 625–632; (b) Takakura, T., Kato, M. and Yamabe, M. (1984) Fluorine-containing segmented polyurethanes for antithrombogenic elastomers. *J. Synth. Org. Chem., Jpn*, **42**, 822–828.

304. Edelman, P. G., Castner, D. G. and Ratner, B. D. (1990) Perfluoropolyether soft segment containing polyurethane: synthesis, characterization and surface properties. *Polym. Prepr. (Am. Chem. Soc., Div. Polym. Chem.)*, **31**, 314–315.

305. (a) Hearn, M. J., Briggs, D., Yoon, S. C. and Ratner, B. D. (1987) Sims and XPS studies of polyurethane surfaces. 2. Polyurethanes with fluorinated chain extenders. *Surf. Interface Anal.*, **10**, 384–391; (b) Yoon, S. C. and Ratner, B. D. (1986) Surface structure of segmented poly(ether urethanes) and poly(ether urethane ureas) with various perfluorochain extenders. An X-ray photoelectron spectroscopic investigation. *Macromolecules*, **19**, 1068–1079.

306. Yoon, S. C., Sung, Y. K. and Ratner, B. D. (1990) Surface and bulk structure of segmented poly(ether urethanes) with perfluoro chain extenders. 4. Role of hydrogen bonding on thermal transitions. *Macromolecules*, **23**, 4351–4356.

307. Takahara, A., Jo, N. J., Takamori, K. and Kajiyama, T. (1990) In *Progress in Biomedical Polymers* (Eds, Gebelein, C. G. and Dunn, R. L.) Plenum Press, New York, pp. 217–228.

308. Kajiyama, T. and Takahara, A. (1991) Surface properties and platelet reactivity of segmented poly(etherurethanes) and poly(etherurethaneureas). *J. Biomater. Appl.*, **6**, 42–71.

309. Yoon, S. C., Ratner, B. D., Ivan, B. and Kennedy, J. P. (1994) Surface and bulk structure of segmented poly(ether urethanes) with perfluoro chain extenders. 5. Incorporation of poly(dimethylsiloxane) and polyisobutylene macroglycols. *Macromolecules*, **27**, 1548–1554.

310. Chen, K. Y. and Kuo, J. F. (2000) Synthesis and properties of novel fluorinated aliphatic polyurethanes with fluoro chain extenders. *Macromol. Chem. Phys.*, **201**, 2676–2688.

311. Tonelli, C., Trombetta, T., Scicchitano, M. and Castiglioni, G. (1995) New perfluoropolyether soft segment containing polyurethanes. *J. Appl. Polym. Sci.*, **57**, 1031–1042.

312. Tonelli, C., Trombetta, T., Scicchitano, M., Simeone, G. and Ajroldi, G. (1996) New fluorinated thermoplastic elastomers. *J. Appl. Polym. Sci.*, **59**, 311–327.

313. Ho, T. and Wynne, K. J. (1992) A new fluorinated polyurethane: polymerisation, characterization and mechanical properties. *Macromolecules*, **25**, 3521–3527.

314. (a) Honeychuck, R. V., Ho, T., Wynne, K. J. and Nissan, R. A. (1992) Preparation and characterization of polyurethanes based on a series of fluorinated diols. *Polym. Mater. Sci. Eng.*, **66**, 521–522; (b) Honeychuck, R. V., Ho, T. and Wynne, K. J. (1993) Preparation and characterization of polyurethanes based on a series of fluorinated diols. *Chem. Mater.*, **5**, 1299–1306.

315. Sbarbati Del Guerra, R., Lelli, L., Tonelli, C., Trombetta, T., Cascone, M. G., Taveri, M., Narducci, P. and Giusti, P. (1994) In vitro biocompatibility of fluorinated polyurethanes. *J. Mater. Sci. Mater. Med.*, **5**, 452–459.

316. Yu, X. H., Okkema, A. Z. and Cooper, S. L. (1990) Synthesis and physical properties of poly(fluoroalkyl ether) urethanes. *J. Appl. Polym. Sci.*, **41**, 1777–1795.

317. Yang, S., Xiao, H. X., Higley, D. P., Kresta, J., Frisch, K. C., Farnham, W. B. and Hung, M. H. (1993) Novel fluorine-containing anionic aqueous polyurethanes. *J. Macromol. Sci., Pure Appl. Chem.*, **A30**, 241–252.

318. Kashiwagi, T., Ito, Y. and Imanishi, Y. (1993) Synthesis and nonthrombogenicity of fluoroalkyl polyether-urethanes. *J. Biomater. Sci. Polym. Ed.*, **5**, 157–166.

319. Tang, Y. W., Santerre, J. P., Labow, R. S. and Taylor, D. G. (1996) Synthesis of surface-modifying macromolecules for use in segmented polyurethanes. *J. Appl. Polym. Sci.*, **62**, 1133–1145.

320. Tang, Y. W., Santerre, J. P., Labow, R. S. and Taylor, D. G. (1997) Use of surface-modifying macromolecules to enhance the biostability of segmented polyurethanes. *J. Biomed. Mater. Res.*, **35**, 371–381.

321. Bonanni, A. P. and Angelo, P., Diisocyanate based laminating resin. US 3,088,934 (assigned to Naval Air Engineering Center) (*Chem. Abstr.*, **59**, 22453 (1963)).

322. Wall, L. A. (1972) *Fluoropolymers*, Wiley Intersc., New York, pp. 201–218.

323. Trischler, F. D., Hollander, J. and Gosnell, R. B. (1967) Preparation of fluorine-containing polyethers. *J. Polym. Sci., Part A: Polym. Chem.*, **5**, 2343–2349.

324. Hirn, B. (1993) Polyéthers fluorés: synthèse et caractérisation. Applications aux polyuréthanes fluorés. PhD Thesis, University of Montpellier.

325. Ho, T., Wynne, K. J. and Malik, A. A. (1995) Polyurethanes based on a fluorinated oligomer. *Polym. Prepr. (Am. Chem. Soc., Div. Polym. Chem.)*, **36**, 737–738.

326. Collet, A., Commeyras, A., Hirn, B. and Viguier, M. (1994) Perfluorinated polyether polyols for use in polyurethanes. French Patent 2,697,256.

327. Cathebras, N. (1997) Polymères uréthanes perfluoroalkylés. Applications: revêtements poly uréthanes, antifoulings non biocides, épaississants associatifs. PhD Thesis, University of Montpellier.

328. Filler, R., O'Brien, J. F., Fenner, J. V. and Hauptschein, M. (1953) Fluorinated esters. II. Diesters of perfluorocarboxylic acids with alcohols and glycols. *J. Am. Chem. Soc.*, **75**, 966–968.

329. Boutevin, B., Dongala, E. B. and Pietrasanta, Y. (1981) Synthèse de polyesters fluorés par polytransestérification. *J. Fluorine Chem.*, **17**, 113–119.

330. Schweiker, G. C. and Robitschek, P. (1957) Condensation polymers containing fluorine. I. Synthesis of linear polyesters from fluorine-containing diols. *J. Polym. Sci.*, **24**, 33–41.

331. (a) Pilati, F., Bonora, V., Nanaresi, P., Munari, A., Toselli, M. and Re, A. (1989) Preparation of poly(ethylene terephthalate) in the presence of telechelic perfluoropolyether. *J. Polym. Sci., Part A: Polym. Chem.*, **27**, 951–962; (b) Messori, M., Toselli, M. and Pilati, F. (1999) Poly-(ε-caprolactone)–poly(fluoroalkene oxide)–poly(ε-caprolactone) block copolymers as surface modifiers of poly(vinyl chloride). *Macromolecules*, **32**, 6969–6976 (and presented at the Fluorine in Coatings IV Conference, Brussels (Belgium), March 3–5, 2001).

332. (a) Pilati, F., Toselli, M., Vallieri, A. and Tonelli, C. (1992) Synthesis of polyesters–perfluoropolyethers block copolymers. *Polym. Bull.*, **28**, 151–157; (b) Pilati, F., Toselli, M., Messori, M., Credali, U., Tonelli, C. and Berti, C. (1998) Unsaturated polyester resins modified with perfluoropolyethers. *J. Appl. Polym. Sci.*, **67**, 1679–1691.

333. (a) Pospiech, D. U., Jehnichen, D., Haussler, L., Voigt, D. and Grundke, K. (1998) Semifluorinated polyesters with low surface energy. *Polym. Prepr. (Chem. Soc., Div. Polym. Chem.)*, **39**, 882–883; (b) Pospiech, D. U., Haussler, L., Eckstein, K., Voigt, D., Jehnichen, D., Gottwald, A., Kollig, W., Janke, A., Grundke, K., Werner, C. and Kricheldorf, H. (2001) Tailoring of polymer properties in segmented block copolymers. *Macromol. Symp.*, **163**, 113–126; (c) Pospiech, D. U., Haussler, L., Eckstein, K., Komber, H., Voigt, D., Jehnichen, D., Friedel, P., Gottwald, A., Kollig, W. and Kricheldorf, H. (2001) Synthesis and phase separation behavior of high performance multiblock copolymers. *High Perform. Polym.*, **13**, 275–292; (d) Grundke, K., Pospiech, D., Kollig, W., Simon, F. and Janke, A. (2001) Wetting of heterogeneous surfaces of block copolymers containing fluorinated segments. *Colloid Polym. Sci.*, **279**, 727–735; (e) Gottwald, A., Pospiech, D., Jehnichen, D., Haussler, L., Friedel, P., Pionteck, J., Stamm, M. and Floudas, G. (2002) Self-assembly and viscoelastic properties of semifluorinated polyesters. *Macromol. Chem. Phys.*, **203**, 854–861; (f) Pospiech, D., Häußler, L., Jehnichen, D., Kollig, W., Eckstein, K. and Grundke, K. (2003) Bulk and surface properties of blends with semifluorinated polymers and block copolymers. *Macromol. Symp.*, **198**, 421–434.

334. Mitsunobu, O. (1981) The use of diethyl azodicarboxylate and triphenylphosphine in synthesis and transformation of natural products. *Synthesis*, **1**, 1–28.

335. Johncock, P. (1970) Fluorinated polymers. German Patent DE 1,954,999 (*Chem. Abstr.*, **73**, 99391 (1970)).

336. Cook, E. W., Fluorinated elastomers containing oxygen in the polymer chain. US Patent 3,391, 118 (*Chem. Abstr.*, **69**, 36886 (1968)).

337. O'Rear, J. G., Griffith, J. R. and Reines, S. A. (1971) New fluorinated epoxies and polymeric derivatives. *J. Paint. Technol.*, **43**, 113–119.

338. Griffith, J. R. (1974) Polymers of 2-(fluorophenyl)-hexafluoro-2-propyl glycidyl ether. Canadian Patent CA 956,398 (assigned to US Navy) (*Chem. Abstr.*, **83**, 28824 (1975)).

339. Brady, R. (1994) Fluorinated polyurethane coatings for unique defence applications. Fluorine in Coatings I Conference, Salford, UK, September 28–30, 1994, Communication No. 7.

340. Guymon, C. A. and Bowman, C. N. (1997) Kinetics analysis of polymerisation rate acceleration during the formation of polymer/smectic liquid crystal composites. *Macromolecules*, **30**, 5271–5278.

341. Guymon, C. A., Hoggan, E. N., Clark, N. A., Rieker, T. P., Walba, D. M. and Bowman, C. N. (1997) Effect of monomer structure on their organization and polymerisation in a smectic liquid crystal. *Science*, **275**, 57–59.

342. Guymon, C. A. and Bowman, C. N. (1998) Formation of polymer stabilized ferroelectric liquid crystal thin films using a fluorinated diacrylate. *Polym. Prepr. (Am. Chem. Soc., Div. Polym. Chem.)*, **39**, 972–973.

343. Steinhauser, N. and Mülhaupt, R. (1994) Preparation and properties of copolyamides-6,6F and copolyamides-4,8F containing oligo(tetrafluoroethylene) segments. *Macromol. Chem. Phys.*, **195**, 3199–3211.

344. Webster, J. A., Butler, J. M. and Morrow, T. J. (1972) Synthesis and properties of imide and isocyanurate-linked fluorocarbon polymers. *Polym. Prepr. (Am. Chem. Soc., Div. Polym. Chem.)*, **13**, 612–616.

345. Boutevin, B. and Hugon, J. P. (1978) Synthèse de polyuréthanes fluorés. *Tetrahedron Lett.*, **2**, 129–137.

346. (a) Yamazaki, N., Matsumoto, M. and Higashi, F. (1975) Reactions of the *N*-phophonium salts of pyridines. XIV. Wholly aromatic polyamides by the direct polycondensation reaction by using phosphites in the presence of metal salts. *J. Polym. Sci., Polym. Chem. Ed.*, **13**, 1373–1380; (b) Yoneyama, M., Yamazaki, K., Kakimoto, M. and Imai, Y. (1999) Synthesis and properties of new semifluoroalkylene-containing aromatic-aliphatic polyamides derived from aromatic diamines and semifluoroaliphatic dicarboxylic acids. *Macromol. Chem. Phys.*, **200**, 2208–2212.

347. Chapman, T. P. and Marra, K. G. (1995) Determination of low critical surface tensions of novel fluorinated poly(amide urethane) block copolymers. 2. Fluorinated soft-block backbone and side chains. *Macromolecules*, **28**, 2081–2085.

348. Zhang, H., Marra, K. G., Ho, T., Chapman, T. M. and Gardella, J. A. Jr. (1996) Surface composition of fluorinated poly(amide urethane) block copolymers by electron spectroscopy for chemical analysis. *Macromolecules*, **29**, 1660–1665.

349. Sasaki, S. and Nishi, S. (1996) In *Polyimides: Fundamentals and Applications* (Eds, Ghosh, M. K. and Mittal, K. L.) Marcel Dekker, New York.

350. (a) Hougham, G., *Fluorine-Containing Polyimides*. In Ref. [3], Chap. 13, pp. 233–276; (b) Ando, S., Matsuura, T. and Sasaki, S., *Synthesis and Properties of Perfluorinated Polyimides*. In Ref. [3], Chap. 14, pp. 277–304; (c) Matsuura, T., Ando, S. and Sasaki, S., *Synthesis and Properties of Partially Fluorinated Polyimides for Optical Applications*. In Ref. [3], Chap. 15, pp. 305–350; (d) Harris, F. W., Li, F. and Cheng, S. Z. D., *Novel Organo-soluble Fluorinated Polyimides for Optical, Microelectronic, and Fiber Applications*. In Ref. [3], Chap. 16, pp. 351–370.

351. Critchley, J. P., Grattan, P. A., White, M. A., Mary, A. and Pippett, J. S. (1972) Perfluoroalkylene-linked aromatic polyimides. I. Synthesis, structure and some general physical characteristics. *J. Polym. Sci., Part A: Polym. Chem.*, **10**, 1789–1807.

352. Webster, J. A. (1974) Ether-linked aryl tetracarboxylic dianhydrides. US Patent 489,008 (*Chem. Abstr.*, **82**, 172298 (1976)).

353. Strepparola, E., Caporiccio, G. and Monza, E. (1984) Elastomeric polyimides from α,ω-bis(aminomethyl) poly[oxy(perfluoroalkylenes)] and tetracarboxylic acids. *Ind. Eng. Chem. Prod. Res. Dev.*, **23**, 600–605.

354. Eastmond, G. C. and Paprotny, J. (1999) The Nature of Segmented Polyimides with Different Aliphatic Sequences, Proceedings of the Fifth European Technical Symposium on Polyimides and High Performance Functional Polymers, May 3–5, 1999, Montpellier, France.

355. Andre, S., Guida-Pietrasanta, F., Rousseau, A., Boutevin, B. and Caporiccio, G. (2000) Synthesis characterization, and thermal properties of anhydride terminated and allyl terminated oligoimides. *J. Polym. Sci., Part A: Polym. Chem.*, **38**, 2993–3003.

356. (a) Bes, L., Boutevin, B., Rousseau, A., Mercier, R., Belleni, B., Decoster, D., Larchanche, J. F. and Vilcot, J. P. (2000) Synthesis and microwave characterizations of crosslinked oligoimide. *Synth. Mater.*, **115**, 251–256; (b) Bes, L., Rousseau, A., Boutevin, B. and Mercier, R. (2001) Syntheses and characterizations of bis(trialkoxysilyl)oligoimides. II. Syntheses and characterizations of fluorinated crosslinkable oligoimide with low refractive indices. *J. Polym. Sci., Part A: Polym. Chem.*, **39**, 2602–2619.

357. Bes, L., Rousseau, A., Boutevin, B., Mercier, R. and Kerboua, R. (2001) Synthesis and properties of novel polyimides containing fluorinated alkoxy side groups. *Macromol. Chem. Phys.*, **202**, 2954–2961.

358. Genies, C., Mercier, R., Sillon, B., Cornet, N., Gebel, G. and Pineri, M. (2001) Soluble sulfonated naphthalenic polyimides as materials for proton exchange membranes. *Polymer*, **42**, 359–367.

359. Piccardi, P., Massardo, P., Modena, M. and Santoro, E. (1973) Addition of carbon tetrachloride to 3,3,4,4-tetrafluorohexa-1,5-diene. *J. Chem. Soc., Perkin I*, 982–988.

360. Piccardi, P., Modena, M. and Santoro, E. (1971) Reactions of 3,3,4,4-tetrafluorohexa-1,5-diene. Part I. Cycloadditions with tetrafluoroethylene and chlorotrifluoroethylene. *J. Chem. Soc. (C)*, 3894–3898.

361. Coe, P. L., Milner, N. E. and Smith, J. A. (1975) Reactions of perfluoroalkylcopper compounds. Part V. The preparation of some polyfluoroalkyl-substituted acids and alcohols. *J. Chem. Soc., Perkin I*, 654–656.

362. Smart, B. E., Feiring, A. E., Krespan, C. G., Yang, Z. Y., Hung, M. H. and Resnick, P. R. (1995) New industrial fluoropolymer science and technology. *Macromol. Symp.*, **98**, 753–767.

363. Piccardi, P., Modena, M. and Cavalli, L. (1971) Reactions of 3,3,4,4-tetrafluorohexa-1,5-diene. Part II. Cyclisation to a four-membered ring in the thermal addition of pentafluoroiodoethane. *J. Chem. Soc.*, 3959–3966.

364. Ameduri, B., Boutevin, B., Kostov, G. K. and Petrova, P. (1999) Synthesis and polymerisation of fluorinated monomers bearing a reactive lateral group. Part 10. Copolymerisation of vinylidene fluoride (VDF) with 5-thioacetoxy-1,1,2-trifluoropentene for the obtaining of a novel PVDF containing mercaptan side-groups. *Des. Monomers Polym.*, **2**(4), 267–285.

365. (a) Arcella, V., Brinati, G., Albano, M. and Tortelli, V. (1996) Fluoroelastomers comprising monomeric units deriving from a bis-olefin. US Patent 5,585,449 (assigned to Ausimont); (b) Arcella, V., Brinati, G., Albano, M. and Tortelli, V. (1997) Fluorinated thermoplastic elastomers having superior mechanical and elastic properties, and the preparation process thereof. US Patent 5,612,419 (assigned to Ausimont); (c) Maccone, P., Apostolo, M. and Ajroldi, G. (2000) Molecular weight distribution of fluorinated polymers with long chain branching, *Macromolecules*, **33**, 1656–1663; (d) Albano, M., Apostolo, M., Arcella, V. and Marchese, E. (2000) Eur. Patent Appl. EP 1031607 (assigned to Ausimont S.p.A.) (*Chem. Abstr.*, **133**, 194044 (2000)); (e) Arcella, V., Ghielmi, A., Apostolo, M. and Abulseme, J. (2002) Crosslinked fluorinated ionomers. Eur. Patent Appl. EP 1167400 A1 (assigned to Ausimont S.p.A.) (*Chem. Abstr.*, **136**, 70270 (2002)).

366. (a) Arrigoni, S., Apostolo, M. and Arcella, V. (2002) Fluoropolymers architecture design: the branching and pseudo living technology. *Trends Polym. Sci.*, **7**, 23–40; (b) Arcella, V., Brinati, G. and Apostolo, M. (1997) New high performance fluoroelastomers. *Chim. Oggi*, **79**, 345–351; (c) Brinati, G. and Arcella, V. (2001) Branching and pseudo-living technology in the synthesis of high performance fluoroelastomers. *Rubber World*, **224**, 27–32.

367. Tatemoto, M., Yamamoto, Y., Yamamoto, K. and Onogi, H. (1989) Modified polytetrafluoroethylene and its preparation. Eur. Patent Appl. EP 302,513 (assigned to Daikin Industries Ltd).

368. Evers, R. C. (1974) Fluorocarbon ether bisbenzoxazole polymers from novel bis(*O*-aminophenols). *Polym. Prepr. (Am. Chem. Soc., Div. Polym. Chem.)*, **15**, 685–687 (*Chem. Abstr.*, **84**, 17775 (1976)).

369. Rosser, R. W., Parker, J. A., De Pasquale, R. J. and Stump, E. C. (1975) Poly(perfluoroalkylene) ethers as high-temperature sealants. *ACS Symp. Ser.*, **6**, 185–198.

370. Caporiccio, G. and Bargigia, G. (1974) Triazine ring-containing fluorinated elastomers. German Patent DE 2,502,505 (assigned to Montefluos) (*Chem. Abstr.*, **83**, 194859 (1975)).

371. Grindahl, G. A., Bajzer, W. X. and Pierce, O. R. (1967) The preparation and coupling of some α-haloperfluoromethyl-*s*-triazines. *J. Org. Chem.*, 603–607.

372. De Pasquale, R. J., Padgett, C. D. and Rosser, R. W. (1975) Highly fluorinated acetylenes. Preparation and some cyclisation reaction. *J. Org. Chem.*, **40**, 810–811.

373. Lau, K. S. and Kelleghan, W. J. (1986) Ethynylphenoxy and ethynylthiophenoxy derivatives of diphenylhexafluoropropane. PCT Int. Appl. WO 8,601,504 (*Chem. Abstr.*, **105**, 61056 (1986)).

374. Baum, K., Cheng, P. G., Hunadi, R. J. and Bedford, C. D. (1988) Polymerisation of fluorinated diacetylenes. *J. Polym. Sci., Part A: Polym. Chem.*, **26**, 3229–3233.

375. Amico, V., Meille, S. V., Corradi, E., Messina, M. T. and Resnati, G. (1998) Perfluorocarbon–hydrocarbon self assembling. 1D infinite chain formation driven by nitrogen···iodine interactions. *J. Am. Chem. Soc.*, **120**, 8261–8262.

376. (a) Metrangolo, P. and Resnati, G. (2001) Halogen bonding: a paradigm in supramolecular chemistry. *Chem. Eur. J.*, **7**, 2511–2519; (b) Metrangolo, P., Pilati, T., Resnati, G. and Stevenazzi, A. (2003) Halogen bonding driven self-assembly of fluorocarbons and hydrocarbons. *Curr. Opin. Colloid Interface Sci.*, **8**, 215–222.

377. (a) Liantonio, R., Logothetis, T. A., Messina, M. T., Metrangolo, P., Moiana, A. and Resnati, G. (2002) 2,2′:6′,2″-Terpyridine as monodentate ligand: halogen bonding driven formation of discrete 2 : 1 aggregates with 1,2,4,5-tetrafluoro-3,6-diiodobenzene. *Collect. Czech. Chem. Commun.*, **67**, 1373–1379; (b) Fontana, F., Forni, A., Metrangolo, P., Panzeri, W., Pilati, T. and Resnati, G. (2002) Perfluorocarbon–hydrocarbon discrete intermolecular aggregates: an exceptionally short N···I contact. *Supramol. Chem.*, **14**, 47–53.

378. (a) Walsh, R. B., Padgett, C. W., Metrangolo, P., Resnati, G., Hanks, T. W. and Pennington, W. T. (2001) Crystal engineering through halogen bonding: complexes of nitrogen heterocycles with organic iodides. *Cryst. Growth Des.*, **1**, 165–174; (b) Valerio, G., Raos, G., Meille, S. V., Metrangolo, P. and Resnati, G. (1999) Halogen bonding in fluoroalkylhalides: a quantum chemical study of increasing fluorine substitution. *J. Phys. Chem. A*, **104**, 1617–1624; (c) De Santis, A., Forni, A., Liantonio, R., Metrangolo, P., Pilati, T. and Resnati, G. (2003) N···Br halogen bonding: 1D infinite chains through the self-assembly of dibromotetrafluorobenzenes with dipyridyl derivatives. *Chem. Eur. J.*, **9**, 3974–3981.

379. Metrangolo, P., Panzeri, W., Recupero, F. and Resnati, G. (2002) ^{19}F NMR study of the halogen bonding between haloperfluorocarbons and heteroatom containing hydrocarbons. *J. Fluorine Chem.*, **114**, 27–33.

380. (a) Corradi, E., Meille, S. V., Messina, M. T., Metrangolo, P. and Resnati, G. (2000) Halogen bonding versus hydrogen bonding in driving self-assembly processes. *Angew. Chem., Int. Ed. Engl.*, **39**, 1782–1792; (b) Messina, M. T., Metrangolo, P., Pappalardo, S., Parisi, M. F., Pilati, T. and Resnati, G. (2000) Interactions at the outer faces of calixarenes: 2D infinite network formation with perfluoroarenes. *Chem. Eur. J.*, **6**, 3495–3503; (c) Liantonio, R., Luzzati, S., Metrangolo, P., Pilati, T. and Resnati, G. (2002) Asilines as new electron donor molecules for halogen bonded infinite chain formation. *Tetrahedron*, **58**, 4023–4029.

381. (a) Messina, M. T., Metrangolo, P., Panzeri, W., Pilati, T. and Resnati, G. (2001) Intermolecular recognition between oxygen-donors and perfluorocarbon iodine acceptors: the shortest O···I non-covalent bond. *Tetrahedron*, **57**, 8543–8552; (b) Chu, Q., Wang, Z., Huang, Q., Yan, C. and Zhu, S. (2001) Fluorine-containing donor–acceptor complex: infinite chain formed by oxygen···iodine interaction. *J. Am. Chem. Soc.*, **123**, 11069–11074.

382. Jay, J. I., Padgett, C. W., Walsh, R. D. B., Hanks, T. W. and Pennington, W. T. (2001) Noncovalent interactions in 2-mercapto-1-methylimidazole complexes with organic iodides. *Cryst. Growth Des.*, **1**, 501–509.

383. (a) Farina, A., Meille, S. V., Messina, M. T., Metrangolo, P. and Resnati, G. (1999) Resolution of racemic 1,2-dibromohexafluoropropane in halogen bonded supramolecular helices. *Angew. Chem., Int. Ed. Engl.*, **38**, 2433 (*Angew. Chem.*, **111**, 2585–2593 (1999)); (b) Grebe, J., Geisler, G., Harms, K. and Denicke, K. (1999) Donor–akzeptor-komplexe von kalogenidionen mit 1,4-diiodtetrafluorbenzol. *Z. Naturforsch. B*, **54**, 77–86; (c) Farnham, W. B., Dixon, D. A. and Calabrese, J. C. (1998) Novel fluorine-bridged polyfluorinated iodine structures. Presence of fluorine as the central atom in a five-center, six-electron bond. *J. Am. Chem. Soc.*, **110**, 8453–8457; (d) Liantonio, R., Metrangolo, P., Pilati, T. and Resnati, G. (2003) Fluorous interpenetrated layers in a three-component crystal matrix. *Cryst. Growth & Des.*, **3**, 355–362; (e) Weiss, R., Schwab, O. and Hampel, F. (1999) Ion-pair strain as the driving force for hypervalent adduct formation between iodide

ions and substituted iodobenzenes: structural alternatives to Meisenheimer complexes. *Chem. Eur. J.*, **5**, 968–975.

384. (a) Messina, T., Metrangolo, P., Panzeri, W., Ragg, E. and Resnati, G. (1998) Perfluorocarbon–hydrocarbon self-assembly. Part 3. Liquid phase interaction between perfluoroalkyl halides and heteroatom containing hydrocarbons. *Tetrahedron Lett.*, **39**, 9069–9076; (b) Cardillo, P., Corradi, E., Lunghi, A., Meille, S. V., Messina, M. T., Metrangolo, P. and Resnati, G. (2000) The N⋯I intermolecular interaction as a general protocol in the formation of perfluorocarbon–hydrocarbon supramolecular architectures. *Tetrahedron*, **56**, 5535–5537.

385. Resnati, G., Personal Communication.

386. Bertani, R., Metrangolo, P., Moiana, A., Perez, E., Pilati, T., Resnati, G., Rico-Lattes, I. and Sassi, A. (2002) Supramolecular route to fluorinated coatings: self-assembly between poly(4-vinylpyridines) and haloperfluorocarbons. *Adv. Mater.*, **14**, 1197–1204.

387. Apostolo, M., Arcella, V., Sorti, G. and Morbidelli, M. (2002) Free radical controlled polymerisation of fluorinated copolymers produced in microemulsion. *Macromolecules*, **35**, 6154–6166.

388. Jo, S. M., Lee, W. S., Ahn, B. S., Park, K. Y., Kim, K. A. and Paeng, I. S. R. (2000) New AB or ABA type block copolymers: atom transfer radical polymerisation (ATRP) of methyl methacrylate using iodine-terminated PVDFs as (macro)initiators. *Polym. Bull. (Berlin)*, **44**, 1–8.

389. (a) Zhang, Z., Ying, S. and Shi, Z. (1998) Synthesis of fluorine-containing block copolymers via ATRP. 1. Synthesis and characterization of PSt–PVDF–PSt triblock copolymers. *Polymer*, **40**, 1341–1345; (b) Zhang, Z., Ying, S. and Shi, Z. (1999) Synthesis of fluorine-containing block copolymers via ATRP. 2. Synthesis and characterization of semifluorinated di- and triblock copolymers. *Polymer*, **40**, 5439–5444.

390. Modena, S., Pianca, M., Tato, M. and Moggi, G. (1989) Radical telomerisation of vinylidene fluoride in the presence of 1,2-dibromotetrafluoroethane. *J. Fluorine Chem.*, **43**, 15–25.

CHAPTER 3

SYNTHESIS, PROPERTIES AND APPLICATIONS OF FLUOROALTERNATED COPOLYMERS

1. Introduction

Copolymers represent a huge branch of polymer science since they can cumulate the properties of each of the comonomers. Many various kinds of copolymers are available and can be classified according to random (or statistic) copolymers, alternated (or alternating) copolymers, block copolymers, graft copolymers, star copolymers and hyperbranched copolymers. The order given here represents decreasing production quantities but also increasing difficulties in synthesis; this is also true for fluorinated copolymers.

Although random copolymers have been reported in many books, reviews, publications and patents, this chapter is devoted to the syntheses, properties and applications of fluorinated alternating copolymers while the following chapters concern those of fluorinated block and graft copolymers. Before describing these particular fluorocopolymers in which at least one of the comonomers bears fluorine atom(s), it is useful to review some theoretical concepts regarding this class of copolymers.

2. Theoretical Concepts

A copolymerisation reaction lies in the competition between the four propagation equations from two monomers [1–4].

If one considers that A and D are the accepting and donating monomers, respectively, with both corresponding reactivity ratios, r_A and r_D, these four equations can be written as follows:

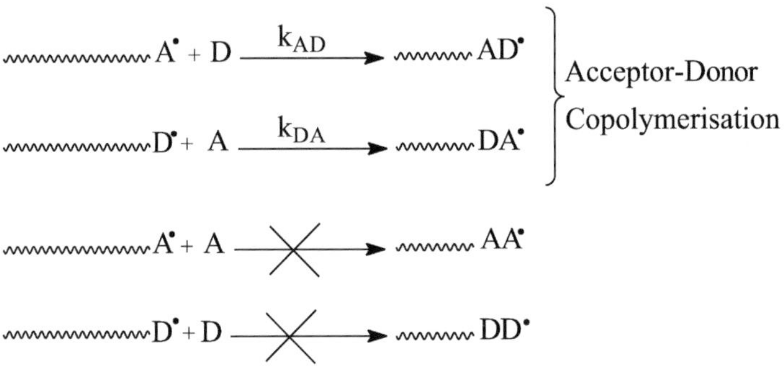

$$\sim\!\sim\!\sim\!\sim A^\bullet + D \xrightarrow{\;k_{AD}\;} \sim\!\sim\!\sim AD^\bullet$$

$$\sim\!\sim\!\sim\!\sim D^\bullet + A \xrightarrow{\;k_{DA}\;} \sim\!\sim\!\sim DA^\bullet$$

Acceptor-Donor Copolymerisation

$$\sim\!\sim\!\sim\!\sim A^\bullet + A \xrightarrow{\quad\times\quad} \sim\!\sim\!\sim AA^\bullet$$

$$\sim\!\sim\!\sim\!\sim D^\bullet + D \xrightarrow{\quad\times\quad} \sim\!\sim\!\sim DD^\bullet$$

This case is ideally reached if both comonomers do not (or poorly) homopolymerise. An example is the radical copolymerisations of vinyl ethers (VEs) or of maleic anhydride. However, this is not typically the case. Therefore, even in known alternating copolymers, short homopolymeric sequences can be produced that may disturb the overall properties of copolymers. This fact will be described later.

2.1. Polarity of Monomers

It has been established that, to favour the alternance [1–3], the groups borne by both comonomers must be of opposed polarity (Scheme 1). Both withdrawing or donating groups enable the macroradical to react more or less easily with the comonomer as shown is Scheme 1.

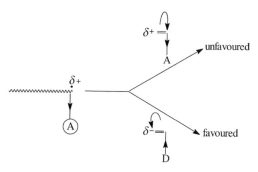

SCHEME 1.   Expected copolymerisation involving acceptor and donor monomers.

2.2. Reactivity Ratios

Alfrey and Price [5] reported an experimental relation to describe the reactivity of two comonomers according to two parameters (Q and e) that take into account the stabilisation by resonance and the polar effects of both monomers, respectively.

Such a system has the advantage of identifying each comonomer by associating both these parameters to represent its activity in copolymerisation:

$$r_A = \frac{k_{AA}}{k_{AD}} = \frac{Q_A}{Q_D} \cdot \exp[-e_A(e_A - e_D)]$$

$$r_D = \frac{k_{DD}}{k_{DA}} = \frac{Q_D}{Q_A} \cdot \exp[-e_D(e_D - e_A)]$$

In order to attempt the synthesis of an alternating copolymer, k_{AA} and k_{DD} are small about k_{AD} or k_{DA}. Hence, r_A and r_D are low and $r_A \times r_D$ product tends to zero. According to both equations above, it is noted that $r_A \times r_D$ is proportional to $\exp[-(e_A - e_D)^2]$. Hence, it is deduced that $|e_A - e_D|$ must be as high as possible. This shows that, once again, the polarity between two monomers must be the most opposed as possible.

Table I lists e and Q values [6–36] of different commercially or synthesised fluorinated monomers bearing one (or more) fluorine atom(s) or fluorinated

TABLE I

e AND Q VALUES OF FLUORINATED MONOMERS

Monomer	e	Q	References
$H_2C=CHCF_3$	0.42	0.130	[6]
	0.72	0.008	[6]
$H_2C=CHC_6H_4-F$	-0.12	0.830	[7]
$H_2C=C(CF_3)OAc$	1.51	0.069	[8]
$H_2C=CFC_6H_5$	-0.70	0.900	[9]
	-2.04	2.49	[10]
$H_2C=CFCONH_2$	1.45	0.540	[11]
$H_2C=C(CF_3)CO_2CH_3$	2.90	0.800	[12]
	2.50	0.740	[13]
$H_2C=C(CF_3)CN$	3.1	2.5	[12]
$H_2C=C(CF_3)C_6H_5$	0.40	0.780	[7, 14]
	0.90	0.430	[15]
$HFC=CH_2$	0.60	0.250	[16]
	0.40	0.008	[17]
	0.72	0.008	[6, 18]
	1.28	0.012	[18]
	-0.05	0.016	[17]
	-0.80	0.010	[19]
$H_2C=CFCF_2OR_F$	0.75	0.030	[20]
$F_2C=CH_2$	0.50	0.015	[14]
	0.40	0.008	[17]
	1.20	0.036	[21]
$F_2C=CCl_2$	2.10	0.041	[18]
$CF_3CF=CFH$ (cis)	2.10	0.002	[19]
$F_2C=CFH$	1.15	0.009	[17]
$F_2C=CHCF_3$	0.31	0.002	[22]
$F_2C=CFCH_2OH$	1.52	0.011	[23]
	2.24	0.075	[24]
$F_2C=CFC_3H_6OH$	1.04	0.008	[23]
$F_2C=CFC_3H_6OAc$	1.14	0.060	[25]
$F_2C=CFC_3H_6SAc$	1.68	0.045	[26]
$F_2C=CFCl$	1.84	0.031	[17]
	1.56	0.026	[6, 7, 14]
	1.48	0.020	[18]
$F_2C=C(CF_3)COF$	2.35	0.011	[21]
$F_2C=C(CF_3)OCOC_6H_5$	2.27	0.081	[27]
$F_2C=CFCO_2CH_3$	2.37	0.035	[28]
	1.20	0.048	[6, 13, 14]
	1.5–2.1	0.01–0.20	[29]
	2.2–2.6	0.04–0.06	[30]
$F_2C=CFOCF_3$	3.01	0.250	[31]
$F_2C=CFOC_3F_7$	2.48	0.031	[31]
$F_2C=CFO(HFP)OC_3F_7$	1.37	0.004	[32]
$F_2C=CFO-R_F-CO_2R$	1.08	0.006	[33]
$F_2C=CFO-R_F-SO_2F$	2.26	0.047	[34]
$F_2C=CFCF_3$	4.09	0.047	[6]
	1.50	0.005	[35]

TABLE I   (*continued*)

Monomer	e	Q	References
$F_2C{=}CF_2$	1.22	0.049	[29]
	1.63	0.032	[13, 35]
	1.84	0.031	[17]
$H_2C{=}CFCH{=}CH_2$	−0.42	2.100	[18]
	0.63	1.880	[6]
$F_2C{=}CF{-}CF{=}CF_2$	0.58	0.820	[6]
	4.09	0.047	[17]
	0.47	0.930	[36]
$F_2C{=}CF{-}C_6H_5$	0.95	0.380	[18]

substituent on the ethylenic carbon atom(s) ranging from fluoroalkenes to more "exotic" monomers.

2.3.   REACTION MECHANISMS

The alternating radical copolymerisation of various comonomers has been known for years, leading to various discussions (rather in disagreement) on the mechanism. However, an invariable property of different pairs of monomers yielding alternated systems concerns the complementarity between the electron donor/electron acceptor behaviour of the couple. Indeed, a highly electron donating monomer (D) should copolymerise in an alternating fashion with a highly electron accepting monomer (A), as shown in Figure 1.

As a matter of fact, three independent theories have been suggested to justify this statement.

Bartlett and Nosaki [37] proposed the first mechanism in which the two monomers form a donor–acceptor complex [DA] that has a higher reactivity than those of free monomers.

This complex can mainly be added to the growing chain. This mechanism was confirmed by several further studies, especially those of Butler [38].

Walling [39] suggested a second theory for which exist electrostatic interactions and differences of polarity between the radical end-chain and the inserting monomer. Consequently, the activation energy of the alternating propagation is lowered about that of the homopropagation.

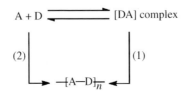

FIG. 1.   Mechanism of A–D copolymerisation.

A third mechanism was proposed by Shirota *et al.* [40] and by Tsuchida and Tomono [41]. In this case, both the free monomers and the charge transfer complex (CTC) participate in the propagation.

These three theories can be summarised as shown in Figure 1. However, many authors, especially Butler [38], discussed the mechanism of the CTC in detail.

The best evidence of the participation of the DA complex deals with the polymerisation rate which is maximum for a monomeric ratio 1 : 1 (i.e., the highest concentration is [DA] complex).

[AD] complex is indeed an equilibrium represented below:

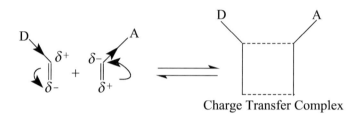

Charge Transfer Complex

These CTCs have been evidenced by various authors, either by NMR spectroscopy [42] or by UV [43].

2.4. KINETICS

Tabata and Du Plessis [44] showed that the copolymerisation rate was maximised when the copolymerisation reaction involves a stoichiometric amount of both donating and accepting monomers, as shown in Figure 2. This was demonstrated with non-fluorinated monomers [e.g. (isoprene; 2,4-dicyanobutene) and (2,4-dicyanobutene, α-methyl styrene) couples] and also for the (chlorotrifluoroethylene (CTFE); vinyl ether) couple.

Hence, various criteria are suggested to give evidence that an alternated copolymer has been obtained. However, the easiest method consists in assessing the composition of copolymers from different feed compositions, and determining at low monomer conversions the monomer composition of the copolymers. The polymer vs. monomer curve that must be obtained is represented by Figure 3. The closer the curve is to the horizontal ($F_A = 0.5$), the more alternated the copolymer (curve (3)). However, in most cases the curve is not close to the horizontal (curves (2) and (1)) and the process becomes a key-parameter to obtain (or not) alternating copolymers.

A nice example of such behaviour is given by styrene/maleïc anhydride copolymers [45] (which have already led to commercially available SMA$^{®}$ copolymers by the Total company). Actually, $Q_A = 0.86$ and $e_A = 3.69$ while $Q_D = 1.0$ and $e_D = -0.8$ (where A and D stand for maleic anhydride and styrene, respectively) leading to a curve very close to ideality (curve (3)).

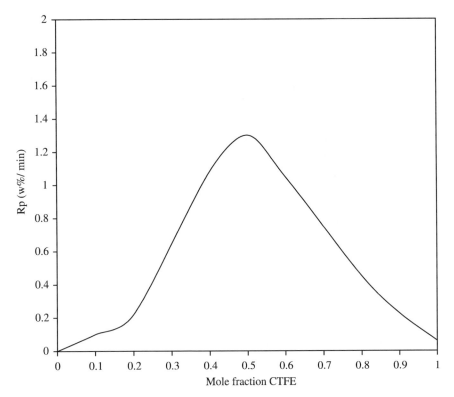

FIG. 2. Relationship between the rate of copolymerisation as a function of the CTFE mol. fraction in monomer mixture (reproduced with permission from Tabata and Du Plessis, *J. Polym. Sci.* (*Part A-1*), **9**, 3425–3435 (1971)).

However, the Total company has succeeded in marketing MA/styrene with the following mole contents: 1 : 1, 1 : 2 and 1 : 3 according to the process of copolymerisation.

This brief overview of the fluorinated alternating copolymers and of the criteria that allow one to predict the compatibility of two or more monomers ought not to belittle the tremendous scientific work which was carried out to control the structure–property relationships. Actually, as will be reported later, slight defects of chaining and/or defects in the alternance can have a marked effect on physical properties.

3. Uses and Applications of Fluorinated Monomers

3.1. INTRODUCTION

This section concerns only fluoromonomers bearing fluorinated atoms (or perfluoroalkyl substituents) directly linked to the ethylenic carbon atoms.

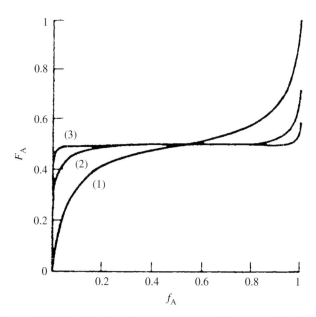

F_IG. 3. Various possibilities of polymer/monomer composition curves for alternating copoly-merisation: (1) poor tendency to the alternance; (2) tendency to the alternating copolymerisation; (3) quasi-ideal alternating copolymerisation.

This part does not consider the acrylic monomers possessing a fluorinated side chain for which a number of surveys have been carried out [46].

One of the key sources of interest in fluoromonomers arises from the presence of these fluorinated electron-withdrawing substituents (or fluorine atom(s)) that makes them good electron acceptors. Hence, these are ideal candidates for the alternating copolymerisation. The most used monomers are the commercially available fluoroalkenes such as tetrafluoroethylene (TFE), vinylidene fluoride (VDF or VF_2), CTFE, hexafluoropropylene (HFP), 3,3,3-trifluoropropene (TFP) and perfluoroalkyl vinyl ether (PAVE) for which the syntheses have been reported [47]; the less utilised ones are hexafluoroisobutylene (HFIB), $\alpha,\beta,\beta,$-trifluoroacrylate, perfluorobutadiene, bromotrifluoroethylene, trifluoroethylene, methyl α-trifluoroacrylic acid and 2,2,2-trifluoroethyl α-fluoroacrylate. To a lesser extent, other monomers have been reported in publications but have not yet been used industrially; e.g., alkyl β,β- and α,β-difluoroacrylate and acetyl- or benzoyloxypentafluoropropene ($F_2C=C(CF_3)OCOR$ with R: CH_3 or C_6H_5) [27].

However, all fluorinated monomers reported above exhibit poor reactivity with other classic non-fluorinated monomers (such as styrenic or acrylic monomers) because they exhibit very low Q value of resonance stabilisation. Hence, the choice of non-fluorinated monomers is limited to those presenting low Q values and negative e values [48].

As mentioned in chapter 1, fluorinated polymers exhibit a unique combination of high thermal stability, chemical inertness (to acids, bases, solvents and petroleum), low dielectric constants and dissipation factors, low water absorptivity, excellent weatherability and a good resistance to oxidation and ageing, low flammability and very interesting surface properties [49, 50].

Therefore, such products are involved in many applications (e.g., aerospace, aeronautics, engineering, optics, textile finishing, stone and leather treatments, microelectronics) in spite of their high price and they are experiencing an increasing market.

However, fluoroplastics exhibit various drawbacks: homopolymers are often crystalline (which cause costly processing) and have a poor solubility to classical organic solvents and this does not enable one to characterise them properly. Furthermore, homopolymers and copolymers are cured with difficulty [46].

One of the solutions to avoid these drawbacks consists in performing a copolymerisation of fluorinated monomers with either hydrogenated or fluorinated comonomers. Actually, these latter derivatives may provide complementary properties such as solubility (provided by a cyclohexyl group), facility to cure (brought by a hydroxy or epoxide function), decrease of crystallinity and also adhesion onto substrates (from carboxylic or phosphonic acids).

The copolymerisation of fluorinated with non-fluorinated monomers has led to various industrial products and are of growing interest [48]. Two main thermoplastics, CTFE/ethylene (E) and E/TFE, have been marketed, particularly by the Solvay-Solexis (formerly Ausimont) and DuPont companies under the HALAR® and TEFZEL® trade names, respectively. In addition, Allied patented CM1 composed of HFIB and VDF comonomers. As for elastomers, the Asahi Glass company markets TFE/propene (P) copolymers (under the AFLAS® trade name). From the 1980s, several companies have proposed novel coatings prepared from CTFE/VE copolymers (such as LUMIFLON® from Asahi Glass [49]), from CTFE/2-allyl oxyethanol/vinyl acetate copolymers (FLUOROBASE® from Ausimont, presently Solvay-Solexis [50]) and a tetrapolymer of VDF/CTFE/TFE/2-allyl oxyethanol from Atofina [51]. Interestingly, the Daikin company produces an original thermoplastic elastomer, chlorine-free ZEFFLE®, based on a similar concept from TFE.

3.2. FLUORINATED ALTERNATING COPOLYMERS FROM CHLOROTRIFLUOROETHYLENE (CTFE)

3.2.1. Introduction

CTFE is the most used fluoroalkene after TFE and VDF in the series of fluorinated monomers. Usually, it is synthesised from dechlorination of 1,2,2-trifluorotrichloroethane in the presence of zinc [46, 47]. This latter was one of the most known chlorofluorocarbons in the 20th century but its production was stopped a few years ago because of ozone depletion.

Indeed, CTFE exhibits a reactivity closer to that of TFE rather than to that of VDF, and can often be substituted to TFE or copolymerised with it to modify the properties of the corresponding polymers. However, the presence of the chlorine atom contributes to decreasing its thermostable performances. However, as in the hydrogenated series of PVC, this chlorine atom brings useful film-forming properties for coatings. This can explain its use in paints such as Lumiflon® marketed by the Asahi company, and for which more information will be supplied later. CTFE has been extensively studied in copolymerisation reactions to obtain alternating copolymers in general, which is the subject of this chapter.

Hence, we propose a non-exhaustive state of the art for all comonomers of CTFE and, above all, those which allow to prepare alternating copolymers. First, it is required to recall Q and e values of CTFE as: $Q_{CTFE} = 0.020, 0.026, 0.031$ and $e_{CTFE} = +1.48, +1.56$ and $+1.84$, respectively (Table I).

Consequently, it can be expected that comonomers showing low Q values, and hence poorly stabilised by resonance, and whose double bonds bear electron-donating group(s) (i.e., negative values of e), lead to alternating copolymers.

Table I takes into account the Q and e values supplied by the Polymer Handbook and by a review [46] and such comonomers can be expected to possess alkyl groups, oxygen or nitrogen atoms directly linked to the double bond. This leads to a description of three main kinds of monomer in this section:

(i) olefins and derivatives: ethylene, propylene and isobutylene (IB);
(ii) VEs and derivatives;
(iii) other monomers.

These three main kinds are reported step by step below.

3.2.2. Olefins and Derivatives

Ethylene–CTFE and vinyl ether–CTFE copolymers are the most representative examples of alternating copolymers based on CTFE.

The syntheses of the former copolymers were pioneered by the DuPont de Nemours company in 1946 [52], but Allied Chemical, in 1974, first commercialised the production of E–CTFE in the USA under the Halar® trademark.

After many economical changes well explained by Stanitis [53], E–CTFE alternating copolymer is presently produced by the Ausimont (now Solvay-Solexis) company.

Two basic publications from Montecatini (ex-Ausimont company) [54, 55] supply the interesting structural and physico-chemical properties of this copolymer.

These studied copolymers were synthesised at different temperatures ranging from -78 to $+60\,°C$ from the radical initiating BEt₃/O₂ system in 1,2-dichlorotetrafluoroethane that enabled the homogenisation of both comonomers [54, 55]. Although the authors did not report that the produced copolymer is a

TABLE II

REACTIVITY RATIOS OF CTFE INVOLVED IN RADICAL COPOLYMERISATION WITH OTHER M MONOMERS

CTFE	Comonomers: r_M (°C)	References
0.012 (0.008)	Vinyl ethyl ether: 0	[44]
0.04	Isobutylene: 0	[56]
0.06	Propylene: 0 (−35 °C)	[56]
0.24	Propylene: 0 (−78 °C)	[56]
0.01	Ethylene: 0.07 (−78 °C)	[54, 55]
0.05	Ethylene: 0.116 (−40 °C)	[54, 55]
0.0096	Ethylene: 0.172 (0 °C)	[54, 55]
0.029	Ethylene: 0.252 (60 °C)	[54, 55]
0.05	Vinyl silane: 0.20 (60 °C)	[74]
0.01	Vinyl acetate: 0.60 (60 °C)	[68]
0.04	Vinyl acetate: 0.68 (30 °C)	[69]
0.014	Vinyl acetate: 0.44 (45 °C scCO$_2$)	[70]
0.001	Styrene: 7 (60 °C)	[68]
0.005	MMA: 75 (60 °C)	[68]
0.01	VC: 2.53 (60 °C)	[72]
0.02	VDC: 17.14 (60 °C)	[72]
0.3	NVP: 0.38 (20 °C)	[71]

highly crystalline white powder, they assessed the reactivity ratios (Table II) which without ambiguity give evidence that this is a highly alternating copolymer. Indeed, $r_1 \times r_2$ product ranges between 7×10^{-3} and 7×10^{-2} whatever the reaction temperature.

It is worth noting that r_{CTFE} is close to zero and much lower than r_E. In addition, both these r_{CTFE} and r_E values increase with the reaction-temperature and show that the tendency deviates from a perfectly alternating structure when the syntheses are carried out at higher temperatures, as evidenced by the composition curves [54].

The authors used the Price equation [5] to determine the molar fractions of alternating sequences of length n ($n = 2$ to 10) in the copolymer. It is worth pointing out that the maximum mole fraction of alternating sequences of length 10 in the copolymer is shifted from 0.925 at −78 °C to 0.475 at 60 °C, confirming the significance of temperature upon the regular structure of the copolymer.

The second part of the study is devoted to physico-chemical characteristics of the copolymers. The "T_m vs. E content" curve clearly shows the influence of the structure on the properties. Indeed, the melting points (T_m) of PE and PCTFE homopolymers are 137.5 and 200–225 °C, respectively, while that of the perfectly alternating E–CTFE copolymer exhibits the highest T_m of 264 °C (when it is prepared at −78 °C) with a crystallinity rate of 60%.

In addition, it can be noted that the copolymers prepared from γ irradiation [56] or by peroxides [52, 57] exhibit lower T_m (ca. 237 °C). Hence, the initiating system has a drastic influence on the structure of the resulting copolymers, and hence on their melting temperatures.

Further studies carried out by Sibilla [57] by X-ray diffraction and various other methods (e.g., IR, NMR) complete these investigations on the structure and the physico-chemical characterisations of these copolymers.

This copolymer has found many applications such as fibres [58] and materials used in dielectrics [59] or resistant to fire [60]. The interesting chapter by Stanitis [53] also mentions many other industrial applications of E–CTFE copolymers.

As for poly(CTFE-*alt*-P) and poly(CTFE-*alt*-IB) copolymers (where P and IB stand for propylene and isobutylene, respectively), most investigations were carried out by Tabata *et al.* [56, 61].

Regarding P–CTFE copolymers, ^{19}F NMR characterisations demonstrated the alternance of the copolymer as also evidenced by the values of the corresponding reactivity ratios (Table II): $r_p = 0$ and $r_{CTFE} = 0.06$ or 0.24 (at -35 and $-78\,°C$, respectively [56]). Indeed, the signal centred at -127 ppm corresponding to the $-CF_2C\underline{F}ClCF_2CFCl$-diad observed by Bovey [62] is used as a good assessment of the deviation of the copolymer from an alternating structure.

From isobutylene, the resulting IB–CTFE copolymer is highly crystalline and alternated ($r_{IB} = 0.04$ and $r_{CTFE} = 0$) [56].

However, it is worth noting that the reactivity is lower than in the previous case which is lower than that of the E–CTFE system. Hence, the industrial choice of the (E, CTFE) pair can be clearly understood.

3.2.3. Vinyl Ethers (VEs)

The pioneering work in this area launched in early 1970s involved the radiation copolymerisation (under ^{60}Co initiation) of CTFE with ethyl vinyl ether (EVE) in liquid phase at 20 and $-78\,°C$ by Tabata *et al.* [44]. In a remarkable way, the composition curve shows the perfect alternance of this system and the assessed values of the reactivity ratios confirm such a behaviour: $r_{CTFE} = 0.012$ (or 0.008) and $r_{EVE} = 0$.

In addition, the "copolymerisation rate vs. the CTFE molar fraction" curve reaches a maximum for a stoichiometric composition (Figure 2) as does that of the intrinsic viscosity vs. the monomer composition [44].

All these data make the CTFE/VE system a very interesting basic example of perfectly alternating copolymers.

We ourselves [63, 64] found it interesting to investigate extensively the theoretical study of that reaction to analyse in a more accurate way the mechanism of achievement of these copolymers. Indeed, they may result either from the homopolymerisation of the CTC [40, 41] or from the alternating copolymerisation of free monomers [39]. Hence, the formation of the CTC was investigated by ^{19}F NMR in the course of the copolymerisation of CTFE with EVE. The evolutions of the chemical shift of the fluorine atom located in *trans* position about that of the chlorine atom for different EVE concentrations enabled us to determine the value of the equilibrium constant K between the free and linked species of both monomers from the Hanna and Ashbaugh equation [42].

Indeed, the high K value $(1.41\, 1\, mol^{-1})$ [64] tends to favour a mechanism via the CTC.

In contrast, we carried out the radical telomerisation of CTFE and 2-acetoxyethyl vinyl ether with a fluorinated mercaptan ($C_6F_{13}C_2H_4SH$) to study the structures of the two first adducts produced. Actually, two monoadducts [63, 64]:

$$C_6F_{13}C_2H_4S-CF_2CFCl-H \quad and \quad C_6F_{13}C_2H_4S-CH_2\underset{\underset{OC_2H_4OAc}{|}}{CH}-H$$

and one diadduct: $C_6F_{13}C_2H_4S-CF_2CFCl-CH_2CH_2OC_2H_4OAc$ were isolated (by distillation) and characterised.

This result clearly gives evidence that the reaction proceeds from free monomers since the expected structure achieved from the CTC of diadduct should be:

$$C_6F_{13}C_2H_4S-CF_2CFCl-\underset{\underset{O-C_2H_4OAc}{|}}{CHCH_3}$$

taking into account the normal polarisation of CTC:

$$F_2\overset{\delta+}{C}=\overset{\delta-}{C}FCl$$

$$H_2\underset{\delta-}{C}=\underset{\delta+}{C}H-OR$$

Hence, it can be suggested that the complex is a monomer reserve, that the equilibrium provides free monomers for the reaction and that the difference of polarisability explains the alternating structure.

Meanwhile, a new type of paint resin [49] consisting of a copolymer from fluoroethylene (TFE or CTFE) and VE was developed and marketed by the Asahi Glass Co. Ltd, under the Lumiflon® trade name [49, 65, 66] and very well documented in Takakura's review [67].

An excellent representation of the alternating structure of these paints is given in Figure 4 [67] showing the alternance of CTFE and VEs. Interestingly, the choice of the substituent or the function borne by the VE enables the copolymer to exhibit original properties: solubility from cyclohexyl group, crosslinking from hydroxy group (in the presence of the melamine, HMM or isocyanato group) or from epoxide, adhesion from carboxylic acid and also with the amido group for the improvement of the mechanical properties. The copolymer has an

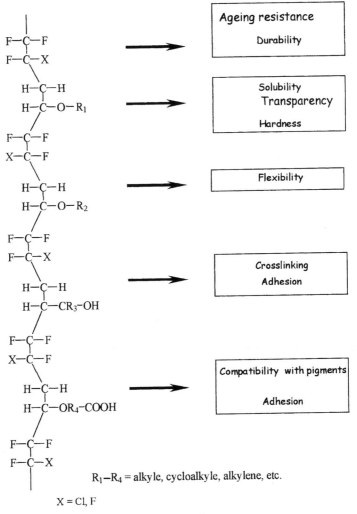

FIG. 4. Molecular structure or design of Lumiflon® (reproduced with permission from Wiley, "Modern Fluoropolymers" (1997) (Ed., Scheirs, J.) Chap. 29, p. 560).

excellent stability to ageing and has found relevant applications in paints for which its guarantee is for 20 years.

3.2.4. Other Monomers Involved in the Copolymerisation of CTFE

Table II lists several known values of r_1 and r_2 reactivity ratios of comonomers successfully used in the copolymerisation of CTFE. As expected, copolymers obtained from styrene and MMA are not alternated. The same behaviour is also observed from vinyl chloride or vinylidene chloride, although their values obtained cannot easily be explained since the stabilisation by resonance of these monomers is close to that of CTFE.

More interesting cases are those of vinyl acetate (VAc), N-vinyl pyrrolidone (NVP) and allylic monomers.

From the radical copolymerisation of CTFE with VAc, the composition curve reported by Thomas and O'Shaughnessy [68] and the assessed values of the reactivity ratios ($r_{CTFE} = 0.01$ and $r_{VAc} = 0.6$) have shown a tendency to the alternance. However, the authors have supplied few data on the curve and it should be useful to study further these first results.

Several decades later, Murray *et al.* [69] determined both reactivity ratios $r_{CTFE} = 0.04$ and $r_{VAc} = 0.68$ under emulsion conditions.

Interestingly, Baradie and Soichet [70] investigated the kinetics of copolymerisation of CTFE and VAc in supercritical CO_2 initiated by diethyl peroxydicarbonate. They assessed both reactivity ratios: $r_{CTFE} = 0.014$ and $r_{VAc} = 0.44$ at 45 °C. These Canadian authors observed that CTFE–VAc copolymers were synthesised in high yields when the CTFE mol% in the feed was lower than 50. However, at high concentrations of CTFE, the gradual consumption of VAc results in lower polymer yields and molar masses, suggesting a poor solubility of CTFE in sc CO_2.

The difference in reactivity ratios measured in CO_2 and emulsion probably reflects differences in solubility of VAc, CTFE and macroradicals in both media. As mentioned by these authors [70], VAc probably solvates the propagating macroradical chain in CO_2, thereby accounting for the higher yields and molar masses attained at higher VAc concentrations. Polymerisation in CO_2 is often considered as a precipitation polymerisation because the monomers are soluble in CO_2, but the macroradical propagating chains are only weakly soluble. Precipitation affects copolymer composition and monomer sequence distribution when accessibility of the reaction site is different for each monomer. This is probably the case in CO_2, where preferential solvation of one monomer is enhanced by differences in the polarity of the solvent and the monomers; the polar monomer is pushed by the non-polar solvent from the solution phase to the polymer phase.

From NVP [71], the results are also interesting. The tendency to the alternance is striking (from both the composition curve and the polymerisation rate vs. the monomer composition) but $r_1 \times r_2$ product is rather high (0.114) and gives evidence of defects of alternance.

The copolymers obtained exhibit interesting properties. First, in contrast to PCTFE, poly(NVP-*alt*-CTFE) is soluble in DMF and THF. Second, their thermal stability is improved compared with that of PNVP thanks to the CTFE base units.

Such characteristics are very often noted on alternated copolymers involving fluoromonomers as for E–CTFE and CTFE–VE copolymers.

Other monomers such as vinyl chloride [72], vinylidene chloride [72], styrene [68] and MMA [68] have also been studied in kinetics of copolymerisation of CTFE.

Figure 5 represents the copolymerisation curves (i.e., the CTFE content in the copolymer vs. that in feed) of all the above CTFE-monomer pairs showing the

Copolymerisation of CTFE

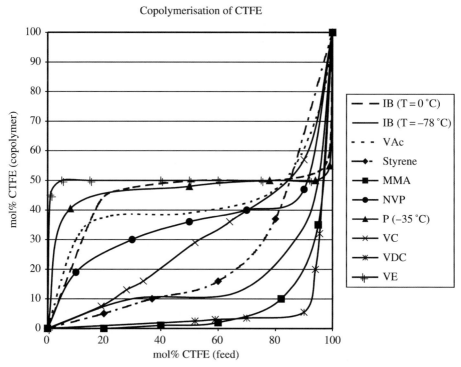

FIG. 5. Monomer–polymer composition curve of the copolymerisations of CTFE with M monomers.

expected case of alternance while others are random. Surprisingly, it is noted that all copolymerisations do not lead to any copolymer containing a CTFE amount higher than 50 mol%.

From allylic comonomers [73], few theoretical studies on the structures have been investigated. However, the analyses of the compositions of copolymers show that the resulting copolymers are not alternated, in spite of the presence of the methylene group adjacent to the double bond. This is probably due to the presence of the electron-withdrawing oxygen atom linked to this methylene group (as for $H_2C=CHCH_2-O-CF_2CF_2H$ [73]).

A particular case investigated by Pike and Bailey [74] concerns the copolymerisation of CTFE with olefins bearing $Si(OR)_x(R')_{3-x}$ groups (or vinyl siloxane, VS). These authors showed that the VS monomers were less reactive than acrylonitrile, styrene and NVP, while with monomers such as vinyl chloride, VAc and CTFE, both r_M and r_{CTFE} were smaller than one. For example, they assessed the values of the reactivity ratios of both comonomers: $r_{CTFE} = 0.05$ and $r_{VS} = 0.2$ at 60 °C, hence giving a copolymer with a high tendency to the alternance. The authors also studied the rate of polymerisation vs. the percentage of VS and they noted that the curve reaches a maximum for 25% of VS in the feed (or monomer mixture).

Finally, the structure of the VS did not appreciably alter the rate of copolymerisation and the higher the percentage of VS in the feed, the lower the molecular weights of the copolymers.

3.2.5. Conclusion

As expected, the overview of these copolymers obtained from CTFE shows that the Alfrey and Price theory [5] (the Q, e scheme) is still up to date and can be successfully applied. Hence, only comonomers with a low Q value and negative value of e can lead to alternating copolymers with CTFE. However, it is required to modulate this conclusion because certain couples are rather surprising.

For example, ethylene (E) perfectly alternates with CTFE (while e_E value is very close to zero); whereas VAc ($e_{VAc} = -0.22$) and allylic monomers ($e < 0$) do not alternate (see Table II). Hence, the theory gives a tendency but the experiment supplies the real behaviour.

Finally, it is necessary to investigate more thoroughly the copolymerisation of certain couples such as (CTFE, VAc) and (CTFE, vinyl olefins and especially fluorinated ones) for which the literature is not extensive.

3.3. COPOLYMERS OF TETRAFLUOROETHYLENE (TFE)

3.3.1. Introduction

For the most part, TFE and CTFE behave rather similarly in homopolymerisation and in copolymerisation. However, several differences have been observed, such as (i) better thermal stability of TFE copolymers and (ii) better performance of CTFE copolymers for coatings. Nevertheless, minor differences are noted in their alternating copolymers which greatly determine their suitability for specific applications.

3.3.2. Ethylene–Tetrafluoroethylene Copolymers

Copolymers of ethylene (E) and tetrafluoroethylene (TFE) have been known since 1946 [52] but E–TFE resins were only commercialised in the 1970s because the pure alternating copolymer exhibits poor resistance to cracking at elevated temperature.

In the 1970s, researchers introduced some other monomers (in 1–10 mol%) having a bulky pendent side group in order to improve the properties of E–TFE resins [75]. Examples are perfluorobutylethylene, hexafluoropropylene, perfluoropropyl vinyl ether and hexafluoroisobutylene.

E–TFE resins are manufactured by DuPont (Tefzel®), Asahi Glass (Aflon®), Daikin (Neoflon®) and Solvay-Solexis (Halon®).

Concerning the synthesis and the respective reactivities of both comonomers, important studies have been carried out by Borsini *et al.* [76]. The reactivity ratios have been assessed at various temperatures:

$$\text{at } -30\,^{\circ}\text{C}, \quad r_E = 0.10 \pm 0.02 \qquad r_{TFE} = 0.013 \pm 0.008$$

$$\text{at } +65\,^{\circ}\text{C}, \quad r_E = 0.14 \pm 0.02 \qquad r_{TFE} = 0.045 \pm 0.007$$

These results show again that an increase in the reaction temperature decreases the degree of alternance. Indeed, these Italian authors represented the molar percentage of alternating sequence vs. the TFE mol. content in feed according to the Price method [77]. They noted that a temperature effect is detectable for the concentration of $(C_2H_4)_n$ sequences in the copolymers obtained at -30 and $+65\,^{\circ}\text{C}$ (for an equimolar ratio 50/50).

For $n = 2$, the values are 1.5 and 3.3%, respectively, and the percentages of alternating sequence worth 97 and 93% for the same experiment, respectively [78].

Kostov and Nikolov [79] revisited this copolymerisation and investigated the influences of the temperature, pressure and solvent composition on the sequence distribution.

All processes of copolymerisation have been described and/or patented for obtaining these copolymers. The process in bulk is difficult (because of the exothermicity) but suspension or emulsion processes are the most commonly used. Sometimes, an inert co-solvent is added such as 1,1,2-trichloro- or 1,2,2-trifluoroethane in order to help the polymerisation that proceeds in the solvent phase. Water allows reduction of the viscosity of the medium and removal of calories. In addition, as for most polymerisations of fluoro-monomers, perfluorinated surfactants are generally used.

Finally, to obtain a high degree of alternance, it is required to start the radical copolymerisation with a feed mol. composition of 75/25 (E/TFE). Actually, most commercialised grades of E–TFE contain a slight excess of TFE to increase the properties. Crosslinking of these resins can be achieved either in a radical way (via peroxides/triallyl isocyanurate system) or by radiochemical routes [80].

The properties of E–TFE resins have been reported in various books and reviews that we do not repeat in this present book. Nevertheless, most of the information is clearly described in Kerbow's review [81].

However, it is worth mentioning a peculiar characteristic published by a Russian team [82] regarding the very high crystallinity of these copolymers and the variation of the melting point (T_m) vs. the monomer composition.

Indeed, T_m reaches a maximum value (of 313 $^{\circ}$C which is very high [76]) for the alternating composition. However, this T_m is usually lower in the industrial grades because of the slight excess of TFE $(T_m = 270^{\circ}\text{C})$. As a comparison, the melting point of PVDF, which is an isomer of E–TFE copolymer, ranges between 158 and 170 $^{\circ}$C [83] only. This difference has been explained by the carbon chain which is in a planar zig-zag orientation for the E–TFE copolymer.

Adjacent $-CF_2-$ groups are on opposite sides of the chain, followed by a similar unit for the adjacent $-CH_2-$ unit. Interpenetration of adjacent chains occurs when electronic interactions cause bulky $-CF_2-$ groups of one chain to nestle into the space above smaller $-CH_2-$ groups of an adjacent chain.

Beside the viscosity characterisation [84], conformation and packing analysis [85] and dynamical–mechanical relaxation [86] of alternated E–TFE, a very interesting comparison of the physico-chemical, mechanical and electrical properties of PVDF, E–TFE and E–CTFE has been given by Kerbow [81] in Table III.

Main applications of these resins are in wire and cable insulation. Other applications, as for those of PVDF resins, concern the tubings, sheets, piping and rod stocks obtained by extrusion.

3.3.3. Copolymerisation of TFE with Propylene

In contrast to E–TFE copolymers, which are thermoplastics, TFE–P copolymers are elastomers, marketed by the Asahi company under the Aflas® trademark. The research has essentially been carried out by Japanese teams and mainly by Tabata from the 1960s [87] up to the 1980s. These elastomers have been discovered thanks to his extensive investigations on radiation-induced copolymerisation. A very interesting chronology of these researches is reported in the article of Machi et al. [88], which reviews these overall works carried out in bulk [88, 89], then in liquid phase [88, 90] and finally in emulsion [88, 91].

These are alternating copolymers [89] and the authors assessed their reactivity ratios: $r_P = 0.10$ and $r_{TFE} = 0.01$ at $-23\,°C$ that were confirmed in 1992 by Kostov and Petrov [92] at higher temperatures (50–$90\,°C$). These authors

TABLE III

COMPARISON OF VARIOUS PROPERTIES OF ETFE, PVDF AND ECTFE COPOLYMERS

Property	Test method	Units	E–TFE	PVDF	E–CTFE
Melting point	DSC	°C	225–270	154–184	236–246
Specific gravity					
Melt			1.3		
Solid			1.72	1.75–1.78	1.7
Tensile strength (23 °C)	D 638	MPa	38–48	36–56	50–65
Elongation (23 °C)	D 638	%	100–350	25–500	150–250
Tensile modulus (23 °C)	D 638	MPa	830	1,340–2,000	
Izod impact strength (23 °C)	D 256	J m^{-1}	No break	160–530	No break
Coefficient of thermal expansion	D 696		9×10^{-5}	$\sim 10^{-4}$	5×10^{-5}
Critical shear rate		s^{-1}	200–10,000		
Processing temperature range		°C	300–345	200–300	260–300
Dielectric constant (1 kHz)	D 150	ε	2.6	7.5–13.2	2.6
Dielectric strength	D 149	kV mm^{-1}	59	260–950	80–90
Dissipation factor (1 kHz)	D 150		0.0008	0.013–0.019	0.0024
Limiting oxygen index	D 2863		30	44	64

Source: Reproduced with permission from Modern Fluoropolymers (Ed., Scheirs, J.) Chapt. 15, p. 305.

initiated bulk copolymerisation by *tert*-butylperoxy benzoate. They deduced Q and e values: $Q_{TFE} = 0.002$; $e_{TFE} = 1.7$ and $Q_P = 0.002$; $e_P = -0.78$, close to those assessed by Naberezhnykh *et al.* [17] and Moggi *et al.* [35] (Table I). Extensive kinetic studies were also investigated and the Bulgarian authors obtained a value of the transfer to the polymer $C_p = 5 \times 10^{-4}$, which is logically assigned to the propylene sequence.

Physical characteristics of the elastomers obtained by this initiation process are similar to those achieved from radiochemical initiation.

In addition, the semi-batch processes describing this type of reaction have been mentioned by Machi *et al.* [88] and these authors have supplied a reaction mechanism starting from an initiation by ˙OH (arising from the action of γ rays onto water).

To enhance the properties of TFE–P copolymers (by an increase of the molecular weights), crosslinking of these macromolecules was accomplished in the presence of 2,5-dimethyl-2,5-di(*t*-butyl peroxy)hexane [88].

Table IV lists several physico-chemical characteristics of Aflas® copolymers (M_n ca. 10^5).

The stability of these fluoroelastomers in acid and alkali media is excellent and even better than that of poly(VDF-*co*-HFP) copolymers [88].

These copolymers have a solubility parameter of 9.5 (or 2×9.5 according to the standards used) which corresponds to a maximum swelling in THF.

Finally, interesting ^{19}F and ^{1}H NMR spectroscopic studies have been carried out by Tabata's group [93] and later in our Laboratory [94]. From different copolymers of variable compositions (achieved out of the alternating conditions), the chemical shifts of each group were attributed, and the authors recalculated the reactivity ratios (according to the Markovian statistic). From the ^{19}F NMR characterisation, the values determined were found close to those assessed above.

More recently, the introduction of a cure site monomer was attempted [95], leading to an original hydroxy side-group containing TFE/E. Although the thermal stability was somewhat affected, the T_gs (-2 to $0\,°C$) were acceptable.

All these investigations clearly show that these copolymers are highly alternating when the copolymerisations have been realised at low temperature;

TABLE IV

PHYSICO-CHEMICAL CHARACTERISTICS OF TFE–P COPOLYMERS

Specific gravity^a: 1.49
$T_g{}^a$: -8 to $-5\,°C$
$T_{dec}{}^a$: $360–450\,°C$
Tensile strength: $130–250\,kg\,cm^{-2}$
Elongation at break: $310–400\%$
Modulus of elasticity at 100% elongation: $24–75\,kg\,cm^{-2}$
Volume resistivity: $10^{17}\,\Omega\,cm^{-1}$

^{a} Uncrosslinked polymer.

however, when the temperature is increased, a non-negligible deviation to the alternance was observed.

3.3.4. Copolymerisation of TFE with Isobutylene (IB)

Radiation-induced copolymerisation of TFE with IB in the liquid state at $-78\,°C$ leads to perfectly alternated copolymers, the structure of which is [96]:

$$-(CF_2-CF_2-CH_2-\underset{\underset{CH_3}{|}}{\overset{\overset{CH_3}{|}}{C}})_n-$$

The copolymer is crystalline and its melting point is about $200\,°C$, which is close to the decomposition temperature when the molecular weight is low. However, the thermal stability can be increased to $300-350\,°C$ for copolymers having high molecular weights. In an investigation which was devoted to the kinetics of copolymerisation of these monomers, Tabata et al. [97] assessed both their reactivity ratio: $r_{TFE} = 0$ and $r_{IB} = 0.2$ and their Q and e parameters:

$$Q_{TFE} = 0.049 \text{ and } Q_{IB} = 0.033$$

$$e_{TFE} = 1.22 \text{ and } e_{IB} = -0.96$$

It is also worth citing a patent deposited by the DuPont company [98] that reports the terpolymerisation of TFE, IB and a hydrophilic monomer (mainly acrylic acid, and also vinyl phosphonic acid). This patent claims comonomer contents up to 5% but in many cases amounts lower than 1% have been indicated. In spite of its application interest, this work is not convincing. However, from these results, it is recalled that if alternating copolymers are obtained, they are highly crystalline and this is rather peculiar when they are compared to copolymers arising from other olefins.

3.3.5. Copolymers of TFE with Vinyl Ethers (VEs)

Hikita et al. [99] accomplished a pertinent investigation on the copolymerisation of TFE with VEs initiated by γ rays. They could establish a decreasing reactivity series of various VE used: Ethyl VE (80) > Methyl VE (42) ~ n-Butyl VE (38) > i-Butyl VE (21) (including their respective yields obtained in the same experimental conditions). Their structure can be represented by:

$$-(CF_2-CF_2-CH_2-\underset{\underset{OR}{|}}{CH})_n-$$

These Japanese authors then investigated their thermal properties as listed in Table V.

TABLE V

GLASS TRANSITION TEMPERATURE (T_g) AND DECOMPOSITION TEMPERATURE (T_{dec}) OF VARIOUS
COPOLYMERS OF TETRAFLUOROETHYLENE AND VINYL ETHER

Comonomer	T_g (°C)	T_{dec} (°C)	
n-Butyl VE	−28	300 (N₂)	230 (air)
i-Butyl VE	−12	–	225
Ethyl VE	−12	–	220
Methyl VE	−12	–	245

Finally, the kinetics of radiation-induced copolymerisation achieved at 27 °C led to the reactivity ratios: $r_{TFE} = 0.005$ and $r_{n\text{-Bu VE}} = 0.0015$ [99] (Table VI). Such values are in excellent agreement with an alternating copolymerisation and these authors brought quite convincing ^{1}H and ^{19}F NMR spectroscopic evidence by comparing the chemical shifts of the characteristic group in the structure of the copolymers with those of the corresponding homopolymers.

More recently, Feiring et al. [100] have studied this same reaction in organic solvents involving VEs bearing bulky (tricyclic and bicyclic) groups. Using VEs containing an adamantyl and a norbonyl group, these American authors obtained alternating copolymers in good yields, soluble in common organic solvents and endowed with high T_gs (ca. 150 °C) which are close to those of the corresponding acrylates.

Furthermore, the decomposition temperatures are very high (400 °C under nitrogen).

Hence, the radical copolymerisation of TFE with VE shows the same behaviour (alternance) as that involving CTFE (Section 3.2.3) and this explains that many companies use TFE and CTFE mixtures and also HFP in copolymerisations with VEs.

3.3.6. Copolymerisation of TFE with Vinyl Acetate (VAc)

The radical copolymerisation of TFE with VAc has been investigated by several authors since there is an emerging interest in obtaining hydrophobic fluoropolymers bearing hydrophilic hydroxyl side groups.

Indeed, the most interesting search applications concern the obtaining of barrier-polymers with an increase of this property/application in aqueous or at least humid medium.

In 1967, Modena et al. [101] carried out this reaction in emulsion at 50 °C initiated by persulfates. They concluded that the copolymer is statistic and that VAc is more reactive than TFE. In contrast, these Italian authors obtained polymers of high molecular weights ($1–2 \times 10^6$ g mol^{-1}) and, for TFE content

TABLE VI

REACTIVITY RATIOS OF TETRAFLUOROETHYLENE (TFE) INVOLVED IN RADICAL COPOLYMERISATION WITH VARIOUS M MONOMERS

M	r_{TFE}	r_M	References
$H_2C=CH_2$	0.10	0.15	[105]
	0.07	0.52	[79]
$H_3CCH=CH_2$	0.06	1.00	[89]
	0.01	0.17	[106]
$CH_3CO_2-CH=CH_2$	−0.009	0.95	[70]
$F_3C-CH=CH_2$	0.12	5.0	[107]
$C_3F_7CH=CH_2$	0.21	2.3	[107, 108]
$FCH=CH_2$	0.06	0.30	[17]
$tr-HFC=CHCF_3$	22	0.18	[107]
$F_2C=CH_2$	3.73	0.23	[109]
	0.28	0.32	[17]
$(F_3C)_2C=CH_2$	0.58	0.09	[107]
$F_2C=CHF$	1.14	0.46	[17]
$F_3CCF=CH_2$	0.37	5.4	[107]
$F_2C=CFCl$	0.75	1.04	[110]
	0.80	1.10	[17]
	1.0	1.0	[18]
$F_2C=CFBr$	0.82	0.24	[110]
$F_2C=CFCH_2OH$	2.47	0.41	[23]
$F_2C=CFC_3H_6OH$	1.57	0.45	[23]
$F_2C=CFC_3H_6OAc$	0.18	0.20	[111]
$F_2C=CFCO_2CH_3$	5.0	0.10	[112]
	3.0	0.15	[29]
$F_2C=CFCF_3$	15	0	[113]
	65	0	[17]
	3.5	0	[114]
$F_2C=CFCF_2OR_FSO_2F$	10	0.3	[115]
$F_2C=CFOCF_3$	1.73	0.09	[116]
$F_2C=CFOC_3F_7$	8.72	0.06	[116]
$F_2C=CFOCF_2CF(CF_3)OC_3F_7$	15.60	0.02	[116]
$F_2C=CFO-R_F-CO_2R$	7.0	0.14	[33, 117]
$F_2C=CFO-R_F-SO_2F$	8.0	0.08	[117, 118]

ranging between 2 and 15%, the resulting copolymers were soluble in organic solvents.

Later, Oxenrider *et al.* [102] revisited this reaction by keeping constant the TFE/VAc molar ratio and they obtained copolymers practically alternated (by monitoring the fluorine content in the copolymers). For that, they found an original process involving a hydrosoluble cosolvent in order to swell the copolymer particles and to allow the copolymerisation in bulk. After hydrolysis, the hydroxyl fluorocopolymer exhibited high crystallinity with a melting point of 205 °C.

Further results were published by Lousenberg *et al.* [103], who carried out such a copolymerisation in supercritical CO_2 and the same group [70] also assessed the reactivity ratios of both comonomers: $r_{TFE} = -0.009$ and $r_{VAc} = 0.95$ at 45 °C in sc CO_2.

To our knowledge, the most recent study was achieved by Feiring *et al.* [100]. As in the case of VEs, these authors used vinyl esters bearing norbornyl and adamantyl bulky groups. The resulting copolymers were not totally alternating but the TFE mol% in the copolymers were ranging between 42 and 47%, which is close to the alternating structure. As above, the T_gs were high (103–149 °C, respectively) and the thermal stability was acceptable ($T_{dec} = 340°C$ which is however, ca. 60 °C lower than that of the copolymers containing VE). These American authors claimed potential applications of the copolymers in coatings or optics such as advanced photoresists for microlithography [104].

In conclusion, from these studies, in contrast to VE, VAc does not lead to alternating copolymers although, in certain cases, the TFE content reached 47 mol%, which is not too far from the alternance.

With regard to the kinetic point of view, it is worth listing the reactivity ratios [17, 70, 79, 89, 107–118] of TFE and its comonomers assessed from their copolymerisations (Table VI).

3.3.7. Application of the Alternating Copolymerisation for Obtaining Particular Tetrapolymers

Kostov *et al.* [119] smartly applied their knowledge to synthesise original materials for membrane applications. Hence, they used an equimolar mixture composed of acceptor monomers (e.g., TFE and $F_2C=CFCO_2CH_3$ (MTFA)) with an ethylenic/propylene blend since these comonomers are well known to lead to alternated copolymers.

Obviously, the fluorinated ester was used as an interesting precursor of hard carboxylic acid because of the environment of this acid function. Moreover, this method allows better dispersal of the hydrophilic side groups (i.e., CO_2H) about the polymeric backbone and improvement in the processability and the mechanical performances of the resulting cation exchange membranes.

The acid content ranged from 0 to 50% and, when propylene was used as the only donor monomer, the T_g varied from −5 to 21 °C, whereas the decomposition temperatures of these polymers (assessed under nitrogen) decreased from 407 to 362 °C.

Interestingly, an alternated copolymer of MTFA and propylene was also achieved. Its formula is:

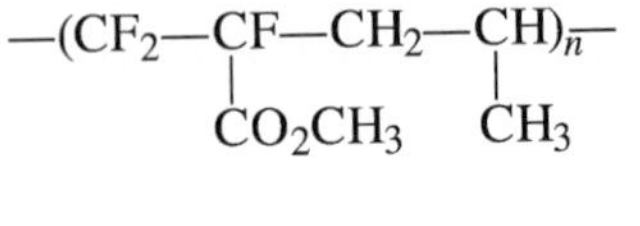

$$-(CF_2-\underset{\underset{CO_2CH_3}{|}}{CF}-CH_2-\underset{\underset{CH_3}{|}}{CH})_{\overline{n}}-$$

Finally, these researchers also prepared tetrapolymers and assessed their physico-chemical and thermal properties before and after the hydrolysis.

Such a method which associates various acceptor and donor monomers enables one to introduce specific and peculiar chemical functions onto the polymeric backbones.

3.3.8. Conclusion

The obtained results are in good agreement with expectation and they show a striking similarity to the results of the radical copolymerisation of TFE and of CTFE. However, to the best of our knowledge, no article reports any synthesis of alternating copolymers containing HFP, yet this monomer bears electron-withdrawing groups such as those borne by TFE or CTFE.

3.4. ALTERNATED FLUOROCOPOLYMERS CONTAINING VINYLIDENE FLUORIDE (VDF)

Although VDF does not react with hydrogenated VEs such as CTFE [67], 1,1-dichlorodifluoroethylene [120], perfluoroacrylonitrile [121], TFE [99, 100] or methyltrifluoroacrylate (MTFA) [122] do, it is able to produce alternating copolymers with three monomers so far: these are 3,3,3-trifluoro-2-trifluoro-methylpropene (or hexafluoroisobutylene, HFIB) [123–126], MTFA [112, 127] and α-(trifluoromethyl) acrylic acid [128]. Unexpectedly, such copolymerisation did not occur via a CTC, in contrast to those of VE with fluoromonomers as reported above.

3.4.1. Hexafluoroisobutylene (HFIB)

The emulsion copolymerisation of HFIB with VDF led to an alternated copolymer [123–126] with the following structure:

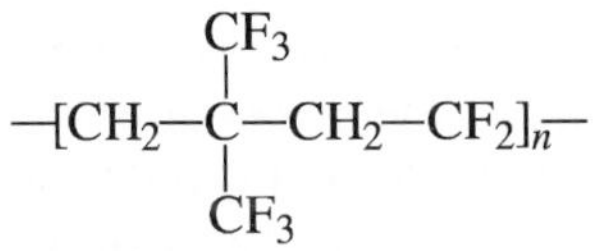

$$-[CH_2-\underset{\underset{CF_3}{|}}{\overset{\overset{CF_3}{|}}{C}}-CH_2-CF_2]_n-$$

This alternance, as with the sense of addition, is still difficult to explain from classic theories of the electronic delocalisation of fluoromonomers and from known rules of alternance.

Nevertheless, the fact is both these comonomers led to alternating copolymers.

Moreover, the exceptional properties of these copolymers can be explained by the protection displayed by both CF_3 side groups onto both adjacent methylenes. Interestingly, such a copolymer has already been industrially produced by Allied Chemical Corporation under the CM1$^{®}$ trademark, but its production seems to have stopped because of the toxicity of HFIB. A relevant article which compares its properties with those of PVDF and PTFE was published by Minhas and Petrucelli [124]. This alternating copolymer is highly crystalline ($T_m = 327\,^\circ C$) and its T_g is 132 °C [125].

It is poorly soluble in organic common solvents but, in contrast to PTFE, it can be melt-processed. Its thermal and chemical stabilities are in the same order as those of PTFE and its critical surface tension is 19.3 dyne cm^{-1} (18.8 and 31.2 dyne cm^{-1} for PTFE and poly(E-alt-CTFE), respectively). Undoubtedly, this is an exceptional copolymer.

3.4.2. Methyltrifluoroacrylate (MTFA)

MTFA does not homopolymerise [13] under radical conditions. However, the copolymerisation of MTFA with VDF was successful, as pioneered by Watanabe [112], although his team performed one experiment only. Recently, its kinetics of copolymerisation were achieved [127], leading to the assessment of the reactivity ratios of both comonomers. Its behaviour has a tendency to the alternation. The contents of both comonomers in the resulting copolymers were assessed by ^{19}F NMR. We found $r_{VDF} = 0.30 \pm 0.03$ and $r_{MTFA} = 0$ at 50 °C when $tert$-butyl peroxypivalate was chosen as the initiator (Table VII). This surprising alternance (arising from two electron-withdrawing monomers) can be explained by the steric hindrance of the ester group in MTFA.

3.4.3. α-(Trifluoromethyl) Acrylic Acid (TFMAA)

This monomer is marketed by the Tosoh F-Tech company.

As for the above fluoromonomer, TFMAA homopolymerises poorly but it copolymerises with VDF with a tendency to the alternation. The kinetics of copolymerisation enabled us to determine both reactivity ratios [128]: $r_{VDF} = 0.33 \pm 0.03$ and $r_{TFMAA} = 0$ at 50 °C, showing a tendency to the alternance. As in the case of HFIB, the reaction was very fast, accompanied by a strong exotherm, and led to high yields. Again, the surprising alternance may come from the steric hindrance of both bulky CO_2H and CF_3 groups in TFMAA.

Table VII lists the reactivity ratios of both VDF and these comonomers compared with those of other fluorinated comonomers, ethylene [98] and VAc [70, 129, 130]. Figure 6 shows the tendency of the copolymerisation of various pairs of comonomers. The polymer/monomer composition curves can be classified into four main groups: (i) that for which $r_{VDF} > 1$ and $r_M < 1$ (case when M is HFP, PAVE, pentafluoropropene and $C_6F_{13}CH=CF_2$), (ii) that for which $r_{VDF} < 1$ and $r_M > 1$ (case of ethylene, VAc, CTFE, TFE, BrTFE and $F_2C=CF(CH_2)_3-XAc$ (X = S or O)), (iii) the case where $r_{VDF} < 1$ and $r_M < 1$ (e.g., $F_2C=CFCH_2OH$, $F_2C=C(CF_3)OCOR$) and, finally, (iv) the case where alternation occurs ($r_{VDF} \sim 0$ and $r_M \sim 0$ for HFIB, MTFA and TFMAA).

TABLE VII

REACTIVITY RATIOS OF VINYLIDENE FLUORIDE (VDF) INVOLVED IN RADICAL COPOLYMERISATION WITH VARIOUS M MONOMERS

Monomer B	r_A	r_B	$r_A r_B$	$1/r_A$	References
$H_2C=CH_2$	0.05	8.5	0.42	20.00	[105]
$H_2C=CHOCOCH_3$	−0.40	1.67	−0.67	−2.5	[70]
	0.50	2.0	1.00	2.0	[130]
$H_2C=C(CF_3)CO_2H$	0.33	0	0	3.03	[128]
$FCH=CH_2$	0.17	4.2–5.5	0.71–0.94	5.88	[131]
	0.20–0.43	3.8–4.9	0.76–2.11	2.33–5.00	[19]
$H_2C=CFCF_2OR_F$	0.38	2.41	0.92	2.63	[20]
$F_2C=CFH$	0.70	0.50	0.35	1.43	[132]
$F_2C=CHCF_3$	9.0	0.06	0.54	0.11	[22]
$F_2C=CHC_6F_{13}$	12.0	0.90	10.80	0.08	[133]
$CFCl=CF_2$	0.73	0.75	0.55	1.37	[109]
	0.17	0.52	0.09	5.88	[134]
$CFBr=CF_2$	0.43	1.46	0.63	2.33	[109]
$CF_2=CF_2$	0.23	3.73	0.86	4.35	[109, 113]
	0.32	0.28	0.09	3.13	[17]
$CF_3-CF=CF_2$	6.70	0	0	0.15	[35]
	2.45	0	0	0.40	[113]
	2.90	0.12	0.35	0.34	[46]
$F_2C=CFOCF_3$	3.40	0	0	0.29	[31]
$F_2C=CFOC_3F_7$	1.15	0	0	0.86	[31]
$F_2C=CFO(HFP)OC_2F_4SO_2F$	0.57	0.07	0.04	1.75	[34]
$CF_2=CFCH_2OH$	0.83	0.11	0.09	1.02	[24]
$CF_2=CF(CH_2)_2Br$	0.96	0.09	0.09	1.00	[130]
$CF_2=CF(CH_2)_3OAc$	0.17	3.26	0.59	5.56	[25]
$F_2C=CF(CH_2)_3SAc$	0.60	0.41	0.25	4.07	[26]
$CF_2=CFCO_2CH_3$	0.30	0	0	3.33	[127]
$F_2C=C(CF_3)COF$	7.60	0.02	0.15	0.13	[21]
$F_2C=C(CF_3)OCOC_6H_5$	0.77	0.11	0.08	1.30	[27]

3.4.4. Conclusion

VDF is known to copolymerise randomly with other monomers. Among the most well known commercially available fluorinated monomers, VDF seems rather atypical since it has led to the least alternating copolymers and it does not induce any acceptor–donor copolymerisation. This behaviour can be explained by the number of electron-withdrawing atoms, which is smaller than those of TFE and CTFE. However, the most surprising results arise from three pairs of monomers: (VDF, $H_2C=C(CF_3)_2$), (VDF, $F_2C=CFCO_2CH_3$) and (VDF, $H_2C=C(CF_3)CO_2H$), which lead to alternating copolymers. Such behaviour is explained with difficulty from the polarity of both comonomers of each pair, yet the explanation of these structures can be given by the presence of bulky groups in the comonomers.

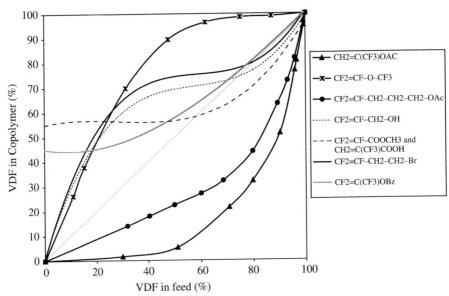

F$_{IG.}$ 6. Monomer–polymer composition curve of the copolymerisations of VDF with various
M monomers.

3.5. S$_{YNTHESIS}$, P$_{ROPERTIES}$ and A$_{PPLICATIONS}$ of F$_{LUORINATED}$ A$_{LTERNATED}$ C$_{OPOLYMERS}$ which D$_{O}$ not C$_{ONTAIN}$ VDF, CTFE and TFE

3.5.1. Introduction

Apart from TFE, CTFE and VDF, several fluoromonomers have attracted the
interest of various researchers who studied their alternating copolymerisations
with other comonomers. This section gives a non-exhaustive list of examples
involving a chlorinated monomer ($F_2C = CCl_2$) or 3,3,3-trifluoropropene, which
have already been used in many telomerisation reactions (Chapter 1): hexafluoroisobutylene, conjugated 1,3-perfluorobutadiene, maleic-type monomers,
α-(trifluoro) methacrylic acid derivatives, α-fluoro acrylates and those containing
an aromatic ring.

3.5.2. 1,1-Dichloro-2,2-difluoroethylene

This monomer has not been well investigated either from the academic or the
industrial point of view. It can be synthesised by dehalogenation of the
corresponding chlorofluoroethane:

$$ClCF_2CCl_3 \xrightarrow[Ac_2O]{Zn} F_2C{=}CCl_2$$

This alkene also contains CFCl=CFCl as a by-product (even up to 20%) arising from the isomer of the precursor alkane.

In the 1950s, various investigations carried out by McBee *et al.* [135] showed that this monomer copolymerised with VAc, styrene and 1,3-butadiene but in a non-alternating way.

Forty-five years later, the copolymerisation of $F_2C=CCl_2$ with VE was studied in our Laboratory [120]. Whatever the feed composition, by elemental analysis of chlorine and fluorine atoms the obtained copolymer exhibited an alternating structure. Furthermore, the yields obtained from various feeding ratios were close to the theoretical ones that corresponded to the stoichiometric composition. The characterisation of the CTC by ^{19}F NMR showed that the equilibrium constant was $0.056\,1\,m^{-1}$ at 20 °C, while that arising from the copolymerisation of CTFE with VE was $1.4\,1\,m^{-1}$ [64].

At 70 °C, this value dropped to $0.018\,1\,m^{-1}$ (ca. one third of the former value) and gives evidence that a low amount of monomers has been involved in the CTC at the reaction temperature. This has been well confirmed by spectroscopical studies of CTFE/VE cotelomers, which gives proof that the mechanism is closer to a mechanism by free monomers than to the CTC.

Nevertheless, such $CF_2=CCl_2$/VE copolymers are highly alternated.

3.5.3. 3,3,3-Trifluoropropene (TFP)

Although this monomer has been used in telomerisation (leading to low molecular weight telomer—see Chapter 1), it has only rarely been used in polymerisation since it does not homopolymerise. However, it is worth noting that it is an industrial reactant for the synthesis of fluorinated silicones (e.g., poly(methyl-3,3,3-trifluoropropyl siloxane), marketed by the Dow Corning company under the Silastic® trade name). The first step is a hydrosilylation reaction [136]:

$$\equiv Si-H + H_2C=CHCF_3 \xrightarrow{\text{``Pt''}} \equiv Si-C_2H_4-CF_3$$

As a matter of fact, this fluoroalkene has been copolymerised with various other comonomers. Tabata's group [137], as experts in the (co)polymerisation initiated by γ rays, investigated its copolymerisation with two different monomers at 20 °C: itaconic acid (IA) and EVE whose formulae are the following:

$$H_2C=C\begin{smallmatrix}\diagup CH_2CO_2H\\[4pt]\diagdown CO_2H\end{smallmatrix} \quad \text{and} \quad H_2C=CH-O-C_2H_5$$

In the former case, the obtained copolymer (recovered as a powder) is not alternated, as shown by the values of the reactivity ratios: $r_{TFP} = 0.55$ and $r_{IA} = 13.0$.

In contrast, the copolymerisation with the latter monomer led to a white grease copolymer and the authors noted a high tendency to the alternance: ($r_{TFP} = 0.3$ and $r_{EVE} = 0.2$) as shown by the copolymer/monomer composition curve and the copolymerisation rate vs. the feed composition [137].

However, this alternating behaviour is less striking than that noted in the copolymerisation with TFE.

3.5.4. Hexafluoroisobutylene (HFIB)

As mentioned in Section 3.4.1, HFIB copolymerises with VDF alternatively. Although highly toxic, this monomer is used to provide other alternating copolymers since it possesses two CF_3 groups borne by the same ethylenic carbon atom. This enables the electronic density of the double bond to be greatly decreased. Hence, it can be expected that the copolymerisation of HFIB with a donating monomer tends to be alternated. In this sub-section, only VAc as comonomer has been chosen.

Extensive investigations on the microstructure of this copolymer were achieved by NMR spectroscopy. The ^{13}C NMR spectra of β carbon atoms in HFIB and VAc gave chemical shifts of 127.5 and 96.2 ppm, respectively [138]. Because of the high difference of these values, the theories of Hatada *et al.* [139] and Herman and Teyssie [140] allow one to predict a high tendency towards the alternance. This is confirmed by the diagram of composition which shows a perfect parallel line to that of the 50 mol% composition of HFIB in the copolymer, whatever the feed composition.

3Furthermore, in contrast to VAc, HFIB does not homopolymerise. The authors determined the probability of VAc chaining and the curves show that, up to compositions higher than 80 mol% in VAc, it is very low. Their study also concerned the assessment of reactivity ratios by using either the terminal or the penultimate model, and they obtained the following results: $r_{11} = 0.0066$ and $r_{21} = 0.0405$; $r_{22} = r_{12} = 0$, where 1 and 2 represent VAc and HFIB, respectively.

With the terminal model, Wu *et al.* [138] obtained $r_2 = 0$ and $r_1 = 0.0214$. After extensive discussion, these authors confirmed that this pair is an excellent example of application of the penultimate model, mainly because of steric considerations.

Interestingly, the authors also investigated the chains conformation from different nuclei-type NMR. They noted that the copolymer is alternating at a "certain degree of stereoregularity" that is related to the polymerisation temperature.

The thermal properties of this copolymer before and after hydrolysis of acetate groups were studied; T_g values ranged between 44 and 63 °C vs. the molecular weights (of 78,000 and 136,000, respectively). Interestingly, the hydrolysis process enabled the T_g to be increased by almost 25 °C.

In addition, these copolymers were totally amorphous and exhibited a high film-forming behaviour. A theoretical determination of the T_g of the hypothetical poly(HFIB) led to values ranging between 80 and 90 °C [138], which is in disagreement with values obtained by the same calculation from VDF/HFIB copolymers (T_g PHFIB $= 197$ °C), as mentioned in the part above [123, 124].

3.5.5. Perfluoro-1,3-butadiene

Perfluoro-1,3-butadiene is an electron-poor monomer which has been known for decades and whose most common synthesis is the following, first pioneered by Haszeldine [141] and then revisited by Bargigia et al. [142]:

$$ICl + F_2C=CFCl \longrightarrow ClCF_2CFClI \xrightarrow{coupling} ClCF_2CFCl-CFCl-CF_2Cl$$

$$Zn \Big| Ac_2O$$

$$CF_2=CF-CF=CF_2$$

Zurkova et al. [143] studied its radical copolymerisation with various monomers. With α-olefins (e.g., ethylene, propylene and isobutylene), these Czech authors noted that the obtained copolymers were not alternating. In contrast, from VEs, alternated copolymers were produced as evidenced by the elemental analysis and the "polymerisation rate vs. feed composition" diagram that reaches a maximum in a classic way for a stoichiometric composition. The characterisation of the microstructure of these copolymers by ^{19}F NMR shows that about 75% is 1,4-dienic units while ca. 25% is a 1,2-perfluorobutadiene unit, which is close to the percentages obtained when 1,3-butadiene is polymerised under radical conditions.

3.5.6. Maleic-Type Monomers

These monomers are obviously in competition with the fluoromonomers because of the electron-withdrawing carbonyl groups adjacent to the double bond. As a matter of fact, two kinds of such a type of monomer bearing fluorine atom(s) or fluorinated group(s) are available: (i) those bearing these groups in the lateral chain and (ii) those in which the fluorine atom(s) is(are) linked to the ethylenic carbons.

3.5.6.1. Maleic Monomers Bearing Side Fluorinated Groups

These monomers are maleimides or bismaleimides of aromatic compounds substituted by fluorinated atoms.

For example, Searle [144, 145] synthesised N-pentafluoromaleimides while Barrales Rienda et al. [146] improved the procedure.

In addition, Cummings and Lynch [147] could obtain fluorinated bismalei-mides such as:

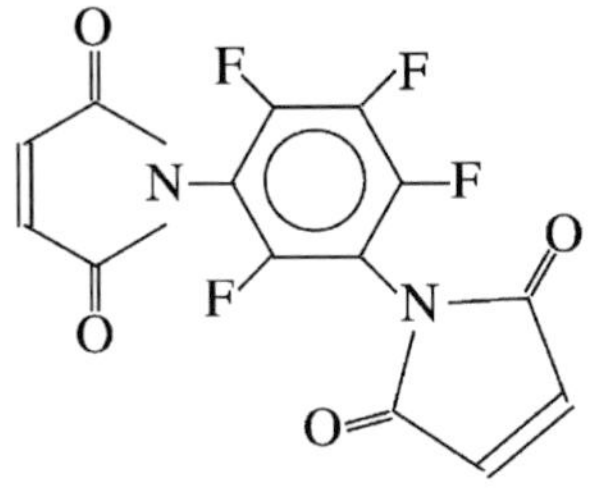

These monomers are known for leading to alternating copolymers with VEs. Very interesting surveys were carried out by Wagener [148] and the obtained copolymers exhibit very good properties at high temperatures.

In addition monomaleimides of structure:

$$X = O \text{ or } S$$

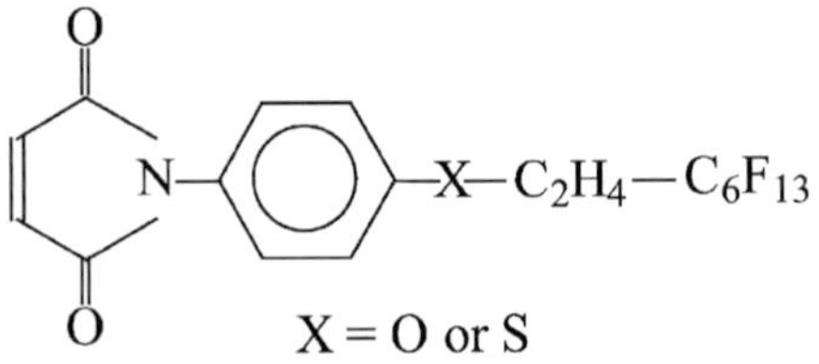

were synthesised in our Laboratory [149].

Their copolymerisations with various VEs (such as $H_2C{=}CH{-}O{-}R$, with R: $CH_2{-}C_6H_5$ and CH_2CH_2Cl) were investigated to check whether the resulting copolymers were alternated by varying the initial monomeric amounts.

This research clearly demonstrated that the alternance was not reached since the maleimides mainly incorporated the copolymers. In addition, the chlorinated VE is more reactive than its aromatic homologues. However, the thermal stabilities of these copolymers are rather high. They start to decompose from 270 °C and the T_gs ranged between 140 and 175 °C when the fluorinated maleimide/VE ratios were varied from 1.1 and 2.8.

3.5.6.2. Maleimides Bearing Fluorine Atoms in the Double Bond

Canessa [150] investigated the copolymerisation of phenyl vinyl ether with:

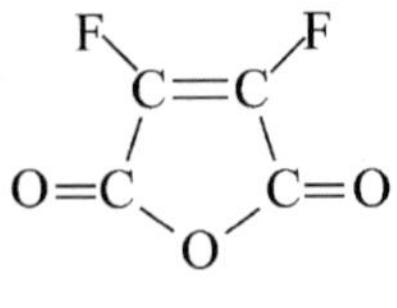

and the corresponding esters.

This author obtained alternated copolymers with rather high molecular weights (up to 70,000), the structures of which were characterised by ^{19}F NMR.

In the same way, Green *et al.* [151] studied the fluorinated aromatic bismaleimides of structure:

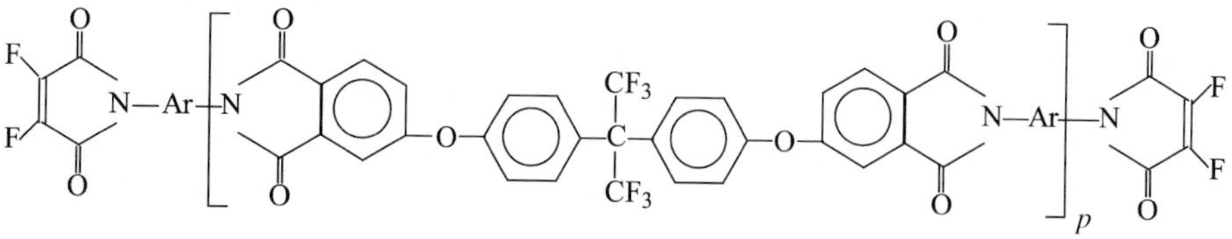

where $p > 3$ and Ar represents an aromatic group.

In 1997, additional information was published by Laschewsky's group [152], who synthesised alternating copolymers from the radical copolymerisation of partially fluorinated maleimides with VEs or with styrene derivatives. The resulting copolymers were very stable against thermal degradation in comparison with conventional vinyl and vinylidene polymers, and they had high T_gs.

3.5.7. Copolymers from Maleates and Maleic Anhydride

The copolymerisation of maleic anhydride with fluorinated (meth)allylic ethers was carried out by Pittman and Wasley [153]:

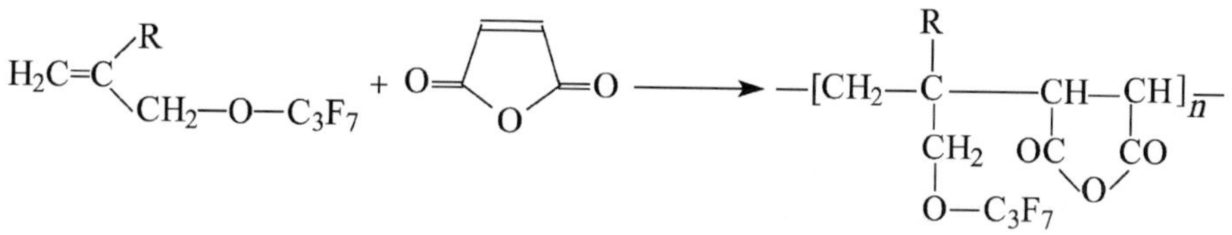

When $R = CH_3$, the obtained copolymers were perfectly alternated and exhibited high molecular weights (M_w ca. 500,000), in contrast to the results arising from the allylic homologue. Interestingly, surface properties were assessed. These authors measured the surface tensions of 16.8 and 15.7 dyne cm^{-1} when $R = CH_3$ and H, respectively, which are lower than that of PTFE ($\gamma_c = 18$ dyne cm^{-1}).

3.5.8. α-Fluoroacrylate Monomers

In 1984, Majumbar et al. [154] investigated the copolymerisation of methyl α-fluoroacrylate with styrene, α-deuterostyrene and β,β-dideuterostyrene. These reactions were initiated either by AIBN at 60 °C or by ethyl aluminium sesquichloride as complexing agent at −15 °C.

By ^{13}C NMR spectroscopy, these authors showed that, in the latter case, the copolymers were ideally alternated in contrast to the classic copolymerisation of the former initiating system. In addition, a fuller survey regarding the ^{1}H, ^{19}F and ^{13}C NMR spectra enabled the authors to propose a value of the coisotacticity parameter of 0.66 for the alternating copolymer. This quite high value (0.34 according to Patnaik and Gaylor [155]) is close to that suggested by the Japanese team of Yokota [156] (0.69) showing a certain discrepancy.

A more recent survey concerns the copolymerisation of 1,3-cyclohexadiene with 1-fluoro-1-cyanoethylene (or α-fluoroacrylonitrile) that led to alternating copolymers [157]. In addition, these copolymers were thermally stable thanks to both bulky substituents. A carefully structural analysis showed that these macro-molecules contained both 1,2- and 1,4-linkages across the cyclohexene unit.

3.5.9. Fluorinated Monomers Containing Aromatic Group

Bernardo-Bouteiller [158] and, above all, Montheard's group [159] investi-gated the copolymerisation of vinylidene cyanide with various styrenic monomers bearing fluorinated pendant groups:

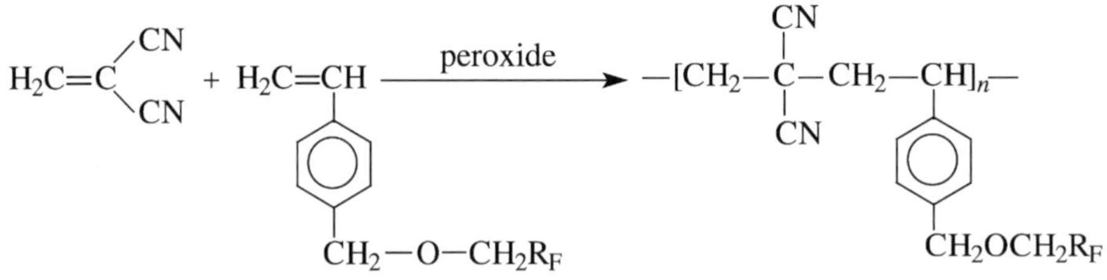

R_F: $CH_2C_4F_9$ or C_7F_{15}

Alternating copolymers were obtained and presented piezoelectric properties [159].

In contrast to an amorphous copolymer bearing a C_4F_9 side group ($T_g = 88\,°C$), the second copolymer (C_7F_{15} as lateral end-group) was crystalline due to the presence of such a longer fluorinated group.

In addition, the former copolymer started to decompose at 200 °C while the latter one was more thermal stable ($T_{dec} = 230\,°C$).

3.5.10. Other Monomers

Other fluorinated monomers bearing nitrile (or cyano) function (e.g., $CF_2=CFCN$) have also led to alternating copolymers [121, 160].

In addition, Aglieto et al. [161] investigated the radical copolymerisation of electron-acceptor α-trifluoromethyl acrylic acid, TFMAA (and its methyl ester, MTFMA) with various electron-donor α-olefins and VEs. Both α-CF3 monomers homopolymerise with difficulty although the literature reports that TFMAA does not homopolymerise [12, 162]; but the authors noted that the acid one was more reactive than its ester homologue. The conversions of both comonomers in copolymerisation were high (79–90%), except for the copolymerisation of TFMAA with VEs. In this last case, starting from an equimolar ratio of both comonomers, the TFMAA content in the copolymer was 80%. However, the copolymerisation of TFMAA and 1-decene or 2-methyl hexene in dioxane initiated by t-butylcyclohexyl peroxydicarbonate led to almost alternating copolymers. Similar results were obtained from the copolymerisation of VEs (Bu, 2-ethylhexyl or menthyl vinyl ethers) with MTFMA in bulk, initiated by AIBN. The overall structure is as follows:

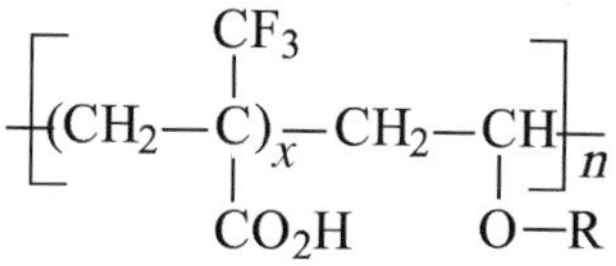

$$\left[(CH_2-\underset{\underset{CO_2H}{|}}{\overset{\overset{CF_3}{|}}{C}})_x-CH_2-\underset{\underset{O-R}{|}}{CH}\right]_n$$

with R: Butyl, 2-ethylhexyl, and (−) menthyl, and $x = 1.0-1.6$.

These authors noted a higher degree of alternation in the case of methyl vinyl ether/MTFMA since the hydrogenated monomer bears a higher electron-donating substituent.

Molecular weights and T_gs of these copolymers obtained were assessed, $\bar{M}_n$ ranging from 20,000 to 50,000. T_gs were in the $22-117\ ^\circ$C range, the highest one being assigned to bulky MTFMA/methyl vinyl ether copolymers.

4. Conclusion

Alternated copolymers should be considered slightly separately from well-controlled architectured polymers. Actually, their syntheses do not require special processes (as in the case of ionic polymerisations) or the manipulation of delicate molecules (such as counter radicals or complex catalysts, as in the case of free radical controlled polymerisations). However, obtaining them requires an excellent knowledge of both the electronic and resonance effects of the chosen comonomers.

Because of the electron-withdrawing effect of the fluorine atom (or of highly fluorinated substituents), fluorinated monomers constitute a basic class of monomers called "acceptor".

In this chapter, we have also quoted monomers, such as ethylene and its derivatives (propylene, butylene), and VEs which have led to industrial and commercially available copolymers for thermoplastics and coating applications.

However, theoretical studies on other monomers have not been widely attempted and, of those that have, these studies have been limited, as in the case of VAc. In addition, associations with other monomers (e.g., hydrophilic ones to counter balance the hydrophobic behaviour arising from the fluorinated monomers) should be further studied.

Finally, it would be interesting to use alternating copolymers to protect various other hydrogenated monomers that are electrodonating located between two fluoromonomers.

Also, for applications in coatings, high value added materials (e.g., for optics) should be studied on account of the other peculiar properties of the fluorinated groups (low refractive index, low dielectric constant and obviously the surface properties).

References

1. Cowie, J. M. (1985) *Alternating Copolymers*, Plenum Press, New York.
2. (a) Hamielec, A. E., MacGregor, J. F. and Pendilis, A. (1989) Copolymerization. In *Comprehensive Polymer Science*, Vol. 3 (Eds, Bevington, J. C., Allen, G., Eastmond, A. L. and Russo, S.) Chap. 2, pp. 17–31; (b) Tirrell, D. A. (1989) Copolymer composition. In *Comprehensive Polymer Science*, Vol. 3 (Eds, Bevington, J. C., Allen, G., Eastmond, A. L. and Russo, S.) Chap. 15, pp. 195–206; (c) Braun, D. and Czerwinski, W. K. (1989) Rates of copolymerization. In *Comprehensive Polymer Science*, Vol. 3 (Eds, Bevington, J. C., Allen, G., Eastmond, A. L. and Russo, S.) Chap. 16, pp. 207–218.
3. (a) Polic, A. L., Duever, T. A. and Pendilis, A. (1998) Case studies and literature review on the estimation of copolymerization reactivity ratios. *J. Polym. Sci., Part A: Polym. Chem.*, **36**, 813–822; (b) Hagiopol, C. and Frangu, O. (2003) Strategies in improving the accuracy of reactivity ratios estimation. *J. Macromol. Sci. Part A, Pure Appl. Chem.*, **A40**, 571–584 and **36**, 813–822.
4. (a) Yamabe, M. (1992) A challenge to novel fluoropolymers. *Macromol. Chem. Macromol. Symp.*, **64**, 11–18; (b) Ameduri, B. and Boutevin, B. (2000) Copolymerisation of fluorinated monomers: recent developments and future trends. *J. Fluorine Chem.*, **104**, 53–62.
5. Alfrey, T. and Price, C. C. (1947) Radical copolymerisation. *J. Polym. Sci., Part A*, **2**, 201–212.
6. Brown, D. W. and Wall, L. A. (1968) Radiation-induced copolymerization of tetrafluoroethylene and 3,3,3-trifluoropropene under pressure. *J. Polym. Sci., Polym. Chem. Ed.*, **6**, 1367–1379.
7. Greenley, R. Z. (1989) Free radical copolymerization reactivity ratios. In *Polymer Handbook*, Vol. II, 3rd ed. (Eds, Brandrup, J. and Immergut, E. H.) Wiley, New York, pp. 181–308 (see also Q and e values for free radical copolymerization of vinyl monomers and telogens. pp. 267–274).
8. Haas, H. C., MacDonald, R. L. and Chiklis, C. K. (1969) α-(Trifluoromethyl)vinyl acetate. II. *J. Polym. Sci., Polym. Chem. Ed.*, **7**, 633–641.
9. Majumdar, R. N. and Harwood, H. J. (1984) Synthesis and NMR characterization of copolymers of α-fluorostyrene with methyl acrylate. *Polym. Sci. Technol. (New Monomers Polym.)*, **25**, 285–309.
10. Knebelkamp, A. and Heitz, W. (1991) Synthesis and properties of poly(α-fluorostyrenes). *Makromol. Chem. Rapid Commun.*, **12**, 597–606.
11. Ueda, M., Shouji, S., Ogata, T., Kamachi, M. and Pittman, C. U. (1984) Radical-initiated homo- and copolymerization of α-fluoroacrylamide "living" radicals in a homogeneous system. *Macromolecules*, **17**, 2800–2804.
12. Iwatsuki, S., Kondo, A. and Harashina, H. (1984) Free radical copolymerization behavior of methyl α-(trifluoromethyl) acrylate and α-(trifluoromethyl) acrylonitrile: penultimate monomer unit effect and monomer reactivity parameters. *Macromolecules*, **17**, 2473–2479.
13. Ito, H., Giese, B. and Engelbrecht, R. (1984) Radical reactivity and Q–e values of methyl α-(trifluoromethyl)acrylate. *Macromolecules*, **17**, 2204–2205.
14. Greenley, R. Z. (1999) Q and e values for free radical copolymerization of vinyl monomers and telogens. In *Polymer Handbook*, Vol. II, 4th ed. (Eds, Abe, A., Bloch, D. R. and Immergut, E. H.) Wiley, New York, pp. 309–378.
15. Ueda, M. and Ito, H. (1988) Radical reactivity of α-trifluoromethylstyrene. *J. Polym. Sci., Part A: Polym. Chem.*, **26**, 89–98.
16. Burkhart, R. D. and Zutty, N. L. (1963) Copolymerization studies. III. Reactivity ratios of model ethylene copolymerizations and their use in Q–e calculations. *J. Polym. Sci., Part A: General Papers*, **1**, 1137–1163.
17. Naberezhnykh, R. A., Sorokin, A. D., Volkova, E. V. and Fokin, A. V. (1974) Radiation copolymerization of fluoroolefins. *Izv. Akad. Nauk SSSR, Ser. Khim.*, **1**, 232–233.
18. Young, L. J. (1961) Copolymerization parameters. *J. Polym. Sci.*, **54**, 411–455.
19. Caporiccio, G. and Sianesi, D. (1970) Polymerization and copolymerization of vinyl fluoride. II. Radical copolymerization. *Chim. Ind.*, **52**, 37–42.
20. Ameduri, B. and Bauduin, G. (2003) Synthesis and polymerization of fluorinated monomers bearing a reactive lateral group. XIV. Radical copolymerization of vinylidene fluoride with

methyl 1,1-dihydro-4,7-dioxaperfluoro-5,8-dimethyl non-1-enoate. *J. Polym. Sci., Part A: Polym. Chem.*, **41**, 3109–3121.

21. Khodzhaev, S. G., Yusupbekova, F. Z. and Yul'chibaev, A. A. (1981) Synthesis of derivatives of perfluoromethacrylic acid and some polymers made of them. *Sbornik Nauchnykh Trudov., Tashk. Gos.*, **667**, 34–48 (*Chem. Abstr.*, **97**, 163545).

22. Usmanov, Kh. U., Yul'chibaev, A. A., Mukhamadaliev, N. and Sarros, T. K. (1975) Radiation copolymerization of 2-hydropentafluoropropylene with vinylidene fluoride. *Izv. Vys. Uch. Zav. Kh.*, **18**, 464–466 (*Chem. Abstr.*, **83**, 28687).

23. Ameduri, B., Bauduin, G., Boutevin, B., Kostov, G., Petrova, P. and Rousseau, A. (1999) Synthesis and polymerization of fluorinated monomers bearing a reactive lateral group. Part 7. Copolymerization of tetrafluoroethylene with ω-hydroxy trifluorovinyl monomers. *J. Appl. Polym. Sci.*, **73**, 189–202.

24. Guiot, J., Ameduri, B. and Boutevin, B. (2002) Synthesis and polymerization of fluorinated monomers bearing a reactive lateral group. XII. Copolymerization of vinylidene fluoride with 2,3,3-trifluoroprop-2-enol. *J. Polym. Sci., Part A: Polym. Chem.*, **40**, 3634–3643.

25. Ameduri, B., Bauduin, G., Boutevin, B., Kostov, G. and Petrova, P. (1999) Synthesis and polymerization of fluorinated monomers bearing a reactive lateral group. Part 9. Copolymerization of vinylidene fluoride with 4,5,5-trifluorovinyl-4-ene pentyl acetate. *Macromolecules*, **32**, 4544–4550.

26. Ameduri, B., Bauduin, G., Boutevin, B., Kostov, G. and Petrova, P. (1999) Synthesis and polymerization of fluorinated monomers bearing a reactive lateral group. Part 10. Copolymerization of vinylidene fluoride (VDF) with 5-thioacetoxy-1,2,2-trifluoropentene for the obtaining of novel PVDF containing mercaptan side groups. *Design Mon. Polym.*, **2**, 189–202.

27. Guiot, J., Ameduri, B., Boutevin, B. and Lannuzel, T. (2003) Synthesis and polymerization of fluorinated monomers bearing a reactive lateral group. 13. Copolymerization of vinylidene fluoride with 2-benzoyloxy pentafluoropropene. *Eur. Polym. J.*, **39**, 887–896.

28. Matsuda, O., Watanabe, T., Tabata, Y. and Machi, S. (1979) Radiation-induced copolymerisation of methyl trifluoroacrylate with propylene. *J. Polym. Sci., Polym. Chem. Ed.*, **17**, 1795–1800.

29. Weise, J. K. (1971) Termonomer induced copolymerization of methyl methacrylate and tetrafluoroethylene. *Polym. Preprints (American Chemical Society, Division of Polymer Chemistry)*, **12**(1), 512–520.

30. Laita, Z., Paleta, O., Posta, A. and Liska, F. (1975) Haloacrylic acids. V. Polymerization of methyl trifluoroacrylate. *Collect. Czech. Chem. Commun.*, **40**, 2059–2062.

31. Ameduri, B., Boucher, M. and Manseri, A. (2001) Radical copolymerization of vinylidene fluoride with perfluorovinyl ethers, Presented at the 15th Winter Fluorine Conference, St Petersburg Beach, Fa, January 2001.

32. Kochkina, L. G., Erokhova, V. A. and Loginova, N. N. (1984) Kinetics of tetrafluoroethylene copolymerization with perfluoroalkyl vinyl ethers of different structure. *Zh. Prikl. Khim.*, **57**, 1126–1128 (*Chem. Abstr.*, **101**, 131147).

33. Miyake, H., Sugaya, Y. and Yamabe, M. (1998) Synthesis and properties of perfluorocarboxylated polymers. *J. Fluorine Chem.*, **92**, 137–140.

34. Ameduri, B., Boucher, M. and Manseri, A. (2002) Elastomères réticulables nitriles fluorosulfonés à base de fluorure de vinylidène présentant une faible Tg et procédés pour leurs préparations, World Demand WO 02/50,142 A1 (assigned to Hydro-Québec).

35. Moggi, G., Bonardelli, P. and Russo, S. (1983) Emulsion polymerization of the vinylidene fluoride–hexafluoropropene system. *Conv. Ital. Sci. Macromol.*, **2**, 405–408.

36. Kobunshi Gakkai Kyojugo (Copolymerisation) (1975) pp. 396–413.

37. Bartlett, P. D. and Nozaki, K. (1946) The polymerization of allyl compounds: III. The peroxide-induced copolymerization of allyl acetate with maleic anhydride. *J. Am. Chem. Soc.*, **68**, 1495–1500.

38. Butler, G. B., Olson, K. G. and Tu, C. L. (1984) Monomer orientation control by donor–acceptor complex participation in alternating copolymerization. *Macromolecules*, **17**, 1884–1887.

39. Walling, C. B., Emorene, R., Wolfstirn, K. B. and Mayo, F. R. (1948) Copolymerization. X. Effect of m- and p-substitution on the reactivity of the styrene double bond. *J. Am. Chem. Soc.*, **70**, 1537–1542.

40. Shirota, Y., Yoshimura, M., Matsumoto, A. and Mikawa, H. (1974) Mechanism of charge-transfer polymerization. VI. Alternating radical copolymerization of *N*-vinylcarbazole with electron-accepting monomers. *Macromolecules*, **7**, 4–11.

41. Tsuchida, E. and Tomono, T. (1971) Mechanism of alternating copolymerization of styrene and maleic anhydride. *Makromol. Chem.*, **141**, 265–298.

42. Hanna, M. W. and Ashbaugh, A. L. (1964) Nuclear magnetic resonance (N.M.R.) study of molecular complexes of 7,7,8,8-tetracyanoquinodimethane and aromatic donors. *J. Phys. Chem.*, **68**, 811–816.

43. Yamashita, Y. (1965) Charge transfer complex assessed by UV investigations. *Macromol. Chem.*, **89**, 205–212.

44. Tabata, Y. and Du Plessis, T. A. (1971) Radiation-induced copolymerization of chlorotrifluoroethylene with ethyl vinyl ether. *J. Polym. Sci., Part A-1*, **9**, 3425–3435.

45. (a) Trivedi, B. C. and Culbertson, B. M. (1982) Random addition copolymerizations. In *Maleic Anhydride*, Chap. 9, pp. 269–305, Plenum, New York; (b) Altering Addition copolymerizations, Chap. 10, pp. 307–458, Plenum, New York.

46. Ameduri, B., Boutevin, B. and Kostov, G. (2001) Fluoroelastomers: synthesis, properties and applications. *Prog. Polym. Sci.*, **26**, 105–187.

47. Ebnesajjad, S. (Ed.) (2000) *Fluoroplastics Melt Processable Fluoropolymers*, Plastic Design Library, Handbook Series, Vol. 2, Worwich, New York.

48. (a) Boutevin, B. and Ameduri, B. (1994) Copolymerization of fluorinated monomers with nonfluorinated monomers: reactivity and mechanisms. *Macromol. Symp.*, **82**, 1–17; (b) Scheirs, J. (1997) *Modern Fluoropolymers*, Wiley, New York; (c) Hougham, G., Cassidy, P., Jonhs, K. and Davidson, T. (1999), *Fluoropolymers*, Kluvert, New York.

49. (a) French Patent 81,15212 (assigned to Asahi Glass Co. Ltd), August 1981; (b) Unoki, M., Kimura, I. and Yamauchi, M. (2001) Solvent soluble fluoropolymers for coatings: chemical structure and weatherability, presented at the "Fluorine in Coatings IV" Conference, Brussels, March 2001, paper 8; (c) Ishida, T., Suzuki, T., Takimoto, Y. and Sasakura, H. (2001) The effect of monomer sequence on the polymer durability in the copolymers of chlorotrifluoroethylene with alkyl vinyl ester or alkyl vinyl ethers, presented at the "Fluorine in Coatings IV" Conference, Brussels, March 2001, paper 12; (d) Masuda, S. (2003) Fluoropolymer for powder coatings, presented at the "Fluorine in Coatings V" Conference, Orlando, January 2003, paper 15.

50. Pozzio, T., Lenti, D. and Masini, L. (1994) Water based fluoroelastomers coatings with high barrier properties, "Fluorine in Coatings I" Conference, Salford, September 28–30, 1994, paper 24.

51. (a) Tournut, C. (1994) New copolymers of vinylidene fluoride. *Macromol. Symp.*, **82**, 99–109; (b) Tournut, C. (1997) Thermoplastic copolymers of vinylidene fluoride. In *Modern Fluoropolymers* (Ed, Scheirs, J.) Wiley, New York, Chap. 31, pp. 577–596.

52. Hanford, E. W. (1946) Copolymers of chlorotrifluoroethylene and olefin hydrocarbons. US Patent 2,392,378 (assigned to Du Pont de Nemours and Co.).

53. Stanitis, G. (1997) ECTFE copolymers. In *Modern Fluoropolymers* (Ed, Scheirs, J.) Wiley, New York, Chap. 27, pp. 525–540.

54. Ragazzini, M., Garbuglio, C., Cevidalli, G. B., Carcano, D. and Minasso, B. (1967) Copolymerization of ethylene and chlorotrifluoroethylene by trialkylboron catalysts-I. *Eur. Polym. J.*, **3**, 129–136.

55. Garbuglio, C., Ragazzini, M., Pilati, O., Carcano, D. and Cevidalli, G. B. (1967) Copolymerization of ethylene and chlorotrifluoroethylene by trialkylboron catalysts-II. Physico-chemical characterization of the copolymers. *Eur. Polym. J.*, **3**, 137–144.

56. Tabata, Y., Ishigure, K., Higaki, H. and Oshima, K. (1970) Radiation-induced copolymerization of fluorine-containing monomers. *J. Macromol. Sci. Chem.*, **A4**, 801–813.

57. Sibilia, J. P., Roldan, G. L. and Chandrasekaran, S. (1972) Structure of ethylene–chlorotrifluoroethylene copolymers. *J. Polym. Sci., Part A*, **10**, 549–559.
58. Robertson, A. B. (1973) Poly(ethylene–chlorotrifluoroethylene) fibers. *Appl. Polym. Symp.*, **21**, 89–100.
59. Khanna, Y. P., Taylor, T. J. and Chandrasekaran, S. (1989) Dielectric properties of Halar, an alternating copolymer of ethylene and chlorotrifluoroethylene. *J. Appl. Polym. Sci.*, **38**, 135–145.
60. (a) Lin, S. C. (2001) Flame resistance of ECTFE powder coatings, presented at the "Fluorine in Coatings IV" Conference, Brussels, March 2001, paper 3; (b) Lin, S. C. (2003) Ignition resistance of ECTFE powder coatings "Fluorine in Coatings V" Conference, Orlando, January 21–22, 2003, paper 16.
61. Ishigure, K., Tabata, Y. and Oshima, K. (1975) Fluorine-19 nuclear magnetic resonance of chlorotrifluoroethylene–propylene alternating copolymer. *Macromolecules*, **8**, 177–181.
62. Tiers, G. V. D. and Bovey, B. A. (1963) Polymer NMR [nuclear magnetic resonance] spectroscopy: VII. The stereochemical configuration of poly(trifluorochloroethylene). *J. Polym. Sci.*, **1**, 833–841.
63. Boutevin, B., Cersosimo, F., Youssef, B. and Kappler, P. (1991) Télomérisation des éthers vinyliques et leur cotelomerisation avec les monomères fluorés. *J. Fluorine Chem.*, **52**, 403–418.
64. Boutevin, B., Cersosimo, F. and Youssef, B. (1992) Studies of the alternating copolymerization of vinyl ethers with chlorotrifluoroethylene. *Macromolecules*, **25**, 2842–2847.
65. (a) Yamabe, M., Massaki, T., Higaki, H., Shinobaru, T., Tanabe, H., Nukayoma, S. and Yokohama, K. (1983) Korrosionsschutz-Beschichtungsverfahren, German Patent 3303824 C2 (assigned to Asahi); (b) Kojima, G. and Yamabe, M. (1984) A solvent-soluble fluororesin for paints, Vol. 42, Yuki Gosei Kagaku Kyokaishi, Japan, pp. 841–849.
66. Yamabe, Y. (1984) New Fluoropolymers Coatings, Organic Coatings: Science and Technology, Vol. 7, Marcel Dekker, New York, pp. 25–42.
67. Takakura, T. (1997) CTFE/vinyl ether copolymers. In *Modern Fluoropolymers* (Ed, Scheirs, J.) Wiley, New York, Chap. 29, pp. 557–564.
68. Thomas, W. M. and O'Shaughnessy, M. T. (1953) Kinetics of chlorotrifluoroethylene polymerization. *J. Polym. Sci.*, **11**, 455–470.
69. Murray, D. L., Harwood, H. J., Shendy, S. M. M. and Piirma, I. (1995) The use of sequence distributions to determine monomer feed compositions in the emulsion copolymerization of chlorotrifluoroethylene with vinyl acetate and vinyl propionate. *Polymer*, **36**, 3841–3848.
70. Baradie, B. and Shoichet, M. S. (2002) Synthesis of fluorocarbon–vinyl acetate copolymers in supercritical carbon dioxide: insight into bulk properties. *Macromolecules*, **35**, 3569–3575.
71. Jim, C., Otsuhata, K. and Tabata, Y. (1985) Radiation-induced copolymerization of *N*-vinylpyrrolidone with monochlorotrifluoroethylene. *J. Macromol. Sci. Chem.*, **A22**, 379–386.
72. Kliman, N., Košinar, M. and Lazár, M. (1959) Copolymerization of trifluorochloroethylene with vinyl chloride and vinylidene chloride. *Chem. Prumysl.*, **9/34**, 668–670.
73. Ohmori, A., Tomihashi, N. and Shimizu, Y. (1984) Fluorine-containing copolymer and composition containing the same. Eur. Patent Sp. 121,934 (assigned to Daikin Kogyo Co. Ltd).
74. Pike, R. M. and Bailey, D. L. (1956) Copolymerization of vinyl siloxanes with organic vinyl monomers. *J. Polym. Sci.*, **22**, 55–64.
75. Carlson, D. P. (1971) Copolymers of ethylene/tetrafluoroethylene and of ethylene/chlorotri-fluoroethylene. US Patent 3,624,250 (assigned to DuPont de Nemours) 30-11-1971.
76. (a) Borsini, G., Modena, M., Nicora, C. and Ragazzini, V. M. (1968) Process for the production of fluorinated polymeric materials. US Patent 3,401,155 (assigned to Montecatini Edison S.p.A.); (b) Modena, M., Garbuglio, C. and Ragazzini, M. (1972) New high temperature thermoplastic material. Tetrafluoroethylene-ethylene copolymers. *J. Polym. Sci., Part B*, **10**, 153–156; (c) Ragazzini, M., Modena, M. and Carcano, D. (1973) Copolimeri tetrafluoroetilene-etilene, sintesi e caratterizzazione. *Chim. Ind.*, **55**, 265–272.
77. Price, F. D. (1962) Copolymerization mathematics and the description of stereoregular polymers. *J. Chem. Phys.*, **36**, 209–218.

78. Modena, M., Garbuglio, C. and Ragazzini, M. (1972) A new high temperature thermoplastic material: tetrafluoroethylene–ethylene copolymer. *Polym. Lett.*, **10**, 153–156.
79. Kostov, G. and Nikolov, A. (1995) Study of radical copolymerization of tetrafluoroethylene with ethylene in bulk. *J. Appl. Polym. Sci.*, **57**, 1545–1555.
80. (a) Lyons, B. J. (1984) The crosslinking of fluoropolymers with ionising radiation: a review. Second International Conference on Radiation Processing for Plastics and Rubbers, Canterbury, UK, March 1984, pp. 1–8; (b) Lyons, B. J. (1997) The radiation crosslinking of fluoropolymers. In *Modern Fluoropolymers* (Ed., Scheirs, J.) Wiley, New York, Chap. 18, pp. 335–347; (c) Forsythe, J. S. and Hill, D. J. T. (2000) Radiation chemistry of fluoropolymers, *Prog. Polym. Sci.*, **25**, 101–136; (d) Dargaville, T. R., George, G. A., Hill, D. J. J. and Whittaker, A. K. (2003) High energy radiation grafting of fluoropolymers. *Prog. Poly. Sci.*, **28**, 1355–1376.
81. Kerbow, D. L. (1997) Ethylene–tetrafluoroethylene copolymer resins. In *Modern Fluoropolymers* (Ed, Scheirs, J.) Wiley, New York, Chap. 15, pp. 301–310.
82. Naberezhnykh, R. A., Sorokin, A. D., Galperin, E., Volkova, E. V. and Simakina, A. (1977) Radiation copolymerization. *Vysokomol. Soedin.*, **19**, 33–37.
83. Seilers, D. A. (1997) PVDF in the chemical process industry. In *Modern Fluoropolymers* (Ed, Scheirs, J.) Wiley & Sons, New York, Chap. 25, pp. 487–506.
84. Wang, Z., Tontisakis, A., Tuminello, W. H., Buck, W. and Chu, B. (1990) Viscosity characterization of an alternating copolymer of ethylene and tetrafluoroethylene. *Macromolecules*, **23**, 1444–1446.
85. Farmer, B. L. and Lando, J. B. (1975) Conformational and packing analysis of the alternating copolymer of ethylene and tetrafluoroethylene. *J. Macromol. Sci. Phys.*, **B11**, 89–119.
86. Guerra, G., De Rosa, C., Iuliano, M., Petraccone, V. and Corradini, P. (1993) Structural variations as a function of temperature and dynamic-mechanical relaxations for ethylene–tetrafluoroethylene and ethylene–chlorotrifluoroethylene alternating copolymers. *Makromol. Chem.*, **194**, 389–396.
87. Tabata, Y., Shibano, H. and Sobue, H. (1964) Copolymerization of tetrafluoroethylene with ethylene induced by ionizing radiation. *J. Polym. Sci., Part A*, **2**, 1977–1986.
88. Machi, S., Matsuda, O., Ito, M., Tabata, Y. and Okamoto, J. (1977) Synthesis of a new fluoroelastomer from tetrafluoroethylene and propylene by radiation processing. *Radiat. Phys. Chem.*, **9**, 403–417.
89. Kojima, G. and Tabata, Y. (1972) Radiation-induced copolymerization of tetrafluoroethylene and propylene: I. Copolymerization in bulk. *J. Macromol. Sci. Chem.*, **A6**, 417–438.
90. Kojima, G. and Tabata, Y. (1973) Radiation-induced copolymerization of tetrafluoroethylene and propylene: 2. Copolymerization in solution. *J. Macromol. Sci. Chem.*, **A7**, 783–793.
91. (a) Matsuda, O., Okamoto, J., Suzuki, N., Ito, M. and Tabata, Y. (1974) Radiation-induced emulsion copolymerization of tetrafluoroethylene with propylene I. *J. Macromol. Sci. Chem.*, **A8**, 775–791; (b) Kojima, G. and Hisasue, M. (1981) Die emulsions copolymerisation von tetrafluorethylen mit propylen bei niedrigen temperaturen. *Makromol. Chem.*, **182**, 1429–1439.
92. (a) Kostov, G. K. and Petrov, P. Chr. (1992) Study of synthesis and properties of tetrafluoroethylene–propylene copolymers. *J. Polym. Sci., Part A: Polym. Chem.*, **30**, 1083–1088; (b) Yamaguchi, K., Hayakawa, N. and Okamoto, J. (1978) Molecular weight characteristics of tetrafluoroethylene–propylene copolymer produced by radiation-induced emulsion copolymerization. *J. Appl. Polym. Sci.*, **22**, 2653–2660.
93. (a) Ishigure, K., Tabata, Y. and Oshima, K. (1972) Nuclear magnetic resonance of fluorine-containing polymer: IV. Fluorine spectra of tetrafluoroethylene–propylene copolymer. *J. Fac. Eng., Univ. Tokyo, Ser. A (Japan)*, **10**, 52–53 (*Chem. Abstr.*, **79**, 66957 (1973)); (b) Kojima, N., Wachi, H., Ishigure, K. and Tabata, Y. (1976) ^{19}F- and ^{1}H-NMR spectra of tetrafluoroethylene–propylene copolymers. *J. Polym. Sci.: Polym. Chem. Ed.*, **14**, 1317–1330.
94. Kostov, G., Bessiere, J. M., Guida-Pietrasanta, F., Bauduin, G. and Petrov, P. (1999) Study of the microstructure and thermal properties of tetrafluoroethylene–propylene elastomers. *Eur. Polym. J.*, **35**, 743–749.

95. Ameduri, B., Boutevin, B., Kostov, G., Bauduin, G., Petrova, P. and Petrov, P. C. (1999) Synthesis and copolymerization of functional fluoromonomers. Part 8. Terpolymerization of tetrafluoroethylene and propylene with fluorinated alcohols. *J. Polym. Sci., Part A: Polym. Chem.*, **37**, 3991–3999.

96. Tabata, Y., Ishigure, K., Oshima, K. and Sobue, H. (1964) Copolymerization of tetrafluoroethylene with isobutene induced by ionizing radiation. *J. Polym. Sci., Part A*, **2**, 2445–2453.

97. Tabata, Y., Ishigure, K. and Oshima, K. (1965) Radiation-induced copolymerization of tetrafluoroethylene with isobutene at low temperature. *Die Makromol. Chem.*, **85**, 91–101.

98. Stilmar, F. B. (1968) Tetrafluoroethylene, isobutylene, carboxylic copolymers. US Patent 3,380, 974 (assigned to E. I. du Pont de Nemours and Company).

99. Hikita, T., Tabata, Y., Oshima, K. and Ishigure, K. (1972) Radiation-induced copolymerization of tetrafluoroethylene with vinyl ethers. *J. Polym. Sci., Polym. Chem. Ed.*, **10**, 2941–2949.

100. Feiring, A. E., Wonchoba, E. R., Fischel, B. E., Thieu, T. V. and Nassirpour, M. R. (2002) Amorphous fluoropolymers from tetrafluoroethylene and bulky vinyl esters or vinyl ethers. *J. Fluorine Chem.*, **118**, 95–98.

101. Modena, M., Borsini, G. and Ragazzini, M. (1967) Vinyl acetate and vinyl alcohol copolymers with tetrafluoroethylene. *Eur. Polym. J.*, **3**, 5–12.

102. Oxenrider, B. C., Long, D. J. and Mares, F. (1991) Copolymerization of vinyl acetate and a fluoromonomer in an aqueous medium. US Patent 5,070,162 (assigned to Allied Signal Inc.).

103. Lousenberg, D. and Shoichet, M. (2000) Synthesis of linear poly(tetrafluoroethylene-*co*-vinyl acetate) in carbon dioxide. *Macromolecules*, **33**, 1682–1685.

104. Fiering, A. E. and Feldman, J. (2000) Photoresist for microlithography, Intern Demand PCT WO 00 17,712 (assigned to DuPont).

105. Sorokin, A. D., Volkova, E. V. and Naberezhnykh, R. A. (1972) Radiation copolymerization of fluoroolefins with ethylene. *Radiat. Khim.*, **2**, 295–297.

106. Petrov, P. Chr. and Kostov, G. K. (1994) Emulsion copolymerization of tetrafluoroethylene and propylene with redox system containing *tert*-butylperbenzoate: II. Polymerization mechanism. *J. Polym. Sci., Part A: Polym. Chem.*, **32**, 2235–2239.

107. Wall, L. A. (1972) *Fluoropolymers*, Wiley Interscience, New York.

108. Brown, D. W., Lowry, R. E. and Wall, L. A. (1970) Radiation-induced copolymerization of tetrafluoroethylene and 3,3,4,4,5,5,5-heptafluoro-1-pentene under pressure. *J. Polym. Sci., Polym. Chem. Ed.*, **8**, 2441–2452.

109. Moggi, G., Bonardelli, P. and Bart, J. C. J. (1984) Copolymers of 1,1-difluoroethene with tetrafluoroethene, chlorotrifluoroethene, and bromotrifluoroethene. *J. Polym. Sci., Polym. Phys. Ed.*, **22**, 357–365.

110. Moggi, G., Bonardelli, P., Monti, C. and Bart, J. C. J. (1985) Copolymers of tetrafluoroethene with chlorotrifluoroethene and with bromotrifluoroethene. *J. Polym. Sci., Polym. Phys. Ed.*, **23**, 1099–1108.

111. Kostov, G., Ameduri, B., Bauduin, G., Boutevin, B. and Stankova, M. (2004). *J. Polym. Sci., Part A: Polym. Chem.*, **42**, 1693–1706.

112. Watanabe, T., Momose, T., Ishigaki, I., Tabata, Y. and Okamoto, J. (1981) Radiation-induced copolymerization of methyl trifluoroacrylate with fluoroolefins. *J. Polym. Sci., Polym. Lett. Ed.*, **19**, 599–602.

113. Bonardelli, P., Moggi, G. and Turturro, A. (1986) Glass transition temperatures of copolymer and terpolymer fluoroelastomers. *Polymer*, **27**, 905–906.

114. Kabankin, A. S., Balabanova, S. A. and Markevich, A. M. (1970) Radical copolymerization of tetrafluoroethylene with ethylene and hexafluoropropylene. *Vysokomol. Soedin. Ser. A*, **12**, 267–272.

115. Kostov, G. K., Kotov, St. V., Ivanov, G. D. and Todorova, D. (1993) Study on the synthesis of perfluorovinyl-sulfonic functional monomer and its copolymerization with tetrafluoroethylene. *J. Appl. Polym. Sci.*, **47**, 735–741.

116. Kochkina, L. G., Erokhova, V. A. and Loginova, N. N. (1984) Kinetics of tetrafluoroethylene copolymerization with perfluoroalkyl vinyl ethers of different structures. *Zh. Prikl. Khim.*, **57**, 1126–1128 (*Chem. Abstr.* **101**, 131147).

117. Seko, M., Ogawa, S. and Kimoto, K. (1982) Perfluorocarboxylic acid membranes and membrane chlor-alkali process developed by Asahi Chemical Industry. In *Perfluorinated Ionomer Membranes,* American Chemical Society Symposium Series 180 (Eds, Eisenberg, A. and Yaeger, H. L.) ACS, Washington, DC, Chap. 15, pp. 365–410.

118. Kato, M., Akiyama, K. and Yamabe, M. (1983) New perfluorocarbon polymers with phosphonic groups. *Reports Res. Lab. Asahi Glass Co., Ltd.*, **33**, 135–141.

119. (a) Kostov, G. K. and Atanasov, A. N. (1991) Ion-exchange copolymers based on tetrafluoroethylene, hexafluoropropylene, and acrylic acid. *J. Appl. Polym. Sci.*, **42**, 1607–1613; (b) Kostov, K., Matsuda, O., Machi, S. and Tabata, Y. (1992) Radiation synthesis of ion-exchange carboxylic fluorine containing membranes. *J. Membr. Sci.*, **68**, 133–140.

120. Beaune, O., Bessiere, J. M., Boutevin, B. and El Bachiri, A. (1995) Study of the alternating radical copolymerization of dichlorodifluoroethylene with ethyl vinyl ether. *J. Fluorine Chem.*, **73**, 27–35.

121. (a) Gotlub, V. (1991) Copolymerization of vinyl ether with cyano fluoromonomers. *J. Appl. Polym. Sci.*, **23**, 412–420; (b) Laita, Z., Paleta, O., Posta, A. and Liska, F. (1977) Polymerization of trifluoroacrylonitrile. *Coll. Czech. Chem. Commun.*, **42**, 1536–1539.

122. Liartaon, I. (1992) Copolymers of vinyl ether with methyl trifluoroacrylate. *J. Polym. Sci.*, **24**, 1201–1210.

123. Chandraskhan, S. and Mueller, M. B. (1972) Copolymers of 3,3,3-trifluoro-2-trifluoromethyl propene and vinylidene fluoride. US Patent, 3,706,723 (assigned to Allied Chem.).

124. Minhas, P. S. and Petrucelli, F. (1977) A new high-performance fluoropolymer that can be readily melt-processed. *Plast. Eng.*, **33**, 60–63.

125. Litt, M. H. and Lando, J. B. (1982) The crystal structure of the alternating co-polymer of hexafluoroisobutylene and vinylidene fluoride. *J. Polym. Sci., Polym. Phys. Ed.*, **20**, 535–552.

126. (a) Froix, M. F., Goedde, A. O. and Pochan, J. M. (1977) An NMR relaxation study of a hexafluoroisobutylene/vinylidene fluoride copolymer (HFIB/VF2). *Macromolecules*, **10**, 778–781; (b) Pochan, J. M., Hinman, D. F., Froix, M. F. and Davidson, T. (1977) Dielectric and dynamic mechanical relaxations in 3,3,3-trifluoro-2-trifluoromethylpropene (hexafluoroisobutylene)/1,1-difluoroethylene alternating copolymers. *Macromolecules*, **10**, 113–116.

127. Souzy, R., Guiot, J., Ameduri, B., Boutevin, B. and Paleta, O. (2003) Unexpected alternating copolymerization of vinylidene fluoride with methyl trifluoroacrylate. *Macromolecules*, **36**, 9390–9395.

128. Souzy, R., Ameduri, B. and Boutevin, B. (2004) Radical copolymerization of α-trifluoromethylacrylic acid with vinylidene fluoride and vinylidene fluoride/hexafluoropropene. *Macromol. Chem. Phys.*, **205**, 476–485.

129. Panchalingam, V. and Reynolds, J. R. (1989) New vinylidene fluoride copolymers: poly(vinyl acetate-*co*-vinylidene fluoride). *J. Polym. Sci., Part C: Polym. Lett.*, **27**, 201–208.

130. Guiot, J. (2003) Nouveaux monomères fluorofonctionnels et leurs copolymérisations avec le fluorure de vinylidène, PhD dissertation, Université de Montpellier.

131. Sianesi, D. and Caporiccio, G. (1968) Radical copolymerisation of vinylidene fluoride with vinyl fluoride. *J. Polym. Sci., Part A-1: Polym. Chem.*, **6**, 335–339.

132. Yagi, T. and Tatemoto, M. (1979) A fluorine-19 NMR study of the microstructure of vinylidene fluoride–trifluoroethylene copolymers. *Polym. J.*, **11**, 429–436.

133. Otazaghine, B. and Ameduri, B. (2000) Radical copolymerization of vinylidene fluoride with 2-hydroperfluoro-1-octene, presented at the 16th International Symposium in Fluorine Chemistry, Durham, UK, July 16–20. Oral communication B-11.

134. Dohany, R. E., Dukert, A. A. and Preston, S. S. (1989) Copolymers of vinylidene fluoride. *Encycl. Polym. Sci. Technol.*, **17**, 532–547.

135. McBee, E. T., Hill, H. M. and Bachman, G. B. (1949) Polymerization of vinylidene fluoride and 1,1-dichloro-2,2-difluoroethylene. *Ind. Eng. Chem.*, **41**, 70–75.

136. (a) Pierce, O. R., Holbrook, G. W., Johannson, O. K., Saylor, J. C., Brown, E. D. and Kim, Y. K. (1960) Synthesis and properties of fluorosilicones. *Ind. Eng. Chem.*, **52**, 783–788; (b) Kim, Y. K. (1971) Original fluorosilicones. *Rubber Chem. Technol.*, **44**, 1350–1362.

137. Otsuhata, K., Jin, C. and Tabata, Y. (1988) Copolymerization of trifluoropropene with itaconic acid and ethyl vinyl ether by using gamma rays. *J. Polym. Sci., Part A: Polym. Chem.*, **26**, 601–608.

138. Wu, C., Brambilla, R. and Yardley, J. T. (1990) Copolymerization of hexafluoroisobutylene and vinyl acetate. *Macromolecules*, **23**, 997–1005.

139. Hatada, H., Nagata, K., Hasegawa, T. and Yuki, H. (1977) Carbon-13 NMR spectra and reactivities of vinyl compounds in ionic and radical polymerizations. *Makromol. Chem.*, **178**, 2413–2420.

140. Herman, J. J. and Teyssie, P. (1978) Determination of vinyl monomers reactivity by carbon-13 nuclear magnetic resonance spectroscopy. *Macromolecules*, **11**, 839–845.

141. Haszeldine, R. N. (1952) Fluoroolefins. Part 1. The synthesis of hexafluoro-1,3-butadiene. *J. Chem. Soc.*, 4423–4431.

142. Bargigia, G., Tortelli, V., Tortelli, C. and Modena, S. (1986) Process for the preparation of perfluoroalkadienes, Eur. Patent Appl., EP 270,956 (assigned to Ausimont).

143. Zurkova, E., Bouchal, K., Chvatal, Z. and Dedek, V. (1987) Radical copolymerization and terpolymerization of 1,1,2,3,4,4-hexafluoro-1,3-butadiene with alkylvinyl ethers, alkenes, and methyl 2,3,3-trifluoroacrylate. *Die Angew Makromol. Chem.*, **155**, 101–110.

144. Searle, N. E. (1949) Synthesis of nu-aryl-maleimides. US Patent 2,444,536 (assigned to Du Pont).

145. Arnold, H. W. and Searle, N. E. (1949) Fly spray. US Patent 2,462,835 (assigned to Du Pont) (*Chem. Abstr.*, **43**, 4421c (1949)).

146. Barrales-Rienda, J. M., Gonzalez Ramos, J. and Sanchez Chaves, M. (1977) Synthesis of *N*-(fluorophenyl)maleamic acids and *N*-(fluorophenyl)maleimides. *J. Fluorine Chem.*, **9**, 293–308.

147. Cummings, W. and Lynch, E. R. (1965) Polyimides. British Patent, 1,077,243 (assigned to Monsanto Chem. Ltd), *Chem. Abstr.*, **68**, 41104j, (1968).

148. Wagener, K. B., Do, C. H., Johnson, M. and Smith, M. A. (1987) Donor acceptor polymerization chemistry as a vehicle to low energy cure matrix resins, Interim report for February 1986 to April 1987 (report # AFWAL-TR-87-4093).

149. Beaune, O., Bessière, J. M., Boutevin, B. and Robin, J. J. (1994) Synthèse de polymères hautement fluorés amorphes et à T_G élevées: copolymérisation de maléimides fluorèes avec des éthers vinyliques. *J. Fluorine Chem.*, **67**, 159–166.

150. Canessa, G. C. (1981) Copolimerización de fenilvinyleter con anhidrido difluormaleico y derivados. Una contribución al mecanismo de la copolimerización alternante mediante espectroscopia de ^{19}F-RMN. *Bol. Soc. Chil. Quim.*, **26**, 25–32.

151. Green, H. E., Dones, R. J. and O'Rell, M. K. (1979) Bis(difluoromaleimide) capped prepolymers and polymers. US Patent 4,173,700 (assigned to TRW Inc.), *Chem. Abstr.*, **92**, 95022v (1979).

152. (a) El-Guweri, M., Hendlinger, P. and Laschewsky, A. (1997) Partially fluorinated maleimide copolymers for Langmuir films of improved stability. Part 1. Synthesis, copolymerization behavior, and bulk properties. *Macromol. Chem. Phys.*, **198**, 401–418; (b) Hendlinger, P., Laschewsky, A., Bertrand, P., Delcorte, A., Legras, R., Nysten, B. and Moebius, D. (1997) Partially fluorinated maleimide copolymers for Langmuir films of improved stability: 2. Spreading behavior and multilayer formation. *Langmuir*, **13**, 310–319.

153. Pittman, A. G. and Wasley, W. L. (1972) Polymers derived from fluoroketones: V. Copolymers of fluoroalkyl ethers and maleic anhydride. *Polym. Lett.*, **10**, 279–286.

154. Majumdar, R. N., Lyn, F. T. and Harwood, H. J. (1984) Synthesis and NMR characterization of the alternating copolymer of methyl α-fluoroacrylate with styrene. *Polym. J.*, **16**, 175–182.

155. Patnaik, B. K. and Gaylord, N. G. (1973) Donor–acceptor complexes in copolymerization. XLIV. Copolymerization of styrene and α-methylstyrene with acrylic and α-substituted acrylic esters in the presence of aluminum compounds. *J. Macromol. Sci. Chem.*, **7**, 1247–1263.

156. Hirabayasi, T. and Yokota, K. (1976) Alternating copolymerization of styrene and methyl α-chloroacrylate with diethylaluminum chloride. *J. Polym. Sci.*, **14**, 45–51.

157. Panchalingam, V. and Reynolds, J. R. (1991) New fluorinated copolymers: poly(1,3-cyclohexadiene-alt-α-fluoroacrylonitrile). *Polym. Prepr. (Am. Chem. Soc., Div. Polym. Chem.)*, **32**(3), 185–186.

158. Bernardo-Bouteiller, V. (1995) Synthèses et caractérisations de nouveaux polystyrènes porteurs de chaines latérales fluorées, PhD dissertation, Université de Paris VI.

159. Montheard, J. P., Zerroukhi, A., Ouillon, I., Boinon, B. and Pham, Q. T. (1995) Copolymerization reactions of vinylidene cyanide with styrenes carrying a fluorocarbon segment: preliminary studies of their microstructures. *Macromol. Rep. (Suppl.)*, **A32**, 1–11.

160. Matsuda, O., Kostov, G. K., Tabata, Y. and Machi, S. (1979) Radiation-induced copolymerization of α,β,β-trifluoroacrylonitrile with α-olefin. *J. Appl. Polym. Sci.*, **24**, 1053–1059.

161. Aglietto, M., Passaglia, E. and Montagnini di Mirabello, L. (1995) Synthesis of new polymers containing α-(trifluoromethyl)-acrylate units. *Macromol. Chem. Phys.*, **196**, 2843–2853.

162. Ito, H., Miller, D. C. and Willson, C. G. (1982) Polymerization of methyl α-(trifluoromethyl) acrylate and α-(trifluoromethyl)acrylonitrile and copolymerization of these monomers with methyl methacrylate. *Macromolecules*, **15**, 915–920.

CHAPTER 4

SYNTHESIS, PROPERTIES AND APPLICATIONS OF FLUORINATED DIBLOCK, TRIBLOCK AND MULTIBLOCK COPOLYMERS

1. Introduction

Macromolecular engineering has become a useful tool for designing well-architectured polymers (telechelics, cycles, networks, dendrimers, hyperbranched, gradient, alternated, star, graft or block copolymers) [1–3]. Since fluorinated telechelic and graft (co)polymers are reviewed in Chapters 2 and 5, respectively, the last kind of fluoropolymers is reported in this section.

Block copolymers [1–10] have received much attention as "novel polymer materials" with several components since they are made of different polymer blocks linked together. The reason for their importance comes from their unique chemical structure that brings new physical and thermodynamical properties related to their solid-state and solution morphologies. Frequently, block copolymers exhibit phase separation producing a dispersed phase consisting of one block type in a continuous matrix of the second block type. Their unusual colloidal and mechanical properties allow modification of solution viscosity, surface activity or elasticity and impact resistance of polymers. Thus, several block copolymers have produced a wide range of materials with tailorable properties depending on the nature and length of blocks. They have found significant applications [1–3, 7–10] such as adhesives and sealants, surface modifiers for fillers and fibres, crosslinking agents for elastomers, additives for resin gelation and hardening, compatibilising agents [9] and stable emulsions of homopolymer blends that can find applications in recovery and recycling plastic waste. In addition, because of their self-organisation behaviour, they are potential candidates as nanostructured materials (e.g., as in the case of polystyrene-*b*-polybutadiene-*b*-polymethyl methacrylate triblock copolymers for the reinforcement of polyvinylidene fluoride [11]).

A number of synthetic methods have been successfully developed for the synthesis of block copolymers. They include polycondensation, anionic, cationic, coordinative, free-radical polymerisations and also mechanochemical synthesis [7, 8, 12].

Block copolymers synthesised by radical polymerisation or telomerisation have been reported [10] and a non-exhaustive list of fluorinated block copolymers was mentioned in recent reviews [13, 14].

We have found it interesting to revisit their syntheses and to supply a more detailed review on such compounds by separating the synthesis, properties and applications of fluorinated diblock copolymers to those of fluorotriblock and multiblock copolymers. First, the strategies of synthesis of fluorodiblock copolymers achieved by traditional radical oligomerisation, telomerisation and polymerisation are presented. The second part reports the modern techniques of radical controlled polymerisation used for obtaining such diblock copolymers: from nitroxide-mediated polymerisations, atom transfer radical polymerisation (ATRP), iodine transfer polymerisation (ITP), reversible addition-fragmentation chain transfer (RAFT), MADIX or other degenerative transfer from methods involving tetraphenyl ethanes for example. Third, these fluorodiblock copolymers can also be produced from methods that combine traditional radical ways and controlled radical systems. In a following part, straightforward preparations of fluorosegmented copolymers that do not use any radical system are mentioned. This basically concerns cationic methods (e.g., cationic polymerisations of fluorinated vinyl ethers or by ring opening polymerisation of ε-caprolactame with fluorinated alcohols) and other different techniques like group-transfer polymerisation, anionic polymerisation, metathesis and then polycondensation. In another section, the preparations, properties and applications of fluorinated triblock copolymers from the same content as above are described. Finally, the synthesis of fluorinated multiblock copolymers, different from those reported in Chapter 2 (fluorotelechelics), their properties and applications are described.

2. Synthesis, Properties and Applications of Fluoroblock Copolymers

2.1. FLUORINATED DIBLOCK COPOLYMERS

This section is divided into five main parts. The first one reports the use of traditional (or uncontrolled) radical polymerisation and telomerisation, while the second topic describes the synthesis of F-diblock copolymers from modern techniques of radical pseudo-living polymerisation as mentioned above. A further step concerns the preparation of fluorinated diblock copolymers by combining the above systems (i.e., classic and pseudo-living techniques) or condensation/radical methods. The following part reports methods of controlled syntheses of F-diblock copolymers that do not involve any radical methods: cationic, anionic, group-transfer polymerisation (GTP) and ring-opening metathesis polymerisation. Finally, the last sub-section gives some examples of F-diblock copolymers achieved from condensation methods.

2.1.1. Synthesis of Fluorodiblock Copolymers by Traditional Radical Oligomerisation, Telomerisation and Polymerisation

2.1.1.1. From Radical Addition of Fluorinated Transfer Agents to Hydrogenated Monomers

This part is devoted to the synthesis and the properties of $C_nF_{2n+1}-C_pH_{2p+1}$ molecules that represent interesting models for polytetrafluoroethylene-*b*-polyethylene, PTFE-*b*-PE, copolymers.

The synthesis of these fluorocarbon–hydrocarbon diblocks have been investigated by many authors and pioneered by Tiers [15] and Brace [16]. Their preparation was achieved in high yield from a two-step scheme, starting from the addition of a perfluoroalkyl iodide onto an ω-unsaturated hydrogenated olefin (that usually does not propagate under radical initiation):

$$C_nF_{2n+1}-I + H_2C=CH(CH_2)_{p-2}H \xrightarrow[\text{(2) H}^-]{\text{(1) rad.}} C_nF_{2n+1}-C_pH_{2p+1}$$

This first step can be initiated photochemically, electrochemically and biochemically and in the presence of organic initiators, metals, salts or metallic complexes [17–20]. The radical initiators are commercially available azo (the most used one is AIBN), peroxydicarbonates, peresters or peroxides.

The second step can be carried out in the presence of either tributyl stannane [21], Zn/HCl or AlLiH$_4$ or the ω-perfluoroiodoalkane can undergo a reductive dehalogenation which was recently reviewed by Gambaretto *et al.* [22].

Many R_F-R_H diblock co-oligomers have been synthesised [23–40] and their properties investigated (Table I). The preparation, properties and uses of these highly fluorinated amphiphiles regarding the structural aspects and their applications in the biomedical fields have been extensively reviewed by Napoli [39]

TABLE I

Synthesis and Properties of Various C_NF_{2N+1}–C_PH_{2P+1} Diblock Co-oligomers

C_nF_{2n+1}–C_pH_{2p+1}	Investigations or properties	Reference
C_6F_{13}–C_pH_{2p+1}, $p = 2-20$	Structural and thermal studies (T_m)	[23]
$C_6F_{13}C_{10}H_{21}$	Stabilisation of PFC/EYP emulsion	[24–26]
C_8F_{17}–C_2H_5	Surface properties	[27]
C_8F_{17}–C_8H_{17}	Gel phase formation	[28]
C_8F_{17}–$C_{12}H_{25}$	Formed reversed micelles in F-solvents	[29]
C_8F_{17}–$C_{16}H_{33}$	Formed reversed micelles in F-solvents	[29, 30]
C_8F_{17}–$C_{16}H_{33}$	Inclusion of β-cyclodextrin	[31]
C_8F_{17}–$C_{18}H_{37}$	Purification by gel phase formation	[32]
C_8F_{17}–C_pH_{2p+1}, $p = 2-20$	Thermal transitions	[23]
$C_8F_{17}CH=CHC_8H_{17}$	Stabilisation of PFC/EYP emulsion	[25]
$C_{10}F_{21}$–C_6H_{13}	Surface properties	[27]
$C_{10}F_{21}$–$C_{10}H_{21}$	Liquid crystal phase	[33, 34]
$C_{10}F_{21}$–C_pH_{2p+1}, $p = 2-20$	Structural and thermal studies (T_m)	[23]
$C_{12}F_{25}$–$(CH_2)_pH$, $p = 2-20$	For $p = 2, 4, 6$, the diblock is tilted	[35]
$C_{12}F_{25}(CH_2)_pH$, $p = 4-14$	Solution and solid-state properties (melting, crystallisation and micelle associations)	[36, 37]
$C_{12}F_{25}$–$C_{20}H_{41}$	Melting and crystallisation behaviours	[38]

T_m, PFC and EYP stand for melting temperature, perfluorocarbon and egg-yolk phospholipids, respectively.

and Krafft and Reiss [25], respectively. Table I displays a non-exhaustive list of various fluorocarbon–hydrocarbon diblocks reported in the literature. These molecules were studied in relation to their interesting properties: surface, melting and crystallisation behaviour, solution and solid-state properties. For example, calorimetric studies of R_F–R_H diblocks with at least four to six methylene units and more than six fluorocarbon groups led to one or even two solid–solid transitions of high cooperativity [23, 33, 36]. The solid–solid transition resulted in a change in the molecular packing with no conformation disordering [35]. It was also shown that the heat of melting was almost independent of the length of the hydrocarbon segment until it reaches more than 14 methylene groups. In contrast, the enthalpy of the solid–solid transition was found to increase with the number of methylene units [35]. As a matter of fact, at elevated temperatures, association in toluene so far regarded as the best solvent for these diblock molecules, was not observed [37]. Regarding the structure, Höpken et al. [37] noted "more complicated structures than spherical or cyclindrical micelles". Furthermore, Nostro et al. [31] obtained original inclusion compounds from C_8F_{17}–$C_{16}H_{33}$ with β-cyclodextrin (CD) in water. Interestingly, this A–B co-oligomer readily penetrated the hydrophobic CD's cavity and produced a fine crystalline powder. The aggregation of several R_F–R_H/CD inclusion compounds yielded the formation of large nanotubes observed by AFM. In addition, their stabilising characteristics and high lipophilicity [41, 42] (for example, these molecular "dowels" [42] or interfacially active compounds [43]) have increased affinity for the perfluorocarbon (PFC)/egg-yolk phospholipids interface. Indeed, their addition into PFC emulsion results in remarkable stabilisation for particles ranging in size from 0.13 to 16 μm since they induce a reduction from 86 to 74 $\overset{\circ}{A}{}^{2}$ per molecule of the area occupied by the phospholipid polar head groups at the interface [25].

These investigations, well carried out by Krafft *et al.* [44], also indicated that differences in stabilisation were noted even for very low sizes of particles between $C_6F_{13}-C_{10}H_{21}$ and a perfluorinated stabilising additive $C_{16}F_{34}$ of comparable molecular weight. In the same review and in another one [45], the interests of R_F-R_H in the development of PFC emulsion as blood substitutes are well documented. As a matter of fact, $C_8F_{17}CH=CHC_8H_{17}$ discovered for the first time in the 1970s, is nowadays involved in many investigations [25] for the preparation of oxygen-carrying PFC emulsions. They are stable enough to allow their heat-sterilisation under standard conditions.

In addition, the same team obtained interesting mixed Langmuir monolayers of 10,12-pentacosadiynoic acid and $C_8F_{17}-C_{16}H_{33}$ [30] as an original contribution compared the pioneer survey of Gaines [46] on the surface activity of R_F-R_H.

Other surveys were recently performed by Gambaretto's group, which demonstrated the low surface tensions obtained [22] and high chemical inertness. This Italian team [28] further optimised the synthesis of $C_8F_{17}-C_8H_{17}$ diblock by modifying the reactant molar ratios, the temperature and the reaction time. Their optimal conditions were achieved at 95 °C with an excess of 1-hexene initiated by AIBN and they noted that the perfluorooctyl iodide was converted quantitatively after 20 min. These authors also studied the dilution ratio and the gel points in various alcohols (methanol, ethanol and propanol) [28, 32].

On the other hand, American, Dutch and German groups [23, 27, 29, 36–38] devoted their attention onto the melting and crystallisation behaviour of these diblock co-oligomers.

These semifluorinated diblocks have already found many applications: stabilisation of blood emulsion [44, 45], contrast agents [47] especially for $C_nF_{2n+1}CH=CIC_pH_{2p+1}$ (obtained by radical addition of perfluoroalkyl iodide onto hydrogenated alkyne), waxes for ski or snowboard soles [48–50], surfactants [51] and self-assembled vesicules with unusual properties [52–54].

2.1.1.2. *From Telomerisation Reaction*

As shown above and especially in Chapter 1, telomerisation reactions are useful key reactions to achieve functional (or perfluorinated) low molecular weight polymers. In this sub-section, two key reactions are considered: (i) those involving a hydrogenated transfer agent for the telomerisation of fluoromonomers (different from fluoroalkenes since such a topic is extensively reported in Chapter 1) and (ii) those including a fluorinated transfer agent (and, in these cases, fluorinated thiol or iodoalkyl telogen) in the telomerisation of hydrogenated or fluorinated monomers.

2.1.1.2.1. Telomerisation of Fluoromonomers with Hydrogenated Telogens. Several telomerisations of fluoro(meth)acrylates [55–59] have already been reported, hence leading to telomers:

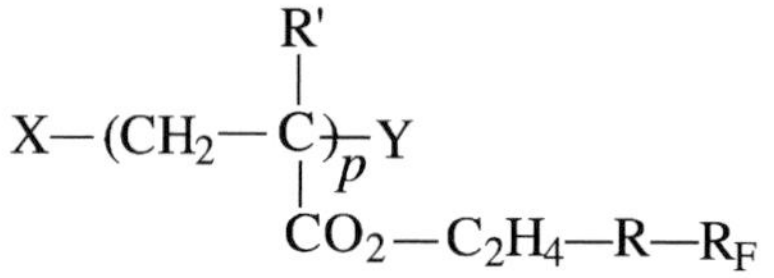

$$X-(CH_2-\underset{\underset{CO_2-C_2H_4-R-R_F}{|}}{\overset{\overset{R'}{|}}{C}})_{\!p}\!-Y$$

(where R represents "—" or NR_1SO_2 and R' = H, CH_3)

The most common transfer agents used were mercaptans [59, 60].

2.1.1.2.2. Telomerisation of Hydrogenated Monomers with Fluorinated Transfer Agents. In this sub-section, non-exhaustive examples of R_F-*b*-poly(M) diblock copolymers are reported. Two various monomers have been selected according to their complementary properties that they can bring from their resulting homo-block obtained: poly(acrylamide) improves surfactant properties of the synthesised R_F-*b*-poly(acrylamide) because of its hydrophilicity, while polybutadiene block enables the R_F-*b*-polybutadiene diblock copolymer to exhibit a certain softness.

2.1.1.2.2.1. Telomerisation of 1,3-Butadiene with Perfluoroalkyl Iodides. Although radical monoaddition of perfluoroalkyl iodides to 1,3-butadiene was investigated by various authors [61–65], we have recently studied the telomerisation of this diene with these transfer agents [66, 67] leading to various C_nF_{2n+1}-oligo(butadiene) (with $n = 6, 8$) whose molecular weights were ranging between 1300 and 3000, as follows:

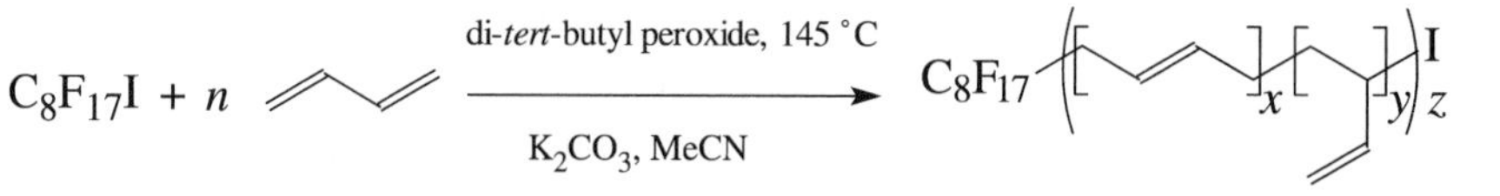

$$C_8F_{17}I + n \;\overset{\text{di-}\textit{tert}\text{-butyl peroxide, 145 °C}}{\underset{\text{K}_2\text{CO}_3, \text{ MeCN}}{\xrightarrow{\hspace{3cm}}}}\; C_8F_{17}\!\left[\left[\right]_{\!x}\left[\right]_{\!y}\right]_{\!z}\!I$$

As expected, the 1,4-butadienic units were in a higher content (ca. 90/10 with respect to 1,2-butadiene ones).

2.1.1.2.2.2. Telomerisation of Acrylamide with Fluorinated Transfer Agents
2.1.1.2.2.2.1. Telomerisation of Acrylamide with Fluorinated Mercaptan or Alkyl Iodides. Among many industrial surveys, it is worth citing various patents deposited by the Ciba-Geigy [68] and Atochem [69, 70] companies. The resulting fluorinated telomers were used as surfactants [51, 71].

TABLE II

FAMILY OF FLUORINATED TELOMERS AND THEIR SURFACE TENSION DEVELOPED BY THE CIBA-GEIGY COMPANY [68]

Fluorinated telomers	n	Surface tension (dyne cm^{-1})		
		0.1%	0.01%	0.001%
$R_FC_2H_4S-(CH_2-\underset{\underset{CONH_2}{\mid}}{CH})_n-H$	15	21	30	62
$R_FC_2H_4S-[CH_2-\underset{\underset{COOCH_2CHCH_2N^{\oplus}(CH_3)_3Cl^{\ominus}}{\mid}}{\overset{\overset{CH_3}{\mid}}{C}}]_n-H$ $\underset{OH}{\mid}$	5	24	–	–
	25	27	–	–
	200	40	–	–
$R_FC_2H_4S-[CH_2-\underset{\underset{CONHC(CH_3)_2CH_2SO_3H}{\mid}}{CH}]_n-H$	5	26	33	53
$R_FC_2H_4S-[CH_2-\underset{\underset{\text{(N-methylpyridinium iodide)}}{\mid}}{CH}]_n-H$ 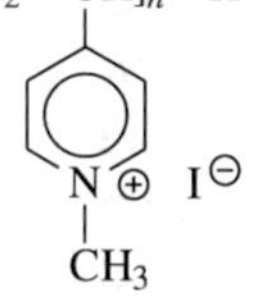	5	21	37	–
	10	22	41	–
$R_FC_2H_4S-[CH_2-\underset{\underset{\text{(pyrrolidone)}}{\mid}}{CH}]_n-H$ 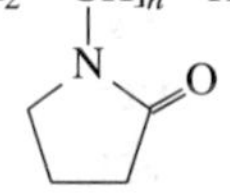	12	19	27	56
	40	20	27	59
	120	21	46	65

The former company has been developing different fluorinated acrylamido telomers but also fluorinated vinyl pyridino and vinyl pyrrolidone telomers, the superficial tensions of which are listed in Table II.

The Atochem's patents [69, 70] concern telomers having the formulae:

$$C_nF_{2n+1}(CH_2)_m-(CH_2-CR_1R_2)_pX$$

where $n = 6-20$, $m = 0, 1$ or 2, $p = 5-1000$, $R_1 = H$ or CH_3, $R_2 = CO_2H$ or $CONR_3R_4$ where R_3 and R_4 stand for H or alkyl or hydroxyalkyl substituents and X represents an I, Cl or Br atom. Regarding the academic investigations, Boutevin et al. [72] synthesised original amphiphilic diblock telomers from the radical telomerisation of acrylamide with fluorinated thiols, perfluoroalkyl iodides initiated by AIBN, as follows:

$$C_nF_{2n+1}I + n\ H_2C=\underset{\underset{CONH_2}{\mid}}{CH} \longrightarrow C_nF_{2n+1}-(CH_2CH)_n-\underset{\underset{CONH_2}{\mid}}{I}$$

$n = 6$ or 8

A wide range of various telomers was produced with $\overline{DP}_n$ varying from 15 to 560 according to the initial reactant molar ratios. The authors also determined the transfer constants to the fluorinated thiol ($C_T = 6500 \times 10^{-4}$), to $C_6F_{13}I$ ($C_T = 550 \times 10^{-4}$). These authors also used $C_6F_{13}C_2H_4I$ as another telogen. However, its transfer constant ($C_T = 40 \times 10^{-4}$) is much lower than that of the corresponding perfluorinated transfer agent.

These compounds exhibit tensioactive properties. For example, the superficial tension at the equilibrium for a telomer of average $\overline{DP}_{35}$ was 24.2 and 18.2 dyne cm^{-1} for concentrations of 100 and 1000 ppm, respectively [72]. Their syntheses were scaled-up and these products have been marketed by Atochem, under the Forafac$^®$ trademark. They can be used as original foams for extinguishers (called fire-fighting foams [73]) or as surfactants for emulsion polymerisation and well summarised in a recent review [74].

However, the redox telomerisation of acrylamide was also investigated, but with fluorinated transfer agents bearing a trichloromethyl end-group [75].

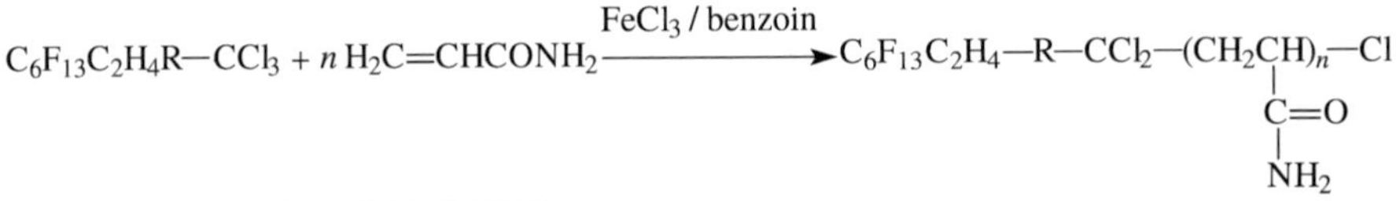

$$C_6F_{13}C_2H_4R-CCl_3 + n\,H_2C=CHCONH_2 \xrightarrow{\text{FeCl}_3\text{ / benzoin}} C_6F_{13}C_2H_4-R-CCl_2-(CH_2CH)_n-Cl$$
$$\begin{matrix} & & C=O \\ & & | \\ & & NH_2 \end{matrix}$$

with R: OCO, CO$_2$CH$_2$, OCOCHClCH$_2$

As above, according to the initial reactant molar ratios, the average $\overline{DP}_n$ ranged between 15 and 115, and it was noted that the lower the initial [FeCl$_3$/benzoin]$_o$/ [acrylamide]$_o$ molar ratio, the higher the yield (up to 93%) and the higher the $\overline{DP}_n$. However, because these fluorinated polyacrylamide telomers exhibit a poor hydrolytic stability, a new generation of amphiphilic diblock cotelomers was supplied [51, 71, 75].

   2.1.1.2.2.2.2.   Telomerisation of Tris(hydroxymethyl) Acrylamide with Fluorinated Mercaptan.   Since 1987, Pucci et al. [76–81] have investigated the telomerisation of acrylamido derivatives (e.g., tris(hydroxymethyl) acryla-midomethane (THAM)) with commercially available mercaptans $CF_3(CF_2)_mC_2H_4SH$ with $m = 5, 7, 9$ [77] or more recently with an original thiol containing an inner fluoro-group, as follows:

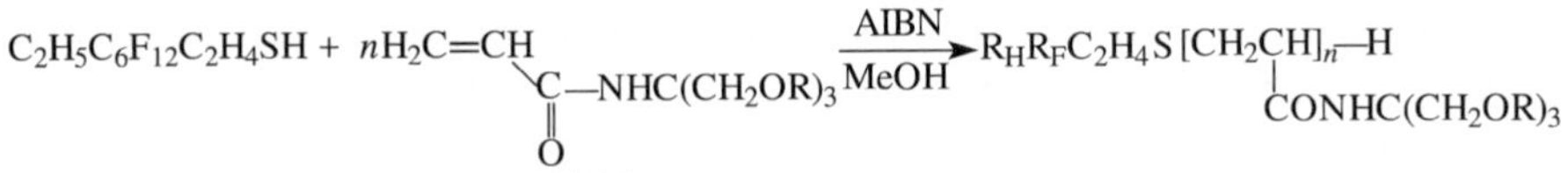

$$C_2H_5C_6F_{12}C_2H_4SH + n\,H_2C=CH \xrightarrow[\text{MeOH}]{\text{AIBN}} R_HR_FC_2H_4S\,[CH_2CH]_n-H$$
$$\begin{matrix} & C-NHC(CH_2OR)_3 & & CONHC(CH_2OR)_3 \\ & \| & \\ & O & \\ & \text{(THAM)} & \\ & & R = H \text{ or Gal} \end{matrix}$$

Indeed, the kinetics of telomerisation showed that the hydroxylated THAM was 50 times more reactive than the galactosylated THAM [80].

Various degrees of telomerisation were achieved depending on the reactant molar ratio and the R substituents (H or galactosylated (Gal) ones) and the yields were higher than 73%. The original fluorinated mercaptan was prepared from 1,6-diiodoperfluorohexane in a three-step procedure [79] as follows:

$$IC_6F_{12}I + H_2C=CH_2 \xrightarrow{\text{rad.}} IC_2H_4-C_6F_{12}-C_2H_4I \xrightarrow[\text{(3) } NH_4OH]{\substack{\text{(1) } SnBu_3H/AIBN \\ \text{(2) } (H_2N)_2CS}}$$

$$C_2H_5-C_6F_{12}-C_2H_4SH$$

These fluorinated telomers were used as non-ionic surfactants in biomedical applications [78], membranes detergents, emulsifiers, or drug carriers [81]. The high content of THAM units brought water solubility while the hydrophobic counter-part supplied the amphiphilic character. Hence, the physico-chemical properties of such surfactants were adjusted according to the biomedicinal requirements [76].

The critical micellar concentrations (cmc) of these original hydrogeno-fluorinated telomers were assessed. It was observed that the cmc of these derivatives (cmc $= 0.35-0.45$ mM) are 10 times higher than those exhibiting $C_8F_{17}C_2H_4$ tail [77] although it remained lower than that of the corresponding hydrogenated compounds. In addition, free energies of micellisation were evaluated from the cmc. They show that micelles form more favourably in the case of ethylfluorocarbon compounds ($C_2H_5C_6F_{12}C_2H_4$ tail; $\Delta G = -19$ kJ) than for hydrocarbon surfactants ($C_{10}H_{21}$ tail; $\Delta G = -16.7$ kJ) and less easily than for their hydrogenated analogues ($C_8F_{17}C_2H_4$ tail; $\Delta G = -25$ kJ). Further investigations on a new generation of hybrid surfactants $C_2H_5-C_6F_{12}-C_3H_6-R$ (where R stands for OSO_3Na or OPO_3X_2) were also achieved [79].

2.1.1.2.2.3. Telomerisation of Acrylic Acid with CTFE Telomers. Interestingly, the bistelomerisation can be a useful tool to synthesise diblock cotelomers. Indeed, the telomers produced in the first telomerisation step act as original transfer agents in the second telomerisation reaction. Hence, amphiphilic cotelomers can be produced when the first block is fluorinated while the second one is hydrophilic. Several examples can be supplied, such as those investigated in our Laboratory. The first one concerns the telomerisation of chlorotrifluoroethylene with carbon tetrachloride [82]. The telomers produced were efficient transfer agents for the redox telomerisation of 2-hydroxyethyl acrylate (HEA). The hydroxyl function was acrylated and these cotelomers were successfully used as original fluorinated photocrosslinkable products [83]. The reaction scheme is as follows:

$$CCl_4 + n\,F_2C=CFCl \longrightarrow CCl_3-(CF_2CFCl)_n-Cl$$

$$CCl_3-(CF_2CFCl)_n-Cl + m\,H_2C=CHCO_2C_2H_4OH \longrightarrow PCTFE\text{-}b\text{-}PHEA$$
$$\text{HEA}$$

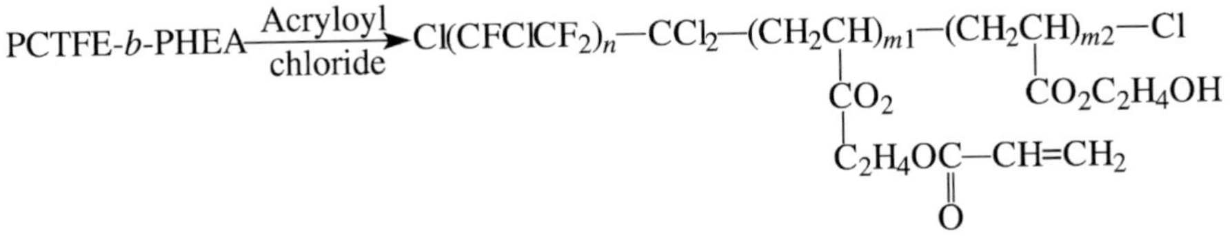

$$\text{PCTFE-}b\text{-PHEA} \xrightarrow[\text{chloride}]{\text{Acryloyl}} Cl(CFClCF_2)_n - CCl_2 - (CH_2CH)_{m1} - (CH_2CH)_{m2} - Cl$$

$$\begin{array}{cc} | & | \\ CO_2 & CO_2C_2H_4OH \\ | & \\ C_2H_4OC - CH = CH_2 & \\ \| & \\ O & \end{array}$$

A similar strategy was used starting from the redox telomerisation of acrylic acid with CCl_3Br. The ω-trichloromethyled oligoacrylic acids thus formed from a redox catalysis acted as potential telogens in the radical telomerisation of a fluorinated acrylate to yield an amphiphilic cotelomer [84]:

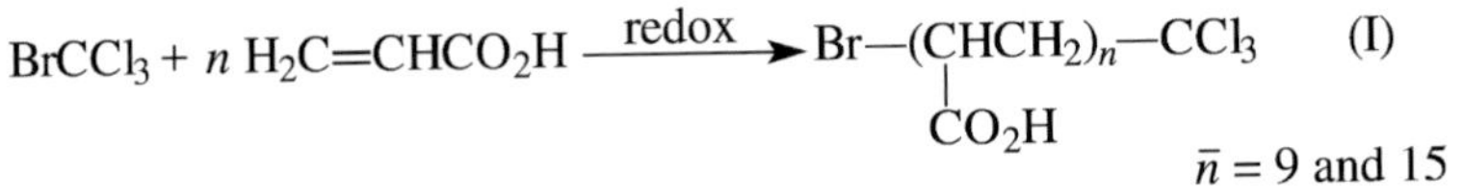

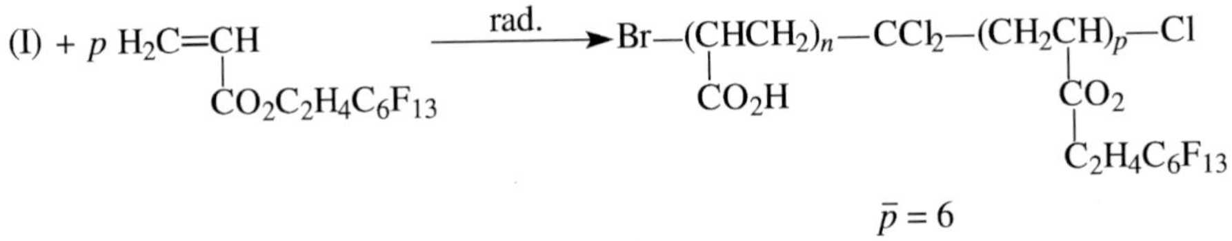

$$BrCCl_3 + n\ H_2C{=}CHCO_2H \xrightarrow{\text{redox}} Br - (CHCH_2)_n - CCl_3 \qquad \text{(I)}$$
$$\begin{array}{c} | \\ CO_2H \end{array}$$
$$\bar{n} = 9 \text{ and } 15$$

$$\text{(I)} + p\ H_2C{=}CH \xrightarrow{\text{rad.}} Br - (CHCH_2)_n - CCl_2 - (CH_2CH)_p - Cl$$
$$\begin{array}{cc} | & | & | \\ CO_2C_2H_4C_6F_{13} & CO_2H & CO_2 \\ & & | \\ & & C_2H_4C_6F_{13} \end{array}$$
$$\bar{p} = 6$$

The synthesis of fluorinated diblock cotelomers was also achieved in a two-step process from CCl_4 or $Cl_3CCF_2CCl_3$ [85] transfer agents. The first step dealt with the redox monoaddition of these telogens onto pentafluorostyrene or onto an allyl perfluoroalkylether, as follows:

$$R - CCl_3 + H_2C{=}CH - R_F \xrightarrow[82-115\ °C]{\text{redox}} R - CCl_2 - CH_2CHCl - R_F$$

$$R : Cl \text{ or } Cl_3CCF_2$$

$$R_F : C_6F_5 \text{ or } CH_2OC_2H_4C_nF_{2n+1}\ (n = 6, 8)$$

This fluorinated allyl ether was prepared by phase transfer catalysis of allyl chloride with 1,1,2,2-tetrahydro perfluorooctanol [86, 87].

In order to generate a telomer that exhibits a reactive CCl_3 end-group, the redox conditions (especially when CuCl was used as the catalyst) enabled us to produce monoadducts mainly, although the $FeCl_3$/benzoïn system produced the first three pentafluorostyrene adducts.

The second step involved these monoadducts as original transfer agents in the telomerisation of various M monomers such as acrylamide, acrylic acid, N-vinyl pyrrolidinone, NVP and MMA:

$$R_F \text{www} CCl_3 + n\ M \longrightarrow R_F \text{www} CCl_2(M)_n Cl$$

Usually, the [initiator or redox catalyst]$_o$/[M]$_o$ and [$R_F$wwwCCl_3]$_o$/[M]$_o$ initial molar ratios were $6 \times 10^{-3} - 3 \times 10^{-2}$ and 0.02–0.04, respectively. The yields ranged from 50 to 100% while the $\overline{DP}_n$ from 28 to 100, according to the above initial molar ratios. Interestingly, the telomerisation of acrylic acid initiated in the presence of a redox system led to an average cumulated degree of telomerisation $(\overline{DP}_n)$ of 34 while, when the reaction was initiated by AIBN, $\overline{DP}_n$ was 23,000. This confirms that the transfer to the telogen is negligible compared to that of the redox catalyst.

The telomerisation of NVP was initiated by AIBN, leading to NVP telomers with a $\overline{DP}_n$ reaching 52–75.

2.1.1.2.3. Telomerisation of Fluoromonomers with Fluorinated Transfer Agent. Numerous examples have been supplied in Chapter 1 (Fluorotelomers) and two key examples regarding the telomerisation of vinylidene fluoride with perfluoropolyether are reported hereafter.

A process involving a C–Br cleavage was used by Marchionni *et al.* [88, 89] who telomerised vinylidene fluoride (VDF) with ω-bromoperfluorinated poly-ethers, PFPE–Br (i.e., Fomblin$^®$ derivatives) in a non-controlled process:

$$\underset{\text{PFPE–Br}}{CF_3(OC_2F_4)_m(OCF_2)_p Br} + n\,VDF \xrightarrow{\text{rad.}} PFPE-b-PVDF$$

The radical initiators were peroxydicarbonates or peroxides. The authors showed that the higher the $[VDF]_o/[PFPE–Br]_o$ (where PFPE stands for perfluoropo-lyether) initial molar ratio, the higher the length of the PVDF block (reaching up to 71 VDF units).

These diblock copolymers exhibited a heterogeneous morphology composed of two amorphous zones assigned to both blocks. That attributed to the perfluorinated polyether block had a T_g of $-143\,°C$ in all the cases, while the second one depended upon the chain length of VDF blocks, hence varying from -82 to $-52\,°C$ for a number of VDF units ranging from 13 to 71.

A similar investigation leading to PFPE-*b*-poly(VDF-*co*-HFP) from the radical cotelomerisation of VDF and HFP with PFPE–I was recently reported [90].

$$\underset{\quad\;\;CF_3\quad\;\;CF_3}{F(CFCF_2O)_n CFI} + n\,VDF + m\,HFP \xrightarrow{\quad DTBP \quad} PFPE\text{-}b\text{-poly(VDF-}co\text{-HFP)}$$

The PFPE–Is were Krytox$^®$ derivatives supplied by the DuPont Company with an average-number molecular weight of 1200–1800. The reaction was carried out in solution in perfluorohexane and the reaction initiated by di*tert*-butyl peroxide (DTBP). A wide range of PFPE-*b*-PVDF and PFPE-*b*-poly(VDF-*co*-HFP) block copolymers were obtained from various ($[VDF]_o + [HFP]_o)/[PFPE–I]_o$, $[initiator]_o/([VDF]_o + [HFP]_o)$ and $[VDF]_o/([VDF]_o + [HFP]_o)$ initial molar ratios. The molecular weight of these (co)telomers assessed by ^{19}F NMR spectroscopy ranged between 2000 and 30,300 [90].

In contrast to the previous example, all these diblock copolymers showed one T_g only, varying from -50 to $-30\,°C$.

2.1.1.3. By Dead-End Polymerisation of Fluoromonomers from Macroinitiators

Radical dead end polymerisation (DEP) [91] enables one to synthesise A−B or A−B−A type copolymers, the A sequence being contributed by the initiator.

This initiator is usually introduced in a high amount and, consequently, oligomers are obtained, as mentioned in Table III that lists the main differences between the DEP and the classical radical polymerisation. Although this concept has led to numerous F-triblock co-oligomers (see Section 2.2.1.1), one interesting example has been reported in the literature.

An original strategy to prepare block copolymers was suggested by the Nippon Oils and Fats Company which commercialised a polyperoxide that enables to initiate the polymerisation of a (fluoro)monomer yielding a diblock copolymer. The macroperoxide was synthesised by condensation of a telechelic diacid chloride with hydrogen peroxide as follows [92]:

$$n \; Cl\!-\!\overset{\displaystyle O}{\overset{\|}{C}}\!-\!R\!-\!\overset{\displaystyle O}{\overset{\|}{C}}\!-\!Cl + H_2O_2 \longrightarrow -[\overset{\displaystyle O}{\overset{\|}{C}}\!-\!R\!-\!\overset{\displaystyle O}{\overset{\|}{C}}\!-\!O\!-\!O]_n\!-$$

The most common one is an oligoester named "ATPPO" in which R represents:

$$-(CH_2)_4\!-\!\overset{\displaystyle O}{\overset{\|}{C}}\!-\!O(C_2H_4O)_3\,\overset{\displaystyle O}{\overset{\|}{C}}(CH_2)_4\!- \quad [92].$$

In this case, the average molecular weight in number is 5700 ($n = 16$) with PDI $= 1.90$.

This Japanese company claimed to prepare original block copolymers from ATPPO in a two-step procedure: first, by the polymerisation of hydrogenated methacrylates (MMA, hydroxyethyl methacrylate) initiated by ATPPO at 70 °C, followed by a further radical polymerisation of a fluorinated acrylate

TABLE III

COMPARISON BETWEEN THE CHARACTERISTICS OF THE DEAD END POLYMERISATION [91] AND THE CLASSICAL RADICAL POLYMERISATION ($t_{1/2}$ MEANS THE HALF-LIFE OF THE INITIATOR)

	Dead end polymerisation	Classical radical polymerisation
Principle	Termination of growing chains by primary recombination	Various possible ways of termination: disproportionation, recombination, transfers, etc.
Initiation	Quick production of radicals	Radicals generated more homogeneously vs. time
Initiator concentration	Usually high (1−15%)	Low (0.1−1.0%)
$t_{1/2}$ Initiator	Low (≈ 20 min)	High (1−10 h)
Chain length	Usually low M_n	Variable M_n
Functionality of polymer	Production of telechelic oligomers from telechelic initiator ($f = 2.0$)	Difficult to control even from a telechelic initiator

(e.g., 1H,1H,2H,2H-perfluorodecyl acrylate, FOA) as follows:

$$\text{ATPPO} + n\,\text{MMA} \longrightarrow \text{PMMA}-(\text{O}-\text{O})_n \xrightarrow{\text{FOA}} \text{PMMA-b-PFOA}$$

Both steps were achieved in nearly quantitative yields. These diblocks copolymers are commercially available as Modiper® trademark.

These block copolymers showed interesting surface properties [93] since the fluorinated chains migrated remarkably to the surface. The surface layer was not formed by the F-segments alone, but consisted of both polymeric segments of block copolymer, resulting in the formation of a "sea-island" structure.

2.1.1.4. By Chemical Changes and Coupling

This section displays several examples of syntheses of F-diblock copolymers by chemical modification of hydrogenated polymers P into C_nF_{2n+1}–P. To our knowledge, the fluorination of hydrogenated block copolymers is not selective enough to obtain readily such halogenated diblock.

First, examples such as R_F–PDMS and R_F–PEO (where PDMS and PEO stand for polydimethylsiloxane and polyethylene oxide, respectively) are supplied. The latter are prepared in a two-step procedure starting from the telomerisation of fluoromonomers.

Then coupling reactions (e.g., esterification of two (fluoro)telomers) are reported.

2.1.1.4.1. Chemical Changes of Non-fluorinated Polymers

2.1.1.4.1.1. Synthesis of R_F-b-Polydimethylsiloxane. The Daikin Company [94] synthesised original fluorinated PDMS in two ways: either from the hydrosilylation [95] of an unsaturated fluoroderivative onto hydrogenodimethylsilyl PDMS, or from a two-step procedure starting from the radical addition of a perfluoroalkyl iodide onto an ω-unsaturated PDMS. Both routes are summarised below:

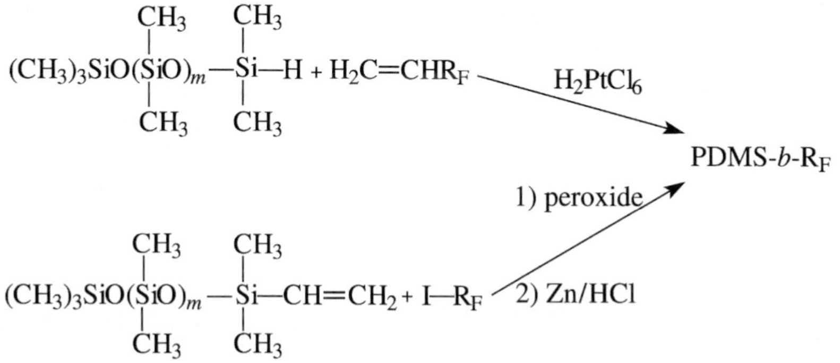

The Japanese researchers used a wide range of fluorinated derivatives in which R_F represents C_nF_{2n+1} ($n = 4, 6, 8, 10, 12, 14, 16$), $C_4F_9[CF_2CF(OC_3F_7)]_p(C_2F_4)_q$

$(p/q = 32/68$ and the number-average molecular weight was 840 or 980), $F(C_4F_8O)_nC_2F_4$ ($n = 26$ or 28) and $C_3F_7O-(C_4F_8O)_{24}-C_2F_4$.

   2.1.1.4.1.2.   Synthesis of R_F-PEO.   Three examples are given below.

   (a) PEO-*b*-poly(F-acrylate) diblock copolymer is synthesised from the telomerisation of $H_2C=CHCO_2C_2H_4N(CH_3)SO_2C_2H_4C_6F_{13}$ with an original poly(ethylene oxide), PEO, mercaptan $HSCH_2CO_2(C_2H_4O)_n-H$ ($n = 14$), initiated from persulfate, peroxides or AIBN, at pH ranging from 2 to 11 and in temperatures in the 40–120 °C range [56, 57, 96].

   Its formula is as follows:

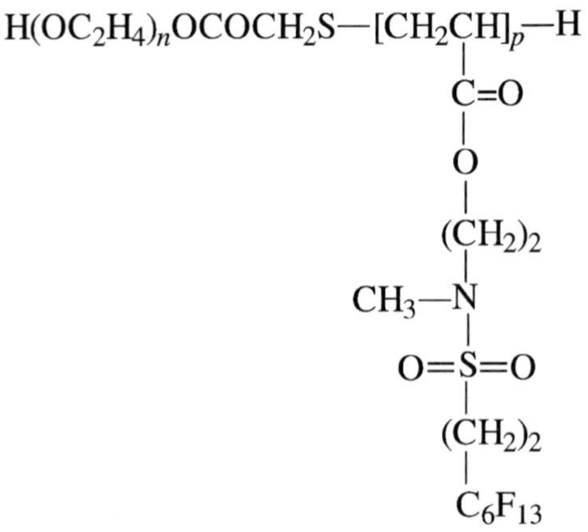

$$
\begin{array}{c}
H(OC_2H_4)_nOCOCH_2S-[CH_2CH]_p-H \\
| \\
C=O \\
| \\
O \\
| \\
(CH_2)_2 \\
CH_3-N \\
| \\
O=S=O \\
| \\
(CH_2)_2 \\
| \\
C_6F_{13}
\end{array}
$$

Its applications concern the textile treatment (for hydrophobic and oleophobic properties) for clothes, furniture, carpets and leather [96].

   This copolymer is formulated with surfactants, various additives and with or without any solvent. Furthermore, thermocondensables and catalysts which favour the good adhesion of the layers onto the textile are also introduced [96]. The formulation is applied as solutions or dispersions by various techniques such as enduction, impregnation, pulverisation, brushing or padding. Such a commercially available product was marketed first by Ugine Kuhlmann, then by Elf Atochem followed by the Atofina Company under the Foraperle® Tradename and now by the DuPont Company.

   (b) A similar strategy was recently used in our laboratory [97] from a two-step procedure. The first one consisted in obtaining a macromolecular mercaptan (from the condensation of thioglycolic acid with a polyethylene oxide (PEO) of molecular weight 2000 or 5000) followed by the telomerisation of 1H,1H,2H,2H-perfluorodecyl acrylate (FDA) as follows:

$$
HSCH_2CO_2H + \underset{PEO}{HO(C_2H_4O)_n-CH_3} \longrightarrow HSCH_2CO_2\underset{PEO-SH}{(C_2H_4O)_nCH_3}
$$

$$
PEO-SH + n\,H_2C=CH\underset{FDA}{CO_2C_2H_4C_8F_{17}} \xrightarrow[C_6H_5CF_3]{AIBN} PEO-b-PFDA
$$

The molecular weight of the fluorinated block ranged between 4900 and 79,400. Interestingly, these fluorodiblock copolymers behave as surfactants or "dispersing aids" (with a fluorinated CO_2-philic block and a PEO hydrophilic block) and enable emulsion of an aqueous solution in liquid CO_2 under supercritical pressure to be achieved. On the other hand, such surfactants can also be useful to realise dispersion polymerisation successfully and could interact with the growing polymer in a thermodynamically favourable monomer.

In addition, the condensation of ω-perfluorinated thiols onto PEO bearing an end-halogen atom was successfully achieved by Cambon's team [98] as follows:

$$C_nF_{2n+1}C_2H_4SH + Cl(C_2H_4O)_mH \xrightarrow{\text{EtONa}} C_nF_{2n+1}C_2H_4S(C_2H_4O)_m-H$$
$$\underset{R_F-PEO}{}$$

$$R_F-PEO \xrightarrow[\text{(3) PEO/EtONa}]{\substack{\text{(1) TosCL} \\ \text{(2) }(NH_2)_2CS}} R'_F-PEO$$

where $n = 6, 8$ and the PEO $= (C_2H_4O)_p$ with $p = 2, 3, 4$ and $R'_F=C_nF_{2n+1}C_2H_4$ $S(C_2H_4O)_pS$. This group has successfully synthesised various fluorinated block copolymers bearing ether or thioether bridges that represent a real exclusivity.

The originality of this research deals with the monodispersity of the diblock derivative, in contrast to all other compounds obtained in other studies.

2.1.1.4.2. Coupling of (Fluoro)telomers by Esterification. In addition to these systems described, above few coupling reactions of two species (one being a fluorinated sequence while the other one a hydrogenated block) have been investigated. One of the rare examples consists in the esterification reaction of two telomers (one of both containing polyfluoroacrylate segment), as follows [58].

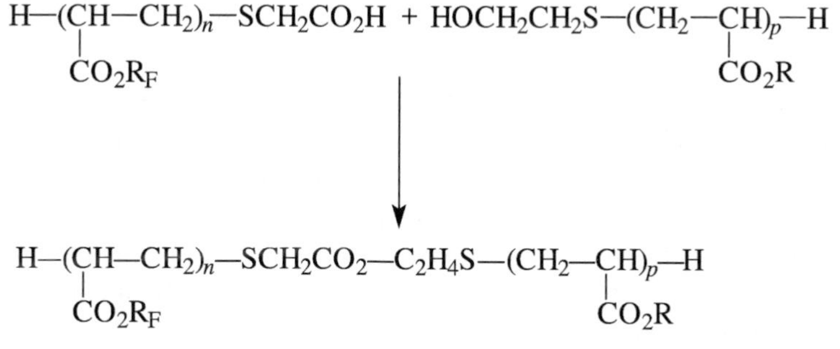

$$H-(\underset{\underset{CO_2R_F}{|}}{CH}-CH_2)_n-SCH_2CO_2H + HOCH_2CH_2S-(CH_2-\underset{\underset{CO_2R}{|}}{CH})_p-H$$

$$\downarrow$$

$$H-(\underset{\underset{CO_2R_F}{|}}{CH}-CH_2)_n-SCH_2CO_2-C_2H_4S-(CH_2-\underset{\underset{CO_2R}{|}}{CH})_p-H$$

2.1.2. Synthesis of Fluorodiblock Copolymers from Controlled Radical Polymerisation

2.1.2.1. Introduction

As mentioned in Section 1, various methods have been used for the preparation of block copolymers. Despite the exceptional amount of attention paid to the prospects of various catalytic systems, radical polymerisation has not lost any of its importance, particularly in this area. Its competitiveness with other methods of conducting polymerisation is attributable to the simplicity of the mechanism and good reproducibility. Actually, the extensive use of free radical polymerisation in practice is well understood when considering the ease of the process, the soft processable conditions of vacuum and temperature, the fact that reactants do not need to be extremely pure and the absence of residual catalyst in the final product. Thus, it can be easily understood that more than 50% of all plastics have been produced industrially via radical polymerisation.

Methods of obtaining block copolymers by radical processes have been developed relatively recently compared with other processes, especially ionic methods. This may be due to the nature of the radical, which is an intermediate with a very short lifetime and a very high non-selective reactivity. These characteristics do not favour a well-controlled architecture, as in the case of living carbanions appearing in anionic polymerisation. However, the recent development of new methods to control the reactivity of radicals and to give a living character to the growing macroradicals offers fascinating new possibilities.

Since the 1990s, the international research activity in the field of controlled/living radical polymerisation (LRP) has grown to a considerable extent [99, 100]. The general principle of the methods reported so far relies on a reversible activation–deactivation process between dormant chains (or capped chains) and active chains (or propagating radicals), as follows:

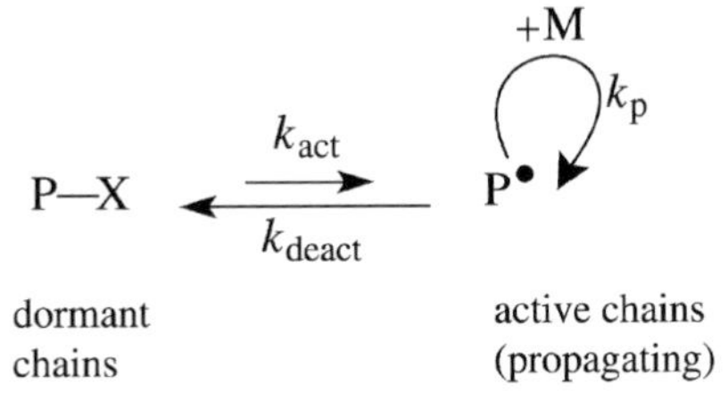

$$\text{P—X} \underset{k_{\text{deact}}}{\overset{k_{\text{act}}}{\rightleftharpoons}} \text{P}^{\bullet} \overset{+M}{\underset{}{\circlearrowright}} k_p$$

dormant chains	active chains (propagating)

The major benefit of such methods is the possibility of tailoring well-defined polymers (telechelic, block, graft, or star copolymers) by convenient radical polymerisations. The oldest techniques are the iniferter method [101–105] (where iniferter stands for INItiation–transFER–TERmination), nitroxide-mediated radical polymerisation (NMP) [106–109], ATRP [110–113], ITP [114] and RAFT [115–120] including macromolecular design via interchange of xanthates (MADIX) [118–120]. Their respective specific mechanisms are briefly summarised below:

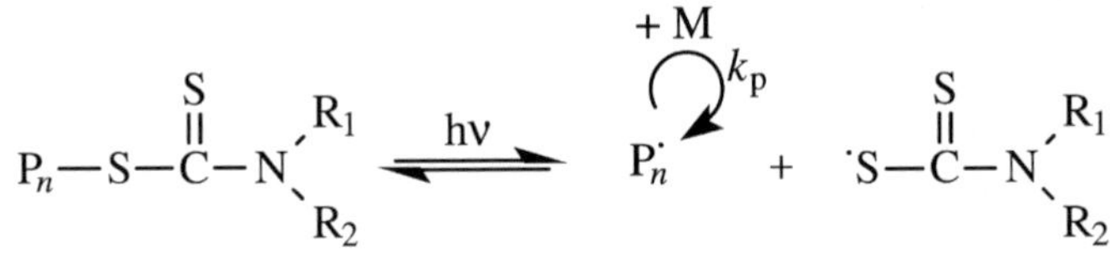

$$P_n\text{—S—}\overset{\overset{\text{S}}{\|}}{\text{C}}\text{—N}\overset{R_1}{\underset{R_2}{}} \overset{h\nu}{\rightleftharpoons} P_n^{\bullet} \overset{+M}{\underset{}{\circlearrowright}} k_p + {}^{\bullet}\text{S—}\overset{\overset{\text{S}}{\|}}{\text{C}}\text{—N}\overset{R_1}{\underset{R_2}{}}$$

(a) Photo-iniferter method

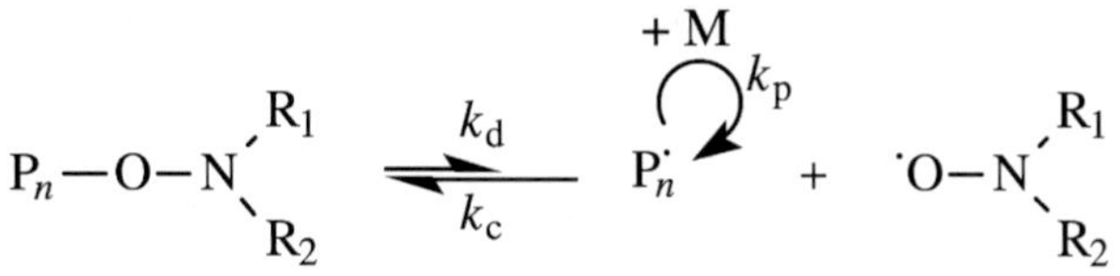

$$P_n-O-N\overset{R_1}{\underset{R_2}{\diagup}} \quad \underset{k_c}{\overset{k_d}{\rightleftharpoons}} \quad P_n^{\cdot}\overset{+M}{\overset{\curvearrowright}{}}{}^{k_p} \quad + \quad {}^{\cdot}O-N\overset{R_1}{\underset{R_2}{\diagup}}$$

(b) NMP

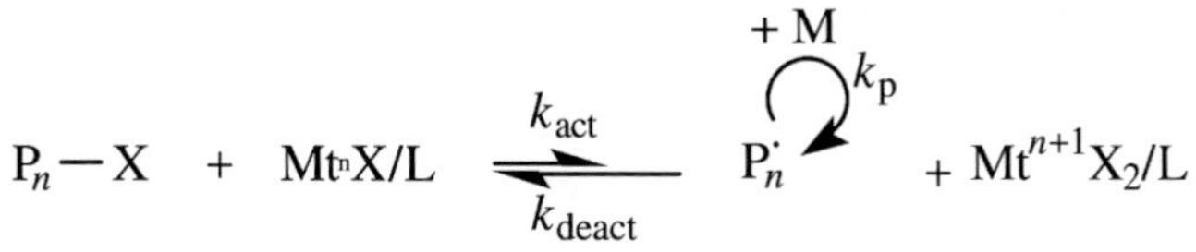

$$P_n-X \quad + \quad Mt^vX/L \quad \underset{k_{deact}}{\overset{k_{act}}{\rightleftharpoons}} \quad P_n^{\cdot}\overset{+M}{\overset{\curvearrowright}{}}{}^{k_p} \quad + \quad Mt^{n+1}X_2/L$$

(c) ATRP

$$P_m-I \quad + \quad P_n^{\cdot}\overset{+M}{\overset{\curvearrowright}{}}{}^{k_p} \quad \underset{k_{ex}}{\overset{k_{ex}}{\rightleftharpoons}} \quad P_m^{\cdot}\overset{+M}{\overset{\curvearrowright}{}}{}^{k_p} \quad + \quad P_n-I$$

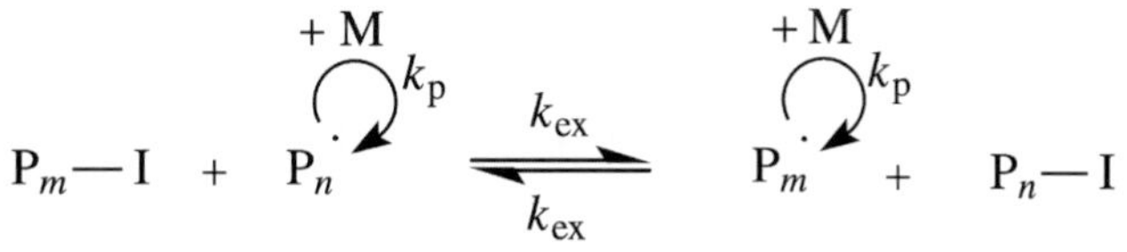

(d) Iodine transfer radical polymerisation (ITP)

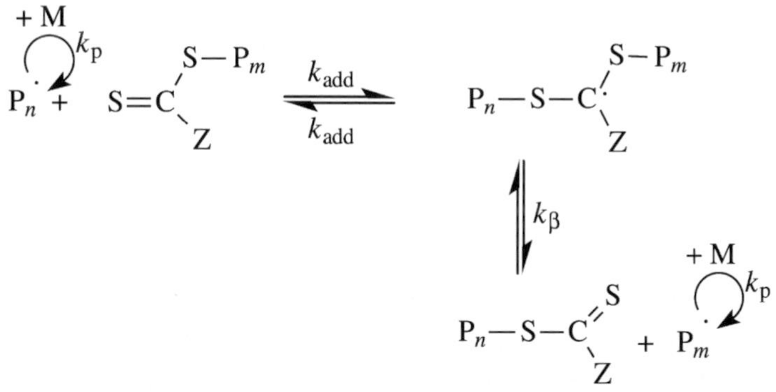

$$P_n^{\cdot}\overset{+M}{\overset{\curvearrowright}{}}{}^{k_p} \quad + \quad S{=}C\overset{S-P_m}{\underset{Z}{\diagup}} \quad \underset{k_{add}}{\overset{k_{add}}{\rightleftharpoons}} \quad P_n-S-C\overset{S-P_m}{\underset{Z}{\diagdown}}{}^{\cdot}$$

$$\updownarrow k_\beta$$

$$P_n-S-C\overset{S}{\underset{Z}{\diagup}} \quad + \quad P_m^{\cdot}\overset{+M}{\overset{\curvearrowright}{}}{}^{k_p}$$

with Z: C_6H_5, OR, SR (R: alkyl group)

(e) Reversible addition–fragmentation chain transfer process (RAFT and MADIX)

In Scheme (a), the example of dithiocarbamates is given for simplicity, but other molecules such as xanthates $R_1SC(S)OR_2$ also fall in this category.

In Scheme (e), for xanthates, some authors sometimes refer to the name MADIX, as used by the Rhodia company for this process.

Fluorinated organic compounds have sometimes been used in LRP to take advantage of electronic or steric effects, to serve as labelling agents or to impart specific properties to the resulting materials such as low surface energy, chemical resistance and solubility in supercritical carbon dioxide, to name a few. Non-exhaustive examples are going to be given below regarding various species involved in these schemes.

2.1.2.2. By Polymerisation of Fluoromonomers Using Alkoxyamines (Nitroxide-Mediated Polymerisation, NMP)

As mentioned above, nitroxides do not behave as initiators but are efficient counter-radicals to trap a macromolecular growing radical $P_n^{\cdot}$ (living species) to lead to a "dormant species" $P_n-O-NR_1R_2$. This polymer has a P_n-O bond that is thermally reversible and hence gives back the $P_n^{\cdot}$ macroradical able to further react onto the monomer in the propagation step.

An attempt at synthesis of fluorinated diblock copolymers was performed in our laboratory by a controlled radical stepwise copolymerisation first involving 2,2,6,6-tetramethyl piperidyloxy (TEMPO) as a counter radical [121, 122]. Then, the introduction of the second block was achieved from NMP of meta/para(1H,1H,3H,3H,4H,4H-2-oxaperfluorododecyl) styrene (FDS) from PS–TEMPO, the molecular weights of which being between 3000 and 10,000 and with a narrow range of polydispersity indices (PDI) (1.10–1.37). Its structure is as follows:

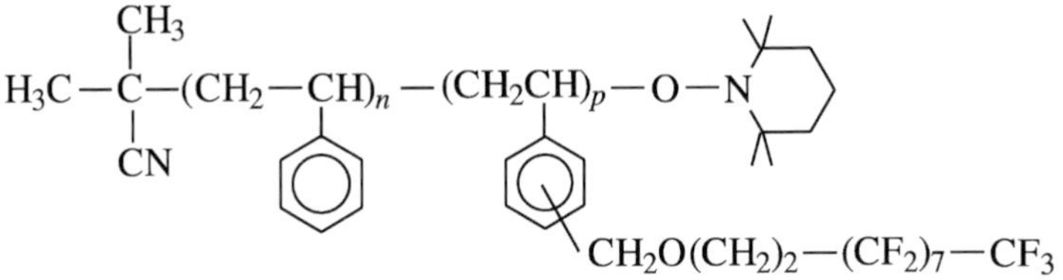

The molecular weight of the PS-*b*-PFDS diblock copolymer was 20,000–80,000.

A similar strategy for the synthesis of poly(styrene)-*b*-poly(1,1,2,2-tetrahydroperfluorodecyl acrylate) (PS-*b*-PFDA) block copolymers was also used by the same authors involving N-*tert*-butyl-1-diethyl phosphono-2,2-dimethyl propyl nitroxide (DEPN) as the counter radical [122]. They were obtained with molar masses ranging from 40,000 to 60,000. Indeed, this phosphonyl persistent radical was suitable to control the polymerisation of both styrenic and acrylate monomers (e.g., *n*-butyl acrylate [122]), in contrast to TEMPO that is mainly limited to control the polymerisation of styrenic monomers such as styrene [121, 122] or chloromethyl styrene [123].

Interestingly, melting transition temperatures were noted ($T_m = 55-59\,°C$ or $72-73\,°C$ for PFDS and PFDA blocks, respectively) brought by the fluorinated block (the fluorinated group may behave as side-chain liquid-crystalline polymers).

Andruzzi *et al.* [124, 125] also investigated similar syntheses recently, using TEMPO as the counter radical.

As a matter of fact, the fluorinated CO_2-philic blocks, if large enough, impart solubility in dense CO_2 to the diblock copolymers, making them useful as original macromolecular surfactants in such a medium since they are able to stabilise the dispersion polymerisation of styrene in supercritical (sc) CO_2.

Indeed, PFDS [121] and PFDA [126] homopolymers were shown to be soluble in dense CO_2 and cloud point measurements of PS-*b*-PFDA block copolymers in dense CO_2 exhibited good solubility at moderate temperatures (25–65 °C) and pressures (120–250 bar).

2.1.2.3.   From Atom Transfer Radical Polymerisation (ATRP)

2.1.2.3.1.   Introduction.   In this part, two kinds of routes for obtaining fluorinated block copolymers by ATRP can be considered: (i) the radical polymerisation of a monomer with a fluorinated telomer or a fluorinated initiator is followed by an ATRP step, and (ii) two consecutive or sequential ATRP steps.

ATRP requires an activated C–X (X is a halogen atom) bond of the initiator and the examples below consider X as Cl, Br or I.

Various halogenated initiators have been involved in ATRP. They may exhibit different C–I, C–Br and C–Cl cleavable bonds to enable successful controlled reactions but those that possess the alter two bonds are mainly used. A few examples are given below.

However, one atypical fluorinated initiator is a simple perfluorinated sulfonyl chloride (using the method pioneered by Percec [110]) from which Fiering *et al.* [127] achieved the original synthesis of ω-perfluorinated polystyrenes by ATRP of styrene in the presence of CuCl and 2,2′-bipyridine (bipy) as coordinating ligand:

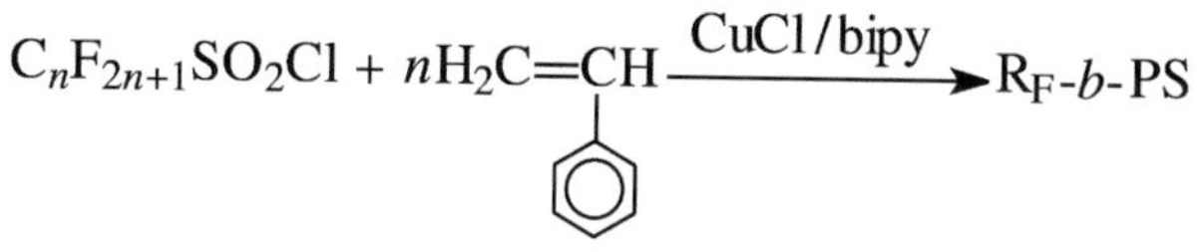

$$C_nF_{2n+1}SO_2Cl + nH_2C=CH \xrightarrow{\text{CuCl/bipy}} R_F\text{-}b\text{-}PS$$

2.1.2.3.2.   From Telomeric Initiators Reactivated by a C–I Cleavable Bond.   Iodoperfluoroalkane $CF_3(CF_2)_2CF_2I$ was used as the initiator in ATRP of MMA with cuprous salt/bipyridine catalysts [128]. However, further experiments with poly(vinylidene fluoride)-capped iodine macroinitiators led to a low initiator efficiency because the propagation rate was much faster than the initiation rate.

Similar results were obtained in our group from $CF_3-(CF_2)_5-I$ as the initiator and CuX/pyridine-2-carbaldehyde *N*-pentylimine as the catalysts (X = Br, I), with or without addition of Cu°, in ATRP of methyl methacrylate and styrene [129].

2.1.2.3.3.   Initiators that Possess a C–Br Cleavable Bond.   One example of synthesis of fluorodiblock copolymers from initiator containing a C–Br reactivable bond was recently achieved by Ming *et al.* [130], who synthesised an ω-perfluorinated ester bearing a C–Br bond. These Dutch authors obtained two different PMMAs by ATRP: one PMMA containing a perfluorinated end-group and one PMMA that did not bear any fluorinated group, whose molar masses, PDI and T_g were 4400 and 3300, 1.3 and 1.3, and 84 and 83 °C, respectively. The synthesis of R_F–PMMA was as follows:

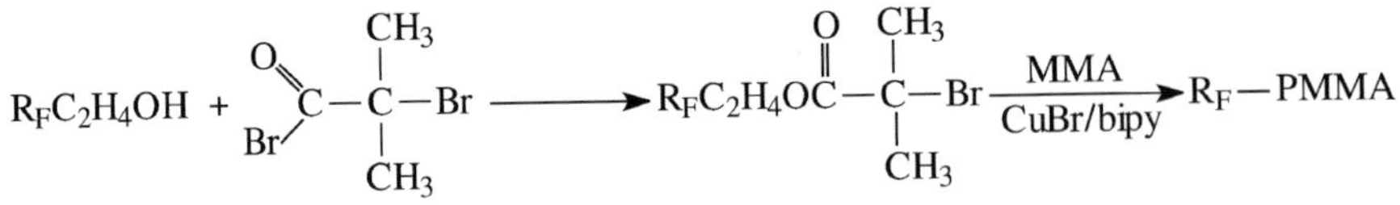

$$R_FC_2H_4OH \; + \; \underset{Br}{\overset{O}{\underset{}{}}}\overset{CH_3}{\underset{CH_3}{C-C-Br}} \longrightarrow R_FC_2H_4O\overset{O}{\overset{\parallel}{C}}-\overset{CH_3}{\underset{CH_3}{C}}-Br \xrightarrow[CuBr/bipy]{MMA} R_F-PMMA$$

$R_F : C_6F_{13} ; C_8F_{17}$

2.1.2.3.4.   From Initiators Possessing a C–Cl Cleavable Bond.   Examples of cleavage of the C–Cl bond are scarce, probably due to its stronger bond dissociation energy (see Section 2.1.2.3.4).

One of them, described by Lim *et al.* [131], involves the ATRP of fluorinated methacrylate (FMA) from a polyethylene oxide macroinitiator bearing a $CH(CH_3)$–Cl end-group, PEO–Cl. The synthesis was achieved in two steps:

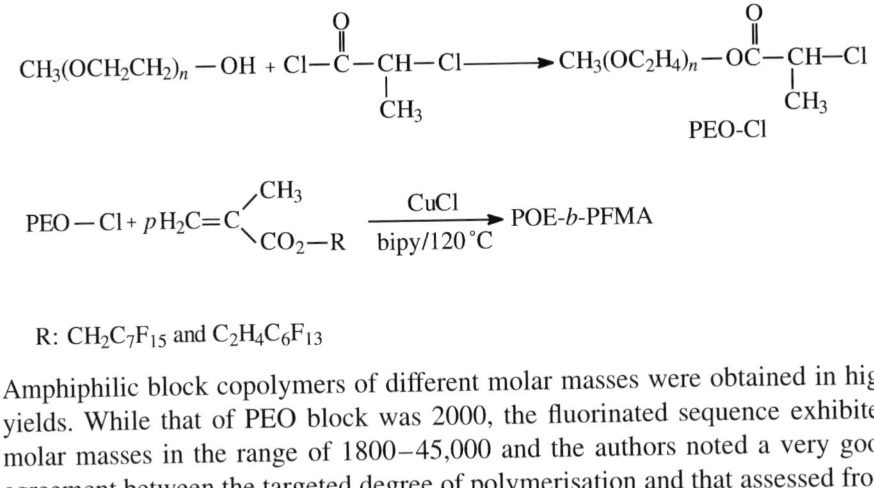

$$CH_3(OCH_2CH_2)_n - OH + Cl-\overset{O}{\overset{\parallel}{C}}-\underset{CH_3}{CH}-Cl \longrightarrow CH_3(OC_2H_4)_n-O\overset{O}{\overset{\parallel}{C}}-\underset{CH_3}{CH}-Cl$$
$$PEO-Cl$$

$$PEO-Cl + p\,H_2C=C\overset{CH_3}{\underset{CO_2-R}{}} \xrightarrow[bipy/120°C]{CuCl} POE\text{-}b\text{-}PFMA$$

R: $CH_2C_7F_{15}$ and $C_2H_4C_6F_{13}$

Amphiphilic block copolymers of different molar masses were obtained in high yields. While that of PEO block was 2000, the fluorinated sequence exhibited molar masses in the range of 1800–45,000 and the authors noted a very good agreement between the targeted degree of polymerisation and that assessed from ^{1}H NMR. In addition, by means of dynamic light scattering (DLS) and TEM, this Korean team observed that these block copolymers self-assembled in chloroform to form micelles and that they showed different morphologies of spheres and rods by changing the copolymer composition. Further, these surfactants lowered the interfacial tension between water and sc CO_2 to stabilise water-in-CO_2 emulsions.

2.1.2.3.5.   Synthesis of Block Copolymers by Stepwise ATRP Processes.   This sub-section reports various examples of syntheses of fluorodiblock copolymers prepared by stepwise or sequential ATRPs in which the first polymer produced (that constitutes the first block) acts as an original initiator involved in a second ATRP step of another monomer. It is composed of two parts, the first reporting the synthesis and properties of *aliphatic* A–B copolymers and the second one describing those of *aromatic* fluorodiblocks. For both, those in which the first block is hydrogenated (whereas the second one is fluorinated) are considered, followed by those in which the fluorinated block is introduced first.

   2.1.2.3.5.1.   Synthesis of Aliphatic Fluorinated A–B Copolymers.   Betts *et al.* [132] obtained amphiphilic block copolymers as follows:

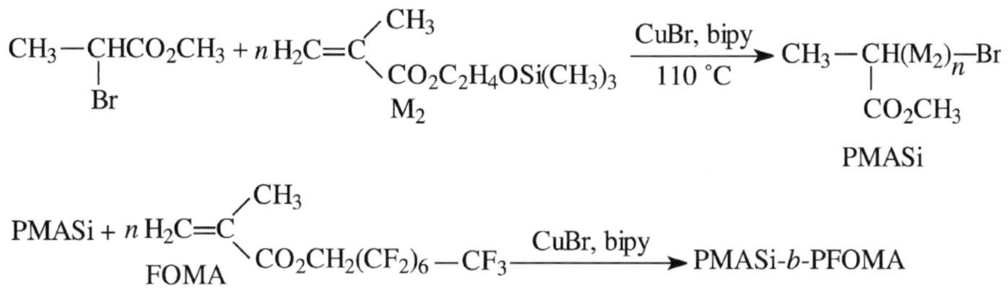

$$CH_3-\underset{\underset{Br}{|}}{C}HCO_2CH_3 + n\,H_2C=C\overset{\diagup CH_3}{\underset{\diagdown CO_2C_2H_4OSi(CH_3)_3}{}} \xrightarrow[110\,°C]{CuBr,\ bipy} CH_3-\underset{\underset{CO_2CH_3}{|}}{C}H(M_2)_{\overline{n}}Br$$

$$\qquad\qquad\qquad\qquad\qquad M_2 \qquad\qquad\qquad\qquad\qquad\qquad\qquad PMASi$$

$$PMASi + n\,H_2C=C\overset{\diagup CH_3}{\underset{\diagdown CO_2CH_2(CF_2)_6-CF_3}{}} \xrightarrow{CuBr,\ bipy} PMASi\text{-}b\text{-}PFOMA$$

$$\qquad\quad FOMA$$

Interestingly, by ATRP, various poly(M_1)-*b*-poly(FOMA) block copolymers were obtained, M_1 representing MMA, 2-hydroxyethyl methacrylate, styrene, *tert*-butyl acrylate and 2-ethylhexyl acrylate with molar masses ranging from 40,000 to 56,000.

   Similar investigations were carried out by Wu *et al.* [133] and Gaynor *et al.* [134, 135], who prepared fluorinated diblock copolymers with improved surface properties.

   Another sequential coATRP was achieved by a Chinese group [136] which was able to prepare poly(acrylate)-*b*-poly(fluoromethacrylate) diblock copolymer and poly(fluoromethacrylate)-*b*-poly(acrylate)-*b*-poly(fluoromethacrylate) triblock copolymers. The first block, PA, was introduced by ATRP of methyl or butyl acrylate in bulk in the presence of monobrominated or dibrominated initiator. The molar masses and PDI ranged from 9700 to 14,000 and from 1.33 to 1.48, respectively.

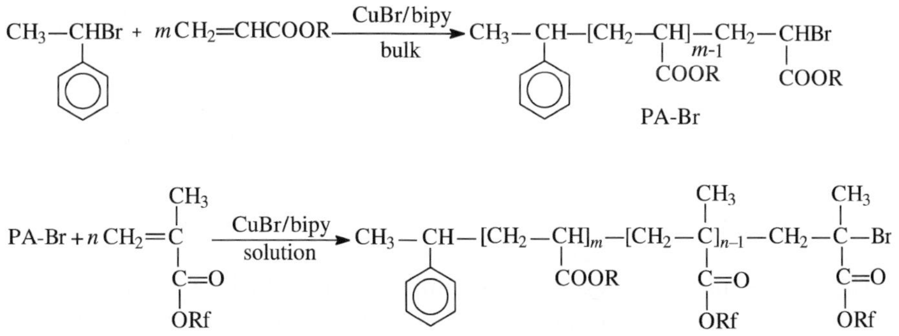

$$CH_3-CHBr + m\,CH_2=CHCOOR \xrightarrow[bulk]{CuBr/bipy} CH_3-CH-[CH_2-CH]_{\overline{m-1}}-CH_2-CHBr$$

$$\qquad\qquad\qquad\qquad\qquad\qquad\qquad\qquad\qquad\qquad\qquad\qquad COOR \qquad COOR$$

$$\qquad\qquad\qquad\qquad\qquad\qquad\qquad\qquad\qquad\qquad PA\text{-}Br$$

$$PA\text{-}Br + n\,CH_2=\underset{\underset{\underset{ORf}{|}}{\underset{C=O}{|}}}{\overset{\overset{CH_3}{|}}{C}} \xrightarrow[solution]{CuBr/bipy} CH_3-CH-[CH_2-CH]_m-[CH_2-\underset{\underset{\underset{ORf}{|}}{\underset{C=O}{|}}}{\overset{\overset{CH_3}{|}}{C}}]_{n-1}-CH_2-\underset{\underset{\underset{ORf}{|}}{\underset{C=O}{|}}}{\overset{\overset{CH_3}{|}}{C}}-Br$$

$$\qquad\qquad\qquad\qquad\qquad\qquad\qquad\qquad\qquad\qquad\qquad\qquad COOR$$

$$R = -CH_3, -CH_2CH_2CH_2CH_3$$

$$Rf = -CH_2CH_2OC_9F_{19}$$

The second or third blocks were introduced in ATRP of fluoromethacrylate in solution, leading to diblock or triblock copolymers in 62–66% yields. Molar masses reached 29,000 with PDI remaining as low as 1.27–1.34.

A similar strategy was successfully applied by Li *et al.* [137], who started from PMMA (or PS) macroinitiators bearing a bromine end-atom to obtain PMMA-*b*-PFOSMA or PS-*b*-PFOSMA diblock copolymers (where FOSMA stands for 2-(*N*-ethyl perfluorooctane sulfonamido)ethyl methacrylate). Molar masses were characterised by laser scattering ($M_\omega = 3.0 \times 10^6$ and 1.2×10^6, respectively) and GPC, and the authors noted the presence of a bimodal distribution suggested as either high molar mass polymers or "conglomerated" macromolecules caused by the difference between both comonomers. This second assumption was confirmed by the determination of the particle size distribution of the copolymer solutions: both PMMA-*b*-PFOSMA and PS-*b*-PFOSMA exhibit the same profile showing two kinds of particles of different sizes (the smaller size being ca. 10 nm while the larger one was about 100 nm). These Chinese authors also studied the thermal stability of the fluorinated diblock copolymers that combined the thermal stabilities of both homopolymers better than that of PFOSMA homopolymer. Indeed, PMMA-*b*-PFOSMA and PS-*b*-PFOSMA started to degrade from 180 and 290 °C, respectively.

Another investigation regarding the synthesis of F-diblock copolymers by sequential ATRP involving initiator with a C–Br bond was achieved by Xia *et al.* [138], who synthesised poly(1,1-dihydroperfluorooctyl acrylate) (PFOA) and poly(1,1-dihydroperfluorooctyl methacrylate) (PFOMA) macroinitiators in sc CO_2 and they found that the more hydrophobic the bipyridine ligand, the higher the fluorinated (meth)acrylate conversion.

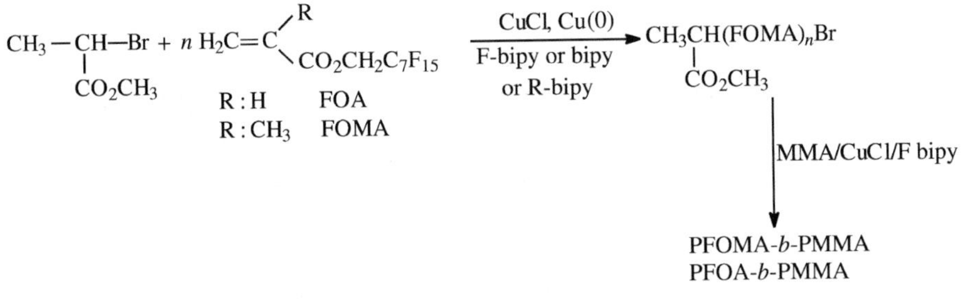

$$CH_3-CH-Br + n\, H_2C=C\begin{smallmatrix}R\\CO_2CH_2C_7F_{15}\end{smallmatrix} \xrightarrow[\text{or R-bipy}]{\substack{\text{CuCl, Cu(0)}\\ \text{F-bipy or bipy}}} CH_3CH(FOMA)_nBr$$

CH₃—CH—Br	
CO₂CH₃	

R : H   FOA
R : CH₃ FOMA

CH₃CH(FOMA)ₙBr
CO₂CH₃

MMA/CuCl/F bipy

PFOMA-*b*-PMMA
PFOA-*b*-PMMA

The second step was also successful when 2-(*N*-dimethylamino)ethyl methacrylate (DMAEMA) was used as monomer [139].

The authors also investigated the solubility of these block copolymers in CO_2 at various temperatures and pressures. Interestingly, at 65 °C and for 5000 psi (i.e., 34.5 MPa), the PFOMA-*b*-PDMAEMA was soluble in that medium.

2.1.2.3.5.2. Synthesis of Aromatic Fluorinated A–B Copolymers. Aromatic fluorodiblocks have also been obtained. Obviously, they can contain PS segments [137] but also blocks in which the fluorine atoms are linked to the aromatic ring, just like in original poly(4-fluorostyrene), poly(2,3,4,5,6-pentafluorostyrene) and poly(4-methoxy-2,3,5,6-tetrafluorostyrene) blocks.

Becker and Wooley [140, 141] synthesised amphiphilic diblock copolymers containing poly(methyl acrylate) and poly(4-fluorostyrene) blocks by ATRP with CuBr and N,N,N',N'',N''-pentamethyldiethylenetriamine (PMDETA) as coordinating ligand, with molar masses and PDI ranging between 8000 and 12,000, and 1.12 and 1.18, respectively. The reaction was as follows:

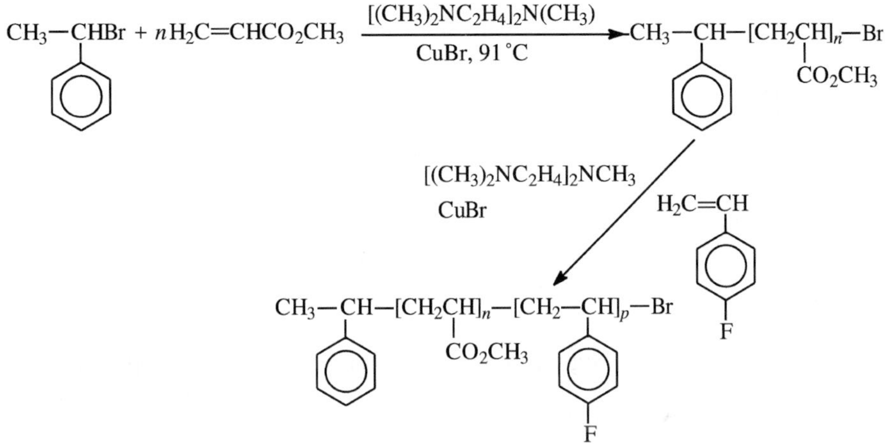

Alternatively, the polymerisation sequence was reversed, in which a bromoalkyl-terminated poly(4-fluorostyrene) was first prepared by ATRP, leading to (4-fluorostyrene) homopolymers endowed with $\overline{DP}_n$ of 24–101. These polymers were utilised as original fluorinated macroinitiators for the ATRP of methyl acrylate to generate poly(methyl acrylate)-b-poly(4-fluorostyrene) of $\overline{M}_n$ and PDI ranging from 8000 to 22,000 and 1.09 to 1.21, respectively. The degree of polymerisation was controlled by the ratio of initiating species to monomer and by the extent of subsequent monomer conversion. These stepwise copolymerisations were monitored by SEC and characterised by IR, ^{1}H, ^{13}C and ^{19}F NMR spectroscopies [141].

Interestingly, the block copolymers were hydrolysed in dioxane under acidic conditions to give poly(4-fluorostyrene)-b-poly(acrylic acid). These diblock copolymers were assembled into micelles and converted into shell-crosslinked nanoparticles (SCKs) in the presence of diamine by covalent stabilisation of the acrylic acid units in the shell. By DLS experiments, these authors noted that the higher the mole ratio of diamine, the lower the average hydrodynamic diameter of the resulting nanoparticle. These SCK nanostructures were characterised by AFM and TEM, showing that the fluorine aryl moieties were located within the core. The ^{19}F tags served as probes for the continued physical and biological characterisation of SCKs [141].

More recently, Hvilsted *et al.* [142] investigated the polymerisation of 2,3,4,5,6-pentafluorostyrene (FS) by ATRP at 110 °C using 1-phenylethyl bromide (PhEBr) as the initiator and CuBr/2,2′-bipyridine as the catalytic system [142, 143]. 96% conversion of the fluoroaryl monomer was achieved in 100 min (leading to a molar mass of 16,000 with $\overline{M}_w/\overline{M}_n \leq 1.2$ and $T_g = 95$ °C) and the polymerisation appeared to perform in a controlled manner. This high rate assumingly reflects the electron-withdrawing character of the fluorine atoms on the phenyl ring. Furthermore, the molar masses increased linearly with monomer conversions. These ranged between 4000 and 52,000 with PDI of 1.08 and 1.50, while the glass transition temperatures were 77 and 101 °C, respectively. Finally, the determined molar masses fit with the theoretical values with relative low polydispersities. The Br-terminated poly(2,3,4,5,6-pentafluorostyrene) (PFS) thus prepared acted as an original macroinitiator for the synthesis of PFS-*b*-PS block copolymers with relatively narrow polydispersities. After 260 min, their molar mass, polydispersity index and T_g were 20,800, 1.32 and 101 °C, respectively.

FS block copolymers can likewise be prepared from a similar PS macroinitiator. From PS macroinitiator obtained in 59% yield after 270 min with $\overline{M}_n = 10,000$, PDI = 1.13 and $T_g = 91$ °C, the PS-*b*-PFS block copolymer prepared after 880 min in 87% had the following characteristics: $\overline{M}_n = 14,200$, PDI = 1.35 and $T_g = 96$ °C. Three main properties of these fluoroaromatic diblock copolymers were investigated by Hvilsted's team. FS homo- and block copolymers gained higher thermal stability compared with PS [143, 144]. The extent of the thermal stability reflects the relative size of the FS block since PFS was more thermally stable than poly(4-methoxy-2,3,5,6-tetrafluorostyrene) which was more stable than PS [144]. T_g of FS homo- and block copolymers with styrene depends on the molar masses up to approximately 16,000, where a final value at around 101 °C was reached. The solubility of these polymers was realised in chlorinated, oxygen-containing or aromatic common organic solvents. Although the solubility of FS-containing polymers, especially that of the homopolymer, is much smaller than that of PS, the materials could still be handled as solutions. In particular, fluorobenzene is a good solvent for the PFS with a somehow reduced solubility for the block copolymers and PS itself [143]. The introduction of 4-methoxy substituent improves the solubility of PTFMS as compared with that of PFS in the investigated solvents, with the exception of fluorobenzene. However, the general solvent resistance is still higher than that of PS.

Solutions of homopolymers as well as block copolymers were expected to be viable for the creation of new low energy, low friction or higher thermal resistance surfaces cast or spin coated on other polymer surfaces.

Other fluorinated but functionalised styrenes such as 4-methoxy-2,3,5,6-tetrafluorostyrene (TFMS) were prepared and polymerised under ATRP conditions [144] with molar masses reaching 17,000 and PDI below 1.3.

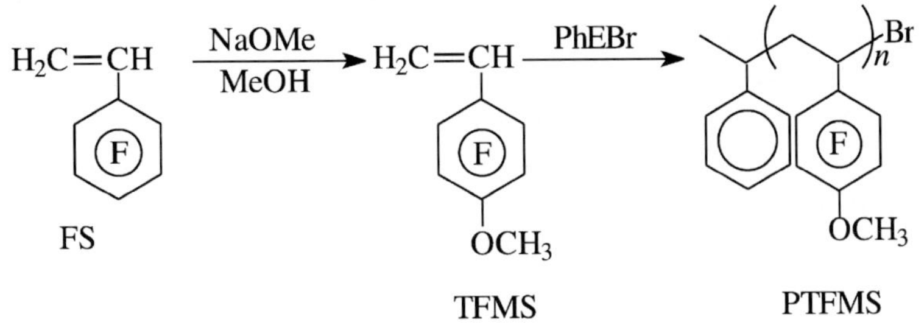

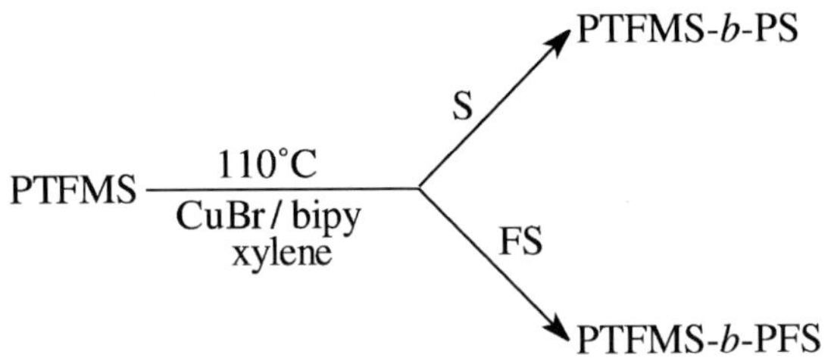

Interestingly, the apparent rate coefficients in ATRP of various styrenes [144, 145], substituted methoxy [144, 145] or fluorinated styrenes [144] were calculated. They show that the five fluorine atoms greatly increase the rate of polymerisation of FS compared with that of styrene, and that the electron-donating methoxy group enhances the polymerisation rate of TFMS even further as compared with that of FS. The decreasing reactivity series in ATRP of various styrenes was established as follows: TFMS > FS > 4-trifluoromethyl styrene > S. Block copolymers with styrene (PTFMS-*b*-PS) or with 2,3,4,5,6-pentafluorostyrene (PTFMS-*b*-PFS) were also synthesised. In the latter case, the molar masses were 12,600 or 13,800 (assessed by SEC or ^{1}H NMR, respectively) with a polydispersity index of 1.20 and a T_g of 95 °C. Again, both types of macroinitiators were employed with success. Optimisation of the hydrolysis of the methoxy groups was also realised [144] and the best conditions were achieved in the presence of boron tribromide. This opened up new ways to functionalise these highly fluorinated block copolymers, since the resulting 4-hydroxy-2,3,5,6-tetrafluorostyrene sites underwent alkylations via the Williamson ether synthesis approach [143, 144]. Hence, original fluorinated block copolymers with azobenzene side chains formed materials exhibiting liquid crystal properties (mesophase of smetic type) or can be potential candidates for optical storage of information through polarisation holography [144].

*2.1.2.4.   From Degenerative Transfer: By Iodine-Transfer Polymerisation,
          Thiuram or RAFT*
*2.1.2.4.1.   Synthesis of Fluorinated Block Copolymers by Iodine-Transfer
Polymerisation (ITP) of Fluoroalkenes*
   2.1.2.4.1.1.   Introduction.   ITP is one of the radical living processes pioneered in the late 1970s by Tatemoto [114, 146–153]. Actually, this process

requires (per)fluoroalkyl iodides because their highly electron-withdrawing (per)fluorinated group R_F provides the lowest level of the CF_2–I bond dissociation energy. The cleavage of the C–I bond, the bond dissociation energy of which amounts to 55 kcal mol^{-1}, can be achieved by various methods [18, 19, 152, 154]. Such a C–I cleavage is possible to a lesser extent for $R_FCH_2CH_2I$ [155].

Based on well-selected monomers, a method has been developed to control polymerisation with alkyl iodides (ITP) which can be easily applied to fluorinated alkenes.

For example, original PFPE-*b*-PMA, PFPE-*b*-PMMA or PFPE-*b*-PAN diblock copolymers (where PFPE, PMA, PMMA, and PAN stand for perfluoropolyether, polymethylacrylate polymethylmethacrylate and polyacrylonitrile, respectively), were achieved in a living process [156]. They were obtained from the ITP of methylacrylate, MMA and acrylonitrile with PFPE–I (of $\overline{M}_n = 4000-5000$), as depicted below:

$$C_3F_7(C_3F_7O)_pC_2F_4I + nH_2C=\overset{\overset{\displaystyle R_1}{|}}{\underset{\underset{\displaystyle R_2}{|}}{C}} \xrightarrow{\text{rad.}} \begin{array}{l} \text{PFPE-}b\text{-PMA} \\ \text{PFPE-}b\text{-PMMA} \\ \text{PFPE-}b\text{-PAN} \end{array}$$

PFPE-I

$$R_1: H, CH_3$$

$$R_2: CO_2CH_3 \text{ or } CN$$

Surprisingly, these authors claimed that ethylene and 1-hexene were required to make these reactions successful, otherwise the syntheses of diblock copolymers failed. Consequently, some copolymerisation of these α-olefins with MA, MMA or AN occurred in high yields and with narrow molecular distribution.

These block copolymers were used as greases and as additives (lubricants) for automobile engine oils.

The same strategy was used by the same author to prepare PFPE-*b*-PS copolymers blended with polystyrene (PS) [114]. This author studied the hydro- and oleophobic properties of this blend and observed that: (i) 1% addition of the block copolymer was sufficient to cover the main surface with PFPE segments and (ii) 10% addition of PFPE itself, although immiscible with PS, has no drastic effect on the surface contact angles in spite of its presence characterised by ESCA measurements.

A similar approach made it possible to obtain a water-soluble structure illustrated as perfluoropolyether-*b*-poly(sodium acrylate) [114, 156].

In addition, one of the best examples led to the synthesis of thermoplastic elastomers (TPE), in which the poly(VDF-*co*-HFP) block is elastomeric (soft segment) while the PTFE sequence is thermoplastic (hard block) [114, 146, 147, 157]:

$$iC_3F_7I + nVDF + pHFP \longrightarrow \underset{\text{poly(VDF-}co\text{-HFP)-I}}{iC_3F_7[(VDF)_x(HFP)]_n-I}$$

$$\text{poly(VDF-}co\text{-HFP)-I} + mTFE \longrightarrow \text{poly(VDF-}co\text{-HFP)-}b\text{-PTFE}$$

Such an ITP concept was named "degenerative transfer" in the mid-1990s and concerned the controlled radical polymerisation of methyl or butyl acrylate [156, 158, 159] or of styrene [158–160] from an iodinated transfer agent.

Iodine-containing compounds have been used for a long time as chain transfer agents in telomerisation [161, 162]. Fluorinated alkyl iodides such as $CF_3-(CF_2)_5-I$ and, to a lesser extent, $CF_3-(CF_2)_5-(CH_2)_2-I$, were reported to be efficient telogens (Chapter 1). However, the living character of the polymerisation carried out in the presence of alkyl iodides was revisited [114, 151–153, 156, 163]. A large variety of branched or linear perfluoroalkyl iodides [114, 146, 147, 156] or α,ω-diiodoperfluoroalkanes [114, 152, 164–168] and even polyiodides [169] have been involved in ITP or in stepwise cotelomerisation of fluoroalkenes [170]. Although the radical telomerisation of fluoroalkenes is reported in Chapter 1, we have decided not to mention it thereafter, but to give non-exhaustive examples of stepwise cotelomerisation followed by ITP of fluoroolefins. This order has been suggested by the increasing chain lengths of the fluorinated products varying from the fluorocotelomers to the fluoropolymers.

2.1.2.4.1.2. Stepwise Cotelomerisation. Various examples of direct cotelomerisation of fluoroalkenes have been supplied in Chapter 1, but here it is worth mentioning that several telomers can act as original transfer agents and hence enable the stepwise or sequential cotelomerisation of fluoroalkenes.

This was thus investigated in order to prepare "living" and well-defined tetrafluoroethylene (TFE) telomers as follows [171]:

$$C_2F_5I + n\,C_2F_4 \rightarrow C_4F_9I \xrightarrow{\text{TFE}} C_6F_{13}I \xrightarrow{\text{TFE}} C_8F_{17}I \rightarrow \text{etc.}$$

Such a living telomerisation can be initiated either thermally or in the presence of radical initiators or redox catalysts [171].

The application of such a process allows the step-by-step synthesis of block co-oligomers, either from fluoroalkyl iodides or α,ω-diiodoperfluoroalkanes. Chambers' investigations or those performed in our laboratory have led to the extensive use of fluoroalkyl iodides with chlorotrifluoroethylene (CTFE) for the preparation of efficient transfer agents X–Y with X = Y = I [172] or Br [173, 174] and also X = I and Y = Cl [175], Br [176] or F [177, 178].

$$X-Y + n\,CF_2{=}CFCl \longrightarrow X(C_2F_3Cl)_n Y$$

Chambers *et al.* [172] obtained difunctional oligomers with iodinated telogens. It has been observed, however, that from ICl or [IF], generated *in situ* from iodine and iodine pentafluoride, monoadducts were obtained as the sole products [177, 178]. Addition of [IF] to CTFE yields CF_3CFClI, which was successfully used as the transfer agent for the telomerisations of CTFE and hexafluoropropene (HFP) [178] or for stepwise cotelomerisations of CTFE/HFP, CTFE/HFP/VDF, HFP/VDF and HFP/trifluoroethylene [170]. All the above cotelomers were successfully end-capped by ethylene, allowing further functionalisation [170].

In addition, the CFClI end-group shows a higher reactivity than that of CF_2I, as Haszeldine reported previously [179].

Furthermore, stepwise cotelomerisations of various commercially available fluoromonomers also led to interesting, highly fluorinated derivatives, as in the following scheme:

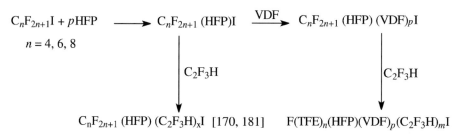

$$C_4F_9I + n\ CH_2{=}CF_2 \longrightarrow C_4F_9(C_2H_2F_2)_nI \xrightarrow{HFP} C_4F_9(VDF)_n(HFP)I\ [170]$$

$$iC_3F_7I + n\ CH_2{=}CF_2 \longrightarrow i\ C_3F_7(C_2H_2F_2)_nI \xrightarrow{HFP} iC_3F_7(C_2H_2F_2)_nCF_2CFICF_3$$
$$[170, 180]$$

$$C_nF_{2n+1}I + pHFP \longrightarrow C_nF_{2n+1}(HFP)I \xrightarrow{VDF} C_nF_{2n+1}(HFP)(VDF)_pI$$
$$n = 4, 6, 8$$

$$\downarrow C_2F_3H \qquad\qquad\qquad\qquad \downarrow C_2F_3H$$

$$C_nF_{2n+1}(HFP)(C_2F_3H)_xI\ [170, 181] \qquad F(TFE)_n(HFP)(VDF)_p(C_2F_3H)_mI$$

All these cotelomers were successfully end-capped with ethylene that offered new ω-functionalised fluorinated compounds [170].

As a model of ITP described below, the stepwise cotelomerisation of fluoroalkenes (such as VDF, trifluoroethylene, hexafluoropropene or chlorotrifluoroethylene) was performed, taking into account that the fluoroiodo end-group of the obtained telomer is capable of being reactivated in a further telomerisation reaction [170]. Interestingly, a large variety of telomers containing various polyfluorinated units exhibit amorphous to semicrystalline behaviour depending on the number of difluoromethylene groups (more than six lead to crystallinity [170]). Branches with trifluoromethyl groups decrease their crystallinity.

2.1.2.4.1.3. Iodine Transfer Polymerisation of Fluoroalkenes. Various fluorinated monomers have been successfully used in ITP. Basic similarities in these living polymerisation systems are found in the stepwise growth of polymer chains with each active species. The active living centre, generally located in the end-groups of the growing polymer, has the same reactivity at any time during polymerisation, even when the reaction is stopped [114, 152]. In the case of ITP of fluoroalkenes, the terminal active bond is always the $C{-}I$ bond originating from the initial iodine-containing chain transfer agent and monomer, as follows:

$$C_nF_{2n+1}{-}I + (p+1)H_2C{=}CF_2 \xrightarrow{R\cdot\ or\ \Delta} C_nF_{2n+1}(C_2H_2F_2)_p{-}CH_2CF_2{-}I$$

Molecular weights were claimed to be in the 30,000–10,000,000 range [114] and yet polydispersities were narrow (1.2–1.3) [114, 151, 152].

Tatemoto *et al.* [114, 151–153] used peroxides as initiators of polymerisation in solution usually involving perfluorinated or chlorofluorinated solvents.

Improvement was also possible by using diiodo- [114, 152, 164–168] and polyiodo-compounds [169].

The pioneering example of diblock copolymers [146, 147] is mentioned in Section 2.1.2.4.1.1.

Several investigations have shown that ITP can also occur in emulsion, solution or, later, by microemulsion [182] processes. This is not extensively described here but several articles and patents from Tatemoto [147, 149, 157, 164], DuPont [155, 165] or Ausimont [166–169] have been reported using ammonium persulfate as the initiator and involving tetrafluoroethylene (TFE), vinylidene fluoride (VDF) and hexafluoropropene (HFP) as monomers (see schemes above).

The first fluoroelastomers produced by ITP which could be peroxide-curable led to commercially available Dai-El® [147, 157] developed by Daikin (mainly by its subsidiary called Dai-Act). Such a polymer is stable up to 200 °C and finds many applications in high technology such as in O-rings and gaskets [149–151], transportation and electronics.

A similar strategy was also developed at the DuPont [155, 165] and Ausimont [166–168] companies in the 1990s leading to original TPE (see Section 2.2.2.2).

Fluorinated alkyl iodide $C_6F_{13}I$ was also used to perform ITP of styrene in aqueous emulsion and miniemulsion [182, 183]. In conventional batch emulsion polymerisation, the efficiency of the transfer agent was low, resulting in molar masses higher than expected. This is explained by a slow rate of diffusion of the hydrophobic perfluorinated transfer agent through the water phase, from the monomer droplets to the active latex particles. This problem was overcome in miniemulsion polymerisation for which the transfer agent was directly located in the polymer particle.

Iodine-transfer polymerisation is one of the few methods that make it possible to control the polymerisation of fluoroalkenes such as TFE or VDF. Because of strong dissociation energy of the CF_2–Br and CF_2–Cl bonds, polymerisations with brominated or chlorinated transfer agents have not been described.

2.1.2.4.1.4. Conclusion. Even though, in 1955, Haszeldine [179] and, in 1957, Hauptschein *et al.* [184, 185] showed a certain living nature of the radical telomerisation of CTFE with $ClCF_2CFClI$, only the present research has attracted many academic or industrial chemists towards this fascinating area. In addition, Dear and Gilbert [186] and Sharp *et al.* [187, 188] performed the telomerisation of various fluoroalkenes with disulfides, but they did not examine the living character of this reaction.

However, ITP and stepwise cotelomerisation of fluoroalkenes have shown a living character, and possible polymerisation by organometallic systems, which is quite successful for non-fluorinated monomers, could be an encouraging route for these haloalkenes.

Nevertheless, ITP is still a very interesting means to obtain diblock and even triblock copolymers (as mentioned in Section 2.2.2.2). The pioneered investigation developed by the Daikin company has also attracted the interest of DuPont and Ausimont since living block copolymers have been achieved. Further, these companies could obtain original fluorinated TPE that could be easily cured and involved in many applications. As a matter of fact, Dai-El® product is the only commercially available fluoroelastomer obtained by controlled radical polymerisation, although DuPont and Ausimont (now Solvay-Solexis) have deposited several patents on similar topics.

2.1.2.4.2. By Iniferter Method.
Three main examples are proposed here, involving thiurams disulfides and alkyl dithiocarbamates.

2.1.2.4.2.1. Thiurams.
A Japanese team [189] synthesised amphiphilic acrylic resins composed of (meth)acrylate esters and fluoroalkyl (meth)acrylates by a two-step procedure involving a thiuram disulfide as follows:

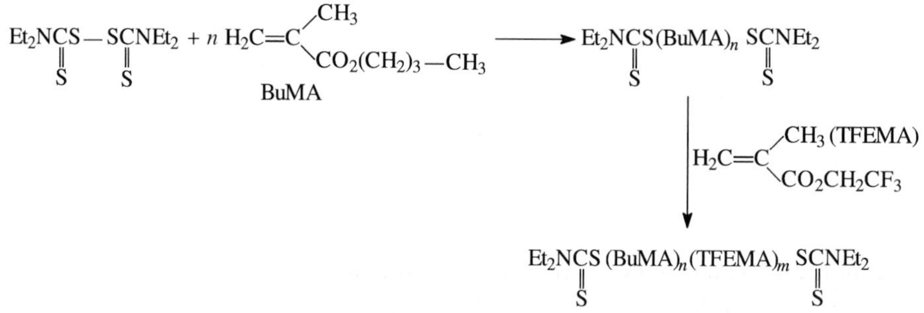

$$Et_2NCS{-}SCNEt_2 + n\ H_2C{=}C \overset{CH_3}{\underset{CO_2(CH_2)_3{-}CH_3}{\diagdown}} \longrightarrow Et_2NCS(BuMA)_n\ SCNEt_2$$

BuMA

$$H_2C{=}C \overset{CH_3\ (TFEMA)}{\underset{CO_2CH_2CF_3}{\diagdown}}$$

$$Et_2NCS\ (BuMA)_n(TFEMA)_m\ SCNEt_2$$

These diblock copolymers were used in formulations for paints that were resistant to acid, heat and weather.

A similar approach was used by Guan and DeSimone [190] in radical photopolymerisation of 1,1-dihydroperfluorooctyl acrylate (FOA) with a telechelic polystyrene synthesised by polymerisation of styrene with tetra-ethylthiuram disulfide, to produce:

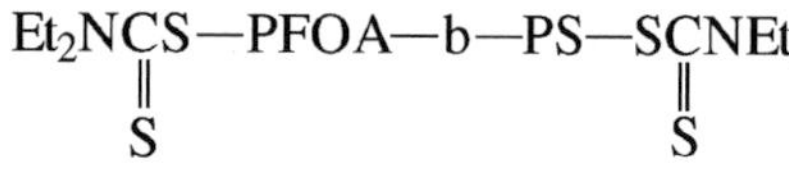

$$Et_2NCS{-}PFOA{-}b{-}PS{-}SCNEt_2$$

Although Otsu [101, 104, 105] has not worked on fluoromonomers, he optimised the concept of pseudo-living polymerisation of (meth)acrylates by using mixtures of thiuram disulfide and dithiocarbamates, R'SCNR$_2$ in order to compensate both:

(i) the loss of ˙SCNR$_2$ radicals (that enable the control of the polymerisation) by transfer or direct initiation and (ii) the cleavage in ⌇⌇SC$-$NR$_2$ that led to non-living species.

DeSimone and co-workers [191] also prepared amphiphilic PDMAEMA-b-PFOA diblock copolymers by two-step controlled radical polymerisations of conventional methacrylates such as 2-(N-dimethylaminoethyl) methacrylate (DMAEMA, M$_1$) with fluoroacrylate from benzyl N,N-diethyldithiocarbamate (under photoinitiation), as follows:

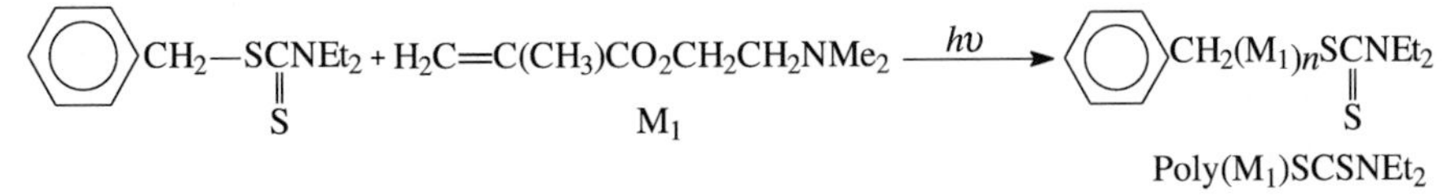

$$\langle\bigcirc\rangle CH_2-\underset{\underset{S}{\parallel}}{S}CNEt_2 + H_2C{=}C(CH_3)CO_2CH_2CH_2NMe_2 \xrightarrow{h\upsilon} \langle\bigcirc\rangle CH_2(M_1)_n\underset{\underset{S}{\parallel}}{S}CNEt_2$$
$$M_1$$
$$Poly(M_1)SCSNEt_2$$

$$Poly(M_1)SCSNEt_2 + \underset{FOA}{H_2C{=}CH-CO_2CH_2(CF_2)_6-CF_3} \rightarrow PDMAEMA\text{-}b\text{-}PFOA$$

This "iniferter" controlled free radical technique made it possible to prepare surfactants used in numerous applications including the stabilisation of dispersion polymerisation of various monomers [192] such as styrene [193] and methyl methacrylate [194, 195] in sc CO_2.

Actually, these authors used such block copolymers containing: (i) a PS segment which readily undergoes partition into the lipophilic core and anchors itself to a growing PS particle and (ii) a poly(fluoroacrylate) (PFOA) block which is readily solubilised in CO_2 and serves to stabilise the growing PS particle through a steric stabilisation mechanism. DeSimone and co-workers [193] also demonstrated that this block copolymeric surfactant stabilised the *in situ* generation of a polymer colloid during the dispersion polymerisation of styrene in CO_2.

More recently, the same team [196] revisited the above reaction and also that involving macroiniferter poly(M_1)SCSNEt$_2$ and 1,1,2,2-tetrahydroperfluorooctyl acrylate (FA) as the monomer to generate PDMAEMA-*b*-PFA diblock copolymer.

2.1.2.4.3. Block Copolymers from the RAFT Process

2.1.2.4.3.1. Dithioesters.   ω-Perfluorinated dithioesters were prepared and successfully used in RAFT process to obtain polymers and block copolymers bearing a fluorinated end-group [197]. Dithioesters **A–D** were synthesised by reacting phenylmagnesium bromide with carbon disulfide, followed by nucleophilic substitution on the appropriate brominated compounds. Dithioester **D** suffered from degradation at 105 °C for 7 days (loss of CS$_2$). Alternatively, dithioesters **A$_1$** and **A$_2$** were successfully produced by a convenient and straight-forward transesterification reaction between the corresponding fluorinated thiol $R_F(CH_2)_2SH$ and a readily commercially available dithioester, (thiobenzoylthio) acetic acid [198]. Dithioesters **E–F** resulted from the direct esterification of the commercially available (thiobenzoylthio)acetic acid with the corresponding fluorinated alcohol [197].

2.1.2.4.3.2. Synthesis of Fluorinated Block Copolymers by RATF.   Poly-merisation of styrene showed the following trend in the control of the polymerisation: **F > E > A**. Dithioester **F**, with the shorter spacer between the ester group and the fluorinated moiety (methylene instead of ethylene spacer), led to a lower polydispersity index, indicating the beneficial effect of the electron-withdrawing fluorine atoms on the activity of the chain transfer agent for this structure [197]. Otherwise, in polymerisation of styrene, methyl methacrylate, ethyl acrylate and of butadiene, dithioester **C** gave the most promising results and a decreasing efficiency series of dithioesters was suggested (**C** > **B** > **E** > **A**), as expected from steric and electronic effects (methyl or phenyl substituents) [197].

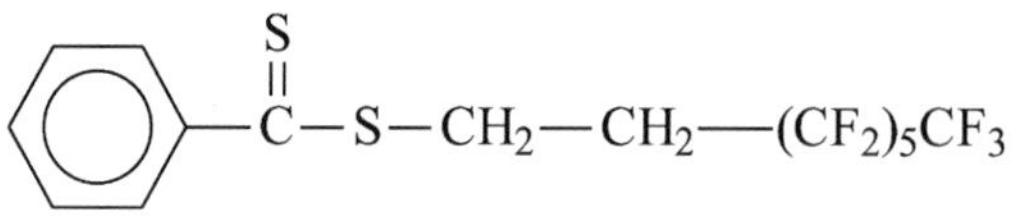

$$\text{C}_6\text{H}_5-\overset{\overset{\text{S}}{\|}}{\text{C}}-\text{S}-\text{CH}_2-\text{CH}_2-(\text{CF}_2)_5\text{CF}_3 \qquad \textbf{A1}$$

$$\text{C}_6\text{H}_5-\overset{\overset{\text{S}}{\|}}{\text{C}}-\text{S}-\text{CH}_2-\text{CH}_2-(\text{CF}_2)_7\text{CF}_3 \qquad \textbf{A2}$$

$$\text{C}_6\text{H}_5-\overset{\overset{\text{S}}{\|}}{\text{C}}-\text{S}-\underset{\underset{\text{CH}_3}{|}}{\text{CH}}-\overset{\overset{\text{O}}{\|}}{\text{C}}-\text{O}(\text{CH}_2)_2(\text{CF}_2)_5\text{CF}_3 \qquad \textbf{B}$$

$$\text{C}_6\text{H}_5-\overset{\overset{\text{S}}{\|}}{\text{C}}-\text{S}-\text{CH}-\overset{\overset{\text{O}}{\|}}{\text{C}}-\text{O}(\text{CH}_2)_2(\text{CF}_2)_5\text{CF}_3 \qquad \textbf{C}$$

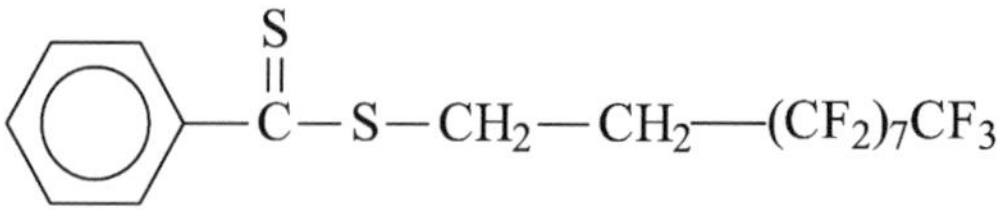

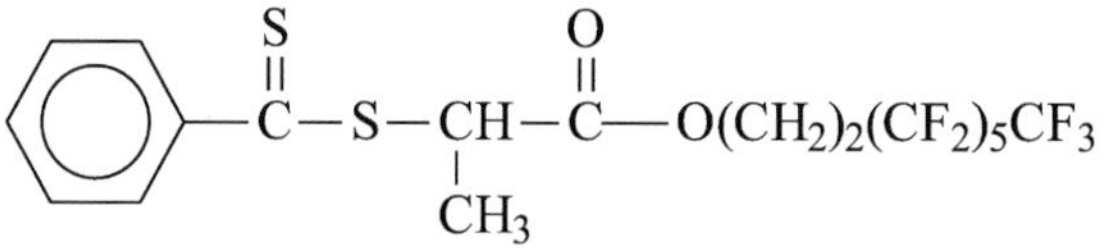

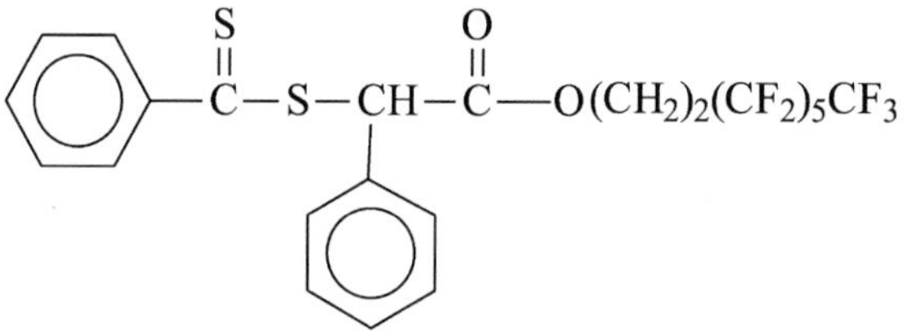

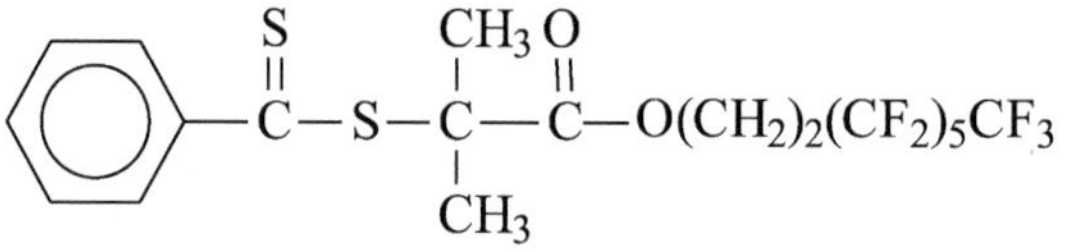

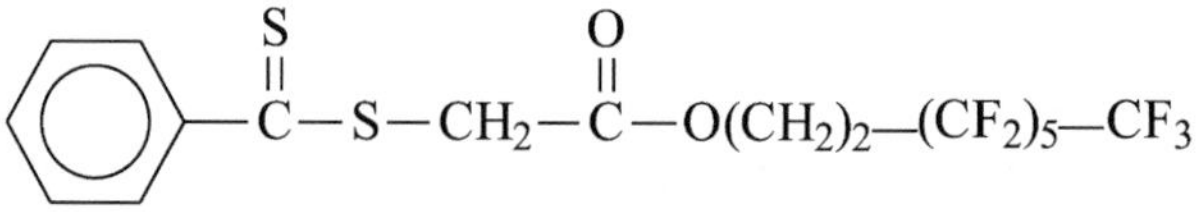

$$\text{C}_6\text{H}_5-\overset{\overset{\text{S}}{\|}}{\text{C}}-\text{S}-\underset{\underset{\text{CH}_3}{|}}{\overset{\overset{\text{CH}_3}{|}}{\text{C}}}-\overset{\overset{\text{O}}{\|}}{\text{C}}-\text{O}(\text{CH}_2)_2(\text{CF}_2)_5\text{CF}_3 \qquad \textbf{D}$$

$$\text{C}_6\text{H}_5-\overset{\overset{\text{S}}{\|}}{\text{C}}-\text{S}-\text{CH}_2-\overset{\overset{\text{O}}{\|}}{\text{C}}-\text{O}(\text{CH}_2)_2-(\text{CF}_2)_5-\text{CF}_3 \qquad \textbf{E}$$

$$\text{C}_6\text{H}_5-\overset{\overset{\text{S}}{\|}}{\text{C}}-\text{S}-\text{CH}_2-\overset{\overset{\text{O}}{\|}}{\text{C}}-\text{OCH}_2\text{CF}_2\text{CF}_2\text{H} \qquad \textbf{F}$$

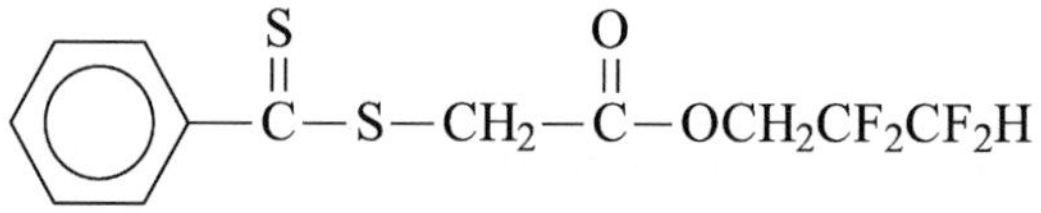

The use of these ω-perfluorinated dithioesters enabled us to achieve a series of various oligomers and polymers bearing a perfluorinated end-group: R_F–PS, R_F–PMMA, R_F–PEA and R_F–PBut. Hence, R_F–PS were synthesised with molar masses and PDI ranging from 2200 to 34,500 and 1.08 to 1.70, respectively, while those of R_F–PMMA were 19,000–130,000 and 1.28–1.65, respectively. $\overline{M}_n$ and PDI of R_F–PBut ranges were 1200–2500 and 1.24–1.50, respectively.

In addition, using these polymeric ω-perfluorinated dithioesters, we synthesised poly(ethyl acrylate)-b-polystyrene of $\overline{M}_n = 32,600$ and a PDI = 1.20 and PMMA-b-PS ($\overline{M}_n = 147,000$ and PDI = 1.17) block copolymers bearing an ω-perfluorinated group [197]:

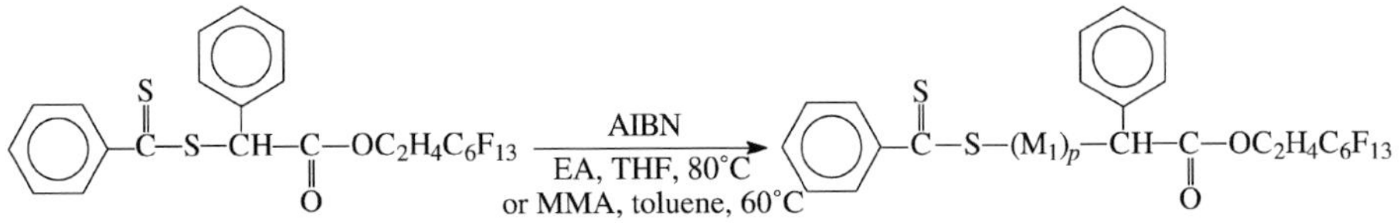

Although the contact angles of these polymers were not assessed, interesting surface properties can be expected.

2.1.2.4.4. Block Copolymers from the MADIX Process. Recently, the Rhodia company has investigated the synthesis of a wide range of controlled polymers and block copolymers by Macromolecular Design via Interchange of Xanthates (MADIX) [118–120], in the presence of the following fluorinated xanthates:

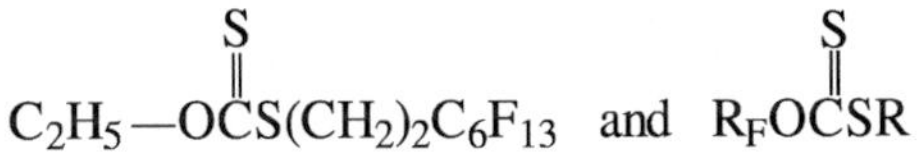

$$C_2H_5 - O\overset{S}{\overset{\|}{C}}S(CH_2)_2C_6F_{13} \quad \text{and} \quad R_FO\overset{S}{\overset{\|}{C}}SR$$

where R_F stands for CF_3-CH_2, C_6F_5 or $(C_2H_5O)_2P(O)CH(CF_3)$ while R represents $CH_2C_6H_5$ or $CH(CH_3)CO_2C_2H_5$. The breaktrough concerns the presence of electron-withdrawing fluorinated groups which activate the xanthates.

2.1.2.5. From Fluorinated Tetraphenyl Ethanes

Tetraphenylethanes represent another class of iniferters (activated by thermal activation) that were pioneered and then extensively investigated by Bledzki and Braun [199]. However, to our knowledge, only one investigation dealing with fluorinated tetraphenyl ethane was realised. Actually, Roussel and Boutevin [200] synthesised a fluorinated tetraphenylethane derivative used in the thermal polymerisation of 2,2,2-trifluoroethyl methacrylate (TFEMA) to yield telechelic (α,ω-difunctional) oligomers (see also Chapter 2). These latter served as efficient macroinitiators to prepare PTFEMA-b-PS block copolymers [200] beside the expected formation of homopolystyrene. The efficiency of the TFEMA oligomers as macroinitiator was 85%. Interestingly, in this special case, the fluorinated substituents were useful in the characterisation of the polymer chain-ends by ^{19}F NMR. The chemical pathway summarising the synthesis of these block copolymers is as follows:

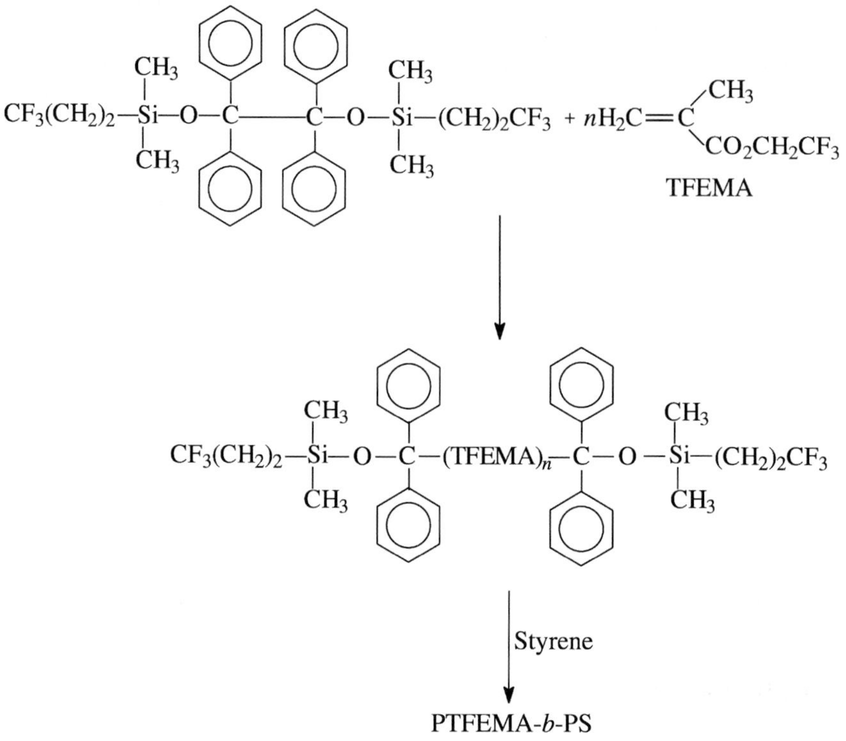

$$CF_3(CH_2)_2-\underset{\underset{CH_3}{|}}{\overset{\overset{CH_3}{|}}{Si}}-O-\underset{}{\overset{}{C}}-\underset{}{\overset{}{C}}-O-\underset{\underset{CH_3}{|}}{\overset{\overset{CH_3}{|}}{Si}}-(CH_2)_2CF_3 \ + \ nH_2C=C\overset{\diagup CH_3}{\diagdown CO_2CH_2CF_3}$$

TFEMA

$$CF_3(CH_2)_2-\underset{\underset{CH_3}{|}}{\overset{\overset{CH_3}{|}}{Si}}-O-\underset{}{\overset{}{C}}-(TFEMA)_n-\underset{}{\overset{}{C}}-O-\underset{\underset{CH_3}{|}}{\overset{\overset{CH_3}{|}}{Si}}-(CH_2)_2CF_3$$

Styrene

PTFEMA-*b*-PS

Starting from TFEMA oligomers of $M_n = 18,000$, the molar masses of the diblock copolymers ranged between 30,500 and 56,400.

2.1.3. Synthesis of Diblock Copolymers by Combination of Different Systems

Condensation reactions followed by telomerisation and esterification of (fluoro)telomers have been reported in Sections 2.1.1.4.1.2 and 2.1.1.4.2 covering the synthesis of R_F-*b*-PEO and poly(fluoroacrylate)-*b*-poly(acrylate), respectively. Nevertheless, it is worth indicating below a special emphasis on two examples of F-diblock copolymers: first, a procedure that takes into account a telomerisation reaction followed by a condensation and then by the dead-end polymerisation [91] of a silicone-containing methacrylate; second, the combination of radical telomerisation and controlled radical polymerisation.

2.1.3.1. Coupling of Telomerisation/Condensation/Dead-End Polymerisation Methods

This exotic process was achieved by Kim *et al.* [201], who succeeded in the synthesis of poly(fluoroacrylate)-*b*-poly(ω-trimethoxysilane methacrylate) block copolymers in three steps. First, the telomerisation of fluorinated acrylates (bearing a C_8F_{17} end-group) with 2-mercaptoethanol led to expected ω-hydroxy fluoroacrylate telomers (PFAM). These telomers were synthesised with targeted molar masses achieved from the assessment of the transfer constant to 2-mercaptoethanol. In optimal cases, the yields were 94%. Then, these hydroxy telomers were condensed with the azo-bis(4-cyanopentanoic acid chloride) (ACPC) to produce an original azo initiator (PFAMI) bearing poly(fluoroacrylate) end-groups in 85% with an average molecular weight in number of 3800. The third step consisted of the polymerisation of a silicon-containing methacrylate (SiMA) initiated by the above macromolecular azo to give a poly(F-acrylate)-b-poly(SiMA) diblock copolymer, depicted in the following scheme:

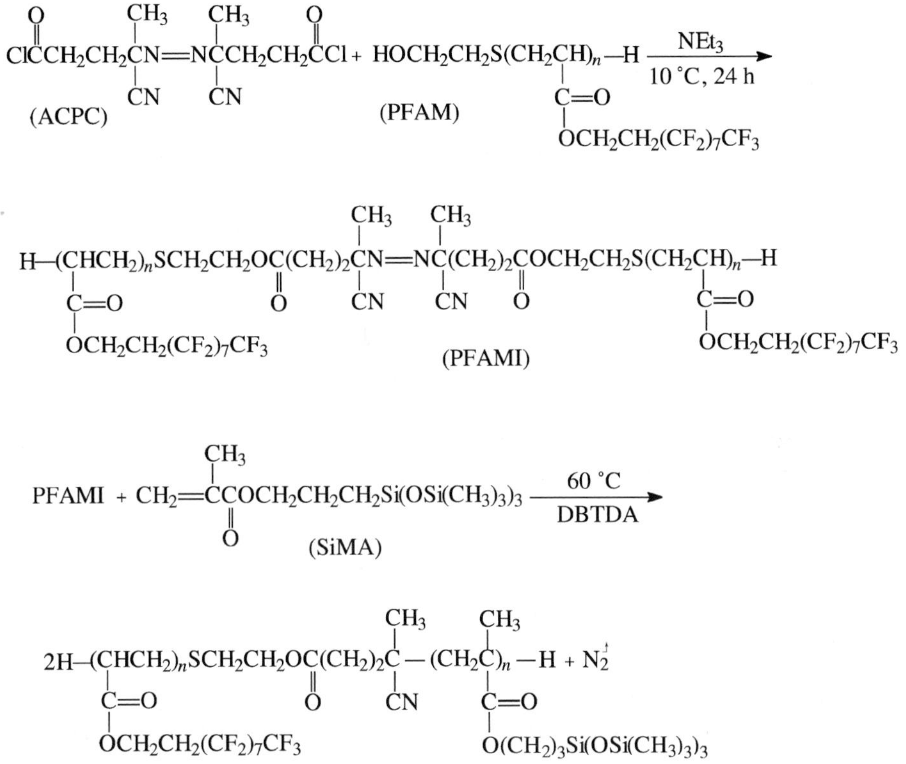

$$\underset{(ACPC)}{ClCCH_2CH_2CN=NCCH_2CH_2CCl} + \underset{(PFAM)}{HOCH_2CH_2S(CH_2CH)_n-H} \xrightarrow[10\,°C,\,24\,h]{NEt_3}$$

$$\underset{(PFAMI)}{H-(CHCH_2)_nSCH_2CH_2OC(CH_2)_2CN=NC(CH_2)_2COCH_2CH_2S(CH_2CH)_n-H}$$

$$PFAMI + \underset{(SiMA)}{CH_2=CCOCH_2CH_2CH_2Si(OSi(CH_3)_3)_3} \xrightarrow[DBTDA]{60\,°C}$$

$$2H-(CHCH_2)_nSCH_2CH_2OC(CH_2)_2C-(CH_2C)_n-H + N_2$$

Surprisingly, the authors did not claim any other by-product, yet it is known that the radical polymerisation of methacrylate leads to a mixture of recombined and disproportionated polymers [201].

However, four different diblock copolymers were prepared in 74–82% yield with molecular weights in number ranging from 4300 to 5400 (and with strange narrow PDI ranging from 1.13 to 1.20). Their surface tensions were rather low: 9.7–13.0 dyne cm^{-1}. These Korean authors used such fluorinated diblock copolymers as coatings on PVC and they noted that the perfluorinated group of the corresponding block migrated to the outermost layer of the surface.

2.1.3.2.   Coupling of Non-living/Living Radical Methods

Other methods of synthesising of fluorodiblock copolymers, taking into account radical controlled polymerisations combined with other processes, are also possible. Two examples can be suggested. First, the condensation reaction involving polyethylene oxide (PEO) bearing a hydroxy end-group as the first step, followed by an ATRP of fluorinated methacrylates (FMA). This is mentioned in Section 2.1.2.3.4 that reports the synthesis of PEO-*b*-PFMA block copolymers from original PEO containing a $-O_2CCH(CH_3)Cl$ terminal group.

The second example concerns the coupling of the telomerisation and the use of the resulting telomers as original initiators for ATRP. Hence, the synthesis of PVDF-*b*-poly(M) diblock copolymers [202] deals with the chlorine transfer radical polymerisation (via a CCl_3 end-group) of various monomers (M) (styrene, MMA, methyl acrylate and butyl acrylate) initiated by VDF telomers ($Cl_3C-PVDF$) with $\overline{DP}_n$ ranging between 5 and 16, as follows:

$$Cl_3C-H + nH_2C{=}CF_2 \xrightarrow{\text{rad.}} \underset{Cl_3C-PVDF}{Cl_3C-(VDF)_n-H}$$
$$\underset{VDF}{}$$

$$Cl_3C-PVDF + pM \xrightarrow{\text{CuCl,bipy}} PVDF\text{-}b\text{-poly(M)}$$

Interestingly, whatever the VDF telomer/M couple, the number-average molecular weight in number ($\overline{M}_n$) of PVDF-*b*-PM increased with the monomer conversion, and experimental values were close to the theoretical ones. In addition, narrow molecular weight distributions and low polydispersities ($\overline{M}_w/\overline{M}_n < 1.2$) were obtained as evidence of a controlled radical polymerisation.

More recently, Jankova *et al.* [203] have used higher molecular weight PVDFs bearing a CCl_3 end-group (with a $\overline{DP}_n$ of ca. 70; $\overline{M}_n \sim 4600$) to successfully achieve the synthesis of original PVDF-*b*-poly(2,3,4,5,6-pentafluorostyrene) block copolymers by ATRP.

2.1.4.   Preparation of Fluorodiblock Copolymers by Living Methods Other Than Radical Ones

Various common methods different from the radical ones have been successfully used for the synthesis of fluorinated diblock copolymers, although few examples are reported: these are cationic, anionic, ring opening metathesis and also GTP and they are discussed below.

2.1.4.1.   Cationic Methods

Two distinct families of block copolymers are considered: (i) those achieved from sequential cationic living polymerisation of vinyl ethers and (ii) the ring opening polymerisation of cyclic esters, amides or oxazolines.

2.1.4.2. Sequential Cationic Polymerisation of (Fluoro)vinyl Ethers

Regarding the former family, although the cationic homopolymerisation of fluorinated vinyl ethers was achieved by various authors [204–206], only four publications report the synthesis of fluorinated block copolymers from the stepwise cationic copolymerisation of hydrogenated and fluorinated vinyl ethers.

Percec and Lee [207], using triflic acid as initiator, synthesised by sequential cationic living copolymerisation of ω-[(4-cyano-4'-biphenyl)oxy] alkyl vinyl ether (VE) and of 1H,1H,2H,2H-perfluorodecyl vinyl ether, fluorinated diblock copolymers endowed with narrow PDI (1.15–1.26):

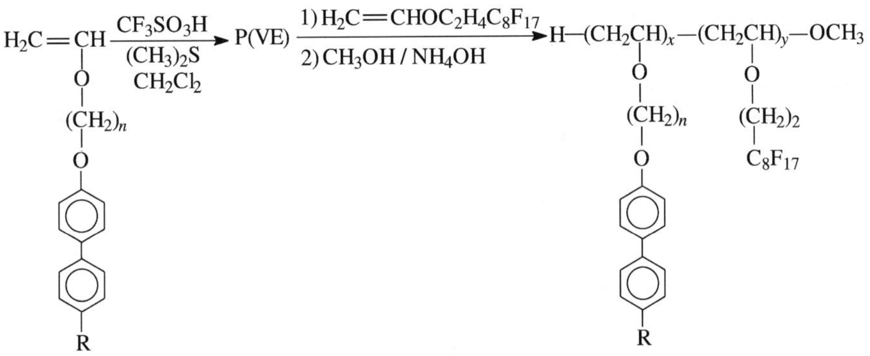

$n = 2, 3, 9$
R = H or CN

It was noted that the weight ratio composition of the block copolymers was in good agreement with the feed weight ratio of each comonomer.

Yields were higher than 80% and molar masses ranged between 5900 and 8200.

Interestingly, the authors showed from DSC analysis and optical polarised microscope that all these block copolymers exhibit microphase separated morphology when the hydrogenated block was located in the liquid crystalline phase and that similar morphology also occurred in certain cases in the melt phase of hydrogenated and fluorinated blocks.

(b) The second survey was successfully achieved by Matsumoto *et al.* [208], who used $ZnCl_2$ as the initiating system in the sequential cationic polymerisation of 2-(acetoxy ethyl) vinyl ether and 2-(2,2,2-trifluoroethoxy) ethyl vinyl ether (TFEOVE). This led to original, well-architectured amphiphilic PHOVE-*b*-PTFEOVE block copolymers after deprotection of the acetyl side groups:

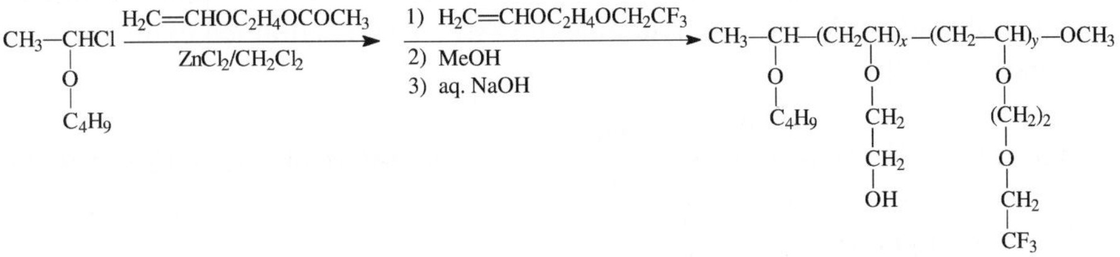

The average degrees of polymerisation x and y were in the ranges of 28–34 and 10–31, respectively, while the PDI were quite low (1.08–1.16).

These copolymers were soluble in water and in toluene and showed interesting surface activity (the surface tensions of the polymeric aqueous solutions decreased to 30 mN m^{-1}). Critical micelle concentrations (cmc) of these block copolymers were found to be around 10^{-5} and 10^{-4} mol l^{-1}. SAXS measurements revealed that the micelle had a core-shell spherical morphology and that the aggregation number increased as the length of the fluorinated block increased. In addition, these authors also showed that these block copolymers had a high ability to solubilise a fluorinated compound compared with non-fluorinated amphiphilic block copolymers.

(c) In a further study [209], the same authors prepared two diblock copolymers; PHOVE-b-PTFEOVE ($x : y = 78 : 23$) and its non-fluorinated homologue (i.e., PHOVE-b-PNBVE, where NBVE stands n-butyl vinyl ether, with $x : y = 74 : 23$). They had the same ratio of degree of polymerisation in the hydrophilic and hydrophobic block, with the same total chain length. The cmc of both these diblocks, assessed by water surface tension, were ca. 1.0×10^{-4} mol l^{-1}. These Japanese authors demonstrated that the micelles of the fluorinated copolymer could solubilise fluorinated compounds more than could non-fluorine-containing polymer micelles. This was shown by solubilisation experiments using a mixture of perfluorinated (e.g., decafluorobiphenyl) and non-fluorinated solubilisates, which revealed that the fluorine-containing polymer micelle could preferentially solubilise the perfluorinated solubilisate. Although the hydrophobic segment of the A–B polymers was not perfluorinated, this copolymer exhibited the unique characteristics of fluorinated materials (named as the "perfluoroalkyl-philic" property) [209].

(d) In addition, Clark et al. [210] synthesised original diblock copolymers containing a poly(fluorinated vinyl ether) block followed by a poly(methyl vinyl ether) in sc CO$_2$. Their molar masses ranged from 16,000 to 18,000, endowed with low PDI (1.10).

They were successfully used for cationic dispersion polymerisation of styrene in sc CO$_2$. The authors observed the formation of PS particles ranging in diameter from 300 nm to 1 μm.

2.1.4.3. Ring-Opening Polymerisation of Cyclic Amides Esters and Oxazoline

Recently, Storojakova et al. [211] have reported the ring opening oligomerisation of ε-caprolactame by ω-hydroxyl TFE-telomers under catalytic amounts of lead oxide and triethylamine or DMF. The reaction led to diblock co-oligomers as follows:

$$H(C_2F_4)_nCH_2OH + m\,HN(CH_2)_5CO \xrightarrow[\text{NEt}_3 \text{ or DMF}]{\text{PbO/260 °C}} H[NH(CH_2)_5CO]_mOCH_2(C_2F_4)_nH$$

with $n = 1$–4 and $m = 2, 3$.

These Russian authors noted that: (i) very low yields of diblock co-oligomers were obtained when the reaction was carried out at a temperature lower than 260 °C; (ii) the higher the TFE unit-number, the lower the yield and, as expected, the higher the thermal stability of the diblock co-oligomers; and (iii) DMF led to lower molecular weights of poly(ε-caprolactame) block but reduced the dehydrofluorination reaction.

On the other hand, Saegusa's group [212] synthesised poly(fluoroalkyloxazo-line)-b-poly(methyloxazoline) copolymers by sequential ring opening polymerisation of (fluoro)oxazolines. Interestingly, this was the first example of a water-soluble fluorine-containing block copolymer, a field of research comprehensively reviewed [213–215].

2.1.4.4. Anionic Methods

Fluorinated A–B copolymers achieved by anionic living polymerisation started to be synthesised quite recently (in the last 5 years).

The literature reports several investigations by Lodge's group [216], Antionnetti's team [217] and carried out in Ober's Laboratory [218] regarding various chemical changes of hydrogenated diblock copolymer (e.g., *quasi-monodispersed* PS-*b*-P(butadiene) or PS-*b*-P(isoprene)) synthesised by sequential anionic polymerisations: for example, the addition of various perfluoroalkyl radicals onto the pendent butadienic or isoprenic double bonds, or the condensation of fluorinated acid chloride onto the pendent hydroxyl groups obtained by oxidative hydroboration.

Although such copolymers exhibit block structures, it is noted that the (per)fluorinated substituents are located in the side chain quite far from the polymeric backbone. Consequently, we have considered that these structures are closer to graft copolymers than to block copolymers and we invite the reader to find more information on that topic in Chapter 5, Section 2.2.2.2. However, it is worth mentioning one typical example in which the fluorinated chain is short enough to consider the resulting macromolecule as a block copolymer. In this typical example, Ren *et al.* [219] first synthesised, by sequential anionic copolymerisation of styrene (S) and isoprene (I_p), PS-*b*-PI$_p$ diblock copolymers ($\overline{M}_n = 6100–61,000$ and PDI $= 1.01–1.05$) on which the addition of difluorocarbene onto the double bonds of the PI$_p$ block was achieved [219]. Difluorocarbene was generated *in situ* from the thermolysis of hexafluoropropylene oxide.

The resulting fluorinated copolymers had the following structure:

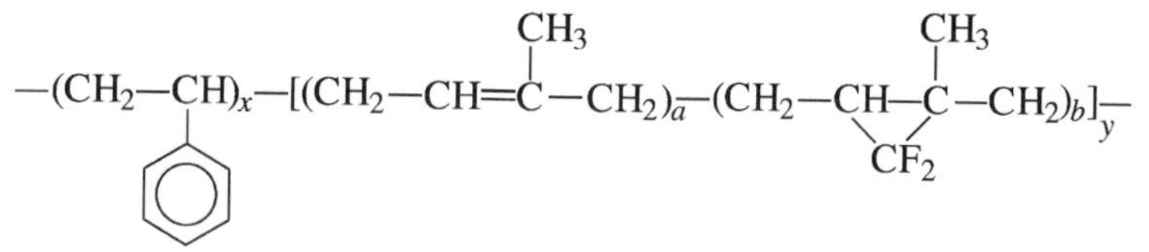

$$-(CH_2-CH)_x-[(CH_2-CH=C-CH_2)_a-(CH_2-CH-C-CH_2)_b]_y-$$

Four examples of synthesis of fluorinated diblock copolymers from direct stepwise anionic polymerisation are given below.

(a) The first one was proposed in 1998 by Sugiyama *et al.* [220], who synthesised poly(2-perfluorobutylethyl methacrylate)-*b*-poly(2-hydroxyethyl methacrylate) amphiphilic block copolymer in high yields by sequential anionic polymerisation of the first monomer, followed by that of the hydroxy-protected second one, initiated by *sec*-butyllithium/1,1-diphenylethylene at $-78\,°C$ in the presence of LiCl under high vacuum conditions. Such a stepwise reaction offered well-controlled copolymers of 23,000 molar masses with a polydispersity of 1.05. The molar content of the fluorinated block was 57%. They were characterised by TEM and XPS and, interestingly, static light scattering measurements of these block copolymers in methanol revealed that these fluorinated products formed micelles. Although these Japanese researchers prepared only one diblock copolymer, more investigations were carried out on a wide range of fluorinated A–B–C triblock copolymers (see Section 2.2.4.2).

(b) Then, in 1999, Hems *et al.* [221] prepared PMMA-*b*-PFMA block copolymers by using a modified version of the screened anionic polymerisation methodology [222]. Two main features were noted: the reactions were carried out at $0\,°C$ whilst still maintaining control of the polymerisation. This resulted from the shielding of the propagating species by a bulky lithium–aluminium alkyl complex at the enoate chain-end. The polymerisation of MMA was living and thus offered the ability to "tune" the length of the PMMA sequence and, further, that of the PFMA block in the copolymers according to specific requirements. In addition, and in an elegant way, the authors added 1,3-bis(trifluoromethyl) benzene as a cosolvent with toluene to enable the solubility of the fluorinated block copolymer in the medium. Starting from molar masses of the PMMA sequence varying from 3000 to 14,000, the authors obtained PMMA-*b*-PFMA diblock copolymers in 83–91% yields having molar masses of 17,000–76,000.

Interestingly, these fluorinated A–B copolymers were successfully used as surfactants in the radical polymerisation of MMA initiated by AIBN in sc CO_2 at 285 bar/70 °C [221]. The reaction was achieved in high yields and led to PMMA with molar masses up to 390,000. The authors noted that the higher the molar masses and the fluorine contents of the fluorodiblock copolymer, the better the surfactant.

(c) Another survey was carried out by Imae's group [223], which also investigated the synthesis of fluorinated diblock copolymers by sequential anionic copolymerisation of MMA (initiated by BuLi, LiCl in THF at $-78\,°C$) and 2-perfluorooctylethyl methacrylate (FMA). The PMMA and PMMA-*b*-PFMA had molar masses and PDI of 41,100 and 78,400, and 1.03 and 1.07, respectively. These Japanese authors studied the self-assemblies of these block copolymers in organic solvents and observed by light scattering that they formed aggregates of 410 molecules with a 306 Å radius in acetonitrile and of 26 molecules of smaller radius (ca. 200 Å) in chloroform.

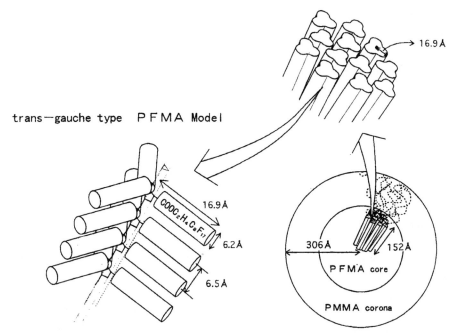

FIG. 1. Schematic representation of *trans–gauche* PFMA chain structure and polymer micelle structure in acetonitrile (reproduced with permission from Imae *et al.*, *Colloids Surf. Phys. Eng. Aspects*, 167, 73–81 (2000)).

Cryo-transmission electronic microscopy and AFM imaging indicated their spherical morphology. These spherical particles were composed of copolymer micelles with a PFMA core of 152 Å radius and a PMMA Corona (Shell) of 154 Å thickness (Figure 1).

The authors also suggested that the copolymer micelles behaved as hard spheres into which acetonitrile molecules penetrated poorly.

(d) More recently, Busse *et al.* [224] carried out the synthesis of amphiphilic block copolymers by stepwise anionic polymerisation of *tert*-butyl methacrylate and 2-(*N*-methyl perfluorobutanesulfonamido) ethyl methacrylate (FBSMA) the second block bringing the hydrophobicity. The hydrophilic block was the sodium salt of poly(methacrylic acid) (PMAA) obtained by thermal annealing at 200 °C, followed by alkaline treatment of the respective poly(*tert*-butyl ester) block. Interestingly, such a thermal treatment did not upset the poly(fluorosulfonamido methacrylate) block and this German group noted that the obtained block copolymers were soluble in water when the PFBSMA content was lower than 10 mol%. Surface tension measurements (STM) allowed these authors to assess the cmc of $67 \, \mathrm{g \, l^{-1}}$ (while it is $2 \, \mathrm{g \, l^{-1}}$ for PFBSMA-*b*-PMAA-*b*-PFBSMA triblock copolymers).

Indeed, from light scattering and SAXS, these authors noted that the formation of micelles dominated in a wide range of concentrations whereas, for triblock copolymers, an aggregation of the micelles occurred. By TEM experiments, they determined the size of the micelles and their inner structure.

2.1.4.5.   By Ring-Opening Metathesis Polymerisation (ROMP)

To our knowledge, only two works deal with the preparation of fluorinated diblock copolymers by ring opening metathesis polymerisation [225]: by researchers at the DuPont company [40] and by Koshravi's group [226]. The authors at DuPont prepared linear polyethylene (PE) bearing perfluorodecyl end-group by ring opening metathesis polymerisation of cyclododecene in the presence of partially fluorinated acyclic olefin (e.g., $C_{10}F_{21}(CH_2)_6-CH=CH_2$) that acted as chain transfer agents, followed by the reduction of the double bond in the backbone. The initiating system was $WCl_6/SnBu_4$. This metathesis reaction is a well-known process that has been extensively studied by this American company [225] for two decades.

Hence, PTFE-*b*-PE block copolymers of various molar masses (3300 and 5100) containing different fluorine contents (12.15 and 7.76%) showed a high hydrophobicity (the advance contact angles in water were 127 and 130 °C, respectively).

The second survey was accomplished by Koshravi *et al.* [227], who reported in an interesting review various fluorinated (co)polymers obtained by ROMP from Schrock alkylidenes initiators (especially complexes of tungsten and molybdenum):

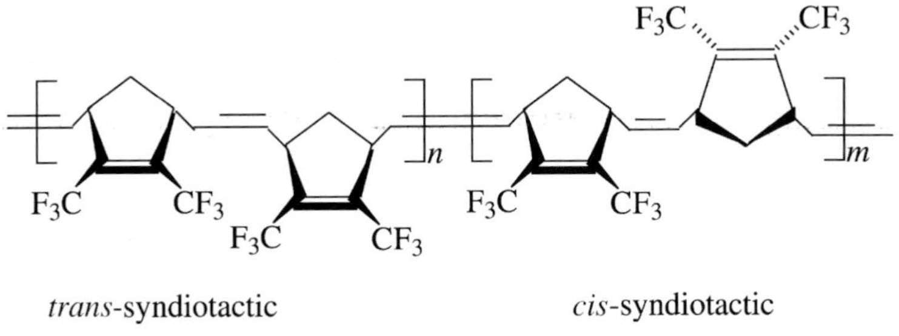

trans-syndiotactic   *cis*-syndiotactic

Interestingly, this group obtained stereoblock copolymers containing *trans*- and *cis*-syndiotactic blocks having molar masses of 73,000 (theoretical 46,000) and $M_w/M_n = 1.16$. The observed polydispersity is remarkably narrow considering the number of steps involved and indicates a very well-defined living polymerisation process.

2.1.4.6.   By Group-Transfer Polymerisation

Few syntheses of fluorodiblock copolymers were achieved from the nucleophilic catalysed GTP [227]. An interesting example was proposed by Krupers *et al.* [228], who carried out the preparation of PMMA-*b*-PFMA in a two-step procedure (where FMA stands for 1H,1H,2H,2H-perfluoroalkyl methacrylate) represented by the following scheme:

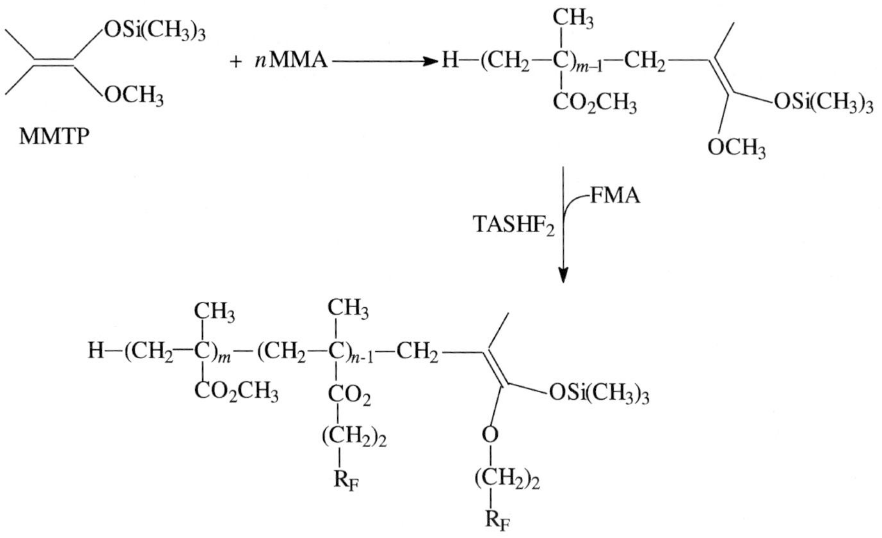

$$\underset{\text{MMTP}}{\overset{\text{OSi(CH}_3)_3}{\diagdown\diagup}}\overset{}{\underset{\text{OCH}_3}{}} + n\,\text{MMA} \longrightarrow \text{H}-(\text{CH}_2-\overset{\text{CH}_3}{\underset{\text{CO}_2\text{CH}_3}{\text{C}}})_{m-1}-\text{CH}_2-\diagup\diagdown\overset{\text{OSi(CH}_3)_3}{\underset{\text{OCH}_3}{}}$$

$$\downarrow \text{TASHF}_2 \overset{\text{FMA}}{\diagup}$$

$$\text{H}-(\text{CH}_2-\overset{\text{CH}_3}{\underset{\text{CO}_2\text{CH}_3}{\text{C}}})_m-(\text{CH}_2-\overset{\text{CH}_3}{\underset{\underset{(\text{CH}_2)_2}{\underset{R_F}{|}}{\text{CO}_2}}{\text{C}}})_{n-1}-\text{CH}_2-\diagup\diagdown\overset{\overset{O}{|}}{\underset{\underset{(\text{CH}_2)_2}{R_F}}{}}\text{OSi(CH}_3)_3$$

In the first step, the polymerisation of MMA was initiated by adding tetrabutylammonium benzoate (TBAB) to a solution containing MMA and 1-methoxy-2-methyl-1-(trimethyl)silyloxy-1-propene (MMTP) as the initiator. Then, the addition of 1H,1H,2H,2H-perfluoroalkyl methacrylate (FMA) (the alkyl group being butyl or hexyl), followed by that of tris(dimethylamino)sulfonium bifluoride (TASHF$_2$), enabled the obtaining of the F-diblock copolymers.

Interestingly, in contrast to long reaction times required for the synthesis of block copolymers by GTP (ca. 1 day [225]), the authors noted that after less than 4 h the polymerisation of PMMA-b-PFMA was completed [229].

While the length of the PMMA block and the PDI were varying from 5000 to 19,100 and from 1.11 to 1.20, respectively, those of the diblock copolymers were in the ranges of 14,000–21,200 and 1.09–2.00, respectively [228–230].

The isolated yields were fair to very high (from 55 to 90%). The micellar morphology was examined by using size exclusion chromatography, TEM, AFM and the measurement of the contact angles. Interestingly, these semi-fluorinated diblock copolymers were used as surfactants for the formation of water-blown closed cell rigid polyurethane foams by the "tube"-foam method [231]. The smallest sizes of the cells obtained were about 100 μm.

Another similar investigation was carried out by Yang $et\ al.$ [232, 233], who first introduced a poly(tetrahydropyranyl methacrylate) (PTHPMA) sequence followed by a poly(fluorinated methacrylate) block containing C$_3$F$_7$ or C$_7$F$_{15}$ side groups. By changing both the length and the nature of the fluorinated side chain, the surface properties were greatly influenced. The block copolymers containing the perfluoroheptyl side chain exhibited the lowest surface tension ever measured (down to 6.7 mN m^{-1}), while those bearing the perfluoropropyl group had a surface tension of about 12 mN m^{-1}. Although the relative molar ratio of both blocks (assessed by ^{1}H NMR) did not upset the surface energy, the authors [232, 233] observed that the surface tension changed drastically when the perfluorinated lateral chain contained a CF_2H end-group, increasing to 18 mN m^{-1} for the three $-CF_2-$ units. After thermal/acid decomposition of the THP labile group, original poly(methacrylic acid)-*b*-poly(fluorinated methacrylate) diblock copolymers were obtained. These copolymers that possess such a photoacid cleavable THP group are of interest for lithography and photoimaging applications [233]. Thermal crosslinking of these acid groups increased the mechanical robustness, leading to better adhesion and surface stability.

2.1.5.   Synthesis of Fluorinated Diblock Copolymers by Condensation

This sub-section concerns A–B block copolymers formed by condensation of two blocks, one being fluorinated. Although Section 2.1.1.4.2 has already mentioned the synthesis, this part will focus on fluorinated poly(ethylene oxide) (PEO). Indeed, these polymers are of growing interest since they are water-soluble, allowing them to act as associative thickeners. Several strategies have been investigated and Hogen-Esch's team has comprehensively studied both the synthesis and the rheological properties of R_F–PEO and R_F–PEO–R_F.

Various esterifications involving functionalised perfluorocarbons and PEO have been reported in the literature.

Trabelsi *et al.* [234] carried out the condensation of ω-perfluorinated acyl chloride onto a low molecular weight ω-hydroxy PEO to lead to C_nF_{2n+1}-oligo(ethylene oxide) ($n = 4$–12). These authors claimed that the fluoro-PEO can be used as components of microemulsions or emulsions of fluorocarbons for biomedical purposes or to form niosomes for the vectorisation of active principles.

One of the key reactions was the condensation at the molten state of PEO monoethylether with a carboxylic acid containing a C_8F_{17} end-group [235] to yield $CH_3O(C_2H_4O)_nCO(CH_2)_2C_8F_{17}$ in which the molar mass of PEO was 5000. The micellar and associative behaviours were 2×10^{-4} and 10^{-3} g ml^{-1}, respectively.

Besides R_F–PEO, fluorocarbon derivatives of water-soluble polymers such as polyacrylamide [236] and (hydroxyethyl) cellulose [237] have also shown hydrophobic intermolecular association of the R_F groups much stronger than their corresponding hydrocarbon derivatives of the same carbon number [238].

Another class of water-soluble polymers with associative properties is hydrophobic ethoxylated urethane (HEUR) polymers which are PEO end-functionalised through urethane linkages with a hydrophobic (e.g., perfluorinated) end-group.

The reaction involved poly(ethylene oxide) monomethyl ether with degrees of polymerisation of 120 and 250 in the presence of an excess of isophorone diisocyanate [239], as follows:

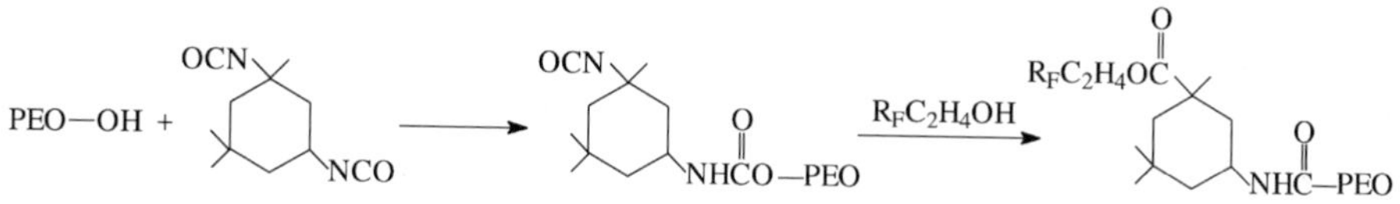

Various similar investigations were achieved and the rheological and associative properties have been reviewed [240].

A further alternative of F-diblock by condensation was achieved by a Dutch group [241] by carrying out the esterification of an excess of adipic acid with 1,4-butanediol. Then, the resulting ω-hydroxy oligomer reacted with a perfluoroacid chloride to yield:

$$C_7F_{15}CO[O(CH_2)_4OCO(CH_2)_4CO]_n(CH_2)_4O-H$$

A careful characterisation by NMR and MALTI-TOF MS enabled these authors to estimate n (ca. 7).

Another synthesis of F-diblock copolymers (by anionic polymerisation) was achieved by Hogen-Hesch's group [242] in 2001. It concerns the polymerisation of N,N-dimethylacrylamide (DMAA) initiated by triphenylmethyl caesium, followed by the reaction of PDMAA monoanion with perfluorooctanoyl chloride as follows:

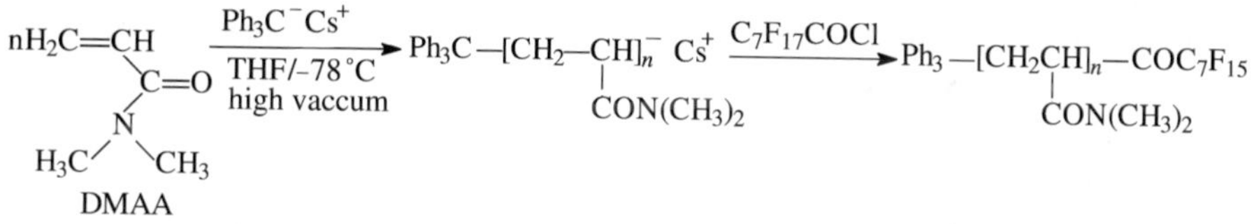

This team showed that the end functionalisation by the fluorinated block was almost quantitative. The values of molar masses (15,000–47,000), obtained by SEC and based on PS standards, were in good agreement with the calculated values assessed from both the initial $[monomer]_o/[initiator]_o$ ratio and the results of elemental analyses.

The character of the living behaviour of this polymerisation was also shown by PDI lower than 1.19.

Interestingly, these American authors [242] claimed that these fluorinated PDMAA block copolymers were the first water-soluble PDMAA reported in the literature. Furthermore, by choosing telechelic PDMAA dianions, these authors obtained C_7F_{15}-b-PDMAA-b-C_7F_{15} triblock copolymers (see Section 3). Xie et al. [242] showed that reduced viscosity measurements were consistent with predominantly dimer association of the perfluorooctanoyl end-functionalised PDMAAS. They also indicated that, below a concentration of about 2 g dl^{-1}, the reduced viscosities of the mono- and bis-R_F-end-functionalised PDMAAs (with a molecular weight of about 47,000) and the isobaric precursor PDMAA are similar, which is consistent with a lack of significant association of the monofluorocarbon-functionalised polymers. Above 2.0 g dl^{-1}, the reduced viscosities increased significantly compared with that of the non-functionalised PDMAA, suggesting association, as indicated by the larger Huggins constant ($k_H = 0.80$). Thus, these authors claimed that the assembly of one-ended polymeric surfactants into star-shaped micelles with aggregation numbers of two or larger seemed possible.

2.1.6. Conclusion

As shown by these various examples, many co-oligomers, cotelomers and copolymers containing a fluorinated block have been synthesised (Table IV). This can first be explained by the comprehensive investigations on the telomerisation reactions involving fluorinated transfer agents (such as $C_nF_{2n+1}I$), which are exploited at the frontier between the traditional radical polymerisations and the controlled radical polymerisation that is a more recent topic. Hence, many surveys led to a wide range of different cotelomers or copolymers containing both a fluorinated sequence and another block bringing complementary properties. Many surveys on semifluorinated R_F-R_H molecular compounds that act as models of block copolymers have been investigated on the structural and thermal (mainly melting) behaviours, on the stabilisation of emulsion, liquid crystals surface properties and micelle associations. These amphiphilic oligomers have been used as coatings of various substrates (PVC or textiles), surfactants, in biomedical applications and as dispersing aids: in most cases, hydrophilic blocks have been obtained, but various sequences have also been supplied, e.g., those used in radical polymerisation of fluoromonomers in sc CO_2 since the fluorinated chains are CO_2-philic.

In addition, various authors have successfully used methods different from the radical polymerisations for synthesising original A–B fluoropolymers. Basically, four methods were relevant: sequential cationic polymerisation of fluorovinyl ethers and the ring opening polymerisations of ε-caprolactame (CL) (leading to oligoTFE-*b*-oligoCL co-oligomers) and, on the other hand, anionic and GTP of fluorinated methacrylates that led to high molecular weight copolymers. Several properties such as the surface tension and, above all, the micelle behaviour at cmc were searched.

Beside the synthesis, properties and applications of fluorinated diblock copolymers, many investigations were conducted on A–B–A or A–B–C fluorotriblock copolymers and these are reported below.

2.2. Synthesis, Properties and Applications of Fluorinated Triblock Copolymers

As for fluorodiblock copolymers, the synthesis of fluorinated A–B–A or A–B–C triblocks is challenging and many investigations have been carried out. Interestingly, A–B–C triblock copolymers show much more complicated

TABLE IV

OVERVIEW OF THE SYNTHESIS OF FLUORINATED DIBLOCK COPOLYMERS ACHIEVED FROM VARIOUS METHODS

Method	Initiator, catalyst or chain transfer	Structure of F-diblock	Reference
Radical			
Iniferter	Thiuram	PBuMA-*b*-PTFEMA	[189]
		PS-*b*-PFOA	[190]
		PDMAEMA-*b*-PFOA	[191]
		PDMAEMA-*b*-PFA	[196]
	Tetraphenyl ethane	PTFEMA-*b*-PS	[200]
RAFT	Dithioester	R_F–PS, R_F–PMMA	[197]
		R_F–PEA, R_F–PBut	[197]
	Xanthate	R_F–PVAc	[118, 119]
ITP	$R_F I$	Poly(fluoroalkene)-*b*-poly(fluoroalkene)	[114, 151, 164–168]
ATRP	CCl$_3$–PVDF telomers	PVDF-*b*-PS, PVDF-*b*-PMMA	[202]
		PVDF-*b*-PEA	[202]
	CCl$_3$–PVDF telomers	PVDF-*b*-PPFS	[203]
Cationic	CF$_3$SO$_3$H	PHVE-*b*-PFVE	[207]
	ZnCl$_2$	PHVE-*b*-PFVE	[208, 209]
	X/scCO$_2$	PFVE-*b*-PHVE	[210]
	HO–PTFE telomers	PTFE-*b*-PCL	[211]

(continued on next page)

TABLE IV (continued)

Method	Initiator, catalyst or chain transfer	Structure of F-diblock	Reference
Anionic	sec–BuLi/1,1-diphenyl-ethylene	PHEMA-b-PFBMA	[220]
	BuLi/LiCl	PMMA-b-PFMA	[221]
		PMMA-b-PFMA	[223]
	1,1-Diphenyl-3-methyl	P(tBuMA)-b-P(2-FBSMA)	[224]
	Pentyllithium/LiCl	PS-b-P(But-R_F), PS-b-P(I_p−R_F)	[217–219]
Metathesis	C_8F_{17}	PTFE-b-PE	[40]
GTP	MMTP	PMMA-b-PFMA	[227–230]
		PTHPMA-b-PFMA	[232, 233]
Condensation	None	PEO-b-R_F	[234, 235, 239]
		PDMAA-b-R_F	[242]

But, EA, FA, FMA, FOA, FBMA, FBSMA, FVE, HVE, I_p, MMTP, PE, PFS, THPMA, TFEMA and VAc stand for 1,3-butadiene, ethyl acrylate, F-acrylate, 1H,1H,2H-perfluoroalkyl methacrylate, 1,1-dihydro-perfluorooctyl acrylate, 2-perfluorobutyl ethyl methacrylate, 2-(N-methyl perfluorobutane sulfonamido) ethyl methacrylate, F-vinyl ether, hydrogenated vinyl ether, isoprene, 1-methoxy-2-methyl-1-trimethylsilyloxy-1-propene, polyethylene, 2,3,4,5,6-pentafluorostyrene, tetrahydropyranyl methacrylate, trifluoroethyl methacrylate and vinylacetate, respectively. The other acronyms are listed in the "Nomenclature and Abbreviations" section at the beginning of the book.

microphase separated structures than diblocks. As in the previous sub-section, this part summarises the different synthesis routes. Five main families can be distinguished: (i) those involving radical oligomerisation and telomerisation or traditional polymerisation (the use of novel α,ω-bisperfluoroinitiators involved in DEP deserves special mention); (ii) by controlled radical polymerisation involving sequential ATRP or ITP; (iii) consecutive combination of methods (such as telomerisation or condensation to prepare a telechelic macroinitiator used in controlled radical polymerisation); (iv) ionic routes (sequential ionic polymerisation or starting from telechelic initiators); and (v) condensation of perfluorinated acid chloride with α,ω-dihydroxy oligomers or polymers. All these alternatives are described below.

2.2.1. By Classical Radical Oligomerisation, Telomerisation and Polymerisation

2.2.1.1. By Oligomerisation of Monomers from Fluoroinitiators (Dead-End Polymerisation [91])

Direct radical polymerisation (or oligomerisation) requires a radical initiator and a monomer and, if the fluorinated initiator $R_F–X–X–R_F$ is adequately chosen, either diblock or triblock copolymers can be produced. The following scheme illustrates this methodology:

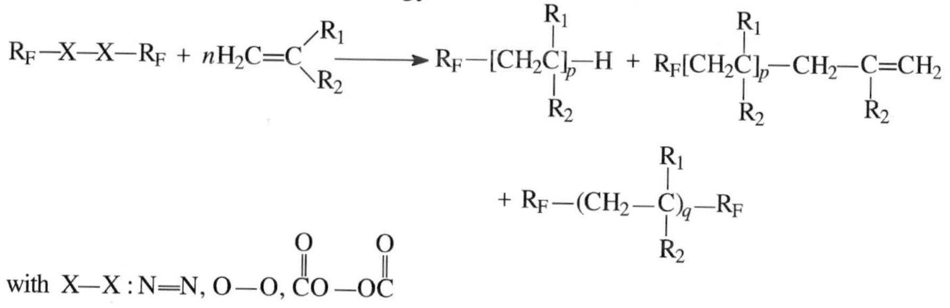

$$R_F–X–X–R_F + nH_2C=C\underset{R_2}{\overset{R_1}{\big\langle}} \longrightarrow R_F–[CH_2C]_p^{R_1}{}_{R_2}–H + R_F[CH_2C]_p^{R_1}{}_{R_2}–CH_2–\underset{R_2}{\overset{R_1}{C}}=CH_2$$

$$+ R_F–(CH_2–\underset{R_2}{\overset{R_1}{C}})_q–R_F$$

with $X–X : N=N, O–O, \overset{O}{\overset{\|}{C}}O–O\overset{O}{\overset{\|}{C}}$

If R_1 and R_2 are not hydrogen atoms, disproportionation occurs whereas combination happens for either $R_1 = R_2 = H$ or $R_1 = H$ and $R_2 = CO_2CH_3$ or C_6H_5.

Obviously, the nature of the monomer (i.e., the tendency of the monomer to undergo recombination and/or disproportionation) is a key-point for the selective formation of triblock copolymers or a mixture of diblock and triblock copolymers.

The thermal decomposition of these initiators present in a higher amount than that used in classical radical polymerisation leads to perfluorinated radicals that initiate the polymerisation of vinyl monomers. In these conditions of DEP [91], the formation of oligomers is usually favoured. Table III (p. 244) compares the concepts and methodologies, the experimental conditions and the nature of the oligomers and polymers obtained by both methods.

Non-exhaustive examples from fluorinated peroxides and from F-azo initiators are provided below.

2.2.1.1.1. From Fluorinated Peroxides. These per- or polyfluorodiacyl peroxides (or bis(per)fluoroalkanoyl peroxide), or perfluoroalkyl peroxydicarbonates, have short half-lives. For example, that of $[C_3F_7OCF(CF_3)CO_2]_2$ is 59 min at 30 °C [243], that of $(CF_3OCO_2)_2$ is 34.4 min at 72 °C and 140 min at 60 °C [244], while that of $[FO_2SC_2F_4OCF(CF_3)CO_2]_2$ is about 79 and 23 min at 30 and 40 °C, respectively [245].

In one example, the synthesis of fluorinated diacyl peroxide was achieved in a two-step process in our laboratory [246–248]:

$$C_8F_{17}-C_2H_4-CO_2H + SOCl_2 \xrightarrow{\text{reflux}} R_F-C_2H_4-COCl \qquad \text{Yield} = 92\%$$

$$R_F-C_2H_4-COCl + H_2O_2 + NaOH \xrightarrow[\text{2h, 25 °C}]{F_{113}} R_F-C_2H_4-\underset{O}{\underset{\|}{C}}-OO-\underset{O}{\underset{\|}{C}}-C_2H_4-R_1$$

$$\text{Yield} = 87\%$$

Similarly, we also carried out the synthesis of another perfluorinated diacyl peroxide from $C_7F_{15}CO_2H$ but the yield of the second step was 40% giving mainly unreacted R_FCOCl [247]. However, its dissociation constant is much higher [243] (Table V).

The most significant investigations have been carried out by Sawada *et al.* [249–261], who have published more than 20 articles on this topic that was reviewed in 1996 [254], 2000 [259] and then in 2003 [261]. Various fluorinated initiators and monomers were used. Fluorinated diacyl peroxides undergo a homolytic scission of the O–O central bond under heating, leading to fluorinated acyloxy radicals that are decarboxylated to generate $R_F^{\cdot}$ radicals [254]:

TABLE V

HALF-LIVES AND DISSOCIATION RATES (AT THE REACTION TEMPERATURE) OF VARIOUS FLUORINATED INITIATORS

Initiator	Half-life (min)	T (°C)	k_d (s^{-1})	Reference
$(CF_3-\underset{O}{\overset{\|}{C}}-O)_2$	140	60	–	[244]
	34	72	–	[244]
$(C_3F_7OCF-\underset{CF_3\ O}{\overset{\|\ \ \ \|}{C}}-O)_2$	59	30	–	[243]
$(R_F-\underset{O}{\overset{\|}{C}}-O)_2$	–	–	3.3×10^{-4} (40 °C)	[243]
$(FO_2SC_2F_4OCF-\underset{CF_3\ O}{\overset{\|\ \ \ \|}{C}}-O)_2$	79	30	–	[245]
	23	40	–	[245]
$(R_F-C_2H_4-\underset{O}{\overset{\|}{C}}-O)_2$	–	–	7.3×10^{-5} (82 °C)	[265]
$C_8F_{17}-Azo$	50	82	2.3×10^{-4} (82 °C)	[265]

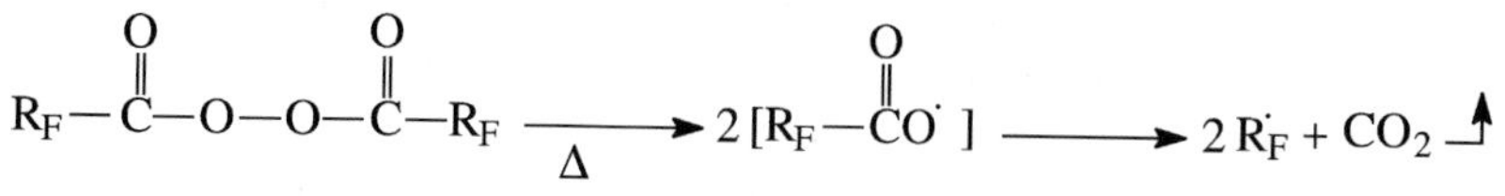

$$R_F-\overset{\overset{\displaystyle O}{\|}}{C}-O-O-\overset{\overset{\displaystyle O}{\|}}{C}-R_F \xrightarrow{\Delta} 2\,[R_F-\overset{\overset{\displaystyle O}{\|}}{C}O^{\cdot}\,] \longrightarrow 2\,R_F^{\cdot} + CO_2 \uparrow$$

These (per)fluorinated radicals react with the monomers leading to novel (per)fluoroalk(ox)yl end-capped co-oligomers.

Many examples concern the radical polymerisation of methacrylates according to the following scheme:

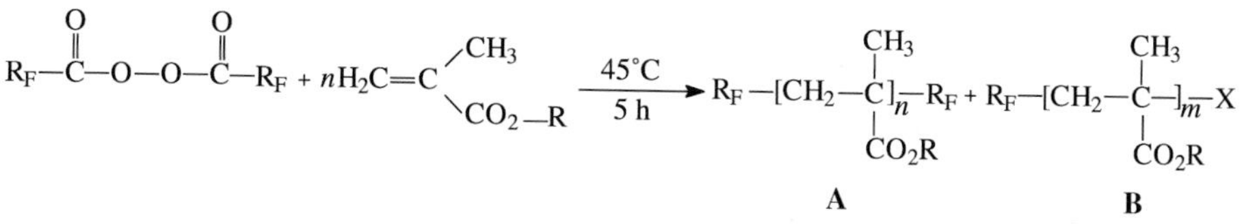

$$R_F-\overset{\overset{\displaystyle O}{\|}}{C}-O-O-\overset{\overset{\displaystyle O}{\|}}{C}-R_F + nH_2C=C\overset{\diagup CH_3}{\diagdown CO_2-R} \xrightarrow[5\,h]{45°C} R_F-[CH_2-\overset{\overset{\displaystyle CH_3}{|}}{\underset{\underset{\displaystyle CO_2R}{|}}{C}}]_{\overline{n}}R_F + R_F-[CH_2-\overset{\overset{\displaystyle CH_3}{|}}{\underset{\underset{\displaystyle CO_2R}{|}}{C}}]_{\overline{m}}X$$

$$\qquad\qquad\qquad\qquad\qquad\qquad\qquad\qquad\qquad\mathbf{A}\qquad\qquad\qquad\qquad\mathbf{B}$$

where R_F substituents stand for C_3F_7, $CF(CF_3)OC_3F_7$, and $CF(CF_3)OCF_2CF(CF_3)OC_3F_7$ [250, 251, 259] while X can be:

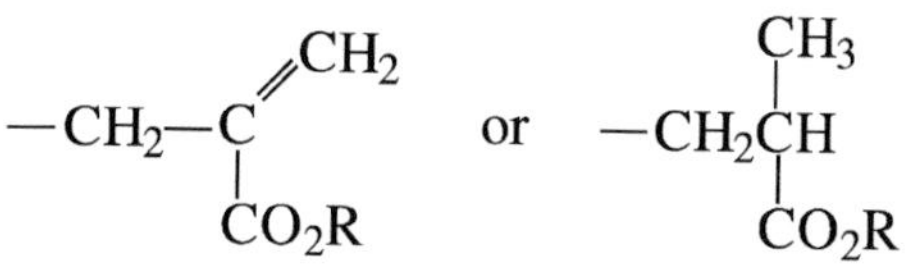

$$-CH_2-C\overset{\diagup CH_2}{\diagdown CO_2R} \qquad or \qquad -CH_2\overset{\overset{\displaystyle CH_3}{|}}{\underset{\underset{\displaystyle CO_2R}{|}}{CH}}$$

A mixture of fluorinated diblock (**B**) and triblock (**A**) co-oligomers or copolymers can be produced, since methacrylates undergo termination either by recombination (leading to triblock copolymers) or by disproportionation (yielding diblock copolymers).

Sawada's team investigated the radical polymerisation of methacrylates containing various key groups or functions (e.g., quaternary ammonium end-group) such as [258]:

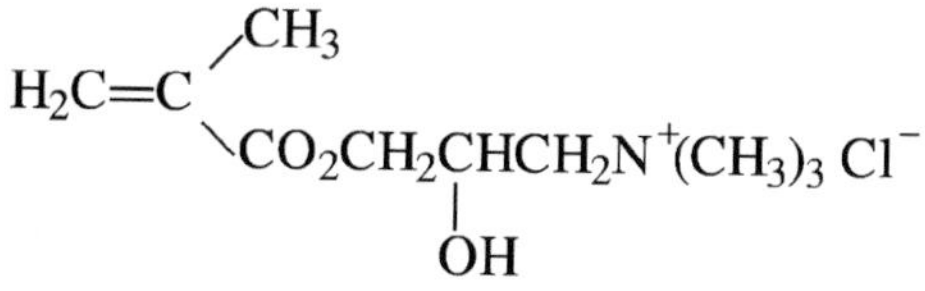

$$H_2C=C\overset{\diagup CH_3}{\diagdown CO_2CH_2\underset{\underset{\displaystyle OH}{|}}{CH}CH_2N^+(CH_3)_3\ Cl^-}$$

and also copolymerised this monomer with MMA or with trimethylvinylsilane [258], obtaining "gel like" triblock copolymers in water, methanol, ethanol, DMF and DMSO [257]. Interestingly, the polymers obtained exhibit acceptable surface properties ($\gamma_c = 15$ mN m^{-1}), a good antibacterial activity and show aggregation of the fluoroalkyl segments.

The same group also introduced an isocyanatoethyl methacrylate 2-butanone oxime adduct (IEM–BO) and N,N-dimethylacrylamide (DMAA) [260]. These fluoroalkoxy end-capped IEM–BO–DMAA co-oligomers, soluble in water and hexane, were able to reduce the surface tension of water to 18 mN m^{-1}. They formed self-assembled molecular aggregates of 17.5 nm size in aqueous solutions, and could interact strongly with ethidium bromide to form a host–guest intermolecular complex. In addition, they bound with functional aromatic moieties such as 5-fluorouracyl (5-FU) and 9-aminoacridine and, with the former, the resulting co-oligomer-bound 5-FU had remarkably strong interaction with oligoDNA [260].

F-triblock copolymers and partially fluorinated polysoaps were also obtained by the same authors by adopting a similar strategy using other hydrophilic monomers such as acrylic acid [252]. Interestingly, these acrylic acid oligomers bearing two fluoroalkyl end-groups were obtained via primary radical termination or radical chain transfer to the peroxide. Molecular weights and PDI ranged between 4600 and 14,800 and 1.43 and 1.82, respectively. Elemental analyses of fluorine confirmed that these oligomers contain two fluoroalkylated end-groups [249]. They were soluble in water, methanol, ethanol and THF [250, 251] and enabled effective reduction of the surface tension of water with a clear breakpoint resembling a cmc at around 15 mM [250]. In addition, various F-triblock copolymers were successfully used as surfactants in the formulation of potential inhibitors of the HIV virus [253, 255] or of anti-human immunode-ficiency viruses [258].

Using the same strategy, Guan *et al.* [245] have recently prepared in moderate to high yields (52–97%) original oligostyrene end-capped by perfluoro[1-(2-fluorosulfonyl)ethoxy] ethyl groups, as follows:

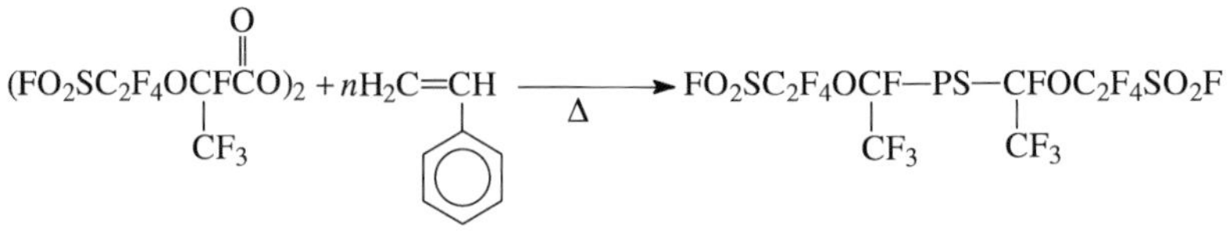

$$(FO_2SC_2F_4OCFCO)_2 + nH_2C=CH \xrightarrow{\Delta} FO_2SC_2F_4OCF-PS-CFOC_2F_4SO_2F$$

Depending on the initial [initiator]$_o$/[styrene]$_o$ molar ratios and on the reaction temperatures, different F-block copolymers were obtained. Their molar masses ranged from 4200 to 8300 while their polydispersities varied from 1.99 to 3.30. These authors have found that the fluoroalkyl end-capped oligo(styrene)s showed rather high contact angles (with water) up to 102° compared with that of polystyrene (95°).

Another interesting result concerning the synthesis of telechelic fluoroelasto-mers by dead end copolymerisation of VDF and HFP was achieved by Rice and Sandberg [262–264]. Diacyl peroxide [262] was also fluorinated and its synthesis was as follows:

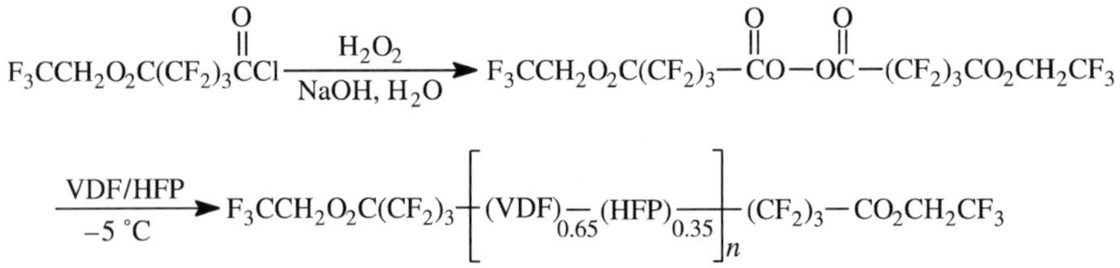

$$F_3CCH_2O_2C(CF_2)_3CCl \xrightarrow[NaOH, H_2O]{H_2O_2} F_3CCH_2O_2C(CF_2)_3-CO-OC-(CF_2)_3CO_2CH_2CF_3$$

$$\xrightarrow[-5\,°C]{VDF/HFP} F_3CCH_2O_2C(CF_2)_3 \left[(VDF)_{0.65}-(HFP)_{0.35} \right]_n (CF_2)_3-CO_2CH_2CF_3$$

The molecular weights of these fluoroelastomers were estimated by vapour pressure osmometry (*ca.* 4000).

2.2.1.1.2.   From Fluorinated Azo Initiators.   A similar strategy was also investigated using fluorinated azo initiators prepared in our laboratory. Four examples are reported here. First, the radical bismonoaddition of fluorinated mercaptans onto α,ω-bisunsaturated azines followed by a nitrilation step led to an original fluorinated hydrazine. Then, bromination and debromination reactions enabled Bessiere *et al.* [246–248] to successfully achieve the synthesis of α,ω-bis(perfluorooctyl) azo initiators:

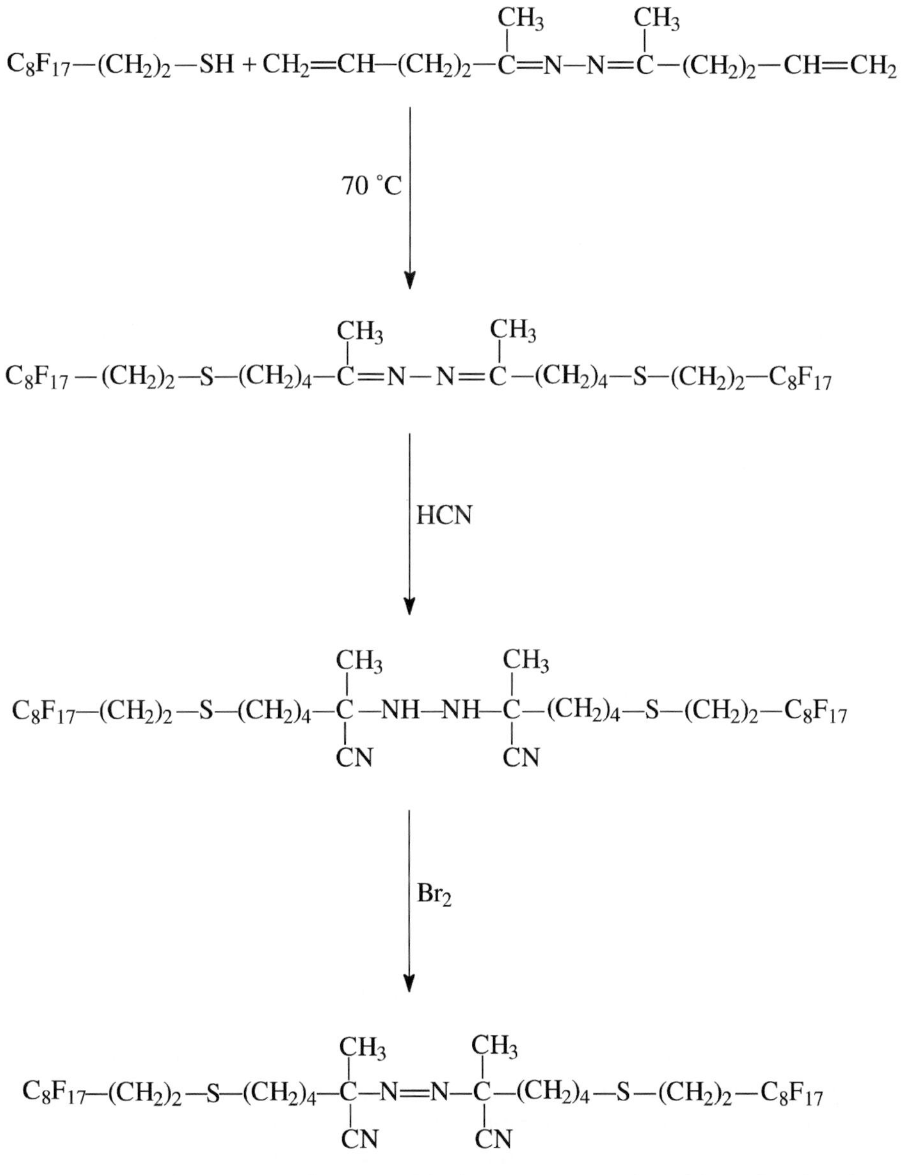

$$C_8F_{17}-(CH_2)_2-SH + CH_2=CH-(CH_2)_2-\overset{\overset{\displaystyle CH_3}{|}}{C}=N-N=\overset{\overset{\displaystyle CH_3}{|}}{C}-(CH_2)_2-CH=CH_2$$

70 °C

$$C_8F_{17}-(CH_2)_2-S-(CH_2)_4-\overset{\overset{\displaystyle CH_3}{|}}{C}=N-N=\overset{\overset{\displaystyle CH_3}{|}}{C}-(CH_2)_4-S-(CH_2)_2-C_8F_{17}$$

HCN

$$C_8F_{17}-(CH_2)_2-S-(CH_2)_4-\overset{\overset{\displaystyle CH_3}{|}}{\underset{\underset{\displaystyle CN}{|}}{C}}-NH-NH-\overset{\overset{\displaystyle CH_3}{|}}{\underset{\underset{\displaystyle CN}{|}}{C}}-(CH_2)_4-S-(CH_2)_2-C_8F_{17}$$

Br$_2$

$$C_8F_{17}-(CH_2)_2-S-(CH_2)_4-\overset{\overset{\displaystyle CH_3}{|}}{\underset{\underset{\displaystyle CN}{|}}{C}}-N=N-\overset{\overset{\displaystyle CH_3}{|}}{\underset{\underset{\displaystyle CN}{|}}{C}}-(CH_2)_4-S-(CH_2)_2-C_8F_{17}$$

Another procedure started from azo-bis(4-cyanopentanoic acid) (ACPA), whose acid chloride derivative reacted onto a fluorinated alcohol, as follows [247, 248]:

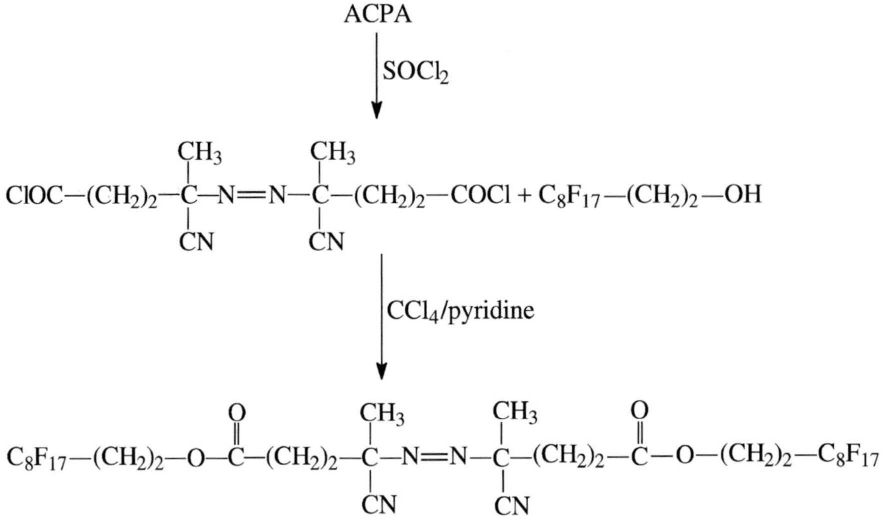

ACPA

$\downarrow$ SOCl$_2$

$$\underset{\overset{|}{CN}}{\overset{\overset{\displaystyle CH_3}{|}}{ClOC-(CH_2)_2-C}}-N{=}N-\underset{\overset{|}{CN}}{\overset{\overset{\displaystyle CH_3}{|}}{C}}-(CH_2)_2-COCl \; + \; C_8F_{17}-(CH_2)_2-OH$$

$\downarrow$ CCl$_4$/pyridine

$$C_8F_{17}-(CH_2)_2-O-\overset{\overset{\displaystyle O}{\|}}{C}-(CH_2)_2-\underset{\overset{|}{CN}}{\overset{\overset{\displaystyle CH_3}{|}}{C}}-N{=}N-\underset{\overset{|}{CN}}{\overset{\overset{\displaystyle CH_3}{|}}{C}}-(CH_2)_2-\overset{\overset{\displaystyle O}{\|}}{C}-O-(CH_2)_2-C_8F_{17}$$

A third alternative concerned the condensation of a fluorinated acid chloride onto an α,ω-dihydroxylazo compound [247]:

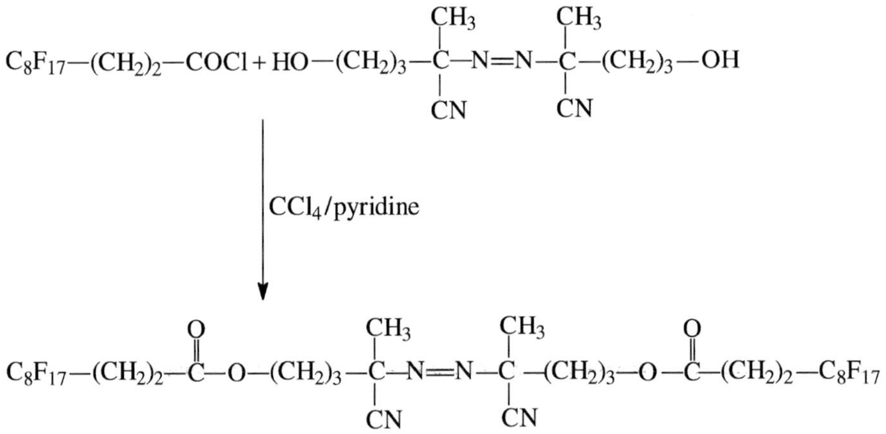

$$C_8F_{17}-(CH_2)_2-COCl + HO-(CH_2)_3-\underset{\overset{|}{CN}}{\overset{\overset{\displaystyle CH_3}{|}}{C}}-N{=}N-\underset{\overset{|}{CN}}{\overset{\overset{\displaystyle CH_3}{|}}{C}}-(CH_2)_3-OH$$

$\downarrow$ CCl$_4$/pyridine

$$C_8F_{17}-(CH_2)_2-\overset{\overset{\displaystyle O}{\|}}{C}-O-(CH_2)_3-\underset{\overset{|}{CN}}{\overset{\overset{\displaystyle CH_3}{|}}{C}}-N{=}N-\underset{\overset{|}{CN}}{\overset{\overset{\displaystyle CH_3}{|}}{C}}-(CH_2)_3-O-\overset{\overset{\displaystyle O}{\|}}{C}-(CH_2)_2-C_8F_{17}$$

Finally, from a similar fluorotelechelic azo initiator (C_8F_{17}–Azo), Lebreton and Boutevin [265] synthesised a block perfluorinated oligo(acrylamide) by radical polymerisation of acrylamide at 82 °C. Surprisingly, the authors also noted the presence of a triblock copolymer while the diblock copolymer only was expected because of the disproportionation of acrylamide [266]. Indeed, the authors assessed the decomposition rate constant ($k_d = 2.33 \times 10^{-4}$ s^{-1} at 82 °C) and the half-life (50 min at 82 °C) of such a fluorinated azo initiator [265] (Table IV). Acetonitrile was chosen as a suitable solvent because of its quite high boiling point (82 °C) and its low transfer constant. The same authors extensively studied the kinetics of radical polymerisation of acrylamide initiated by C_8F_{17}–Azo as follows:

$$[HO_2CC_2H_4-\overset{\overset{\displaystyle CH_3}{|}}{\underset{\underset{\displaystyle CN}{|}}{C}}-N]_{\overline{2}} + HOC_2H_4C_8F_{17} \xrightarrow{DCCI} [C_8F_{17}C_2H_4O\overset{\overset{\displaystyle O}{\|}}{C}C_2H_4-\overset{\overset{\displaystyle CH_3}{|}}{\underset{\underset{\displaystyle CN}{|}}{C}}-N]_{\overline{2}} \quad (70\%)$$

$$C_8F_{17}\text{-Azo}$$

$$C_8F_{17}\text{-Azo} + n\,H_2C{=}CH\overset{\overset{\displaystyle O}{\|}}{C}NH_2 \xrightarrow{82\,°C} C_8F_{17}C_2H_4O\overset{\overset{\displaystyle O}{\|}}{C}C_2H_4\overset{\overset{\displaystyle CH_3}{|}}{\underset{\underset{\displaystyle CN}{|}}{C}}(AA)_nR$$

$$\mathbf{AA}$$

where R designates H, an unsaturated group, or $\overset{\overset{\displaystyle CH_3}{|}}{\underset{\underset{\displaystyle CN}{|}}{C}}-C_2H_4-\overset{\overset{\displaystyle O}{\|}}{C}-O-C_2H_4-C_8F_{17}$ group.

Although the radical polymerisation proceeded heterogeneously due to the unsolubility of poly(acrylamide), the following kinetic law, deviating a great deal from the classical law, was supplied: $R_{pi} \sim [C_8F_{17}{-}Azo]_o^{0.76}[AA]_o^{1.06}$ for: $0.05\% < [C_8F_{17}{-}Azo]_o/[AA]_o < 1.00\%$ and $R_{pi} \sim [C_8F_{17}{-}Azo]_o^{0.29}[AA]_o^{1.95}$ for: $1\% < [C_8F_{17}{-}Azo]_o/[AA]_o < 7\%$ showing that primary radical termination predominated in this latter case.

Hence, it was possible to obtain triblock copolymers (Table VI) that were separated from diblock copolymers by solvent fractionation. Their molar masses ranged from 11,600 to 54,000.

Interestingly, associative properties of the triblock copolymer were noted [267], as shown by the shape of the GPC chromatogram in various solvents, although acrylamide undergoes disproportionation [266]. Indeed, size exclusion chromatography of these oligo(acrylamide)s was carried out in three eluents: water, water/acetonitrile 90/10 (v/v) and water/acetonitrile 70/30 (v/v). The elution of the polymer was strongly influenced by the composition of the carrier solvent. These R_F-*b*-oligo(acrylamide)-*b*-R_F hydrophobically-modified water-soluble polymers associated in water and formed aggregates of high hydrodynamical volumes [267]. The associative characteristic of the triblock structure was confirmed by a rheological study (showing a Newtonian behaviour over a wide range of shear rates).

Interestingly, these triblock oligo(acrylamide) copolymers were used in associative thickeners in painting formulations [268] and in fire-fighting aqueous formulations [73].

TABLE VI

AMOUNT OF FLUORINATED DIBLOCK AND TRIBLOCK COPOLYMERS ACCORDING TO THE [INITIATOR, $C_8F_{17}{-}Azo]_o$/[ACRYLAMIDE, AA]$_o$ INITIAL MOLAR RATIOS [265]

$[C_8F_{17}{-}Azo]_o/[AA]_o$ (%)	Diblock copolymer (%)	Triblock copolymer (%)
1	50	50
3	30	70
5	23	77
7	17	83

Another interesting alternative was proposed by Inoue *et al.* [269], who extensively investigated the surface characteristics of poly(fluoroalkylsiloxane)-*b*-PMMA-*b*-poly(fluoroalkylsiloxane) triblock copolymers of different molecular weights and containing various fluorinated side groups. These block copolymers were synthesised according to a similar strategy as above from the DEP of methyl methacrylate (MMA) initiated by poly(fluoroalkylsiloxane)-containing azo initiator, as follows:

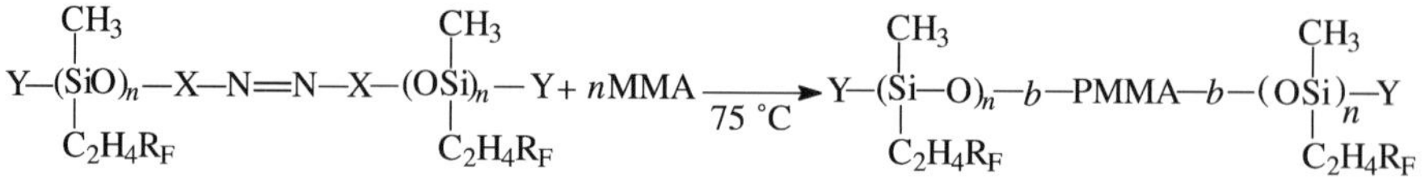

$$
\underset{\underset{C_2H_4R_F}{|}}{Y-(SiO)_n}-X-N=N-X-\underset{\underset{C_2H_4R_F}{|}}{(OSi)_n}-Y + nMMA \xrightarrow[75\,°C]{} \underset{\underset{C_2H_4R_F}{|}}{Y-(Si-O)_n}-b-PMMA-b-\underset{\underset{C_2H_4R_F}{|}}{(OSi)_n}-Y
$$

R_F: CF_3, C_6F_{13}, C_8F_{17}

Y: $-[Si(CH_3)_2O]-C_3H_6NH_2$; X: variable spacer

Cast films were prepared on which XPS, ESCA and contact angles were assessed. A relationship between surface segregation behaviour and adhesion performance was established. A significant effect of fluoroalkyl groups onto the surface accumulation, not only in silicon but also fluorine atoms, was observed for a relatively long fluoroalkyl side chain that enhanced the hydrophobicity and the oleophobicity.

2.2.1.2.   By Telomerisation

As reported in Section 2.1.1.1 for the synthesis of model R_F-R_H diblock copolymers, telomerisation and, to a lesser extent, monoaddition reaction is a powerful tool to enable the preparation of model $R_F-R_H-R_F$ or $R_H-R_F-R_H$ triblocks. The choice of an adequate transfer agent is crucial to allow one to obtain the expected fluoroblock(s). For instance, straightforward syntheses can be achieved as in the examples below:

$$
n\,R_FI + H_2C=CH-R_H-CH=CH_2 \longrightarrow R_F-R_H-R_F
$$

$$
IR_FI + n\,H_2C=CH-R_H \longrightarrow R_H-R_F-R_H
$$

In addition, sequential telomerisation is also possible when the transfer agent is adequately chosen. Indeed, the intermediate telomers which are produced act as original transfer agents, since one (or both) extremity(ies) is (are) reactivable to allow a further reaction.

In those cases, telomers that exhibit a CFX$-$I (where X represents H, F atoms or a CF_3 group) are suitable and particularly $I(CF_2)_x-I$. The latter have been successfully used in iodine transfer polymerisation (ITP).

As a matter of fact, other methods involving transfer steps are also useful to favour the production of triblock copolymers.

They require xanthates or dithioesters that contain a specific cleavable group such as:

$$\text{\textasciitilde\textasciitilde\textasciitilde CYZ} \overset{\displaystyle S}{\underset{\displaystyle \downarrow}{\underset{\|}{\text{\textbardbl SCW—R}}}}$$

where Y, Z stand for H, CH_3, C_6H_5 and W represents "_" (in the case of dithioester) or an oxygen atom (case of xanthate) to introduce fluorinated or non-fluorinated polymeric units in a controlled way.

Hence, the chronology of this sub-section is as follows: after reporting various syntheses dealing with radical monoaddition (Section 2.2.1.2.1), stepwise cotelomerisation (Section 2.2.1.2.2) is going to demonstrate how A–B–A, B–A–B and A–B–C are synthesised. Then, other preparations of fluorinated triblock copolymers achieved by controlled radical polymerisation are depicted in further sections. They concern ATRP (Section 2.2.2.1), ITP (Section 2.2.2.2), RAFT and MADIX (Section 2.2.2.3).

2.2.1.2.1.   Synthesis of Models of F-Triblocks from Radical Monoadditions.   As mentioned in Section 2.1.1 that reports the synthesis and the properties of $R_F–R_H$ semi-fluorinated diblock copolymers achieved by radical monoaddition, models such as $C_nF_{2n+1}–C_pH_{2p}–C_nF_{2n+1}$ fluorotriblock co-oligomers have also been described in the literature. Non-exhaustive examples are $C_4F_9–(CH_2)_n–C_4F_9$ (where $n = 1–4$) [16, 19, 20], $C_{10}F_{21}(CH_2)_8C_{10}F_{21}$ [37], $C_{12}F_{25}–(CH_2)_p–C_{12}F_{25}$ (where $p = 8$ and 10) [37], and $C_nF_{2n+1}(CH_2)_6CH=CH(CH_2)_6C_nF_{2n+1}$ (where $n = 4$ or 10) [40].

The first three models were obtained by radical bismonoaddition of an excess of perfluoroalkyl iodide onto a non-conjugated diene followed by the selective reduction of the iodine atoms, although cyclic derivatives were also produced:

$$C_nF_{2n+1}I + H_2C=CH-(CH_2)_x-CH=CH_2 \xrightarrow[\text{(2) H}^-]{\text{(1) rad. init.}} C_nF_{2n+1}-(CH_2)_{x+4}-C_nF_{2n+1}$$

The synthesis of these F-triblock molecules and their morphological and structural aspects were recently reviewed by Napoli [39].

2.2.1.2.2.   Stepwise Cotelomerisation.   This sub-section reports an elegant method in which the telomer produced acts as an original transfer agent for a further telomerisation reaction.

α,ω-Diiodofluoroalkanes allow access to well-defined block cotelomers. Tortelli and Tonelli [270] and Baum and Malik [271, 272] performed the synthesis of such telechelic products by heating iodine crystal with tetrafluoroethylene. Such halogenated reactants were successfully involved in telomerisation of VDF [273] or hexafluoropropene (HFP) [270–272, 274, 275] (see Chapter 1 dealing with the fluorinated telomers) to form α,ω-diiodo PVDF-*b*-PTFE-*b*-PVDF or α,ω-diiodo HFP-*b*-PTFE-*b*-HFP triblock cotelomers [276] as interesting models of fluoro A–B–A terpolymers. These products are potential starting materials for FKM-type multiblock cotelomers [274] (e.g., Viton®, Dai-el®, Tecnoflon®, Fluorel®, etc.) as shown in the following examples:

$$I(VDF)_p(TFE)_n(VDF)_qI + mHFP \longrightarrow I(HFP)_x(VDF)_p(TFE)_n(VDF)_q(HFP)_yI$$

$$I(HFP)(TFE)_n(HFP)_tI + qVDF \longrightarrow I(VDF)_z(HFP)(TFE)_n(HFP)_t(VDF)_\omega I$$

Further, these novel α,ω-diiodides underwent functionalisations [276], especially for the preparation of original fluorinated nonconjugated dienes [277] (utilised in the preparation of hybrid fluorosilicones [278] which show excellent properties at low and high temperatures), or in original fluorinated polyesters and polyurethanes from fluorinated α,ω-diols [272] (see Section 3.3 and Chapter 2 on fluorotelechelics).

2.2.2. Synthesis of F-Triblocks by Controlled Radical Polymerisation

Three sub-sections now report the preparation of fluorinated triblocks by controlled (or pseudo-living) radical polymerisation: (i) by sequential ATRP of (fluoro)monomers; (ii) by ITP of fluoroalkenes (essentially C_2–C_3 gaseous olefins); and (iii) by reversible addition fragmentation transfer (RAFT). Several examples are given below.

2.2.2.1. By Sequential ATRP of (Fluoro)monomers

Zhang et al. [279] prepared two series of triblock copolymers from sequential ATRP of styrene (or copolymerisation of styrene with a silicon-containing methacrylate (SiMA)) and methacrylate (FNEMA) starting from a telechelic dibromino end-capped polystyrene, as follows:

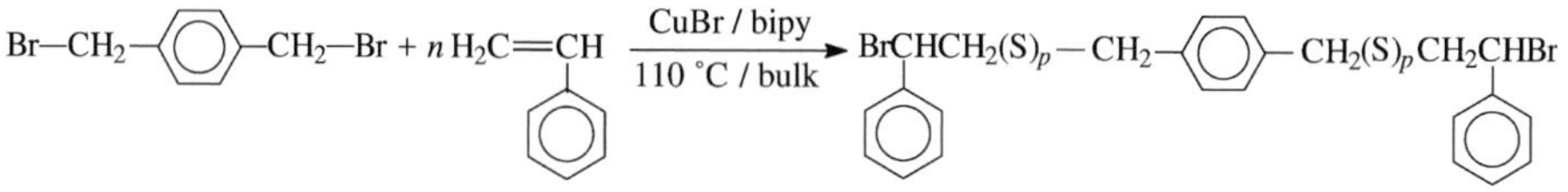

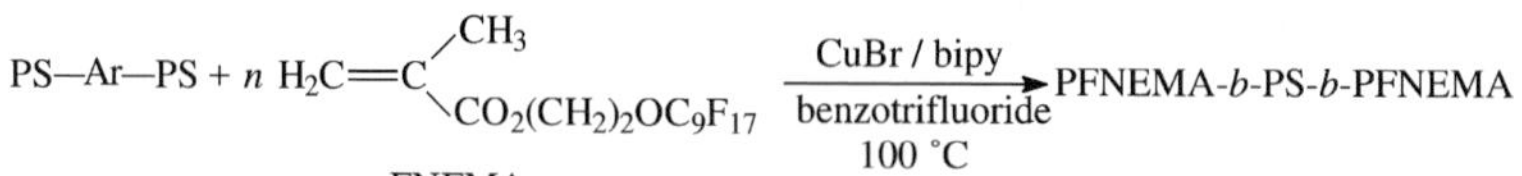

The molar masses and the PDI of the α,ω-dibromo polystyrenic initiator obtained in the first step and of the triblock copolymer were 6200 and 13,400, and 1.37 and 1.49, respectively.

From a similar strategy, this Chinese group [279] could achieve the poly(EGMAFO)-b-poly(S-co-SiMA)-b-poly(EGMAFO) triblock copolymers using ethylene glycol mono-methacrylate mono-perfluorooctanoate (EGMAFO) as the fluorinated monomer. The molar mass obtained was 14,300, starting from the telechelic dibromopoly(S-co-SiMA) whose $\overline{M}_n$ was 6400.

2.2.2.2. Synthesis of Fluorinated A–B–A by Iodine Transfer Polymerisation (ITP)

Although ITP has already been described in Section 2.1.2.3.1 for the preparation of fluorinated diblock copolymers, the use of efficient iodoinitiators is also a very interesting method for the preparation of triblock copolymers. The most relevant results were achieved from $I(C_2F_4)_nI$ ($n = 4$ or 6 mainly, which are thermally stable, and also because IC_2F_4I undergoes β-scission to generate iodine and tetrafluoroethylene (TFE) under radical conditions [273, 280]). These telechelic diiodides (the syntheses of which are comprehensively reported in Chapters 1 and 2) are obtained from the telomerisation of TFE with molecular I_2. These telomers are commercially available from the Daikin and Asahi Glass companies but are also produced in a pilot plant at the Solvay-Solexis Company.

They are interesting precursors to many fluorotelechelic derivatives, nonconjugated dienes, as well as a wide range of novel materials or fluoropolycondensates (see Chapter 2). For example, they easily react with $H_2C{=}CHSiCl_3$ under peroxidic initiation to produce

$$I(CHCH_2)_n{-}R_F{-}(CH_2CH)_m{-}I$$
$${\overset{|}{SiCl_3}}\phantom{)_n{-}R_F{-}(CH_2}{\overset{|}{SiCl_3}}$$

(where $n + m \geq 2$) triblock copolymers [281]. These fluoropolymers were easily moisture curable and led to original transparent, tough, oil- and water-repellent films with good acid resistance (even at 100 °C). They are useful as sealants, coupling agents, finishes and release coatings.

However, one of the most exciting investigations concerns the preparation of thermoplastic elastomers (TPEs). These A–B–A fluorocopolymers are phase segregated with an amorphous zone and crystalline domains, and they combine various advantages inducing a wide range of temperatures of service.

These fluorinated triblock copolymers are synthesised in a two-step procedure: a novel α,ω-diiodofluoropolymer (usually an elastomeric I–Soft–I block) is produced from the first step and can be further successfully utilised as an original initiator in a second reaction as follows:

$$I{-}R_F{-}I + n\,M_1 + m\,M_2 \xrightarrow{\text{rad.}} I[(M_1)_x(M_2)_y]_z{-}R_F{-}[(M_1)_x(M_2)_y]_z{-}I$$
$$\phantom{I{-}R_F{-}I + n\,M_1 + m\,M_2 \xrightarrow{\text{rad.}} I[(M_1)_x(M_2)_y]_z}_{\text{I–Soft–I}}$$

$$I{-}Soft{-}I + p\,M_3 + q\,M_4 \longrightarrow$$

$$poly(M_3\text{-}co\text{-}M_4)\text{-}b\text{-}poly(M_1\text{-}co\text{-}M_2)\text{-}b\text{-}poly(M_3\text{-}co\text{-}M_4)$$

The process can be emulsion, suspension, microemulsion or solution. Various companies have already shown much interest in this research.

Daikin's pioneering works from 1979 [114, 146, 152] were confirmed at the DuPont Company [165] and at Ausimont (now Solvay-Solexis) [166–168].

A *quasi*-exhaustive summary of these investigations is given in Table VII, where M_1, M_2, M_3 and M_4 (fluoro)alkenes can be equally well chosen among TFE, VDF, hexafluoropropylene (HFP), perfluoromethyl vinyl ether (PMVE),

TABLE VII

IODINE TRANSFER POLYMERISATION OF FLUOROALKENES FOR THE SYNTHESIS OF FLUORINATED HARD-*b*-SOFT-*b*-HARD TRIBLOCK COPOLYMERS

Soft block[a]	% Comonomers in soft block	Hard block	Soft/hard ratio (wt%)	T_g (°C)	T_m (°C)	Reference
$I[(VDF)_xHFP]_yI$	n.g.	PVDF	n.g.	n.g.	160	[151]
$I[(VDF)_xHFP]_yI$	n.g.	Poly(E-*alt*-TFE)	n.g.	n.g.	220	[151]
$I[(CTFE)_x(VDF)_y]_zI$	45/55	Poly(E-*co*-CTFE)	85/15	− 6	247	[164]
$I[(CTFE)_x(VDF)_y]_zI$	45/55	Poly(E-*co*-TFE)	90/10	− 8	252	[164]
$I[(VDF)_x(FVA)_y]_zI$	n.g.	PVDF	n.g.	n.g.	n.g.	[114]
$I[(TFE)_xP]_yI$	55/45	Poly(E-*co*-TFE)	80/20	− 1	267	[165]
$I[(VDF)_xHFP(TFE)_y]_zI^{b,c}$	56/19/25	PVDF	80/20	− 15 to − 12	165	[166, 167]
$I[(VDF)_xHFP(TFE)_y]_zI$	35/40/25 (wt)	Poly(E-*alt*-TFE)	n.g.	− 8	222	[151]
$I[(VDF)_xHFP(TFE)_y]_zI$	50/30/20	Poly(E-*co*-HFP-*co*-TFE)	85/15	n.g.	n.g.	[282]
$I[(VDF)_xHFP(TFE)_y]_zI^{b}$	54/21/25	Poly(E-*alt*-TFE)	80/20	− 13	266	[166]
$I[(VDF)_xPMVE(TFE)_y]_zI^{b}$	62/19/19	PVDF	80/20	− 30	160	[166, 168]
$I[(VDF)_xPMVE(TFE)_y]_zI$	73/17/10	Poly(E-*co*-TFE)	72/28	− 33	254	[165]
$I[(VDF)_xPMVE(TFE)_y]_zI^{b}$	n.g.	Poly(E-*co*-TFE)	n.g.	− 15 to − 13	266	[166, 167]
$I[(VDF)_xPMVE(TFE)_y]_zI^{b}$	57/23/20	poly(E-*co*-TFE-*co*-PMVE)	75/25	n.g.	180	[168]
$I[(TFE)_xP(VDF)_y]_zI$	n.g.	Poly(E-*co*-TFE)	85/15	− 13	262	[165]
$I[(TFE)_xE(PMVE)_y]_zI$	45/19/36	Poly(E-*co*-TFE)	71/29	− 16	245	[165]

[a]VDF, HFP, TFE, CTFE, PMVE, FVA, E and P represent vinylidene fluoride, hexafluoropropene, tetrafluoroethylene, chlorotrifluoroethylene, perfluoromethyl-vinylether, perfluoro vinylacetic acid, ethylene and propylene, respectively.

[b]Also tetrapolymerisation with less than 1 mol% of $H_2C=CH-C_6F_{12}-CH=CH_2$.

[c]Also tetrapolymerisation with less than 1 mol% of $H_2C=CH-C_6F_{12}-CH=CH_2/H_2C=CHC_6F_{12}C_2H_4I/IC_2H_4C_6F_{12}C_2H_4I$ mixture.

n.g. stands for not given.

chlorotrifluoroethylene (CTFE), ethylene (E) and propylene (P). Typical combinations of comonomers by pair, trio or quartet enabled the authors to obtain both soft and hard segments. In the case of the preparation of blocks by the above copolymerisation of two olefins, the authors were able to synthesise soft or hard segments by varying the amounts of comonomer and taking into account their reactivity ratios (comprehensively reviewed in Ref. [12]). It is recalled that a poly(VDF-co-HFP) copolymer containing less than 15 mol% HFP is a thermoplastic (or hard block), while that having an HFP content higher than 15 mol% exhibits elastomeric properties [12]. More information is supplied in Table I of Chapter 2, which lists various elastomers based on VDF, HFP, TFE, CTFE, PMVE and P.

Usually, the central elastomeric segments have molar masses of 30,000 or more, while those of the plastomeric (or hard) blocks are higher than 10,000, although Yagi et al. [164] has already claimed molar masses of TPE up to 1,000,000.

As a consequence, original TPE composed of soft and hard segments (as for the well-known PS-b-PBut-b-PS or SBS triblock) were produced in a living fashion by sequential ITP. They were endowed with a negative glass transition temperature, T_g, and high melting temperatures, T_m, sometimes higher than 250 °C (Table VII). Figure 2 represents Dai-el® (from Daikin) or Tecnoflon TPE supplied by Ausimont in which both amorphous and crystalline zones are seen.

Tatemoto [282] used a mixture of VDF/HFP/TFE and $I-(CF_2)_4-I$ in the presence of bis(trichloroperfluorohexanoyl) peroxide in 1,2,2-trifluorotrichloroethane to obtain a soft $I(HFP)_a(VDF)_b(TFE)_cI$ diiodide containing a 30 : 50 : 20 molar ratio which exhibited a number-average molecular weight of 3300 with a polydispersity index of 1.27.

The hard segment can be composed of PVDF or PTFE, or a combination of VDF (in high quantity) with HFP or CTFE, or copolymers of TFE with E, PMVE,

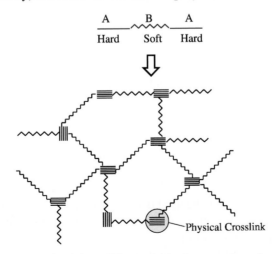

FIG. 2.   Schematic structural morphology of thermoplastic elastomers (reproduced with permission from, and courtesy of, M. Apostolo et al. (2002) Trends Polym. Sci., **7**, 23–40 [283]).

TABLE VIII

EFFECT OF THE NATURE OF THE TRANSFER AGENT IN IODINE TRANSFER POLYMERISATION OF
FLUOROALKENES

Transfer agent	Fluoroalkenes	M_n	PDI	Reference
None	VDF/HFP	–	5.0	[283]
i-C_5H_{12} or i-$C_{10}H_{22}$	VDF/HFP	–	5.0	[114, 152]
EtOAc	VDF/HFP	–	2.9	[283]
i-C_3F_7I	VDF/HFP	140,000	2.7	[114, 152]
IC_4F_8I	VDF/HFP	120,000	2.3	[284]
IC_4F_8I	VDF/HFP	143,000	1.7	[114, 152]
IC_4F_8I	VDF/HFP	–	1.5	[114]
$IC_6F_{12}I$	VDF/HFP	–	1.7	[283]
$H_2C=CHC_6F_{12}CH=CH_2$/IC_4F_8I	VDF/HFP	–	1.3	[284]
Triazinetriiodide	VDF/HFP/TFE	116,000	1.9	[169]

M_n and PDI stand for molar mass and polydispersity index, respectively.

VDF or with HFP, or terpolymers TFE/E with P, isobutylene, HFP, pentafluoropropene, PMVE, 3,3,3-trifluoropropene or hexafluoroisobutylene.

For example, the hard sequence can be composed of three comonomers (E/HFP/TFE in a 43 : 8 : 49 molar ratio in the plastomeric block [282]). A similar approach was also used with a central soft block composed of E/HFP/VDF in 20 : 30 : 50 molar ratio [150]. Additional advantages of using ethylene as the comonomer are: (i) its low cost; (ii) the livingness of the polymerisation [156]; and (iii) a further key advantage is the fact that this monomer reduces the base sensitivity of VDF-containing copolymers [165, 283].

In addition, the use of α,ω-diiodofluoroalkanes favours the control of the (co)polymerisation. Table VIII lists various transfer agents used in the ITP of fluoroalkenes. Although not all the reactions have been carried out in the same experimental conditions, it is observed that polydispersities of the final TPE as low as 1.73 could be achieved from diiodides. Interestingly, when a fluorinated nonconjugated diene [166, 168, 283–285] was involved in terpolymerisation of fluoroalkenes, the PDI was even lower (1.4). This improved control of ITP, called "pseudo living and branching technology" [283–285], was extensively studied at Ausimont (now Solvay-Solexis S.A.) and enabled the production of competitive TPE. This is the case, for example, for Tecnoflon® FTPE XPL.

Table VII also lists the T_gs and T_ms of these fluorinated TPE. Even when PMVE was added in the soft block, the T_g could not be significantly lowered. Further, the introduction of a small amount of nonconjugated diene slightly lowered the T_g and increased the T_m by a few degrees [166–168].

Most of these TPEs could be used from − 30 to 250 °C [151] and can be used in continuous service at 200 °C (in contrast to conventional elastomers that decompose at this temperature).

Interestingly, the TPEs can be crosslinked by the peroxide/triallyl (iso)-cyanurate system [286] or by gamma (^{60}Co) or beta rays [168] which confer improved thermal stability, a high resistance to vapour, UV, chemical agents, ozone and hydrocarbons, and better mechanical properties (tensile modulus, compression set) than those characteristic of crosslinked fluoroelastomers. Tatemoto [114] compared the thermal decomposition of three TPEs and supplied the following order: poly(E-*alt*-TFE)-*b*-poly(VDF-*co*-HFP)-*b*-poly(E-*alt*-TFE); $T_{dec} = 345\,°C <$ poly(E-*alt*-CTFE)-*b*-poly(VDF-*co*-HFP)-*b*-poly(E-*alt*-CTFE); $T_{dec} = 379\,°C <$ PVDF-*b*-poly(VDF-*co*-HFP)-*b*-PVDF; $T_{dec} = 410\,°C$.

Regarding their rheological properties, the above TPEs can easily be processed as gums.

Commercially available Dai-els® [151] (Figure 3) exhibit very interesting properties such as a high specific volume (1.90), a high melting point (160–220 °C), a high thermostability (up to 380–400 °C), a refractive index of 1.357 and good surface properties ($\gamma_c \sim 19.6–20.5\,mN\,m^{-1}$). These characteristics offer excellent resistance against aggressive chemicals and strong acids, fuels, oils and ozone. In addition, their tensile moduli are close to those of cured fluoroelastomers.

Interestingly, poly(E-*co*-M)-*b*-poly(VDF-*co*-CTFE)-*b*-poly(E-*co*-M) triblock copolymers (where M stands for TFE or CTFE) exhibit smaller transmittance than PMMA does in the UV range of 200–300 nm, while they had substantially the same transmittance as that of PMMA in the visible range [164]. Hence, these A–B–A block copolymers can cut the UV light which is dangerous for eyes and can be successfully used in optics and contact lenses. Other applications concern the compatibilisers for polymer alloys and blends which comprise a hydrocarbon

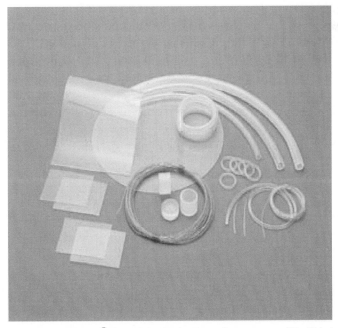

FIG. 3.   Various items of Dai-el® thermoplastic elastomer (by courtesy of Dr Shimizu, Dai Act).

polymer and a fluorine-containing polymer, a dispersant for a fluorine-containing paint, a carrier for electrophotography and an electrostatic charge-adjusting or fusion-preventing agent for toner particles.

Further, numerous applications in high-tech areas such as aerospace, aeronautics and engineering or in optics [164], arising from their outstanding properties, are listed in Table IX.

These TPEs are useful as O-rings [286], hot melts [287], pressure-sensitive adhesives, tough transparent films [281], sealants having good chemical and ageing resistance, protective coatings for metals [281] or even for denture plates [287] (since Dai-el® T530 is claimed to show neither erythrocyte hemolysis nor cytotoxicity and does not generate any harmful substances).

In addition, binary blends of TPE with fluoroelastomers [282] or with thermoplastics [288], and ternary blends of two TPEs and one thermoplastics [289] have also been investigated. They can find applications as pressure-sensitive adhesives [282], sealants [282], electrical devices (e.g., conductors) [288], fuel hoses for engines, [289], O-rings [289] and shaft seals [289].

2.2.2.3. Synthesis of F-Triblock Copolymers by RAFT and MADIX Techniques

Fluorinated dithioesters (mentioned in Section 2.1.2.4.3.1) are also efficient chain transfer agents and enabled us to synthesise fluorinated block copolymers [197].

The syntheses of C_8F_{17}-b-PS-b-PM triblock copolymers (where M stands for MMA, ethylacrylate (EA) or 1,3-butadiene) have recently been achieved by sequential RAFT copolymerisations of styrene and M monomers [197]. For example, we prepared: (i) C_8F_{17}-b-PEA-b-PS copolymer of $\overline{M}_n = 32,600$ and PDI = 1.20 from a C_8F_{17}-b-PEA having $\overline{M}_n$ of 14,200 and a PDI of 1.14, and (ii) a C_8F_{17}-b-PMMA-b-PS of $\overline{M}_n = 147,000$ and PDI = 1.36 from a C_8F_{17}-b-PMMA of $\overline{M}_n = 23,000$ and PDI = 1.20, as follows:

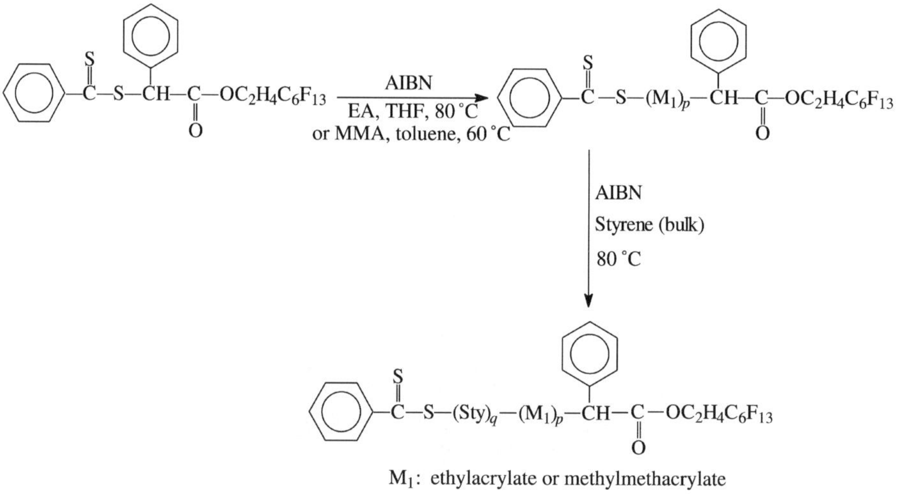

M$_1$: ethylacrylate or methylmethacrylate

TABLE IX

FIELDS OF APPLICATIONS AND ITEMS FROM THERMOPLASTIC ELASTOMERS [114, 152, 164]

	Moulding	Weather and chemical resistance	Transparency	Elasticity	Safety	Vulcanisability	Mechanical properties	Applications
Medical, biochemical	X	X	X	X	X			Tubing, rubber stoppers, medical materials
Physical and chemical equipment	X	X	X	X		X		Tubing, seals, pipettes
Semiconductor industry	X	X	X	X	X			Tubing and seals
Chemical industry		X		X		X		Anticorrosion paints, diaphragms, hoses, linings
Food industry	X	X	X	X	X			Hoses, sealing materials
Electronics industry	X	X		X			X	Wiring, sealing material
Civil engineering	X	X		X			X	Sealing materials, films, non-flammable materials
Textile industry				X			X	Coatings, filaments, fibres, water-repellant treatments
Optics			X		X			Lenses
Other fields			X	X			X	Adhesives (hot melts)

A few years ago, the Rhodia company [119, 120] claimed an original process of controlled radical polymerisation of various monomers (styrene, alkyl acrylate and, interestingly, vinyl acetate that has never been radically polymerised in a controlled way) using xanthates as potential chain transfer agents. These xanthates,

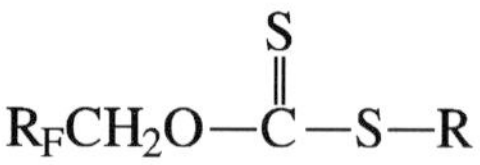

$$R_FCH_2O-\overset{\overset{\displaystyle S}{\|}}{C}-S-R$$

[with R=CH(CH$_3$)CO$_2$Et; CH(CO$_2$CH$_3$)$_2$, C(CH$_3$)R$_1$R$_2$ with (R$_1$, R$_2$): (CH$_3$, C$_6$H$_5$) or (CO$_2$Et, SC$_6$H$_5$), and R$_F$: CF$_3$, C$_3$F$_7$ or C$_6$F$_{13}$CH$_2$] were synthesised in a two-step procedure from the addition of a fluorinated alcohol (R$_F$CH$_2$OH) to CS$_2$ and then sodium hydride, followed by reaction with an alkyl halide (RX).

These xanthates enabled a wide range of diblock copolymers bearing a fluorinated end-group (see Section 2.1.2.4.4) to be produced by sequential radical polymerisation, with a total conversion of xanthates. Such copolymers have the following structure:

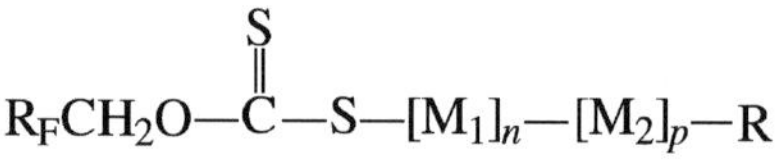

$$R_FCH_2O-\overset{\overset{\displaystyle S}{\|}}{C}-S-[M_1]_n-[M_2]_p-R$$

with M$_1$: styrene, alkyl (ethyl, butyl or t-butyl) acrylate; M$_2$: alkyl (methyl, ethyl, t-butyl) acrylate, vinyl acetate.

Evidence for control of the polymerisation reaction was provided by: (i) a *quasi*-linear "$\overline{M}_n$ vs. monomer conversion" relationship, in good agreement with the theoritical curve, and (ii) a lower polydispersity index (PDI) at higher monomer conversion. Conversely, such behaviour of polymerisations of the same monomers in identical conditions but involving hydrogenated xanthates was not observed. The molar masses and the PDI were in the ranges of 5000–14,000 and 1.45–1.66, respectively.

A few examples of the synthesis of fluoro A–B–C triblock copolymers by MADIX are reported in a patent assigned to Rhodia [119], concerning the preparation of R$_F$–(M$_1$)$_n$–(M$_2$)$_m$ where R$_F$, M$_1$ and M$_2$ are CF$_3$ (or C$_6$F$_5$), VAc, and S or MMA, respectively.

2.2.2.4. *Conclusion*

Various methods of controlled radical polymerisation are useful for the synthesis of fluorotriblock copolymers: sequential ATRP, two-step ITP, stepwise RAFT and MADIX polymerisations. Interestingly, the number of publications and patents has increased over recent years.

Although ITP was pioneered in the late 1970s, more interest is noted from the mid-1990s when pseudo-living radical controlled polymerisation emerged. This explains why many investigations have been carried out and also why Dai-el$^{®}$ is commercially available, although few scientific articles (most of them from Tatemoto) have been published. The disadvantage arises from the cost of

α,ω-diiodoperfluoroalkanes, industrially produced by Japanese companies and in a pilot plant by Solvay-Solexis. Nevertheless, ITP can be adapted to methacrylate and styrenic monomers and particularly to fluoroalkenes, for which ITP is the only method that enables controlled radical polymerisations. Another advantage concerns the versatility of the process, which can occur in solution, suspension, emulsion or microemulsion.

More recent investigations carried out at Ausimont/Solvay-Solexis from the branching and pseudo-living technology have produced very interesting fluoropolymers (e.g. Tecnoflon® FTPE XPL).

Other methods such as RAFT and MADIX require complex chain transfer agents (dithioesters or xanthates, respectively) and sometimes the copolymers obtained may retain the strong smell from these sulfur-containing derivatives. Nevertheless, both approaches are complementary in term of the chosen monomers.

Finally, sequential ATRP seems attractive although the excess cuprous halide and ligand used (bipyridine is highly toxic) is a limitation, even after several work-ups and purifications. Nevertheless, experimental evidence supports pseudo-living polymerisation as a method that can be well adapted to styrenic and, especially, (meth)acrylic monomers.

2.2.3. Synthesis of Fluorotriblock Copolymers by Combining Different Methods

This sub-section summarises several routes to prepare F-triblocks by using the sequential mixed systems. They can combine either condensation methods or classical radical oligomerisation or telomerisation, or may involve controlled radical polymerisation (mainly ATRP). Hence, non-exhaustive examples are supplied in the following order:

    (i) condensation then telomerisation;
    (ii) telomers that are then condensed;
    (iii) coupling of telomerisation/ATRP;
    (iv) condensation followed by ATRP.

In most cases, A–B–A triblocks are expected and the nature of their precursors (usually telechelic or α,ω-difunctional derivatives, see Chapter 2) is crucial for their syntheses.

Beside them, one different approach was suggested by Schulte et al. [290], who prepared

$$C_nF_{2n+1}\!-\!\!\!\langle\bigcirc\rangle\!-\!C_nF_{2n+1}$$

from the Ullman coupling reaction between an excess of $C_nF_{2n+1}I$ (where $n = 6$, 7, 8, 10, 12) and 1,4-diiodobenzene. This American group studied the effect of a stiff phenyl unit and cooperative motion between fluorocarbon segments by Raman spectroscopy and calorimetry. By DSC, the authors noted low-temperature phase transition in $C_7F_{15}-C_6H_4-C_7F_{15}$, suggesting that its low-temperature crystal structure was similar to that of its longer chain analogues at room temperature.

2.2.3.1. Synthesis of Triblock Copolymers by Coupling Condensation and Telomerisation Methods

This route enables one to synthesise F-triblocks in a simple two-step procedure where the first step allows the synthesis of the transfer agent by condensation.

One of the best examples is a poly[F-sulfonamido(meth)acrylate]-b-poly(ethylene oxide)-b-poly[F-sulfonamido(meth)acrylate] terpolymer marketed by the 3M company under the Scotchguard® trademark [55, 96]. It was prepared by telomerisation of fluorinated (meth)acrylate with a telechelic polyethylene oxide (PEO) dithiol as follows:

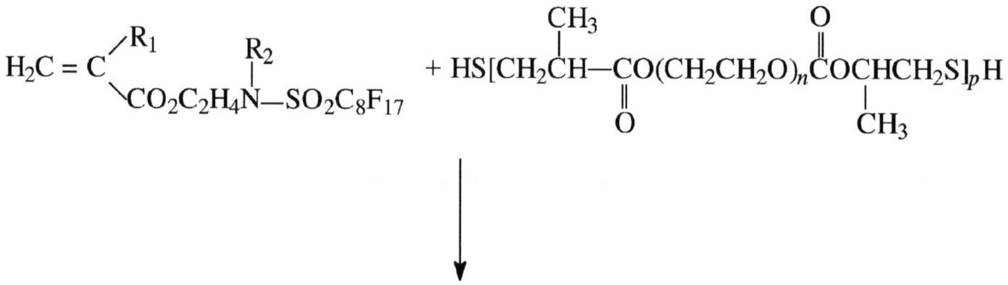

$$H_2C = C \overset{R_1}{\underset{CO_2C_2H_4N-SO_2C_8F_{17}}{}} \quad + \quad HS[CH_2\overset{CH_3}{\underset{\underset{O}{\parallel}}{CH}}-CO(CH_2CH_2O)_n\overset{O}{\overset{\parallel}{C}}OCHCH_2S]_pH$$

$$\downarrow$$

poly[F(meth)acrylate]-b-PEO-b-poly[F(meth)acrylate]

with R_1: H or CH_3, R_2: CH_3 or C_2H_5, $n = 4$ and $p \leq 100$.

Stochguard® product is regarded as a "hybrid" coating that exhibits both antispot and soil-release properties thanks to its hydrophobic and oleophobic fluorinated sulfonamido-acrylic sequences and a hydrophilic block composed of ethylene tetroxide (ETO). It usually contains two lateral sequences of three acrylic base-units separated by a range of 10–100 of ETO units. These copolymers also have good mechanical properties (durability of coatings [291]) on various substrates [292–294] and resistance to washings [96, 295].

Because of recent surveys on bioaccumulation of fluorine in human blood, this product has been withdrawn from the market. Nowadays, Foraperle®, Zonyl®, Lezanova® and Asahi Guard® (to name a few) produced by the Atofina (now DuPont), DuPont, Daikin and Asahi companies, respectively, are the most important chemicals and mainly used for textile treatment and finishes.

2.2.3.2. Condensation of Telomers

An example of synthesis of A–B–A type triblocks by coupling telomerisation and condensation methods was achieved by Akemi et al. [296]. These were triblock polyurethanes to prepare anticoagulants of blood. These Japanese authors obtained them from a telechelic diisocyanato polystyrene telomer (this B block was generated from the telomerisation of styrene with bis(4-isocyanato-phenyl) disulfide) followed by its condensation with ω-hydroxy- or amino-telomers of 3,3,4,4,5,5,6,6,6-nonafluorohexyl acrylate (HQ-PFHA leading to A block), as follows:

$$HQCH_2CH_2SH + nH_2C=CHCO_2C_2H_4C_4F_9 \longrightarrow HQC_2H_4S-(CH_2-CH)_n-H$$

Q: O or NH

(FHA)

HQ-PFHA $\quad \overset{|}{C}O_2$

$\overset{|}{C}_2H_4C_4F_9$

$$OCN-\langle\bigcirc\rangle-S-PS-S-\langle\bigcirc\rangle-NCO$$

HQ-PFHA $\longrightarrow$ PFHA-*b*-PS-*b*-PFHA

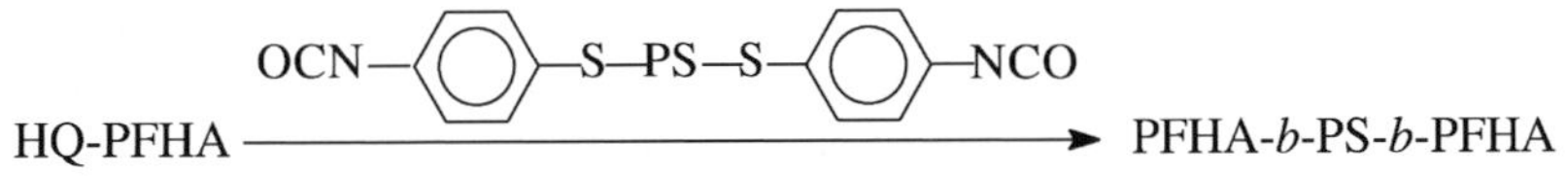

These authors supplied the kinetics monitoring of the reaction by showing the potential adjustment of the number of fluoroacrylate units.

2.2.3.3. Synthesis of Fluorinated A–B–A Terpolymers by the Coupling of Classic and Controlled Radical Techniques

As mentioned in Section 2.1.3.1 for the synthesis of A–B fluorocopolymers, an elegant preparation of fluorotriblocks can be achieved by ATRP of hydrogenated monomers (e.g., styrenic or (meth)acrylic derivatives) from fluorinated telechelic telomers that act as the original initiator:

$$X-R_F-X + H_2C=C\overset{R_1}{\underset{R_2}{}} \xrightarrow{\text{CuX/Ligand}} R_H-R_F-R_H$$

Conversely, the literature reports another strategy that requires a telechelic initiator (e.g., a polyether X–PEO–X) obtained by condensation of commercially available dihydroxy compounds as efficient precursors of A–B–A copolymers:

$$X-PEO-X + nH_2C=C\overset{R_1}{\underset{CO_2(CH_2)_x-R_F}{}} \xrightarrow{\text{CuX/L}} PFMA-b-PEO-b-PFMA$$

FMA

Both these routes are described below.

2.2.3.3.1. From ATRP of Hydrogenated Monomers Initiated by Telechelic Dihalogenated Fluorotelomers

2.2.3.3.1.1. Telomers that Possess C–I Cleavable Bonds. α,ω-Diiodi-nated PVDF telomers were synthesised to initiate the ATRP of MMA, leading to a PMMA-*b*-PVDF-*b*-PMMA triblock copolymer [128] as follows:

$$I-(CF_2)_4-I + nH_2C=CF_2 \longrightarrow I-PVDF-I \xrightarrow[\text{CuI/bipy}]{\text{MMA}} PMMA-b-PVDF-b-PMMA$$

The kinetic plots (ln ($[M]_0/[M]$) vs. time) showed nearly a linear behaviour, the $\overline{M}_n$ (assessed by NMR) of block copolymers linearly increasing with conversion.

In the same way, original α,ω-diiodo-PVDFs were successfully used to prepare PMMA-b-PVDF-b-PMMA and PHEA-b-PVDF-b-PHEA triblock copolymers to be involved in solutions deposited onto a trough to prepare Langmuir–Blodgett films [297].

Unfortunately, the ^{19}F and ^{1}H NMR characterisations of I–PVDF–I are not accurate enough to insure that ⌇⌇⌇CH$_2$CF$_2$I end-groups only have been produced to be sure that triblock copolymers were obtained. Actually, it is known that the propagation of VDF leads to defects of chaining (i.e., head-to-head and above all tail-to-tail chainings). As a consequence, transfer reaction onto the latter isomer should lead to ⌇⌇⌇CF$_2$CH$_2$I which should be inactive in the ATRP of MMA.

In addition, the syntheses of the triblock copolymers are in contradiction with the investigations of Tatemoto [156], who found that the yield of the PMA-b-R$_F$-b-PMA block copolymer was poor from the ITP of methylacrylate (MA), even more reactive than MMA.

2.2.3.3.1.2. Telomers Containing C–Br Cleavable Bonds. On the other hand, involving the cleavage of the C–Br bond, regarded as stronger than the C–I one but weaker than C–Cl bond, two processes have been used to prepare triblock copolymers.

The first one is based on the controlled radical polymerisation of styrene by bromine transfer [298] from α,ω-dibrominated derivatives Br–PVDF–Br, considered as "macroinitiators" [299]. These dihalogenated compounds were prepared by radical telomerisation of vinylidene fluoride with 1,2-dibromotetrafluoroethane [300], as follows:

$$\text{BrCF}_2\text{CF}_2\text{Br} + n\text{VDF} \xrightarrow{\text{peroxide}} \text{BrR}_\text{F}(\text{VDF})_n\text{Br}$$
$$\text{Br–PVDF–Br}$$

$$\text{Br-PVDF-Br} + p\text{H}_2\text{C=CH} \xrightarrow[\text{CuBr/bipy}]{\text{ATRP}} \text{Br(Sty)}_x\text{R}_\text{F}(\text{VDF})_n(\text{Sty})_y\text{Br}$$
$$\text{PS-}b\text{-PVDF-}b\text{-PS}$$

The Chinese authors showed that molecular weights of the polymers increased linearly with styrene conversion, and that polydispersities were low, suggesting the formation of a triblock copolymer in a living fashion. However, these articles [298, 299] do not report any information on: (i) the presence or absence of PVDF oligomers or polymers as by-products generated from the direct addition of the radicals (arising from the peroxidic initiators) onto VDF and (ii) the nature of the end-groups of the VDF telomers that should contain both –CF$_2$CH$_2$Br and –CH$_2$CF$_2$Br extremities but the authors did not report which was the more reactive end-group for a successful ATRP of styrene.

The second example reports that, under similar ATRP conditions and from a "telechelic" ethylene bis(2-bromopropionate), poly(FOA) (where FOA stands for

1H,1H-perfluorooctyl acrylate) bearing two bromo end-groups was obtained with $\overline{M}_{n(NMR)} = 12,000$. This diester acted also as a macroinitiator for obtaining PMMA-*b*-PFOA-*b*-PMMA triblock copolymer that contained 49 MMA units [138].

   2.2.3.3.2. By Coupling Condensation/ATRP of Fluoromonomers. In 2002, by ATRP of fluorinated methacrylates, Busse *et al.* [301] synthesised amphiphilic block copolymers of poly(ethylene oxide) (PEO) and poly(fluoromethacrylate) (PFMA) which were able to form stable micelles. In the case of the synthesis of the triblock copolymers, the telechelic macroinitiator was produced from the condensation of α,ω-dihydroxy PEOs of various molar masses (ranging from 6000 to 20,000) with 2-bromopropionyl bromide [302]. Then, the reaction is as follows:

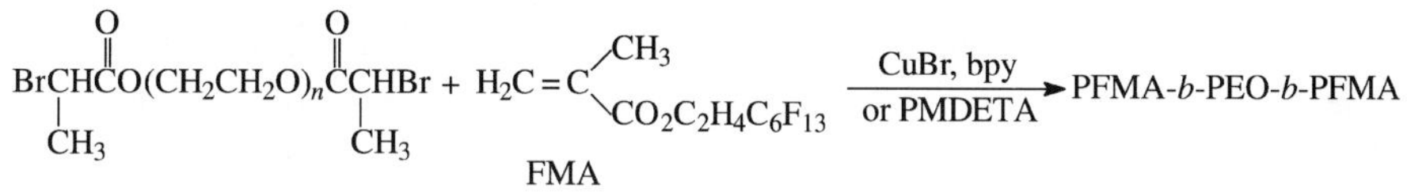

Starting from PEO of 6000, 10,000 and 20,000, the authors obtained triblock copolymers containing a wt% of PFMA ranging from 20 to 60, 5 to 18 and 4 to 62, respectively. Almost all these fluoroblock copolymers exhibit PDI varying from 1.1 to 1.8.

   Interestingly, these authors studied the bulk properties by X-ray scattering (SAXS and WAXS), DSC and polarised light microscopy (PLM) [302], and the self-association of these fluorocopolymers in aqueous solution using STM, DLS and transmission electron microscopy (TEM) [303]. SAXS data showed morphology of hexagonal packed cylinders in melt and lamellar morphology in solid for the triblock copolymer produced from PEO of 20,000 containing 24 wt% of PFMA. In addition, a reduction of the crystallinity of PEO was observed, accompanied with a decrease of its melting temperature that was enhanced for longer PFMA blocks. By PLM, the authors noted that the PFMA content strongly affected the spherulitic texture and its growth-rate in the copolymers.

   First, STM showed a clear inflection point from the surface tension vs. concentration plots [302]. These authors assigned the concentration corresponding to the inflection point to the cmc and they observed that the higher the fluorine content of the triblock copolymers up to 11 wt.% PFMA (solubility limit), the lower the cmc value.

   Second, from DLS investigations carried out on various samples above, the cmc revealed: (i) the existence of small aggregates (micelles) and single chains for F-diblock copolymer solution and (ii) large clusters as the dominant species in the F-triblock copolymer solutions.

   These clusters were interpreted as loose network-like structures formed because of an intermicellar interaction through bridges. In addition, the micelle size was resistant to elevated temperatures [302, 303] although a non-negligible increase in apparent hydrodynamic radius was noted when the concentration was increased, while both temperature and concentration affected the large clusters (especially in concentrated solutions).

Third, TEM observations after depositing the aqueous solutions to carbon-coated copper grids indicated that the initial concentration of samples has a strong influence on the morphology of the aggregate formed: the German team [303] observed single micelles, fibrous networks and even films at high concentrations. Finally, the addition of colloidal gold particles into the solutions led to typical covering by the fluoroblock copolymer.

The same strategy of synthesis was realised by Korean researchers [131], who synthesised PFOMA-b-PEO-b-PFOMA triblocks with controlled molar masses and narrow polydispersities from telechelic PEO bearing chlorine atoms at both ends of the polymeric chain. This Cl–PEO–Cl macroinitiator was prepared by the condensation of telechelic dihydroxy-PEO with 2-chloropropionyl chloride while the triblock arose from ATRP of 1H,1H-perfluorooctyl methacrylates (FOMA) as follows:

$$HO\text{--}PEO\text{--}OH + Cl\text{--}\overset{\overset{\displaystyle O}{\|}}{C}\text{--}\underset{\underset{\displaystyle CH_3}{|}}{CHCl} \longrightarrow ClCH\text{--}\overset{\overset{\displaystyle O}{\|}}{CO}\text{--}PEO\text{--}\overset{\overset{\displaystyle O}{\|}}{OC}\text{--}\underset{\underset{\displaystyle CH_3}{|}}{CHCl}$$

$$\underset{\displaystyle CH_3}{|}$$

Cl–PEO–Cl

$$Cl\text{--}PEO\text{--}Cl + n\,H_2C = C\overset{\displaystyle CH_3}{\underset{\displaystyle CO_2CH_2C_7F_{15}}{<}} \xrightarrow{\text{CuCl / bipy}} PFOMA\text{-}b\text{-}PEO\text{-}b\text{-}PFOMA$$

FOMA

As in the case above, these A–B–A triblocks exhibit amphiphilic properties and self-assemble in chloroform to promote micelles, evidenced by DLS and TEM. Although the authors did not assess the cmc, they observed different morphologies of spheres and rods by changing the copolymer composition.

Their differences, compared with those of the previous study (Busse et al. [224] noted that their triblocks organised into hexagonally packed cylinder or into lamellae), might be due to the nature of the fluoromethacrylate that exhibits a longer fluorinated chain and a shorter spacer between the perfluorinated end-group and the ester function. Besides, as previously shown in the course of study of the kinetics of radical homopolymerisation of fluorinated methacrylates [304], the longer this spacer, the greater the reactivity of the monomers.

Furthermore, these novel surfactants lower the interfacial tension between water and supercritical carbon dioxide to stabilise water in carbon dioxide emulsions.

2.2.3.4. Miscellaneous

Another interesting study, carried out by Li et al. [305], concerns the use of a PS-b-PDMS diblock copolymer prepared by anionic polymerisation to obtain a triblock copolymer. The diblock macromolecule contained a hydroxyl end-group, chemically esterified in a first stage to yield an ω-bromo-PDMS-*b*-PS (PS-*b*-PDMS-Br) which acted as an original initiator in the ATRP of perfluorooctylsulfonamido methacrylate (FOSMA), in the presence of CuCl/pentamethyldiethylenetriamine (PMDETA), as described by:

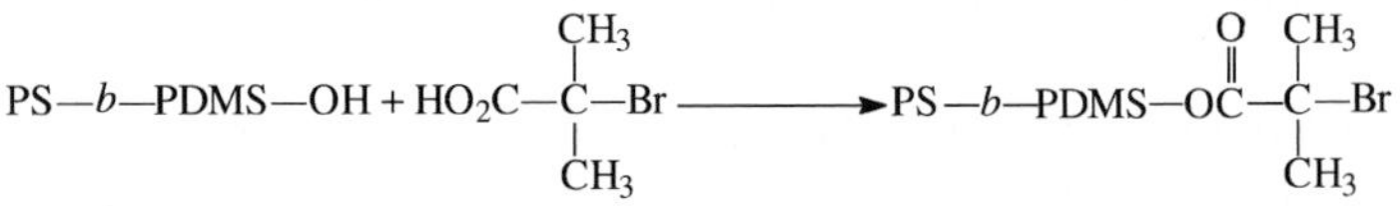

$$PS-b-PDMS-OH + HO_2C-\underset{\underset{CH_3}{|}}{\overset{\overset{CH_3}{|}}{C}}-Br \longrightarrow PS-b-PDMS-O\underset{\underset{CH_3}{|}}{\overset{\overset{O\ CH_3}{||\ \ \ |}}{C-C}}-Br$$

PS-b-PDMS-Br

$$PS-b-PDMS-Br + n\,H_2C=C\overset{CH_3}{\underset{\underset{\underset{Et}{|}}{CO_2C_2H_4NSO_2C_8F_{17}}}{\diagup}} \xrightarrow{\ CuCl\ /\ PMDETA\ } PS-b-PDMS-b-PFOSMA$$

FOSMA

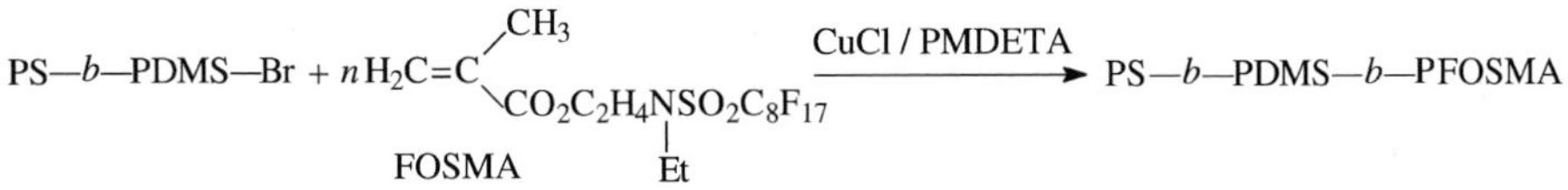

2.2.4. Synthesis of Fluorinated Triblock Copolymers by Living Methods Different from the Radical Ones

As mentioned in Section 2.1.4, specific living polymerisations other than the radical controlled polymerisation are interesting means to prepare fluorotriblocks copolymers. They consist of cationic and anionic polymerisations and of ring opening metathesis.

2.2.4.1. Cationic Polymerisation

In contrast to the synthesis of fluorodiblock copolymers achieved from sequential cationic living polymerisation of vinyl ethers, no preparation of triblock copolymer was attempted from this route.

However, the ring opening cationic polymerisation of oxazolines and of ε-caprolactone successfully led to fluorinated A−B−A terpolymers.

2.2.4.1.1. From the Cationic Polymerisation of Oxazoline. Water-soluble amphiphilic surfactants were synthesised by German teams [306] from the living cationic polymerisation of 2-methyl-2-oxazoline, initiated by trifluoromethane sulfonic acid-1-ethylperfluorocetyl ester. Their structures were:

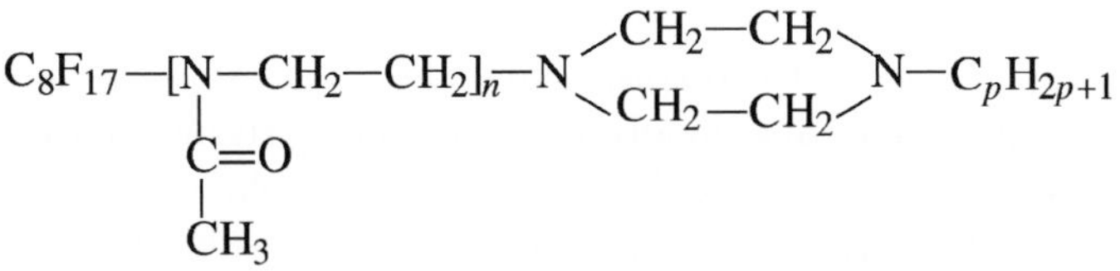

$$C_8F_{17}-[N-CH_2-CH_2]_n-N\overset{\diagup CH_2-CH_2\diagdown}{\underset{\diagdown CH_2-CH_2\diagup}{}}N-C_pH_{2p+1}$$
$$\underset{\underset{CH_3}{|}}{\overset{|}{C}=O}$$

where $n = 25-30$ and $p = 6-18$ (even numbers).

Average degrees of polymerisation (n) assessed from ^{1}H NMR, end-group analyses of both the piperidine (that was used to terminate a small part of the polymer mixture) and from the lipophilic features were in good agreement. They corresponded closely to the values expected from the initial [monomer]$_o$/ [initiator]$_o$ molar ratio (n was ranging between 25 and 30 whatever the chain length of the hydrogenated end-group) and the PDI showed a narrow distribution (PDI = 1.13–1.20).

The association behaviour (free and aggregated fluorinated end-group) was investigated by ^{19}F NMR and showed that the degree of aggregation increased linearly with the polymer concentration.

2.2.4.1.2.   From the Ring Opening Polymerisation of Oxygen-Containing Cyclic Monomers.   Another way to prepare fluorinated A–B–A copolymers consists of starting from fluorinated telechelic diols that act as original initiators in the cationic polymerisation of cyclic oxygen-containing monomers (ε-caprolactone and 1,3-dioxolane).

2.2.4.1.2.1.   From ε-Caprolactone.   To our knowledge, the only work was carried out by Pilati *et al.* [307–310], dealing with the ring opening polymerisation of ε-caprolactone (CL) in bulk, initiated by different metal oxides, from telechelic dihydroxy perfluoropolyethers (PFPE), as follows:

$$\text{HO–PFPE–OH} + \underset{\text{CL}}{\text{ε-caprolactone}} \xrightarrow[110-120\ °C]{\text{bulk Met}_x\text{O}_y} \text{PCL-}b\text{-PFPE-}b\text{-PCL}$$

These perfluoropolyethers were Fomblin® derivatives, formely marketed by Ausimont but now by the Solvay-Solexis company. They exhibit molecular weights ranging from 1100 to 3400 and T_gs as low as $-140\ °C$. The [CL]$_o$/[HO–PFPE–OH]$_o$ initial molar ratio was in the 4:1 to 200:1 range, hence leading to triblock copolymers with molar masses of 3600–12,000.

Interestingly, these authors investigated the influence of the nature of the metallic initiator onto the ε-caprolactone conversion and supplied the following decreasing series [309]:

initiator:

$$\text{Bu}_2\text{SnO} > \text{Ti(OBu)}_4 \gg \text{Ca(OAc)}_2 > \text{Al(O}i\text{Pr)}_3 > \text{Sm(OAc)}_3 \gg \text{Sb}_2\text{O}_3$$

CL conv.(%):

    36–100   18–100   5–23   2–15   0–8   0–2

2.2.4.1.2.2.   From 1,3-Dioxolane.   In addition, such fluorotelechelic PFPE diols ($\overline{M}_n = 2000$) were also involved in the polymerisation of 1,3-dioxolane (DIOX) initiated by triflic acid to successfully achieve the synthesis of PDIOX-b-PFPE-b-PDIOX [311] triblock copolymers with amphiphilic properties. After fractionation, the authors recovered various fractions of triblock copolymers whose molar masses were 4600, 11,000 and 20,000.

2.2.4.2.   Anionic Polymerisation

Sugiyama *et al.* [220] prepared A–B–C triblock copolymers composed of three immiscible sequences: poly[2-(perfluorobutyl) ethyl methacrylate], poly (*tert*-butyl methacrylate) and poly(2-hydroxyethyl methacrylate). These triblocks containing a polyfluoroalkyl methacrylate segment were quantitatively synthesised by sequential anionic living polymerisation of 2-(perfluorobutyl) ethyl methacrylate, (*tert*-butyl methacrylate) and 2-(trimethylsilyloxy)ethyl methacrylate initiated by *sec*-BuLi/1,1-diphenylethylene/LiCl catalyst. By changing the addition order and the molar ratios of the three above monomers in the feed, a wide range of polymers having different sequence orders and segment lengths were produced. Their molar masses and PDI ranged between 19,000–61,000 and 1.03–1.08, respectively.

By characterising these A–B–C copolymers by means of TEM, XPS and static light scattering, the authors observed that the sequential order of the blocks significantly affects the microphase separation, the surface construction of the polymer films and the micelle formation in selective solvents.

More recently, Busse *et al.* [224] prepared PFBSMA-*b*-PtBMA-*b*-PFBSMA triblock copolymers by stepwise anionic polymerisation of *tert*-butyl methacrylate initiated by potassium naphthenate followed by that of (2-*N*-methyl perfluorobutylsulfonamido) ethyl methacrylate (FBSMA).

These triblock copolymers exhibit various amounts of FBSMA (varying from 1 to 38), different ranges of molar masses (from 49,000 to 72,400) and narrow polydispersities (PDI varied from 1.09 to 1.19). These German researchers observed that an aggregation of the micelle occurred in contrast to the fluorodiblocks for which the formation of micelles dominates in a wide range of concentrations. In addition, from light scattering and TEM, these authors demonstrated that these triblock copolymers could cover gold particles.

2.2.4.3.   From Metathesis

As in the case of the synthesis of fluorodiblock copolymers, the only example obtained from metathesis was successfully carried out by McLain *et al.* [40]. This group from DuPont obtained PTFE-*b*-PE-*b*-PTFE triblocks from the ring opening metathesis polymerisation of cyclododecene in the presence of $R_F(CH_2)_6CH=CH(CH_2)_6R_F$ (where R_F is C_4F_9 or $C_{10}F_{21}$) followed by the reduction of the double bonds in the backbone. Hence, $C_4F_9-(CH_2)_{260}-C_4F_9$, $C_{10}F_{21}(CH_2)_{130}C_{10}F_{21}$ and $C_{10}F_{21}(CH_2)_{270}C_{10}F_{21}$ were obtained for which their melt surface tensions were investigated. These American authors noted that the third triblock copolymer was the most efficient at lowering the melt surface tension.

2.2.4.4.   Conclusion

Ionic (cationic and anionic) and metathesis polymerisations also offer smart ways to synthesise fluorotriblock copolymers. However, to our knowledge, the literature does not report any surveys from GTP, in contrast to the synthesis of fluorodiblock copolymers (Section 2.1.4.4). Nevertheless, these methods have been used for a limited number of monomers such as oxazolines, ε-caprolactone, 1,3-dioxolane, methacrylates and cyclododecene.

In addition, these processes are quite demanding and require specific conditions (e.g., high purity of the monomer and extreme experimental conditions such as a deep vacuum) providing yet another limitation to produce such fluorinated triblock copolymers on an industrial scale.

2.2.5. Synthesis of Fluorinated Triblock Copolymers by Condensation

Various condensation reactions have been reported in the literature and can be summarised in a non-exhaustive order as follows:

(i)   esterification between an OH and COX (X = Cl or OH) functions;
(ii) condensation of a chlorine atom and a hydroxyl group;
(iii) addition of an acid chloride onto an anion to terminate an anionic polymerisation.

In examples below, the first two kinds of condensation reactions have been motivated by macromolecules containing a central poly(ethylene oxide) (PEO) block and both perfluoroalkyl end-groups. Indeed, one of the most interesting properties of these block copolymers concerns their associative characteristics [312]. Such a behaviour has highlighted these polymers because of their original and specific rheological properties [313, 314]. They can be explained by the incompatibility between the different groups (hydrophobic and hydrophilic) that confer to these macromolecules a more or less pronounced antagonist character according to the strength of repulsion (such as steric or electrostatic effects). Upon application of shear, the network-like structures, held together by the relatively weak hydrophobic associations, are disrupted, giving rise to typical pseudo-plastic behaviour. However, upon removal of shear, the network structures reform and the viscosities of the polymer solutions completely recover their original low shear values. There is a growing interest in the interpretation and the understanding of the structure-properties relationship of the associative polymers from a good knowledge and the characterisation of the structure of these polymers.

This is why water-soluble associative polymers [312–314] have found increasing application as rheology modifiers (e.g., surfactants). In aqueous solutions, these polymers exhibit viscosifying properties and are named "associative thickeners".

Similar methodologies of syntheses started from dihydroxy telechelic PEO of narrow polydispersity index (1.07–1.18) esterified with perfluoroalkyl chloride [235] or ω-perfluoroalkyl acid [315]. Indeed, in the first case, Boschet *et al.* [235] used a wide series of PEO of molar masses between 300 and 35,000. Several attempts were made. Although condensation of activated telechelic dialcoholate PEO with $C_8F_{17}C_2H_4COCl$ was unsuccessful in solution, the reaction carried out in bulk led to C_8F_{17}ww PEO ww C_8F_{17}. The yield was higher than 90% when the reaction was carried out in molten PEO. The fluorinated acid chloride was successfully obtained from commercially available 3-perfluorooctylpropanoic acid:

$$C_8F_{17}C_2H_4CO_2H \xrightarrow[\text{2) } 110°C/3h]{\text{1) } PCl_5/RT} C_8F_{17}C_2H_4COCl$$

$$\downarrow HO-PEO-OH$$

$$C_8F_{17}-b-PEO-b-C_8F_{17}$$

For the triblock copolymer containing a PEO central block of 10,000, the cmc, aggregation concentration and critical concentration were 7×10^{-5}, 3×10^{-3} and 8×10^{-2} g ml^{-1}, respectively.

For the latter case, Hartman *et al.* [315] carried out the esterification, efficiently catalysed by dicyclohexylcarbodiimide (DCCI).

$$HO-PEO-OH + 2C_8F_{17}(CH_2)_{10}-CO_2H \xrightarrow{DCCI}$$

$$C_8F_{17}(CH_2)_{10}-CO_2-PEO-OCO(CH_2)_{10}-C_8F_{17}$$

The fluorinated nonadecanoic acid was prepared from commercially available perfluorooctyl iodide (see Chapter 2) in a two-step procedure [315].

Interestingly, NMR spectroscopy revealed a micellar form below the aggregation concentration while viscosimetric measurements could detect intramolecular associations. Hence, a flower-like structure was deduced.

Besides, longer F-triblock copolymers were synthesised, in which the hydrophobic end-groups were linked to the hydrophilic central block by an isophorone diurethane spacer, hence leading to hydrophobically modified ethoxylated urethane (HEUR). Winnick's group [316] suggested the following synthesis:

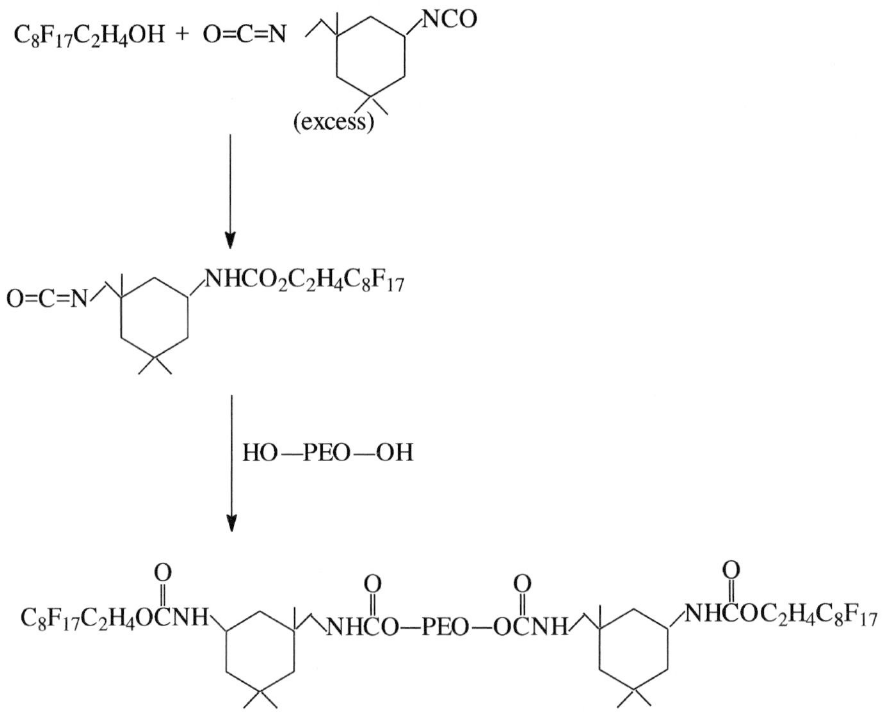

These Canadian authors noted that the steady-state shear viscosity increased strongly with the polymer concentration and was enhanced when changing the fluorinated chain length from C_8F_{17} to C_6F_{13} (the length of C_6F_{13} and C_8F_{17} end-groups are about 0.65 and 0.91 nm, respectively [317], while at 30 °C, the hydrodynamic radius of an unmodified PEO chain with $M_w = 35,000$ is 5.7 nm [318]. These authors observed a structure built up of flower-like micelles connected by bridging PEO chains to form a three-dimensional network. They could only form when the entropic penalty of looping was overcome by the insolubility of the chain-end so that, in dilute solution, the aggregated system preferred a structure with the end-group confined to a common core rather than free in solution. In aqueous solution, the polymers exhibited unusual rheological properties depending on the fluorocarbon length, the polymer concentration and the shear rate (frequency) [316]. Like most other associative polymers, their O-shear viscosity increased with the concentration.

In addition, this team also monitored the dilution and, by ^{19}F NMR parameters (e.g., ^{19}F transverse relaxation and ^{19}F chemical shift), they could distinguish aggregated and non-aggregated chain-ends. This enabled them to follow the formation of an infinite network [319].

An alternative leading to almost the same molecule was proposed by Cathebras et al. [320], proceeding first by the condensation of a telechelic diol PEO ($\overline{M}_n = 6000$; 10,000 or 20,000) with an excess of isophorone diisocyanate. Experimental conditions were recently improved to prevent polycondensation [321]. The resulting α,ω-diisocyanato isophorone urethane PEO was added to an excess of $C_8F_{17}-(CH_2)_{11}-OH$ to yield various C_8F_{17}-b-PEO-b-C_8F_{17} triblock copolymers. The degree of functionalisation was determined by ^{19}F NMR. The linear viscoelasticity of aqueous solutions of these F-HEUR copolymers was investigated as a function of the polymer concentration [320], temperature [320], intermediate chain length [320, 321], degree of functionalisation [321] and presence of a surfactant (SDS) [321]. In addition, Serreno et $al.$ [322] reported that these F-HEURs self-assemble in solution into star-like flowers in the dilute regime and, using small-angle neutron scattering, they investigated their form and structure factors.

Compared with fully hydrogenated end-capped PEO, the relaxation times of F-HEUR were found to be 10^3 times larger. The associated activation energy ($40\,k_BT$) was about twice that of the non-fluorinated homologue.

Using the above methodology, Zhou et $al.$ [323] also investigated the associative behaviour of these F-HEURs in aqueous solutions by NMR and fluorescent probe techniques. Various probes (such as pyrene derivatives (even fluorinated)) were chosen and perfluorooctanoyl pyrene was effective and informative in the study of the association of triblock R_F-b-PEO-b-R_F as a result of its good affinity to the fluorocarbon domains.

Another strategy that enabled Boschet et $al.$ [324] to synthesise novel A–B–C copolymers containing a PEO block was based on the sequential anionic copolymerisation of styrene and ethylene oxide, followed by an end-capping of $C_8F_{17}C_2H_4COCl$ (prepared in two steps from $C_8F_{17}C_2H_4I$). These PS-b-PEO-b-C_8F_{17} were obtained in high yields and were well characterised by various analytical methods. They exhibited narrow PDI (1.1–1.2), with molar masses of the PS-b-PEO ranging between 7500 and 14,500. These authors [325] also investigated the rheological properties of the triblock copolymers and showed that, below the aggregation concentration, competition occurred between the micellar associations of C_8F_{17} and PS. Above the aggregation concentration, hydrophobic associations were assigned to C_8F_{17} end-groups, suggesting that the associative behaviour was linked to a binding phenomenon. Adding β-cyclodextrin (which allowed the formation of inclusion complexes with hydrophobic or amphiphilic species) or methanol broke the associations between C_8F_{17} groups. As a consequence, these F-triblocks exhibit rheological properties similar to those of the non-associative precursors, hence confirming the associative behaviour of these fluorinated A–B–C copolymers.

Another example of $R_F-(M)_n-R_F$ triblocks was suggested by a Dutch team [241] which synthesised polybutylene adipate oligomers, followed by a "partial fluorination" by esterification to yield

$$C_7F_{15}-\overset{\displaystyle O}{\overset{\|}{C}}-O(CH_2)_4-O-[\overset{\displaystyle O}{\overset{\|}{C}}(CH_2)_4\overset{\displaystyle O}{\overset{\|}{C}}O(CH_2)_4O]_n-\overset{\displaystyle O}{\overset{\|}{C}}-C_7F_{15}$$

in small amount in addition to the formation of diblock co-oligomers (see Section 2.1.5). Evidence for the presence of both series of products was obtained from NMR, MALDI-TOF MS and GPC, although it was not possible to accurately quantify their relative amounts.

Another example concerning the synthesis of a fluoroblock copolymer of fluorinated silicone and poly(ethylene oxide) was recently reported in the literature [326]: 3,3,3-trifluoropropyldimethylchlorosilane was first condensed with a poly(fluorosiloxane) yielding an ω-CF$_3$ alkyl poly(fluorosiloxane) (CF$_3$–FSi). An ω-hydroxy PEO was condensed with one chlorine atom of a dimethylchlorosilane leading to PEO–Si(CH$_3$)$_2$Cl. The condensation of CF$_3$–FSi and PEO–Si(CH$_3$)$_2$Cl led to the fluoroblock copolymer. Such a macromolecule acted as a novel surfactant stabilising monodispersed particles of poly(styrene-co-acrylamide) in water-in-fluorosilicone oil emulsions.

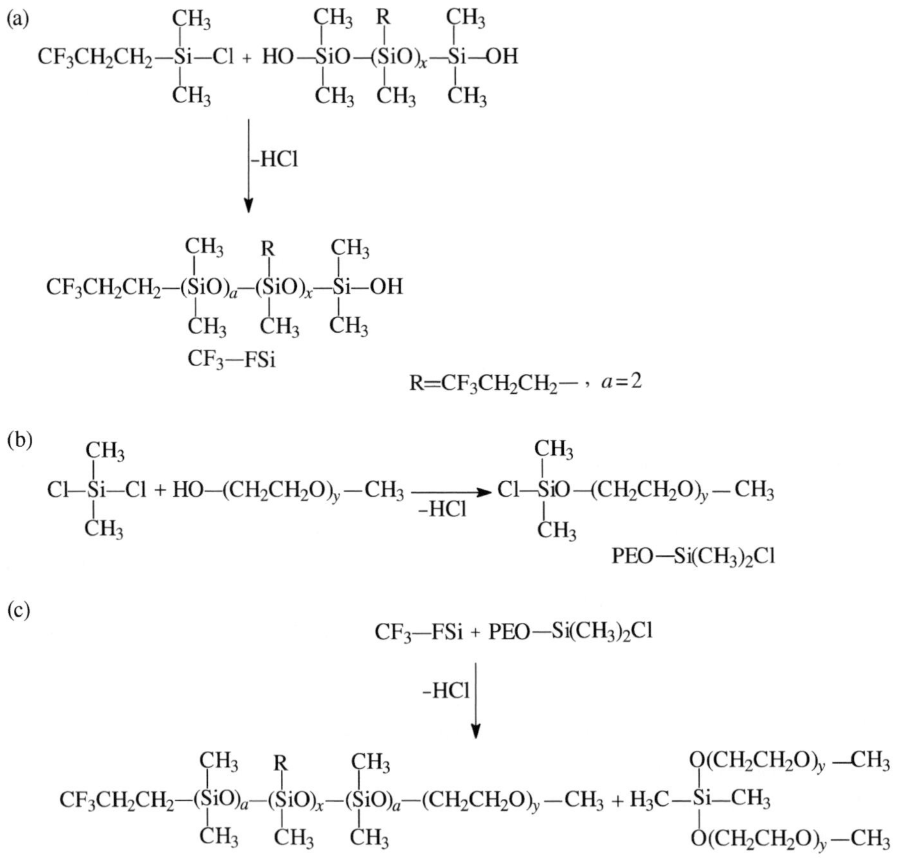

(a)

$$CF_3CH_2CH_2-\underset{\underset{CH_3}{|}}{\overset{\overset{CH_3}{|}}{Si}}-Cl \;+\; HO-\underset{\underset{CH_3}{|}}{\overset{\overset{CH_3}{|}}{Si}O}-(\underset{\underset{CH_3}{|}}{\overset{\overset{R}{|}}{Si}O})_x-\underset{\underset{CH_3}{|}}{\overset{\overset{CH_3}{|}}{Si}}-OH$$

$\downarrow$ -HCl

$$CF_3CH_2CH_2-(\underset{\underset{CH_3}{|}}{\overset{\overset{CH_3}{|}}{Si}O})_a-(\underset{\underset{CH_3}{|}}{\overset{\overset{R}{|}}{Si}O})_x-\underset{\underset{CH_3}{|}}{\overset{\overset{CH_3}{|}}{Si}}-OH$$

CF$_3$—FSi

R=CF$_3$CH$_2$CH$_2$—, a=2

(b)

$$Cl-\underset{\underset{CH_3}{|}}{\overset{\overset{CH_3}{|}}{Si}}-Cl \;+\; HO-(CH_2CH_2O)_y-CH_3 \xrightarrow{-HCl} Cl-\underset{\underset{CH_3}{|}}{\overset{\overset{CH_3}{|}}{Si}O}-(CH_2CH_2O)_y-CH_3$$

PEO—Si(CH$_3$)$_2$Cl

(c)

CF$_3$—FSi + PEO—Si(CH$_3$)$_2$Cl

$\downarrow$ -HCl

$$CF_3CH_2CH_2-(\underset{\underset{CH_3}{|}}{\overset{\overset{CH_3}{|}}{Si}O})_a-(\underset{\underset{CH_3}{|}}{\overset{\overset{R}{|}}{Si}O})_x-(\underset{\underset{CH_3}{|}}{\overset{\overset{CH_3}{|}}{Si}O})_a-(CH_2CH_2O)_y-CH_3 \;+\; H_3C-\underset{\underset{O(CH_2CH_2O)_y-CH_3}{|}}{\overset{\overset{O(CH_2CH_2O)_y-CH_3}{|}}{Si}}-CH_3$$

Another strategy of synthesis [242] deals with the anionic polymerisation of N-dimethylacrylamide (DMAA) initiated by caesium naphthalemide or

1,4-dicaesio-1,1,4,4-tetraphenyl-butane followed by the reaction of PDMAA dianions with perfluorooctanoyl chloride, as follows:

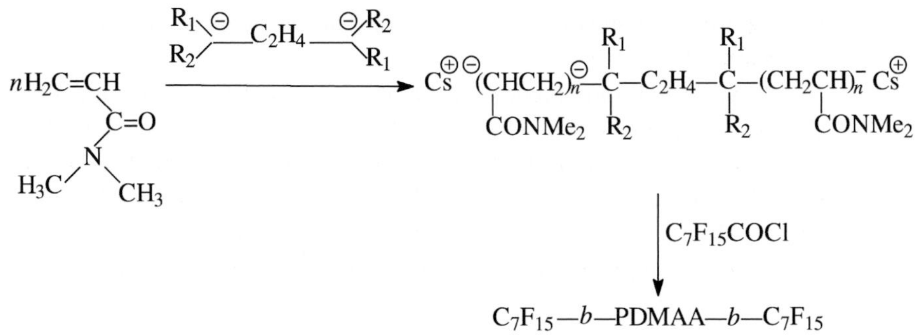

where (R_1, R_2) couple stands for (Ph, Ph) or (H, $CONMe_2$).

Molar masses ranged between 49,000 and 150,000 with an end-functionality close to 2.0 and narrow PDI (1.08–1.33).

These American researchers [242] showed that the presence of oligo(ethylene oxide) spacers between the hydrophobic end-groups led to micellar bridging.

Hence, a wide variety of fluorotriblock copolymers can also be produced in different ways, the last step being a condensation reaction. Indeed, the products containing a PEO block show interesting rheological properties and have found various applications as surfactants and thickeners.

2.2.6. Conclusion

As for the fluorinated diblock derivatives, a wide range of possibilities of F-triblock oligomers, telomers and copolymers has been reported. Various investigations concern the DEP of hydrogenated monomers M initiated by fluorinated diacyl peroxides or azos, hence leading to $R_F-(M)_n-R_F$ triblock co-oligomers. Both basic and applied researches were developed: the assessment of the dissociation rates and half-lives of these initiators were given and several oligomers have already found applications in medicine (thanks to antibacterial properties and their strong interactions with oligoDNA), as surfactants, associative thickeners and fire-fighting agents.

However, the most interesting breakthrough arises from α,ω-diiodoperfluoroalkanes that are involved in three key reactions which exploit the weak CF_2-I bond:

(i) By reacting with hydrogenated monomers M under telomerisation conditions, these diiodides are interesting precursors for $(M)_n-R_F-(M)_n$ triblocks for which many bismonoadditions were achieved (leading to semifluorinated $R_H-R_F-R_H$ models).

(ii) The stepwise telomerisation of commercially available fluoroalkenes enables one to obtain original longer and highly fluorinated α,ω-diiodo(per)fluoroalkanes. They can be easily functionalised, hence leading to a new generation of telechelic derivatives, as also described in Chapter 2.

(iii) ITP of various fluoroalkenes led to copolymers still bearing iodinated groups, the presence of which is exploited to prepare novel triblock copolymers. Interestingly, this procedure yields "living" iodinated fluoro(co)polymers and nowadays Daikin, Solvay-Solexis and DuPont prepare their own TPE by ITP. These high value added materials have found many applications in various fields (aerospace, engineering, optics...).

Besides the radical systems (ITP, RAFT, MADIX), various routes such as cationic, anionic and metathesis have produced well-defined triblock copolymers.

Various authors also synthesised such A−B−A or A−B−C fluoropolymers by combining several techniques and interesting materials have been obtained.

Finally, several investigations involving condensation reactions have produced R_F−PEO−R_F or R_F−U−PEO−U−R_F (where U represents a urethane bridge) endowed with interesting rheological and, above all, associative properties.

2.3. SYNTHESIS, PROPERTIES AND APPLICATIONS OF FLUORINATED MULTIBLOCK COPOLYMERS

Because numerous examples of preparations, characteristics and applications of fluorinated multiblock and fluorosegmented copolymers have been reported in section 3 of Chapter 2 (fluorotelechelics), we do not want to describe them again. However, we have found it of interest to complete the above section by including the syntheses, properties and applications of multiblock copolymers in which one block is fluorinated while the second is siliconated, and having the following chemical structure:

$$-\left[R_F-(\underset{\underset{R}{|}}{\overset{\overset{CH_3}{|}}{Si}}-O)_x\right]_n$$

These are strictly alternating copolymers on which investigations have been carried out by many authors since the 1970s. The companies that have been the most involved in this research are Dow-Corning, Shin-Etsu Corporation and, to a lesser extent, the Daikin company (via its collaboration with our laboratory).

These studies can be divided into three main parts according to their general structures. Two of the structures involve telechelic fluorinated oligomers, in which the fluorinated block is located within the polymeric backbone. The third one is devoted to fluorinated silicones which bear fluorinated C_nF_{2n+1} side groups.

These three different kinds of structures can be written as shown in the following figures.

The first one is:

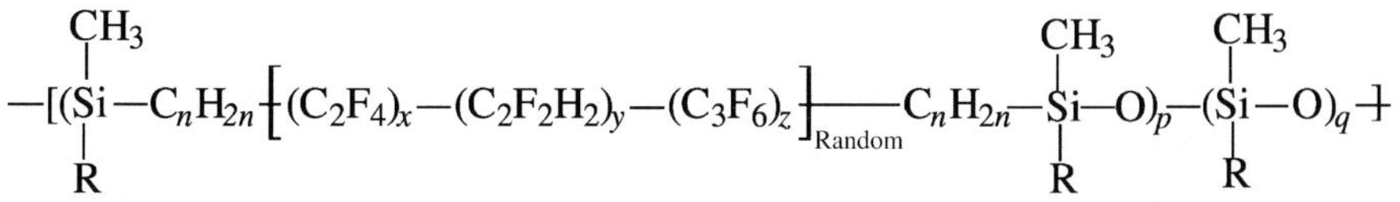

$$-[(\overset{\overset{\text{CH}_3}{|}}{\underset{\underset{\text{R}}{|}}{\text{Si}}}-C_nH_{2n}\underbrace{[(C_2F_4)_x-(C_2F_2H_2)_y-(C_3F_6)_z]}_{\text{Random}}-C_nH_{2n}-\overset{\overset{\text{CH}_3}{|}}{\underset{\underset{\text{R}}{|}}{\text{Si}}}-O)_p-(\overset{\overset{\text{CH}_3}{|}}{\underset{\underset{\text{R}}{|}}{\text{Si}}}-O)_q+$$

In this first series, fluorinated copolymers containing the three types of commercially available fluoroalkenes alternate with polysiloxane chains (with various side groups) and both kinds of chain are separated by a C_nH_{2n} spacer (usually C_2H_4 or C_3H_6). To the best of our knowledge, no products from this series have been marketed.

The structure of the multiblock fluorosilicones of the second series can be summarised as:

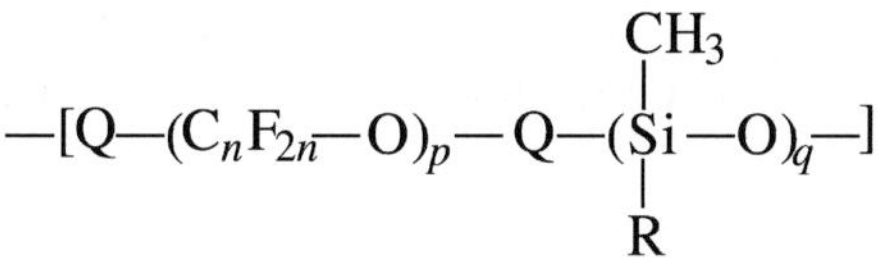

$$-[Q-(C_nF_{2n}-O)_p-Q-(\overset{\overset{\text{CH}_3}{|}}{\underset{\underset{\text{R}}{|}}{\text{Si}}}-O)_q-]$$

Here, a (per)fluorinated polyether (such as Fomblin[®] or Krytox[®]-type) alternates with short-length (or molecular) siloxane chains and Q is a more complex spacer than in the previous series. This kind of fluorosilicone is commercialised by the Shin-Etsu Company under the tradename Sifel[®].

The general formula of the silicon-containing fluorinated multiblock of the third series can be written as follows:

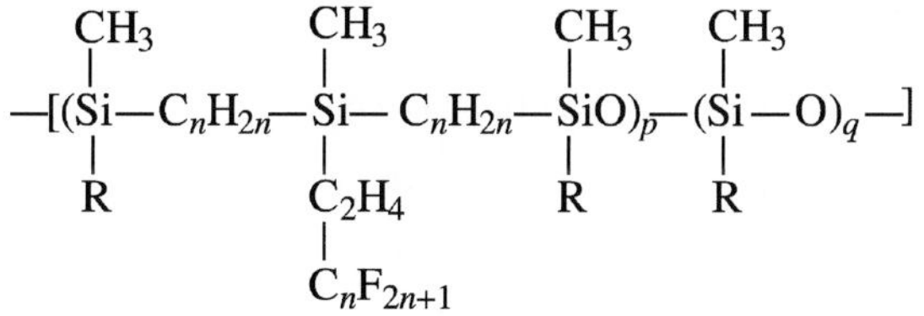

$$-[(\overset{\overset{\text{CH}_3}{|}}{\underset{\underset{\text{R}}{|}}{\text{Si}}}-C_nH_{2n}-\overset{\overset{\text{CH}_3}{|}}{\underset{\underset{\underset{\underset{C_nF_{2n+1}}{|}}{C_2H_4}}{|}}{\text{Si}}}-C_nH_{2n}-\overset{\overset{\text{CH}_3}{|}}{\underset{\underset{\text{R}}{|}}{\text{SiO}}})_p-(\overset{\overset{\text{CH}_3}{|}}{\underset{\underset{\text{R}}{|}}{\text{Si}}}-O)_q-]$$

In this series, there is an alternance of a polysilane, in which one of the silicon atoms bears a fluorinated side chain, with a classical siloxane.

We describe below the reported syntheses with as much care as possible since the literature exclusively indicates patents. Hence, it is sometimes required to "imagine" intermediates or reactions that are not quoted in the traditional literature.

2.3.1. Syntheses Involved in the First Series

The first step deals with the preparation of the telechelic nonconjugated diene containing fluoroalkenes with the following formula:

$$H_2C=CH-(CH_2)_a-[(C_2F_4)_x-(C_2F_2H_2)_y-(C_3F_6)_z-]_{stat}-(CH_2)_a-CH=CH_2$$
$$a = 0 \text{ or } 1$$

The syntheses of these dienes have been reported in previous chapters (1 and 2) and we recall both basic reactions (in monounsaturated series) [277, 327]:

$$\text{for } a = 0 : R_F-X + CH_2=CH_2 \longrightarrow R_F-CH_2-CH_2-X \xrightarrow{^-OH} R_F-CH=CH_2$$

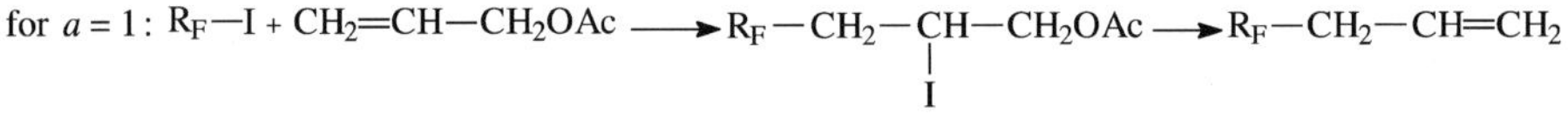

$$\text{for } a = 1 : R_F-I + CH_2=CH-CH_2OAc \longrightarrow R_F-CH_2-\underset{\underset{I}{|}}{CH}-CH_2OAc \longrightarrow R_F-CH_2-CH=CH_2$$

Kim [328] pioneered the syntheses of these dienes in the 1970s mainly using TFE as fluoroalkene and choosing $a = 0$.

Several years ago, we introduced different fluoroalkenes [273, 275, 278], starting from $I-C_nF_{2n}-I$ (with $n = 2, 4, 6$) since we wished to induce disorder to lower the T_gs of the resulting oligomers. Indeed, as seen below, T_gs obtained by Kim [328] were too high ($T_g = -27\,°C$) for any application in aeronautics.

Several generations of products have been prepared from these fluorinated nonconjugated dienes, symbolised by $H_2C=CH-R_F-CH=CH_2$.

2.3.1.1. First Generation from Kim [329, 330]

The scheme describing the synthesis of hybrid fluorosilicone is as follows:

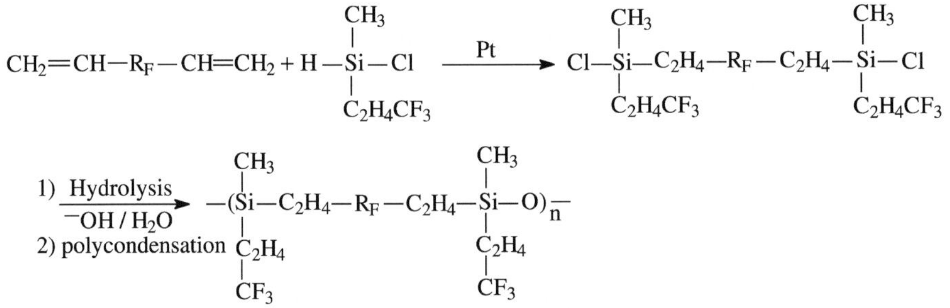

As mentioned above, R_F represents C_nF_{2n} mainly. For $n = 1$, the T_g of the fluorosilicone is $-38\,°C$ while for $n > 1$, the T_g is ca. $-27\,°C$. In contrast, their thermal stabilities are excellent under air or nitrogen and are much higher than those of commercially available poly(3,3,3-trifluoropropyl methyl siloxane) of structure

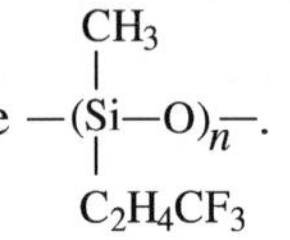

$$-(\underset{\underset{C_2H_4CF_3}{|}}{\overset{\overset{CH_3}{|}}{Si}}-O)_n-.$$

However, the T_gs are too high except for when $n = 1$ which is unfortunately the least thermal stable fluorosilicone. It is worth noting that the $-CH_2CH_2CF_3$ side chain is essential; otherwise, as in the case where a methyl side group is introduced, the silicone undergoes a crystallisation which is obviously unfavourable with regard to obtaining elastomers.

To lower the T_g values, Kim [331] synthesised original nonconjugated dienes which contain vinylidene fluoride units as follows:

$$Br-C_2F_4-Br + nH_2C=CF_2 \longrightarrow Br \left[C_2F_4-(CH_2-CF_2)_n \right] Br \xrightarrow{\substack{(1)C_2H_4 \\ (2)KOH}}$$

$$CH_2=CH \left[C_2F_4-(CH_2-CF_2)_n \right] CH=CH_2$$

Hybrid fluorosiloxanes were prepared from these nonconjugated dienes and the T_g value was $-34\,°C$ for $n = 2$. In addition, this was an improvement without affecting the thermal stability.

Finally, Kim [332] synthesised several copolymers and introduced siloxane groups by co-hydrolysis of hybrid α,ω-dichlorosiloxanes with elementary α,ω-dichlorosilanes.

2.3.2. Second Generation from Investigations Carried Out in Our Laboratory [278, 333–336]

This second generation of fluorinated hybrid silicones stems from difficulties that occurred in the synthesis of the chlorofluorinated silane used by Kim. This precursor of fluorosilicone was obtained by disproportionation according to the following equilibrium:

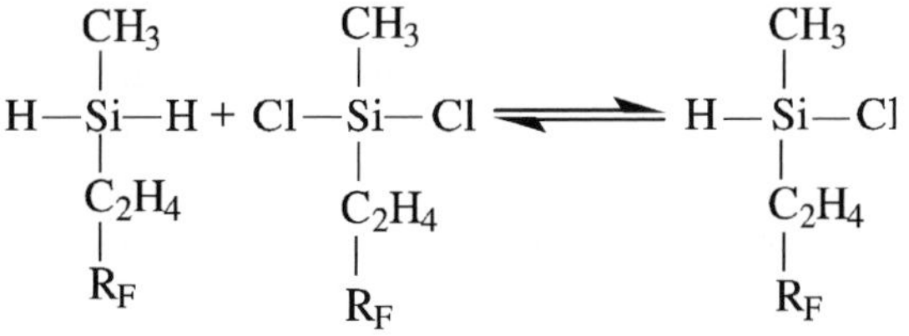

$$
\underset{\substack{|\\ C_2H_4 \\ |\\ R_F}}{\overset{\substack{CH_3 \\ |}}{H-Si-H}} +
\underset{\substack{|\\ C_2H_4 \\ |\\ R_F}}{\overset{\substack{CH_3 \\ |}}{Cl-Si-Cl}}
\rightleftharpoons
\underset{\substack{|\\ C_2H_4 \\ |\\ R_F}}{\overset{\substack{CH_3 \\ |}}{H-Si-Cl}}
$$

Unfortunately, such an equilibrium is shifted toward the reactants when the length of the R_F group is increased (i.e., from CF_3 to C_6F_{13}). Difficulties of separation (by distillation) also increase with this chain length.

In addition, as the T_gs observed by Kim were rather high, we have found another method involving reactants with longer and more complex fluorinated chains, and using $\underset{\substack{|\\ C_2H_4R_F}}{\overset{\substack{CH_3 \\ |}}{H-Si-H}}$ directly in the following hydrosilylation reaction:

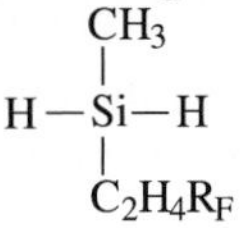

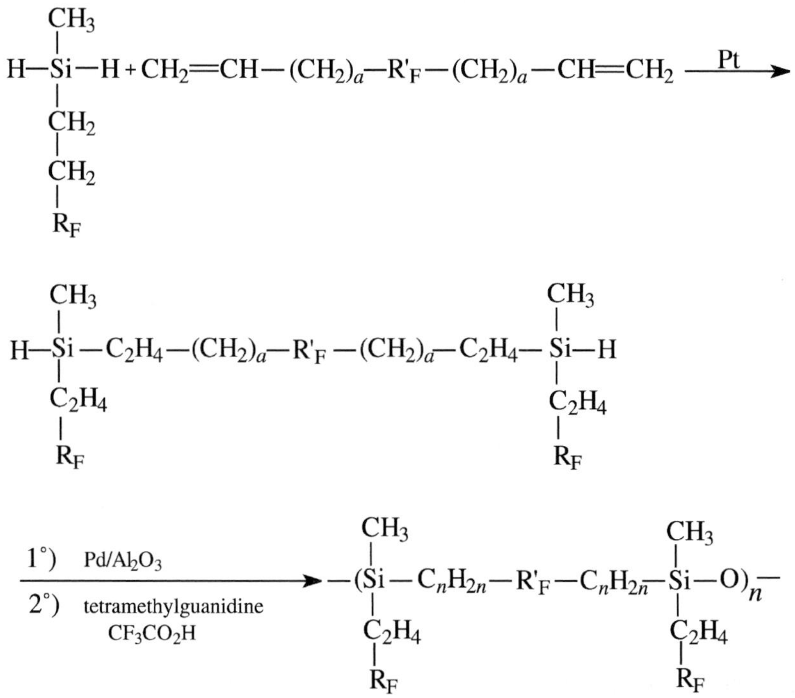

$$
\begin{array}{c}
\text{CH}_3 \\
| \\
\text{H}-\text{Si}-\text{H} + \text{CH}_2\!=\!\text{CH}-(\text{CH}_2)_a-\text{R}'_F-(\text{CH}_2)_a-\text{CH}\!=\!\text{CH}_2 \xrightarrow{\ \text{Pt}\ } \\
| \\
\text{CH}_2 \\
| \\
\text{CH}_2 \\
| \\
\text{R}_F
\end{array}
$$

$$
\begin{array}{c}
\text{CH}_3 \qquad\qquad\qquad\qquad\qquad\qquad\quad \text{CH}_3 \\
| \qquad\qquad\qquad\qquad\qquad\qquad\qquad\quad | \\
\text{H}-\text{Si}-\text{C}_2\text{H}_4-(\text{CH}_2)_a-\text{R}'_F-(\text{CH}_2)_a-\text{C}_2\text{H}_4-\text{Si}-\text{H} \\
| \qquad\qquad\qquad\qquad\qquad\qquad\qquad\quad | \\
\text{C}_2\text{H}_4 \qquad\qquad\qquad\qquad\qquad\qquad \text{C}_2\text{H}_4 \\
| \qquad\qquad\qquad\qquad\qquad\qquad\qquad\quad | \\
\text{R}_F \qquad\qquad\qquad\qquad\qquad\qquad\quad\ \text{R}_F
\end{array}
$$

1°) Pd/Al₂O₃

2°) tetramethylguanidine
     CF₃CO₂H

$$
\begin{array}{c}
\qquad\qquad\qquad \text{CH}_3 \qquad\qquad\qquad\qquad \text{CH}_3 \\
\qquad\qquad\qquad | \qquad\qquad\qquad\qquad\qquad\ | \\
\longrightarrow -(\text{Si}-\text{C}_n\text{H}_{2n}-\text{R}'_F-\text{C}_n\text{H}_{2n}-\text{Si}-\text{O})_{\!n}- \\
\qquad\qquad\qquad | \qquad\qquad\qquad\qquad\qquad\ | \\
\qquad\qquad\qquad \text{C}_2\text{H}_4 \qquad\qquad\qquad\qquad \text{C}_2\text{H}_4 \\
\qquad\qquad\qquad | \qquad\qquad\qquad\qquad\qquad\ | \\
\qquad\qquad\qquad \text{R}_F \qquad\qquad\qquad\qquad\ \text{R}_F
\end{array}
$$

$\text{R}_F = -\text{C}_x\text{F}_{2x+1}$ with   x = 1, 4, 6

or $\text{CF}_3-\text{CFCl}-\text{CF}_2-\underset{\underset{\displaystyle \text{CF}_3}{|}}{\text{CF}}-$, $\text{C}_4\text{F}_9-\text{CF}_2-\underset{\underset{\displaystyle \text{CF}_3}{|}}{\text{CF}}-$

$a = 0$ or 1

where R'_F represents a fluorinated group composed of a combination of TFE, VDF and HFP base-units.

Using a Pd/Al₂O₃ catalytic system enabled us to achieve the hydrolysis of Si−H end-groups into silanols, while tetramethylguanidine/trifluoroacetate was an excellent catalyst for polycondensation of the silanol functions.

New hybrid fluorinated siloxanes endowed with low T_gs and good thermostabilities could be obtained (Table X) by introducing the above modifications.

This table gives the thermal properties of the fluorosilicones obtained. The influence of the key group on these properties is as follows:

1. The longer the spacer between the fluorinated central group and the silicon atom, the higher the T_g and the less thermostable the fluorosilicon is in air, which agrees with a previous investigation performed on model fluorosilicones.

2. Substituting CH_3 by $CF_3C_2H_4$ as the fluorinated side group on the polymeric backbone increases the T_g but a lateral $C_4F_9C_2H_4$ group lowers the T_g of the fluorosilicone containing a $CF_3C_2H_4$ group and slightly decreases its thermostability in air.

3. The fluorinated central group in the polymeric backbone has a very pronounced effect on both the T_g and T_{dec}. The introduction of CF_3 side groups arising from the HFP base units lowers the T_g (e.g., from $-18\,°C$ for fluorosilicones where $R = CF_3C_2H_4$, $R' = C_6F_{12}$ and $x = 1$ to $-34\,°C$ for $R = CF_3C_2H_4$, $R' = HFP/C_4F_8/HFP$ and $x = 1$), but does not improve thermostability.

2.3.2.1. Third Generation from Investigations Carried Out in Our Laboratory [337]

This new generation implies easier syntheses and the preparation of polymers with lower and lower T_g.

First, we have avoided both complex preparations of silanes bearing fluorinated side chains and the use of $H(R_1)Si(R_2)H$ which requires $AlLiH_4$. Second, the lengths of the Si–O–Si sequences (that confer the softness, since PDMS and PHMS exhibit T_gs of -123 and $-128\,°C$, respectively) were increased. However, the length must be carefully adjusted because of the poor thermostability of siloxanes that generate cyclic arrangements from intramolecular reactions at high temperatures [338].

As a consequence, the siloxane chain should have less than five silicon atoms and the last two must be linked to a carbon atom to avoid the formation of a cyclosiloxane with three grafts.

TABLE X

THERMAL DATA FOR HYBRID F-SILICONE HOMOPOLYMERS

$\begin{array}{cc} CH_3 & CH_3 \\ \mid & \mid \\ HO\text{-}[Si\text{-}R'\text{-}SiO]_n\text{-}H \\ \mid & \mid \\ R & R \end{array}$		DSC (10 °C min^{-1})			TGA (5 °C min^{-1})	
		T_g	T_m	T_c	$T_{50\%}$ (N$_2$)	$T_{50\%}$ (air)
$R = CH_3$	$R' = C_2H_4C_6F_{12}C_2H_4$	-53	26	-11	470	380
	$R' = C_3H_6C_6F_{12}C_3H_6$	-40	25	-27	465	330
$R = C_2H_4CF_3$	$R' = C_2H_4C_6F_{12}C_2H_4$	-28			490	410
	$R' = C_3H_6C_6F_{12}C_3H_6$	-18			465	360
$R = C_2H_4C_4F_9$	$R' = C_2H_4C_6F_{12}C_2H_4$	-42			490	360
	$R' = C_3H_6C_6F_{12}C_3H_6$	-29			470	310
$R = CH_3$	$R' = C_3H_6/HFP/C_4F_8/HFP/C_3H_6$	-49			425	300
$R = C_2H_4CF_3$	$R' = C_3H_6/HFP/C_4F_8/HFP/C_3H_6$	-34			445	310
$R = C_2H_4C_4F_9$	$R' = C_3H_6/C_2F_4/VDF/HFP/C_3H_6$	-47			420	315

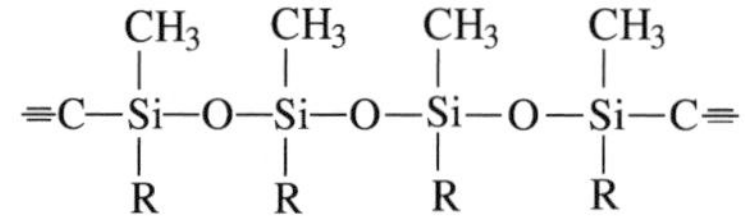

$$\overset{\displaystyle CH_3}{\underset{\displaystyle R}{|}}\ \overset{\displaystyle CH_3}{\underset{\displaystyle R}{|}}\ \overset{\displaystyle CH_3}{\underset{\displaystyle R}{|}}\ \overset{\displaystyle CH_3}{\underset{\displaystyle R}{|}}$$

$\equiv C-Si-O-Si-O-Si-O-Si-C\equiv$ chain is thus the most appropriate.

The synthesis was carried out by polyaddition of fluorinated dienes with α,ω-dihydroxysiloxanes of structure:

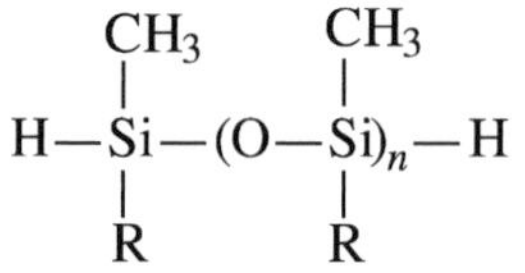

$$H-\underset{\underset{\displaystyle R}{|}}{\overset{\overset{\displaystyle CH_3}{|}}{Si}}-(O-\underset{\underset{\displaystyle R}{|}}{\overset{\overset{\displaystyle CH_3}{|}}{Si}})_n-H \qquad n = 1, 2, 3$$

The formed products exhibit the structure below:

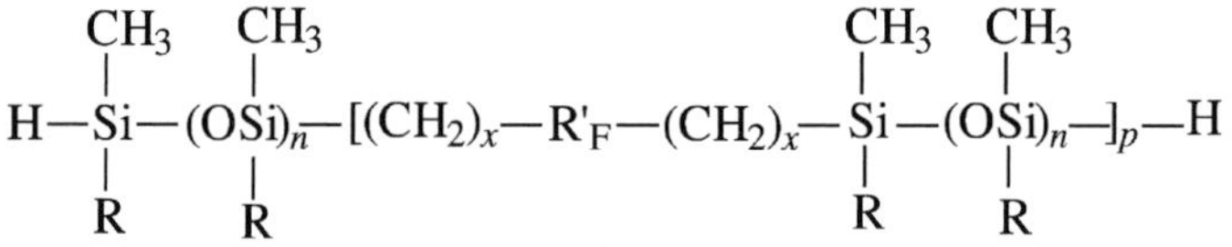

$$H-\underset{\underset{\displaystyle R}{|}}{\overset{\overset{\displaystyle CH_3}{|}}{Si}}-(OSi)_{\overline{n}}-[(CH_2)_x-R'_F-(CH_2)_x-\underset{\underset{\displaystyle R}{|}}{\overset{\overset{\displaystyle CH_3}{|}}{Si}}-(OSi)_n-]_p-H$$

with $x = 2$ and 3; $n = 1, 2, 3$ and R'_F groups are the same as above.

Table XI sums up the characteristics and results for fluorosiloxanes of various structures.

2.3.2.2. Fourth Generation from Specific Polymer's Research [339]

The process patented by Specific Polymer is based on the polycondensation of dichlorosilane with a dihydroxysiloxane (L_2OH) known for years, but the synthesis of L_2OH has recently been patented by Dow-Corning [340]. In a simple synthesis, the dichlorosilane contains a fluorinated chain as above, and the silicon atoms bear a methyl group attached to each silicon atom to simplify synthesis.

This easier synthesis allowed perfect preservation of the length of the siloxane spacer (four silicon atoms) between the fluorinated chain. The general scheme is the following:

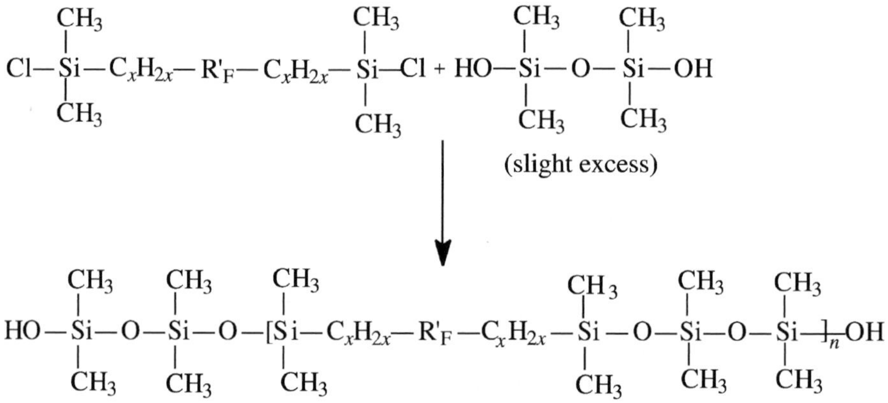

$$Cl-\underset{\underset{\displaystyle CH_3}{|}}{\overset{\overset{\displaystyle CH_3}{|}}{Si}}-C_xH_{2x}-R'_F-C_xH_{2x}-\underset{\underset{\displaystyle CH_3}{|}}{\overset{\overset{\displaystyle CH_3}{|}}{Si}}-Cl + HO-\underset{\underset{\displaystyle CH_3}{|}}{\overset{\overset{\displaystyle CH_3}{|}}{Si}}-O-\underset{\underset{\displaystyle CH_3}{|}}{\overset{\overset{\displaystyle CH_3}{|}}{Si}}-OH$$

(slight excess)

$$HO-\underset{\underset{\displaystyle CH_3}{|}}{\overset{\overset{\displaystyle CH_3}{|}}{Si}}-O-\underset{\underset{\displaystyle CH_3}{|}}{\overset{\overset{\displaystyle CH_3}{|}}{Si}}-O-[\underset{\underset{\displaystyle CH_3}{|}}{\overset{\overset{\displaystyle CH_3}{|}}{Si}}-C_xH_{2x}-R'_F-C_xH_{2x}-\underset{\underset{\displaystyle CH_3}{|}}{\overset{\overset{\displaystyle CH_3}{|}}{Si}}-O-\underset{\underset{\displaystyle CH_3}{|}}{\overset{\overset{\displaystyle CH_3}{|}}{Si}}-O-\underset{\underset{\displaystyle CH_3}{|}}{\overset{\overset{\displaystyle CH_3}{|}}{Si}}]_n-OH$$

These bis(silanol)s can be modified (into α,ω-nonconjugated dienes or into bis(SiH) derivatives or blocked) and their molecular weights can easily be

TABLE XI

THERMAL DATA FOR HYBRID F-SILICONE RANDOM COPOLYMERS

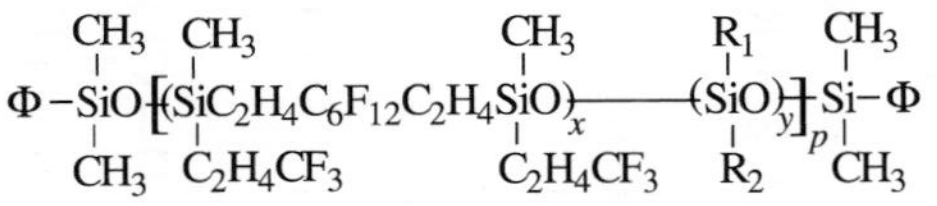

$$\underset{CH_3}{\overset{CH_3}{\Phi-SiO}}\underset{C_2H_4CF_3}{\overset{CH_3}{\Big[SiC_2H_4C_6F_{12}C_2H_4SiO\Big]_x}}\underset{R_2}{\overset{R_1}{(SiO)}}\underset{CH_3}{\overset{CH_3}{\Big]_p}Si-\Phi}$$

	x	y	p	T_g (°C)	$T_{-50\%}$ (N$_2$)
		(from ^{1}H NMR)		DSC (5 °C min^{-1})	TGA (10 °C min^{-1})
R$_1$=CH$_3$ R$_2$ = C$_2$H$_4$C$_6$F$_{13}$	1	0.68	10.5	−40	475
R$_1$ = C$_2$H$_4$CF$_3$ R$_2$ = C$_2$H$_4$C$_6$F$_{13}$	1	0.65	8.8	−42	460
R$_1$ = CH$_3$ R$_2$ = C$_2$H$_4$CF(CF$_3$)C$_5$F$_{11}$	1	0.63	9.4	−44	470
R$_1$ = CH$_3$ R$_2$ = C$_2$H$_4$CF(CF$_3$)CF$_2$CFClCF$_3$	1	0.6	10.3	−46	460

regulated, in contrast to those of fluorosiloxanes of the third generation. Parallel to these investigations on the use of fluorinated polyolefinic chains, the Shin-Etsu company has studied the introduction of perfluoropolyethers as reported below.

2.3.3. Synthesis of Fluorosiloxanes Containing Perfluoropolyethers (PFPE) for the Second Series

There are four principal commercially available fluorinated polyethers. These are [341]:

(i) FOMBLIN® (formerly produced by Ausimont and now by Solvay-Solexis): $(CF_2O)_x-(C_2F_4O)_y$

(ii) KRYTOX® (from the DuPont Company): $-(CF_2-\underset{\underset{CF_3}{|}}{CFO})_n-$

(iii) AFLUNOX® (from Unimatec Corporation, formerly Nippon Mektron):

$-(CF_2\underset{\underset{CF_3}{|}}{CFO})_m$

(iv) DEMNUM® (from Daikin): $-(CF_2CF_2CF_2O)_p-$

Control of the functionalisation of these chains is not easy and, in spite of much research, few of the studies have led to industrial development of fluorosiloxanes, although that marketed by the Shin-Etsu company under the Sifel® tradename is a notable exception.

Conceptually, the synthesis concerns the polyaddition of telechelic dienes containing a fluorinated polyether with fluorinated or non-fluorinated compounds possessing at least two Si–H end-groups. Both can be represented by the following structures:

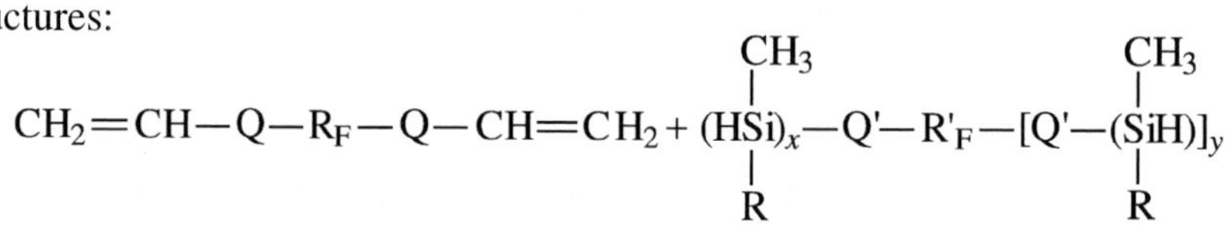

$$CH_2{=}CH-Q-R_F-Q-CH{=}CH_2 + (HSi)_x-\underset{\underset{R}{\overset{\overset{CH_3}{|}}{|}}}{}Q'-R'_F-[Q'-(SiH)]_y \quad \underset{\underset{R}{\overset{\overset{CH_3}{|}}{|}}}{}$$

where Q and Q' are the varied spacers; x: 1, 2, 3; R_F: perfluoropolyether; and y: 0, 1 or 2.

In a recent patent, Shin-Etsu [342] has reported original structures such as:

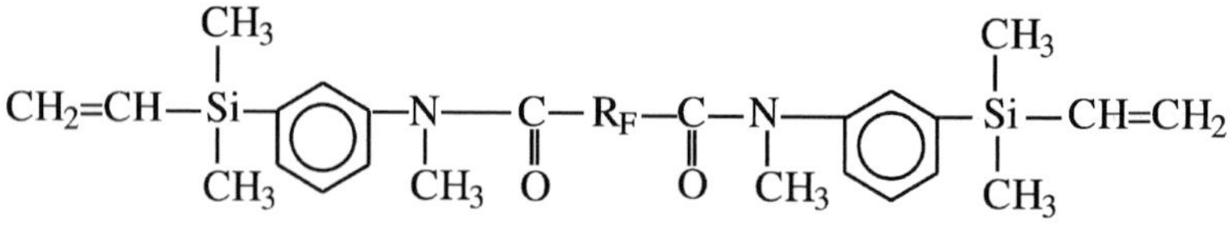

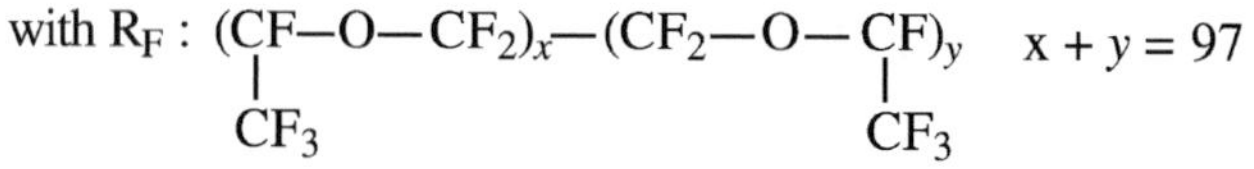

$$CH_2{=}CH-\underset{\underset{CH_3}{|}}{\overset{\overset{CH_3}{|}}{Si}}-\bigcirc-\underset{\underset{CH_3}{|}}{N}-\underset{\underset{O}{\|}}{C}-R_F-\underset{\underset{O}{\|}}{C}-\underset{\underset{CH_3}{|}}{N}-\bigcirc-\underset{\underset{CH_3}{|}}{\overset{\overset{CH_3}{|}}{Si}}-CH{=}CH_2$$

with R_F : $(\underset{\underset{CF_3}{|}}{CF}-O-CF_2)_x-(CF_2-O-\underset{\underset{CF_3}{|}}{CF})_y \quad x+y=97$

In addition, various crosslinking agents have been claimed:

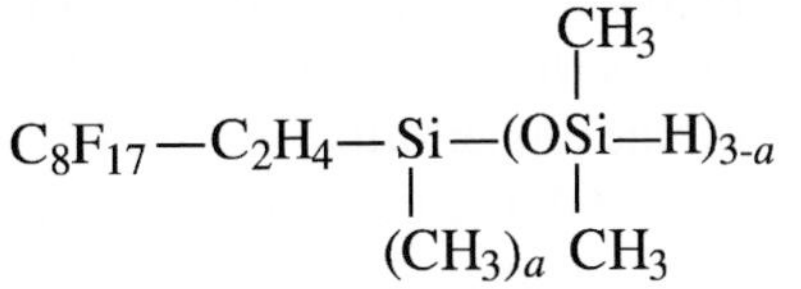

$$C_8F_{17}-C_2H_4-\underset{\underset{(CH_3)_a}{|}}{\overset{\overset{CH_3}{|}}{Si}}-(O\underset{\underset{CH_3}{|}}{Si}-H)_{3-a} \quad \text{with } a = 0 \text{ and } 1$$

$$\text{and } (H-\underset{\underset{R}{|}}{\overset{\overset{CH_3}{|}}{Si}}-O)_3Si-C_2H_4-(CF_2)_5-C_2H_4-Si(O-\underset{\underset{R}{|}}{\overset{\overset{CH_3}{|}}{Si}}-H)_3$$

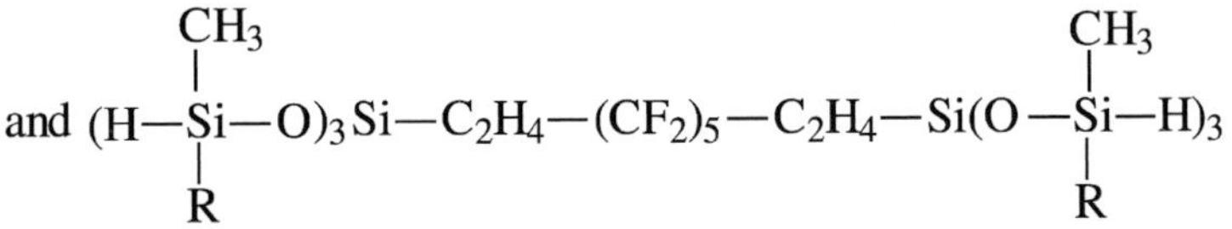

In this patent, the authors emphasise the need to proceed in three steps in order to obtain elastomers endowed with very good properties.

Moreover, the above novel crosslinking agents, reported by the same company [343], were synthesised according to an original method summarised below:

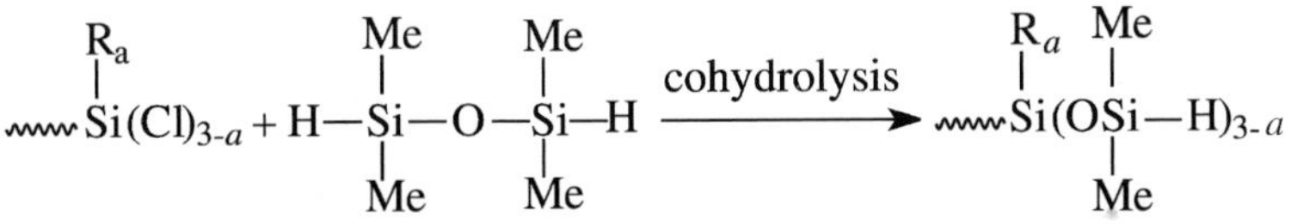

$$\underset{\text{www}}{}\overset{\overset{R_a}{|}}{Si}(Cl)_{3-a} + H-\underset{\underset{Me}{|}}{\overset{\overset{Me}{|}}{Si}}-O-\underset{\underset{Me}{|}}{\overset{\overset{Me}{|}}{Si}}-H \xrightarrow{\text{cohydrolysis}} \underset{\text{www}}{}\overset{\overset{R_a \ Me}{| \ |}}{Si}(O\underset{\underset{Me}{|}}{Si}-H)_{3-a}$$

In a more recent patent [344], another variety of novel crosslinkers has been claimed and seems to contribute to the improvement of properties at high temperatures, as explained by Uritani [345].

The structures of these compounds are reported as follows:

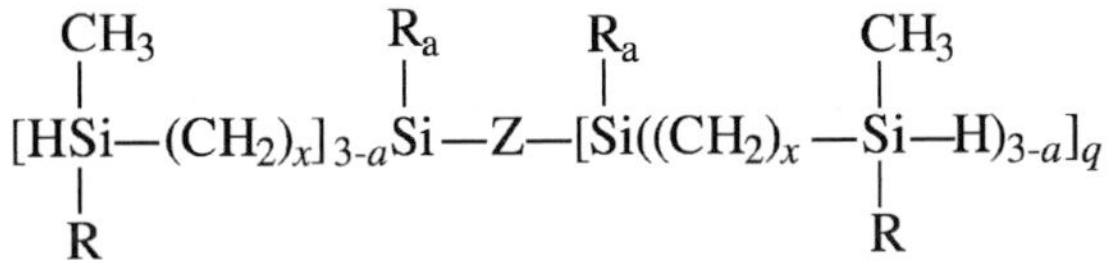

$$[H\underset{\underset{R}{|}}{\overset{\overset{CH_3}{|}}{Si}}-(CH_2)_x]_{3-a}\overset{\overset{R_a}{|}}{Si}-Z-[Si(\overset{\overset{R_a}{|}}{}(CH_2)_x-\underset{\underset{R}{|}}{\overset{\overset{CH_3}{|}}{Si}}-H)_{3-a}]_q$$

$q = 0$ or 1; $a = 0, 1, 2$; $x = 1, 2, 3, 4$; Z represents various fluorinated or non-fluorinated spacers.

Interestingly, the presence of

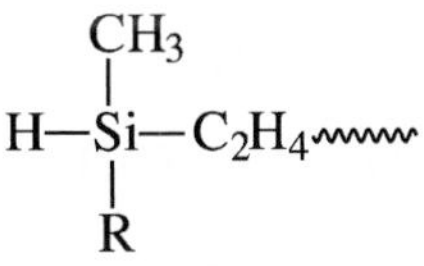

$$H-\underset{\underset{R}{|}}{\overset{\overset{CH_3}{|}}{Si}}-C_2H_4\text{www}$$

groups instead of

$$H-\underset{\underset{R}{|}}{\overset{\overset{CH_3}{|}}{Si}}-O\text{www}$$

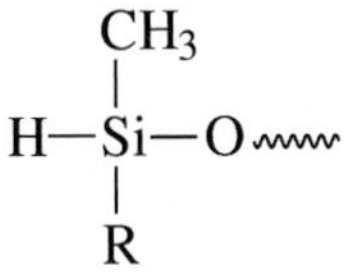

would represent an innovative means to improve the thermal stability of elastomers. In this patent, the comparison between Sifel® and classic commercially available Viton® E, high performance Viton® GLT and a fluorosilicone is supplied. This shows the superiority of Sifel® over its competitors with regard to T_g values (ca. $-60\,°C$), the swelling properties in various polar and non-polar solvents and the overall mechanical properties including the compression set.

2.3.4.   Synthesis and Procedures of the Third Series

This family of fluorosilicones mainly concerns Kobayashi's work for the Dow-Corning Toray Silicone Company.

Kobayashi's main idea dealt with the polyaddition of a dihydrogenosilane bearing a fluorinated side chain with various molecular or macromolecular siloxanes [346]:

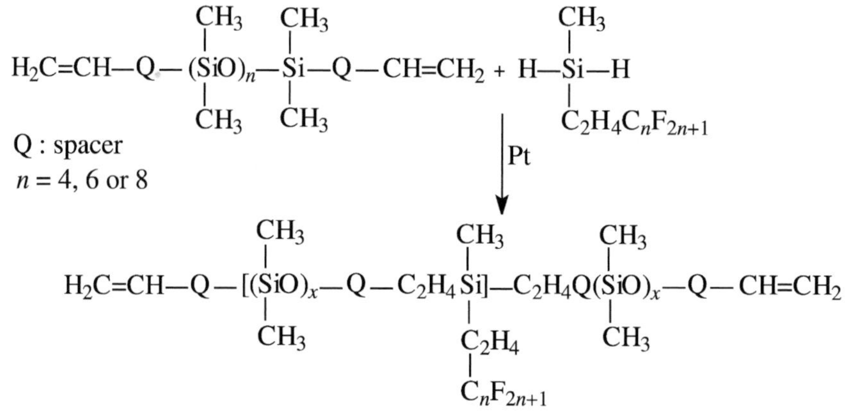

$$H_2C=CH-Q-(SiO)_n-\underset{\underset{CH_3}{|}}{\overset{\overset{CH_3}{|}}{Si}}-Q-CH=CH_2 + H-\underset{\underset{C_2H_4C_nF_{2n+1}}{|}}{\overset{\overset{CH_3}{|}}{Si}}-H$$

Q : spacer
n = 4, 6 or 8

$$\Big\downarrow Pt$$

$$H_2C=CH-Q-[(SiO)_x-Q-C_2H_4\underset{\underset{C_2H_4}{|}}{\overset{\overset{CH_3}{|}}{Si}}]-C_2H_4Q(SiO)_x-Q-CH=CH_2$$

with Q:– and x = 1.

One of the structures most used by Kobayashi is the following synthon [347]:

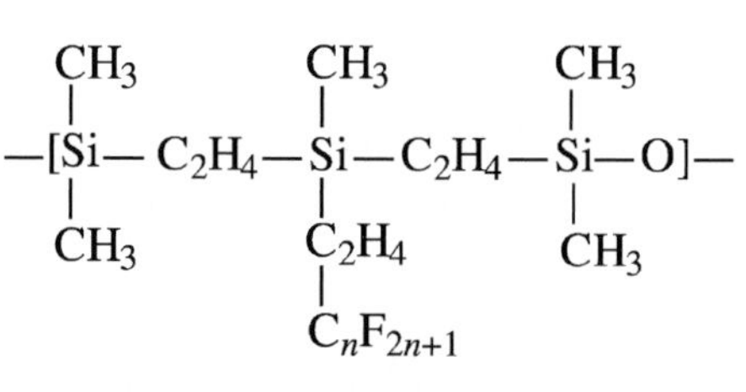

$$-[\underset{\underset{CH_3}{|}}{\overset{\overset{CH_3}{|}}{Si}}-C_2H_4-\underset{\underset{C_2H_4}{|}}{\overset{\overset{CH_3}{|}}{Si}}-C_2H_4-\underset{\underset{CH_3}{|}}{\overset{\overset{CH_3}{|}}{Si}}-O]-$$

Such an intermediate was simply synthesised from an excess chlorovinyl silane:

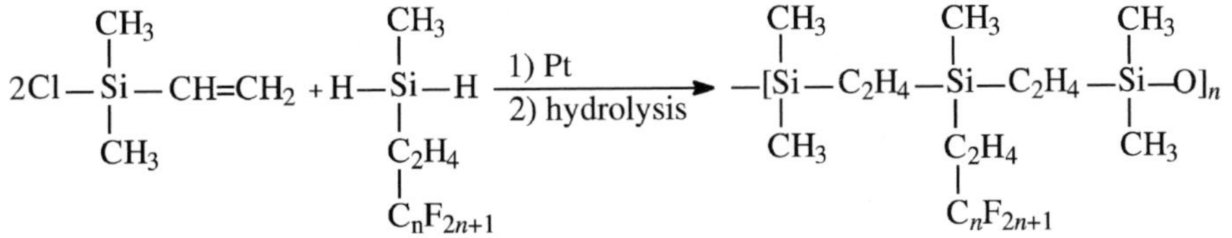

$$2Cl-\underset{\underset{CH_3}{|}}{\overset{\overset{CH_3}{|}}{Si}}-CH=CH_2 \; + \; H-\underset{\underset{\underset{C_nF_{2n+1}}{|}}{C_2H_4}}{\overset{\overset{CH_3}{|}}{Si}}-H \xrightarrow[\text{2) hydrolysis}]{\text{1) Pt}} \; -[\underset{\underset{CH_3}{|}}{\overset{\overset{CH_3}{|}}{Si}}-C_2H_4-\underset{\underset{\underset{C_nF_{2n+1}}{|}}{C_2H_4}}{\overset{\overset{CH_3}{|}}{Si}}-C_2H_4-\underset{\underset{CH_3}{|}}{\overset{\overset{CH_3}{|}}{Si}}-O]_n$$

In a recent patent, this Japanese author [348] has also reported the preparation of a siloxane block copolymer by redistribution in the presence of potassium hydroxide:

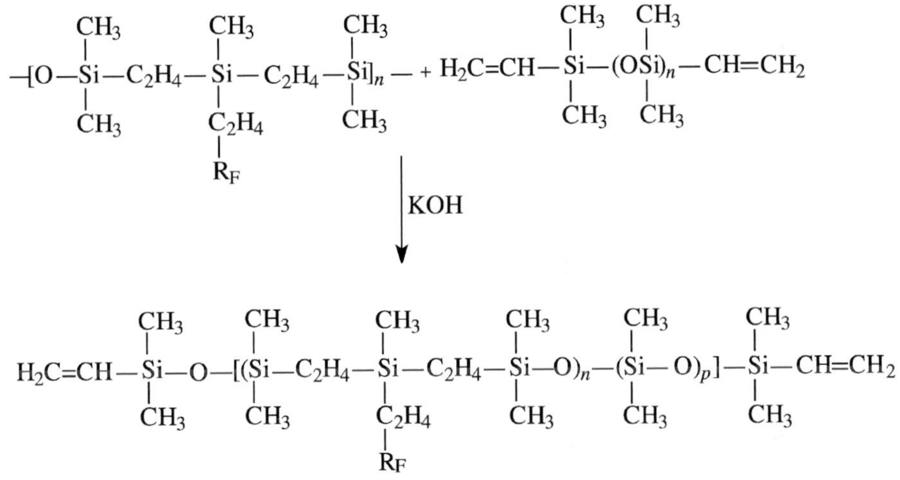

$$-[O-\underset{\underset{CH_3}{|}}{\overset{\overset{CH_3}{|}}{Si}}-C_2H_4-\underset{\underset{\underset{R_F}{|}}{C_2H_4}}{\overset{\overset{CH_3}{|}}{Si}}-C_2H_4-\underset{\underset{CH_3}{|}}{\overset{\overset{CH_3}{|}}{Si}}]_n- \; + \; H_2C=CH-\underset{\underset{CH_3}{|}}{\overset{\overset{CH_3}{|}}{Si}}-(O\underset{\underset{CH_3}{|}}{\overset{\overset{CH_3}{|}}{Si}})_n-CH=CH_2$$

$$\Big\downarrow KOH$$

$$H_2C=CH-\underset{\underset{CH_3}{|}}{\overset{\overset{CH_3}{|}}{Si}}-O-[(\underset{\underset{CH_3}{|}}{\overset{\overset{CH_3}{|}}{Si}}-C_2H_4-\underset{\underset{\underset{R_F}{|}}{C_2H_4}}{\overset{\overset{CH_3}{|}}{Si}}-C_2H_4-\underset{\underset{CH_3}{|}}{\overset{\overset{CH_3}{|}}{Si}}-O)_n-(\underset{\underset{CH_3}{|}}{\overset{\overset{CH_3}{|}}{Si}}-O)_p]-\underset{\underset{CH_3}{|}}{\overset{\overset{CH_3}{|}}{Si}}-CH=CH_2$$

In all the syntheses proposed by Kobayashi, the fluorinated substituent is found in a lateral position about the hybrid siloxane chain which does not contain any fluorinated chain. Although these products are claimed for use in resins, elastomers and primers of adhesion, the author has mainly been researching surface properties brought by the fluorinated side group. Actually, it is well known that these fluorinated lateral chains reduce the surface tension, as shown by Pitman [349]. Kobayashi and Owen [350] have developed methods of STM on these products, although neither value of T_g nor any mechanical property of these elastomers has been reported.

Finally, it is also worth citing the interesting investigation recently carried out by Cassidy's team [351, 352] which has synthesised original "poly(silanes-siloxanes)" bearing hexafluoroisopropylidene groups. The hydrosilylation reaction, in the presence of Karstedt catalyst (Pt/divinyltetramethyldisiloxane (Pt-DVTMD)), was achieved in sc CO_2, which is original:

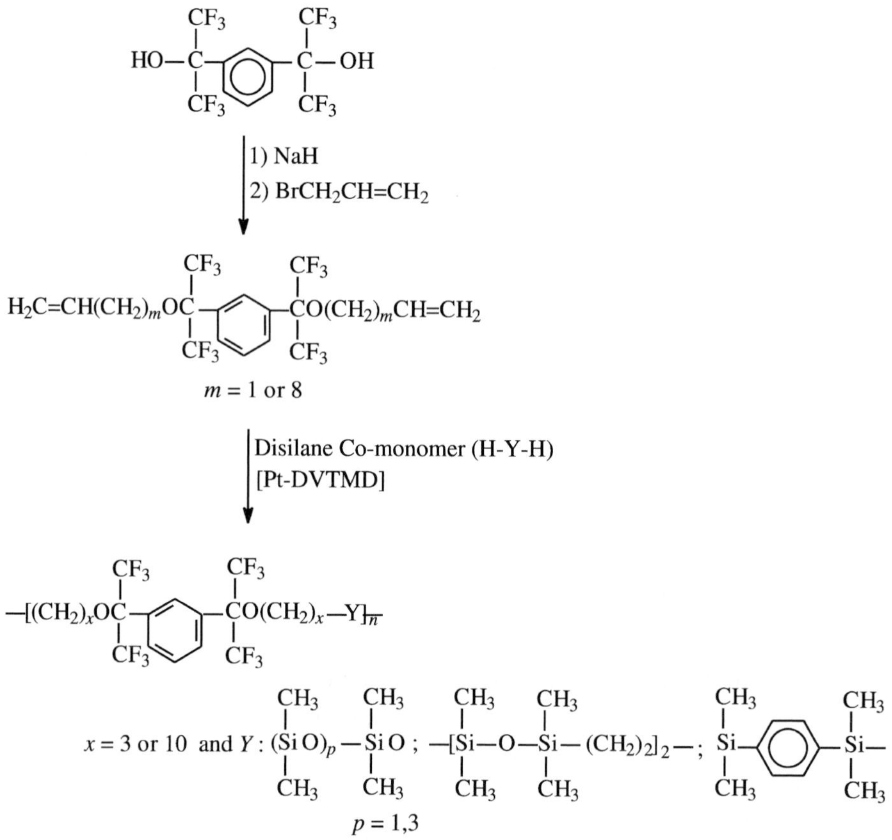

The molecular weights of these hybrid copolymers obtained in sc CO_2 were notably higher than those synthesised in organic solvents such as benzene. They also exhibit high thermal stability ($T_{10\%} = 360\,°C$ in air). Further, hydrolysis and thermal curing of these polymers led to materials endowed with higher molecular weights which retain low T_g values [352].

2.3.5. Conclusion

The tremendous amount of work carried out in the preparation of siloxane-fluorinated chain block copolymers was driven by the need to synthesise high-performance elastomers. Indeed, the term "high performance" implies the ability to withstand very low temperatures ($T_g \leq -60\,°C$), preserve elastomeric properties in the space and resist at high temperatures [353]. For example, from Mach 2.05 (as for Concorde) to Mach 2.4, the requirement for these materials is to preserve a good thermal stability from 110 to 177 °C for 60,000 h (i.e., 6 years and 10 months).

Meeting such a challenge is not trivial. In addition, the copolymers must be endowed with good mechanical properties (elastic strain and return) but also show excellent swelling-resistance whatever the nature of the solvent (polar or non-polar). All these numerous and impressive investigations are not yet conclusive and several decades might still be needed to reach a satisfactory goal.

3. Conclusion

This chapter summarises various routes to the synthesis of fluorinated diblock, triblock and multiblock copolymers.

Many investigations have been carried out on the preparations of fluorodiblocks and more recent interests are focused on those of triblock copolymers. However, to the best of our knowledge, no methods for direct fluorination have been reported in the literature, probably because of the high efficiency of fluorination.

It is observed that, except for a few examples of recent growing interest (sequential cationic, anionic, metathesis, GTP, or condensation), most of the strategies deal with radical initiations. Dead-end polymerisation of hydrogenated monomers with fluorinated peroxides or azo, and telomerisation leading mainly to oligomers and telomers have been widely studied, while more recent and elegant alternatives are emerging, yielding higher molar mass macromolecules. Examples are provided by the controlled radical polymerisation of fluoromonomers (especially (meth)acrylates or styrenic derivatives) in ATRP or those of hydrogenated monomers with a wide range of fluorinated initiators (dithiocarbamates, xanthates, tetraphenylethanes, alkoxyamines, fluorinated alkyl halides and dithioesters). These systems ensure a pseudo-living character and a better control of the polymerisation. In some cases, the presence of fluorine atoms significantly improved the living process, as in the case of xanthates where fluorine atoms in the alkoxy moiety allow tuning of reactivity in the MADIX process. Although the RAFT technique is attractive for the polymerisation of styrene, MMA and 1,3-butadiene, MADIX technology seems more convenient for vinyl acetate, which cannot be polymerised by other living techniques. The presence of heteroatoms could also further extend the potential of LRP. For instance, phosphorus- and silicon-based compounds could be very useful in the future in the search for new LRP processes. As an illustration, the thiophosphorus compounds ($>P(S)S-$) widely used in agriculture could be efficient reversible transfer agents. In other cases, the fluorinated compounds are mainly used to impart specific properties to the resulting materials (in particular, interesting surface properties, low cohesive energy, solubility properties for supercritical carbon dioxide applications and fluorine labeling). The polymerisation of fluorinated (meth)acrylic and styrenic monomers can be well controlled by the iniferter method, NMP, ATRP and by RAFT, with essentially the same restrictions as for their non-fluorinated homologues (for instance, polymerisation of methacrylics cannot be controlled by NMP because of disproportionation reactions). However, not all monomers have been tested (e.g., few investigations have been carried out on pentafluorostyrene and its derivatives). In contrast, the most appropriate method for living radical (co)polymerisation (LRP) of fluorinated alkenes (such as VDF, TFE and CTFE or their combinations with HFP, PMVE or TrFE) is ITP in which the iodine atom acts as an interesting reversible transfer site. The versatility of ITP enabled various companies involved in fluorine chemistry (e.g., Daikin, DuPont and Ausimont (now Solvay-Solexis)) to produce original TPE endowed with soft and hard segments.

However, the characteristics and properties of the polymers claimed in many patents contrast with those of published articles which might be expected to have a better understanding of the control of these sequential radical polymerisations. For example, few articles report the molar masses vs. monomer conversion relationship in contrast to the LRP of S, MMA and other hydrogenated monomers. Hence, future research will undoubtedly aim at improving such knowledge of ITP polymerisations.

Modelling the bond dissociation energy between the fluorinated unit and the end-groups introduced by dithiocarbamates, tetraphenylethanes, nitroxides, alkyl halides (Cl, Br, I), xanthates and dithioesters deserves closer investigation. In addition, the characterisation of higher-fluorinated polymers is still a challenge, especially by size exclusion chromatography because of the lack of appropriate fluorinated solvents and standards. This chapter has also shown that several methods of preparation of fluorinated block copolymers by LRP are available. Lastly, LRP in supercritical carbon dioxide is an attractive approach since some fluoropolymers are soluble in such a medium, which makes it possible to achieve quite high molecular weights (especially due to the absence of transfer to the solvent).

Nevertheless, the nature and the length of the blocks are crucial to the tuning of desired properties that the presence of R_F block(s) drastically modifies. Indeed, many properties (micellar morphology, melting, structural behaviours and surface properties) have been investigated for model R_F–R_H or R_H–R_F–R_H copolymers, and applications have already been found in many fields such as medicine (stabilisation of emulsions and oxygen carrier compounds), blends, elastomeric sealants or gaskets, surfactants, compatibilisers, textile finishings, coatings and dispersing agents.

Hence, the synthesis of further fluoroblock copolymers is still a challenge and it is understandable why increasing interest has been shown over the last 5 years.

References

1. Mishra, M. K. (1995) *Macromolecular Engineering*, Plenum Press, New York.
2. Sawamoto, M. (1991) Modern cationic vinyl polymerization. *Prog. Polym. Sci.*, **16**, 111.

3. Davis, K. A. and Matyjaszewski, K. (2002) Statistical, gradient, block and graft copolymers by controlled/living radical polymerizations. *Adv. Polym. Sci.*, **159**, 2–166.
4. Noshay, A. and Mc Grath, J. E. (1977) *Block Copolymers: Overview and Critical Survey*, Academic Press, New York.
5. Riess, G., Hurtrez, C. and Bahadur, P. (1985) Block copolymers. In *Encyclopedia of Polymer Science and Technology*, Vol. 2 (Ed, Kroschwitz, J. I.) Wiley, New York, pp. 324–374.
6. Hazer, B. (1989) Synthesis and characterization of block copolymers. In *Handbook of Polymer Sciences and Technology* (Ed, Cheremsinoff, N. P.) Marcel Dekker, New York, Chap. IV, p. 133.
7. Cowie, J. M. G. (1989) Block and graft copolymers. In *Comprehensive Polymer Science*, Vol. 3(3) (Eds, Allen, G., Bevington, J. C., Eastmond, A. L. and Russo, A.) Pergamon, Oxford, p. 33.
8. Quirk, R. P. and Kim, J. (1994) Block copolymers. In *Encyclopedia of Advanced Materials*, Vol. 1 (Eds, Bloor, D., Brook, R. J., Flemings, M. C. and Mahajan, S.) Pergamon, Oxford, pp. 280–298.
9. Datta, S. and Lohse, D. J. (1996) *Polymeric Compatibilizers*, Henser, New York.
10. Ameduri, B., Boutevin, B. and Gramain, P. (1997) Synthesis of block copolymers by radical polymerization and telomerization. *Adv. Polym. Sci.*, **127**, 87–142.
11. Court, F., Liebler, L., Mourran, A., Navarro, C. and Royackkers, V. (1998) Compositions à base de résines thermoplastiques semi-crystallines à tenue mécanique et thermique améliorées, leur procédé de préparation et leurs utilisations. PCT/FR Patent 98.02635 WO 99/29772 (assigned to Atofina).
12. Ameduri, B., Boutevin, B. and Kostov, G. (2001) Fluoroelastomers: synthesis, properties and applications. *Prog. Polym. Sci.*, **26**, 105–187.
13. (a) Hadjichristidis, N. (2002) *Block Copolymers: Synthetic Strategies, Physical Properties, and Applications*, Wiley Interscience, New York; (b) Lodge, T. P. (2003) Block copolymers: past successes and future challenges. *Macromol. Chem. Phys.*, **204**, 265–273.
14. Lacroix-Desmazes, P., Ameduri, B. and Boutevin, B. (2002) Use of fluorinated organic compounds in living radical polymerizations. *Collect. Czech. Chem. Commun.*, **67**, 1383–1415.
15. Tiers, G. V. D. (1962) Some free radical-catalyzed additions of perfluoralkyl iodides to olefins. *J. Org. Chem.*, **27**, 2261–2266.
16. Brace, N. O. (1962) Addition of polyfluoroalkyl iodides to unsaturated compounds and products produced thereby, US Patent 3,145,222 (assigned to DuPont de Nemours) 31-07-1962.
17. Huang, W.-Y. (1994) Synthesis of fluorinated monomers trough sulfinato-dehalogenation and related reactions. *Macromol. Symp.*, **82**, 67–75.
18. Kotora, M., Hajek, M., Kvicala, J., Ameduri, B. and Boutevin, B. (1993) Rearrangement of 2-iodo-3-perfluoroalkyl-1-propyl acetate to 1-iodo-3-perfluoroalkyl-2-propyl acetate. *J. Fluorine Chem.*, **64**, 259–268 (and references herein).
19. Brace, N. O. (1999) Syntheses with perfluoroalkyl radicals from perfluoroalkyl iodides. A rapid survey of synthetic possibilities with emphasis on practical applications. Part one: alkenes, alkynes and allylic compounds. *J. Fluorine Chem.*, **93**, 1–25.
20. Brace, N. O. (1999) Syntheses with perfluoroalkyl iodides. Part II. Addition to nonconjugated alkadiene; cyclization of 1,6-heptadiene, and of 4-substituted 1,6-heptadienoic compounds: *bis*-allyl ether, ethyl diallylmalonate, *N,N'*-diallylamine and *N*-substituted diallylamines; and additions to homologous *exo*-and *endo*cyclic alkenes, and to bicyclic alkenes. *J. Fluorine Chem.*, **96**, 101–127 (and references herein).
21. Neumann, W. J. (1987) Tri-*n*-butyltin hydride as reagent in organic synthesis. *Synthesis*, 665.
22. Gambaretto, G., Conte, L., Fornasieri, G., Zarantonello, C., Tonei, D., Sassi, A. and Bertani, R. (2003) Synthesis and characterization of a new class of polyfluorinated alkanes: tetrakis (perfluoroalkyl)alkane. *J. Fluorine Chem.*, **121**, 57–63 (and references herein).
23. Rabolt, J. F., Russell, T. P. and Twieg, R. J. (1984) Structural studies of semifluorinated *n*-alkanes. 1. Synthesis and characterization of $F(CF_2)_n(CH_2)_mH$ in the solid state. *Macromolecules*, **17**, 2786–2794.

24. Cornélus, C., Krafft, M. P. and Riess, J. G. (1994) About the mechanism of stabilization of fluorocarbon emulsions by mixed fluorocarbon/hydrocarbon additives. *J. Colloid Interface Sci.*, **163**, 391–394.

25. Cornélus, C., Krafft, M. P. and Riess, J. G. (1994) Improved control over particle sizes and stability of concentrated fluorocarbon emulsions by using mixed fluorocarbon/hydrocarbon molecular dowels. In *Artif. Cells, Blood Subs. Immobil. Biotechnol.* (Guest Editor, Riess, J. G.), Vol. 22, 1183–1191; see also pp. 1267–1272.

26. Reiss, J. G. and Weers, J. G. (1996) Emulsions for biomedical uses. *Curr. Opin. Colloid Interface Sci.*, **1**, 652–659.

27. Sheiko, S., Lerman, E. and Moller, M. (1996) Self-dewetting of perfluoroalkyl methacrylate films on glass. *Langmuir*, **12**, 4015–4024.

28. Napoli, M., Conte, L. and Guerrato, A. (2001) Synthesis of $F(CF_2)_8(CH_2)_8H$ and gel phase formation from its solution in homogeneous alcohols. *J. Fluorine Chem.*, **110**, 47–58.

29. Turberg, M. P. and Brady, J. E. (1988) Semifluorinated hydrocarbons: primitive surfactant molecules. *J. Am. Chem. Soc.*, **110**, 7797–7801.

30. Wang, S., Lunn, R., Krafft, M. P. and Lebanc, R. M. (2000) One and a half layers? Mixed Langmuir monolayer of 10,12-pentacosadiynoic acid and a semifluorinated tetracosane. *Langmuir*, **16**, 2882–2886.

31. Lo Nostro, P., Santoni, I., Bonini, M. and Baglioni, P. (2003) Inclusion compound from a semi-fluorinated alkane and β-cyclodextrine. *Langmuir*, **19**, 2313–2317; (b) Lo Nostro, P. (2003) Aggregates from semifluorinated *n*-alkanes: how incompatibility determines self-assembly. *Curr. Opin. Colloid Interface Sci.*, **8**, 223–226.

32. Napoli, M., Conte, L. and Gambaretto, G. (1997) Diblock semifluorinated *n*-alkanes purification by gel phase formation in organic solvents: $C_8F_{17}C_{18}H_{37}$ in alcohols. *J. Fluorine Chem.*, **85**, 163–167.

33. Viney, C., Russell, T. P., Depero, L. E. and Twieg, R. J. (1989) Transitions to liquid crystalline phases in a semifluorinated alkane. *Mol. Cryst. Liq. Cryst.*, **168**, 63–82.

34. Mahler, W., Guillon, D. and Skoulios, A. (1985) Smectic liquid crystal from (perfluoro-decyl)decane. *Mol. Cryst. Liq. Cryst., Lett. Sect.*, **2**, 111–119.

35. Russell, T. P., Rabolt, J. F., Twieg, R. J., Siemens, R. L. and Farmer, B. L. (1986) Structural characterization of semifluorinated *n*-alkanes. 2. Solid–solid transition behavior. *Macromolecules*, **19**, 1135–1143.

36. Pugh, C., Hopken, J. and Möller, M. (1988) Amphiphilic molecules with hydrocarbon and fluorocarbon segments. *Polym. Prep. (Am. Chem. Soc., Div. Polym. Chem.)*, **29**(1), 460–461.

37. Hopken, J., Pugh, C., Richtering, W. and Möller, M. (1988) Melting, crystallization, and solution behavior of chain molecules with hydrocarbon and fluorocarbon segments. *Makromol. Chem.*, **189**, 911–925.

38. Hopken, J. and Möller, M. (1992) On the morphology of (perfluoroalkyl)alkanes. *Macromolecules*, **25**, 2482–2489.

39. Napoli, M. (1996) Diblock and triblock semifluorinated *n*-alkanes: preparations, structural aspects and applications. *J. Fluorine Chem.*, **79**, 59–69.

40. McLain, S. J., Sauer, B. B. and Firment, L. E. (1996) Surface properties and metathesis synthesis of block copolymers including perfluoroalkyl-ended polyethylenes. *Macromolecules*, **29**, 8211–8219.

41. Riess, J. G. and Postel, M. (1992) Stability and stabilization of fluorocarbon emulsions destined injection. *Biomater., Artif. Cells, Immobil. Biotechnol.*, **20**, 819–829.

42. Riess, J. G., Sole-Violan, L. and Postel, M. (1992) A new concept in the stabilization of injectable fluorocarbon emulsions: the use of mixed fluorocarbon–hydrocarbon dowels. *J. Disp. Sci. Technol.*, **13**, 349–355.

43. Meinert, H., Fackler, R., Knoblich, A., Mader, J., Reuter, P. and Rohlke, W. (1992) On the perfluorocarbon emulsions of second generation. *Biomater., Artif. Cells, Immobil. Biotechnol.*, **20**, 805–818.

44. Krafft, M. P., Riess, J. G. and Weers, J. G. (1998) The design and engineering of oxygen-delivering fluorocarbon emulsions. In *Submicron Emulsion in Drug Targeting and Delivery* (Ed, Benita, S.) Harwood Academic Publishers, Toronto, pp. 235–333, Chap. 10.

45. (a) Krafft, M. P. and Riess, J. G. (1998) Highly fluorinated amphiphiles and colloidal system, and their applications in the biomedical field. A contribution. *Biochimie*, **80**, 489–514; (b) Krafft, M. P. and Riess, J. G. (2003) Fluorinated colloids and interfaces. *Curr. Opin. Colloid Interface Sci.*, **8**, 213–214; (c) Krafft, M. P., Chittofrati, A. and Riess, J. G. (2003) Emulsions and microemulsions with a flurocarbon phase. *Curr. Opin. Colloid Interface Sci.*, **8**, 251–258.

46. Gaines, G. L. (1991) Surface activity of semifluorinated alkanes $F(CF_2)_m(CH_2)_nH$. *Langmuir*, **7**, 3054–3056.

47. Sanchez, V., Greiner, J. and Riess, J. G. (1995) Highly concentrated 1,2-*bis*(perfluoroalkyl) iodoethene emulsions for use as contrast agents for diagnosis. *J. Fluorine Chem.*, **73**, 259–264.

48. Traverso, E. and Rinaldi, A. Preparation of (perfluoroalkyl)alkanes as agents affording sliding characteristics, Eur. Patent 444,752 (04-09-1991) and US Patent 5,202,041 (assigned to Enichem Synthesis S.p.A.) 13-04-1993.

49. Gambaetto, G. P. (1998) Solid lubricants and process for preparing it, US Patent 4,724,093 (assigned to Enichimia Secondaria).

50. Karydas, A. (1999) Fluorinated lubricants for polyethylene snow sliders such as skis, sleds and snowboards, US Patent 5,914,298 (22-06-1999).

51. Kissa, E. (1994) *Fluorinated Surfactant: Synthesis, Preparations, Applications*, Marcel Dekker, New York.

52. Reiss, J. G. (2001) Oxygen carriers ("Blood substitutes") Raison d'Etre Chemistry, and some physiology. *Chem. Rev.*, **101**, 2797–2919.

53. Reiss, J. G. (2002) Blood substitutes and other biomedical applications of fluorocarbon colloids. *J. Fluorine Chem.*, **114**, 119–126.

54. Brace, N. O. (2003) "Single tail" and "twin tail" 3-(perfluoroalkylethanethia) alkylsuccinic anhydrides give R_F segmented diacids, amic acids, imides, esters and salts. Unusual ^1H NMR of the succinamic acids. *J. Fluorine Chem.*, **121**, 33–53.

55. Bolstad, A. N., Township, A., County, W., Sherman, P. O. and Smith, S. (1962) Process for preparing block and graft copolymers of perfluoroalkyl monomer and resulting product, US Patent 3,068,187 (assigned to 3M).

56. Dessaint, A. and Perronin, J. French Patents 2,328,070 (1977) and 2,442,861 (1980) (assigned to Produits Chimiques Ugine Kuhlmann).

57. Foulletier, L. and Lalu, J. P. Nouveaux composes organiques fluorès, French Patents 1,532,284 (1968); 1,588,865 (1970); 1,604,039 (1970); 2,031,650 (1980) (assigned to Produits Chimiques Ugine Kuhlmann).

58. Hayashi, T., Matsuo, M. and Kawakami, S. (1980) Improvement of polymers, Jap. Patent Kokai Tokkyo Koho, 80.99924 assigned to Asahi Glass Co. Ltd, (*Chem. Abstr.*, **94**, 31291m).

59. Bonardi, C. (1991) Fluoroacrylic telomers and their use as waterproofing and oilproofing products on diverse substrates, Eur. Patent Appl. EP 426, 530 (assigned to Elf Atochem) (*Chem. Abstr.*, **115**, 161338z).

60. Boutevin, B., Rigal, G., El Asri, M. and Lakhlifi, T. (1997) Emulsion telomerisation of styrene with mercaptans. *Eur. Polym. J.*, **33**, 277–286.

61. Brace, N. O. (1961) Addition of polyfluoroalkyl iodides to unsaturated compounds, US 3,145, 222 (assigned to E. I. Du Pont de Nemours and Co.) 23-02-1961 (*Chem. Abstr.* (1964) **61**, 10589g).

62. El Soueni, A., Tedder, J. M. and Walton, J. C. (1981) The reactions of trifluoromethyl and trichloromethyl radicals with conjugated and methylene-interrupted dienes and enynes. *J. Fluorine Chem.*, **17**, 51–56.

63. Umemoto, T., Kuriu, Y. and Nakayama, S. I. (1982) Novel oxyperfluoroalkylation. *Tetrahedron Lett.*, **23**, 4101–4103.

64. Huang, W.-Y. and Zhang, H.-Z. (1991) Studies on sulfinatodehalogenation. XIX. Reaction of perfluoroalkyl iodides and perfluoroalkanesulfonyl bromides with conjugated dienes. *Chin. J. Chem.*, **9**, 76–82.

65. Huang, W.-Y. and Zhang,, H.-Z. (1992) The reaction of perfluoroalkanesulfonyl halides. IX. The syntheses of bisperfluoroalkyl butadienes and perfluoroalkylated dienoates. *Chin. J. Chem.*, **10**, 544–551.

66. Lebreton, P., Ameduri, B., Boutevin, B., Corpart, J. M. and Juhue, D. (2000) Radical telomerization of 1,3-butadiene with perfluoroalkyl iodides. *Macromol. Chem. Phys.*, **201**, 1016–1024.

67. Lebreton, P., Ameduri, B., Boutevin, B., Corpart, J. M. and Cot, D. (2002) Radical telomerization of 1,3-butadiene with perfluoroalkyl iodides in the presence of potassium carbonate. *J. Polym. Sci., Part A, Polym. Chem.*, **40**, 3743–3756.

68. Kleiner, E. K. and Falk, R. A. (1980) Oligomers with perfluoroalkyl end-groups that contain mercapto group, process for their preparation and their use as surface-active substances and as additives in fire-extinguishing compositions, Eur. Patent Appl. 80 19584 (assigned to Ciba Geigy).

69. Boutevin, B., Lantz, A. and Pietrasanta, Y. (1985) Fluorinated telomers with hydrophilic groups: their process of preparation and their use, Eur. Patent Appl. 0,189,698 (assigned to Elf Atochem) 18-12-1985.

70. Boutevin, B., Lantz, A. and Pietrasanta, Y. (1985) Fluorinated telomers with CCl₃ end-groups; their preparation and that of diblock cotelomers there from, Eur. Patent Appl. 0,188,952 (assigned to Elf Atochem) 18-12-1985.

71. Holmberg, K., Jonsson, B., Kronberg, B. and Lindman, B. (2002) *Surfactants and Polymers in Aqueous Solution*, 2nd ed., Wiley, New York.

72. Boutevin, B., Mouanda, J., Pietrasanta, Y. and Taha, M. (1986) Synthesis of block cotelomers involving a perfluorinated chain and a hydrophilic chain. II. Use of fluorinated telogens with iodine or thiol end groups. *J. Polym. Sci., Part A, Polym. Chem.*, **24**, 2891–2903.

73. Pabon, M., Broder, P., Morrillon, E. and Corpart, J. M. Extinguishing composition, Eur. Patent Appl. 1,013,311 (assigned to Elf Atochem) 28-06-2000.

74. Pabon, M. and Corpart, J. M. (2002) Fluorinated surfactants: synthesis, properties and effluent treatment. *J. Fluorine Chem.*, **114**, 149–156.

75. Boutevin, B., Diaf, O., Pietrasanta, Y. and Taha, M. (1986) Synthesis of block cotelomers involving a perfluorinated chain and a hydrophilic chain. Part 1. Use of fluorinated telogens with trichloromethyl end-groups. *J. Polym. Sci., Part A, Polym. Chem.*, **24**, 3129–3137.

76. Pavia, A. A., Pucci, B., Riess, J. G. and Zarif, L. (1992) New perfluoroalkyl telomeric nonionic surfactants: synthesis, physicochemical and biological properties. *Makromol. Chem.*, **193**, 2505–2517.

77. Giusti, F., Mansis, S. and Pucci, B. (2002) Effect of protic interactions on the kinetics of polymerization of tris(hydroxymethyl)acrylamidomethane (THAM) and its derivatives. *New J. Chem.*, **26**, 1724–1732.

78. Pucci, B., Maurizis, J. C. and Pavia, A. A. (1991) Telomers and cotelomers of biological and biomedical interest. IV. Telomers from tris(hydroxymethyl)acrylamidomethane. New nonionic amphiphilic agents. *Eur. Polym. J.*, **27**, 1101–1106.

79. Chaudier, Y., Zito, F., Barthélémy, P., Stroebel, D., Améduri, B., Popot, J. L. and Pucci, B. (2002) Synthesis and preliminary biochemical assessment of ethyl-terminated perfluoroalkylamine oxide surfactants. *Bioorg. Med. Chem. Lett.*, **12**, 1587–1590.

80. Zarif, L., Riess, J. G., Pucci, B. and Pavia, A. A. (1993) Biocompatibility of alkyl and perfluoroalkyl telomeric surfactants derived from THAM. *Biomater. Artif. Cells Immobil. Biotechnol.*, **21**, 597–608.

81. Contino-Pépin, C., Maurizis, J. C. and Pucci, B. (2002) Amphiphilic oligomers: a new kind of macromolecular carrier of antimitotic drugs. *Curr. Med. Chem. Anticancer Agent*, **2**, 645–665.

82. Boutevin, B. and Pietrasanta, Y. (1975) Telomerisation of chlorotrifluoroethylene with carbon tetrachloride. *Eur. Polym. J.*, **12**, 219–226.

83. Boutevin, B., Maliszewicz, M. and Pietrasanta, Y. (1982) Synthesis of photocrosslinkable compounds. 4. Esterification of acrylic acid with 2-hydroxyethyl acrylate telomers. *Makromol. Chem.*, **183**, 2333–2340.

84. Boutevin, B., Pietrasanta, Y. and Parisi, J. P. (1987) Synthesis of photocrosslinkable telomers-8-grafting of acrylic acid onto glycidyl methacrylate telomers with thiols. *Makromol. Chem.*, **188**, 1621–1629.

85. Boutevin, B. and Youssef, B. (1987) Synthesis of block cotelomers involving a perfluorinated chain and a hydrophilic chain. 3. Synthesis and use of fluorinated telogens with a trichloromethyl end-group. *J. Polym. Sci., Part A, Polym. Chem.*, **25**, 3025–3033.

86. Boutevin, B., Boileau, S., Lantz, A. and Pietrasanta, Y. (1984) Procédé de préparation d'oléfines fluorées et nouvelles oléfines résultantes, French Patent 09,662 (assigned to Atochem) 20-06-1984.

87. Boutevin, B., Youssef, B., Boileau, S. and Garnault, A. M. (1987) Synthesis of allylic ethers and thioethers by phase transfer catalysis. *J. Fluorine Chem.*, **35**, 399–406.

88. Marchionni, G. and Viola, G. T. (1988) Peroxide-curable fluoroelastomers containing vinylidene fluoride telomers, Eur. Patent Appl. 251,285 A2 (assigned to Ausimont) 07-01-1988.

89. Moggi, G., Modena, S. and Marchionni,, G. (1990) A vinylidene fluoride block copolymer containing perfluoropolyether groups. *J. Fluorine Chem.*, **49**, 141–149.

90. Gelin, M. P. and Ameduri, B. (2003) Fluorinated block copolymers containing poly(vinylidene fluoride) or poly(vinylidene fluoride-*co*-hexafluoropropylene) blocks from perfluoropolyethers: synthesis and thermal properties. *J. Polym. Sci., Part A, Polym. Chem.*, **41**, 160–171.

91. (a) Tobolski, A. V. (1960) Dead end polymerization. *J. Am. Chem. Soc.*, **80**, 5927–5929; (b) Tobolski, A. V., Rogers, C. E. and Brickman, R. D. (1960) Dead end polymerization, Part II. *J. Am. Chem. Soc.*, **82**, 1277–1280.

92. Oshibe, Y., Hishigaki, H., Ohmura, H. and Yamamoto, T. (1989) Preparation of block and graft copolymers and their applications. VIII. Preparation of fluorine-containing block copolymers and their solubility characteristics. *Kobunshi Ronbunshu*, **46**, 81–87 (*Chem. Abstr.* 111, 39960 (1989)) and IX. Surface properties of fluorine-containing block copolymers, *Kobunshi Ronbunshu*, 46, 89–96 (1989) (*Chem. Abstr.* 111, 8131 (1989)).

93. Yamamoto, T., Aoshima, K., Moriya, Y., Suzuki, N. and Oshibe, Y. (1991) New manufacturing processes for block and graft copolymers by radical reaction. *Polymer*, **32**, 19–28.

94. Koji, F., Yuji, Y. and Tatemoto, M. (1988) Novel iodine-containing compounds and preparation thereof, Eur. Patent Appl. 0,272,698 (assigned to Daikin) 29-06-1988.

95. Marciniek, B. (1992) *Comprehensive Handbook on Hydrosilylation*, Pergamon Press, Oxford.

96. Boutevin, B. and Pietrasanta, Y. (1989) *Les Acrylates et Polyacrylates Fluorés*, Erec, Paris.

97. Richard, J., Deschamps, F., Lacroix-Desmazes, P. and Boutevin, B. (2003) Stabilisation de dispersion de substances hydrosolubles ou hydrophiles dans le dioxyde de carbone à pression supercritique, French Patent 0,305,108 (assigned to Ethypharm/CNRS).

98. Lampin, J. P., Cambon, A., Szoni, F., Delpuech, J. J., Serratrice, G., Thiollet, G. and Lafosse, L. (1985) Nouveaux Composés fluorés non ioniques, leur procédé d'obtention et leurs applications en tant que tensio-actifs, Eur. Patent Appl. 0,165,853 (assigned to Institut National de Recherche Chimique Appliquée) 28-05-1985.

99. Matyjaszewski, K. (2000) Controlled/living radical polymerization: progress in ATRP, NMP and RAFT. ACS Symp. Ser., 768, American Chemical Society, Washington, DC.

100. Matyjaszewski, K. (2002/2003) Controlled/Living Radical Polymerization: Progress in ATRP, NMP and RAFT ACS Symp. Ser. 2002, 824 and ACS Symp. Ser. 2003, 854, American Chemical Society, Washington, DC, pp. 570–585, Chap. 39.

101. Otsu, T. and Yoshida, M. (1982) Role of initiator–transfer agent–terminator (iniferter) in radical polymerizations: polymer design by organic disulfides as iniferters. *Makromol. Chem., Rapid Commun.*, **3**, 127–132.

102. Otsu, T., Yoshida, M. and Tazaki, T. (1982) A model for living radical polymerization. *Makromol. Chem., Rapid Commun.*, **3**, 133–137.

103. Sebenik, A. (1998) Living free-radical block copolymerization using thio-iniferters. *Prog. Polym. Sci.*, **23**, 875–912.

104. Otsu, T. and Matsumoto, A. (1998) Controlled synthesis of polymers using the iniferter technique: developments in living radical polymerization. *Adv. Polym. Sci.*, **136**, 75–134.

105. Otsu, T. (2000) Iniferter concept and living radical polymerization. *J. Polym. Sci., Part A, Polym. Chem.*, **38**, 2121–2132.

106. Georges, M. K., Veregin, R. P. N., Kazmaier, P. M. and Hamer, G. K. (1993). Narrow molecular weight resins by a free-radical polymerization process. *Macromolecules*, **26**, 2987–2988.

107. Veregin, R. P. N., Georges, M. K., Kazmaier, P. M. and Hamer, G. K. (1993). Free radical polymerizations for narrow polydispersity resins: electron spin resonance studies of the kinetics and mechanism. *Macromolecules*, **26**, 5316–5322.

108. Bertin, D., Boutevin, B. and Destarac, M. (1998) Controlled radical polymerization with nitroxyl stable radicals. In *Polymer and Surfaces: A Versatile Combination* (Ed, Hommel, D.) Research Signpost, Trivandrum, India, p. 47.

109. Hawker, C. J., Bosman, A. W. and Harth, E. (2001) New polymer synthesis by nitroxide mediated living radical polymerizations. *Chem. Rev.*, **101**, 3661–3718.

110. Percec, V. and Barboiu, B. (1995) "Living" radical polymerization of styrene initiated by arenesulfonyl chlorides and CuI(bpy)nCl. *Macromolecules*, **28**, 7970–7975.

111. Matyjaszewski, K. and Wang, J. S. (1995) Controlled/"living" radical polymerization. Atom transfer radical polymerization in the presence of transition-metal complexes. *J. Am. Chem. Soc.*, **117**, 5614–5618.

112. Matyjaszewski, K. and Xia, J. (2001) Atom transfer radical polymerization. *Chem. Rev.*, **101**, 2921–2990.

113. Kotani, Y., Kamigato, M. and Sawamoto, M. (1998) Living random copolymerization of styrene and methyl methacrylate with a Ru(II) complex and synthesis of ABC-type "block random" copolymers. *Macromolecules*, **31**, 5582–5587.

114. Tatemoto, M. (1996). Iodine transfer polymerization. In *Polymeric Materials Encyclopedia*, Vol. 5 (Ed, Salamone, J. C.) CRC Press, Boca Raton, FL, pp. 3847–3862.

115. Chiefari, J., Chong, Y. K., Ercole, F., Kristina, J., Jeffery, J., Le, T. P. T., Mayadunne, R. T. A., Meijs, F., Moad, C. L., Moad, G., Rizzardo, E. and Thang, S. H. (1998) Living free-radical polymerization by reversible addition-fragmentation chain transfer: the RAFT process. *Macromolecules*, **31**, 5559–5566.

116. Rizzardo, E., Chiefari, J., Mayadunne, R. T. A., Moad, G. and Thang, S. H. (2000) Synthesis of defined polymers by reversible addition-fragmentation chain transfer: the RAFT process. *ACS Symp. Ser.*, **768**, 278–286.

117. Moad, G., Chiefari, J., Chong, Y. K., Kristina, J., Mayadunne, R. T. A., Pastma, A., Rizzardo, E. and Thang, S. H. (2000) Living free radical polymerization with reversible addition-fragmentation chain transfer (the life of RAFT). *Polym. Int.*, **49**, 993–999.

118. (a) Charmot, D., Corpart, P., Adam, H., Zard, S. Z., Biadatti, T. and Bouhadir, G. (2000) Controlled radical polymerization in dispersed media. *Macromol. Symp.*, **150**, 23–31; (b) Destarac, M., Charmot, D., Franck, X. and Zard, S. Z. (2000) Dithiocarbamates as universal reversible addition-fragmentation chain transfer agents. *Macromol. Rapid Commun.*, **21**, 1035–1039.

119. (a) Destarac, M., Bzducha, W., Taton, D., Gauthier-Gillaizeau, I. and Zard, S. Z. (2002) Xanthates as chain transfer agents in controlled radical polymerization (MADIX): structural effect of the *O*-alkyl group. *Macromol. Rapid Commun.*, **23**, 1049–1054; (b) Chapon, P., Mignaud, C., Lizarraga, G. and Destarac, M. (2003) Automated parallel synthesis of MADIX (co)polymers. *Macromol. Rapid Commun.*, **24**, 87–91.

120. Destarac, M., Charmot, D., Zard, S. and Franck, X. (2000) Synthesis method for polymers by controlled radical polymerisation using halogenated xanthates, WO 00/75207 A1 (assigned to Rhodia) 26-05-2000.

121. Lacroix-Desmazes, P., Young, J. L., Taylor, D. K., DeSimone, J. M. and Boutevin, B. (2002) Eighth Meeting on Supercritical Fluids, Proceedings (ISASF), April 14–17, 2002, Bordeaux, 1, 241 [ISBN 2-905267-34-8].

122. Lacroix-Desmazes, P., Boutevin, B., Taylor, D. and DeSimone, J. M. (2002) Synthesis of fluorinated block copolymers by nitroxide-mediated radical polymerization for supercritical carbon dioxide applications. *Polym. Prep. (Am. Chem. Soc., Div. Polym. Chem.)*, **43**(2), 285–286.

123. Lacroix-Desmazes, P., Delair, T., Pichot, C. and Boutevin, B. (2000) Synthesis of poly(chloromethylstyrene-b-styrene) block copolymers by controlled free-radical polymerization. *J. Polym. Sci., Part A, Polym. Chem.*, **38**, 3845–3854.

124. Andruzzi, L., Chiellini, E., Galli, G., Li, X., Seok, S. H. and Ober, C. K. (2002) Engineering low surface energy polymers through molecular design; synthetic route to fluorinated polystyrene-based block copolymers. *J. Mater. Chem.*, **12**, 1684–1692.

125. Galli, G., Andruzzi, L., Chiellini, E., Li, X., Ober, C. K., Hexemer, A. and Kramer, E. J. (2003) Bulk and surface structures of polystyrene-based semifluorinated block copolymers for low surface energy coatings. Proceedings of the Fluorine in Coatings V Conference, Orlando, January 21–22, 2003, paper #4.

126. Mawson, S. and DeSimone, J. M. (1995) Formation of poly(1,1,2,2-tetrahydroperfluorodecyl acrylate) submicron fibers and particles from supercritical carbon dioxide solutions. *Macromolecules*, **28**, 3182–3189.

127. Feiring, A. E., Wonchoba, E. R., Davidson, F., Percec, V. and Barboiu, B. (2000) Fluorocarbon-ended polymers: metal catalyzed radical and living radical polymerizations initiated by perfluoroalkylsulfonyl halides. *J. Polym. Sci., Part A, Polym. Chem.*, **38**, 3313–3321.

128. Jo, S. M., Lee, W. S., Ahn, B. S., Park, K. Y., Kim, K. A. and Paeng, I. S. R. (2000) New AB or ABA type block copolymers: atom transfer radical polymerization (ATRP) of methyl methacrylate using iodine-terminated PVDFs as macroinitiators. *Polym. Bull., Berlin*, **44**, 1–9.

129. Lacroix-Desmazes, P. and Boutevin, B. Unpublished results.

130. Ming, W., Van de Grampel, R. D., Gildenpfennig, A., Snijder, A., Brongersma, H. H., Van de Linde, R. and De With, G. (2003) Thermodynamically controlled surface segregation of perfluoroalkyl-end-capped PMMA as investigated by low energy ion scattering (LIES). *Polym. Mater. Sci. Engng*, **88**, 517–518.

131. Lim, K. T., Lee, M. Y., Moon, M. J., Lee, G. D., Hong, S. S., Johnston, K. P. and Dickson, J. L. (2002) Synthesis and properties of semifluorinated block copolymers containing poly(ethylene oxide) and poly(fluorooctyl methacrylates) via atom transfer radical polymerization. *Polymer*, **43**, 7043–7049.

132. Betts, D. E., Johnson, T., Anderson, C. and DeSimone, J. M. (1997) Controlled radical polymerization methods for the synthesis of non-ionic surfactants for CO_2. *Polym. Prep. (Am. Chem. Soc., Div. Polym. Chem.)*, **38**(1), 760–761.

133. Wu, B., Li, X., Jiao, J., Wu, P. and Han, Z. (2001) Synthesis and characterization of the diblock copolymer of PMMA and PFLUWET. *Huadong Ligong Daxue Xuebao*, **27**, 60–65 (*Chem. Abstr.* 135, 77164, 2001).

134. Gaynor, S., Edelman, R. and Matyjaszewski, K. (1997) From step growth to living free radical polymerization: the synthesis of ABA block copolymers of vinyl and step growth polymers by ATRP. *Polym. Prep. (Am. Chem. Soc., Div. Polym. Chem.)*, **38**(1), 703–704.

135. Gaynor, S., Edelman, R. and Matyjaszewski, K. (1997) Step-growth polymers as macroinitiators for "living" radical polymerization: synthesis of ABA block copolymers. *Macromolecules*, **30**, 4241–4247.

136. Ying, S. K., Zhang, Z. B., Wang, S. R. and Shi, Z. Q. (1999) Novel fluorinated block copolymers for the construction of low energy surfaces. *Polym. Prep., (Am. Chem. Soc., Div. Polym. Chem.)*, **40**, 1051–1052.

137. Li, Y., Zhang, W., Yu, Z. and Huang, J. (2000) The synthesis of fluorine-containing diblock copolymers and their properties. *Polym. Prep. (Am. Chem. Soc., Div. Polym. Chem.)*, **41**(1), 202–203.

138. Xia, J., Johnson, T., Gaynor, S. G., Matyjaszewski, K. and DeSimone, J. (1999) Atom transfer radical polymerization in supercritical carbon dioxide. *Macromolecules*, **32**, 4802–4809.

139. Jonhson, T. and DeSimone, J. M. (1998) Synthesis of hydrophilic hydrocarbon/fluorocarbon block copolymers using controlled free radical techniques. *Polym. Prep. (Am. Chem. Soc., Div. Polym. Chem.)*, **39**(2), 824–825.

140. Becker, M. and Wooley, K. (2000) Radical-based preparation of block copolymers containing fluorine tags: tools for detailed analysis of nanostructured materials. *Polym. Prep. (Am. Chem. Soc., Div. Polym. Chem.)*, **41**(2), 1328–1329.

141. Becker, M., Remsen, E. E. and Wooley, K. (2001) Diblock copolymers, micelles and shell-crosslinked nanoparticules containing poly(4-fluorostyrene): tools for detailed analyses of nanostructured materials. *J. Polym. Sci. Part A, Polym. Chem.*, **39**, 4152–4166.

142. Hvilsted, S., Borkar, S., Abildgaard, L., Georgieva, V., Siesler, H. W. and Jankova, K. (2002) Polymers and block copolymers of fluorostyrenes by ATRP. *Polym. Prep.*, **43**(2), 26–27.

143. Jankova, K. and Hvilsted, S. (2003) Preparation of poly(2,3,4,5,6-pentafluorostyrene) and block copolymers with styrene by ATRP. *Macromolecules*, **36**, 1753–1758.

144. Hvilsted, S., Borkar, S., Siesler, H. W. and Jankova, K. (2003) Novel fluorinated polymer materials based on 2,3,5,6-tetrafluoro-4-methoxystyrene. In *Advances in Controlled /Living Radical Polymerization*, (Ed, Matyjaszewski, K.) ACS Symposium Series 854, American Chemical Society, Washington, DC, pp. 236–249, Chap. 17.

145. Qiu, J. and Matyjaszewski, K. (1997) Polymerization of substituted styrenes by atom transfer radical polymerization. *Macromolecules*, **30**, 5643–5648.

146. Tatemoto, M. and Nakagawa, T. Segmented polymers, German Patent 2,729,671 (assigned to Daikin Kogyo Co., Ltd) 12-01-1978 (*Chem. Abstr.* (1978) **88**, 137374m).

147. Tatemoto, M. and Nakagawa, T. Segmented polymers containing fluorine and iodine and their production, US Patent 4,158,678 (assigned to Daikin) 19-06-1979.

148. Tatemoto, M. Recent studies on fluoroelastomers of vinylidene fluoride copolymers, The First Regular Meeting of Soviet-Japanese Fluorine Chemists, Tokyo, February 15–16, 1979.

149. Tatemoto, M., Furukawa, Y., Tomoda, M., Oka, M. and Morita, S. Fluorine-containing polymer and composition containing the same, Eur. Patent Appl. EP 14,930 (assigned to Daikin) (*Chem. Abstr.* (1980) **94**, 48603).

150. Tatemoto, M. and Morita, S. (1981) Liquid fluorine-containing polymer and a process for the production thereof, Eur. Patent Appl. EP 27,721 (assigned to Daikin) (*Chem. Abstr.* **95**, 170754).

151. Tatemoto, M. (1985) Fluorinated thermoplastic elastomers. *Int. Polym. Sci. Technol.*, **12**, 85–91 (translated into English from *Nippon Gomu Kyokaishi*, 57, 761–767 (1984)).

152. Oka, M. and Tatemoto, M. (1984) Vinylidene fluoride-hexafluoropropylene copolymer having terminal iodines. *Contemp. Top. Polym. Sci.*, **4**, 763–769.

153. (a) Tatemoto, M. (1990) Fluoroelastomers. *Kagaku Kogyo*, **41**, 78–87 (*Chem. Abstr.* 114, 8081); (b) Tatemoto, M. (1992) Fluoroelastomers. *Kobunshi Ronbunshu*, **49**, 765–771 (*Chem. Abstr.* 118, 22655).

154. Fossey, J., Lefort, D. and Sorba, J. (1995) *Free Radicals in Organic Chemistry*, Wiley, New York.

155. Hung, M. H. (1993) Iodine-containing chain transfer agents for fluoroolefin polymerization, US Patent, 5231154 (assigned to DuPont) 27-07-1993.

156. Yutani, Y. and Tatemoto, M. (1991) Process for preparing polymer. Eur. Patent Appl. 0,489,370 A1 (assigned to Daikin).

157. Tatemoto, M., Tomoda, M. and Ueda, Y. (1980) Co-crosslinkable mixture, Ger. Patent DE 2,940,135 (assigned to Daikin) (*Chem. Abstr.* **93**, 27580).

158. Matyjaszewski, K., Gaynor, S. G., Greszta, D., Mardare, D., Shigemoto, T. and Wang, J. S. (1995) Unimolecular and bimolecular exchange reactions in controlled radical polymerization. *Macromol. Symp.*, **95**, 217–228.

159. Gaynor, S. C., Wang, J. S. and Matyjaszewski, K. (1995) Controlled radical polymerizations: the use of alkyl iodides in degenerative transfer. *Macromolecules*, **28**, 2093–2100.

160. Goto, A., Ohno, K. and Fukuda, T. (1998) Mechanism and kinetics of iodide-mediated polymerization of styrene. *Macromolecules*, **31**, 2809–2815.
161. Starks, C. M. (1974) *Free Radical Telomerization*, Academic Press, New York.
162. Ameduri, B. and Boutevin, B. (1997) Telomerisation reactions of fluoroalkenes. In *Current Topics in Chemistry* (Ed, Chambers, R.D.), 192, 165–233.
163. Bak, P. I., Bidinger, G. P., Cozens, R. J., Klich, P. R. and Mayer, L. A. (1994) Pseudo-living radical polymerization, Eur. Patent Appl. 0,617,057 A1 (assigned to Geon).
164. Yagi, T., Tsuda, N., Noguchi, T., Sakaguchi, K., Tanaka, Y. and Tatemoto, M. (1990) Fluorine-containing block copolymers and artificial lens comprising the same. Eur. Patent Appl. 0,422, 644 A2, (assigned to Daikin) 11-10-1990.
165. Carlson, D. P. Fluorinated thermoplastic elastomers with improved base stability. Eur. Patent Appl. 0,444,700 A2, and US Patent 5,284,920 (assigned to E.I. DuPont de Nemours and Co.), 01-03-1991 and 08-02-1994.
166. Arcella, V., Brinati, G., Albano, M. and Tortelli, V. (1995) Fluorinated thermoplastic elastomers having superior mechanical and elastic properties. Eur. Patent Appl. 0,661,312 A1 (assigned to Ausimont S.p.A.) 05-07-1995.
167. Arcella, V., Brinati, G., Albano, M. and Tortelli, V. (1995) New fluorinated thermoplastic elastomers having superior mechanical and elastic properties, and preparation process thereof. Eur. Patent Appl. 0,683,186 A1 (assigned to Ausimont S.p.A.) 09-05-1995.
168. Gayer, U., Schuh, T., Arcella, V. and Albano, M. (2002) Fluorinated thermoplastic elastomers and articles therefrom. Eur. Patent Appl. EP 1,231,239 A1 (assigned to Ausimont S.p.A.) 14-08-2002.
169. Wlassics, I., Rapallo, G., Apostolo, M., Bellinzago, N. and Albano, M. (1999) Fluoroelastomers. Eur. Patent Appl. EP 0,979,832 A1 (assigned to Ausimont S.p.A.) 29-07-1999.
170. Balague, J., Ameduri, B., Boutevin, B. and Caporiccio, G. (2000) Controlled step-wise telomerization of vinylidene fluoride, hexafluoropropene and trifluoroethylene with iodofluorinated transfer agents. *J. Fluorine Chem.*, **102**, 253–268.
171. Bauduin, G., Boutevin, B., Bertocchio, R., Lantz, A. and Verge, C. (1998). Etude de la télomérisation de C_2F_4 avec les iodures de perfluoroalkyle. Partie I. Telomerisation radicalaire. *J. Fluorine Chem.*, **90**, 107–118.
172. Chambers, R. D., Greenhall, M. P., Wright, A. P. and Caporiccio, G. (1995) A new telogen for telechelic oligomers of chlorotrifluoroethylene. *J. Fluorine Chem.*, **73**, 87–92.
173. Paleta, O., Posta, A. and Frantisek, C. (1970) Preparation of fluoroacetic acids based on trifluorochloroethylene. *Collect. Czech. Chem. Commun.*, **35**, 1302–1311.
174. Rogozinski, M., Shorr, L. M., Hasman, U. and Ader-Barcas, D. (1968) Bromine telomerization. *J. Org. Chem.*, **33**, 3859–3866.
175. Ameduri, B., Boutevin, B., Kostov, G. and Petrova,, P. (1995) Novel fluorinated monomers bearing reactive side groups. Part 1. Preparation and use of $ClCF_2CFClI$ as the telogen. *J. Fluorine Chem.*, **74**, 261–267.
176. Ameduri, B., Boutevin, B., Kostov, G. and Petrova, P. (1999) Synthesis and polymerization of fluorinated monomers bearing a reactive lateral group. Part 5. Radical addition of iodine monobromide to chlorotrifluoroethylene to form a useful intermediate in the synthesis of 4,5,5-trifluoro-4-penten-1-ol. *J. Fluorine Chem.*, **93**, 117–127.
177. (a) Chambers, R. D., Musgrave, W. K. R. and Savory, J. (1961) Mixtures of halogens and halogen polyfluorides as effective sources of the halogen monofluorides in reactions with fluoroolefins. *J. Chem. Soc.*, 3779–3783; (b) Chambers, R. D. and Musgrave, W. K. R. (1961) The addition of "halogen monofluorides" to fluoroolefins. *Proc. Chem. Soc.* 113–116.
178. Amiry, M. P., Chambers, R. D., Greenhall, M. P., Ameduri, B., Boutevin, B., Caporiccio, G., Gornowicz, G. A. and Wright, A. P. (1993) The peroxide-initiated telomerization of chlorotrifluoroethylene with perfluorochloroalkyl iodides. *Polym. Prep. (Am. Chem. Soc. Div. Polym. Chem.)*, **34**, 411–412.

179. Haszeldine, R. N. (1955) Fluoroolefins. IV. Synthesis of polyfluoroalkanes containing functional groups from chlorotrifluoroethylene, and the short-chain polymerization of olefins. *J. Chem. Soc.*, 4291–4293.

180. Apsey, G. C., Chambers, R. D., Salisbury, M. J. and Moggi, G. (1988) Polymer chemistry. Part 1. Model compounds related to hexafluoropropene–vinylidene fluoride elastomer. *J. Fluorine Chem.*, **40**, 261–270.

181. Balague, J., Ameduri, B., Boutevin, B. and Caporiccio, G. (1995) Synthèse de télomères fluorés. Partie II. Télomérisation du trifluoroéthylène avec des iodures de perfluoroalkyle. *J. Fluorine Chem.*, **73**, 237–245.

182. Lansalot, M., Farcet, C., Charleux, B., Vairon, J. P. and Pirri, R. (1999) Controlled free-radical miniemulsion polymerization of styrene using degenerative transfer. *Macromolecules*, **32**, 7354–7360.

183. Butté, A., Storti, G. and Morbidelli, M. (2000) Miniemulsion living free radical polymerization of styrene. *Macromolecules*, **33**, 3485–3492.

184. Haupstchein, M., Braid, M. and Fainberg, A. H. (1961) Addition of iodine halides to fluorinated olefins. I. The direction of addition of iodine monochloride to perhaloolefins and some related reactions. *J. Am. Chem. Soc.*, **83**, 2495–2498.

185. Haupstchein, M., Braid, M. and Lawlor, F. E. (1957) Thermal syntheses of telomers of fluorinated olefins. I. Perfluoropropene. *J. Am. Chem. Soc.*, **79**, 2549–2554.

186. Dear, R. E. A. and Gilbert, E. E. (1974) Telomerization of bis(trifluoromethyl)disulfide with polyfluoroolefins. *J. Fluorine Chem.*, **4**, 107–112.

187. Sharp, D. W. A. and Miguel, H. T. (1978) Fluorine-containing dithioethers. *Isr. J. Chem.*, **17**, 144–152.

188. Haran, G. and Sharp, D. W. A. (1972) Photochemically initiated reactions of bis(trifluoromethyl) disulfide with olefins. *J. Chem. Soc. Perkin Trans.*, **4**, 34–42.

189. Kageishi, K., Yasuda, S. and Kishi, N. Jpn Kokai Tokyo Koho JP 62,230,811 (assigned to Atom Chemical Paint Co,Ltd., Japan) 10-09-1987.

190. Guan, Z. and DeSimone, J. M. (1994) Fluorocarbon-based heterophase polymeric materials. 1. Block copolymer surfactants for carbon dioxide applications. *Macromolecules*, **27**, 5527–5532.

191. Kassis, C. M., Steehler, J. K., Betts, D. E., Guan, Z., Romack, T. J., DeSimone, J. M. and Linton, R. W. (1996) XPS studies of fluorinated acrylate polymers and block copolymers with polystyrene. *Macromolecules*, **29**, 3247–3253.

192. Kendall, J. L., Canelas, D. A., Young, J. L. and DeSimone, J. M. (1999) Polymerizations in supercritical carbon dioxide. *Chem. Rev.*, **99**, 543–563.

193. Canelas, D. A., Betts, D. E. and DeSimone, J. M. (1996) Dispersion polymerization of styrene in supercritical carbon dioxide: importance of effective surfactants. *Macromolecules*, **29**, 2818–2821.

194. Lepilleur, C. and Beckman, E. J. (1997) Dispersion polymerization of methyl methacrylate in supercritical CO_2. *Macromolecules*, **30**, 745–756.

195. Giles, M. R., Griffiths, R. M. T., Agular-Ricardo, A., Silva, M. M. C. G. and Howdle, S. W. (2001) Fluorinated graft stabilizers for polymerization in supercritical carbon dioxide: the effect of stabilizer architecture. *Macromolecules*, **34**, 20–25.

196. Arnold, M. E., Nagai, K., Spontak, R. J., Freeman, B. D., Leroux, D., Betts, D. E., DeSimone, J. M., DiGiano, F. A., Stebbins, C. K. and Linton, R. W. (2002) Microphase-separated block copolymers comprising low surface energy fluorinated blocks and hydrophilic blocks: synthesis and characterization. *Macromolecules*, **35**, 3697–3707.

197. Lebreton, P., Ameduri, B., Boutevin, B. and Corpart, J. M. (2002) Use of original ω-perfluorinated dithioesters for the synthesis of well-controlled polymers by reversible addition-fragmentation chain transfer (RAFT). *Macromol. Chem. Phys.*, **203**, 522–537.

198. Severac, R., Lacroix-Desmazes, P. and Boutevin, B. (2002) Reversible addition-fragmentation chain-transfer (RAFT) copolymerization of vinylidene chloride and methyl acrylate. *Polym. Int.*, **51**, 1117–1122.

199. Bledzki, A. and Braun, D. (1981) Initiation of polymerization with substituted ethanes. 1. Polymerization of methyl methacrylate with 1,1,2,2-tetraphenyl-1,2-diphenoxyethane. *Makromol. Chem.*, **182**, 1047–1058.

200. Roussel, J. and Boutevin, B. (2001) Control of free-radical polymerization of 2,2,2-trifluoroethyl methacrylate (TEMA) by a substituted fluorinated tetraphenylethane-type INITER. *J. Fluorine Chem.*, **108**, 37–45.

201. Kim, D. K., Lee, S. B., Doh, K. S. and Nam, Y. W. (1999) Synthesis of block copolymers having perfluoroalkyl and silicone-containing side chains using diazo macroinitiator and their surface properties. *J. Appl. Polym. Sci.*, **74**, 1917–1926.

202. Destarac, M., Matyjaszewski, K., Silverman, E., Ameduri, B. and Boutevin, B. (2000) Atom transfer radical polymerization initiated with vinylidene fluoride telomers. *Macromolecules*, **33**, 4613–4616.

203. Jankova, K., Hvilsted, S., Ameduri, B. and Boutevin, B. (in preparation).

204. Choi, W. O., Sawamoto, M. and Higashimura, T. (1988) Living cationic homo- and copolymerization of vinyl ethers bearing a perfluoroalkyl pendant group. *Polym. J.*, **20**, 201–210.

205. Hopken, J., Moller, M., Lee, M. and Percec, V. (1992) Synthesis of poly(vinyl ether)s with perfluoroalkyl pendant groups. *Macromol. Chem.*, **193**, 275–284.

206. Vandooren, C., Jerome, R. and Teyssie, P. (1994) Living cationic polymerization of 1H, 1H, 2H, 2H-perfluorooctyl vinyl ether. *Polym. Bull.*, **32**, 387–393.

207. Percec, V. and Lee, M. (1992) Molecular engineering of liquid crystal polymers by living polymerization. 23. Synthesis and characterization of AB block copolymers based on ω-[(4-cyano-4'-biphenyl)-oxy]alkyl vinyl ether, 1H,1H,2H,2H-perfluorodecyl vinyl ether, and 2-(4-biphenyloxy)ethyl vinyl ether with 1H,1H,2H,2H-perfluorododecyl vinyl ether. *JMS Pure Appl. Chem.*, **A29**, 723–740.

208. Matsumoto, K., Kubota, M., Matsuoka, H. and Yamaoka, H. (1999) Water-soluble fluorine-containing amphiphilic block copolymer: synthesis and aggregation behavior in aqueous solution. *Macromolecules*, **32**, 7122–7127.

209. Matsumoto, K., Mazaki, H., Nishimura, R., Matsuoka, H. and Yamaoka, H. (2000) Perfluoroalkyl-philic character of poly(2-hydroxyethyl vinyl ether)-*block*-poly[2-(2,2,2-trifluoroethoxy)ethyl vinyl ether] micelles in water: selective solubilization of perfluorinated compounds. *Macromolecules*, **33**, 8295–8300.

210. Clark, M. R., Kendall, J. L. and DeSimone, J. M. (1997) Cationic dispersion polymerisation in liquid carbon dioxide. *Macromolecules*, **30**, 6011–6014.

211. Storojakova, H. A., Efanova, E. I. and Rachimov, A. I. (2002) Reaction of polyfluoroalcohol telomers with ε-caprolactame. *J. Appl. Chem. (Russ.)*, **75**, 1749–1751.

212. Miyamoto, M., Aoi, K. and Saegusa, T. (1989) Poly[(acylimino)alkylene] block copolymers having perfluoroalkyl hydrophobic blocks. Synthesis and surfactants properties. *Macromolecules*, **22**, 3540–3543.

213. (a) Zhang, Y. X., Da, H. A., Hogen-Esch, T. E. and Bulter, G. B. (1991) In *Water Soluble Polymers: Synthesis, Solution Properties and Appreciation*, (Eds, Shalaby, S. W., McCormik, C. L. and Butler, G. B.) ACS Symposium Series 467, American Chemical Society, Washington, DC, pp. 159–185; (b) Halperin, A., Tirell, M. and Lodge, T. P. (1992) Tethered chains in polymer microstructures. *Adv. Polym. Sci.*, **100**, pp. 31–71.

214. Bock, J., Varadaraj, R., Schultz, D. N. and Maurer, J. J. (1994) In *Macromolecular Complexes in Chemistry and Biology. Solution Properties of Hydrophobically Associating Water-Soluble Polymers* (Eds, Dubin, P., Bock, J., Davis, R., Schulz, D. N. and Thies, C.) Springer, Berlin, pp. 33–51.

215. Tuzar, Z. and Kratochvil, P. (1993) In *Surface and Colloid Science*, Vol. 15 (Ed, Matjevic, E.) Plenum Press, New York, pp. 1–83, Chap. 1.

216. Ren, Y., Lodge, T. P. and Hillmyer, M. A. (2001) A simple and mild route to highly fluorinated model polymers. *Macromolecules*, **34**, 4780–4787.

217. Oestreich, S. and Antionnetti, M. (1999) Novel fluorinated block copolymers. Synthesis and applications. In *Fluoropolymers: Synthesis*, Vol. 1 (Eds, Hougham, G., Cassidy, P. E., Johns, K. and Davidson, T.) Plenun Publ., New York, pp. 151–166, Chap. 10.

218. Boker, A., Reihs, K., Wang, J., Sadler, R. and Ober, C. K. (2000) Selectively thermally cleavable fluorinated side chain block copolymers: surface chemistry and surface properties. *Macromolecules*, **33**, 1310–1320.

219. (a) Ren, Y., Lodge, T. P. and Hillmyer, M. A. (1998) A new class of fluorinated polymers by a mild, selective, and quantitative fluorination. *J. Am. Chem. Soc.*, **120**, 6830–6831; (b) Ren, Y., Lodge, T. P. and Hillmyer, M. A. (2000) Synthesis, characterization, and interaction strengths of difluorocarbene-modified polystyrene–polyisoprene block copolymers. *Macromolecules*, **33**, 866–876.

220. Sugiyama, K., Hirao, A. and Nakahama, S. (1998) Synthesis and characterization of ABC triblock copolymers containing poly(perfluoroalkyl methacrylate) segments. *Polym Prep. (Am. Chem. Soc. Div. Polym. Chem.)*, **39**(2), 839–840.

221. Hems, W. P., Yong, T. M., Van Nunen, J. L. M., Cooper, A. I., Holmes, A. B. and Griffin, D. A. (1999) Dispersion polymerisation of methyl methacrylate in supercritical carbon dioxide—evaluation of well-defined fluorinated AB block copolymers as surfactants. *J. Mater. Chem.*, **9**, 1403–1407.

222. David, T. P., Haddleton, D. M. and Richards, S. N. (1994) Controlled polymerization of acrylates and methacrylates. *Rev. Macromol. Chem. Phys.*, **34**, 243–249.

223. Imae, T., Tabuchi, H., Funayama, K., Sato, A., Nakamura, T. and Amaya, N. (2000) Self-assemblies of block copolymers of 2-perfluorooctylethyl methacrylate and methyl methacrylate. *Colloid Surf. A: Physicochem. Engng Aspects*, **167**, 73–81.

224. Busse, K., Kressler, J., VanEck, D. and Horing, S. (2002) Synthesis of amphiphilic block copolymers based on *tert*-butyl methacrylate and 2-(-N-methylperfluorobutanesulfonami-do)ethyl methacrylate and its behavior in water. *Macromolecules*, **35**, 178–184.

225. Eleuterio, S. (1991) Scientific discovery and technological innovation: an eclectic odyssey into olefin metathesis chemistry. *J. Macromol. Sci. Chem.*, **A28**, 907–915.

226. Sogah, D. Y., Hertler, W. R., Webster, O. W. and Cohen, G. M. (1987) Group transfer polymerization. Polymerization of acrylic monomers. *Macromolecules*, **20**, 1473–1488.

227. (a) Feast, W. J. and Khosravi, E. (1999) Synthesis of fluorinated polymers via ROMP: a review. *J. Fluorine Chem.* **100**, 117–125; (b) Khosravi, E. (2002) Novel polymeric materials via ROMP using well-defined initiators. In *Ring Opening Metathesis Polymerisation and Related Chemistry* (Eds, Khosravi, E. and Szymanska-Buzar, T.) Kluwer Academic Publishers, Dordrecht, pp. 133–141.

228. (a) Krupers, M. J., Cabelo, F. M. E. and Moller, M. (1996) Surface active semifluorinated diblock copolymers prepared by group transfer polymerization. *Macromol. Symp.*, **102**, 99–106; (b) Krupers, M. J. and Moller, M. (1997) Synthesis and characterization of semifluorinated polymers via group transfer polymerization. *J. Fluorine Chem.*, **82**, 119–124.

229. Krupers, M. J. and Moller, M. (1997) Semifluorinated diblock copolymers. Synthesis, characterization and amphiphilic properties. *Macromol. Chem. Phys.*, **198**, 2163–2179.

230. Krupers, M. J., Sheiko, S. S. and Moller, M. (1998) Micellar morphology of a semifluorinated diblock copolymer. *Polym. Bull.*, **40**, 211–222.

231. Krupers, M. J., Bartelink, C. F., Grunhauer, H. J. M. and Möller, M. (1998) Permeation of rigid polyurethane foams with semifluorinated diblock copolymeric surfactants. *Polymer*, **39**, 2049–2053.

232. Yang, S., Wang, J., Ognio, K., Sundararajan, N. and Ober, C. K. (1999) Synthesis and characterization of micropatternable low surface energy methacrylic block copolymers. *Polym. Prep. (Am. Chem. Soc. Div. Polym. Sci.)*, **40**(1), 416–417.

233. Yang, S., Wang, J., Ogino, K., Valiyaveettil, S. and Ober, C. K. (2000) Low-surface-energy fluoromethacrylate block copolymers with patternable elements. *Chem. Mater.*, **12**, 33–40.

234. Trabelsi, H., Geribaldi, S., Cambon, A., Lowe, K. C. and Edwards, C. M. Esters per(poly)fluorés poloxyéthyles et leurs précurseurs. Monaco Patent MC 2418, 05-02-1999.

235. Boschet, F., Branger, C., Margaillan, A. and Condamine, E. (2002) Synthesis, characterization and aqueous behaviour of a one-ended perfluorocarbon-modified poly(ethylene glycol). *Polymer*, **43**, 5329–5334.

236. Zhang, H., Hogen-Hesch, T. E., Boschet, F. and Margaillan, A. (1998) Complex formation of β-cyclodextrin- and perfluorocarbon-modified water-soluble polymers. *Langmuir*, **14**, 4972–4977.

237. Hwang, F. S. and Hogen-Hesch, T. E. (1993) Fluorocarbon modified water-soluble cellulose derivatives. *Macromolecules*, **26**, 3156–3160.

238. Hogen-Hesch, T. E. and Amis, E. J. (1995) Hydrophobic association in perfluorocarbon-containing water-soluble polymers. *Trends Polym. Sci.*, **3**, 98–104.

239. Zhang, H., Pan, J. and Hogen-Hesch, T. E. (1998) Synthesis and characterization of a one-ended perfluorocarbon-functionalized derivatives of poly(ethylene glycol)s. *Macromolecules*, **31**, 2815–2821.

240. Berret, J. F., Calvet, D., Collet, A. and Viguier, M. (2003) Fluorocarbon associative polymers. *Curr. Opin. Colloid Interface Sci.*, **8**, 296–306.

241. Ming, W., Lou, X., Van de Grampel, R. D., Van Dongen, J. L. J. and van der Linde, R. (2001) Partial fluorination of hydroxyl end-capped oligoesters revealed by MALDI-TOF mass spectrometry. *Macromolecules*, **34**, 2389–2393.

242. Xie, D., Tomczak, S. and Hogen-Hesch, T. E. (2001) Synthesis and hydrophobic association of poly(*N,N*-dimethylacrylamide) end-functionalized with perfluorocarbon and hydrocarbon groups. *J. Polym. Sci., Part A, Polym. Chem.*, **39**, 1403–1418.

243. Zhao, C. X., Zhou, R. M., Pan, H. Q., Qu, Y. L. and Jiang, X. K. (1982) Synthesis of fluorinated diacyl peroxides. *J. Org. Chem.*, **47**, 2009–2018.

244. Burgos Paci, M. A., Arguelo, G. A., Garcia, P. and Willner, H. (2003) Kinetics of the thermal decomposition of bis(trifluoromethyl) peroxydicarbonate, $CF_3OC(O)OOC(O)OCF_3$. *Int. J. Chem. Kinet.*, **35**, 15–19.

245. Guan, C. L., Chen, L., Deng, C. H. and Zhao, C. X. (2003) Synthesis and characterization of perfluoro[1-(2-fluorosulfonyl)ethoxy] ethyl end-capped styrene oligomers. *J. Fluorine Chem.*, **119**, 97–100.

246. Bessiere, J. M., Boutevin, B. and Loubet, O. (1993) Synthesis of a perfluorinated azo initiator. Application to the determination of the recombination/disproportionation rate during the polymerisation of styrene. *Eur. Polym. J.*, **31**, 673–677.

247. Loubet, O. (1991) Synthèse, caractérisations et applications d'amorceurs de polymérisation radicalaire de type diazoique, PhD Dissertation, University of Montpellier.

248. Bessiere, J. M., Boutevin, B. and Loubet, O. (1993) Determination of kinetic parameters for isothermal decomposition of azo initiators of polymerisation by differential scanning calorimetry. *Polym. Bull.*, **30**, 545–549.

249. Sawada, H., Minoshina, Y. and Nakajima, H. (1992) Reactions of acrylic acid with fluoroalkanoyl peroxides—the formation of acrylic acid oligomers containing two fluoroalkylated end groups. *J. Fluorine Chem.*, **65**, 169–176.

250. Sawada, H. (1993) Synthesis of perfluoro-oxa-alkylated compounds by the use of perfluoro-oxa-alkanoyl peroxides and their applications. *J. Fluorine Chem.*, **61**, 253–272.

251. Sawada, H., Minoshima, Y. and Nakajima, H. (1993) Reactions of acrylic acid with fluoroalkanoyl peroxides—the formation of acrylic acid oligomers containing two fluoroalkylated end-groups. *J. Fluorine Chem.*, **65**, 169–173.

252. Sawada, H., Sumino, E., Oue, M., Baba, M., Kira, T., Shigeta, S., Mitani, M., Nakajima, H., Nishida, M. and Moriya, Y. (1995) Synthesis and surfactant properties of novel acrylic acid oligomers containing perfluoro-oxa-alkylene units: an approach to anti-human immunodeficiency virus type-1 agents. *J. Fluorine Chem.*, **74**, 21–26.

253. Sawada, H., Ohashi, A., Oue, M., Baba, M., Abe, M., Mitani, M. and Nakajima, H. (1995) Synthesis and surfactant properties of fluoroalkylated acrylic acid co-oligomers containing dimethylsilicone segments as potential inhibitors of HIV. *J. Fluorine Chem.*, **75**, 121–129.

254. Sawada, H. (1996) Fluorinated peroxides. *Chem. Rev.*, **96**, 1779–1808.

255. Sawada, H., Kita, H., Yoshimizu, M., Kyokane, J., Kawase, T., Hayakawa, Y. and Yoshino, K. (1997) Direct aromatic fluoroalkylations of poly(*p*-phenylene) with fluoroalkanoyl peroxides: an approach to highly soluble fluorinated conducting polymers. *J. Fluorine Chem.*, **82**, 51–54.

256. Sawada, H., Tanimura, T., Katayama, S. and Kawase, T. (1997) Aggregation of fluoroalkyl units: synthesis of gelling fluoroalkylated end-capped oligomers containing hydroxy segments possessing metal ion binding and releasing abilities. *Chem. Commun.*, 1391–1392.

257. Sawada, H., Tanimura, T., Katayama, S., Kawase, T., Tomita, T. and Baba, M. (1998) Synthesis and properties of gelling fluoroalkylated end-capped oligomers containing hydroxy segments. *Polym. J.*, **30**, 797–804.

258. Sawada, H. (2000) Architecture and applications of novel self-assembled aggregates of fluoroalkyl-end-capped oligomers. *J. Fluorine Chem.*, **101**, 315–324.

259. Sawada, H. (2000) Chemistry of fluoroalkanoyl peroxide, 1980–1998. *J. Fluorine Chem.*, **105**, 219–220.

260. Sawada, H., Ikeno, K. and Kawase, T. (2002) Synthesis of amphiphilic fluoroalkoxy end-capped cooligomers containing oxime-blocked isocyanato segments: architecture and applications of new self-assembled fluorinated molecular aggregates. *Macromolecules*, **35**, 4306–4313.

261. Sawada, H. (2003) Novel self-assembled molecular aggregates formed by fluoroalkyl end-capped oligomers and their application. *J. Fluorine Chem.*, **121**, 111–130.

262. Rice, D. E. and Sandberg, C. L. (1969) Carboxy- and ester-terminated copolymers of vinylidene fluoride and hexafluoropropene. US Patent 3,457,245 (assigned to 3M).

263. Rice, D. E. (1969) Bis(ω-substituted perfluoroacyl)peroxides, US Patent 3,461,155 (assigned to 3M).

264. Rice, D. E. and Sandberg, C. L. (1971) Functionally-terminated copolymers of vinylidene fluoride and hexafluoropropene. *Polym. Prep. (Am. Chem. Soc. Div. Polym. Sci.)*, **12**, 396–397.

265. Lebreton, P. and Boutevin, B. (2000) Primary radical termination and unimolecular termination in the heterogeneous polymerization of acrylamide initiated by a fluorinated azo-derivative initiator: kinetic study. *J. Polym. Sci., Part A, Polym. Chem.*, **38**, 1834–1843.

266. Mukhopadhyay, S., Mitra, B. B. and Palit, S. R. (1971) Estimation of endgroups in acrylamide polymers by the reverse dye partition technique. *Makromol. Chem.*, **141**, 55–61.

267. Lebreton, P., Boutevin, B., Gramain, P. and Corpart, J. M. (1999) Synthesis, characterization and associative properties of triblock and diblock perfluorinated poly(acrylamide)s. *Polym. Bull.*, **43**, 59–66.

268. (a) Boutevin, B., Lebreton, P., Collette, C. and Garcia, G. (1996) Polymère fluoré triblock et polysaccharide. French Patent 96 03,532 (assigned to Elf Atochem) and Corpart, J. M., Collette, C., Boutevin, B., Ciampa, R. (1998) Water-soluble triblock polymers forming associations, US Patent 5,798,421 (assigned to Elf Atochem).

269. Inoue, H., Matsumoto, A., Matsukawa, K. and Ueda, A. (1990) Surface characteristics of fluoroalkylsilicone-poly(methyl methacrylate) block copolymers and their PMMA blends. *J. Appl. Polym. Sci.*, **40**, 1917–1938.

270. Tortelli, V. and Tonelli, C. (1990) Telomerization of tetrafluoroethylene and hexafluoropropene: synthesis of diiodoperfluoroalkanes. *J. Fluorine Chem.*, **47**, 199–210.

271. Baum, K. and Malik, A. A. (1994) Difunctional monomers based on perfluoropropylene telomers. *J. Org. Chem.*, **59**, 6804–6809.

272. Baum, K., Archibald, T. G. and Malik, A. A. (1993) Polyfluorinated, branched-chain diols and diisocyanates, and fluorinated polyurethanes prepared therefrom, US Patent 5,204,441, (assigned to Fluorochem).

273. Manseri, A., Ameduri, B., Boutevin, B., Chambers, R. D., Caporiccio, G. and Wright, A. P. (1995) Synthesis of fluorinated telomers. Part 4. Telomerization of vinylidene fluoride with commercially available α,ω-diiodoperfluoroalkanes. *J. Fluorine Chem.*, **74**, 59–67.

274. Manseri, A., Ameduri, B., Boutevin, B., Chambers, R. D., Caporiccio, G. and Wright, A. P. (1996) Unexpected telomerization of hexafluoropropene with asymmetrical halogenated telechelic telogens. *J. Fluorine Chem.*, **78**, 145–150.

275. Boulahia, D., Manseri, A., Ameduri, B., Boutevin, B. and Caporiccio, G. (1999) Synthesis of fluorinated telomers. Part 6. Telomerization of hexafluoropropene with α,ω-diiodoperfluoroalkanes. *J. Fluorine Chem.*, **94**, 175–182.

276. Ameduri, B. and Boutevin, B. (1999) Use of telechelic fluorinated diiodides to obtain well-defined fluoropolymers. *J. Fluorine Chem.*, **100**, 97–135.

277. Manseri, A., Ameduri, B., Boutevin, B. and Caporiccio, G. (1997) Synthesis of telechelic dienes from fluorinated α,ω-diiodoalkanes. II. Divinyl and diallyl compounds containing tetrafluoroethylene, vinylidene fluoride and hexafluoropropene moieties. *J. Fluorine Chem.*, **81**, 103–113.

278. Ameduri, B., Boutevin, B., Caporiccio, G., Guida-Pietrasanta, F. and Ratsimihety, A. (1999) Use of fluorinated telomers in the preparation of hybrid silicones. In *Fluoropolymers: Synthesis*, Vol. 1 (Eds, Hougham, G., Davidson, T., Cassidy, P. and Johns, K.) Plenum, New York, Chap. 5, pp. 67–80.

279. (a) Zhang, Z. B., Shi, Z. K. and Ying, S. (1998) Synthesis of semifluorinated triblock copolymers by atom transfer radical polymerization (ATRP). *Polym. Prep. (Am. Chem. Soc. Polym. Div.)*, **39**, 820–821; (b) Zhang, Z. B., Ying, S. K., Hu, Q. H. and Xu, X. D. (2002) Semifluorinated ABA triblock copolymers: synthesis, characterization, and amphiphilic properties. *J. Appl. Polym. Sci.*, **83**, 2625–2633.

280. Brace, N. O. (1962) Radical addition of iodoperfluoroalkanes to vinyl and allyl monomers. *J. Org. Chem.*, **27**, 3033–3038.

281. Tatemoto, M. (1989) Preparation and uses of silicon-containing polymers, Eur. Patent 343,526, (assigned to Daikin Ind. Ltd) 29-11-1989.

282. Tatemoto, M. (1990) Fluororubber blend compositions for adhesives and sealants. Eur. Patent 399,543 (assigned to Daikin Ind. Ltd) 28-11-1990.

283. Arrigoni, S., Apostolo, M. and Arcella, V. (2002) Fluoropolymers architecture design: the branching and pseudo-living technology. *Trends Polym. Sci.*, **7**, 23–40.

284. Arcella, V., Brinati, G. and Apostolo, M. (1997) New high performance fluoroelastomers. *Chim. Oggi*, **79**, 345–351.

285. Brinati, G. and Arcella, V. (2001) Branching and pseudo-living technology in the synthesis of high performance fluoroelastomers. *Rubber World*, **224**, 27–32.

286. Albano, M., Tortelli, V., Brinati, G. and Arcella, V. (2001) Peroxide curable fluoroelastomers, particularly suitable for manufacturing O-rings, US Patent 5,625,019 (assigned to Ausimont).

287. Tatemoto, M. and Yagi, T. Thermoplastic fluoroelastomer denture base, Eur. Patent 268,157 (assigned to Daikin Ind. Ltd) 25-05-1988.

288. Cheng, T. C., Mehan, A. K., Weber, C. J., Schwartz, L. and Taft, D. D. Polymeric blends, Eur. Patent 0,524,700 A1 (assigned to Raychem Corp.) 27-01-1993.

289. Brinati, G. and Arcella, V. Fluorinated thermoplastic elastomers, Eur. Patent 0,924,257 A1 (assigned to Ausimont S.p.A.) 15-12-1997.

290. Schulte, A., Hallmark, V. M., Tweig, R., Song, K. and Rabolt, J. F. (1991) Structural studies of semifluorinated *n*-alkanes. 4. Synthesis and characterization of $F(CF_2)_nC_6H_4(CF_2)_nF$ by conventional and Fourrier Raman spectroscopy. *Macromolecules*, **24**, 3901–3905.

291. Fielding, H. C. (1979) In *Organofluorine Chemicals and Their Industrial Applications* (Ed, Banks, R. E.) The Society of Chemical Industry, London, pp. 214–238, Chap. 11.

292. Ziska, D. (1965) Oil and water repellent finishes of natural and manmade-fibres fabrics. *Prace Inst. Wlokiennicktwa*, **15**, 73–85 (*Chem. Abstr.* 66, 19696r (1967)).

293. Pearson, W. M. and Crosby, T. S. (1970) Air-permeable protective materials, US Patent 3,502, 537 (assigned to Bondina BDA Ltd).

294. Smith, H. A. (1971) Cloth treating process and composition, US Patent 3,558,549 (assigned to Dow Chemical Co.).

295. (a) Sherman, P. O., Smith, S. and Johannessen, B. (1969) British Patent 1,215,861 (1967) and Textile characteristics affecting the release of soil during laundering. II. Fluorochemical soil-release textile finishes. *Textile Res. J.*, **39**, 449–456; (b) Sherman, P. O. and Smith, S. Oleophobic and hydrophobic or hydrophilic polymers for textile finishing. French Patent 1,562,070 (1969), US Patents 3,574,791 and 3,728,151, (assigned to 3M).

296. Akemi, H., Aoyagi, T., Shinohara, I., Okamo, T., Kakaota, K. and Sakurai, Y. (1986) Synthesis of a new antithrombogenic block copolymer containing fluorinated segments: poly(nonafluoro-hexylacrylate-b-styrene). *Makromol. Chem.*, **187**, 1627–1638.

297. Drouinaud, R., Ritcey, A. and Ameduri, B. (in preparation).

298. (a) Ying, S., Zhang, Z. and Shi, Z. (1998) Synthesis of ABA triblock copolymers containing both fluorocarbon and hydrocarbon blocks for compatibilizers. *Polym. Prep. (Am. Chem. Soc., Div. Polym. Chem.)*, **39**(2), 843–844; (b) Zhang, Z., Ying, S. and Shi, Z. (1999) Synthesis of fluorine-containing block copolymers via ATRP. 1. Synthesis and characterization of PSt–PVDF–PSt triblock copolymers. *Polymer*, **40**, 1341–1345.

299. Zhang, Z., Ying, S. and Shi, Z. (1999) Synthesis of fluorine-containing block copolymers via ATRP 2. Synthesis and characterization of semifluorinated di- and triblock copolymers. *Polymer*, **40**, 1341–1344.

300. Modena, S., Pianca, M., Tato, M. and Moggi, G. (1989) Radical telomerization of vinylidene fluoride in the presence of 1,2-dibromotetrafluoroethane. *J. Fluorine Chem.*, **43**, 15–24.

301. Busse, K., Hussain, H., Budde, H., Horing, S. and Kressler, J. (2002) Micelle formation of perfluorinated triblock copolymer in water. *Polym. Prep. (Am. Chem. Soc., Div. Polym. Chem.)*, **43**(2), 366–367.

302. Hussain, H., Budde, H., Höring, S., Busse, K. and Kressler, J. (2002) Synthesis and characterization of poly(ethylene oxide) and poly(perfluorohexylethyl methacryalte) containing triblock copolymers. *Macromol. Chem. Phys.*, **203**, 2103–2112.

303. Hussain, H., Busse, K. and Ressler, J. (2003) Poly(ethylene oxide) and poly(perfluorohexylethyl methacrylate) containing amphiphilic block copolymers: associative properties in aqueous solution. *Macromol. Chem. Phys.*, **204**, 936–946.

304. Guyot, B., Ameduri, B., Boutevin, B., Melas, M., Viguier, M. and Collet, A. (1998) Kinetics of homopolymerization of fluorinated acrylates—5—influence of the spacer between the fluorinated chain and the ester group. *Macromol. Chem. Phys.*, **199**, 1879–1885.

305. Li, Y., Zhang, W., Wang, J., Zhang, Y. and Huang, J. (2002) The synthesis of fluorine and silicone containing triblock copolymers, Proceedings of the International Conference on the Polymer Synthesis, Warwick (UK), July 29–August 1, p. 124.

306. Weberskirch, R., Preuschen, J., Spiess, H. W. and Nuyken, O. (2000) Design and synthesis of a two compartment micellar system based on the self association behavior of poly(*N*-acylethyleneimine) end-capped with a fluorocarbon and a hydrocarbon chain. *Macromol. Chem. Phys.*, **201**, 995–1007.

307. Pilati, F., Toselli, M., Vallieri, A. and Tonelli, C. (1992) Synthesis of polyesters-perfluoropolyethers block copolymers 3. Use of various telechelic perfluoropolyethers. *Polym. Bull.*, **28**, 151–157.

308. Toselli, M., Pilati, F., Fusari, M., Tonelli, C. and Castiglioni, C. (1994) Fluorinated poly(butylene terephthalate): preparation and properties. *J. Appl. Polym. Sci.*, **54**, 2101–2106.

309. (a) Pilati, F., Toselli, M., Messori, M., Priola, A., Bongiovanni, R., Malucelli, G. and Tonelli, C. (1999) Poly(ε-caprolactone)-poly(fluoroalkylene oxide)-poly(ε-caprolactone) block copolymers. 1. Synthesis and molecular characterization. *Macromolecules*, **32**, 6969–6976; (b) Pilati, F., Toselli, M., Messori, M., Priola, A., Bongiovanni, R., Malucelli, G. and Tonelli, C. (2001) Thermal surface properties. *Polymer*, **43**, 1771–1779.

310. Messori, M., Toselli, M., Pilati, F. and Fabbri, P. (2001) Poly(ε-caprolactone)-poly(fluoroalkylene oxide)-poly(ε-caprolactone) block copolymers as surface modifiers of poly(vinyl chloride), Fluorine in Coatings IV Conference, Brussels, March 5–7, paper #11.

311. Djebar, A., Reibel, L. and Franta, E. (1996) Synthesis of triblock copolymers made of a fluorinated central block and two outside poly(1,3-dioxolane) blocks. *Macromol. Symp.*, **107**, 219–226.

312. Glass, J. E. (1994) *Polymer in Aqueous Media: Performance Through Association*, American Chemical Society, Washington, DC.

313. DeBons, F. E. and Braun, R. W. (1995) Polymer flooding: still a viable improve oil recovery technique, Eighth European IOR Symposium, Vienna, Austria.

314. Shalaby, W., Mac Cormick, C. L. and Butler, G. B. (1991) *Water Soluble Polymers: Synthesis, Properties and Applications*, American Chemical Society, Washington, DC.

315. Hartman, P., Collet, A. and Viguier, M. (1999) Synthesis and characterization of model fluoroacylated poly(ethylene oxide). *J. Fluorine Chem.*, **95**, 145–151.

316. Xu, B., Li, L., Yekta, A., Masoumi, Z., Kanagalingam, S., Winnick, M. A., Zhang, K., MacDonald, P. and Menchen, S. (1997) Synthesis, characterization and rheological behavior of polyethylene glycol end-capped with fluorocarbon hydrophobes. *Langmuir*, **13**, 2447–2456.

317. Brandrup, J. and Immergut, E. H. (1992) *Polymer Handbook*, Wiley, New York.

318. Devanand, K. and Selser, J. C. (1991) Asymptotic behavior and long-range interactions in aqueous solutions of poly(ethylene oxide). *Macromolecules*, **24**, 5943–5947.

319. Preuschen, J., Menchen, S., Winnick, M. A., Heuer, A. and Spiess, H. W. (1999) Aggregation behavior of a symmetric, fluorinated, telechelic polymer system studied by 19F NMR relaxation. *Macromolecules*, **32**, 2690–2695.

320. Cathebras, N., Collet, A., Viguier, M. and Berret, J. F. (1998) Synthesis and linear viscoelasticity of fluorinated hydrophobically modified ethoxylated urethanes. *Macromolecules*, **31**, 1305–1311.

321. Calvet, D., Collet, A., Viguier, M., Berret, J. F. and Serero, Y. (2003) Perfluoroalkyl end-capped poly(ethylene oxide). Synthesis, characterization, and rheological behavior in aqueous solution. *Macromolecules*, **36**, 449–457.

322. Serreno, Y., Aznar, R., Porte, G. and Berret, J. F. (1998) Associating polymers: from "flowers" to transient networks. *Phys. Rev. Lett.*, **81**, 5584–5587.

323. Zhou, J., Zhuang, D., Yuan, X., Jiang, M. and Zhang, Y. (2000) Association of fluorocarbon and hydrocarbon end-capped poly(ethylene glycol)s: NMR and fluorescence studies. *Langmuir*, **16**, 9653–9661.

324. Boschet, F., Branger, C. and Margaillan, A. (2003) Synthesis and characterization of PS-block-PEO associative water-soluble polymers. *Eur. Polym. J.*, **39**, 333–339.

325. Boschet, F. (2001) Synthèse et caractérisation de nouveaux poly(oxyde d'éthylène)s associatifs fluorés, PhD Dissertation, Université de Toulon et du Pays du Var.

326. Yi, G. R., Manoharan, V. N., Klein, S., Brzezinska, K. R., Pine, D. J., Lange, F. F. and Yang, S. M. (2002) Monodisperse micrometer-scale spherical assemblies of polymer particles. *Adv. Mater.*, **14**, 1137–1140.

327. Ameduri, B., Boutevin, B., Nouiri, M. and Talbi, M. (1995) Synthesis and properties of fluorosilicon-containing polybutadienes by hydrosilylation of fluorinated hydrogenosilanes—Part 1—preparation of the silylation agents. *J. Fluorine Chem.*, **74**, 191–198.

328. Kim, Y. K. (1972) Siloxane elastomers. US Patent 3,542,830 (assigned to Dow-Corning).

329. Kim, Y. K., Loree, L. A. and Pierce, O. R. (1972) Fluorinated silicone, US Patent 3,647,740 (assigned to Dow-Corning) 07-03-1972.

330. Riley, M. O. and Kim, Y. K. (1977) The synthesis of fluoroether–fluorosilicone hybrid polymers. *J. Fluorine Chem.*, **10**, 85–110.

331. Kim, Y. K. (1974) Low temperature fluorosilicone compositions, US Patent 3,818,064 (assigned to Dow-Corning) 18-06-1974.

332. Riley, M. O., Kim, Y. K. and Pierce, O. R. (1978) Synthesis and property comparisons of random and alternating hybrid fluorocarbon-fluorosilicone copolymers. *J. Polym. Sci., Polym. Chem. Ed.*, **16**, 1929–1941.

333. Boutevin, B., Guida-Pietrasanta, F., Ratsimihety, A. and Caporiccio, G. (1996) Method for preparing a hybrid organodisilanol and polymers thereof, US Patent 5,527,933 (18-06-1996) (assigned to Dow Corning).

334. Boutevin, B., Caporiccio, G., Guida-Pietrasanta, F. and Ratsimihety, A. (1997) Synthesis, characterization and thermal properties of a new fluorinated silalkylene-polysiloxane. *Main Group Metal Chem.*, **20**, 133–139.

335. Boutevin, B., Caporiccio, G., Guida-Pietrasanta, F. and Ratsimihety, A. (1997) Hybrid fluorinated silicones. Synthesis and thermal properties of new hybrid homopolymers. *Recent Res. Dev. Int. Polym. Sci.*, **1**, 241–248.

336. (a) Boutevin, B., Caporiccio, G., Guida-Pietrasanta, F. and Ratsimihety, A. (1998) Hybrid fluorinated silicones. Part II. Synthesis of homopolymers and copolymers. Comparison of their thermal properties. *Macromol. Chem. Phys.*, **199**, 61–70; (b) Boutevin, B., Caporiccio, G., Guida-Pietrasanta, F. and Ratsimihety, A. (2003) Poly-silafluoroalkyleneoligosiloxanes: a class of fluoroelastomers with low glass transition temperature. *J. Fluorine Chem.*, **124**, 131–138.

337. Boutevin, B., Caporiccio, G., Guida-Pietrasanta, F. and Ratsimihety, A. (1999) Fluoroelastomers for a wide temperature range of applications. Eur. Patent Appl. 1,097,958 A, (assigned to Daikin) 3-11-1999.

338. Thomas, T. H. and Hendrick, T. C. (1969) Thermal analysis of poly(dimethylsiloxanes). I. Thermal degradation in controlled atmospheres. *J. Polym. Sci.*, **A7**, 537–549.

339. Boutevin, B., Boutevin, G. and Loubat, C. Nouveau procédé de synthese de silicone fluoré hybride. French Patent under deposit (assigned to Specific Polymer).

340. Okawa, T. (2002) Preparation of 1,3-dihydroxytetramethyldisiloxane, Eur. Patent Appl. 1,146, 048 A2 (assigned to Dow-Corning Toray Silicone) 14-04-2002.

341. Scheirs, J. (1997) Perfluoropolyethers: synthesis, characterization and applications. In *Modern Fluoropolymers* (Ed, Scheirs, J.) Wiley, New York, pp. 435–486, Chap. 24.

342. Osawa, Y., Sato, S. and Matsuda, T. (1999) Crosslinkable fluororubber compositions, their manufacture and molding property, Eur. Patent Appl. E.P. 0,967,251 A_1 (assigned to Shin-Etsu) 29-12-1999.

343. Shoda, M., Akagawa, K. and Tomofuji, T. (1997) Infrared solid-state image sensing apparatus, US Patent 5,635,738 (assigned to Shin-Etsu) 30-06-1997.

344. Ato, S., Arai, M., Osawa, Y. and Sato, M. (2003) Curable fluoro polyether rubber compositions and aircraft sealing parts made therefrom, Eur. Patent 1,288,243 A1 (assigned to Shin-Etsu) 05-03-2003.

345. Uritani, P. Liquid perfluoroether elastomers. "High performance elastomers" Conference, November 13–14, 2002, Koln (Germany), paper #14.

346. Kobayashi, H. Fluorine-containing organosilicon compounds and their preparation, Eur. Patent Appl. 0,690,088 A1 (assigned to Dow-Corning Toray) 03-01-1996.

347. Kobayashi, H. Organic fluorosilicone alternating copolymer and its manufacturing method, Eur. Patent 0,665,270 A1 (assigned to Dow-Corning Toray) 02-08-1995.

348. Kobayashi, H. Fluorine-containing organosilicon copolymers, Eur. Patent 0,702,048 A1 (assigned to Dow-Corning Toray) 20-03-1996.

349. Pittman, A. G. (1972) Surface properties of fluorocarbon polymers. In *Fluoropolymers*, Vol. XXV (Ed, Wall, L. A.) Wiley, New York, pp. 419–449, Chap. 13.

350. Kobayashi, H. and Owen, M. J. (1995) Surface properties of fluorosilicones. *Trends Polym. Sci.*, **3**, 330–335.

351. Green, J. W., Rubal, M. J., Osman, B. M., Welsch, R. L., Cassidy, P. E., Fitch, J. W. and Blanda, M. T. (2000) Silicon-organic hybrid polymers and composites prepared in supercritical carbon dioxide. *Polym. Adv. Technol.*, **11**, 820–825.

352. Zhou, H., Venumbaka, S. R., Fitch, J. W. III and Cassidy, P. E. (2003) New poly(silanes-siloxanes) via hydrosilation in supercritical CO_2 and subsequent crosslinking. *Macromol. Symp.*, **192**, 115–121.

353. Hergenrother, P. M. (1996) Polymeric materials for high-speed civil transports. *Trends Polym. Sci.*, **4**, 104–105.

Chapter 5

SYNTHESIS, PROPERTIES AND APPLICATIONS OF FLUORINATED GRAFT COPOLYMERS

1. Introduction

As in the case of block copolymers (Chapter 4), graft copolymers are well-defined copolymers that have already demonstrated relevant properties and hence have been used in many applications (emulsifiers for plastics, hot melt, adhesives, IEMs, impact resistance additives, etc.).

Although hydrogenated graft copolymers are numerous and have led to many contributions in various books [1, 2], reviews [3–5], patents and publications, the number of articles regarding the syntheses, properties and applications of fluorinated graft copolymers has been increasing over the last 5–10 years.

It is well known that heterogeneous (i.e., two phases or more) graft copolymers tend to show the properties of both (or more) the polymeric backbone and the oligomeric or polymeric grafts rather than averaging the homopolymer properties. Not only graft copolymers have properties different from those of homopolymers, random copolymers and polyblends, but the properties of graft copolymers themselves differ depending upon the compositions and the arrangement of the homopolymer sequences. Nevertheless, the overall properties of graft copolymers are similar to those of block copolymers (the fluorinated ones have been reported in Chapter 4). A two-phase graft copolymer is expected to exhibit good miscibility of each homopolymer and have the desirable property of being located at the interface of both phases. Therefore, graft copolymers have attracted peculiar attention (i) in the field of surface modification of various polymers and also (ii) in the improvement of the miscibility of immiscible homopolymers [1, 3–5]. Indeed, the size of the phases of both homopolymers is reduced, enabling better compatibility.

Fluorine-containing graft copolymers having suitable miscibility for the surfaces to be treated have important applications as additives for modifying surface properties in the fields of coatings, adhesives, films, fibres and mouldings because of their miscibility and extremely low surface energy. However, random copolymers containing fluoropolymers are more widely used than graft copolymers as surface modification agents at present because there are many difficulties in preparing well-defined graft copolymers exhibiting both good miscibility and excellent surface properties.

We have found it useful to divide this chapter into five major parts.

Section 1 reports the synthesis and properties of fluorinated graft copolymers obtained by direct co/terpolymerisation of monomers with macromonomers. Section 2 deals with the chemical change of oligomers or polymers. This chemical modification can be carried out by direct reaction of an oligomeric species onto a reactive group. Section 3 describes the telomerisation of M_2 monomer with a macromolecular transfer agent. Section 4 consists in synthesising a macroinitiator (bearing the initiator group as a side function) and using it for the polymerisation of a M_2 monomer. The macroinitiator can be prepared by an activation of P_1 homopolymer. Section 5 concerns the transfer to the polymer in the course of the polymerisation of M_2 monomer.

Basically, the different ways leading to graft copolymers can be summarised as follows, M_2 monomer bringing P_2 grafts in the corresponding reaction:

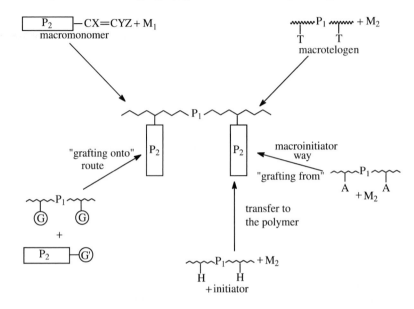

T : transfering group
A : initiator group

Each route is detailed more fully in the separate sections which follow, illustrated by various examples.

2.   Synthesis, Properties and Applications of Fluorinated Graft Copolymers

2.1.   FROM THE MACROMONOMER METHOD

2.1.1.   Introduction

This is a useful method for synthesising graft copolymers arising from the macromonomer technique. A macromonomer is an oligomer or a telomer (see Chapter 1) with an end-group able to polymerise or copolymerise with comonomers to form comb-type graft copolymers. Such a strategy is well known and extensively described in the excellent chapter by Chujo and Yamashita [5] in which many routes have been suggested, initiated by ionic and free radical polymerisations [5–8]. The resulting macromonomer can be (co)polymerised with a backbone-forming comonomer using free radical, cationic, anionic or insertion polymerisation methods as presented below.

In this sub-section, we report the methods for preparing fluorine-containing graft copolymers from fluorinated macromonomers (F-macromonomers) and, first of all, we think it necessary to list several kinds of these F-macromonomers. The precursors of fluorinated macromonomers have been prepared either by electrochemical fluorination (ECF) (especially from the Simon's procedure [9]) or by telomerisation [10–12]. These processes differ on the number of carbon atoms in the perfluorinated chain (even and odd numbers are obtained for the telomerisation and ECF, respectively). Although this latter reaction is extensively reported in Chapter 1, it is worth summarising it as in Scheme 1 (first step).

The telomerisation of tetrafluoroethylene (TFE) with C_2F_5I leads to $C_nF_{2n+1}I$ (n as an odd integer). C_2F_5I is produced from the addition of [IF] (generated from the reaction between IF_5 and I_2) onto TFE. This process is nowadays industrially scaled-up by Daikin, Asahi, Du Pont (which recently bought the corresponding branch of Atofina) and Solvay-Solexis (formerly Ausimont). These perfluoroalkyl iodides are very interesting reactants efficient in both radical additions onto unsaturated derivatives or in telomerisation of (fluoro)monomers (see Chapter 1) and are precursors of many intermediates or products [13]. The radical ethylenation of $C_nF_{2n+1}I$ is rather selective, offering $C_nF_{2n+1}C_2H_4I$ in high yields. This latter can be modified into various functional TFE-containing intermediates as alcohols, acids and amines reported by Brace [14] or as $R_FCH=CH_2$ [15].

The oxidation of $R_FC_2H_4I$ into the hydroxyl homologues was carried out by many authors and this process is also industrially scaled-up, although a non-negligible amount of the corresponding $R_FCH=CH_2$ is produced. Other syntheses of fluorinated alcohols are well documented in Chapters 1 and 2 and in Ref. [13].

Interestingly, tremendous possibilities of various F-macromonomers containing an oligo(TFE) chain and hence graft copolymers have been offered (Scheme 1) and are briefly mentioned below.

2.1.2. Fluorinated Epoxides and Oxetanes Containing TFE Base Units

The fluorinated epoxides [16–18] and oxetanes [19–23], the ring opening polymerisations of which have already been mentioned in Chapter 2 (F-telechelics), led to fluorinated polyethers of structure:

$$-[CH_2-O-\underset{\underset{R_F}{|}}{\overset{\overset{R_1}{|}}{C}}-(CH_2)_x]_m-$$

where $x = 0$ (from fluoroepoxide) or 1 (from oxetane).

$R_1 = H$ (from epoxide), CH_2Cl or $CH_2-O-C_2H_4C_pF_{2p+1}$ (from oxetane $p = 6, 8$) and $R_F = CH_2C_pF_{2p+1}$ (from epoxide) or $CH_2-O-C_2H_4C_pF_{2p+1}$ (from oxetane).

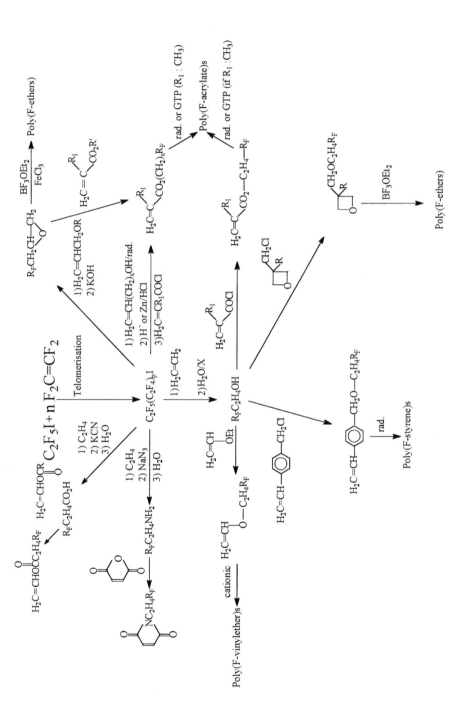

SCHEME 1. Various routes to obtain fluorinated epoxides, oxetanes, vinyl ethers, vinyl esters, styrenics, maleimides, and (meth)acrylates from tetrafluoroethylene (TFE) telomers (R_F stands for C_nF_{2n+1}, $n = 4, 6, 8, 10$).

2.1.3. Fluorinated Styrenic Macromonomers Containing TFE Base Units

The ω-perfluorinated side chain has improved the surface properties and this was clearly seen by Moller's group [24, 25] from fluorinated styrenes containing an ether [24, 25] or thioether [25] bridge in the spacer between the perfluorinated chain-end and the aromatic ring. $H_2C=CH$—⬡—$CH_2XC_2H_4-C_nF_{2n+1}$ (with $n = 6$, 8 and $X = O$ or S). Both these bridges do not influence the polymerisability of these monomers, and the properties of the resulting polymers were investigated in terms of free surface energy ($\gamma = 11-17$ mN m^{-1}) and thermal properties (the thermal stability was worse than that of polystyrene (PS)). These PS bearing fluorinated lateral groups can find a wide range of applications in coatings, optical fibres, perm selective membranes, gels for GPC (or SEC), as explained in Ref. [25], or as original surfactants for polymerisations in supercritical (sc) CO_2 [26].

In addition, although PS could be chain functionalised by one, two, three or four perfluorooctyl groups and their surface energy investigated [27], the breakthrough of these fluorinated styrenes lies in their *pseudo* living radical polymerisation controlled by Tempo [26, 28] to achieve block copolymers bearing lateral fluorinated chains (more information is supplied in Chapter 4, Section 2.1.2.2.).

Moreover, the hydrosilylation of these fluorostyrenes with silanes was successfully achieved [29].

2.1.4. Fluorinated Vinyl Ethers Containing TFE Base Units

The synthesis [30] and the cationic polymerisation of vinyl ethers containing a perfluorinated chain-end ($H_2C=CHO(CH_2)_x-C_nF_{2n+1}$ with $x = 2$, 3 [30–36] or 11 [25a]; $n = 6$, 8) was accomplished by various authors [31–36] and also led to interesting block copolymers (see Chapter 4, Sections 2.1.4.1.1 and 2.2.4.1) by sequential copolymerisation. In addition, these fluorinated monomers were successfully used as surface modifiers (in 1–2 wt% amount) in the cationic photopolymerisation of telechelic hydrogenated vinyl ethers, leading to hydrophobic crosslinked films for which the water contact angles were higher than 105° [37].

2.1.5. Fluorinated (Meth)acrylates Containing TFE Base Units

Undoubtedly, fluorinated acrylates and methacrylates have drawn much interest because (i) they can easily be (co)polymerised under mild conditions (including UV initiation), and (ii) the resulting (co)polymers exhibit very interesting properties and have found relevant applications in the protection of various substrates such as textile, leather, stone, concrete, plaster, metals, optical fibres, etc., thanks to the perfluorinated end-group of the lateral chain that brings good surface properties [13]. The syntheses of fluorinated (meth)acrylates have been reviewed [13, 38–40]. However, there are still many researches dealing with the nature of the Sp spacer between the perfluorinated chain-end and the ester group:

$$H_2C=CH-\overset{\overset{\text{O}}{\|}}{C}-O-Sp-C_nF_{2n+1}$$

This spacer can consist of polymethylenic groups $(CH_2)_p$ with $p = 1, 2, 3, 6, 11$ [41–44]; $CH_2CH_2SCH_2CH_2$ [44, 45], $(CH_2)_pCH(CH_3)SC_2H_4$ ($p = 1, 2, 9$) [45], $CH(CH_3)(CH_2)_qSC_2H_4$ ($q = 1, 2, 9$) [45]; it can contain secondary alcohol [46, 47], morpholino group [48], ethylene oxide base units [47, 49], hexafluoroisopropylidene groups [50] and various aromatic groups leading either to liquid crystal properties [51] inducing crystallisation or structuration [45, 52] containing sulfonamide bridge [53] or triple bonds [54, 55].

Other interesting series of 2-{[bis(fluoroalkyl)phosphoryl]oxy} ethyl acrylates, their radical polymerisation tests of oil and water repellency and fire retardency have recently been reported by Timperley et al. [40].

In addition, the synthesis of α-fluoroacrylates [56, 57], their polymerisation under traditional [57] and controlled [58] methods, their properties and applications have been reviewed [56], just like those of α-trifluoromethyl acrylic acid [59], α-acetoxy [60] and α-propioxy acrylates [60c].

Fluorinated telechelic diacrylates and their polymerisation (also reported in Chapter 2) have been summarised by Turri et al. [61] and have been used for various applications including liquid crystals [62], surface modifiers [63] and optical fibres [64]. Other fluorinated polyacrylates have shown interesting low-dielectric constants for electronic applications [65]. Furthermore, acrylates bearing cyclic fluorinated [66] or SF_5 [67] end-groups have also been obtained.

The kinetics of homopolymerisation of various fluoro(meth)acrylates [48, 60c, 68] and that of copolymerisation [69, 70] were investigated. Indeed, we demonstrated that: (i) fluoroacrylates are more reactive in homopolymerisation than fluorinated methacrylates as in hydrogenated series; (ii) the longer the spacer between the fluorinated chain and the ester group of the acrylate, the faster its homopolymerisation; and (iii) when the fluorinated group is far enough from the double bond, the length of the (per)fluorinated chain has almost no influence on the reactivity [48, 60c, 68].

Other non-exhaustive copolymerisations [71–76] and cotelomerisations of fluoroacrylates [77, 78] were also achieved.

Moreover, interesting properties of poly(fluoroacrylate)s at the nanoscale were also studied [79, 80]. Further interesting works on the ionic, group transfer polymerisation and controlled radical polymerisation (CRP) of fluoro(meth)acrylates have been mentioned in Chapter 4 (Sections 2.1.2.3, 2.1.4.4, 2.2.3.1, 2.2.3.3, 2.2.3.3.2 and 2.2.4.2.1) just like the CRP of α-fluoroacrylates [58].

From an industrial point of view, many investigations on the radical polymerisation of fluorinated (meth)acrylates were carried out by the major industrial companies (DuPont, 3M, Asahi, Daikin, Atofina, etc.).

Other scarce fluorinated monomers have also been synthesised, such as vinyl cyclopropanes [81], vinyl esters [82], maleimides [83] and phenyl maleimides [84].

Hence, a wide range of various macromonomers containing TFE base units have been prepared and their polymerisation studied in the appropriate routes: a cationic method for vinyl ethers and oxetanes, radical initiation for styrenic, (meth)acrylates, α-fluoro, α-trifluoromethyl acrylates and α-alkoxy acrylates, and GTP for methacrylates, as reported in Chapter 4.

Actually, these fluorinated monomers possess TFE units and we have found it interesting to report monomers containing other fluoroalkene units, displayed below.

2.1.6. Fluorinated Macromonomers which do not Contain any TFE Units

Usually, most fluorinated macromonomers possess oligo(TFE) grafts. Indeed, non-exhaustive examples below cite macromonomers bearing different fluorinated side groups such as oligo(VDF), perfluoropolyether grafts, CTFE and mixtures of fluoroalkenes.

Regarding the first case, two possibilities have been reported.

The first one starts from the telomerisation of VDF with CF_3I [85] followed by chemical modification up to obtaining the corresponding oligo(VDF)-containing acrylate [86]:

$$CF_3I + n\,H_2C\!=\!CF_2 \xrightarrow{} CF_3(VDF)_nI \xrightarrow[3)\ H_2C=CHCOCl]{\substack{1)\ H_2C=CHCH_2OH \\ 2)\ SnBu_3H}} H_2C\!=\!CHCO_2(CH_2)_3(VDF)_nCF_3$$

$$\underset{VDF}{}$$

The second case of VDF macromonomer, starting from the telomerisation (see Chapter 1) of VDF with $BrCF_2CFClBr$ [87], is original since it bears (i) a perfluorovinylic group which is reactive in copolymerisation with other fluoroalkenes [88], hence allowing the resulting copolymer to preserve interesting properties, and (ii) a bromo end-atom that enables the produced copolymer to be crosslinked by a peroxide/triallyl isocyanurate system [88]:

$$BrCF_2CFClBr \xrightarrow[2)\ Zn]{1)\ VDF} F_2C\!=\!CF(VDF)_n\!-\!Br$$

In addition, styrenic or acrylic monomers containing VDF units have also been synthesised [89, 90]:

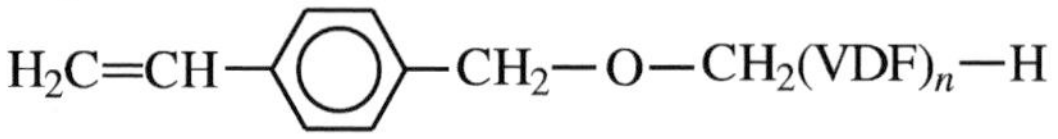

$$H_2C=CH-\langle\bigcirc\rangle-CH_2-O-CH_2(VDF)_n-H \qquad [89]$$

$$H_2C=C\overset{\displaystyle R}{\underset{\displaystyle \underset{\displaystyle O}{\overset{\parallel}{C}}-O-CH_2CH_2-O-\overset{\displaystyle O}{\underset{\displaystyle OH}{\overset{\parallel}{P}}}-(CH_2CF_2)_n-H}{}} \qquad [89,90]$$

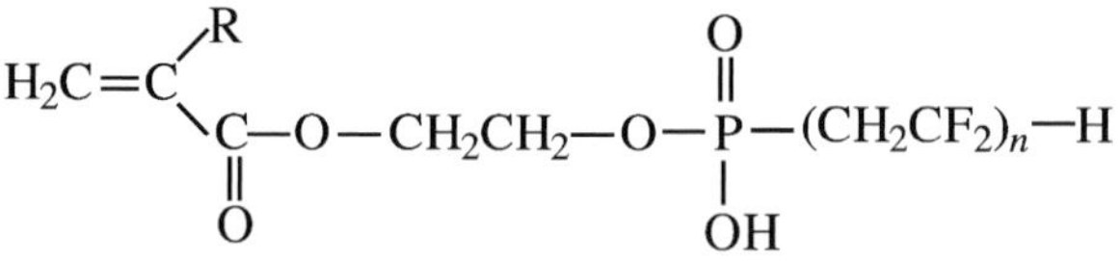

R: H, CH$_3$

from precursors obtained by telomerisation of VDF with methanol [91] or diethyl hydrogenphosphonate [92], respectively.

More recently, the direct terpolymerisation of vinylidene fluoride (VDF) with hexafluoropropylene (HFP) and allylamido perfluoropolyethers (PFPE, Krytox® in that case) was accomplished, leading to original poly(VDF-co-HFP)-g-PFPE graft copolymers [93]:

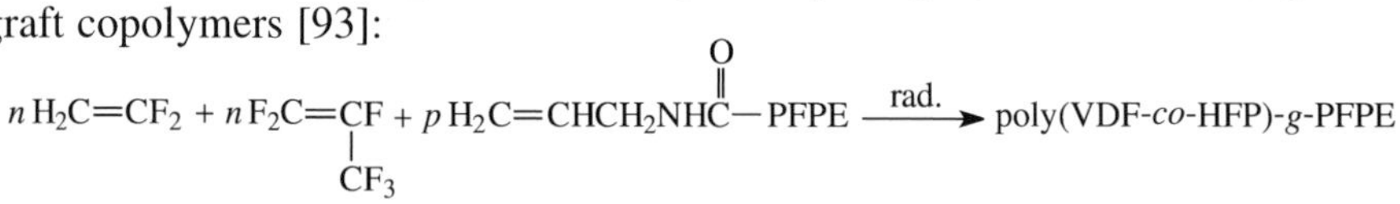

$$n\,H_2C=CF_2 + n\,F_2C=\underset{\displaystyle CF_3}{\overset{\displaystyle |}{C}}F + p\,H_2C=CHCH_2NH\overset{\displaystyle O}{\overset{\parallel}{C}}-PFPE \xrightarrow{\text{rad.}} \text{poly(VDF-co-HFP)-g-PFPE}$$

Although the molecular weights of these original highly fluorinated graft copolymers were not assessed, it was surprising to note that no allylic transfer occurred.

In addition, various strategies to introduce PFPE (especially from Fomblin® or Krytox® oligomers) as grafts have also been investigated [61, 94, 95].

In addition, a wide variety of monomers containing chlorotrifluoroethylene (CTFE), achieved from the telomerisation of CTFE [96] (see Chapter 1), were prepared. They were used either as monomers [97] for polycondensation [98] or as precursors of acrylates [13, 99].

2.1.7. Copolymerisation of Fluorinated Macromonomers with Fluoroalkenes

As mentioned in Chapter 3 (Section 2.2.5), vinyl ethers do not homopolymerise under radical conditions but copolymerise easily in the presence of adequate electron-withdrawing comonomers. An interesting example (that has drawn much academic and industrial interest) concerns the copolymerisation of TFE with a perfluorovinyl ether containing, or not, hexafluoropropylene oxide units and bearing a carboxylic or sulfonyl fluoride end-group:

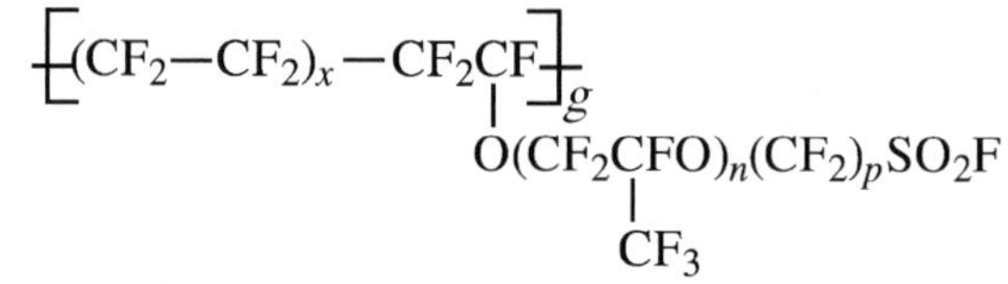

$$\left[(CF_2-CF_2)_x-\underset{\displaystyle \underset{\displaystyle CF_3}{\overset{\displaystyle |}{O(CF_2CFO)_n(CF_2)_pSO_2F}}}{\overset{\displaystyle |}{CF_2CF}}\right]_g$$

x = 6–10

n = 0, 1, 2

p = 1–5

The hydrolysis of sulfonyl fluoride group led to the corresponding sulfonic acid derivative used for membranes in chlor-alkaly, IEMs for electrolysis or for fuel cell applications. The best known copolymers commercially available since 1962 are Nafion® and Flemion® [100–114] (from DuPont and Asahi Glass Co., respectively) when $p = 2$ and $n = 1$, although Tosflex® ($n = 0, 1$ and $p = 1$ or 5), for which the functional end-group is an anion-exchange unit [102], is produced from Tosoh Co. Ltd. In addition, Dow® and Hyflon® Ion H (from Dow Chemical and Solvay-Solexis, respectively) when $n = 0$ [103] and Aciplex® (from Asahi Chemicals Co.) when $n = 2$ [100, 103, 105–112] have also led to potential membranes for these applications [100–106]. Three very interesting reviews on such copolymers for obtaining proton exchange membranes (PEMs) devoted to fuel cells have recently been published by Doyle and Rajendran [101], by Arcella *et al.* [103] and by Li *et al.* [105]. In the current research on fuel cell membranes, efforts have been concentrated on PEMs with lower methanol permeabilities [115] and lower cost than Nafion® (>780 US \$ m^{-2}) [116, 117]. However, it has recently been reported that new solution cast Nafion® membranes (NR-111) would allow the cost to drop to 50 US \$ m^{-2} with volume production of ~2 million m^2 per year [118].

For similar strategical fuel cell applications, other monomers with longer chain lengths [119–121] were also used in the copolymerisations with TFE, and original studies have concerned the use of phosphonate or phosphonic acid end-groups [113, 122–124].

Nevertheless, similar investigations were carried out when VDF replaced TFE leading to various patents [125–127] and also further works report the terpolymerisation involving HFP [128, 129], brominated [130] and nitrile [131] fluorinated comonomers. Interestingly, the cure site monomers bearing Br or CN groups induced a crosslinking reaction that made the resulting network insoluble in all solvents.

In contrast, fluorinated monomers containing a CO$_2$H end-group did not have the same fate as that of the sulfonic acid derivatives [104–108, 113, 132, 133] because of the weaker acidity of the ionic group, although various studies were developed [100, 133]. Nevertheless, kinetics of copolymerisation of TFE with F$_2$C=CFO–R$_F$–CO$_2$H [133] and that of VDF with H$_2$C=CFCF$_2$OCF(CF$_3$)CF$_2$ OCF(CF$_3$)CO$_2$CH$_3$ [134] were investigated.

2.2. FROM THE CHEMICAL MODIFICATION OF HYDROGENATED POLYMERS BY FLUORINATED DERIVATIVES: THE "GRAFTING ONTO" METHOD

2.2.1. Introduction

Various routes have been investigated by several authors to enable the grafting of fluorinated derivatives onto polymeric backbones. This section reports the chemical modification of polydienes (e.g., polybutadiene and polyisoprene and also their block copolymers), poly(hydrogenomethyl siloxanes), polysulfones, polyparaphenylenes, polyphosphazenes, polyvinylalcohols and other copolymers, by different approaches. This strategy can be realised according to both the nature of the polymer and the accessibility of the fluorinated derivative to react onto the key function of these hydrogenated polymers.

2.2.2. Chemical Modifications of Polydienes

Chemical changes of polydienes can be achieved by different ways: (i) radical grafting of perfluoroalkyl iodides or fluorinated mercaptans onto polybutadiene (PBut), polyisoprene (PIp) and block copolymers containing PBut or PIp blocks or (ii) condensation reactions (Table I).

2.2.2.1. Chemical Modification of Polybutadienes by Radical Addition of Perfluoroalkyl Iodides or Fluorinated Mercaptans

The Thiokol company [135] studied the radical addition of perfluoroalkyl iodides ($C_nF_{2n+1}I$ with $2 < n < 20$) onto hydroxy telechelic polybutadienes containing either 80% of 1,4-poly(butadienic) double bonds or 90% of 1,2-poly(butadienic) double bonds. Such α,ω-dihydroxy oligomers (average molecular weights are $\bar{M}_n = 1200$ or 2400) are commercially available products supplied by the Atofina or Nippon Soda companies, respectively. The French group realises the radical dead-end polymerisation (DEP) of 1,3-butadiene initiated by hydrogen peroxide while the Japanese company prepares these oligomers by anionic initiation. Hence, the 1,4/1,2-unit contents differ for both products: these are 80/20 and 10/90, respectively.

The authors noted that the content of fluorine varied according to both the nature of the solvent and the temperature for the same duration and for the same initiator (dibenzoyl peroxide) [135]. This company claimed to prepare fluorinated macrodiols which possess a fluorine content ranging between 6 and 22%.

In 2001, Ren et al. [136] revisited this reaction on poly(butadiene) containing a high amount of 1,2-butadienic units (since the authors prepared these well-defined oligomers of narrow polydispersity by anionic polymerisation). The addition of R_FI onto these lateral unsaturations was initiated by triethylborane. The authors hence prepared a wide range of fluorinated PButs with more than 95% consumption of the double bonds, and with a good preservation of the narrow molecular weight distribution. From identification of the by-products, the authors also suggested a cyclic rather than an open-chain addition mechanism and that five-member ring structures were formed in the course of the radical addition of R_FI onto 1,2-PBut. These authors also added R_FI onto the double bond of 1,2-PBut block in the PS-b-1,2-PBut block copolymer (synthesised by sequential anionic copolymerisation of styrene and 1,3-butadiene (see below)).

By introducing a perfluorinated side group, the authors noted an increase of the T_gs of 75 °C and, after removal of iodine, the resulting fluorinated comb-like polymers showed an increase of the thermostability of about 100 °C and also

TABLE I

GRAFTING OF FLUORINATED REACTANTS ONTO POLY(ENE)S

Poly(ene)	F-Reactants	Experimental conditions	Property/Application	References
PBut	R_FI	BEt_3	Surface	[136]
	$R_FC_2H_4SH$	AIBN	Hydrophobic coatings	[139]
	Difluorocarbene	From HFPO, 170–185 °C	Surface, thermal properties	[153]
HTPBut	R_FI	DBP	Oleophobic/hydrophobic coatings	[135]
	$R_FC_2H_4SH$	Photochemical	Viscosity, surface	[138]
	$R_FC_2H_4SH$	Peroxide or azo	Viscosity, surface	[140]
	$R_FC_2H_4Si(CH_3)_2H$	H_2PtCl_6/iPrOH	Thermal stability	[141]
PI_p	Difluorocarbene	From HFPO, 185 °C	Surface, thermal stability	[154]
PBut-b-PMMA	R_FI	DBP	Structuration	[145]
	$R_FC_2H_4SH$	AIBN	Structuration	[145]
PS-b-PBut	R_FCOCl	9-BBN/H_2O_2/pyridine	Surface	[146]
	R_FCOCl	MCPBA/Bu_4NCl	Surface	[146]
	R_FI	BEt_3	Surface, thermal properties	[136]
	$R_FC_2H_4Si(CH_3)_2H$	Karstedt catalyst	Surface	[158]
PBut-b-PS-b-PMMA	R_FI	DBP	Structuration	[145]
	$R_FC_2H_4SH$	AIBN	Structuration	[145]
PS_d-b-PI_p	Difluorocarbene	From HFPO, 185 °C	Morphology, order–disorder	[154]
	$C_6F_{13}C_2H_4Si(CH_3)_2H$	Karstedt cata/110 °C	Surface, segregation	[149, 158]
	C_3F_7COCl	9-BBN/H_2O_2/NaOH	Surface, segregation	[149]
PS-b-PI_p	R_FCOCl	9-BBN/H_2O_2/NaOH	Thermal stability, surface, $\theta_{adv} = 122°$	[150]
	$(CF_3)_2CHCOCl$	9-BBN/H_2O_2/NaOH	Surface	[159]
	$R_FC_2H_4O$–C_6H_4–COCl	9-BBN/H_2O_2/NaOH	Structuration, surface, morphology	[161]

(continued on next page)

TABLE I (continued)

Poly(ene)	F-Reactants	Experimental conditions	Property/Application	References
	tBOC–CH$_2$C$_8$F$_{16}$CH$_2$OCO C$_2$H$_4$COCl	9-BBN/H$_2$O$_2$/NaOH	Surface	[151]
	NCO–C$_6$H$_4$–CF$_3$	9-BBN/H$_2$O$_2$/NaOH	Surface, thermal stability	[159]
	(R$_{FH}$)$_3$CC$_2$H$_4$COCl	9-BBN/H$_2$O$_2$/NaOH	Liquid crystal	[157]
	C$_6$F$_{13}$C$_2$H$_4$Si(CH$_3$)$_2$H	H$_2$PtCl$_6$, 80 °C	Phase segregation, surface	[152]
	Difluorocarbene	From HFPO, 185 °C	Surface, morphology, order–disorder transitions	[154]
Isocitronellene/propene copolymer	C$_6$F$_{13}$I	Radicals	Morphology	[162]

9-BBN, 9-borabicyclo [3-3-1] nonane; tBOC, $tert$-butoxycarbonyl; But, 1,3-butadiene; HFPO, hexafluoropropylene oxide; I$_p$, isoprene; HTPBut, hydroxytelechelic poly(butadiene); MCPBA, metachlorperbenzoic acid. R$_F$, C$_n$F$_{2n+1}$ (n = 6, 8, 10); R$_{FH}$, R$_F$(CH$_2$)$_p$CO$_2$C$_3$H$_6$; S$_d$, deuterated styrene; DBP, dibenzoyl peroxide.

improved surface properties (their critical surface tensions were ca. 14–16 mN m^{-1}).

Another strategy to synthesise telechelic polybutadienes bearing fluorinated side groups (Table I) deals with the radical addition of fluorinated mercaptans onto hydroxytelechelic polybutadienes (HTPBut).

Addition of CCl_3–$R_{F,Cl}$ onto the double bonds of HTPBut under redox catalysis failed [137] while the radical grafting of fluorinated thiols was successfully achieved whatever the photochemical initiation [138] or that carried out in the presence of organic initiators [139, 140], as follows:

$$HO(CH_2CH{=}CHCH_2)_x \underset{\underset{\substack{|\\CH=CH_2}}{}}{(CH_2CH)_y}{-}OH + R_FC_2H_4SH \xrightarrow[\text{or UV}]{\text{rad.}} HO(C_4H_6)_a \underset{\underset{\underset{C_2H_4R_F}{|}}{S}}{(CH_2CH_2CHCH_2)_b} \underset{\underset{\underset{C_2H_4R_F}{|}}{\underset{S}{|}}{C_2H_4}}{(CH_2CH)_c}OH$$

In the latter case, the radical grafting in the presence of azo compounds (especially AIBN [139]), peroxyesters [140] or peroxides [140] was optimised in terms of yields and selective additions onto 1,4- or 1,2-butadienic units. Indeed, the addition of fluoro mercaptans occurred onto the 1,2-units first, until saturation of all of them, followed by the grafting onto the 1,4-units [140]. This enabled the synthesis of a wide range of fluorinated polybutadienic standards of various fluorine contents (5–58%) and whose physico-chemical characteristics (refractive indexes, viscosities, glass transition and decomposition temperatures) and surface properties were investigated [140]. As expected, these characteristics showed that the higher the fluorine content, the lower the surface tension that could be as low as 8.7 dyne cm^{-1} (when the fluorine content was 57.5%, which represents an almost quantitative grafting).

Figure 1 represents the evolution of the T_gs of virgin and grafted HTPBut vs. the degree of grafting for both commercially available HTPButs. It is noted first that the difference of T_g values of these starting materials arises from their

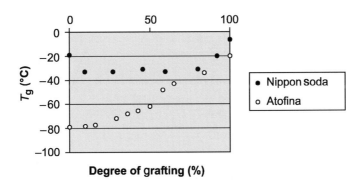

FIG. 1. Glass transition temperature vs. the degree of grafting of $C_6F_{13}C_2H_4SH$ onto hydroxytelechelic polybutadienes supplied by Nippon Soda (full circles) or by Atofina (empty circles) (reproduced with permission from B. Ameduri *et al.*, *J. Polym. Sci., Part A: Polym. Chem.*, 31, 2069–2080 (1993)).

microstructures. Those of the grafted HTPBut from the Nippon Soda company remained constant (ca. $-30\,^{\circ}C$), whereas those of the fluorinated Atofina HTPBut increased from -79 to $-20\,^{\circ}C$. This difference of behaviour is assigned to the difference of microstructure, since the softness of the polybutadiene is related to the content of the 1,4-units. Actually, because the 1,2-units are more reactive than the 1,4-ones with respect to the fluorinated mercaptan, at low degree of grafting the T_g is not affected since the 1,4-units have not reacted yet. However, as soon as the thiol is added onto the 1,4-unit (i.e., at 90 and 20% for the Atofina and Nippon Soda HTPBut products, respectively), the T_g increased drastically [140].

However, the thermal stability of these fluorooligobutadienic derivatives, probably because of the thermooxidative $S-CH_2$ bond and the thermal cyclisation of 1,2-butadienic unit, was not improved about that of virgin HTPBut [140]. Nevertheless, the improvement of the thermal stability (Figure 2) was achieved on original silicon-containing fluoro HTPBut synthesised from the hydrosilylation reaction of the double bonds of HTPBut [141] with fluorinated dimethyl hydrogenosilanes $C_nF_{2n+1}(CH_2)_x-Si(CH_3)_2H$ ($n = 6,\ 8;\ x = 2,\ 3$) [15].

Interestingly, these thermograms show the slight loss of thermal stability of HTPBut grafted with fluorinated thiols (probably because of the CH_2SCH_2 weak point) in contrast to the good thermal behaviour of virgin HTPBut (that makes cycles under heating as observed by Golub [142] and Chiantore et al. [143]). However, that of the HTPBut grafted with the fluorinated silane demonstrates an improvement of the thermal stability.

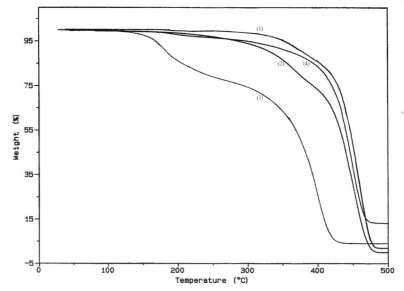

FIG. 2. TGA thermograms of hydroxytelechelic polybutadiene from Atofina ($M_n = 2700$ (1) or 1200 (2), grafted with $C_6F_{13}C_2H_4SH$, d.o.g. $= 25\%$ (3) or with $C_6F_{13}C_2H_4Si(CH_3)_2H$, d.o.g. $= 30\%$ (4).

These fluorinated silanes were obtained in a four-step procedure from perfluoroalkyl iodides as follows [15]:

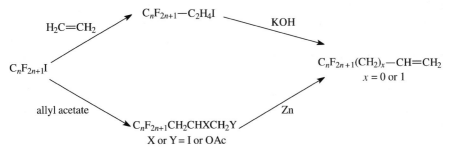

As a matter of fact, the radical addition of $C_nF_{2n+1}I$ onto allyl acetate leads to a $R_FCH_2\dot{C}HCH_2OCOCH_3$ radical that undergoes a thermal rearrangement into $R_FCH_2CH(OAc)\dot{C}H_2$, hence yielding a mixture of both $R_FCH_2CHICH_2OAc/R_FCH_2CH(OAc)CH_2I$ isomers [144]. Nevertheless, both isomers were successfully "deiodoacetylated" and gave $R_FCH_2CH=CH_2$ in high yields [15].

These fluoroolefins are relevant precursors of fluorinated silanes obtained in good yields as follows:

$$R_F(CH_2)_x-CH=CH_2 \ + \ H-\underset{\underset{CH_3}{|}}{\overset{\overset{CH_3}{|}}{Si}}-Cl \ \xrightarrow{H_2PtCl_6} \ R_F(CH_2)_{x+2}-\underset{\underset{CH_3}{|}}{\overset{\overset{CH_3}{|}}{Si}}-Cl \ \xrightarrow{AlLiH_4} \ R_F(CH_2)_{x+2}-\underset{\underset{CH_3}{|}}{\overset{\overset{CH_3}{|}}{Si}}-H$$

$$x = 0, 1$$

2.2.2.2. Addition (or Condensation) of Fluoroderivatives onto Block Copolymers Containing Polybutadienic or Polyisoprenic Blocks

2.2.2.2.1. Grafted Copolymers Containing P(But) Blocks.
From the same strategy, Hilton [145] more recently achieved the radical grafting of R_FI and $R_FC_2H_4SH$ onto 1,2-units of PBut blocks in PBut-*b*-PMMA and PBut-*b*-PS-*b*-PMMA block copolymers (their molar masses and PDIs were in the ranges of 98,000–170,000 and 1.20–1.29, respectively). Beside the optimisation of the grafting accomplished by modifying the experimental conditions and the nature of the solvents and of the initiators, the author identified the by-products and side reactions. The reactions were monitored by ^1H, ^{13}C and ^{19}F NMR and infrared spectroscopies, and by iodine titration. These resulting graft copolymers were characterised by SEC and DSC that showed, as above, an increase of the T_g of the modified block while those of the PS and PMMA were unchanged. Solubility parameters, self-assembly and organisation examined at the nanoscopic scale of these fluorinated polymers were assessed and the key changes were highlighted, namely the increased surfactancy of the modified block copolymers which underwent self-assembly in solution to make very stable micelles. This micellar structure (a core–shell structure composed of a fluorinated core and a hydrocarbon shell was identified) is carried over into the solid state to give

spherical morphologies for the modified copolymers in contrast to their lamellar or cylindrical morphology prior to modification. The properties of structuration in solution were confirmed by visual aspects of solutions at different concentrations of these fluorinated copolymers, showing that the increase of the copolymer concentration intensified the colouration (proved by UV measurements) and increased the viscosity.

The following example was suggested by Oestreich and Antonietti [146], who allowed poly(styrene)-*b*-(polybutadiene) block copolymers (SB) containing 90% of 1,2-butadiene units in the polybutadiene block (prepared by anionic polymerisation) to be fluorinated via two-step procedures from the pendant double bonds of the butadienic block. Indeed, this technology leads rather to "grafted" segments than to "block copolymers" as suggested by these authors. The first method consists in achieving a hydroboration in the presence of 9-bicyclononyl borane (9-BBN) followed by an oxidation that generated hydroxy side groups. Esterification of these functions with perfluorinated carboxylic acid chlorides introduced fluorinated side groups in the butadienic blocks (Table I). The second route deals with the epoxidation of 1,2-double bonds in the presence of metachloroperbenzoïc acid (MCPBA) followed by the ring-opening of the oxirane by the same above fluoroacid chlorides. Both methods are summarised in the following scheme, where only one 1,2-pendant double bond of the butadienic block is presented (R_F represents C_7F_{15}):

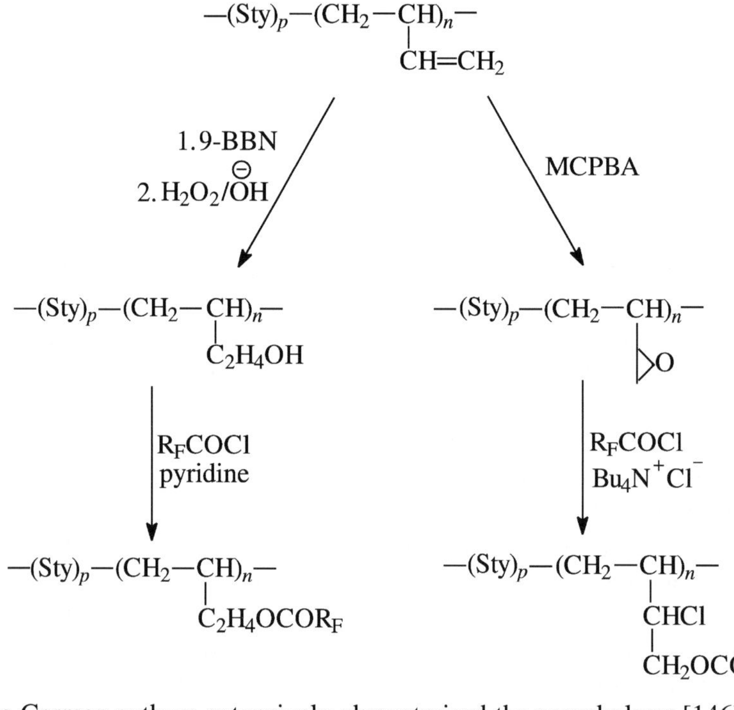

$$-(Sty)_p-(CH_2-CH)_n-$$
$$\overset{|}{C}H=CH_2$$

$$\begin{array}{c} 1.\ 9\text{-BBN} \\ \ominus \\ 2.\ H_2O_2/OH \end{array} \diagdown \qquad \diagdown \text{MCPBA}$$

$$-(Sty)_p-(CH_2-CH)_n- \qquad\qquad -(Sty)_p-(CH_2-CH)_n-$$
$$\overset{|}{C_2H_4OH} \qquad\qquad\qquad\qquad\qquad \triangleright O$$

$$\begin{array}{c} R_FCOCl \\ pyridine \end{array} \qquad\qquad\qquad \begin{array}{c} R_FCOCl \\ Bu_4N^+Cl^- \end{array}$$

$$-(Sty)_p-(CH_2-CH)_n- \qquad\qquad -(Sty)_p-(CH_2-CH)_n-$$
$$\overset{|}{C_2H_4OCOR_F} \qquad\qquad\qquad\qquad \overset{|}{C}HCl$$
$$\qquad\qquad\qquad\qquad\qquad\qquad\qquad\qquad \overset{|}{C}H_2OCOR_F$$

These German authors extensively characterised the morphology [146] of these original well-architectured copolymers and studied their surface properties [146] and dielectric relaxation [147]. By small angle X-ray scattering measurements, they noted that these fluorinated materials, containing a 2 : 1 volume fraction of PS to fluorinated PBut, consisted of a hexagonally packed cylindrical morphology with PS comprising the matrix and the fluorinated PBut making up the cylinders. The contact angle measurements led to surface energies as low as 14.2 mN m^{-1} [146], showing the beneficial action of the perfluorinated end-groups in the lateral chains.

According to the authors, these fluorinated copolymers can act as efficient steric stabilisers for the dispersion polymerisation in solvents with ultra low cohesion energy density [148]. The dispersion particles are effectively stabilised by these amphiphiles.

Interestingly, these materials strongly adhered to glass plates, thanks to pronounced amphiphilicity where the PS blocks were oriented to the glass plates.

In addition, these well-defined fluorinated copolymers were involved in gas permeation. The measurements revealed a remarkable gas selectivity for separating oxygen and nitrogen as well as separating CO_2 from volatile hydrocarbons at a very high flow.

The Ciba-Geigy Company [139] also grafted fluorinated thiols onto 1,2-butadienic double bonds of poly(butadiene-*co*-styrene) and poly(butadiene-*co*-acrylonitrile) copolymers. This company claimed that the yields of that grafting were improved (even being quantitative in optimised conditions) in the presence of solvent (heptane or MEK).

2.2.2.2.2. Grafted Copolymers Containing P(Ip) Blocks. The first pathway above was revisited by Ober's group [149–152], which introduced polyisoprene blocks instead of PBut sequences, and extensively investigated the surface properties of the resulting fluorocopolymers. Films prepared from blends of PS and "monodisperse poly(styrene-*b*-semifluorinated side chain)", which are indeed fluorinated graft copolymers, revealed a high fluorine enrichment at the surface that originated from strong surface segregation of the fluorinated PI_p block (lamellae with periods of about 430 Å). Polymeric surfaces were highly hydrophobic (e.g., the advancing water contact angles reached 122° for perfluorodecanoyl side chain [149]). In addition, X-ray photon spectroscopy (XPS) measurements revealed that the surface of spin cast films of these graft fluorinated PS-*b*-PI_p copolymers was comprised of a layer of fluorinated PI_p sequences up to a depth of 10 nm from the surface [152].

Although the fluorinated side chain was very short to consider that the obtained polymers were fluorinated graft copolymers, Lodge *et al.* [153, 154] succeeded in adding difluorocarbenes (generated in situ from thermal activation of hexafluoropropylene oxide at 170–185 °C) onto the pendant double bonds of the polyisoprenic sequences. Other older studies dealing with the addition of difluorocarbenes or chlorofluorocarbenes (arising from different sources such as the thermolysis of (trifluoromethyl) phenyl mercury or sodium monochlorodifluoroacetate) onto 1,2-polybutadienes or poly(1,1-dimethyl-1-sila-*cis*-pent-3-ene) [155] are well summarised by Reisinger and Hillmyer [156] but these results are not reported in this book.

Various possibilities for "graft" derivatives containing perfluorinated end-groups onto the double bonds of polybutadiene and polyisoprene (as oligomers or in block copolymers) were attempted. These graftings greatly upset the thermal and surface properties [157–161] of the original PBut and PI$_p$ macromolecules, since both the T_gs and the water contact angles were increased when the degree of grafting increased.

2.2.3.   Chemical Modifications of Siloxanes: Synthesis of Silicones Bearing Fluorinated Side Groups

The chemistry and technology of fluorosilicones are very broad and still of growing interest owing to their unique properties such as thermal and oxidative stability, low surface tension, excellent dielectric properties, inertness and moisture resistance [163]. Many applications have been found [164], especially as elastomers, shafts, sealants, high performance coatings, for lithography, defoaming agents, surfactants and wear-resistant surface coatings.

In 1985, Boutevin and Pietrasanta [164] reviewed the strategies of synthesis of fluorosilicones, their properties and their applications. Dow Corning produces ca. 95% of fluorosilicones, mainly Silastics$^{®}$ that bear a short fluorinated side chain. Its synthesis, pioneered by Kim *et al.* [165], consists in a two-step procedure:

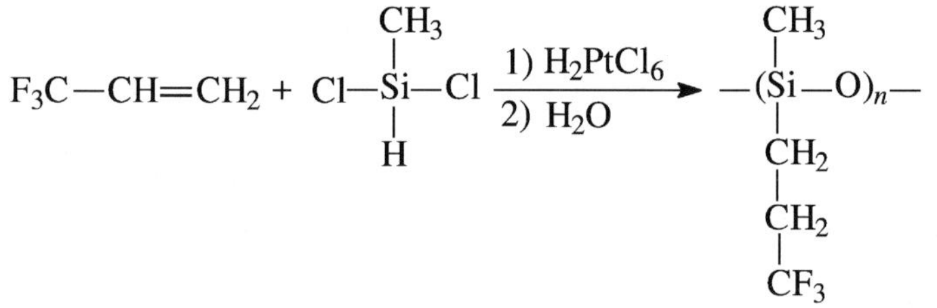

$$F_3C-CH=CH_2 + Cl-\underset{\underset{H}{|}}{\overset{\overset{CH_3}{|}}{Si}}-Cl \xrightarrow[\text{2) H}_2\text{O}]{\text{1) H}_2\text{PtCl}_6} -\underset{\underset{\underset{\underset{CF_3}{|}}{CH_2}}{\overset{CH_2}{|}}}{\underset{|}{\overset{\overset{CH_3}{|}}{(Si}}-O)_n}-$$

Besides this industrial preparation, many investigations devoted to the synthesis of silicones bearing a longer lateral fluorinated group were carried out by various authors, as already reported in pertinent reviews [156, 164, 166]. Besides the radical addition of fluorinated mercaptans onto polymethylsiloxanes bearing vinyl groups or the condensation of fluorinated carboxylic acid onto polymethylsiloxane possessing hydroxy side groups [164, 166], it is worth summarising investigations on the hydrosilylation of fluorinated ω-unsaturated compounds onto poly(methylhydrogeno siloxane) (PMHS).

This reaction usually proceeds in the presence of a platinum catalyst, called Speier or Karstedt catalyst (H$_2$PtCl$_6$/iPrOH or Pt/divinyltetramethyldisiloxane (DVTMDS), respectively) as reported by Marciniek [167].

In addition, di-*tert*-butyl peroxide has also been efficient for such a reaction, making more selective the addition of the $H_2C=$ end-group onto the silicon atom and cobalt complex has also been used [168, 169].

Many surveys concerning the chemical changes of polysiloxanes by introducing fluorinated grafts have been conducted by various authors. These surveys involve the hydrosilylation of unsaturated ω-perfluoroalkyl derivatives onto PMHS according to the following reaction:

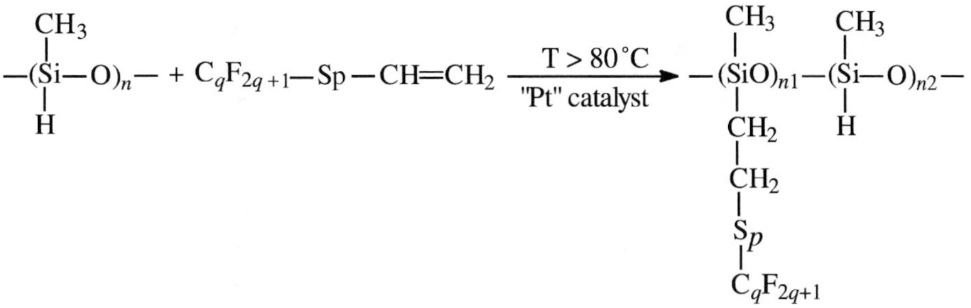

$$
\begin{array}{c}
CH_3 \\
| \\
-(Si-O)_n- \ + \ C_qF_{2q+1}-Sp-CH=CH_2 \\
| \\
H
\end{array}
\xrightarrow[\text{"Pt" catalyst}]{T > 80\,^{\circ}C}
\begin{array}{cc}
CH_3 & CH_3 \\
| & | \\
-(SiO)_{n1}-(Si-O)_{n2}- \\
| & | \\
CH_2 & H \\
| \\
CH_2 \\
| \\
Sp \\
| \\
C_qF_{2q+1}
\end{array}
$$

where n is an integer ($q = 6, 8, 10$ or 12) knowing that a chain length longer than C8 induces crystallinity, and Sp stands for a spacer which is mentioned in Table II.

This reaction leads to high yields and sometimes difficulty arises from the availability of the fluorinated reactant, although a few of them can be easily prepared in two steps from perfluoroalkyl iodides [15].

A wide range of fluorinated compounds were already reported in a well-documented review [166]. However, we do not want to cite them here, only to mention more recent investigations.

The spacer can exhibit various chain lengths (Table II), can be of different nature, aliphatic [168, 170–185] and aromatic [186, 187], and also can contain different functions such as ether [178–181, 185], ester [183, 184, 186] and amido [182] bridges. Interesting properties were shown, such as resistance to solvents, liquid crystalline properties [186], improved surface properties [168, 174, 182–185], fluorescence [180] and lower refractive indices [178]. Numerous applications of these fluorosilicones were found, such as membranes for volatile organic compounds [175], biomaterials [177], liquid crystals [186], surface agents, high performance optical materials [178], release paper [188] and antifouling coatings [179].

Even the presence of a short perfluorinated end-group (e.g., C_3F_7 [170, 171]) upsets the surface properties since the hydrophobicity of the resulting fluorinated PHMS increases (its surface energy drops from 22.6 to 11.5 mJ m^2 [170]).

Many reactions involving different fluorinated olefins, catalysts of hydrosilylation and hydrogenomethyl siloxanes or tetramethyl-tetrahydrocyclotetrasiloxane (D$_4$H) have been carried out. Indeed, the hydrosilylation of a fluorinated olefin with D$_4$H cyclic monomer was pioneered in 1990 as a model of hydrosilylation reaction (catalysed by H_2PtCl_6) involving $C_nF_{2n+1}(CH_2)_mOCH_2$

TABLE II

SYNTHESIS OF FLUORINATED GRAFT COPOLYMERS FROM HYDROSILYLATION REACTION OF FLUORINATED MONOMER, R_F–SPACER–CH=CH$_2$, ONTO POLY(METHYLHYDROGENOSILOXANE)

R_F	Spacer	Properties/use	References
C$_3$F$_7$	None	Structure, surface	[168–171]
C$_4$F$_9$	None	Mechanistic survey	[172]
C$_4$F$_9$	None	Gel from crosslinking	[173]
C$_4$F$_9$	None	Surface tension	[174]
C$_8$F$_{17}$	None	Surface	[168]
C$_8$F$_{17}$	CH$_2$	Selectivity to gas/membranes	[175]
C$_8$F$_{17}$	CH$_2$	Refractive indices, surface tension, thermal decomposition, dielectric constant	[176]
HCF$_2$CF$_2$	C$_2$H$_4$OCH$_2$	Biomaterial	[177]
C$_6$F$_{13}$	CH$_2$OCH$_2$	Photocrosslinkable siloxanes (use of acrylic side groups) optical applications	[178]
C$_n$F$_{2n+1}$ ($n = 6, 8$)	C$_2$H$_4$OCH$_2$	Resistance to biofouling	[179]
C$_6$F$_{13}$	C$_2$H$_4$OCH$_2$	Energy transfer studies (use of fluorescence dies)	[180]
C$_n$F$_{2n+1}$ ($n = 6, 8$)	C$_2$H$_4$OCH$_2$	Photocrosslinkable/gas permeation membrane	[181]
C$_8$F$_{17}$	CONRC$_3$H$_6$ (R=H or CH$_3$)	Textile treatment	[182]
C$_n$F$_{2n+1}$	C$_2$H$_4$OCO(CH$_2$)$_8$	Surface properties	[183]
C$_n$F$_{2n+1}$	(CH$_2$)$_{10}$CO$_2$CH$_2$	Surface properties	[184]
C$_n$F$_{2n+1}$	(CH$_2$)$_{10}$CO$_2$(CH$_2$)$_2$	Surface properties	[184]
C$_n$F$_{2n+1}$; $n = 8, 10, 12$	(CH$_2$)$_p$OCH$_2$; $p = 4, 6, 10$	Surface properties	[185]
C$_n$F$_{2n+1}$; $n = 4, 6, 8$	C$_6$H$_4$–X–C$_6$H$_4$OCO(CH$_2$)$_8$ variable X	Liquid crystals	[186]
Fluoroaromatic	CH$_2$	Higher T_gs (-40 °C)	[187]

$CH=CH_2$. This latter compound was obtained in a three-step procedure as follows [185]:

$$C_nF_{2n+1}I + H_2C=CH(CH_2)_{m-2}OH \xrightarrow[\text{2) SnBu}_3\text{H, AIBN}]{\text{1) AIBN}} C_nF_{2n+1}(CH_2)_m\text{—OH}$$

$$C_nF_{2n+1}(CH_2)_mOH + BrCH_2CH=CH_2 \xrightarrow[\text{phase transfer catal.}]{\text{NaOH, TBAH}}$$

$$C_nF_{2n+1}(CH_2)_mOCH_2CH=CH_2$$

The third step was achieved via a phase transfer catalysed reaction in the presence of tetrabutylammonium hydrogensulfate (TBAH) [189]. Hence, these fluorinated D_4H cyclic monomers, obtained in high yields (IR spectra showed that $-Si-H$ bonds were consumed after a short reaction time), were polymerised by anionic ring opening polymerisation. Siloxane bearing an ω-perfluorinated side chain and with controlled molecular weights was thus obtained [185].

The same strategy was recently achieved by Furukawa et al. [190] using shorter fluorinated substituents.

The hydrosilylation of this D_4H model was extended to that of PMHS (having an average degree of polymerisation of 35) choosing $C_8F_{17}(CH_2)_{10}OCH_2CH=CH_2$ and led to $-[(C_8F_{17}C_{10}H_{20}OC_3H_6)Si(CH_3)O]_{35}-$ [185]. However, even after long reaction times, the $Si-H$ groups could not be quantitatively reacted with the fluorinated allyl derivative. According to the authors, such a behaviour might be caused by changes in the accessibility of the remaining $Si-H$ groups, due to the association of already bound fluorinated residues, although these bonds may crosslink.

Similarly, Bracon et al. [186] carried out the hydrosilylation with the same α,ω-bis(trimethyl silyl) PMHS for novel aromatic ω-perfluorinated monomers bringing a side liquid crystalline chain:

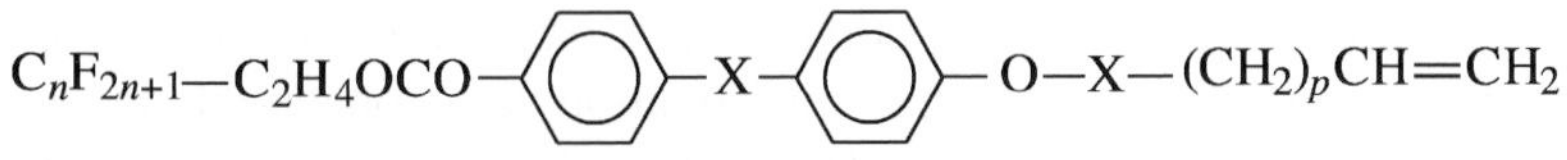

$$C_nF_{2n+1}\text{—}C_2H_4OCO\text{—}\langle\bigcirc\rangle\text{—}X\text{—}\langle\bigcirc\rangle\text{—}O\text{—}X\text{—}(CH_2)_pCH=CH_2$$

with n : 4, 6 or 8   X : "—" or OCO and p = 3 or 10

The resulting grafted PMHS exhibited high smectogen properties.

Olander et al. [177] interestingly introduced the $Si-H$ group onto the surface of PDMS by argon microwave plasma, which, after subsequent hydrosilylation of an allyl tetrafluoropropylether, led to silicone biomaterials with hydrolytically stable structures covalently bound to the surface.

Shorter spacers between the silicon atom and the terminal perfluoroalkyl group, such as in $C_6F_{13}C_2H_4OC_3H_6$, were also laterally introduced [178].

With the same strategy, Pham-Van-Cang et al. [180] synthesised original fluorinated siloxanes labelled with fluorescent dyes by step-wise hydrosilylation of 1H,1H,2H,2H-perfluorooctyl allyl ether and 9-phenanthryl-methyl allyl ether or 9-anthryl methyl ether onto poly(methylhydrogeno siloxane) ($M_n = 2200$, $DP_n = 35$). The resulting comb-like structure polymer was as follows:

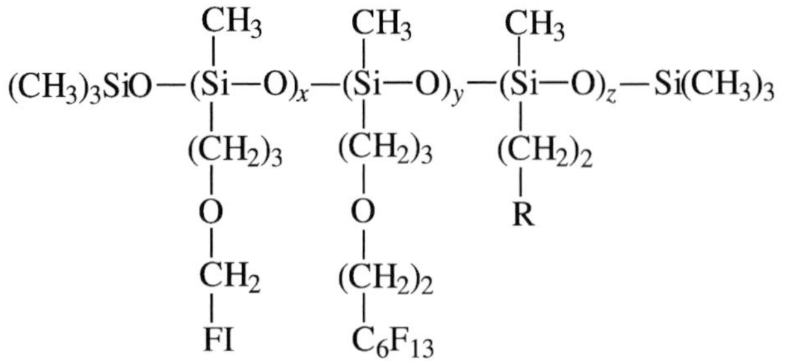

$$
\begin{array}{ccccc}
& CH_3 & CH_3 & CH_3 & \\
& | & | & | & \\
(CH_3)_3SiO-(Si-O)_x-(Si-O)_y-(Si-O)_z-Si(CH_3)_3 \\
& | & | & | & \\
& (CH_2)_3 & (CH_2)_3 & (CH_2)_2 & \\
& | & | & | & \\
& O & O & R & \\
& | & | & & \\
& CH_2 & (CH_2)_2 & & \\
& | & | & & \\
& FI & C_6F_{13} & &
\end{array}
$$

where FI and R represent a fluorescent moiety (9-anthryl or 9-phenanthryl) or other groups ($CH_2OC_8H_{17}$ or $Si(CH_3)_3$), respectively. The different polymers were mutually immiscible liquids labelled with chromophores appropriate for energy transfer studies. Thus, the authors could investigate the kinetics of energy transfer by the assessment of the fluorescence decay profiles. They also noted that the quenching was diffusion controlled.

More recent investigations accomplished in our Laboratory [175, 178, 181] have dealt with the sequential hydrosilylation of ω-perfluoroallyl (ether) monomers and allyl acrylate. The strategy was to carry out a partial first hydrosilylation, allowing unreacted hydrogenosiloxane moieties to react in a subsequent step with the allyl group of allyl methacrylate [175, 178] (Scheme 2) or of allyl urethane acrylate [188]. These original comb-like fluorinated siloxanes bearing lateral acrylate functions were photocrosslinked under UV irradiation, leading to an insoluble network used as original membranes for gas permeation for the purification of natural gas [175], for optics [178] and for release paper [188]. In the first application, permeabilities and selectivities towards hexane, ethyl acetate, ethanol and methylene chloride [175] were tested, while a recent investigation has concerned those of the carbon dioxide/methane gas pair [181]. The other case [178] dealt with the formation of coatings with tunable refractive indices, ultraviolet transparency and Yttrium Aluminium Garnet (YAG) laser resistance with unchanged thermal properties [178].

However, another alternative was suggested by Boileau's group [184] to achieve the synthesis of original fluorosilicone bearing a perfluorinated end-group with a C_{14} spacer prepared by condensation.

As far as the mechanistic point of view is concerned, Furukawa's team investigated the hydrosilylation of PMHS with $C_8F_{17}CH=CH_2$ [172], $C_8F_{17}CH_2CH_2OCH_2CH=CH_2$ [172] and $C_8F_{17}CH_2CH=CH_2$ [176], catalysed by $H_2PtCl_6\cdot6H_2O$. In the first two cases, the authors noted the formation of $C_7F_{15}CF=CHCH_3$ and $C_8F_{17}C_2H_4OCH=CHCH_3$ as by-products, and they suggested a reaction mechanism. However, in the third case, the

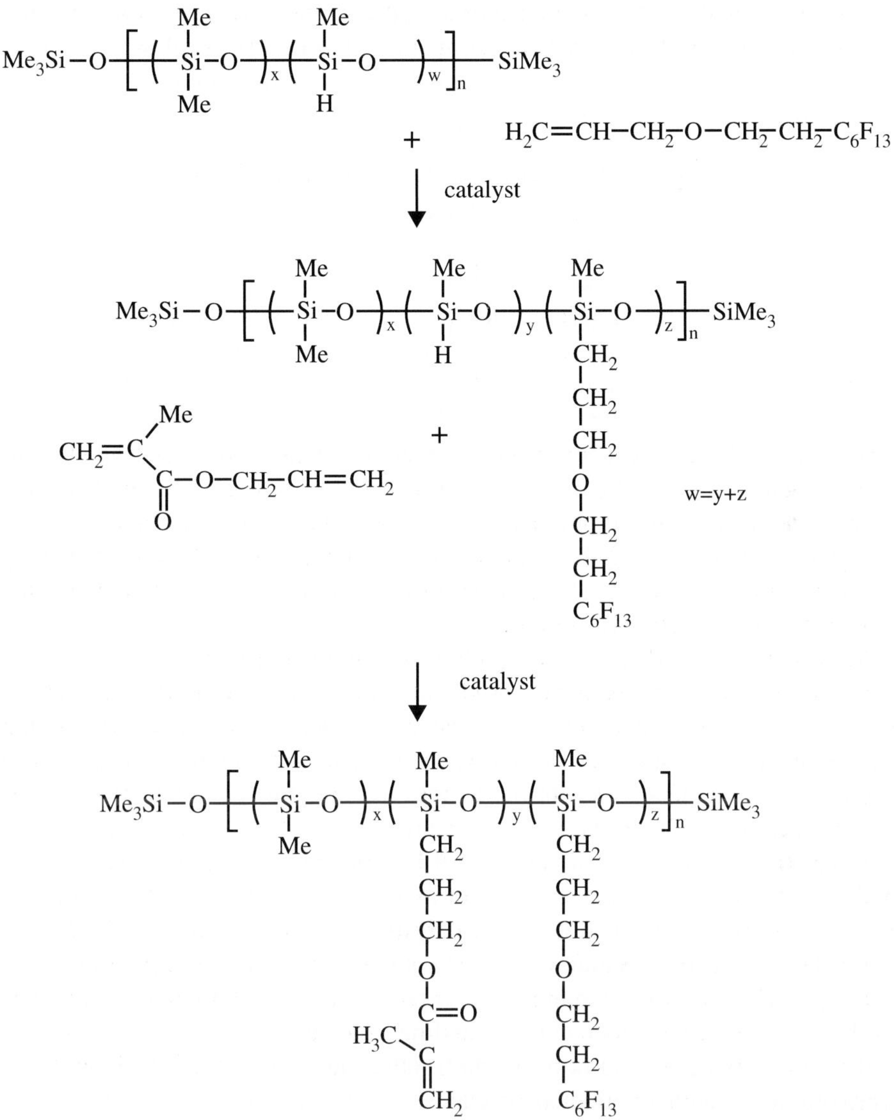

$$Me_3Si-O-\left[\left(-\underset{\underset{Me}{|}}{\overset{\overset{Me}{|}}{Si}}-O-\right)_x\left(-\underset{\underset{H}{|}}{\overset{\overset{Me}{|}}{Si}}-O-\right)_w\right]_n-SiMe_3$$

$$+ \quad H_2C{=}CH{-}CH_2{-}O{-}CH_2{-}CH_2{-}C_6F_{13}$$

↓ catalyst

$$Me_3Si-O-\left[\left(-\underset{\underset{Me}{|}}{\overset{\overset{Me}{|}}{Si}}-O-\right)_x\left(-\underset{\underset{H}{|}}{\overset{\overset{Me}{|}}{Si}}-O-\right)_y\left(-\underset{\underset{CH_2}{|}}{\overset{\overset{Me}{|}}{Si}}-O-\right)_z\right]_n-SiMe_3$$

with side chain: CH₂–CH₂–CH₂–O–CH₂–CH₂–C₆F₁₃

$$CH_2{=}\underset{\underset{\underset{O}{\|}}{\overset{|}{C}}-O-CH_2-CH{=}CH_2}{\overset{Me}{\overset{/}{C}}} \quad + \quad$$

w = y + z

↓ catalyst

$$Me_3Si-O-\left[\left(-\underset{\underset{Me}{|}}{\overset{\overset{Me}{|}}{Si}}-O-\right)_x\left(-\underset{\underset{CH_2}{|}}{\overset{\overset{Me}{|}}{Si}}-O-\right)_y\left(-\underset{\underset{CH_2}{|}}{\overset{\overset{Me}{|}}{Si}}-O-\right)_z\right]_n-SiMe_3$$

left side chain: CH₂–CH₂–CH₂–O–C(=O)–C(H₃C)(=CH₂)

right side chain: CH₂–CH₂–CH₂–O–CH₂–CH₂–C₆F₁₃

SCHEME 2. Synthesis of fluorinated UV-curable polydimethylsiloxanes.

reaction was successful and various properties of the resulting fluorinated polysiloxanes were investigated:

(i) the thermal decomposition started from 245 °C;

(ii) DSC analysis indicated that the C_8F_{17} side chain induced crystallisation and that, in certain cases, a T_g up to 19 °C was observed;

(iii) refractive indices were assessed and the lowest value was 1.3469;

(iv) the dielectric constant values were increasing with the degree of grafting;

(v) liquid and solid surface tensions were determined and surface free energies of 16–21 mN m^{-1} were obtained [176]. The same group [182] later used these fluorosilicones as oleophobic and hydrophobic coatings for polyesters.

A recent investigation concerns the hydrosilylation of 1,3-bis(2-hydroxy-hexafluoro-2-propyl) benzene with PMHS, in the presence of the Karstedt catalyst, leading to a silicone bearing fluoroaromatic moieties [187]. As expected, it was found that the higher the degree of grafting (d.o.g.), the higher the T_g, reaching a value of $-41\,°C$ for 47% d.o.g.

Another interesting survey was achieved by Mera and Morris [173], who carried out the hydrosilylation of PMHS with $C_4F_9CH=CH_2$ in supercritical (sc) CO_2 using Karstedt catalyst. Their results were compared to those obtained in subcritical conventional solvents (i.e., toluene) and the authors noted that the rates of hydrosilylation were dependent on the solvents, the reactant concentrations and the temperature. For a given set of conditions, rates in sc CO_2 were slightly slower than those obtained in subcritical toluene. Increasing the reaction temperature increased the rate of hydrosilylation in sc CO_2. The authors also observed that this reaction proceeded slower in dilute subcritical toluene solution in contrast to concentrated subcritical toluene.

Although phase separation occurred for all reactions in subcritical toluene, this did not occur in sc CO_2. However, the formation of a gel, that did not occur under any conditions in subcritical toluene, was noted in all reactions performed in sc CO_2. According to the authors, the gel could arise from crosslinking through residual Si–H groups.

Obtaining silicones which bear lateral fluorinated chains has attracted much interest. Although the same strategy was employed (hydrosilylation reaction yet in the presence of different catalysts), a wide range of unsaturated fluorinated molecules containing various spacers were synthesised, leading to different applications (membranes, liquid crystals, fluorescent silicones, biomaterials, release paper and antifouling coatings).

Besides polydienes and polysiloxanes, other polymeric backbones have also been modified with fluorinated reactants.

The sub-sections below include various methodologies of syntheses of fluorinated comb-like copolymers: esterification, etherification by nucleophilic substitution and ionic addition.

2.2.4. Synthesis of Fluorinated Graft Copolymers by Condensation

Although esterification reaction has been mentioned in Section 2.2.2.2 for the chemical modification of polydienes, the four non-exhaustive examples below also require this reaction to obtain fluorinated graft copolymers containing modified poly(methylvinylether), PMMA, poly(styrene-*co*-acrylamide) and polysulfones.

2.2.4.1. Chemical Modification of Poly(methylvinylether-alt-maleic anhydride)

The thermal ring opening reaction of poly(methylvinylether-*alt*-maleic anhydride) copolymers (or Gantrez®) with 1H,1H,2H,2H-perfluorohexanol or perfluorooctanol led to an original fluorinated comb-like structure copolymer in 60% yield [191]:

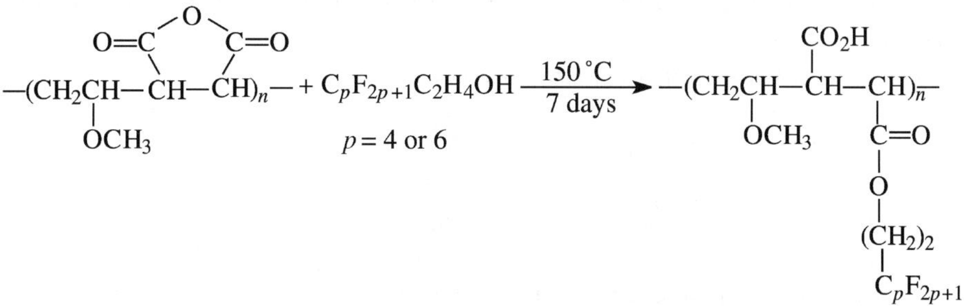

$$-(CH_2CH-CH-CH)_n- + C_pF_{2p+1}C_2H_4OH \xrightarrow[\text{7 days}]{150\,^{\circ}C} -(CH_2CH-CH-CH)_{\overline{n}}$$

(with anhydride ring O=C–O–C=O, OCH₃ substituent, p = 4 or 6; product bears OCH₃, CO₂H and C=O–O–(CH₂)₂–C_pF_{2p+1} graft)

These well-architectured copolymers were successfully used as stabilisers for the dispersion polymerisation of MMA in sc CO_2, leading to PMMA in high yields, with $M_n = 65{,}000-173{,}400$, PDI = 2.1–3.1 and narrow particle size distributions (2.6–3.3 μm).

Actually, the increased length of the fluorinated graft chain has clearly improved the effectiveness of the stabiliser and also led to a decrease in the particle size. In addition, the balance between the length of the grafted chain and hydrocarbon content of the polymer backbone was particularly important. For example, similar investigations using poly(maleic anhydride-*alt*-1-octadecene) grafted with $C_4F_9C_2H_4OH$ led to PMMA which did not form particles even at a 5% loading [192].

2.2.4.2. Modification of PMMA

Original stabilisers for the same polymerisation in sc CO_2 were obtained by Lepilleur and Beckman [193], who condensed acid chloride terminated poly(perfluoropropylene oxide) with poly(methyl methacrylate-*co*-hydroxyethyl methacrylate) copolymer.

The general formula was as follows:

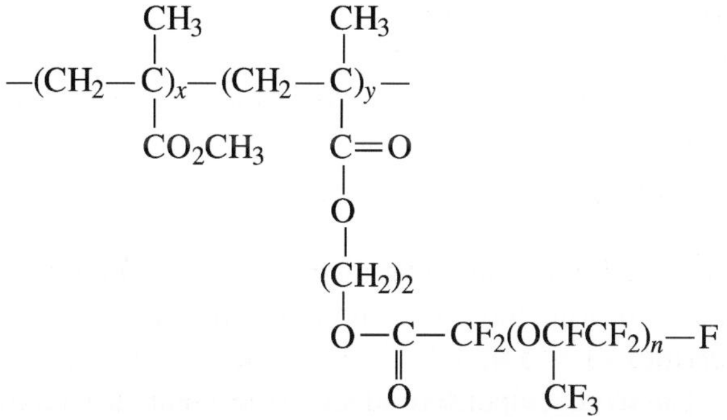

$$-(CH_2-\underset{CO_2CH_3}{\overset{CH_3}{C}})_x-(CH_2-\underset{\substack{C=O\\|\\O\\|\\(CH_2)_2\\|\\O-\underset{O}{\overset{\|}{C}}-CF_2(OCFCF_2)_n-F\\CF_3}}{\overset{CH_3}{C}})_y-$$

The authors mentioned that a careful balance between the size of the anchoring group (backbone length) and the amount of soluble component (graft chain length) was necessary but not sufficient to achieve the best stabilisation.

2.2.4.3. Chemical Modification of Poly(styrene-co-acrylamide)

Similarly, condensation of heptafluorobutyric anhydride onto the hydroxy pendant groups of poly(styrene-*co*-acrylamide)-*g*-poly(ethylene oxide) copolymers led to original fluorinated comb-like copolymers [194]. The molecular weights of the starting materials ranged between 10,000 and 80,000, while those of poly(ethyleneoxide), PEO, grafts were 2500–5000 (i.e., 29–55 wt% of PEO content).

Indeed, the overall statistical structure is:

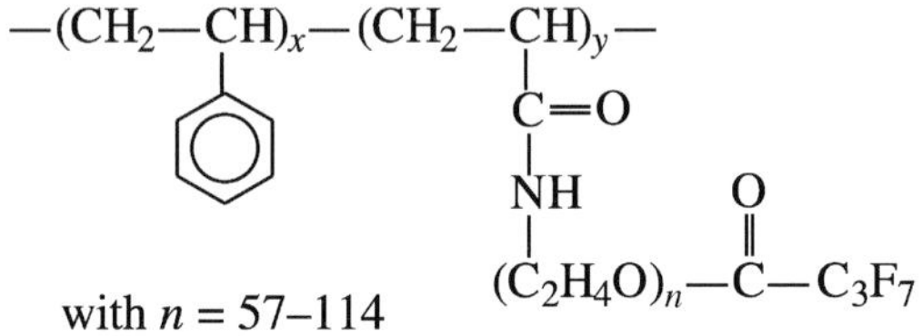

$$-(CH_2-CH)_x-(CH_2-CH)_y-$$

with $n = 57-114$

$(C_2H_4O)_n-\overset{O}{\overset{\|}{C}}-C_3F_7$

Jannasch [194] observed that in PS-*g*-PEO characterised by XPS, PS dominated the surface of the copolymer and that the PEO content decreased with decreasing the sampling depth. In contrast, the surface region of the corresponding fluorinated copolymers was dominated by PEO, and the PEO content increased with decreasing sampling depth. The fluorine content of fluorinated PEO also increased with decreasing sampling depth, indicating that the fluorinated chain ends were segregated at the outermost surface (of few angstroms).

2.2.4.4. Chemical Modification of Polysulfones

Pospeich's group [195] first synthesised polysulfones by condensation of 4,4'-bis(hydroxyphenyl) valeric acid, bisphenol A and 4,4'-bis(chlorophenyl)sulfone. Then, the condensation of $HO(CH_2)_{10}C_{10}F_{21}$ onto the side acid function led to original polysulfones bearing a $-C_{10}H_{20}-C_{10}F_{21}$ lateral group:

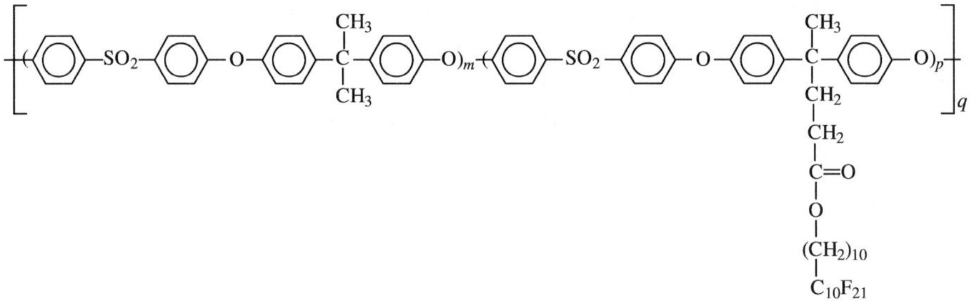

These polymers containing randomly distributed semifluorinated side chains were soluble in chloroform, and possessed molar masses of 36,200–62,300 and broad polydispersities (3.3–5.9).

Interestingly, the oxydecylperfluorodecyl side chain favoured self-organisation of the semifluorinated side chains and hence improved surface properties. Indeed, the contact angles were high (115–122°), making such materials quite hydrophobic, more so than the unmodified polysulfone.

2.2.5. Chemical Modification of Polymers by Nucleophilic Substitution

2.2.5.1. Chemical Modification of Polyparaphenylenes

Another original approach was suggested by Le Ninivin [196], who synthesised novel polyparaphenylenes (PPPs) bearing 2-sulfonic acid-1,1,2,2-tetrafluoroethyl side groups. They were obtained from the nucleophilic substitution of *para*(1-sulfonic acid-1,1,2,2-tetrafluoro)phenolate onto poly(*p*-fluorobenzoyl-1,4-phenylene) in 95% yields. With 1.2 eq. excess of sulfonated phenolate, the degree of sulfonation was 95%, leading to hydrosoluble graft copolymer, whereas a 0.4 eq. of this fluorinated derivative led to 41% degree of sulfonation.

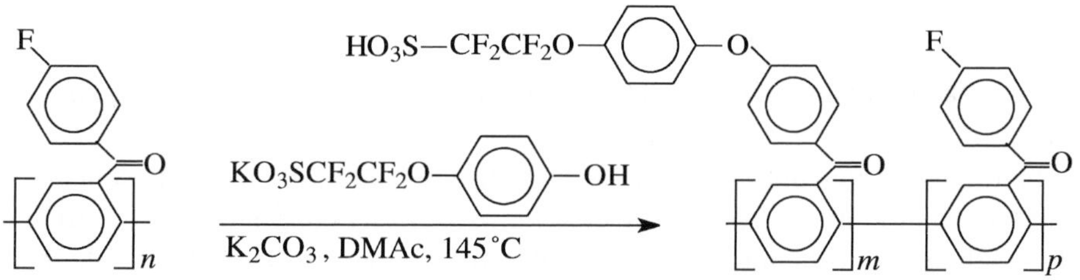

These amorphous polymers ($T_g = 156\,°C$) have fair to high molecular weights (in eq. PS): $\bar{M}_n = 26,000$ and $\bar{M}_w = 65,000$. Interestingly, the electron-withdrawing C_2F_4 adjacent group enables the sulfonic acid to show an enhanced acidic character. By small angle X-ray scattering measurements, the author observed that, from a certain amount of grafting, organisation occurred. The French author could assess a distance of 3 nm between the ionic clusters, the ionic part being far enough from the rigid polyparaphenylene backbone.

From these fluorinated PPPs, this author elaborated original membranes endowed with a good thermal stability (these films were stable in air up to 310 °C), interesting electrochemical properties, very low methanol crossover (its intrinsic permeability to methanol is even lower than that of Nafion® for the same thickness), high ionic exchange capacity (1.3 meq H^+ g^{-1}) and satisfactory conductivity (8.5 mS cm^{-1}) for a thickness of 40 μm. All these relevant characteristics show that these original membranes are potential candidates for direct methanol fuel cells [197].

2.2.5.2. Chemical Modification of Polyphosphazenes

Fluorinated polyphosphazenes [198], rather scarce in contrast to the hydrogenated homologues [199], have been synthesised from the nucleophilic substitution of the halogen atoms in polyphosphazenes by fluorinated alcoholates:

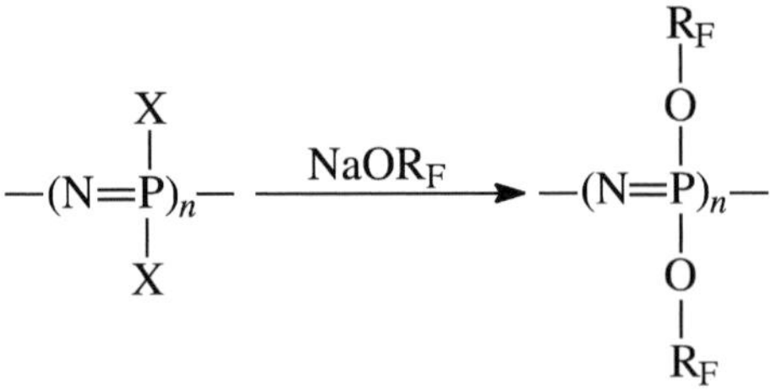

$$
\begin{array}{c}
X \\
| \\
-(N{=}P)_n{-} \\
| \\
X
\end{array}
\quad\xrightarrow{\ \text{NaOR}_F\ }\quad
\begin{array}{c}
R_F \\
| \\
O \\
| \\
-(N{=}P)_n{-} \\
| \\
O \\
| \\
R_F
\end{array}
$$

with R_F: CF_3CH_2, CF_3CFHCH_2.

These fluorinated inorganic–organic hybrid polymers, although bearing short fluorinated grafts, preserve the molecular parameters of the polyphosphazene precursors. The resulting grafted polymers are film- and fibre-forming materials that have hydrophobic and non-adhesive properties.

2.2.6. Modification of Polyvinylalcohol by Ionic Addition of Fluoroalkenes

Feiring and Wonchoba [200] carried out the ionic addition (in the presence of a catalytic amount of potassium *tert*-butoxide) of commercially available fluoroalkenes (TFE and perfluoropropyl vinyl ether) to poly(vinyl alcohol) (PVOH) ($\bar{M}_w = 125,000$ and 100% hydrolysed). The statistic structure of the resulting graft copolymers was as follows:

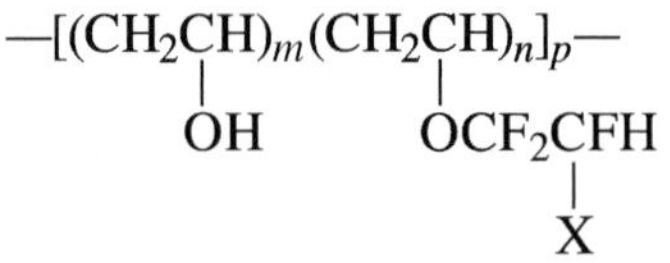

$$
\begin{array}{c}
-[(CH_2CH)_m(CH_2CH)_n]_p- \\
\ \ \ \ \ \ \ \ |\ \ \ \ \ \ \ \ \ \ \ \ \ \ \ | \\
\ \ \ \ \ \ \ \ OH\ \ \ \ \ \ \ OCF_2CFH \\
\ | \\
\ X
\end{array}
$$

with $X = F$ or OC_3F_7.

The degrees of grafting were ranging from 33 to 78%.

Although the grafts were short since the conditions did not allow the propagation of these fluoroolefins, they strongly modified the properties of PVOH: the comb-like copolymers produced were amorphous (T_gs up to 65 °C were assessed), soluble and endowed with a low-refractive index (1.3852).

2.2.7. Conclusion

A wide range of fluorinated grafted copolymers have been proposed and synthesised from various simple reactions (direct thermal grafting, nucleophilic substitutions, condensation). The fluorinated group or the ion-exchange function brings the desired property, and potential applications as membranes for fuel cell, oil and water repellent materials, surfactants for radical polymerisations in sc CO_2 and original amorphous and soluble polymers for optics have already been proposed.

2.3. From the Telomerisation of Monomers with Macrotelogens

In contrast to the synthesis of block copolymers (or block cotelomers) obtained from the telomerisation of monomers with telomeric transfer agents, to the best of our knowledge no graft copolymers have been reported in the literature. The difficulty may arise from the synthesis of a macrotelogen that exhibits transferring side-groups.

However, we suggest an elegant example of fluorinated polymer bearing mercapto groups located in a lateral position about the polymeric fluorinated backbone, although several hydrogenated macromolecules possessing these thiol functions in the side chain have already been synthesised [201, 202].

It is well known that thiol-ene systems are readily used to vulcanise hydrogenated elastomers, but this process has scarcely been used to cure fluoroelastomers because of the difficult availability of mercapto side groups in fluoropolymers. Nevertheless, we used an original trifluorovinyl ω-thioacetate monomer able to copolymerise with VDF [203] and the resulting copolymer was hydrolysed to generate elastomers bearing mercapto lateral groups. In addition, a crosslinking reaction was performed in the presence of non-conjugated dienes, as follows:

$$F_2C=CFCH_2CH=CH_2 + HSCOCH_3 \xrightarrow{\text{UV}} \underset{\text{FSAc}}{F_2C=CFC_3H_6SCOCH_3}$$

$$FSAc + F_2C=CH_2 \xrightarrow{\text{rad.}} \text{poly(VDF-}co\text{-FSAc)}$$

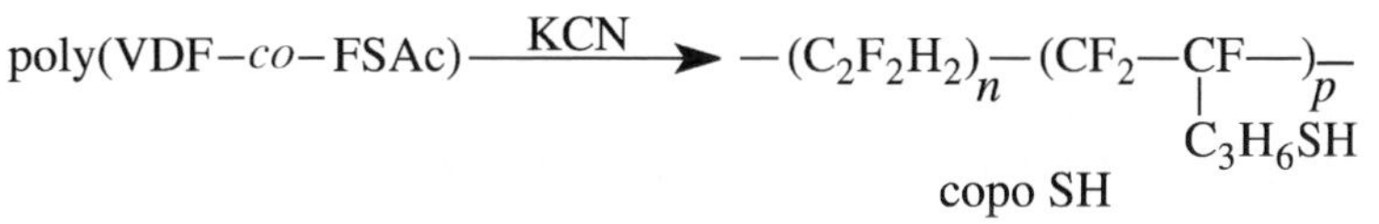

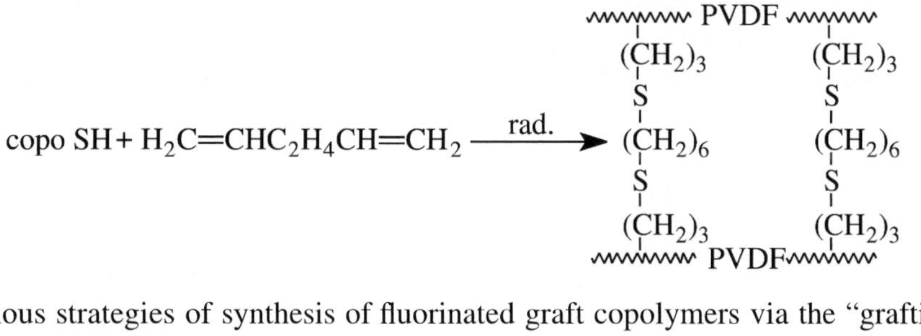

$$\text{poly(VDF-}co\text{-FSAc)} \xrightarrow{\text{KCN}} -(C_2F_2H_2)_n-\underset{\underset{C_3H_6SH}{|}}{(CF_2-CF-)_p}$$

copo SH

$$\text{copo SH} + H_2C=CHC_2H_4CH=CH_2 \xrightarrow{\text{rad.}}$$

$$\begin{array}{cc}
\text{wwww PVDF wwww} \\
| \qquad\qquad | \\
(CH_2)_3 \qquad (CH_2)_3 \\
| \qquad\qquad | \\
S \qquad\qquad S \\
| \qquad\qquad | \\
(CH_2)_6 \qquad (CH_2)_6 \\
| \qquad\qquad | \\
S \qquad\qquad S \\
| \qquad\qquad | \\
(CH_2)_3 \qquad (CH_2)_3 \\
\text{wwww PVDF wwww}
\end{array}$$

Various strategies of synthesis of fluorinated graft copolymers via the "grafting onto" technique have been possible. They can be simple radical addition of R_FI or $R_FC_2H_4SH$ onto the pendant ω-unsaturated groups of poly(butadiene) or poly(isoprene), hydrosilylation of α-unsaturated ω-perfluorinated derivatives onto PHMS or condensation reactions. However, in the second case, the difficulty may arise from the preparation of the fluorinated monomer to graft onto the polysiloxanes. Interestingly, the wide variety of the nature of the polymeric backbones deserves to be mentioned since poly(diene), polysiloxane, polysulfone, polyparaphenylene, poly(methylvinylether-*alt*-maleic anhydride), PMMA, poly(S-*co*-EO) and polyphosphazene have been chemically modified.

2.4. SYNTHESIS OF FLUORINATED GRAFT COPOLYMERS BY THE "GRAFTING FROM" METHOD

2.4.1. Introduction

This sub-section describes the synthesis of original fluorinated graft copolymers mainly achieved from the introduction of grafts brought by macroinitiators, the initiating function being located in the side groups. These macroinitiators can be obtained via two different routes: (i) from the co- or terpolymerisation of fluoromonomers, one of them containing an initiating species which does not participate in the (co)polymerisation; and (ii) from the activation of the fluoropolymers under ozone, plasma, swift heavy ions (SHIs), X-rays or electron beam.

Then these macroinitiators enable one to initiate the radical (co)polymerisations of various monomers leading to graft copolymers. Such a process starting from the initiating functions borne by the polymer is named the "grafting from" technique.

2.4.2. Synthesis of Macroinitiators by the Chemical Method

An original concept of synthesis of fluorografted copolymers was suggested by Japanese companies. First, Oshibe *et al.* (for Nippon Oils and Fats Company) [204] prepared PMMA-*g*-PFDA graft copolymers from the dead-end polymerisation (DEP) [205] of 1H,1H,2H,2H-perfluorodecyl acrylate (FDA) initiated by a PMMA macroperoxide (PMMA-In). This macroperoxide was obtained by radical polymerisation of a methacrylate bearing a peroxide group at low temperature. Each polymeric chain has ca. 1.3 O−O bonds which correspond to about 40% of the O−O bonds in the used PMMA-In. The reaction was suggested as follows:

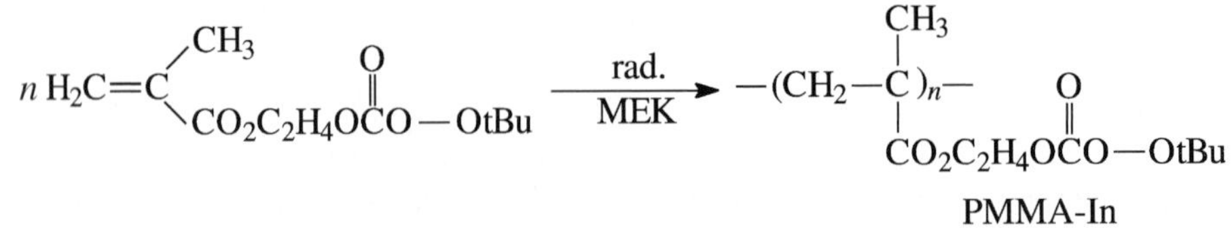

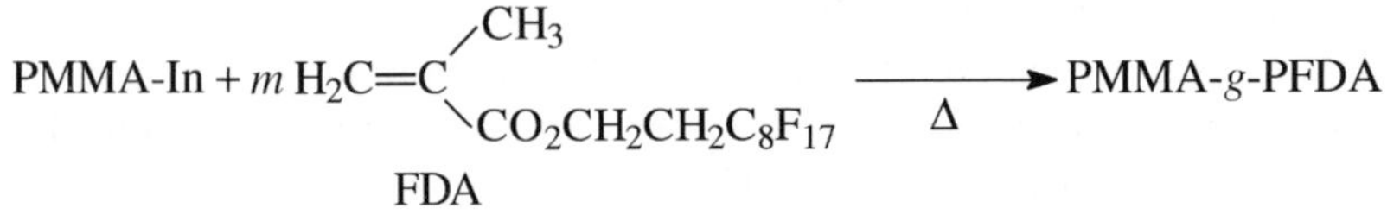

$$n \; H_2C=C \overset{CH_3}{\underset{CO_2C_2H_4OCO-OtBu}{\diagup}} \quad \xrightarrow[\text{MEK}]{\text{rad.}} \quad -(CH_2-\overset{CH_3}{\underset{CO_2C_2H_4OCO-OtBu}{C}})_n-$$

PMMA-In

$$PMMA\text{-}In + m \; H_2C=C \overset{CH_3}{\underset{CO_2CH_2CH_2C_8F_{17}}{\diagup}} \quad \xrightarrow{\Delta} \quad PMMA\text{-}g\text{-}PFDA$$

FDA

Although graft copolymers were expected, the authors also claimed the synthesis of block copolymers since they noted that several of the O–O bonds were surprisingly located at the end of the polymeric chain.

These copolymers were obtained as dispersions with particle sizes of 0.05–0.10 μm that were also affected by the ratio of the monomer feed and the nature of the solvent.

The second example was suggested by the Central Glass Company Ltd [206]. It consists of a two-step procedure in which the first one generates fluorinated macroinitiators. This was realised by carrying out a suspension terpolymerisation of two fluoroalkenes and an allylic monomer (bearing a peroxycarbonate group at low temperature) to enable the reaction of the hydrogenated double bond of *tert*-butyl peroxyallyl carbonate (TBPAC) without decomposing the peroxycarbonate function, as follows:

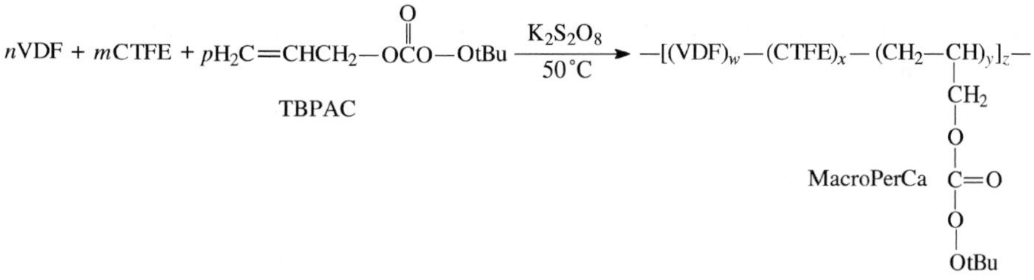

$$n\text{VDF} + m\text{CTFE} + p\text{H}_2\text{C}{=}\text{CHCH}_2{-}\text{O}\overset{\text{O}}{\overset{\|}{\text{C}}}\text{O}{-}\text{OtBu} \xrightarrow[50\,°C]{\text{K}_2\text{S}_2\text{O}_8} -[(\text{VDF})_w{-}(\text{CTFE})_x{-}(\text{CH}_2{-}\text{CH})_y]_z-$$

TBPAC   MacroPerCa

with side chain: CH$_2$–O–C(=O)–O–OtBu

$$\text{MacroPerCa} + q\text{VDF} \xrightarrow{95\ °C} \text{poly(VDF-}co\text{-CTFE)-}g\text{-PVDF}$$

In the various examples, VDF, CTFE and HFP were chosen as the fluoroalkenes to give massic yields of the macroperoxycarbonate (MacroPerCa) of about 60–80%.

By DSC analysis at low temperatures, the T_gs of poly(VDF-*co*-CTFE-*co*-TBPAC) and poly(VDF-*co*-HFP-*co*-TBPAC) were −21 and −19 °C, respectively. The presence of the peroxycarbonate groups was demonstrated (i) by an exothermic peak (assessed by DSC) at 160–180 °C attributed to the peroxycarbonate decomposition, and (ii) by the content of the active oxygen of the terpolymer (0.042 and 0.063%, respectively).

However, neither the molecular weights nor the contents of the three monomers in the terpolymer were given. In addition, the obtaining of high molecular weight polymers may be questionable since, in the course of the terpolymerisation, allylic transfer (arising from the TBPAC) should occur and hence should lead to oligomers.

The preparation of the fluorografted copolymer, composed of a fluoroelastomeric backbone and fluorinated thermoplastic grafts, was achieved from the polymerisation of VDF in solution initiated by the macroperoxycarbonate at 95 °C. These thermoplastic elastomers obtained in high massic yields (81–91%) exhibit a melting point of 155–160 °C.

Unfortunately, no information on the presence (or absence) of PVDF produced from the direct addition of the *tert*-butoxy (or methyl) radical onto VDF was given as reported in a previous study [207].

Conversely, these Japanese authors [208] synthesised poly(CTFE-*co*-TBAPC)-*g*-poly(VDF-*co*-HFP) grafted copolymers in high yield showing a low T_g ($-19\ °C$). In that case, the polymeric backbone consists of a thermoplastic PCTFE copolymer while the grafts are elastomeric.

Interestingly, all those fluorografted copolymers were kneaded and press-shaped at 200 °C, leading to semi-transparent, elastic and flexible sheets that were compared to those prepared from commercially available fluoroelastomers. These original thermoplastic elastomers exhibit much better mechanical properties (break strength and elongation at break).

Such a strategy was also successfully applied at the Nippon Oil and Fats Company Ltd [209], where hydrogenated graft copolymers such as PVAc-*g*-PS and PMMA-*g*-PS were prepared.

2.4.3. Synthesis of F-Graft Copolymers from the Activation of Fluoropolymers Followed by Grafting

2.4.3.1. Introduction

Many studies have been devoted to the synthesis of fluorinated graft copolymers involving a first step of activation of the fluoropolymer. Although such halogenated polymers are resistant to certain irradiations, they can be activated (called pre-irradiation) on their surface or even in bulk by various systems, producing trapped radicals (under vacuum) or peroxides or hydroperoxides (in the presence of air or O_2). As a consequence, the activated fluoropolymer acts as an original macroinitiator able to initiate the polymerisation of a M monomer, hence yielding a graft copolymer [210–216]. This graft copolymer can also be obtained from a simultaneous irradiation/grafting method [210–213].

These reactions can be briefly sketched as follows:

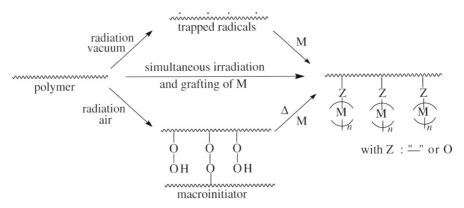

Various fluoropolymers have been activated by different techniques [210–217]: thermally, corona, ozone, SHIs, plasma, electron beam or X-rays. The order of the presentation of these different methods of activation has been motivated by the increasing efficiency of the radiation. For example, ozone smoothly activates fluoropolymers in contrast to electron beam or X-ray radiations which irradiate all the bulk supplying peroxides in high amount. The absorbed radiation dose is defined as the amount of energy imparted to matter. The most recent unit of radiation is Gray (Gy), which corresponds to 1 J kg^{-1}, or 100 rad or 10^4 erg g^{-1} [210–213, 216]. For larger doses, kilogray (kGy) is used. The dose rate is, therefore, defined as the absorbed dose per time unit and is represented as Gy min^{-1}.

The structure of the final grafted copolymer generally depends upon the ratio between the rate of polymerisation and the rate of diffusion of the functional monomer into the fluoropolymeric matrix and, once polymerisation is started, the upper layer with the grafted copolymer usually promotes the diffusion of functional monomer into the lower layer of the fluoropolymeric matrix.

Indeed, one of the breakthroughs of these surveys concerns the grafting of hydrophilic monomers onto these fluoropolymers to obtain, in certain cases, less hydrophobic fluoropolymers, ion-exchange membranes (IEMs) for various applications or interesting polymers for biomedical uses. In both first cases, an improved processing of the F-polymer was noted.

This sub-section deals with the synthesis of fluorinated graft copolymers by the "grafting from" method, i.e., the fluorinated polymers have been first activated by ozonisation or by various other methods (plasma, thermal, γ-ray or electrons) (first step), followed by a subsequent reaction of grafting (second step).

In each step, the fluorinated graft copolymers prepared from different kinds of F-homopolymers (e.g., PTFE, PVDF and PCTFE) and also those arising from their different families of F-copolymers (poly(VDF-co-HFP), poly(VDF-co-TFE), poly(TFE-co-HFP) (or FEP), ETFE and poly(TFE-co-PAVE) (or PFA)) are reported.

2.4.3.2. By Thermal Activation: Chemical Modification of Polyolefins

In this sub-section, we consider first the synthesis and the properties of graft copolymers starting from the activation of hydrogenated polymers followed by the grafting of fluorinated derivatives. The second part concerns the synthesis of fluoropolymer-g-poly(M) graft copolymers from the irradiation of a fluoropolymer followed by the grafting of hydrogenated monomers on the resulting macromolecular initiator.

First, let us consider the grafting of fluorinated derivatives onto hydrogenated polyolefins, and especially onto polyethylene and polypropylene, mentioned in the literature.

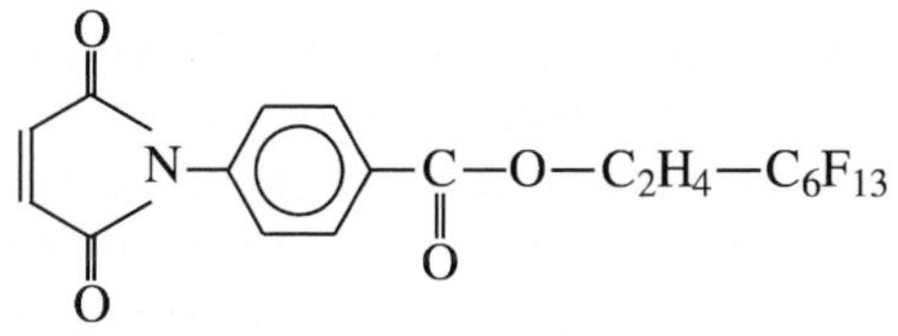

The grafting of fluorinated maleamic acid and maleimide onto high density polyethylene in the molten state was optimised [84b]. This latter derivative, was more reactive than the former one, and the presence of a peroxidic initiator improved the grafting: in the best conditions, a grafting yield of 80% was achieved.

Another strategy was suggested by the DuPont Company [218] to graft $H_2C=CHC_2F_4OC_2F_4SO_2F$ (FM) onto polyethylene in the presence of di-*tert*-butyl peroxide, leading to the corresponding PE-*g*-PFM graft copolymer endowed with a $T_m = 115\ ^\circ C$ and a 10% wt. loss by TGA at 380 °C under nitrogen.

Indeed, this work was inspired from investigations carried out at the Japanese Atomic Energy Research Institute [219], which concerned the preparation of PE-*g*-PGMA graft copolymers (where GMA means glycidyl methacrylate).

Regarding the grafting onto polypropylene (PP), four non-exhaustive examples have been considered. The first one deals with the addition of bis(1,1,2,2-tetrahydroperfluoroalkyl) fumarate in the presence of dicumyl peroxide. The authors noted a fluorinated monomer conversion of 28% [220].

Grafting poly(*N*-butyl perfluorooctanesulfonamidoethyl acrylate) (PBuFO-SEA) was also achieved at the molten state of PP (180–220 °C) using a peroxidic initiator [221] leading to PP-*g*-PBuFOSEA graft copolymers with 63–100% of BuFOSEA conversion.

The proof of the grafting was afforded by IR (presence of the carbonyl absorbance of the poly(F-acrylate) moiety), by hexadecane contact angle measurements (0, 47 and 19–23° for PP, PTFE and PP-*g*-PBuFOSEA, respectively) and by ESCA (that showed the increase of the F : C ratio for higher degrees of grafting). In fact, the surface energies assessed from PP, PTFE and PP-*g*-PBuFOSEA were 32, 22 and 25 dyne cm^{-1}, respectively [221], showing good surface properties of these graft copolymers.

The same strategy was also successfully investigated by Tan *et al.* [222], who grafted 1-heptadecafluorooctylethyl acrylate onto PP at 180 °C in the presence of a peroxide. 52% conversion of fluoroacrylate was reached. Free radical grafting of fluorinated acrylates onto two grades of PP of molecular weights ranging from 74,000 to 307,000 was achieved by introducing styrene as a comonomer. This aromatic comonomer plays the double role (i) to help and improve the grafting of the fluoroacrylic monomers, and (ii) to limit side reactions frequently observed when free radicals are used (indeed, the radical conditions make the polypropylene suffering from severe chain degradation by β-scission [223]). The reaction was carried out in the molten state of PP in the presence of 1,3-bis(*tert*-butylperoxy-isopropyl)benzene (DIPP) at 180 °C.

These authors suggested a radical grafting mechanism in which they observed that styrene(s) reacted first with PP macroradicals and the resulting macroradical terminated by a styryl end-unit initiated a further propagation of the fluoroacrylate. They also noted that the lower the concentration of the free radical generator, the higher the contribution of styrene.

The amounts of grafted fluoroacrylate (FA) and FA conversions were assessed by FTIR and they increased with the increase of DIPP initiator content. However, the highest FA conversion value was 52%, corresponding to a weight composition of 100/5/5/1 for PP/FA/S/DIPP, while this value dropped to 37% in the absence of styrene.

The interest of this survey was to improve the surface properties of PP by introducing perfluorinated side-groups.

Niyogi and Denicola [224] recently claimed the synthesis of PP-g-PVDF graft copolymer from the thermal addition of VDF onto PP in the presence of di(4-*tert*-butylcyclohexyl) peroxy dicarbonate. The PVDF graft content was estimated to be 9.4%.

2.4.3.3. Synthesis of Graft Copolymers from the Ozone-Activation of the Polymer

Ozone, commonly written O_3, is a cheap gas (and quite soluble in fluorinated solvents) and is nowadays well known because of environmental concerns. It is quite useful in industrial fields such as the sterilisation of water, the whitening of wood pulp and the treatment of polymers to get rid of any colour that arises after melt-moulding or sintering [225].

The ozonisation (or ozonation) of polymers has been investigated by many authors. It has allowed the activation of a wide range of polymers, mainly polyolefins (polyethylene, polypropylene, PVC) but also PS, poly(dienes), PDMS, polyurethanes and finally copolymers. All these data can be found in an interesting recent review [226]. In addition, the modification of natural polymers using ozone was also achieved: for example, Yoshida *et al.* [227] prepared thermostable materials based on ozonised lignin, while Platz [228] investigated the decrosslinking of vulcanised rubbers.

The process of ozonisation is a rather simple procedure since polymers can be activated in the solid state without any solvent or any purification, and thus many applications have been developed starting from this reagent: for example (i) the modification of polymeric surfaces by introducing functions that bring complementary properties such as adhesion, hydrophilicity or dyeing, and (ii) the synthesis of graft copolymers after grafting various monomeric units onto a saturated polymeric backbone.

Indeed, the composition of ozone is complex as a result of the coexistence of various chemical species. First, the molecule is quite dipolar, used for its particular oxidative properties. This gas is most often produced from cold plasma based on the flow of air or pure oxygen through electrodes. Oxygen is then decomposed into different species such as negative ions (e.g., O^-, O_2^-, O_3^-, O_4^-), positive ions (e.g., O^+, O_2^+, O_3^+, O_4^+) and neutral species (e.g., O, O_2, O_3), which behave as more or less excited states and are present when ozone is generated.

This gas is relatively stable up to 70 °C and can decompose into ionic species or other excited species when increasing the temperature, O_2^* being the first excited state of oxygen, for example.

Polymers that react with ozone yield peroxides and hydroperoxides, and various functions such as acids or hydroxy groups and, to a less extent, ketones and lactones. Those containing unsaturations (e.g., polydienes) are the most sensitive to ozone and their treatment quickly leads to poorly stable ozonides [229], hence producing a rapid decrease of their molecular weights that generates α,ω-difunctional or telechelic oligomers.

From a mechanistic point of view, Razumovski et al. [230] were the pioneers in the understanding of the peroxidation process of various polymers (PE, PS, polydienes). Basically, the mechanism can be summarised in the following scheme, starting from PVDF:

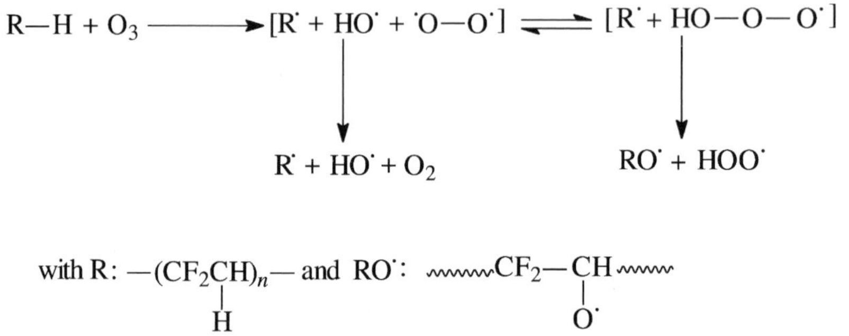

$$R{-}H + O_3 \longrightarrow [R^\cdot + HO^\cdot + {}^\cdot O{-}O^\cdot] \rightleftharpoons [R^\cdot + HO{-}O{-}O^\cdot]$$

$$R^\cdot + HO^\cdot + O_2 \qquad\qquad RO^\cdot + HOO^\cdot$$

with R: $-(CF_2CH)_n-$ and $RO^\cdot$: $\sim\!\!\sim\!\!\sim CF_2{-}CH\sim\!\!\sim\!\!\sim$
with the H below R's CH and $O^\cdot$ below the second CH

In the case of the ozonisation of PP [226, 229], when the treatment time ranged between 1 and 6 h, the content of peroxide and hydroperoxide could be assessed and varied in the following ranges: $0.3{-}1.6 \times 10^{-5}$ mol g^{-1} of PP and $0.25{-}5.95 \times 10^{-5}$ mol g^{-1} of PP, respectively.

Ozonisation is an original means for activating polymers, hence leading to the formation of novel macroperoxides and macrohydroperoxides. This process has been successfully carried out on polyethylene, polypropylene and PVC [231]. For example, PE-g-poly(fluoroacrylate) graft copolymers were synthesised by ozonisation of low-density polyethylene (LDPE) followed by bulk grafting of $H_2C{=}CHCO_2\,C_2H_4C_6F_{13}$ [232]. The corresponding copolymer contained 30 wt% of fluoroacrylate (FA).

The same article reports the preparation of more hydrophilic PE-g-PAAc copolymers by grafting acrylic acid (AAc) onto ozonised LDPE followed by the ring opening of a fluorinated epoxide (R_F: C_6F_{13} or C_8F_{17}) onto these PE-g-PAAc, as follows [232]:

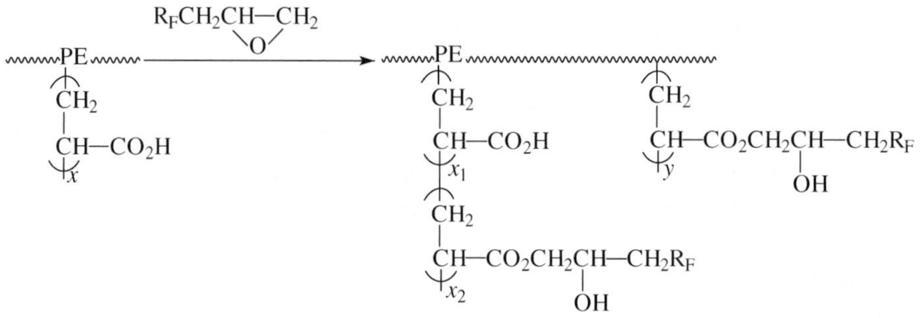

As for the ozonisation of PVDF, slight difficulties have been encountered and, even under hard conditions, little degradation of this polymer was noted [226, 233].

Nevertheless, previous investigations carried out in our laboratory have shown that the activation of PVDF with ozone led to macroinitiators containing mainly peroxides (90%) and hydroperoxides (10%), as observed from the ozonisation of hydrogenated polymers. The content of hydroperoxides decreased as the temperature of ozonisation was increased [234]. The former initiating species were titrated by diphenyl picryl hydrazyl or by iodometry [233], while the latter ones were evidenced according to the method of Zeppenfeld [235]. Indeed, only hydroperoxides in acidic medium can oxidise ferrous ions. Furthermore, the contents of the different peroxide species were assessed. For example, when the ozonisation was achieved in bulk (at 40–80 °C for 12–48 h), the content of peroxides was $1.7–5.5 \times 10^{-5}$ mol g^{-1}, while it was higher (6.3–7.1×10^{-5} mol g^{-1}) when the process occurred in solution (DMF at 25–50 °C for 30–60 min) [234]. In addition, the authors determined the overall dissociation rate constant, k_d, of ozonised PVDF and found it to be abnormally high. For example, k_d was 10.33×10^{-4} s^{-1} at 100 °C and 8.8×10^{-4} s^{-1} at 90 °C. Furthermore, we reported the activation energy and the Arrhenius coefficient of the initiators in the ozone-treated PVDF to be 39 and 5.8 kJ mol^{-1}, respectively, hence deducing the half-life for the decomposition of the macroperoxides of 10 min at 100 °C [234, 236].

Although PVDF is more difficult to activate, obtained species have the same reactivity as those generated from organised PE or PVC.

The second step concerns the grafting, in bulk or in solution, of an oligomer or polymer of MMA, styrene, glycidyl methacrylate (GMA) and AAc. In all cases, the degrees of grafting, assessed from elemental analyses, were higher for processes carried out in bulk (e.g., 5–8 mol% from solution grafting of MMA vs. about 17–29 mol% from bulk grafting). This value increased as the temperature of grafting was lowered. The same tendencies were noted for styrene, GMA and AAc that could be grafted in 16, 50 and 12 mol%, respectively.

Nevertheless, in all cases, adapted experimental recipes were proposed to remove the corresponding homopolymers (that were obtained in ca. 40% yield) [236]. Actually, the grafting reaction was monitored by the torque values of blends throughout the mixing process. This procedure also showed that the graft copolymers could not be obtained if PVDF was not activated by ozone.

These PVDF-g-PMMA, PVDF-g-PS, PVDF-g-PAAc and PVDF-g-PGMA copolymers were tested in various applications. First, the mechanical behaviour of the PVDF/PS blend was greatly improved by adding PVDF-g-PS copolymer. Indeed, this led to a 2.5-fold increase in the break-stress for a 50/50 wt% blend [236].

Second, the adhesion of PVDF to a glassy–epoxy composite was achieved by using PVDF-g-PAAc and PVDF-g-PGMA copolymers (Table III). In both cases, a great increase of the peeling strength was noted, showing that cohesive-type breaks in PVDF were produced [236]. Third, the introduction of hydrophilic

TABLE III

SYNTHESIS OF PVDF-*g*-PM GRAFT COPOLYMERS FROM THE OZONE TREATMENT OF PVDF FOLLOWED BY THE GRAFTING OF M MONOMER

Grafted M monomer	Conditions of ozonation	Properties of the graft-copolymer	Applications	References
AAc (d.o.g. 12%)	Ozonation in bulk powdered PVDF	Adhesion, hydrophilicity	Adhesive compatibiliser	[234, 236]
AAc (d.o.g. 5–100%)	$[O_3] = 0.027$ g l^{-1}/15 min grafting: 60 °C	Good thermal stability, uniform surface composition	pH-sensitive MF membrane	[237, 238]
APHOS grafting 30–60 °C/6–24 h (d.o.g. 10–30%)	Ozonation in bulk powdered PVDF	Adhesion	Coating, adhesion onto steel	[233]
DMAEMA	As above	Hydrophilicity	Amphiphilic polymers	[234]
GMA (d.o.g. 50%)	As above	Adhesion	Adhesive of polymers	[234, 236]
MMA 70–100 °C/1–14 h (d.o.g. 7–29%)	As above	Hydrophilicity	Amphiphilic polymers	[234, 236]
NIPAAM	Ozonation in bulk	Hydrophilicity	T sensitive membrane	[245]
NIPAAM/AAc	Ozonation in bulk	Hydrophilicity	Porous membrane	[245]
NIPAAM/BuMA	Ozonation in bulk	Hydrophilicity	Porous membrane	[245]
NIPAAM grafting in NMP	Ozonation on film	Hydrophilicity, morphology	MF membrane	[246]
PEOMA	Grafting: 100 °C/5–9 min	High ionic conductivity, electrochemical stability	Li ion rechargeable batteries	[241]
Styrene	Ozonation in bulk powdered PVDF	Higher hydrophobic PS	Compatibiliser	[236]
4-Vinyl pyridine, 55 °C	Ozonation on film	Electrochemical stability	Ultrafiltration membrane	[247]

AAc, acrylic acid; APHOS, diethyl phosphonate ethyl acrylate; DMAEMA, dimethylaminoethylmethacrylate; GMA, glycidyl methacrylate; MMA, methyl methacrylate; NIPAAM, *N*-isopropylacrylamide; PEOMA, polyethylene glycol methacrylate; MF, microfiltration; d.o.g., degree of grafting.

groups (brought by acrylic acid, MMA or GMA) enabled PVDF to be less hydrophobic [236].

Grafting AAc onto activated PVDF by ozone was also successfully accomplished by Ying et al. [237, 238] as precursors of original microfiltration membranes. These authors noted that the graft concentration increased with the AAc monomer concentration and 100% grafting occurred when a 6-fold excess of AAc versus PVDF was used. The microstructures and compositions of these PVDF-g-PAAc copolymers were characterised by FTIR, XPS, elemental analysis and thermogravimetric analysis [237]. The membranes resulting from these well-defined copolymers were stable up to 200 °C, this temperature corresponding to the degradation of PAAc. In addition, the flow of the aqueous solution through these PVDF-g-PAAc microfiltration (MF) membranes exhibited a strong and reversible dependence on pH solution in the pH range of 1–6. As a consequence, these new graft copolymers seem to be promising material for producing MF membranes with well-controlled pore size, uniform surface composition and pH-sensitive properties [237].

Other applications of this investigation consisted in achieving two goals: (i) the immobilisation of glucose oxidase onto microporous MF membranes made of PVDF-g-PAAc [238], and (ii) prospects for obtaining hollow fibre membranes.

Similar strategies were utilised in our Laboratory to graft phosphonated (meth)acrylates onto ozonised PVDF [233], leading to well-architectured copolymers having the following structure:

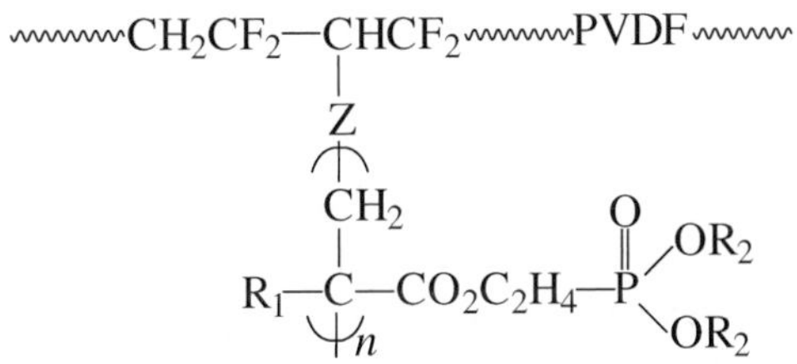

$$\text{~~~~CH}_2\text{CF}_2\text{—CHCF}_2\text{~~~~PVDF~~~~}$$

$$\text{CH}_2$$

$$R_1\text{—C—CO}_2\text{C}_2\text{H}_4\text{—P}\overset{\displaystyle O}{\underset{\displaystyle OR_2}{\overset{\displaystyle OR_2}{}}}$$

R$_1$: H or CH$_3$ and R$_2$: H or C$_2$H$_5$

Z : O or "—"

Phosphorus brings various advantages such as fire retardancy [239], complexation, anticorrosion and adhesion (especially from phosphonic acid function) [240], and this last property was relevantly taken into account for accomplishing a graft copolymer able to be applied onto galvanised steel. In addition, it was observed that the diester phosphonic acrylate was grafted at a higher rate than the homologue acid derivative. Indeed, these experiments show a neat improvement of the adhesive behaviour of the PVDF coating with regard to a similar copolymer containing carboxylic acid functions.

Another relevant investigation was accomplished by Liu et al. [241], who prepared PVDF-g-PPEOMA graft copolymers in two steps.

PEOMA   stands   for   polyethyleneoxide   methacrylate   of   formula:

$$H_2C=C \begin{matrix} \diagup CH_3 \\ \diagdown C-(OCH_2CH_2)_{4.5}-OCH_3 \\ \parallel \\ O \end{matrix}$$

First, ozone treatment of PVDF was carried out at 25 °C with an ozone concentration fixed at 0.027 g l^{-1} in the O$_2$/O$_3$ mixture, leading to peroxides, the concentration of which varied from 5×10^{-5} to 13×10^{-5} mol g^{-1} while the treatment time was ranged between 5 and 15 min, respectively (as above, these concentrations were assessed from reaction with DPPH). These authors noted the increase of peroxide concentration by (i) the decrease of water contact angle of the cast film (from acetone solution) varying from 128 to 115°, and (ii) the increase of the [O]/[C] ratio (determined from the corrected O$_{1s}$ and C$_{1s}$ XPS core-level spectral area ratio) from 0.003 to 0.010.

Second, these PVDF containing peroxide initiated the polymerisation of PEOMA at 100 °C in NMP from [PEOMA]/[PVDF] weight ratios ranging between 1 : 1 and 6 : 1. The structures of the resulting PVDF-g-PPEOMA graft copolymers were investigated by FTIR and XPS showing, as expected, that the graft concentration increased with increasing PEOMA macromonomer concentration [241].

Nanoporous membranes were prepared by the phase inversion method and their surfaces were characterised by XPS. The authors noted that the surface [C]/[F] ratio was higher than the corresponding bulk [C]/[F] ratio assigned to the enrichment of the hydrophilic components from the grafted PEOMA at the outermost surface.

In addition, various electrochemical properties of these membranes were studied. These materials exhibited (i) very high liquid electrolyte uptake capacity (these membranes could absorb 74 wt% of liquid electrolyte—higher than the best ones which absorb 65% in the presence of inorganic fillers [242]); (ii) a high ionic conductivity of 1.6×10^{-3} S cm^{-1} at 30 °C; (iii) a satisfactory transference number of 0.15, characteristic of that in polymer electrolytes or concentrated solutions [243]; and (iv) electrochemical stability. These properties enabled the membranes to have good prospects for applications in lithium ion rechargeable batteries.

This work is original since, to our knowledge, PVDF-g-PPEOMA was synthesised for the first time [244].

Another graft copolymer, achieved from the ozone treated PVDF followed by the grafting of N-isopropylacrylamide (NIPAAM), was accomplished by Iwata et al. [245]. Indeed, PNIPAAM is soluble in water but has a lower critical solution temperature (LCST) around 31–33 °C. These PVDF-g-PNIPAAM graft copolymers were designed as porous membranes with flow-rate retardation. The flow of pure water through the grafted membrane varied by more than 10 times between the temperatures above and below the LCST. Unfortunately, the thickness of graft chains could not be assessed since no accurate data for the pore size and density were available.

A similar approach was investigated by Ying et al. [246], who could achieve the grafting of NIPAAM in N-methyl-2-pyrrolidinone solution and the resulting materials were used for microfiltration membranes.

The same team [247] also attained another objective (especially pH sensitive-ultrafiltration membranes) and succeeded in grafting 4-vinyl pyridine onto activated PVDF from ozone. Interestingly, a well-controlled pore size and a uniform surface were achieved.

2.4.3.4. Swift Heavy Ions (SHI)

In contrast to γ-radiation (see Section 2.4.3.6.1), which leads to a homogeneous modification, the advantage of SHI grafting is that only small regions are modified. Thus, in the case of hydrophilic monomers grafted onto fluoropolymers, the surface may present hydrophilic (corresponding to the modified areas) and hydrophobic (corresponding to the unmodified zones) microdomains of different sizes, depending on the absorbed dose and grafting yield.

SHIs have energies ranging between few MeV amu^{-1} (where amu stands for atomic mass unit) and about 100 MeV amu^{-1} [248]. They interact with matter mainly by inelastic collision processes, leading to ionisation of the target atoms and excitations of the target electrons. Nearly all the SHI energy is lost in the electronic stopping power range.

Table IV supplies several examples of the behaviour of electronic massic stop (S_e) assigned to the natures of both the irradiation and the fluoropolymers (PVDF and poly(VDF-co-HFP)).

For a radiation emitted by ions, it is noted that the lighter the ion (O < Ar < Sn), the lower the S_e value.

Nevertheless, even for a light ion, the energy lost by the ion in the bulk of the fluoropolymer is ca. 1000 times higher than that lost by the electrons in the same polymer.

Compared with the usual ionising radiations (γ-rays or electrons), SHI are known to induce in the irradiated polymer specific groups such as alkynes [249] and allenes [250]. Moreover, the yield of chemical defects resulting from chain scission is enhanced.

TABLE IV

ELECTRONIC MASSIC STOP (S_E) VS. THE NATURE OF THE FLUOROPOLYMER TARGET AND THAT OF THE RADIATION (AMU STANDS FOR ATOMIC MASS UNIT)

Radiation	PVDF (25 μm)	Poly(VDF-co-HFP) (20 μm)
Electron beam (8 MeV)	1.8×10^{-3}	1.8×10^{-3}
^{18}O (11.1 MeV amu^{-1})	2.4	2.4
^{40}Ar (9.1 MeV amu^{-1})	16.9	16.2
^{112}Sn (9.3 MeV amu^{-1})	59.9	64.5

In addition, Le Bouedec *et al.* [251] investigated the irradiation of PVDF films with different SHIs (Sn, Kr, Ar, O) in the electronic stopping power range (4.3– 11.8 MeV amu^{-1}). SHI irradiation induces a large degree of excitation and ionisation along the ion path. By means of FTIR, the authors studied the decrease of crystallinity and the formation of unsaturations: alkynes and double-bonds either isolated or conjugated.

Guilmeau *et al.* [252] studied the kinetics and characterisation of radical- induced grafting of styrene onto E–TFE copolymers at 1.5 MeV in air. The influence of grafting of styrene at 50–80 °C was investigated. By in situ ESR, the decay of peroxy radicals was recorded under various heating and grafting conditions, and this suggested that they reacted rapidly with styrene; but do not initiate the grafting process. To better understand the radiation grafting of styrene onto poly(VDF-*co*-HFP), Aymes Chodur *et al.* [253] also used SEC to study the effect of irradiation of this polymer and the evolution of the molecular weights of the styrene-grafted chains as a function of the grafting time. The authors noted that (i) the gamma irradiation induced both scission and crosslinking of the chains of the fluoropolymer, and that (ii) the styrene-grafted chains increased with grafting time and the asymmetry of the chromatographic peak also increased. These grafted chains could not be regarded as homopolystyrene, which has higher molecular weights. FTIR and transmission and internal reflection spectroscopy were used by Aymes Chodur *et al.* [254] to calculate the PS grafting yield at different depths. In the case of PVDF only, a PS gradient in the thickness was observed which is higher in the case of a γ-ray irradiation than in that of a SHI irradiation. The authors suggested that the PS diffusion was accelerated in the latent track formed after the heavy ions irradiation.

Indeed, different kinds of ionising radiations induce differences in the structure of the grafted films. Whatever the radiation type, the grafting yield was higher in PVDF than in its copolymers, arising from the difference of crystallinity. The PS layer formation was observed by SEM. At small grafting yields, the PVDF spherulites are covered with a network of PS that tends to form a continuous layer when increasing the grafting yield whereas, in the case of copolymer, PS seems to penetrate faster in the bulk so that the formation of the superficial layer is delayed [254].

Extensive investigations on the irradiation of PVDF (α and β forms) and VDF or TFE copolymers by SHIs were accomplished by Betz's team [250, 251, 254–268]. Optimisations of conditions of activation [250, 257] were achieved in terms of the irradiation parameters (absorbed doses, electronic stopping power) [258], polymer substrate factors (bulk, film, thickness, processing), the nature of the ions (Xe, O, C, Ar, Kr, Sn, Pb) [250, 251, 257, 265] and graft polymerisation parameters (effect of the temperature and degree of grafting, use of various monomers—MMA [260], styrene [248, 250, 258, 262], acrylic acid [261] and post-functionalisation of the polymeric grafts (into polystyrene sulfonate [262–264] or PS aspartic acid sulfonamide [262–264])) (Table V).

TABLE V

Synthesis of PVDF-*g*-PM from the Activation of PVDF with Swift Heavy Ions Followed by Grafting M Monomer

M Monomer	Studies	Properties	Applications	References
AAc	FTIR	Immobilisation of amino derivative and peptide	Biomedical	[261]
MMA	–	Ferroelectric		[260, 270]
S	Different conditions of dose rate > 10 kGy FTIR, TIRS	Antithrombogenic	Cardiovascular prostheses	[250, 263]
			Artificial hearts	[266]
S	XPS at each step	Anticoagulative	Heparine-like biomaterials	[264, 267]
S	FTIR at each step, sulfonation of S, addition of aspartic acid sulfonamide	Anticoagulative	Vascular prostheses	[262]
			Blood compatible materials	[248]
S	Influence of dose rate and of type of ions; kinetics	Characterisation (M_n, PDI)	Blood compatible biomaterials	[252, 258]

AAc, acrylic acid; MMA, methyl methacrylate; S, styrene; FTIR, Fourier transform infrared; TIRS, transmission and internal reflection spectroscopy; PDI, polydispersity index.

The authors noted that, because of the high electronic stopping power of the particles used (from 2 to 80 MeV cm^2 mg^{-1}), the grafting initiated by SHIs is different from that initiated with γ-rays [249, 255]. Furthermore, the values of the grafting yield and the molar mass distribution varied, depending on the nature of the ion used [255].

The chain length of the grafts and their polydispersities were also studied [252, 253, 255, 258] by this French group.

The characterisations at each step were achieved by FTIR [249, 254, 259, 262, 265], DSC [260], SAXS [256], WAXS [256], SEM [254, 256, 264, 266], AFM [264] and XPS [267], showing (i) the change of crystallinity of PVDF [260], even leading to an amorphisation of the resulting graft copolymers, and (ii) the modification of the surface chemical composition [267, 268] of the fluorinated (co)polymer.

Many irradiations of PVDF (α- and β-forms) have been investigated by numerous authors. Indeed, this second form is presently of considerable interest because of novel ferroelectric and piezoelectric applications in various fields: captors, sensors and detectors [269].

This β-form can be obtained when α-PVDF form is mechanically stretched or rolling processed, and it melts before it undergoes the ferroelectric-to-paraelectric transition. Betz *et al.* [265, 268] accomplished the irradiation of this β-PVDF by different SHIs (O, Kr, Sn) with doses of 9.9–35.1 kGy in the electronic stopping power range (6.2–7.8 MeV amu^{-1}). Alkyl and peroxy radicals were identified [257] from the activation of both α- and β-form PVDF while only peroxy radicals remain in the use of poly(VDF-*co*-TrFE) [257].

β-PVDF-*g*-PMMA was synthesised by the same authors [270], who extensively studied the ferroelectrical properties at different temperatures (220–440 K). They noted that the dielectric constant of these graft copolymers decreased with increasing the grafting yields.

Dapoz *et al.* [262, 267] enabled a radiografting of PS onto films of PVDF [262] and poly(VDF-*co*-HFP) copolymers [262] activated by SHIs at different doses in order to prepare biomaterials [267]. Then, a chlorosulfonation reaction followed by sulfonamidation and by hydrolysis enabled these authors to functionalise these PVDF-*g*-PS or poly(VDF-*co*-HFP)-*g*-PS graft copolymers [262]. The nature of the functional groups located in the bulk or on the surface was remarkably achieved by transmission FTIR [262], ATR and XPS [267], respectively. Indeed, in case of low dose rates (<100 kGy) to initiate the grafting of styrene, no irradiation effect was noted from FTIR [258a] nor from XPS [262]. Kinetics of radiografting of styrene in both virgin (co)polymers [267b] showed that (i) the higher the irradiation dose, the greater the initial rate and the degree of grafting, and (ii) the initial rate was 10 times lower when the copolymer was irradiated compared with that noted for the activation of PVDF. Both the characteristic frequencies and the F_{1s}, C_{1s}, O_{1s}, S_{2p} and Cl_{2p} were supplied and they justified several features:

(a) besides the expected rays assigned to PS grafts, the authors noted ketone and carboxylic acid functions arising from a slight oxidation;
(b) they could assess the amount of PS on the surface;
(c) a high degree of functionalisation of PS graft was achieved (ranging from 80 to 96%);
(d) in the course of the (chloro)sulfonation, a mixture of SO_2Cl and SO_3H was observed that varied according to the chlorosulfonic acid concentration (a major $ClSO_2$ concentration ca. 3–4 times higher was shown);
(e) the functionalisation occurred mainly in para position but also in ortho/para, although the ratio of monosubstituted and disubstituted PS was assessed with difficulties.

The applications of these PVDF-*g*-PSS and especially poly(VDF-*co*-HFP)-*g*-PSS (where PSS stands for sulfonated polystyrene) concern original antithrombogenic biomaterials (or "heparine-like" materials) [262, 267a].

2.4.3.5. By Plasma Activation

Modifications of polymer surfaces using microwave and radio frequency gas plasma techniques are advantageous because usually minimal altering of bulk properties is anticipated [271, 272]. For that reason, plasma surface reactions, including film and non-film forming processes, have become attractive in numerous applications, ultimately altering such surface properties as polymer biocompatibility [273], adhesion [274], hydrophilicity [275], friction [276] and other surface properties.

Many surveys concern the modification of polymeric surface by plasma activation [277]. This is an original way to introduce specific groups (maleic anhydride [278], amino derivatives [279]) and a few examples are given below.

Baquey *et al.* [280] activated expanded PTFE (ePTFE) by cold plasma, hence modifying its surface. In a second step, radio frequency glow discharge (RFGD) containing AAc followed by several steps of functionalisation by oligopeptides (containing Arg-Gly-Asp-Cys (RGDC) sequence) led to PTFE-*g*-oligopeptide. Known for their role in mediating the adhesion of cells to the extracellular matrix, these oligopeptides led to original graft copolymers as novel biomaterials potentially used as endothelialised vascular prostheses or artificial implants.

The preparation of these ePTFE-*g*-oligopeptides is as follows:

$$\text{ePTFE} \xrightarrow[\text{2) acrylic acid}]{\text{1) Ar/O}_2 \text{ RFGD}} \text{ePTFE-}g\text{-PAAc} \xrightarrow[\text{4) peptide immobilisation}]{\text{3) NH}_2\text{-PEO}} \text{ePTFE-}g\text{-P(EG-peptide)}$$

Yu *et al.* [281] synthesised original PTFE-*g*-PGMA graft copolymers (where GMA means glycidyl methacrylate) from the argon plasma pretreatment of PTFE followed by UV-induced graft copolymerisation with GMA. Interestingly, the epoxide groups of the GMA chains in PTFE-*g*-PGMA were cured and led to a highly crosslinked interphase leading to a strong adhesion of the epoxy resin to

PTFE (a 90°-peel adhesion strength above 15 N cm^{-1} was observed). In addition, the epoxy resin samples obtained from delamination of the epoxy/PTFE-g-PGMA composite showed a lower rate of moisture sorption.

Furthermore, the same team [282] prepared composite films made of polyimide (PI) and poly(TFE-co-HFP), FEP, or of PI and PTFE by a three-step procedure: (i) first, the fluorinated film was argon-plasma treated (with a glow discharge generated at an Ar pressure of 0.6 torr for 1–120 s and a plasma power of 35 W) and it was subsequently exposed to the atmosphere for ca. 10 min to enable the reaction of the surface activated species with oxygen atmosphere to form peroxides and hydroperoxides. (ii) The second step concerned the UV-induced (with a 1000 W high pressure Hg lamp) graft polymerisation of GMA at 28 °C for 1 h, initiated by the F-polymers (FP) containing the (hydro)peroxides. PGMA homopolymer was separated by extraction from the FP-g-PGMA. (iii) A poly(amic acid) of poly(pyromellitic dianhydride-4,4′-oxydianiline) reacted thermally with the epoxy group of the FP-g-PGMA, and even up to 200 °C to enable to complete the imidisation process for the formation of PI/FP-g-PGMA laminates. Interestingly, that reaction occurred at temperatures lower than those conventionally used (ca. 300 °C) for the synthesis of PI and below the decomposition temperature of PGMA homopolymer (>250 °C) [283]. The synthesis of PI/fluoropolymer composite can be summarised by Scheme 3.

The thickness of the PI film on fluoropolymer was 50 μm and the T-peel adhesion strength was 6.5–7.0 N cm^{-1} compared with negligible adhesion strength for a blend of PI and non-grafted FP.

The Singaporian authors characterised the surface of these blends. First, XPS showed that both PI/FEP-g-PGMA and PI/PTFE-g-PGMA composite films delaminated by cohesive failure inside the FEP and PTFE films, respectively. Furthermore, very high static water contact angles (up to 145°) were noted, showing highly hydrophobic laminates produced.

In addition, a Korean team [284] succeeded in preparing PVDF-g-PAAc graft copolymers by grafting AAc onto the surface of a PVDF membrane using Ar plasma polymerisation technique. The degree of grafting depended on the plasma exposition time (less than 3 min) and the evidence of the grafting reaction was shown by both (i) X-ray photon spectroscopy, and (ii) attenuated total reflectance FTIR. These authors noted that, for an argon plasma at 30 W, the grafting rate decreased after the maximum rate was observed at a 30 s exposure. The surface characterisation by XPS of a PVDF membrane after a 30 s plasma exposure and subsequently grafted with AAc showed the greatest O_{1s}/F_{1s} area ratio in the spectra. Its degree of grafting and degree of polymerisation (ranging from 32 to 94) were the highest among the graft membranes. Such original pH-sensitive membranes were used for the permeation of riboflavin through it and the authors observed a decrease in permeability of riboflavin at pH 4–5.

Rafik et $al.$ [285] could achieve deposits of poly(acrylic acid), PAAc and poly(perfluorobutyl ethylene) (PFBE) onto PVDF by plasma polymerisation of

Step 1: Plasma Pretreatment of Fluoropolymer

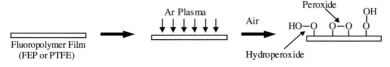

Step 2: UV-Induced Graft Copolymerization of GMA

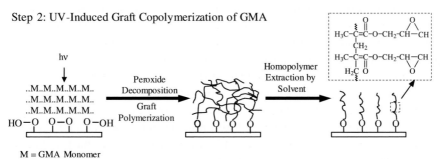

M = GMA Monomer

Step 3: Thermal Imidization of Poly(amic acid) onto GMA Graft-Copolymerized
Fluoropolymer Surfaces

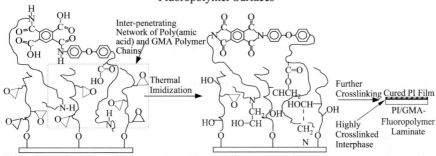

SCHEME 3. Diagram illustrating the synthesis of composite film made of FEP-g-P(Imide) or PTFE-g-P(Imide) graft copolymers (reproduced with permission from Yang *et al.* (2003) *J. Fluorine Chem.*, **119**, 151–160, courtesy of E. T. Kang).

both these monomers onto activated PVDF. This technique of polymerisation was efficient since, after 5 min, the kinetics of polymerisation was not modified. As expected, the nature of the introduced monomers brought specific properties to initial PVDF. For example, PAAc made hydrophilic PVDF, as evidenced by the values of the water contact angles (that decreased from 95° for initial PVDF to 10° after only 10 min plasma polymerisation while the surface energy increased from 25.1 to 75 mN m^{-1}), scanning electron microscopy and FTIR.

In contrast, the authors noted the decrease of the surface tension (from 25.1 to 20.0 mN m^{-1}) for PVDF and PVDF coated with PPFBE, respectively. The thickness of the PPFBE was 7 μm.

These modified PVDFs were used as microfiltration membranes and showed similar filtration properties: flow of 5000 L day^{-1} m^{-2} and retention rate of 45% for dextran solution ($M = 2 \times 10^{6}$).

Such results were obtained after a treatment of 1 h and even 1 min from PVDF coated with PAAc and with PPFBE, respectively.

2.4.3.6. Activation by Electron Beam and Gamma Ray

2.4.3.6.1. Introduction.   Radiation-induced grafting can also be used for the synthesis of graft copolymers [210, 216]. This technology can be realised either from the grafting of hydrogenated (or, to a lesser extent, fluorinated) monomers onto a pre-irradiated fluorinated polymeric backbone or from the grafting of fluorinated monomers onto a hydrogenated activated polymer.

For this method, a polymer endowed with the required mechanical, chemical or thermal properties is irradiated with electron beams or γ-rays (usually emitted from various radioactive isotopes: ^{60}Co obtained by beaming ^{59}Co with neutrons in a nuclear reactor, and ^{137}Cs which is a product of fission of ^{235}U) [210, 216].

^{60}Co is a manufactured metal that emits two γ-rays of 1.17 and 1.33 MeV and one β-ray. ^{137}Cs is a powder that has two modes of disintegration, either a γ-ray and a β-ray or only one β-ray. ^{60}Co has a half-life of 5.25 years while that of ^{137}Cs is 30 years. In spite of a shorter half-life, ^{60}Co is the more often used since it is less expensive, easier to handle and commercially available in greater quantities [207, 213].

Interestingly, there is no risk of induced radioactivity from these radioisotopes.

As for electrons, two cases can be considered and their power of penetration depends upon their energy. β-Rays arising from an isotopic source have a low power of penetration. In contrast, in an accelerator, the energy conferred to the electrons can reach several MeV. The flow of accelerated electrons provides only corpuscular rays having a sufficient penetrating power and satisfactory energies to be used industrially.

On the whole, electron beam enables the activation on the surface. Hence, the polymer to be activated is first processed into thin films. These electrons are usually produced from a cathode (e.g., of lanthanide) and they are introduced in a magnetic field which accelerates them and regulates the direction of the beam.

In contrast, γ-irradiation is efficient in the bulk of the substrate and thicker films can thus be treated. The irradiation causes free radical centres to be formed in the polymeric matrix.

2.4.3.6.2. Types of Radiations.   Different types of high-energy radiation are available to be used for the grafting process [210–213, 216], although crosslinking may occur as notably reported by Forsythe and Hill [217] (for example, the radiation–chemical effects in PVDF showed that the crosslinking proceeds mainly through alkyl macroradicals [286]). This radiation may be either electromagnetic radiation, such as X-rays and gamma rays, or charged particles, such as beta particles and electrons. The most widely used gamma radiation source is the ^{60}Co source which emits radiation of 1.17 and 1.33 MeV (mean value 1.25 MeV). This is because of the ease of its preparation, low cost and fairly long half-life of 5.3 years. Another gamma radiation source which could compete with ^{60}Co is ^{137}Cs. This isotope is a fission product from the nuclear-energy plants and emits radiation of lower energy (0.66 MeV) than that of ^{60}Co.

Both types of gamma radiation sources are available for irradiation [216]. Such sources offer a wide area for irradiation of samples and are, therefore, used on an industrial scale. These "cavity-type" sources are more compact and are widely used on a laboratory scale.

On the other hand, synchrotron also offers the possibility to produce radiation over a broad range of the energy spectrum [287]. Photon energies higher than 0.1 MeV would be interesting for irradiation of fluoropolymers.

A large number of linear accelerators are available which produce electrons of energies in the range of 1–6 MeV [210, 216].

From the grafting point of view, the basic difference in both types of radiation lies in terms of the high penetration power of the electromagnetic radiation [287b]. Charged particles lose energy almost continuously through a large number of small energy transfers while passing through matter. However, photons tend to lose a relatively large amount of intensity by interaction with matter. The advantage of the electromagnetic radiation such as gamma rays is that the fraction of photons which do not interact with a finite thickness of matter is transmitted with their original energy and direction (exponential attenuation law). Hence, the dose rate of radiation may be easily controlled by the use of a suitable attenuator without influencing the photon energy, which is a very important aspect in radiation-initiated polymerisation [210]. Lead, with its half-thickness value of 1.06, is the most widely used attenuator for gamma radiation due to its lower required thickness as compared with other attenuators.

Radical generation throughout the film thickness is necessary. Fortunately, all the kinds of radiation discussed above have sufficient energy to penetrate the bulk of fluorinated films, which are usually of 25–200 μ thickness. This is the advantage of the high-energy radiation grafting over UV-induced photochemical grafting in the presence of a sensitiser; the latter cannot penetrate the bulk of the polymeric film. As a result, the grafting involving UV-radiation remains confined virtually to the surface layers of a film, leaving behind a bulk of the film unmodified [8a]. At the same time, UV-induced grafting needs a photosensitiser which, if left behind in the film matrix, may act as an impurity. Photochemical grafting is, therefore, interesting for applications only where surface properties of a polymer need to be altered.

2.4.3.6.3. Methods of Radiation Grafting. The purpose of the use of gamma rays or an electron beam is to generate radicals in the grafting system. The following three different methods may be used for the radiation-initiated grafting process [210, 216]: (a) simultaneous radiation grafting, (b) pre-irradiation in air (hydroperoxide method) and (c) pre-irradiation in vacuum (trapped-radicals method).

In simultaneous radiation grafting, the polymer and the monomer are exposed to radiation at the same time. The radicals are, therefore, generated in both the polymer and monomer units. A chemical reaction of the monomer with the polymer backbone radical initiates the grafting reaction [216]. Alternatively, a two-step grafting procedure may be adopted. In the first step, the polymer is exposed to radiation which leads to the formation of radicals on the macromolecular chain. If the irradiation is carried out in air, radicals react with oxygen, leading to the formation of peroxides and hydroperoxides (hydroperoxide method). When in contact with a monomer, the irradiated polymer initiates grafting by thermal decomposition of hydroperoxides.

In the absence of air, these macromolecular radicals remain trapped in the polymer matrix and initiate the grafting in the presence of a monomer (trapped radicals method). Simultaneous radiation grafting is, therefore, a single-step process while the pre-irradiation method involves two steps [210, 216].

The sensitivity of a polymer towards radiolytic degradation is an important factor before selecting the irradiation dose. A number of studies have been devoted to the structural changes and mechanism of degradation in fluoropolymers by high-energy irradiation (see Section 2.4.3.6.4). The extent of degradation largely depends on the chemical nature of polymers. Among fluorinated polymers, PVDF and PTFE have been widely used for the grafting of various monomers (see sections below). The latter polymer has a very low resistance to high-energy radiation which causes degradation even for low doses of gamma radiation [212]. However, the poly(tetrafluoroethylene-co-hexafluoropropylene) copolymer (FEP) shows a much higher resistance against gamma radiation (see Section 2.4.3.6.4.2.2.2). Therefore, it is possible to graft under a wide range of radiation doses without much influencing the inherent mechanical properties.

Zhen [288] studied the lifetime of fluorinated radicals at room temperature obtained after irradiation of PTFE, PCTFE and poly(TFE-co-HFP) powders and found that they were higher than 5 years.

In contrast, those assessed after irradiation of PVDF and ETFE were about 1 year.

The free radicals can undergo several reactions, such as combination to form crosslinks [214, 217], chain scission and recombination or disproportionation, and elimination with the formation of double bonds [210]. In the presence of vinyl monomers, the free radical centres can initiate graft copolymerisation [216]. The matrix polymer to be grafted can be in the form of films, membranes, fibres, powders or beads. The irradiation and grafting can be carried out in one, two or more steps.

When a membrane is exposed to irradiation, in certain cases, a (per)fluorination step is accomplished to increase its chemical stability; however, the price of the membrane increases drastically with the degree of fluorination.

This sub-section is divided into two parts: (i) carried out reporting investigations onto the activation of fluorinated homopolymers followed by the grafting and (ii) starting from the irradiation of fluorinated copolymers and continued by the grafting step.

2.4.3.6.4. Synthesis, Properties and Applications of F-Graft Copolymers by Activation of F-Polymers and Grafting from them. Several examples of syntheses of fluorinated graft copolymers, their properties and applications are supplied in this section and we have found it worthwhile considering two main parts: (a) the polymers prepared from the irradiation by electron beam [289, 290] or gamma rays of fluorinated homopolymers (e.g., PVDF, PTFE and PCTFE), and (b) copolymers such as (i) those containing VDF base units, and (ii) those having TFE base units (e.g., FEP, ETFE and PFA), followed by grafting. Both sub-sections are described below.

2.4.3.6.4.1. Activation of Fluorinated Homopolymers

2.4.3.6.4.1.1. Synthesis of PVDF-g-PM Graft Copolymers from the Irradiation of PVDF Followed by Grafting. This sub-section supplies non-exhaustive examples of synthesis, properties and applications of graft copolymers starting from the activation of PVDF. Table VI lists the conditions of irradiation and grafting, leading to PVDF-*g*-PM graft copolymers.

2.4.3.6.4.1.1.1. Synthesis and Properties of PVDF-g-PMAAc. Several authors have chosen to irradiate PVDF in order to graft various M monomers for the preparation of PVDF-*g*-PM graft copolymers. In 1976, Machi *et al.* [291] synthesised poly(fluoroalkene)-*g*-poly(acetoxy styrene) and poly(fluoroalkene)-*g*-poly(acrylic acid) after irradiation of poly(fluoroolefin) films by 20 Mrad electron beam, followed by immersion in acetoxystyrene or AAc. These graft copolymers led to IEM applications. These membranes exhibit an electric resistance of 6.2 Ω cm^{-2} compared, for example, with a film that does not contain any AAc (10–15 Ω cm^{-2}). Four years later, Blaise [292] prepared PVDF-*g*-PMAAc graft copolymers by activating PVDF powder by electron beam (0.8 Mrad). This irradiated PVDF was heated (60 °C) in a solution of methacrylic acid (MAAc) leading to a graft copolymer containing 9.8% of acid functions that show good adhesion onto metals (Fe, Al), while unmodified PVDF was easily removed when deposited onto these metals. The same author [293] used a similar strategy to graft poly[2-(dimethylamino)ethyl methacrylate] onto irradiated PVDF (at 3 Mrad gamma rays). After treatment of the corresponding films with aqueous propylene oxide at pH 4, the membranes reached an electrical resistance of 1.4 Ω cm^{-2}. The patent also claims that the treatment with 4N HCl at 80 °C for 24 h led to a graft copolymer hydrochloride endowed with a resistance of 1.4 Ω cm^{-2}.

In 1998, the synthesis of PVDF-*g*-PMAAc was reported by Mascia and Hashim [294] according to a two-step procedure, as follows:

$$PVDF \xrightarrow{15\ kGy} act.\ PVDF \xrightarrow[MAAc]{H_2C=C\begin{smallmatrix}CH_3\\CO_2H\end{smallmatrix}} PVDF\text{-}g\text{-PMAAc}$$

First, the activation of PVDF powder (Kynar® 7201) was achieved by irradiation of PVDF with γ-rays at 15 kGy, followed by a subsequent treatment in aqueous

TABLE VI

Synthesis, Properties and Application of PVDF-g-PM after Irradiation (Other than Ozone and SHI) of PVDF Followed by the Grafting of M Monomer

Nature of activation	Monomer	Conditions	Properties	Applications	References
e-beam (20 Mrad)	4-Acetoxy styrene, AAc	Bulk grafting 150 μm film	Electrical resistance: 6.2 Ω cm^{-2}	Ion-exchange membrane	[291]
^{60}Co 10 kGy	DMAA	d.o.g. = 17%	Adhesion onto electrode	Electrolyte for Li batteries	[322]
e-beam (25–200 kGy)	S	N$_2$, 175 kV	Good conductivity, good thermal stability	PEM	[298–304]
e-beam (25–200 kGy)	DVB		Electrical persist., electrical resistance 14 Ω cm^{-2}	PEM	[305, 306]
^{60}Co (200 kGy)	VBC		Positive zeta potential	Membrane for ultrapure water	[296]
^{60}Co (6.3 Mrad)	VBC then amination	60–70 °C/1 week (d.o.g. 26%)	Poor prop. after amination	Alkaline AEM	[324, 325]
γ-rays	MAAc	N$_2$/6 h	9.8% acid	Protection of Fe, Al	[293]
γ-rays (3 Mrad)	S	Grafting of 25 mm film /N$_2$	Electrical resistance 1.4 Ω cm^{-2}	Membranes	[311]
e-beam (15 Mrad)	S then sulfonation	d.o.g. = 18–30%	Conductivity = 30 mS cm^{-1}, IEC = 1.7 meq mol^{-1}	IEM	[323]
γ-rays	S		Permeability	Reverse osmosis	[297]
γ-rays (15 kGy)	MMAc	Grafting in water	Hydrophilicity	Compatibiliser	[294]
γ-rays (15 kGy)	Olefin ionomer	Grafting in water		Compatibiliser	[295]
ATRP	PEOMA	CuCl/ligand	Amphiphilic, fouling resistance	Asymmetric membrane for oil/water separation	[310]
e-beam	MASi		Good impact and cold resistance		[326]

AA, acrylamide; AAc, acrylic acid; AEM, anion-exchange membrane; DMAA, dimethylacrylamide; DMAEM, 2-(dimethylamino)ethyl methacrylate; MASi, silane methacrylate; MAAc, methacrylic acid; PMA, diethoxyethyl phosphonate methacrylate; S, styrene; VBC, vinylbenzylchloride; PEM, proton exchange membrane; PEOMA, polyethyleneoxide methacrylate.

solution of MAAc. Unexpectedly, this monomer allowed higher grafting yields than those observed in the case of acrylic acid. From a structural point of view, these PVDF-*g*-PMAAc exhibited a heterogeneous structure assigned to the presence of grafts made of oligo(MAAc) side groups. The objective of that research was to supply compatibilisers of non-miscible PVDF/nylon 6 blends in which the grafted oligo(MAAc)s were compatible with nylon 6. This was evidenced by FTIR that also revealed an enhanced compatibilisation which occurred when zinc acetyl acetonate (ZnAcAc) was introduced into the blend, hence making the co-continuous phases thinner and thinner. Indeed, PVDF-*g*-PMAAc and the amino end-group of the polyamide chain enabled the compatibilisation which was improved by the complexation of zinc cations with amido groups.

Further investigation was carried out by Valenza *et al.* [295], who have shown that PVDF/polyolefin ionomer (POI) blends were successfully achieved by adding such graft copolymers. In contrast to the above survey, these authors accomplished the neutralisation of the acid function to produce the corresponding polymeric zinc salt. In addition, the compatibilisation was achieved by adding ZnAcAc in the PVDF-*g*-PMAAc/PVDF/POI blends. Interestingly, this led to the formation of co-continuous phases that contained encapsulated particles of different size and in various locations.

2.4.3.6.4.1.1.2. Synthesis of PVDF-g-Polyvinyltriphenylphosphonium Bromide. Jan *et al.* [296] investigated the modification of microporous PVDF membrane filters by the incorporation of a positive charged group (such as phosphonium bromide) onto the surface. First, the PVDF was irradiated by [60]Co γ-rays, followed by the grafting of vinyltriphenylphosphonium bromide (VTPPB) leading to PVDF-*g*-PVTPPB graft copolymers. Their electrical characteristics were measured (a positive zeta potential in the pH range from 4 to 9.3 was noted) and the compatibility of these charge-modified membranes with ultrapure water was also investigated.

Interestingly, the density of positive charge sites was assessed to be approximately five times larger than that of unmodified PVDF. In addition, these membranes released less than 1 ppb of total organic carbon into ultrapure water showing potential applications in deionised water systems.

2.4.3.6.4.1.1.3. Synthesis, Properties and Applications of PVDF-g-PS and Functionalised PVDF-g-PS Graft Copolymers. The synthesis, properties and applications of functionalised PVDF-*g*-PS have attracted many investigations. One of the oldest ones was achieved in 1981 by Vigo *et al.* [297] for the preparation of membranes for reverse osmosis, starting from asymmetric PVDF films obtained by casting and gelation technique then modified by radiochemical grafting with styrene, followed by a sulfonation step.

The authors investigated the influences of the evaporation time, grafting, temperature and nature of the solvents.

These membranes exhibited permeabilities of ca. 2000 L m^{-2} day^{-1} and NaCl rejections of ca. 70%.

However, the most pertinent surveys were realised by a Scandinavian team, a Swiss Institute, a French group and an English Laboratory.

(a) *Results from Sundholm's team* [298–308]

These Finnish authors [298–308] synthesised and then fully characterised original sulfonated PVDF-*g*-PS copolymers in a three-step procedure: first, they realised the irradiation of porous films of PVDF by electron beam, followed by the grafting of styrene and, in a final step, the sulfonation of the aromatic ring was achieved in the presence of chlorosulfonic acid.

Sundholm's group [298] carried out the activation of PVDF under nitrogen atmosphere at an acceleration voltage of 175 kV. After the PVDF was irradiated with doses of 25, 50, 100 and 200 kGy, the membranes were immediately immersed into neat styrene and kept for 0.5, 1, 2 and 4 h at room temperature under nitrogen [298]. As expected, it was observed that the higher the dose rate and the longer the grafting time, the higher the degree of grafting (d.o.g.).

Interestingly, the authors claimed that no homopolystyrene was formed. Then, the PVDF-*g*-PS membranes were immersed into chlorosulfonic acid in methylene chloride for about 2–12 min, leading to a degree of sulfonation of 11–71% [298]. However, to achieve 95–100% of sulfonation, a double $ClSO_3H$ concentration was required for a 2-h reaction time. In these conditions, the authors did not observe any trace of chlorine atom in the polyvinylidene fluoride-*graft*-poly(styrene sulfonic acid) (PVDF-*g*-PSSA) membrane. This procedure was summarised by the following reaction scheme:

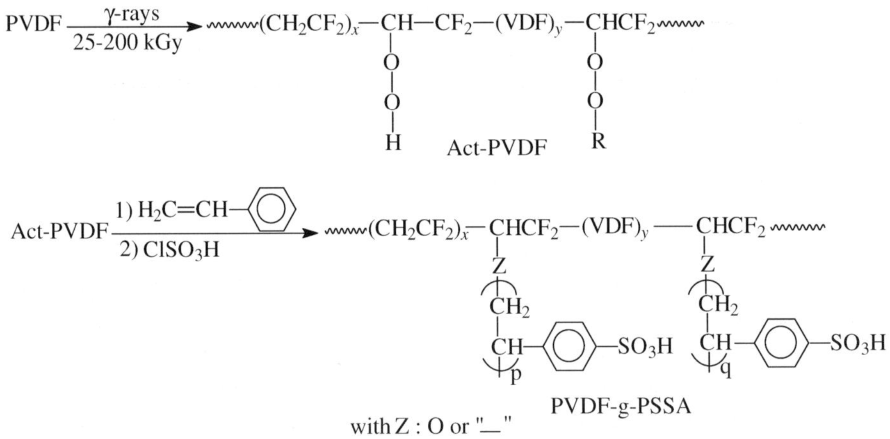

These authors brought relevant evidence of the structure [298, 299], thermal behaviour [298–300] and conductivities (up to 120 mS cm^{-1} at room temperature) [298–301] close to those of Nafion$^®$. These PVDF-*g*-PSSA membranes were stable up to 370 °C under air atmosphere and to 270 °C in a highly oxidising atmosphere [299, 300] (from 340 °C, PS grafts started to decompose [300]).

The grafting reaction of styrene, initiated in the amorphous regions and at the surfaces of the crystallites in the semi-crystalline PVDF matrix [302], was quite efficient. These authors observed a high degree of grafting (50–86%) and they also noted that the grafts were formed both from C–H and C–F branch sites of PVDF. Nevertheless, the presence or absence of C–O bond (that could arise from the formation of peroxide) was not mentioned.

Sundholm's team showed that the sulfonation step was realised in high yields (up to 100%) and that this reaction occurred mainly in the para position of the phenyl ring. As expected, the crystallinity content of PVDF decreased with the degree of grafting and the sulfonation steps, and the authors explained such a behaviour by the efficient penetration of the grafts in the crystallites and branching linked to the grafts [299, 302]. Both the crystallinity and the size of the crystallites of PVDF decreased in the grafting reaction. A further decrease in crystallinity was observed in the sulfonation reaction. Residual crystallinity of PVDF was quite high (10–20%), even in membranes which underwent severe reaction conditions. These Finnish authors assumed that the grafting took place in the amorphous regions of PVDF [299].

These original films were characterised by Raman [298] and NMR [303] spectroscopies, wide angle X-ray scattering (WAXS) and small angle X-ray scattering (SAXS) [298, 299, 304].

Tests of swelling in various solvents and in water and results of conductivity measurements (by impedance spectroscopy) indicated that these films were potential PEMs for fuel cell applications [301, 305]. Indeed, the higher the content of sulfonic acid functions, the higher the conductivity, and these values were enhanced when the crystallinity content was decreasing [299].

Interestingly, in the case of fully sulfonated membranes (for a degree of grafting of ca. 100%), conductivities up to 120 mS cm^{-1} (at room temperature) were assessed. This high value confirms Scherer's statement [216] on the improved conductivity brought by suitable aromatic sulfonic acid function for PEMs.

Sundholm's research was well summarised in a review article [301] where the description and characterisation of novel polymer electrolyte membranes for low temperature fuel cells are reported.

In addition, these membranes made of PVDF-g-PSSA graft copolymers were crosslinked by divinylbenzene and/or bis(vinylphenyl)ethane [302, 305, 306] and were compared with non-crosslinked membranes. The authors observed that the ion conductivity of the crosslinked membranes was lower than that of the non-crosslinked ones (caused by the inefficient sulfonation of the crosslinked materials and also by a low water uptake at low degree of grafting).

This Finnish team [307] also characterised by confocal Raman spectroscopy the fuel cell tested PVDF-g-PSSA membranes and they also noted that the crosslinked membranes (in the present of the same above crosslinking agents) did not undergo any degradation as noted on the non-crosslinking ones.

In addition, this same group performed the synthesis of PVDF-g-PVBC graft copolymers (also reported in Section 2.4.3.6.4.1.1.4) where VBC stands for vinyl benzyl chloride according to the same strategy [308]:

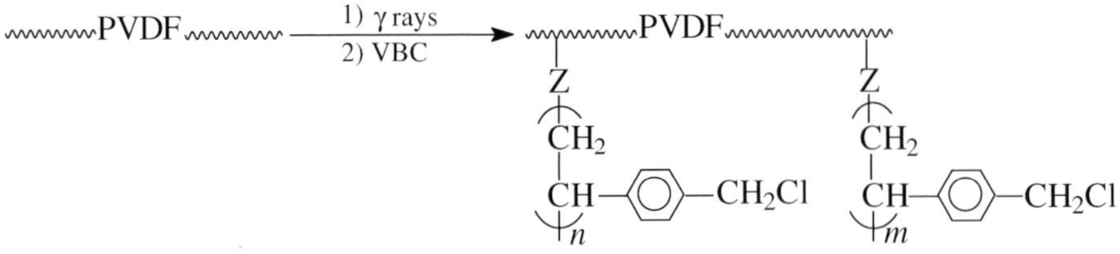

PVDF-g-PVBC

with Z : O or "—"

These PVDF-g-PVBC copolymers acted as suitable macroinitiators via their chloromethyl side groups in the atom transfer radical polymerisation (ATRP) (see Section 2.1.2.2 in Chapter 4) of styrene with the copper bromide/bipyridine catalytic system leading to controlled PVDF-g-[PVBC-g-PS] graft copolymers, as follows:

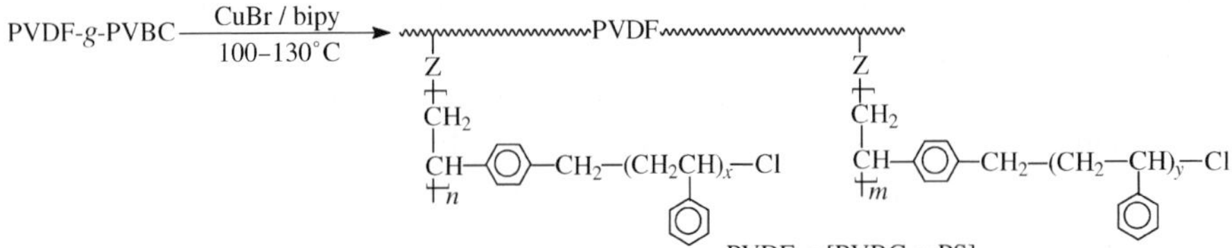

PVDF-g-[PVBC-g-PS]

The degree of styrene grafting increased linearly with time, hence indicating first-order kinetics, regardless of the polymerisation temperature, and it reached 400% after 3 h at 120 °C [308]. As a matter of fact, such a high d.o.g. could not be possible with conventional uncontrolled radiation-induced grafting methods because of termination reactions. The polystyrene grafts were sulfonated, hence leading to well-architectured PVDF-g-[PVBC-g-PSSA] graft copolymers for designing proton-exchange membranes for fuel cell applications. The highest conductivity measured for these membranes was 70 mS cm^{-1}, which is of interest since this value is slightly lower than that of Nafion$^{®}$. SEM/energy-dispersive X-ray results showed that the membranes had to be grafted through the matrix with both PVBC and PS to become proton conducting after sulfonation. These Finnish authors also synthesised original PVDF-g-[PVBC-g-(PS-b-PtBuA)] membranes from the sequential ATRP of styrene and tert-butyl acrylate (tBuA) with PVDF-g-PVBC initiator [308].

   This work is in good agreement with that previously achieved by Liu et al. [309], who prepared original PS-g-PFA by ATRP of 2,2,3,3,4,4,5,5-octa-fluoropentyl acrylate (FA) initiated by a poly(S-co-VBC) catalysed by CuCl/bipyridine. Still on the topic of controlled radical polymerisation (CRP), Mayes et al. [310] claimed the synthesis of PVDF-g-PPEOMA graft copolymers by ATRP of poly(ethylene oxide) monomethacrylate from PVDF. Although this synthesis is quite surprising, the authors mentioned that these comb polymers were used for water filtration in which the hydrophilic domains provide a pathway for water transport.

(b) *Surveys carried out by Scherer*

Similar investigations were also extensively carried out at the Paul Scherer Institute [311, 312].

PVDF was activated by γ-rays from a ^{60}Co source (dose of 20 kGy at 1.33 MeV with a dose rate of 5.9 kGy h^{-1} at room temperature in air). The grafting of styrene occurred at 60 °C. The PVDF-*g*-PS graft copolymers were sulfonated in PVDF-*g*-PSSA as a useful IEM for electrochemical applications [311, 312].

More investigations by this institute were devoted to the synthesis, properties and applications of ETFE-*g*-PM and FEP-*g*-PM graft copolymers reported in Sections 2.4.3.6.4.2.2.1 and 2.4.3.6.4.2.2.2, respectively.

(c) *Results obtained by Porte-Durrieu and Bacquey*

Another relevant survey was carried out by Porte-Durrieu and Baquey's team, which synthesised PVDF-*g*-PS and functionalised PVDF-*g*-PS for biomedical applications [266] (Table V).

Indeed, a conversion from materials into biomaterials requires appropriate preparation protocols and surface processing. Biomaterials are non-living materials used in a medical device, intended to interact with biological systems [313]. They should take part in the making of a device designed for diagnostic purposes, or of a tissue or organ substitute, or of an assistance or replacement device. Obviously, they must be biocompatible and demonstrate at the same time structural properties which fit well with the expected function and surface properties that allow the establishment of a healthy relationship at the material–tissue interface. Considering the replacement of small diameter arteries, this double requirement implies the design of biomaterials which combine structural properties warranting a good mechanical behaviour and surface properties contributing to the prevention of thrombogenesis. However, for numerous applications, it is often difficult to find materials which would fulfil both requirements, and the smartest strategy consists in the selection of a material which satisfies the first condition and which could have its surface modified in order to endow it with the second set of properties.

To this end, Porte-Durrieu and Baquey's team has chosen to use PVDF which exhibits suitable mechanical properties, i.e., lower tensile modulus than those of PET and PTFE (used currently as vascular grafts) and that has been subjected to a special treatment called "heparin-like" [254, 263, 264, 266, 267a, 314].

Previous studies [315–319] have reported that heparin-like materials can be prepared by reacting α-amino acids with grafted chlorosulfonated PS. Unfortunately, this modified PS is too stiff to be used as vascular substitutes. Another approach was also used, consisting of introducing the useful groups (sulfonate, sulfonamide and carboxylic groups) on PVDF-*g*-PS. In this way, interestingly, the French authors used and compared several methods for the activation of PVDF: SHI ($E > 1$ MeV amu^{-1}) and gamma rays.

This procedure was summarised by the following reaction that occurred to the PS grafts:

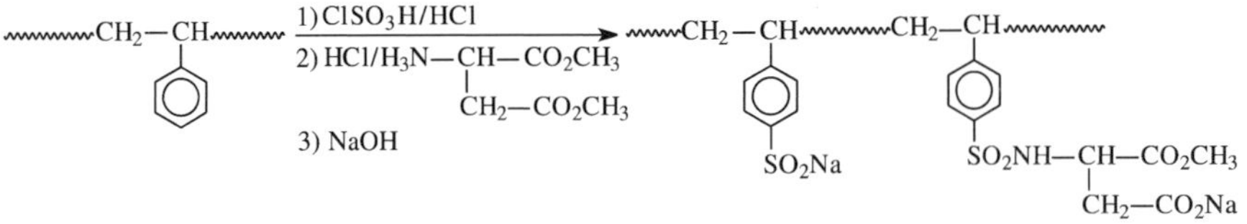

It is known that ionising radiation, mainly electrons, gamma rays or X rays, lose their energy randomly in a polymeric substrate and the subsequently grafted domains are homogeneously distributed. The authors have chosen two methods of activation of PVDF: from SHIs and from gamma rays.

(c1) *From Swift Heavy Ions (SHI)*

Ionising particles such as accelerated ions induce a continuous trail of excitations and ionisations through the solid that leads to the formation of a latent track. The deposited energy is highly localised along the ion path and heterogeneous grafting occurs.

The advantages of such ionising radiation are: (i) low absorbed doses are sufficient to induce significant grafting and (ii) only small regions (the latent tracks) are modified and, thus, the properties of the initial material are not modified [254, 263, 264, 266, 267b]. The size of these microdomains was dependent on the absorbed dose and the grafting yields [254, 263, 264, 266, 267b].

This team also investigated the assessment of the compositions of the graft copolymers by the determination of sulphur, sodium, fluorine and carbon percentages obtained from scanning electron microscopy combined with energy-dispersive X-ray analysis (EDXA) [266]. As expected, and whatever the method of radiation, the higher the grafting content, the lower the fluorine content.

Moreover, when the PS grafted PVDF graft was functionalised, the amount of fluorine also decreased while those of sodium and sulfur increased. Further, these authors noted that the functionalisation seemed to increase the roughness of the surface and its area.

PVDF-*g*-PS graft copolymers and their functionalised homologues showed circular "islands" of different sizes. By EDXA measurements, these authors observed that the amount of fluorine was directly dependent on the presence of these islands [266].

In addition, FTIR and transmission and internal reflection spectroscopy were used to determine the PS yields of grafting at different depths [254]. A PS gradient in the thickness was observed and it was higher in the case of a γ-ray initiation than in that of SHIs [254]. At low grafting yields, PVDF spherulites were covered by a network of PS that tended to form a continuous layer when increasing the grafting yield.

X-ray photoelectron spectroscopy was used to analyse the modification of the surface chemical composition in each step. Concerning the SHIs grafting of styrene onto PVDF, the XPS analysis revealed no fundamental differences. After the chlorosulfonation, a loss of 60–70% of PS was observed due to a reorganisation of the macromolecular chains and/or a loss of homopolymer. 18% of PS present at the PVDF surface was functionalised [254].

The French team investigated the exchange of adsorbed proteins on the surface because of the importance of such events in their bioactivity. The rates of exchange were greater for hydrophilic surfaces than for hydrophobic ones, suggesting a stronger binding or an alteration of the conformation in the latter case. It was observed that, after functionalisation, whatever the irradiated polymer and the ionising irradiation used to induce grafting, fibrinogen and albumin exchange was in the same proportion of about 50–70% for a grafting yield of about 11% [266].

This first technique is known to induce in polymers specific groups such as alkynes [249] and alkenes [250], and it smoothly irradiates the polyfluoroolefin, regarded as a more flexible technique that generates less radicals than other methods of irradiation. Hence, small regions were altered. The size of these microdomains was dependent on the absorbed dose and the grafting yields.

(c2) *From γ-rays*

Conversely, γ-rays enabled PVDF to be irradiated in all the bulk, hence supplying radicals in a higher amount for a homogeneous modification of this fluoropolymer whatever the absorbed dose range [254, 263, 264].

At low absorbed dose (10 kGy), the grafting yield is always higher after a γ-irradiation than after a SHIs one [254]. At the absorbed dose of 10 kGy, the grafting yield is now comparable whatever the radiation type. In the case of SHIs grafting, the grafted domains are restricted to the latent tracks at the beginning of grafting [254].

At the absorbed dose of 100 kGy, samples can be considered as homogeneously irradiated, as in the case of a γ-irradiation resulting in similar grafting yields [254].

In contrast to Klimova *et al.* [320], who investigated the macroradical decay in irradiated PVDF by ESR, these French authors characterised the surface topography by scanning electron microscopy (SEM), atomic force microscopy (AFM), energy dispersive X-rays (EDXA), FTIR/ATR and XPS.

SEM and AFM micrographs of PVDF-*g*-PS using SHIs radiation show circular "islands" from 0.07 to 0.38 μm (for a PS grafting yield of 2.5 and 19.1%, respectively). In the case of gamma radiation, these French authors observed a homogeneous distribution regardless of the grafting yield [254, 263, 264, 266].

Furthermore, preliminary results [314] concerning the evaluation of mechanical properties of PVDF multifilament yarns grafted with PS induced by γ-irradiation with respect to synthesis parameters were obtained. The bulk properties of the materials were analysed by tensile tests. 8 kGy is a high enough dose to initiate significant changes on energy to break, without any change on the Young modulus. Then, limiting the absorbed dose ($D < 8$ kGy) and the grafting time ($1 < t < 2$ h) were necessary to limit any deleterious effect on the PVDF-*g*-PS yarns tensile properties [314].

Moreover, it is known that thrombosis induced by foreign surfaces involves a series of events beginning with the deposition of a protein layer at the blood–artificial interface followed by the adhesion of platelets and possibly leukocytes. When a biomaterial is in contact with blood or plasma, in general, rapid protein adsorption takes place. The composition of the adsorbed protein layer and the reversibility of the adsorption process depend mainly on characteristics of the surface (roughness, interfacial free energy, hydrophobicity and hydrophilicity and the ionic nature of the surface) and flow parameters. Then, adsorption of plasma proteins such as albumin and fibrinogen plays a very important role since the type and conformational state of the adsorbed proteins influences the subsequent events leading to the thrombus formation. Synthetic foreign materials acquire thrombogenic or non-thrombogenic characteristics only after first interacting with blood, and the event which dominates the transformation from an inert polymer into an active surface takes place during the interaction of plasma proteins with the surface.

The authors carried out fibrinogen and albumin adsorption from plasma on these materials, quantitatively assessed using ^{125}I labelled proteins. They showed that, even with an increase in the hydrophilicity of the polymers as a result of functionalisation, the amount of adsorbed proteins on these surfaces increased [266]. The authors showed that, when the samples were functionalised, the amount of proteins adsorbed was greater when the samples were prepared by gamma radiation than when they were prepared by SHIs radiation.

Similar investigations on the polymer grafting for enhancement of biofunctional properties of medical and prosthetic surfaces were also achieved by Golberg *et al.* [321].

Grafting acrylate (Acr) onto PVDF was also accomplished by Kronfli's team [322] for original lithium batteries, and the PVDF-*g*-PAcr graft copolymer led to an improvement in the adhesion of composite electrode to current collectors to an increase of electrolyte solvent uptake and to an increase of the range of solvents for PVDF homopolymers at room temperature. Graphite–LiCoO$_2$ cells containing such modified PVDF-*g*-PM showed good rate performance and stable cycle life. An improved process of preparing microporous membranes for solid electrolytes for Li cells by casting the graft copolymer solution was achieved by the same group.

Using the same strategy, Flint and Slade [323] synthesised PVDF-*g*-PSSA, and the physical and electrochemical properties of the resulting membrane were investigated. These membranes were involved as an electrode membrane assembly used in fuel cell conditions. The conductivity of a membrane (containing 30% grafts) was 30 mS cm^{-1} with an overall resistance of 6.5 Ω cm^{-2}. The grafting occurred in styrene at 80, 90 and 100 °C.

Interestingly, the authors compared the physical and electrochemical properties of their PVDF-*g*-PSSA with those of FEP-*g*-PSSA and Nafion$^{\circledR}$.

2.4.3.6.4.1.1.4.   Synthesis, Properties and Applications of PVDF-g-poly(vinyl benzyl chloride) PVDF-g-PVBC.   Recently, Danks *et al.* [324] synthesised PVDF-*g*-PVBC graft copolymers from the activation of two grades of PVDF by γ-rays ^{60}Co at a dose of 6.3 Mrad at 70 °C for 1 week followed by the grafting of vinyl benzyl chloride (VBC or chloromethylstyrene). According to the grade, the degree of grafting was either 38 or 54%, while the average thickness was 54 μm. Then, the chloromethyl end-groups of the grafts underwent an amination reaction into benzyl trimethylammonium (BTMA) hydroxide [324] or BTMA chloride (BTMAC) [325] for anion-exchange membrane applications.

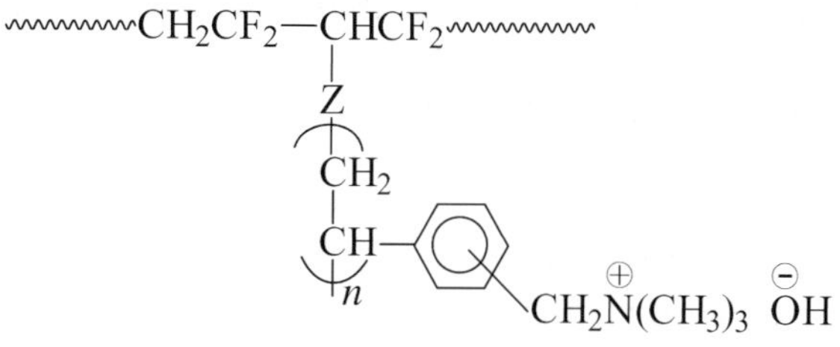

$$\sim\sim\sim\sim\sim CH_2CF_2-CHCF_2\sim\sim\sim\sim\sim$$

$$\begin{array}{c} Z \\ | \\ CH_2 \\ | \\ CH- \end{array}$$

$$CH_2\overset{\oplus}{N}(CH_3)_3 \ \overset{\ominus}{OH}$$

with Z : O or "—"

The originality of the work lies in the ability of these quaternary ammonium side groups to conduct hydroxide ions in order to use this graft copolymer as a novel membrane for alkaline fuel cells for portable applications. Unfortunately, this last step led to too brittle PVDF-g-PBTMAC graft copolymers because of degradation of the polymer backbone by the expected dehydrofluorination. Hence, the resulting material which possessed lowered ion exchange capacity (IEC = 0.31 and 0.71 meq g^{-1}) was unsuitable for use as membranes for fuel cells or electrochemical devices.

That is why these British authors decided to apply the same strategy to a perfluorinated copolymer, a poly(TFE-co-HFP) or FEP copolymer, which is not base-sensitive and they could obtain a membrane endowed with an IEC of 1.0 meq g^{-1} (Section 2.4.3.6.4.2.2.2).

2.4.3.6.4.1.2. Synthesis, Properties and Applications of PTFE-g-PM and PCTFE-g-PM Graft Copolymers. PTFE is the most known commercially available fluoropolymer produced by DuPont (Teflon®), Dyneon (formerly Hoescht, Hostaflon®), Solvay-Solexis (Algoflon®), Daikin (Polyflon®) and ICI (Fluon®). The sensitivity of PTFE to ionising radiation has been known since the early 1950s [210].

Indeed, irradiated PTFE can still be used for grafting polymerisation several years after it has been exposed [327]. The obvious advantages of this method are clear: cost and easy availability of PTFE, functional vinyl monomers and irradiation, as well as the simplicity of the procedure. Radiolysis of PTFE has been extensively studied in various books or reviews [210–213, 327–329].

Lunkwitz *et al.* [330] compared different methods of activation powdered PTFE, FEP and PFA (electron beam and γ-irradiation) in order to modify the hydrophobicity of these fluoropolymers. The first technique induced the formation of carboxylic acid fluoride groups at the surface (identified by FTIR) which were easily hydrolysed into carboxyl groups by air moisture. The second method mainly caused degradation of PTFE and the concentration of carboxyl groups was lower. The increase of the radiation dose led to an increase of the concentration of functional groups in the micropowders of PTFE and favoured an increase of the surface free energy, hence decreasing the oil and water repellencies of PTFE. Application of this modified PTFE can be found in brine electrolysis.

The presence of these polar functions was exploited by Lappan's group [330, 331] to introduce γ-amino propyltriethoxysilane. The authors also noted that the modification of FEP and PFA via these techniques yielded higher contents of carboxyl groups than those assessed in activated PTFE.

As a matter of fact, irradiation of fluorinated polymers and, in this present case, PTFE has drawn much interest in four main topics:

   (i) crosslinking of these halogenated polymers, especially for fluoropolymer wire and cable insulation [214, 217, 328, 329];
   (ii) preparation of low molecular weight—fillers and lubricants from PTFE production scrap [328, 329];
   (iii) improvements of wetting and adhesive properties [330];
   (iv) synthesis of graft copolymers [210, 216, 327] for many applications: membranes, emulsifiers, surfactants, biomaterials, compatibilisers and processing aids.

Usually, undesirable chain scission of fluoropolymer occurred during pre-irradiation processing of fluoropolymer base film. However, that involving ETFE is minimised, especially in combination with electron beam irradiation under inert atmosphere [332].

Many investigations have been carried out on the synthesis of graft copolymers starting from the activation of PTFE or ETFE.

In the first case, most results are summarised in Table VII which lists the different activation conditions, the nature of the monomer ranging from acrylic acid [333–336], methyl trifluoroacrylate (MTFA) [337], styrene [343, 346, 347], vinyl pyridine [338–341] and, to a lesser extent, N-vinyl pyrrolidone (NVP) [342]. Furthermore, the properties and the applications of the resulting graft copolymers are also listed in Table VII. Usually, decreases of both the crystallinity content and melting point of PTFE are noted.

In contrast to investigations on the grafting of PVDF, less work has been very thoroughly carried out on the irradiation of PTFE (Table VII). In the 1960s, Chapiro [210] pioneered that topic and could prepare PTFE-g-PAAc and PTFE-g-PVP (where AAc and VP stand for acrylic acid and 4-vinyl pyridine, respectively), and could also graft both monomers onto Teflon® films. The resulting semipermeable membranes exhibited excellent mechanical and chemical properties (against corrosive media), even in the swollen state.

The second main group who extensively developed researches on the grafting of fluoropolymers (mainly PTFE) was Kostov's team [334, 335, 340, 341] which could reproduce Chapiro's results. This Bulgarian team observed that the T_g and the enthalpies of melting and of crystallisation were decreasing when the d.o.g. of AAc or VP were increasing. Acrylate complexes with ferric ions were obtained

TABLE VII

SYNTHESIS, PROPERTIES AND APPLICATIONS OF PTFE-g-PM AND PCTFE-g-PM* GRAFT COPOLYMERS

Irradiation	Grafted reactant	Property	Application	References
e-beam or γ-ray	$H_2NC_3H_6Si(OEt)_3$	Less hydrophobic PTFE	Brine electrolysis	[330, 331]
γ-ray (in air)	AAc	Chemical inertness, mechanical properties	Membrane	[333]
^{60}Co, 50 kGy, RT	AAc	Ionomer	Enzyme biosensor	[334]
^{60}Co, 100 kGy, RT	AAc (d.o.g. 8–46%)	Hydrophilicity	Membrane	[335]
^{60}Co	AAc	Hydrophilicity	Membrane	[336]
^{60}Co, 1–70 Mrad	MTFA	Chemical resistance, electrical resistance	IEM	[337]
γ-ray (air)	4-VP (then quaternisation)	Chemical inertness, mechanical properties	Membrane	[338]
γ-ray (air)	4-VP	Solvent effects	Membrane	[339]
^{60}Co, 1–35 kGy	4-VP (d.o.g. 0.7–13%) (then quaternisation)	High transport number, high diffusion coefficient, mechanical properties, low specific electrical resistance	AEM	[340]
^{60}Co, 10–50 kGy	4-VP (d.o.g. 2.5–50%)	$T_g = 60\ ^{\circ}C$	AEM	[341]
^{60}Co, 5–30 kGy	NVP (radiochemical grafting)	–	–	[342]
e-beam	Styrene (then sulfonation)	Electrochemical stability	Fuel cell electrolyte	[343]
^{60}Co 1.3–15.0 kGy h^{-1}	Styrene (then sulfonation)	Water uptake, ion-exchange capacity, good thermal stability, chemical, mechanical stability	Proton conducting membrane	[344–347]
Radiografting	Vinyl pyridinium	Ionic conductivity	Membrane	[348]
* e-beam 3.5 kGy	Styrene	–	Membrane	[349]

AAc, acrylic acid; AEM, anion-exchange membrane; IEM, ion-exchange membrane; MTFA, methyl trifluoroacrylate; NVP, N-vinylpyrrolidone; 4-VP, 4-vinyl pyridine.

to prove by Mossbauer spectra the ionic cluster phase structure of the grafted layer [335]. The resulting graft copolymers were used for ionomers, IEMs and enzyme biosensors [334].

In 2000, interesting surveys were also carried out by Nasef *et al.* [344–346] on the synthesis of PTFE-*g*-PSSA copolymers for PEMs, for which striking structural investigations and thermal [344], chemical and mechanical stabilities [344–346] have been carried out (Table VII). Kinetic approaches were also attempted [334, 341].

More recently, a Chinese team [347] investigated a similar strategy for the preparation of fluorinated membranes for fuel cells (thickness: 50 μm). Although these authors could graft styrene onto irradiated PTFE with grafting rates ranging from 0.95 to 35.4%, the grafted sulfonic acid membranes obtained from d.o.g. of 12–21% exhibited good humidity, swelling performance and chemical exposure resistance.

These four teams studied the effect of grafting and the experimental conditions on the d.o.g., the kinetics of grafting and the thermal, chemical and membranous properties of the grafted films.

As for graft copolymers arising from the activation of PCTFE, to the best of our knowledge only one study was reported in the literature, by Zhao [349], who could prepare PCTFE-*g*-PS. Interestingly, the authors could assess the amount of radicals. For a dose of 3.5 kGy, the radical concentration was 5.36×10^{17} eq g^{-1} after 2760 days (and 3.99×10^{17} eq g^{-1} after 3450 days) at room temperatures, while for a dose of 14 kGy, it was 9.15×10^{17} eq g^{-1} after 3390 days.

2.4.3.6.4.2.   Activation of Fluorinated Copolymers Followed by the Grafting of Various Monomers

2.4.3.6.4.2.1.   Synthesis and Properties of Copolymers Based on VDF

*2.4.3.6.4.2.1.1.   From Poly(VDF-*co*-HFP).* Various interesting studies dealing with the preparation, properties and applications of poly (VDF-*co*-M$_1$)-*g*-PM$_2$ graft copolymers (where M$_1$ represents HFP, TFE or trifluoroethylene, TrFE, respectively) have also been reported.

Only polystyrene (which underwent further functionalisation) [248, 252, 263–268] and fluorosilicone [350, 351] as PM$_2$ grafts could be introduced as lateral polymeric chains onto the activated fluoropolymeric backbone. A summary of these researches is listed in Table VIII.

In contrast to the graft copolymers above, the lack of PM$_2$ base unit to graft may arise from the poor degree of grafting, as explained by Porte-Durrieu *et al.* [263, 264, 266]. Indeed, the diffusion of styrene in the polymer bulk is faster in poly(VDF-*co*-HFP) than in PVDF, thus limiting the accumulation of PS on the surface, which is consistent with the absence of a grafting gradient in the case of poly(VDF-*co*-HFP).

*2.4.3.6.4.2.1.2.   Activation of Poly(VDF-*co*-TFE) Copolymers Followed by Grafting* An interesting topic concerns the compatibilisation of poly(VDF-*co*-TFE) with fluorosilicone since, under normal conditions, the blend exhibits a rather coarse morphology and poor interfacial adhesion (Table VIII).

TABLE VIII

SYNTHESIS, PROPERTIES AND APPLICATIONS OF POLY(VDF-co-F-ALKENE)-g-PM GRAFT COPOLYMERS [OR CROSSLINKED POLY(VDF-co-F-ALKENE) COPOLYMERS] BY GRAFTING [OR NOT] A POLY(M) ONTO ACTIVATED POLY(VDF-co-F-ALKENE)

F-polymer	Activation	Grafted reactant	Property	Application	References
Poly(VDF-co-HFP)	γ-rays ^{137}Cs, 3 kGy h^{-1} 0.66 MeV amu^{-1}, air	Styrene (60 °C, 16 h) then sulfonation	Heparine like	Cardiovascular prostheses	[263]
Poly(VDF-co-HFP)	^{36}Ar ion 10.8 MeV amu^{-1} (O$_2$ atm)	Styrene (60 °C, 16 h) then sulfonation	Heparine like	Artificial heart	[264, 266]
Poly(VDF-co-TFE)	^{60}Co, 15 kGy	F-silicone (80 °C/24 h)	Peroxide crosslinking	Compatibiliser	[350]
Poly(VDF-co-TFE)	^{60}Co, 15 kGy	F-silicone (80 °C/24 h)	Good mechanical properties, decrease of T_m (PVDF), increase of T_g (F-silicone)	PVDF as reinforcing agent of F-silicone	[351]
Poly(VDF-co-TrFE)	Swift Sn ion 503 kGy, 5.5 MeV amu^{-1} (vacuum)	None	Crosslinking	–	[257, 268, 270]
Poly(VDF-co-TrFE)	Swift O ion 10 MeV amu^{-1} (O$_2$)	None	Crosslinking	–	[248, 267, 268]
Poly(VDF-co-TrFE)	Swift Kr ion 6.2 MeV amu^{-1} (O$_2$)	None	Crosslinking	–	[248, 268]
Poly(VDF-co-TrFE)	Corona	None	–	Biomaterials	[352]

HFP, hexafluoropropylene; TFE, tetrafluoroethylene; TrFE, trifluoroethylene.

Indeed, Mascia *et al.* [350] could use poly(VDF-*co*-TFE) as a reinforcing agent, and these British authors [351] also pre-grafted multifunctional free radical crosslinking agents on the fluoropolymeric chains before being blended with the fluorosilicone elastomer. A mixture of triallyl isocyanurate and diallyl phthalate was found to be particularly useful using γ-radiation to promote grafting reactions (enhanced by a small amount of dibenzoyl peroxide) [350]. In addition, they also succeeded in synthesising poly(VDF-*co*-TFE)-*g*-fluorosilicone elastomers [351], hence achieving better low temperature ductility without excessively reducing the melting point and the degree of crystallinity of poly(VDF-*co*-TFE) and the T_g of the F-silicone.

2.4.3.6.4.2.2.   Activation of TFE Copolymers Followed by Grafting

*2.4.3.6.4.2.2.1.   From Poly(E-*co*-TFE) or ETFE.*   Copolymers containing ethylene and TFE, called ETFE copolymers are commercially available macromolecules produced by DuPont (Tefzel®), Asahi Glass (Aflon®), Daikin (Neoflon®), Dyneon (Hostaflon®) and, in the past, Ausimont (Halon®) [328, 353]. Most grades contain between 50 and 65 mol% of TFE. Additional information on their synthesis can be found in Chapter 3, Section 3.3.2.

Many surveys on the synthesis of ETFE-*g*-poly(M) were carried out involving mainly acrylic acid [354–357], ethyl acrylate [358], methyl acrylate [358], dimethylaminoethyl methacrylate (DMAEMA) [359], *N,N*-methylene-bis-acrylamide [360], NVP [359], 1-vinyl imidazole [359] and styrene [361–367] that was used for the preparation of IEM. To improve the IEC, some of ETFE-*g*-PAAc membranes were sulfonated [354a], hence leading to both sulpho- and carboxyl groups that increased the IEC values (up to 1.75 meq g^{-1}). Some IEMs were especially suitable for fuel cell applications after sulfonation of ETFE-*g*-PS graft copolymers [362–365] (Table IX).

Various relationships [356] between the membrane properties and composition, IEC and their respective fuel cell performance are reported [356, 357]. Most work in this area has been carried out by Horsfall and Lowell [356, 357, 363] or at the Paul Scherer Institute [312, 332, 365, 366]. The former team [364] showed that oxygen diffusion coefficient and permeability increased when decreasing equivalent weight, while the proton conductivity also increased with the water content. These laboratories also noted that such new types of membranes perform well in comparison with Nafion® standard [363]. Some membranes were found to have stable resistivities, at high current density and high power density, greater than 1 A cm^{-2} [356].

Besides all the electrochemical measurements, Scherer's group also investigated the surface properties and has reported the increasing total surface energy series, as follows:

ETFE film – irradiated film < grafted film < sulfonated membrane.

Besides PEMs for fuel cells, these ETFE-*g*-PM have led to other interesting applications such as ion exchanges for dialysis, electrodialysis [354] and

TABLE IX

ETFE-g-PM GRAFT COPOLYMERS: SYNTHESIS, PROPERTIES AND APPLICATIONS

Irradiation	Grafted monomer	Property	Application	References
Electron beam	AAc (d.o.g. 8–54%)	Good electrochemical, mechanical, thermal and chemical stabilities	IEM for electrodialysis	[354]
Electron beam	AAc	Hydrophilicity	Metal recovery membrane	[355]
^{60}Co	AAc	Mechanical, chemical, thermal stabilities	IEM (fuel cell)	[356]
^{60}Co variable dose rates	AAc (d.o.g. 5–44%)	Hydrophilic membrane	IEM	[357]
e-beam 31.9 kGy	EA (d.o.g. 462%)	Thermal stability	–	[358]
e-beam 25.4 kGy	MA (d.o.g. 300%)	Thermal stability	–	[358]
^{60}Co	DMAEMA			[359]
^{60}Co	MBAA, styrene			[360]
^{60}Co	NVP			[359]
^{60}Co	1-Vinyl imidazole			[359]
e-beam	Styrene or DVB then sulfonation and quaternisation	Stability	Acid recovery by dialysis membrane electrodialysis membrane	[361]
e-beam	Styrene then sulfonation	Reduction of MeOH crossover	Membrane for DMFC	[362]
^{60}Co 12.7 kGy h^{-1}	Styrene then sulfonation (d.o.g. 5–50%)	Hydrophilic membranes	IEM	[363]
^{60}Co 12.7 kGy h^{-1}	Styrene then sulfonation		Solid polymer electrolyte	[364]
	Styrene then sulfonation	Fuel cell performance	IEM	[356]
γ-ray (in O$_2$) 5.9 kGy h^{-1}	Styrene (then sulfonation)	IEC = 1.9–2.0 meq g^{-1}, low methanol coefficient diffusion	IEM for fuel cell	[312, 365, 366]
^{60}Co	Styrene (then sulfonation)		Membranes for DMFC	[367]
γ-ray or e-beam	Styrene		IEM for fuel cell (FC)	[368]
e-beam	Substituted TFS		IEM for FC	[369]
e-beam	Substituted TFV naphthylene		IEM for FC	[369]

AAc, acrylic acid; EA, ethyl acrylate; IEM, ion-exchange membrane; MA, methyl acrylate; DVB, divinyl benzene; DMAEMA, dimethylaminoethyl methacrylate; MBAA, N,N-methylene-bis-acrylamide; TFS, trifluorostyrene; TFV, trifluorovinyl; NVP, N-vinyl pyrrolidone.

electro-electrodialysis [361] and cation-exchange membranes in the recovery of metals (Ag^+, Hg^{2+}, Co^{2+}, Cu^{2+}) [355].

Interestingly, this group compared the performances of the membranes prepared from PVDF-g-PSSA, ETFE-g-PSSA and FEP-g-PSSA and found that the one arising from ETFE polymer exhibited better mechanical properties because of: (i) the higher molecular weight of ETFE; (ii) a better compatibility of ETFE with the graft component; and (iii) the reduced extent of radiation-induced chain scission occurring on ETFE [312, 332, 365, 366].

2.4.3.6.4.2.2.2. Activation of Poly(TFE-co-HFP) or FEP Followed by Grafting. FEP (or fluorinated ethylene–propylene) resin is a linear, semi-crystalline fluoroplastic containing TFE and HFP base units [328, 370]. Interestingly, the introduction of a trifluoromethyl side group disrupts the chain packing and results in a decrease in crystallinity (that of PTFE is 98% compared with 70% for FEP). The molar percentage of HFP may vary depending on the grade of the resin, but typically contains 8–9 mol%. DuPont and Ensinger Companies [328] are main producers of this fluoropolymer.

Early investigations on the radiation chemistry of FEP were carried out by Florin and Wall [371]. It is admitted [217, 328, 370] that FEP undergoes a degradation when it is irradiated in vacuum below its T_g (80 °C).

Hill *et al.* [217] have thoroughly reviewed the influence and the intensity of the dose on the activation, crosslinking or degradation of FEP. They explained that from zero strength time (ZST) measurements an initial dose of about 160 kGy induced first a crosslinking followed by chain scissioning. Additional information on the lifetime of radicals, mechanisms of degradation, effect of UV, VUV irradiations or those from high energy electrons and X-rays has been supplied in their interesting reviews [217].

Irradiation of FEP resin in the presence of air resulted in embrittlement of the FEP film and a dramatic loss in mechanical properties.

By e-beam irradiation of finely divided FEP resin (containing 9 mol% of HFP), Lovejoy *et al.* [372] found that CF_4 was the major gaseous radiolytic product. This gas was generated from the recombination of $\cdot CF_3$ and $F\cdot$ radicals arising from the radiochemical cleavage of the CF_3 side groups.

Further, Rosenberg *et al.* [373] investigated higher dose (2000 kGy) γ-radiolysis of FEP and found it to undergo chain scission when irradiated at room temperature.

However, FEP has been an interesting matrix for grafting various monomers such as acrylic acid [357, 374–376], MMA [377], styrene [343, 363], divinyl benzene [382, 383] and vinylbenzyl chloride (VBC) [324, 384] (Table X).

In addition, useful crosslinking agents such as divinyl benzene [382, 383] or *N,N*-methylene-bis-acrylamide (MBAA) [360] have been used to enhance the properties of the membranes.

As above, most investigations have been developed on the synthesis of FEP-g-PS by Horsfall and Lowell [356, 363, 364], Nasef *et al.* [378, 379], Slade's team [324, 325, 384] and by Scherer's group [332, 365, 366, 380–382] as precursors of

TABLE X

Synthesis, Properties and Applications of FEP-g-poly(M) Graft Copolymers from Irradiation of FEP Followed by Grafting with M Monomer

Irradiation conditions	Grafted monomer	Properties	Applications	References
γ-rays	AAc followed by pyrrole	Morphology, surface, surface resistance: $104\ \Omega\ cm^{-2}$	Composite	[374]
γ-rays	AAc, kinetics of grafting (d.o.g. 20–73%)	Morphology, swelling, electrical conductivity, mechanical properties	CEM	[375, 376]
^{60}Co, 10–30 kGy	AAc; 70 °C, optimisation of d.o.g., variable [AAc]	Homogeneous membrane, EWC = 50–65%, IEC vs. d.o.g., electrol. resistivity vs. d.o.g.	IEM	[357]
^{60}Co	AAc	Hydrophilic membranes	CEM pervaporation	[375, 376]
e-beam or γ-rays	MBAA (crosslinker)	Oxidative stability	PEM for f.c.	[360]
e-beam or γ-rays	MMA	Electrochem. properties	Membrane	[377]
^{60}Co	S then sulf., kinetics of d.o.g., optimisation of d.o.g.	Good mechanical stability	PEM	[378]
^{60}Co 10–100 kGy	S then sulf.	Microstructure, morphology	PEM	[380–382]
^{60}Co, 23 °C, 10–100 kGy	S then sulf., optimisation of d.o.g., [S] = 20–80% v/v	EWC vs. d.o.g., IEC vs. d.o.g.	IEM	[363]
^{60}Co, 6 Mrad	S then sulf., 60 °C/15 h d.o.g. = 19%	Thermal stability, IEC = 1.39 meq mol^{-1}, water uptake: 68%	PEM	[379, 382]
γ-rays	S then sulf.	Electrochemical properties	FC electrolyte	[343]
^{60}Co (low radiation dose) 2–20 kGy	DVB (crosslinker)	Electrochemical properties	Membranes	[382, 383]
γ-rays (^{60}Co)	VBC 1 week, 40–70 °C (d.o.g. 3–30%)	Thermal stability IEC, morphology conductivity = 20 mS cm^{-1}	Low T DMFC Alkaline AEM	[324] [384]

AAc, acrylic acid; AEM, anionic-exchange membrane; CEM, cation-exchange membrane; DMFC, direct methanol fuel cell; DVB, divinyl benzene; EWC, equilibrium water content; IEC, ion-exchange capacity; IEM, ion-exchange membrane; MBAA, N,N-methylene-bis-acrylamide; S, styrene; sulf., sulfonation; VBC, vinylbenzylchloride.

original FEP-*g*-PSSA (after sulfonation of the aromatic rings). Interestingly, these well-architectured FEP-*g*-PSSA copolymers were rather stable (the desulfonation occurred from 200 °C) and were used as original PEMs.

Actually, Rouilly *et al.* [380a] showed that, for a d.o.g. higher than 30%, specific resistivity values as low as 2 Ω cm (at 20 °C) were measured, lower than that of Nafion® 117 (12 Ω cm).

Experimental parameters were established for maximising the d.o.g. and even the optimal treatment duration for maximum amination of FEP-*g*-PVBC graft copolymers was assessed [384].

However, the most interesting work was achieved by Horsfall and Lovell [356, 357], who made some correlations between the d.o.g., the IEC and the equilibrium water content, even starting from various fluorinated (co)polymers such as PVDF, PTFE, FEP, ETFE and PFA.

Danks *et al.* [324, 325] have compared the chemical and thermal behaviours of PVDF-*g*-PVBC and FEP-*g*-PVBC. They showed that the former graft copolymer underwent a dehydrofluorination in the functionalisation step (amination), but this step was successfully achieved from the FEP-*g*-PVBC graft copolymer in which ammonium sites were introduced for potential membranes in alkaline fuel cell applications [324, 384]. Although the IEC decayed more rapidly at 100 °C, amino FEP-*g*-PVBC was stable at 60 °C, even for at least 120 days. The conductivities of these graft copolymers were $0.01-0.02$ S cm^{-1} at room temperature, which is satisfactory for use as membranes in fuel cells.

2.4.3.6.4.2.2.3.   From Poly(TFE-co-perfluoropropylvinylether), Poly(TFE-co-PPVE) or PFA.   Copolymers of TFE and perfluoroalkyl vinyl ether (PAVE, $CF_2=CFOC_nF_{2n+1}$; $n = 1$ and mainly $n = 3$) are called PFA [328, 370, 385]. The synthesis of these comonomers have been reported [370]. These copolymers are commercially available from DuPont or Chemfab companies [328]. As expected, the melting point of PFA, ranging from 305 to 315 °C (according to the PAVE content), is lower than that of PTFE.

A small amount of PAVE is required to reduce crystallinity (and does so more efficiently than HFP because they possess longer side chains), hence leading to a tough polymer [328].

In contrast to huge works concerning the synthesis of graft copolymers starting from the irradiation of ETFE or FEP, a few research studies have been completed on the activation of PFA. Except for a few studies dealing with the crosslinking of PFA by electron beam [386], to the best of our knowledge three main groups have worked on the synthesis of PFA-*g*-PM comb-like copolymers by irradiation of PFA followed by grafting (Table XI).

First, Okada *et al.* [387] activated the surface of PFA copolymers by excimer laser irradiation and then carried out a KrF laser-induced graft reaction of poly(acrylic acid) (PAAc) onto PFA. Interestingly, the contact angle of poly(TFE-*co*-PPVE) decreased from 106 to 40–60°.

Hence, this team made wettable the surface of PFA in this original poly(TFE-*co*-PAVE)-*g*-PAAc graft copolymer. This wettability is demonstrated

TABLE XI

SYNTHESIS, PROPERTIES AND APPLICATIONS OF PFA-g-PM GRAFT COPOLYMERS FROM THE ACTIVATION OF PFA FOLLOWED BY GRAFTING

Activation conditions	Grafted reactant	Property	Application	References
Kr laser pulses 22–67 mJ cm^{-2} pulse^{-1} irrad. En = 100–200 J cm^{-2}	AAc	Surface : wettability	–	[387]
e-beam (100–1200 kGy)	S	Electrochemical, surface and mechanical properties	Electrolyte for fuel cells	[388–392]
Irradiation grafting up to 500 kGy	S	Thermal stability, structure, crystallinity	–	[393–396]

by (i) the values of the water contact angles of PFA that neatly decreased from 106 to 40–60°, (ii) the extensive loss of fluorine atoms assessed by XPS, and (iii) an increase of oxygen atoms in addition to the presence of a C_{1s} line shape, similar to that of PAAc.

The second team [388–392] investigated the irradiation of PFA followed by the grafting of styrene to achieve PFA-g-PS, which was then sulfonated to yield PFA-g-PSSA. These authors optimised the conditions of activation/grafting in terms of monomer concentrations, irradiation doses and dose rates, and different solvents were chosen in the grafting process (CH_2Cl_2 was found to greatly enhance the grafting process) [388]. As in cases above, the higher the d.o.g., the greater the monomer concentrations until it reached the highest styrene concentration of 60 vol%.

A major application of this original well-architectured copolymer deals with polymer electrolyte membrane for fuel cells. Indeed, Nasef et al. [389–392] extensively studied the morphology and various electrochemical, thermal and physico-chemical properties of these membranes. These properties were dependent upon the d.o.g.

The morphology of the membranes plays an important role in the chemical degradation of the membranes [391]. These authors also studied the physico-chemical properties (swelling behaviour, IEC, hydratation number, ionic conductivity and water uptake) which were increasing when the d.o.g. was increasing, just like the thermal and chemical stabilities [390]. These membranes were also characterised by FTIR, X-ray diffraction, XPS [391, 392] and by mechanical property measurements (tensile strength and elongation percentage). They noted that the degree of crystallinity decreased with an increase of grafting and that both the above mechanical properties decreased when the d.o.g. increased [389]. In addition, XPS was used by the authors to monitor the membrane degradation after a fuel cell test. Indeed, an oxidation degradation took place in the PFA-g-PSSA membranes during the fuel cell test and was mainly due to the chemical attack on the tertiary H of α-carbon in PS side chains [389, 391, 392].

However, the drawback of this work concerns the poor chemical stability when these membranes were used in electrochemical applications.

A similar strategy was also chosen by Cardona et al. [393], who optimised the grafting of styrene by varying the experimental conditions [394–396]. They observed that the d.o.g. increased on increasing the radiation dose up to 500 kGy, then stabilised above this dose, but the grafting yield decreased with an increase in the dose rate. As complementary studies to those above, these Australian authors [394, 396] investigated by micro-Raman spectroscopy the penetration depth of the graft and deeply studied the thermal behaviours (in terms of T_g, T_m and T_{dec} of PFA-g-PS containing various perfluoropropyl vinyl ether contents [393]). As expected, they noted first the thermal degradation of the PS graft and, in a second step, that of the PFA backbone.

Finally, these authors compared the radiation grafting of styrene onto PFA copolymer and onto polypropylene (PP) and the influence of styrene concentration, type of solvent, dose rate and irradiation rate on the grafting yield [396]. They observed that, although the d.o.g. was higher in PP chain than in PFA, for the same d.o.g., the penetration depth of the grafted PS into the substrate was higher in PFA than in PP substrate. In both polymers, the crystallinity was drastically affected: the grafting yield processes and the degree of crystallinity decreased with the grafting dose [396].

    2.4.3.6.4.3.   Activation of Other Fluorinated Copolymers.   Although other ways to enable the surface modifications of polymers by grafting are also possible [215], a few examples of grafted fluorinated polymer are reported here.

    Apart from the excimer KrF laser irradiation of PFA followed by grafting of AAc, accomplished by Okada *et al.* [387] as mentioned above, Vasilets *et al.* [397] suggested the simultaneous action of vacuum ultraviolet irradiation and atomic oxygen for the graft polymerisation of a mesogenic monomer onto various fluorine-containing polymeric materials. The resulting two-layer structure was endowed with the combined physico-mechanical properties of the polymer-supporting film and the optical characteristics of the anisotropic liquid crystal layer.

2.4.3.6.5.   Conclusion.   A wide range of different fluorinated polymer-*g*-poly(M) graft copolymers has been synthesised from the activation of fluoropolymers (mainly by ozone, plasma, SHIs, gamma or electrons) followed by grafting. Their properties (mechanical, physico-chemical, thermal and membranous) have been studied. The quality of these well-architectured fluorinated copolymers depends upon the nature of the activation of the fluoropolymer and upon a good compromise in generating as many radicals as possible without affecting the properties of the fluoropolymeric backbone, and it is still a challenge. Both basic and applied researches have been carried out, just like useful investigations on the assessment of the concentration of macroradicals (peroxides or hydroperoxides) and their half-lives, and specific properties for several applications (membranes, compatibilisers, biomaterials, etc.), respectively.

    Interestingly, different states (powders and films) and different types of fluoropolymers were activated: except for PCTFE, for which few surveys have been carried out, most starting polymers are PVDF, PTFE, ETFE, FEP and PFA.

    Various articles have reported the comparison of membranes prepared in similar conditions from the above (co)polymers. For example, taking into account two fluoropolymers isomers (PVDF and ETFE), Slade *et al.* [324, 325, 384] noted that the performance of PVDF was diminished after amination reaction of PVDF-*g*-PVBC which led to dehydrofluorination. Conversely, ETFE was a better candidate that offered potential membranes for alkaline fuel cells. The comparison of grafting monomers with various polymeric substrates and the performances of the resulting graft copolymers were achieved by different teams [298, 312, 356, 363]. Correlations between the d.o.g., IEC and equilibrium water content were shown to be independent of the chemical nature of the fluorinated starting polymer [356b].

    Grafting the required monomer to introduce the function properly has also been closely investigated by many authors, hence offering a wide range of well-architectured copolymers involved in many applications such as biomaterials (e.g., artificial hearts and cardiovascular prostheses), protection of substrates (e.g., metals), pH-sensitive membranes, membranes for purification of water and for fuel cells.

2.5. Synthesis of Fluorinated Graft Copolymers by Transfer to the Polymer

Kihara et al. [398] proposed an interesting way to synthesise an original hydrogenated graft copolymer from the graft polymerisation onto polystyrene bearing thioallyl groups. The process is depicted in the following scheme:

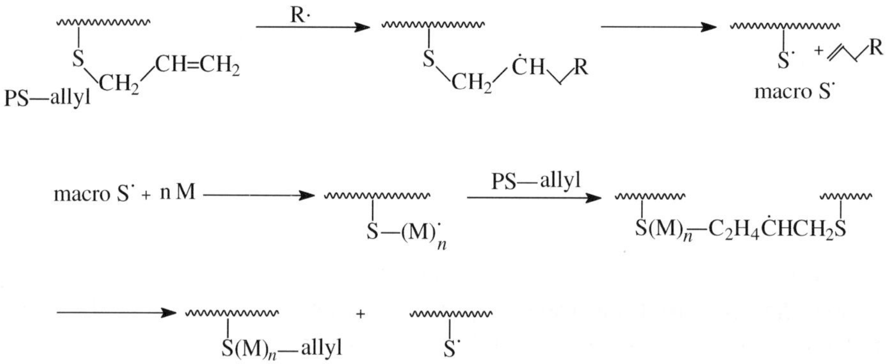

In addition, Corner [399] underlined the transfer to the polymer in his interesting review that describes all the possibilities of transfers in radical polymerisation of hydrogenated monomers.

However, in fluorinated series, few examples have been reported in the literature and, to our knowledge, in the case of the copolymerisation of TFE with propylene which produces alternating copolymers (see Chapter 3), extensive kinetics of this reaction was investigated by Kostov et al. [400]. The authors could assess the value of the transfer constant to the polymer, $C_p = 5 \times 10^{-4}$, which was logically assigned to the propylene sequence.

2.6. Synthesis of Fluorinated Graft Copolymers by Other Methods

Apart from the five main routes above, particular cases of syntheses of fluorinated graft copolymers can be suggested, including methods such as (i) the dead-end polymerisation (DEP) of monomers with fluorinated grafted azo initiators, (ii) controlled radical polymerisation such as reversible addition fragmentation chain transfer (RAFT) or atom transfer radical polymerisation (ATRP) of various monomers involving an initiator bearing a fluorinated lateral chain, and (iii) polycondensation reactions.

These various means have led to fluorinated graft-block hybrid copolymers, described below.

2.6.1. Synthesis of Fluorinated Hybrid Graft-Block Copolymers by Dead-End Polymerisation (DEP)

As mentioned in Chapter 4 (Sections 2.1.1.3 and 2.2.1.1), DEP [205] was explained and enabled the synthesis of block copolymers. An interesting example is proposed by Inoue et al. [401], who extensively investigated the surface characteristics of poly(fluoroalkylsiloxane)-b-PMMA-b-poly(fluoroalkylsiloxane) triblock copolymers of different molecular weights and containing various fluorinated side groups. These block copolymers were synthesised according to the DEP of methyl methacrylate (MMA) initiated by poly(fluoroalkylsiloxane)-containing azo initiator, as follows:

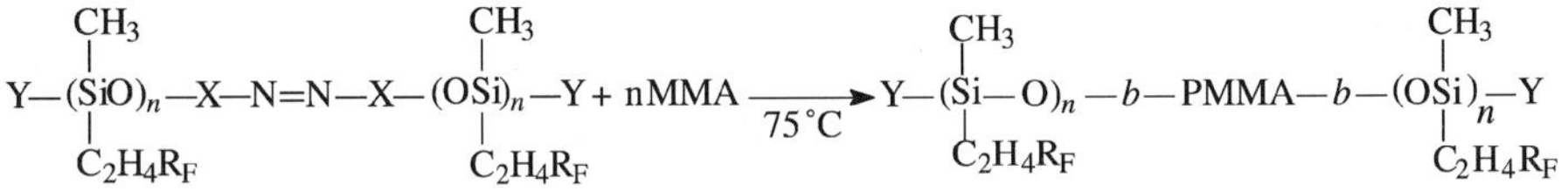

$$Y-(\underset{\underset{C_2H_4R_F}{|}}{\overset{\overset{CH_3}{|}}{SiO}})_n-X-N=N-X-(\underset{\underset{C_2H_4R_F}{|}}{\overset{\overset{CH_3}{|}}{OSi}})_n-Y + nMMA \xrightarrow{75\,°C} Y-(\underset{\underset{C_2H_4R_F}{|}}{\overset{\overset{CH_3}{|}}{Si}}-O)_n-b-PMMA-b-(\underset{\underset{C_2H_4R_F}{|}}{\overset{\overset{CH_3}{|}}{OSi}})_n-Y$$

R_F: CF_3, C_6F_{13}, C_8F_{17}

Y: $-[Si(CH_3)_2O]-C_3H_6NH_2$; X: variable structures

Although it is known that MMA undergoes disproportionation, the authors claimed the synthesis of a triblock copolymer.

Nevertheless, cast films were prepared on which XPS, ESCA and contact angles were assessed. A relationship between surface segregation behaviour and adhesion performance was established. A significant effect of fluoroalkyl groups onto the surface accumulation, not only in silicon but also in fluorine atoms, was observed for a relatively long fluoroalkyl side chain that enhanced the hydrophobicity and the oleophobicity.

2.6.2. Synthesis of Hybrid Graft-Block Copolymers by Controlled Radical Polymerisation

2.6.2.1. By ATRP

A symmetrical fluorinated difunctional brominated initiator $Br-R-C(Et)(R_F)-Z-Br$, with $Z = -C(CH_3)_2CO_2CH_2-$ and $R_F = -CH_2O(CH_2)_3S(CH_2)_2(CF_2)_7-CF_3$ (Br–InR$_F$), was prepared by condensation of 2-bromo-2-methylpropionyl bromide and a fluorinated $HOCH_2C(C_2H_5)(R_F)CH_2OH$ diol [402, 403]:

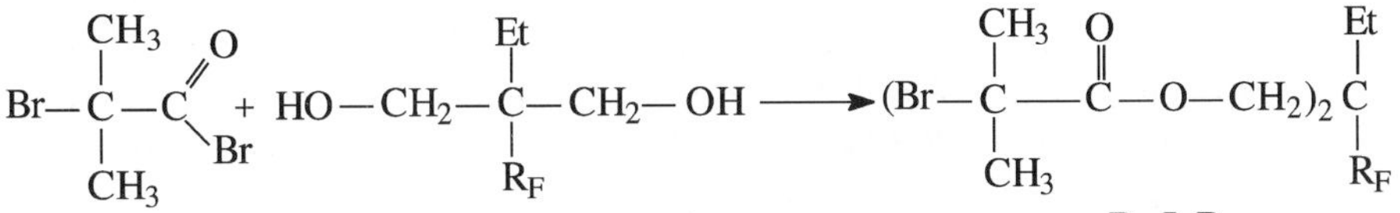

$$Br-\underset{\underset{CH_3}{|}}{\overset{\overset{CH_3}{|}}{C}}-\overset{\overset{O}{\|}}{C}\diagdown_{Br} + HO-CH_2-\underset{\underset{R_F}{|}}{\overset{\overset{Et}{|}}{C}}-CH_2-OH \longrightarrow (Br-\underset{\underset{CH_3}{|}}{\overset{\overset{CH_3}{|}}{C}}-\overset{\overset{O}{\|}}{C}-O-CH_2)_2\underset{\underset{R_F}{|}}{\overset{\overset{Et}{|}}{C}}$$

Br-InR$_F$

where R_F: $CH_2O(CH_2)_3S(CH_2)_2(CF_2)_7-CF_3$

This initiator bearing a fluorinated side chain was successfully used in the ATRP of MMA in the presence of a CuBr/ligand system leading to a PMMA having a perfluorinated side chain and that exhibits a $\bar{M}_n = 26{,}000$ and PDI = 1.2. Such perfluorinated lateral groups containing a $CH_2OC_3H_6SC_2H_4$ spacer between the polymeric backbone and the C_8F_{17} end-group brought improved surface properties.

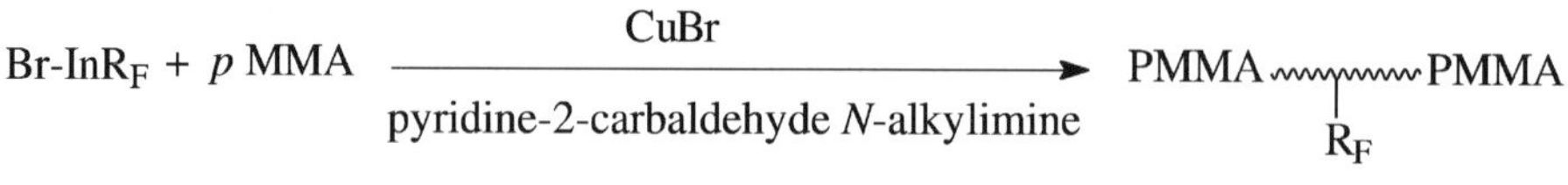

$$Br\text{-}InR_F + p\ MMA \xrightarrow[\text{pyridine-2-carbaldehyde }N\text{-alkylimine}]{\text{CuBr}} PMMA\text{\textasciitilde\textasciitilde\textasciitilde}\underset{R_F}{|}PMMA$$

2.6.2.2. By RAFT

In contrast to various syntheses of block copolymers carried out by the RAFT process, and reported in Chapter 4 (Sections 2.1.2.4.3 and 2.2.3), little research has been attempted for the preparation of fluorinated graft copolymers. To the best of our knowledge, the only example that has also led to hybrid graft-block copolymers has been suggested by Laschewsky's group [404–406]. The authors first synthesised a fluorinated styrene bearing an ω-perfluoroethyl group and a piperazinium spacer that they copolymerised with a hydrophilic vinyl benzyl trimethyl ammonium chloride in the presence of benzyldithiobenzoate according to the RAFT technique. This random copolymer bearing a dithioester end-group was involved in a further copolymerisation of the hydrogenated styrene homologue with the same hydrophilic monomer to lead to a diblock copolymer bearing a fluorinated graft:

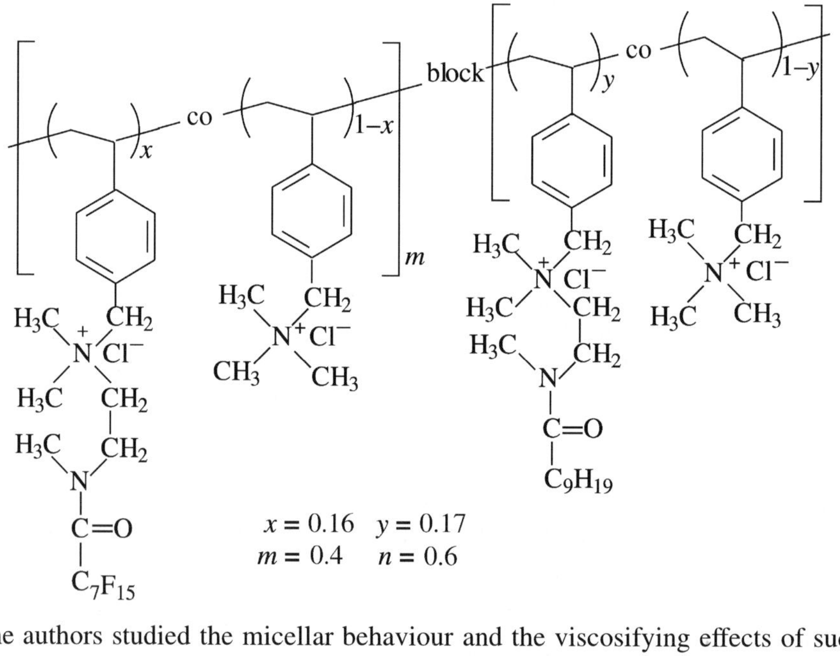

$x = 0.16 \quad y = 0.17$
$m = 0.4 \quad n = 0.6$

The authors studied the micellar behaviour and the viscosifying effects of such hybrid copolymers and they found that these macromolecules acted somewhere in between "well-behaved" polysoaps and associative thickeners. In addition, they noted their ability to hydrophobically self-organise, together with the possibility of preparing microphase-separated amphiphilic hydrocarbon–fluorocarbon block copolymers [404].

In addition, the same group [405] also prepared other polysoaps bearing the same fluorinated grafts but by polyaddition of hydrophobically modified telechelic diamine bearing either a hydrocarbon or a perfluorocarbon hydrophobic side chain with 1,4-dibromobut-2-ene. These amphiphilic hybrid copolymers were studied as polymeric surfactants, especially with respect to their hydrophobic association in aqueous solution. The authors noted a microphase separation into hydrocarbon-rich and fluorocarbon-rich hydrophobic domains, thus yielding multicompartment micelles. Furthermore, these macromolecules showed strong surface activity and very high surface dilatational elasticity [406].

2.6.3. Synthesis of Fluorinated-Graft Multiblock Copolymers by Polycondensation

Original multiblock copolymers bearing fluorinated side groups of different chain lengths were successfully achieved by Pospiech's group [195, 407–410] from the polycondensation of 3-fluoroalkyloxy isophthalic acid (IPA) with 1,4-dihydroxybenzene (or hydroquinone, HQ). The fluorinated isophthalic acid was obtained from the Mitsunobu reaction between 3-hydroxy isophthalic acid with ω-perfluoroalkyl alcohols of various chain lengths, shown below [410]. These ω-perfluorinated side groups improved the surface properties.

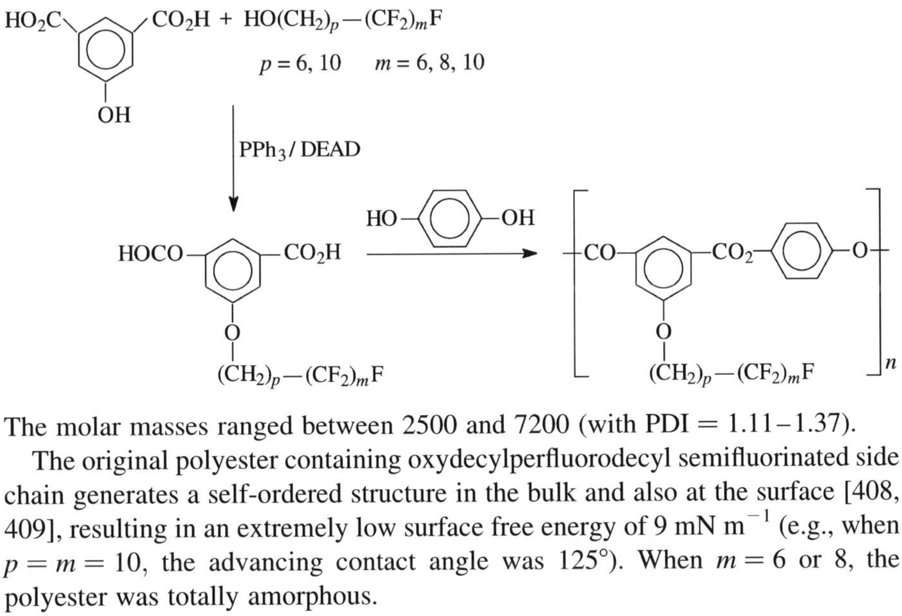

The molar masses ranged between 2500 and 7200 (with PDI $= 1.11$–1.37).

The original polyester containing oxydecylperfluorodecyl semifluorinated side chain generates a self-ordered structure in the bulk and also at the surface [408, 409], resulting in an extremely low surface free energy of 9 mN m^{-1} (e.g., when $p = m = 10$, the advancing contact angle was 125°). When $m = 6$ or 8, the polyester was totally amorphous.

In addition, this same German group [195, 409] also prepared a series of 10 segmented block copolymers containing polysulfone (PSU) and aromatic polyester containing a long semifluorinated (SF) group:

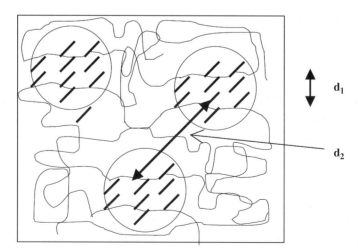

They carefully characterised these materials in terms of amounts of HQ–IPA–HQ, PSU–IPA–HQ and PSU–IPA–PSU triads: according to the stoichiometry, they noted that the first one was generally in high excess, from 82 to 90%, while 3, 25 and 72% content was also observed. This German team also assessed their molecular weights ranging from 13,000 to 44,000 (with PDI of 2.13–3.33).

Further, these authors investigated the structural behaviour of these materials: phase-separated block copolymers of PSU and SF polyester have both the glass transition PSU as indication of an amorphous matrix as well as the thermal transitions of the SF polyester. Based on the results obtained by DSC, SAXS and TEM, Pospiech *et al.* [409] proposed a structural model showing a random distribution of well-ordered SF domains within an amorphous matrix (Figure 3). Interestingly, d_1 is exactly the layer distance obtained in the pure SF–PES system while the larger scattering maximum d_2 represents the periodic distance between the SF–PES domain in the PSU matrix.

These F-polymers show a tendency of self-organisation of SF side chains even stronger than that of randomly distributed SF side groups (see Section 2.2.4), because of the greater interaction of the fluorinated groups, leading to a greater

FIG. 3. Structural model of segmented PSU–SF–PES grafted copolymers indicating the periodic distances obtained in the SAXS measurements (reproduced with permission and courtesy from D. Pospiech *et al.* (2003) *Macromol. Symp.*, **198**, 421–434).

ordered structure in the bulk. In addition, their surface properties show extremely high water contact angles (115–122°), i.e., low surface energies, at high concentrations of SF side chains providing a clearly more hydrophobic behaviour than the unmodified polysulfones. These polymers have been involved in blends [410].

2.6.4. Conclusion

This sub-section reports various strategies of synthesis of original fluorinated graft-block copolymers and their properties.

These well-architectured copolymers can be obtained either from non-controlled (DEP of monomer with suitable azo macroinitiator bearing fluorinated side-groups) or controlled radical polymerisation (RAFT and ATRP), or by polycondensation. The originality and the availability of starting materials (fluorinated azo for DEP, initiator for ATRP, monomers for RAFT or isophthalic acid for polycondensation) are key points in the achievement of these hybrid well-defined fluorocopolymers. Micellar behaviour and surface properties were investigated since the perfluorinated end-groups of the grafts bring higher water contact angles and hence lower surface tensions.

3. Conclusion

This chapter provides an up-to-date review of the syntheses, properties and applications of fluorinated graft copolymers. Four main routes have been successfully used to produce these fluoropolymers of controlled architecture.

The key points are summarised hereafter:

(1) A macromonomer is chosen that can be adequately (co)polymerised: for example (i) (meth)acrylates and styrenics can be (co)polymerised under radical or photochemical conditions; (ii) methacrylates by GTP; (iii) vinyl ethers, oxetanes and epoxides with cationic systems. In fact, the first fluorinated monomers and, more recently, fluorinated oxetanes are commercially available.

However, some of the resulting homopolymers have several important drawbacks, particularly their insolubility in all organic common solvents except in fluorinated ones or in sc CO_2. This is reflected in the growing current interest focussed on the use of this "green" technology.

On the other hand, more copolymerisations have been studied since the second comonomer (which is not necessarily fluorinated) brings complementary properties such as solubility, hydrophilicity, adhesion, crosslinking, enhanced ion exchange capacity or conductivity, improved hardness or mechanical properties, increasing the glass transition temperature or decreasing the degree of crystallinity.

(2) Another alternative which has also led to many academic and industrial surveys concerns the "grafting onto" technique. Most routes to chemical

modifications concern the grafting of a fluorinated reactant onto a hydrogenated oligomer or polymer. This can be achieved by esterification, radical grafting of difluorocarbenes (although only short fluorinated grafts are formed), of fluorinated mercaptans or of perfluoroalkyl iodides onto lateral double bonds of polydienes, hydrosilylation of poly(hydrogenomethyl siloxanes) with various vinylic or (*pseudo*) allylic ω-perfluorinated monomers, and other specific modifications.

(3) A third possibility consists in using macroinitiators. They can be synthesised via different alternatives:

(i) either from the polymerisation of a monomer that contains an initiating function,

(ii) or from a polymer bearing various lateral hydroperoxides or peroxides located all along the polymeric backbone.

(a) In the former case, the experimental conditions must be chosen in order to defavour reaction of the monomer by its peroxide or azo groups, hence leading to a macromolecule bearing various initiating side functions. So far, to our knowledge, only monomers that possess an unsaturation and initiating functions have been synthesised. Hence, these monomers react first at their double bond. The best example of starting monomer is *tert*-butyl peroxyallyl carbonate leading to commercially available fluorothermoplastic elastomer Cefral® graft copolymer.

Nevertheless, it would be interesting to imagine a monomer which generates a macroinitiator by cationic or GTP techniques in the first step, followed by a radical copolymerisation of (an)other monomer(s) to achieve original F-graft copolymers.

(b) In the latter case, the polymer that contains various initiating functions can be prepared by the activation of this initial polymer. Indeed, various techniques of irradiation are possible. Among electron beam, γ-rays, SHIs, plasma, ozone and, to a lesser extent, thermal and redox system, the first three have been extensively used onto various polyfluoroolefins (especially PVDF and PTFE) and onto fluorinated copolymers (particularly poly(VDF-*co*-HFP), ETFE, FEP and PFA copolymers).

Extensive investigations were carried out by Sundholm's, Scherer's, Slades's, Betz's Baquey and Porte-Durrieu's and Horsfall and Lowell's teams which optimised the grafting conditions by varying the dose rates, the radiation times of γ-rays, electron beam or SHIs, hence finding interesting relationships. Furthermore, various authors could assess by titration the hydroperoxides and peroxide contents. Although these contents are still low, further investigations are required to increase these amounts of initiators. The second step following the activation of the polymer in order to complete the "grafting from" technique requires the grafting of a monomer (usually styrenic, (meth)acrylic, acrylic acid, vinyl pyridine, vinyl benzyl chloride (VBC)) and, in this case too, many efforts to optimise the degree of grafting were attempted by changing the grafting time, the temperature and other experimental parameters. Further steps were also achieved to introduce key functions (sulfonic acid or ammonium groups) for IEMs.

(4) Although many results concerning the transfer to the polymer in hydrogenated series have been reported in the literature, to our knowledge, few similar strategies have been described for synthesising fluorinated graft copolymers.

These fluorinated graft copolymers have already found potential applications in many fields, such as antithrombogenic biomaterials (cardiovascular prostheses and artificial hearts); compatibilisers for the protection of substrates (e.g., by adhesion onto steel), pH-sensitive membranes or involved in the purification of water (recovery of metals K^+, Hg^{2+}, Co^{2+}, Cu^{2+}) or in fuel cells.

Finally, except (i) the ATRP of poly(ethylene oxide) monomethacrylate from PVDF [310], (ii) the ATRP of styrene with PVDF-*g*-PVBC copolymers [308] and (iii) the RAFT process recently developed by Kang's team [411] to obtain controlled PVDF-*g*-PPEOMA, PVDF-*g*-[PVBC-*g*-PSSA] and PVDF-*g*-PEO graft copolymers, respectively, all these four possibilities reported above have led to non-living graft copolymers, in contrast to the processes of synthesising *pseudo*living fluorinated block copolymers. To our knowledge, these are the only three examples on the preparation of controlled F-graft copolymers mentioned in the literature. However, a recent investigation was carried out in our laboratory regarding the synthesis of *pseudo*living fluorinated graft copolymers from the ATRP of various monomers (hence leading to "living" grafts) initiated from macromolecules bearing specific brominated groups [412].

In addition, regarding the microstructure of these graft copolymers, the real structure of the link between the polymeric backbone and the grafts deserves to be elucidated.

Presumably, an oxygen atom should be the bridge, as in the case of grafted copolymers arising from the irradiation of polyethylene, but no neat characterisation has so far demonstrated that statement. Actually, such an ether link should make stable the graft to the polymeric backbone in contrast to a carbon–carbon bond between both of them, although several investigations have shown that the graft decomposed first, under heating.

Moreover, almost no information has been given on the chain length of the oligomeric or polymeric grafts. Thus, there is potential for scientists to conduct further investigations which should produce worthwhile results.

References

1. Noshay, A. and McGrath, J. E. (1977) *Block and Graft Copolymers: Overview and Critical Survey*, Academic Press, New York.
2. Andrade, J. D. (1985) *Surface and Interfacial Aspects of Biomedical Polymers*, Vol. 1, Plenum Press, New York, p. 373.
3. Rempp, P. F. and Lutz, P. J. (1989) Synthesis of graft copolymers. In *Comprehensive Polymer Science*, Vol. 6 (Eds, Allen, G., Bevington, J. C., Eastmond, A. L. and Russo, S.) Pergamon Press, Oxford, pp. 403–421, Chap. 12.
4. Cowie, J. M. G. (1989) Block and graft copolymers. In *Comprehensive Polymer Science*, Vol. 3(3) (Eds, Allen, G., Bevington, J. C., Eastmond, A. L. and Russo, A.) Pergamon Press, Oxford, pp. 33.

5. Chujo, Y. and Yamashita, Y. (1989) Macromonomers. In *Telechelic Polymers: Synthesis and Applications* (Ed, Goethals, E. J.) CRC Press, Boca Raton, FL, pp. 163–179, Chap. 8.
6. (a) Tsukahara, Y., Kohno, K., Inoue, H. and Yamashita, Y. (1985) Surface modification of polymer solids by graft copolymers. *Nippon Kagaku Kaishi*, 1070–1078 (*Chem. Abstr.*, **103**, 24217); (b) Imae, T. (2003) Fluorinated polymers. *Curr. Opin. Colloids Interf. Sci.*, **8**, 307–314.
7. Reghunadhan Nair, C. P. (1992) Macromonomers via free radical chain transfer; some aspects of their syntheses and kinetic prediction of functionality. *Eur. Polym. J.*, **28**, 1527–1532.
8. (a) Yagci, Y. and Schnabel, W. (1990) Light-induced synthesis of block and graft copolymers. *Prog. Polym. Sci.*, **15**, 551–601; (b) Moad, G. (1999) The synthesis of polyolefin graft copolymers by reactive extrusion. *Prog. Polym. Sci.*, **24**, 81–142.
9. (a) Simons, J. H. (1950) *Fluorine Chemistry*, Vol. 1, Academic Press, Orlando, FL, pp. 414; (b) Abe, T. and Nagase, S. (1982) Electrochemical fluorination (Simons process) as a route to perfluorinated organic compounds of industrial interest. In *Preparation, Properties, and Industrial Applications of Organofluorine Compounds* (Ed, Banks, R. E.) Ellis Horwood Ltd, Chichester, pp. 19–43, Chap. 1.
10. Starks, C. M. (1974) *Free Radical Telomerization*, Academic Press, New York.
11. Gordon, R. and Loftus, J. E. (1989) Telomerization. In *Encyl. Polym. Sci. Technol.*, 16 (Eds, Kirk, R. E. and Othmer, D. F.) Wiley, New York, pp. 533–572.
12. Ameduri, B. and Boutevin, B. (1997) Telomerisation reaction of Fluoroalkenes. In *Topics Curr. Chem.*, 192 (Ed, Chambers, R. D.) pp. 165–235.
13. Boutevin, B. and Pietrasanta, Y. (1989) *Les Acrylates et Polyacrylates Fluorés*, Erec, Paris.
14. (a) Brace, N. O. (1999) Syntheses with perfluoroalkyl radicals from perfluoroalkyl iodides. A rapid survey of synthetic possibilities with emphasis on practical applications. Part one: alkenes, alkynes and allylic compounds. *J. Fluorine Chem.*, **93**, 1–25 (and references therein); (b) Brace, N. O. (1999) Syntheses with perfluoroalkyl iodides. Part II. Addition to non-conjugated alkadiene; cyclization of 1,6-heptadiene, and of 4-substituted 1,6-heptadienoic compounds: bis-allyl ether, ethyl diallylmalonate, *N,N'*-diallylamine and *N*-substituted diallylamines; and additions to homologous *exo*- and *endo*cyclic alkenes, and to bicyclic alkenes. *J. Fluorine Chem.*, **96**, 101–127 (and references therein).
15. Ameduri, B., Boutevin, B., Nouiri, M. and Talbi, M. (1995) Synthesis and properties of fluorosilicon-containing polybutadienes by hydrosilylation of fluorinated hydrogenosilanes. Part 1. Preparation of the silylation agents. *J. Fluorine Chem*, **74**, 191–198.
16. (a) Coudures, C., Pastor, R., Szonyi, S. and Cambon, A. (1984) Synthèse de F-alkyl oxiranes. *J. Fluorine Chem.*, **24**, 93–104; (b) Chaabouni, B. M., Baklouti, A., Szonyi, S. and Cambon, A. (1990) Nouvelles méthodes de préparation des F-alkyl oxiranes à partir des acétates de bromo-2 F-alkyl-2-éthyle et des bromo-2 F-alkyl-2-éthanol. *J. Fluorine Chem.*, **46**, 307–315.
17. (a) Evans, F. W. and Litt, M. H. (1968) Fluorinated epoxy compounds and polymers. US Patent 3,388,078 (assigned to Allied Chem. Co.); (b) Collet, A., Commeyras, A., Hirn, B. and Viguier, M. (1994) Perfluorinated polyether polyols for use in polyurethanes. French Patent 2,697,256.
18. Ameduri, B., Boutevin, B., Kostov, G. and Petrova, P. (1999) Synthesis and polymerization of fluorinated monomers bearing a reactive lateral group. Part 6. Synthesis of trifluorovinyl epoxide and its 1,2-diol. *J. Fluorine Chem.*, **93**, 139–144.
19. Vakhlamova, L. S. H., Kashkin, A. V., Krylov, A. I., Kashkina, G. S. and Sukhinin, V. S. (1978) Synthesis of fluorine-containing derivatives of oxacyclobutane. *Zh. Vses. Khim.*, **23**, 357–359 (*Chem. Abstr.*, **89**, 110440 (1978)).
20. Ameduri, B., Boutevin, B. and Karam, L. (1993) Synthesis and properties of poly(3-chloromethyl-3-(1,1,2,2-tetrahydroperfluorooctyloxy) methyl oxetane). *J. Fluorine Chem.*, **65**, 43–50.
21. Cook, E. W. and Landrum, B. F. (1965) Synthesis of partially fluorinated oxetanes. *J. Heterocycl. Chem.*, **2**, 327–328.
22. (a) Malik, A. A., Manser, G. E., Archibald, T. G., Duffy-Matzner, J. L., Harvey, W. L., Grech, G. J. and Carlson, R. P. (1996) Mono-substituted fluorinated oxetane monomers. Int. Demand

WO 96/21657 (assigned to Aerojet General Corp.); (b) Fujiwara, T., Makal, W., Wynne, K. J. (2003) Synthesis and characterization of novel amphiphilic telechelic polyoxetanes. *Macromolecules*, **36**, 9383–9389.

23. (a) Thomas, R., Kausch, C. M., Medsker, R. E., Russell, V. M., Malik, A. A. and Kim, Y. (2002) Surface activity and interfacial rheology of a series of novel architecture, water dispersible poly(fluorooxetane)s. *Polym. Prepr. (Am. Chem. Soc., Div. Polym. Chem.)*, **43**(2), 890–891; (b) *Fluorine in Coatings V Conference*, Orlando, January 21–22, 2003, paper 3; (c) Weinert, R. J., Robbins, J. E., Medsker, R. and Woodland, D. (2003) Surface improvements in coatings using polyfluorooxetane modified polyesters. *Fluorine in Coatings V Conference*, Orlando, January 21–22, 2003, paper 13.

24. Höpken, J. and Möller, M. (1992) Low-surface energy polystyrene. *Macromolecules*, **25**, 1461–1467.

25. (a) Bernardo-Bouteiller, V. (1995) Synthèse et caractérisations de nouveaux polystyrènes porteurs de chaines latérales fluorées. PhD Dissertation, Université de Paris VI; (b) Bouteiller, V., Garnault, A. M., Teyssié, D., Boileau, S. and Möller, M. (1999) Synthesis, thermal and surface characterisation of fluorinated polystyrenes. *Polym. Int.* **48**, 765–772.

26. (a) Lacroix-Desmazes, P., Young, J. L., Taylor, D. K., DeSimone, J. M. and Boutevin, B. (2002) *8th Meeting on Supercritical Fluids, Proceedings (ISASF)*, April 14–17, (2002), Bordeaux, 1, 241 [ISBN 2-905267-34-8]; (b) Lacroix-Desmazes, P., Boutevin, B., Taylor, D. and DeSimone, J. M. (2002) Synthesis of fluorinated block copolymers by nitroxide-mediated radical polymerization for supercritical carbon dioxide applications. *Polym. Prepr. (Am. Chem. Soc., Div. Polym. Chem.)* **43**(2), 285–286.

27. Hirao, A., Koide, G. and Sugiyama, K. (2002) Synthesis of novel well-defined chain-end- and in-chain-functionalized polystyrenes with one, two, three, and four perfluorooctyl groups and their surface characterization. *Macromolecules*, **35**, 7642–7651.

28. (a) Andruzzi, L., Chiellini, E., Galli, G., Li, X., Kang, S. H. and Ober, C. K. (2002) Engineering low surface energy polymers through molecular design: synthetic routes to fluorinated polystyrene-based block copolymers. *J. Mater. Chem.*, **12**, 1684–1692; (b) Andruzzi, L., Hexemer, A., Li, X., Ober, C. K., Kramer, E. J., Galli, G. and Chiellini, E. (2002) Surface control using polymer brushes produced by controlled radical polymerization. *Polym. Prepr. (Am. Chem. Soc., Div. Polym. Chem.)*, **43**(2), 76–77.

29. Boutevin, B., Pietrasanta, Y. and Youssef, B. (1989) Synthèses de polysiloxanes fluorés. Partie 1. Hydrosilylation d'oléfines fluorées allyliques et styréniques. *J. Fluorine Chem.*, **31**, 57–63.

30. Boutevin, B., Pietrasanta, Y. and Youssef, B. (1989) Synthèse d'éthers vinyliques à chaine latérale fluorée. *J. Fluorine Chem.*, **44**, 395–402.

31. Choi, W. O., Sawamoto, M. and Higashimura, T. (1988) Living cationic homo- and copolymerization of vinyl ethers bearing a perfluoroalkyl pendant group. *Polym. J.*, **20**, 201–210.

32. Hopken, J., Moller, M., Lee, M. and Percec, V. (1992) Synthesis of poly(vinyl ether)s with perfluoroalkyl pendant groups. *Macromol. Chem.*, **193**, 275–284.

33. Percec, V. and Lee, M. (1992) Molecular engineering of liquid crystal polymers by living polymerization. 23. Synthesis and characterization of AB block copolymers based on ω-[(4-cyano-4(-biphenyl)-oxy]alkyl vinyl ether, 1H,1H,2H,2H-perfluorodecyl vinyl ether, and 2-(4-biphenyloxy)ethyl vinyl ether with 1H,1H,2H,2H-perfluorododecyl vinyl ether. *J. Macromol. Sci., Pure Appl. Chem.*, **A29**, 723–740.

34. Vandooren, C., Jérôme, R. and Teyssié, P. (1994) Living cationic polymerization of 1H, 1H, 2H, 2H-perfluorooctyl vinyl ether. *Polym. Bull.*, **32**, 387–393.

35. Matsumoto, K., Kubota, M., Matsuoka, H. and Yamaoka, H. (1999) Water-soluble fluorine-containing amphiphilic block copolymer: synthesis and aggregation behavior in aqueous solution. *Macromolecules*, **32**, 7122–7127.

36. Matsumoto, K., Mazaki, H., Nishimura, R., Matsuoka, H. and Yamaoka, H. (2000) Perfluoralkyl-philic character of poly(2-hydroxyethyl vinyl ether)-*block*-poly[2-(2,2,2-trifluoroethoxy)ethyl

vinyl ether] micelles in water: selective solubilization of perfluorinated compounds. *Macromolecules*, **33**, 8295–8300.

37. Bongiovanni, R., Sangermano, M., Malucelli, G., Priola, A., Leonardi, A., Ameduri, B., Policcino, A. and Recca, A. (2003). Fluorinated vinyl ethers as new surface agents in the photocationic polymerization of vinyl ether resins. *J. Polym. Sci., Part A: Polym. Chem.*, **41**, 2890–2897.

38. Wakselman, C. (1994) Fluoroacrylic esters and related monomers: preparation and use as synthons. *Macromol. Symp.*, **82**, 77–87.

39. (a) Guery-Rubio, C., Viguier, M. and Commeyras, A. (1995) Surface tension of perfluoroalkyl methacrylic copolymers. *Macromol. Chem. Phys.*, **196**, 1063–1075; (b) Mélas, M. (1995) Alcools perfluoroalkylés et polyacrylates derivés: synthèse, étude structurale et propriétés de surface. PhD dissertation, University of Montpellier.

40. Timperley, C. M., Arbon, R. E., Bird, M., Brewer, S. A., Parry, M. W., Sellers, D. J. and Willis, C. R. (2003) Bis(fluoroalkyl)acrylic and methacrylic phosphate monomers, their polymers and some of their properties. *J. Fluorine Chem.*, **121**, 23–31.

41. Park, J., Lee, S.-B., Choi, C. K. and Kim, K.-J. (1996) Surface properties and structure of poly (perfluoroalkylethyl methacrylate). *J. Colloid Interface Sci.*, **181**, 284–288.

42. Thomas, R. R., Anton, D. R., Graham, W. F., Darmon, M. J., Sauer, B. B., Stika, K. M. and Swartzfager, D. G. (1997) Preparation and surface properties of acrylic polymers containing fluorinated monomers. *Macromolecules*, **30**, 2883–2890.

43. Pan, G., Tse, A. S., Kesavamoorthy, R. and Asher, S. A. (1998) Synthesis of highly fluorinated monodisperse colloids for low refractive index crystalline colloidal arrays. *J. Am. Chem. Soc.*, **120**, 6518–6524.

44. (a) Ameduri, B., Bongiovanni, R., Malucelli, G., Pollicino, A. and Priola, A. (1999) New fluorinated acrylic monomers for the surface modification of UV-curable systems. *J. Polym. Sci., Part A: Polym. Chem.*, **37**, 77–86; (b) Ameduri, B., Bongiovanni, R., Lombardi, V., Pollicino, A., Priola, A. and Recca, A. (2001) Effect of the structural parameters of a series of fluoromonoacrylates on the surface properties of cured films. *J. Polym. Sci., Part A: Polym. Chem.*, **39**, 4227–4235.

45. (a) Pées, B. (2000) Synthèses et polymérisation de monomères acryliques perfluorés. Caractérisations physico-chimiques de leurs polymères. PhD Dissertation, University of Metz; (b) Pées, B., Cahuzac, A., Sindt, M., Ameduri, B., Paul, J. M., Boutevin, B. and Mieloszynski, J. L. (2001) Synthesis of fluoro-substituted acrylic monomers bearing a functionalized lateral chain. Part 1. Preparation of sulfur-containing monomers. *J. Fluorine Chem.* **108**, 133–142; (c) Pées, B., Sindt, M., Paul, J. M. and Mieloszynski, J. L. (2002) Synthèses et caractérisations physico-chimiques de polyacrylates et ω-perfluorooctyl-alkyle: effet pair–impair. *Eur. Polym. J.* **38**, 921–931; (d) Pées, B., Paul, J. M., Oget, N., Sindt, M. and Mieloszynski, J. L. (2003) Synthesis of fluoro-substituted monomers bearing a functionalised lateral chain. Part 2. Preparation of sulfoxides and sulfones containing monomers. *J. Fluorine Chem.* **124**, 139–146.

46. Cirkva, V., Ameduri, B., Boutevin, B. and Paleta, O. (1997) Chemistry of [(perfluoroalkyl) methyl]oxiranes. Regioselectivity of ring opening with O-nucleophiles and the preparation of amphiphilic monomers. *J. Fluorine Chem.*, **84**, 53–59.

47. Kvicala, J., Dolensky, B. and Paleta, O. (1997) Fluoroalkyl derivatives of protected glycerol by nucleophilic substitution. Fluorine-containing amphiphilic mono- and bis-methacrylates. *J. Fluorine Chem.*, **85**, 117–125.

48. Guyot, B., Ameduri, B. and Boutevin, B. (1995) Synthesis and polymerization of novel fluorinated morpholino acrylates and methacrylates. *J. Fluorine Chem.*, **74**, 233–240.

49. Gavelin, P., Jannasch, P. and Wesslén, B. (2001) Amphiphilic polymer gel electrolytes. I. Preparation of gels based on poly(ethylene oxide) graft copolymers containing different ionophobic groups. *J. Polym. Sci., Part A: Polym. Chem.*, **39**, 2223–2232.

50. (a) Sreenivasulu Reddy, V., Lunceford, B. D., Cassidy, P. and Fitch, J. (1997) Synthesis and characterization of new silicon-containing fluoroacrylate monomers and polymers: 3. *Polymer*, **38**, 703–706; (b) Sreenivasulu Redi, V., Bruma, M., Fitch, J. and Cassidy, P. (1999) Polyacrylates containing the hexafluoroisopropylidene function in the pendant groups. In *Fluoropolymers*, Vol. 1 (Eds, Hougham, G., Cassidy, P. E., Johns, K. and Davison, T.) Kluwer Academic, New York, pp. 3–9, Chap. 1.

51. (a) Bracon, F., Guittard, F., Taffin-de-Givenchy, E. and Geribaldi, S. (2001) Preparation and thermotropic properties of partially fluorinated acrylates. *Cryst. Liq. Cryst. Sci. Technol., Sect. A: Mol. Cryst. Liq. Cryst.*, **365**, 263–270; (b) Andruzzi, L., d'Apollo, F., Galli, G. and Gallot, B. (2001) Synthesis and structure characterization of liquid crystalline polyacrylates with unconventional fluoroalkylphenyl mesogens. *Macromolecules*, **34**, 7707–7714; (c) Corpart, J. M., Girault, S. and Juhue, D. (2001) Structure and surface properties of liquid crystalline fluoroalkyl polyacrylates: role of the spacer. *Langmuir*, **17**, 7237–7244.

52. Hartman, P. (2002) Relations structure-propriétés de surface dans les polymères acryliques perfluoroalkylés. PhD Dissertation, University of Montpellier.

53. Bernett, M. K. and Zisman, W. A. (1962) Wetting properties of acrylic and methacrylic polymers containing fluorinated side chains. *J. Phys. Chem.*, **66**, 1207–1208.

54. Marra, K. and Chapman, T. (1996) Poly(fluoroalkyl) acrylates with diacetylene-containing side chains. *Polym. Prepr. (Am. Chem. Soc., Polym. Div.)*, **37**, 286–287.

55. Robert, A., Lopez-Calahorra, F. and Ameduri, B. (2004) Radical polymerisation of 1H,1H,2H, 2H-perfluoro-3,5-alkyldiynol and 1H,1H-perfluoro-2,4-alkyldiynol acrylates and methacrylates: a new family of fluorinated polymers. *Macromol. Chem. Phys.*, **205**, 223–229.

56. (a) Shimizu, T., Tanaka, Y., Kutsumizu, S. and Yano, S. (1994) Ordered structures of poly(1H,1H,2H,2H-perfluorodecyl α-substituted acrylates. *Macromol. Symp.*, **82**, 173–184; (b) Chuvatkin, N. N. and Panteleeva, I. Yu. (1997) Polyfluoroacrylates. In *Modern Fluoropolymers* (Ed Scheirs, J.) Wiley, New York, pp. 191–206, Chap. 9.

57. Boutevin, B., Rousseau, A. and Bosc, D. (1992) Accessible new acrylic monomers and polymers as highly transparent organic materials. *J. Polym. Sci., Part A: Polym. Chem.*, **30**, 1279–1286.

58. Otazaghine, B., Boutevin, B. and Lacroix Desmazes, P. (2003) Controlled radical polymerization of *n*-butyl α-acrylate. I. Use of atom transfer radical polymerization as the polymerization method. *Macromolecules*, **35**, 7634–7641 (and *Polym. Prepr., Am. Chem. Soc. Polym. Div.*, **44** (2), 683–684 (2003)).

59. (a) Castelvetro, V., Aglietto, M., Ciardelli, F., Chiantore, O. and Lazzari, M. (2001) Design for fluorinated acrylic-based polymers as water repellent intrinsically photostable coating materials for stone (Eds, Castner, D. G. and Grainger, D. W.) Proceedings of the American Chemical Society Symposium Series 787, Fluorinated surfaces, Coatings, and Films, Washington, DC, pp. 129–142, Chap. 10; (b) Souzy, R., Ameduri, B. and Boutevin, B., (2004) Radical copolymerization of α-trifluoromethylacrylic acid with vinylidene fluoride and vinylidene fluoride/hexafluoropropene. *Macromol. Chem. Phys.*, **205**, 476–485.

60. (a) Bessière, J. M., Boutevin, B., El Bachiri, A. and Harfi, A. (1993) Synthesis and polymerization of fluorinated acrylic monomers substituted in α position. Part II. Synthesis and polymerization of 2-perfluorohexyl ethyl α-acetoxyacrylate. *Makromol. Chem.*, **194**, 2939–2947; (b) El Bachiri, A., Bessière, J. M. and Boutevin, B. (1993) Synthèse et polymérisation de monomères fluorés acryliques substitués en position α. Partie III. Synthèse et polymérisation de l'α-(heptadecafluoroundécanoloxy) acrylate d'éthyle. *J. Fluorine Chem.*, **65**, 91–100; (c) Guyot, B., Ameduri, B., Boutevin, B. and Sideris, A. (1995) Synthèse et polymérisation de monomères fluorés acrylates substitués en position α. Partie IV. Synthèses et polymérisations de l'α-acétoxy acrylate de 2-perfluorooctyl éthyle et de l'α-propioxyacrylate de 2-perfluorooctyl éthyle. *Macromol. Chem. Phys.*, **196**, 1875–1886.

61. Turri, S., Scicchitano, M., Marchetti, R., Sanguineti, A. and Radice, S. (1999) Structure–property relationships of coatings based on perfluoropolyether macromers from PFPE.

In *Fluoropolymers: Properties*, Vol. 2 (Eds, Hougham, G., Cassidy, P. E., Jonhs, K. and Davidson, T.) Plenum Press, New York, pp. 145–169.

62. Hoyle, C. E., Mathias, L. J., Jariwala, C. and Sheng, D. (1996) Photopolymerization of a semifluorinated difunctional liquid crystalline monomer in a smectic phase. *Macromolecules*, **29**, 3182–3187.

63. Priola, A., Bongiovanni, R., Malucelli, G., Pollicino, A., Tonelli, C. and Simeone, G. (1997) UV-curable systems containing perfluoropolyether structures: synthesis and characterization. *Macromol. Chem. Phys.*, **198**, 1893–1907.

64. Barraud, J. Y., Gervat, S., Ratovelomanana, V., Boutevin, B., Parisi, J. P. and Cahuzac, A. (1993) Materiau polymère de type polyuréthane acrylate pour revêtement de fibre optique ou pour ruban à fibres optiques. Eur. Patent Appl. 93 0,400,880 (assigned to Alcatel).

65. Hu, H. S. W. and Griffith, J. R. (1999) Synthesis and structure property relationships of low-dielectric-constant fluorinated polyacrylates. In *Fluoropolymers: Properties*, Vol. 2 (Eds, Hougham, G., Cassidy, P. E., Jonhs, K. and Davidson, T.) Plenum Press, New York, pp. 167–180.

66. Hayakawa, Y., Terasawa, N., Hayashi, E. and Abe, T. (1998) Synthesis of novel polymethacrylates bearing cyclic perfluoroalkyl groups. *Polymer*, **39**, 4151–4154.

67. Gard, G. L., Winter, R., Nixon, P. G., Hu, Y.-H., Holcomb, N. R., Grainger, D. W. and Castner, D. G. (1998) Recent advances in organosulfur fluorine chemistry—new fluorinated surfaces/ films containing the SF_5 group. *Polym. Prepr. (Am. Chem. Soc., Polym. Div.)*, **39**(2), 962–963.

68. (a) Guyot, B., Ameduri, B., Boutevin, B., Mélas, M., Viguier, M. and Collet, A. (1998) Kinetics of homopolymerization of fluorinated acrylates. 5. Influence of the spacer between the fluorinated chain and the ester group. *Macromol. Chem. Phys.*, **199**, 1879–1885; (b) Guyot, B., Ameduri, B. and Boutevin, B. (1999) Cinétique de polymérisation radicalaire de (méth)acrylates à chaîne latérale fluorée. *Macromol. Chem. Phys.*, **200**, 2111–2123.

69. Guyot, B., Boutevin, B. and Ameduri, B. (1996) Etude de la copolymérisation de monomères acryliques fluorés avec le méthacrylate de morpholinoéthyle. *Macromol. Chem. Phys.*, **197**, 937–949.

70. Saïdi, S., Guittard, F. and Geribaldi, S. (2002) Monomers reactivity ratios of fluorinated acrylates–styrene copolymers. *Polym. Int.*, **51**, 1058–1062.

71. Yamashita, Y., Tsukahara, Y., Ito, K., Okada, K. and Tajima, Y. (1981) Synthesis and application of fluorine containing graft copolymers. *Polym. Bull.*, **5**, 335–340.

72. Niwa, M. and Higashi, N. (1988) Specific copolymerization behavior of a fluorocarbon amphiphilic macromonomer. *Macromolecules*, **21**, 1193–1194.

73. (a) Van der Grinten, M. G. D., Clough, A. S., Shearmur, T. E., Bongiovanni, R. and Priola, A. (1996) Surface segregation of fluorine-ended monomers. *J. Colloids Interface Sci.*, **182**, 511–515; (b) Morita, M., Ogisu, H. and Kobu, M. (1999) Surface properties of perfluoroalkyl-ethyl acrylate/*n*-alkyl acrylate copolymers. *J. Appl. Polym. Sci.*, **73**, 1741–1749.

74. Belfield, K. D. and Abdel-Sadek, G. G. (2002) Homo- and co-fluorinated acrylic ester polymers: synthesis, characterization and coating onto silicon surfaces. *Polym. Prepr. (Am. Chem. Soc., Div. Polym. Chem.)*, **43**(1), 588–589.

75. Park, I. J., Lee, S. B. and Choi, C. K. (1998) Surface properties of the fluorine-containing graft copolymer of poly((perfluoroalky)ethyl methacrylate)-*g*-poly(methyl methacrylate). *Macromolecules*, **31**, 7555–7558.

76. de Crevoisier, G., Fabre, P., Liebler, L., Tencé-Girault, S. and Corpart, J. M. (2002) Structure of fluorinated side-chain smectic copolymers: role of the copolymerization statistics. *Macromolecules*, **35**, 3880–3888.

77. Bannai, N., Yasumi, H. and Hirayama, S. (1986) Fluorine-containing monomer and process for producing the same. US Patent 4,580,981 (assigned to Kureha Kagaku Kogyo Kabushiki).

78. Vanhoye, D., Loubet, O., Ballot, E., Boutevin, B., Legros, R. and Ameduri, B. (1996) Graft copolymers from trifluoroethyl methacrylate-containing telomers for curable coating compositions. Eur. Patent Appl. EP 714,916 A1 (assigned to Elf-Atochem).

79. Volkov, V. V., Plate, N. A., Takahara, A., Kajiyama, T., Amaya, N. and Murata, Y. (1992) Aggregation state and mesophase structure of comb-shaped polymers with fluorocarbon side groups. *Polymer*, **33**, 1316–1320.

80. de Crevoisier, G., Fabre, P., Corpart, J. M. and Liebler, L. (1999) Swichable tackiness and wettability of a liquid crystalline polymer. *Science*, **285**, 1246–1249.

81. Galli, G., Gasperetti, S., Bertolucci, M., Gallot, B. and Chiellini, F. (2002) New architecture of liquid-crystalline polymers from fluorinated vinylcyclopropanes. *Macromol. Rapid Commun.*, **23**, 814–818.

82. Boutevin, B., Pietrasanta, Y. and Sideris, A. (1982) Synthèse et polymérisation de monomères chlorofluorés allyliques, vinyliques et acryliques. *J. Fluorine Chem.*, **20**, 727–732.

83. Beaune, O., Bessière, J. M., Boutevin, B. and Robin, J. J. (1992) Synthesis of maleimides monomers having a perfluoroaliphatic side chain and study of their copolymerization with vinylethers. *Polym. Bull.*, **26**, 605–612.

84. (a) Beaune, O., Bessiere, J. M., Boutevin, B. and Robin, J. J. (1994) Synthèse de polymères hautement fluorés amorphes et à T$_g$ élévées: copolymérisation de maléimides fluorées avec des éthers vinyliques. *J. Fluorine Chem.*, **67**, 159–166; (b) Boutevin, B., Lusinchi, J. M., Pietrasanta, Y. and Robin, J. J. (1995) Grafting of fluorinated oligomers onto high density polyethylene (HDPE) in the molten state. *J. Fluorine Chem.*, **73**, 79–82.

85. Ameduri, B., Ladavière, C., Boutevin, B. and Delolme, F. First Maldi Tof spectrometry of vinylidene fluoride telomers endowed with low-defect-chaining, submitted to *Macromolecules*.

86. Montefusco, F., Bongiovanni, R., Priola, A. and Ameduri, B. (2003) Synthesis of new vinylidene fluoride-containing acrylic monomers and their use as surface modifiers in photopolymerized coatings. *Fluorine in Coatings V Conference*, Orlando, January 21–22, 2003, paper 25, submitted to Macromolecules.

87. Ameduri, B., Boutevin, B. and Kostov, G. (2002) Synthesis and copolymerization of vinylidene fluoride (VDF) with trifluorovinyl monomers. Telomers of VDF with 1-chloro-1,2-dibromotri-fluoroethane as precursors of bromine-containing copolymers. *Macromol. Chem. Phys.*, **203**, 1763–1771.

88. Ameduri, B., Boutevin, B. and Kostov, G. (2001) Fluoroelastomers: synthesis, properties and applications. *Prog. Polym. Sci.*, **26**, 105–187.

89. Duc, M. (1997) Télomérisation du fluorure de vinylidène; applications à la synthèse d'oligomères fluorés fonctionnels. PhD Dissertation, University of Montpellier.

90. Duc, M., Ameduri, B. and Boutevin, B. (2001) Synthesis of polymerizable surfactants containing vinylidene fluoride, Presented at Workshop on *Recent Advances in Fluorinated Surfactants*, Avignon (France), January 26–27, 2001.

91. Duc, M., Ameduri, B., Boutevin, B., Kharroubi, M. and Sage, J. M. (1998) Telomerization of vinylidene fluoride with methanol; elucidation of the reaction process and mechanism by a structural analysis of the telomers. *Macromol. Chem. Phys.*, **92**, 77–84.

92. Duc, M., Ameduri, B. and Boutevin, B. (2001) Radical telomerization of vinylidene fluoride with diethyl hydrogenophosphonate. Characterization of the first telomeric adducts and assessment of the transfer constants. *J. Fluorine Chem.*, **112**, 3–12.

93. Gelin, M. P. and Ameduri, B. (2003) Synthesis of an original poly(vinylidene fluoride-*co*-hexafluoropropylene)-*g*-perfluoropolyether graft copolymer. *J. Fuorine Chem.*, **119**, 53–58.

94. Casazza, E., Mariani, A., Ricco, L. and Russo, S. (2001) Synthesis, characterization, and properties of a novel acrylic terpolymer with pendant perfluoropolyether segments. *Polymer*, **43**, 1207–1214.

95. Sheiko, S. S., Slangen, P.-J., Krupers, M., Mourran, A. and Möller, M. (2001) Wetting behaviour of thin films of polymethacrylates with oligo(hexafluoropropene oxide) side-chains (Eds, Castner, D. G. and Grainger, D. W.) *Proceedings of the American Chemical Society Symposium Series 787*, Fluorinated surfaces, Coatings, and Films, Washington, DC, pp. 71–82, Chap. 6.

96. (a) Boutevin, B. and Pietrasanta, Y. (1973) Télomérisation du monochlorotrifluoroéthylène avec le tétrachlorure de carbone. *Tetrahedron Lett.*, **12**, 887–894; (b) Boutevin, B. and Pietrasanta, Y.

(1975) Télomerisation par catalyse rédox. IV. Télomérisation du chlorotrifluoroéthylène avec le tétrachlorure de carbone. *Eur. Polym. J.*, **12**, 219–226; (c) Boutevin, B., Cals, J. and Pietrasanta, Y. (1975) Télomerisation par catalyse rédox. V. Télomérisation du chlorotrifluoroéthylène avec le bromotrichlorométhane par catalyse rédox. *Eur. Polym. J.*, **12**, 225–232.

97. (a) Battais, A., Boutevin, B., Hugon, J. P. and Pietrasanta, Y. (1980) Synthèse de diols fluorés à partir de dérivés des télomères du chlorotrifluoroéthylène et du tétrachlorure de carbone. *J. Fluorine Chem.*, **16**, 397–405; (b) Ameduri, B., Boutevin, B., Lecrom, C. and Garnier, L. (1992) Synthesis of halogenated monodispersed telechelic oligomers. III. Bistelomerization of allyl acetate with telogens which exhibit α,ω-di(trichloromethyled) end groups. *J. Polym. Sci., Part A: Polym. Chem.*, **30**, 49–60.

98. (a) Boutevin, B., Hugon, J. P. and Pietrasanta, Y. (1977) Synthèse de polyamides fluorés. *C. R. Acad. Sci., Sér. C*, 599–605; (b) Boutevin, B., Dongala, B. and Pietrasanta, Y. (1981) Synthèse de polyesters fluorés par polytransestérification. *J. Fluorine Chem.*, **17**, 113–120; (c) Boutevin, B., Hugon, J. P. and Pietrasanta, Y. (1981) Synthèse et caractérisation de polyuréthanes fluorés. Partie I. Etude des molécules modèles. *Eur. Polym. J.*, **17**, 729–735.

99. Boutevin, B., Pietrasanta, Y., Rousseau, A. and Laroze, A. (1986) Synthesis and polymerization of fluorinated chlorinated and deuterated vinyl carbonates. *Polym. Bull.*, **16**, 391–400.

100. Grot, W. G. (1994) Perfluorinated ion exchange polymers and their use in research and industry. *Macromol. Symp.*, **82**, 161–172.

101. Doyle, M. and Rajendran, G. (2003) Perfluorinated membranes. In *Handbook of Fuel Cells: Fundamentals, Technology and Applications; Fuel Cell Technology and Applications*, Vol. 3 (Eds, Vielstich, W., Lamm, A. and Gasteiger, H.) pp. 351–395, Chap. 30, Wiley, New York.

102. Dunsch, L., Kavan, L. and Weber, J. (1990) Perfluoro anion-exchange polymeric films on glassy carbon electrodes. *J. Electroanal. Chem.*, **280**, 313–325.

103. Arcella, V., Ghielmi, A. and Tommasi, G. (2003) High performance perfluoropolymer films for membranes. *Ann. NY Acad. Sci.*, **984**, 226–244.

104. (a) Eisenberg, A. and Kim, J. S. (1998) *Introduction to Ionomers*, Wiley, New York; (b) Eisenberg, A. and Yaeger, H. L. (1982) Perfluorinated ionomer membranes. In *American Chemical Society Symposium Series 180*, ACS, Washington, DC.

105. Li, Q., He, R., Jensen, J.O. and Bjerrum, N. J. (2003) Approaches and recent development of polymer electrolyte membranes for fuel cells operating above 100 °C. *Chem. Mat.*, **15**, 4896–4915.

106. Trainham, J. A. III (1993) US Patent 5,411,641 (assigned to E.I. Du Pont de Nemours and Company).

107. Yeager, H. L. and Gronowski, A. A. (1995) Membrane applications. In *Ionomers. Synthesis, Structure, Properties, and Applications* (Eds, Tant, M. R., Mauritz, K. A. and Wilkes, G. L.) New York, pp. 333–364.

108. Gibbard, H. F. (2002) Batteries versus fuel cells: the application of a disruptive technology. In *Proceedings of the Eyes for Fuel Cells Conference on Fuel Cells for Portable Applications*, Boston, USA.

109. Cooley, G. E. and D'Agostino, V. F. (1995) Cation exchange membranes and method for the preparation of such membranes. US Patent 5,626,731 (assigned to National Power PLC).

110. Moya, W. (1998) Surface modified polymeric substrate and process. US Patent 6,179,132 (assigned to Millipore Corporation).

111. Naylor, T. and De, V. (1996) Polymer membranes–materials, structures and separation performance. *Rapra Rev. Rep.*, **8**, 39–45.

112. Olah, G. A., Iyer, P. S. and Prakash, G. K. S. (1986) Perfluorinated resinsulfonic acid (Nafion-H®) catalysis in synthesis. *Synthesis*, **7**, 513–531.

113. Ukihashi, H., Yamabe, M. and Miyake, H. (1986) Polymeric fluorocarbon acids and their applications. *Prog. Polym. Sci.*, **12**, 229–270.

114. Heitner-Wirguin, C. (1996) Recent advances in perfluorinated ionomer membranes: structure, properties and applications. *J. Membr. Sci.*, **120**, 1–33.

115. Guo, Q., Pintauro, P. N., Tang, H. and O'Connor, S. (1999) Sulfonated and crosslinked polyphosphazene-based proton-exchange membranes. *J. Membr. Sci.*, **154**, 175–181.

116. Kerres, J. A. (2001) Development of ionomer membranes for fuel cells. *J. Membr. Sci.*, **185**, 3–27.

117. Yu, J., Yi, B., Xing, D., Liu, F., Shao, Z. and Fu, Y. (2003) Development of novel self-humidifying composite membranes for fuel cells. *J. Power Sources*, **124**, 81–89.

118. Curtain, D. E. (2002) High volume low cost manufacturing process for nafion membranes. Abstracts for Fuel Cell Seminar 2002, Palm Springs, FL, November 18–21, p. 834.

119. Kostov, G. K., Kotov, St. V., Ivanov, G. D. and Todorova, D. (1993) Study of the synthesis of perfluorovinyl-sulfonic functional monomer and its copolymerization with tetrafluoroethylene. *J. Appl. Polym. Sci.*, **47**, 735–741.

120. DesMarteau, D. D. (1995) Copolymers of tetrafluoroethylene and perfluorinated sulfonyl monomers and membranes made therefrom. US Patent 5,463,005 (assigned to Gas Research Institute).

121. Hung, M. H., Farnham, W. B., Feiring, A. E. and Rozen, S. (1999) Functional Fluoromonomers and Fluoropolymers. In *Fluoropolymers: Synthesis*, Vol. 1 (Eds, Hougham, G., Cassidy, P. E., Johns, K. and Davidson, T.) Plenum Press, New York, pp. 51–66, Chap. 4.

122. Kotov, S. V., Pedersen, S. D., Qiu, W., Qiu, Z.-M. and Burton, D. J. (1997) Preparation of perfluorocarbon polymers containing phosphonic acid groups. *J. Fluorine Chem.*, **82**, 13–19.

123. Kato, M., Akiyama, K. and Yamabe, M. (1983) New perfluorocarbon polymers with phosphonic groups. *Rep. Res. Lab. Asahi Glass Co., Ltd*, **33**, 135–142.

124. Burton, D. J. and Sprague, L. G. (1989) Allylations of [(Diethoxyphosphinyl) difluoromethyl] zinc bromide as a convenient route to 1,1-difluoro-3-alkenephosphonates. *J. Org. Chem.*, **54**, 613–617.

125. Feiring, A. E., Doyle, C. M., Roelofs, M. G., Farnham, W. B., Bekiarian, P. G. and Blau, H. A. K. (1999) Substantially fluorinated ionomers. World Demand WO 99/45,048 (assigned to E. I. du Pont de Nemours and Company).

126. Doyle, C. M., Farnham, W. B., Feiring, A. E., Morken, P. A. and Roelofs, M. G. (2000) Fluorinated ionomers and their uses. US Patent, 6,025,092 (assigned to E. I. du Pont de Nemours and Company).

127. Ameduri, B., Armand, M., Boucher, M. and Manseri, A. (2001) Elastomères fluorosulfonés à faible Tg à base de fluorure de vinylidène et ne contenant ni du tétrafluoroéthylène, ni de l'hexafluoropropène, ni de groupements siloxane. World Demand WO 01/49,757 (assigned to Hydro-Québec).

128. Connolly, D. J. and Gresham, W. F. (1966) Sulfo derivatives of perfluorovinyl ether monomers. US Patent 3,282,875 (assigned to Du Pont).

129. Ameduri, B., Armand, M., Boucher, M. and Manseri, A. (2001) Elastomères fluorosulfonés à faible Tg à base d'hexafluoropropène et ne contenant ni du tétrafluoroéthylène, ni de groupements siloxane. World Demand WO 01/49,760 A1 (assigned to Hydro-Québec).

130. Ameduri, B., Armand, M., Boucher, M. and Manseri, A. (2001) Elastomères réticulables fluorés bromosulfonés à base de fluorure de vinylidène présentant une faible Tg et procédés pour leurs préparations. World Demand WO 01/96,268 A2 (assigned to Hydro-Québec).

131. Ameduri, B., Boucher, M. and Manseri, A. (2002) Elastomères réticulables nitriles fluorosulfonés à base de fluorure de vinylidène présentant une faible Tg et procédés pour leurs préparations. World Demand WO 02/50,142 A1 (assigned to Hydro-Québec).

132. Yamabe, M. (1992) A challenge to novel fluoropolymers. *Macromol. Symp.*, **64**, 11–18.

133. Miyake, H., Sugaya, Y. and Yamabe, M. (1998) Synthesis and properties of perfluorocarboxylated polymers. *J. Fluorine Chem.*, **92**, 137–140.

134. Ameduri, B. and Bauduin, G. (2003) Synthesis and polymerization of fluorinated monomers bearing a reactive lateral group. XIV. Radical copolymerization of vinylidene fluoride with methyl 1,1-dihydro-4,7-dioxaperfluoro-5,8-dimethyl non-1-enoate. *J. Polym. Sci., Part A: Polym. Chem.*, **41**, 3109–3121.

135. Villa, L. and Iserson, H. (1974) Modified polybutadiene. German Patent 2,325,561 (assigned to Thiokol Chem. Corp.), *Chem. Abstr.*, **81**, 38326w.

136. Ren, Y., Lodge, T. P. and Hillmyer, M. A. (2001) A simple and mild route to highly fluorinated model polymers. *Macromolecules*, **34**, 4780–4787.

137. Benoit, P. (1989) Addition redox de polyhalogénoalcanes sur le polybutadiène hydroxy téléchélique. MSc (DEA) Dissertation, University of Montpellier.

138. Boutevin, B., Hervaud, Y. and Nouiri, M. (1990) Addition de thiols fluorés sur les polybutadiènes hydroxy téléchéliques par voie photochimique. *Eur. Polym. J.*, **26**, 877–885.

139. Muller, J. K. F. (1984) Perfluoroalkyl substituted polymers, Eur. Patent Appl. EP 115,253 (assigned to Ciba-Geigy), *Chem. Abstr.*, **102**, 47471.

140. Ameduri, B., Boutevin, B. and Nouiri, M. (1993) Synthesis and properties of fluorinated telechelic macromolecular diols prepared by radical grafting of fluorinated thiols onto hydroxyl-terminated poly(butadienes). *J. Polym. Sci., Part A: Polym. Chem.*, **31**, 2069–2080.

141. Nouiri, M. (1995) Synthèse et caractérisation de nouveaux polybutadiènes hydroxytéléchéliques fluorosoufrés et fluorosiliciés. Moroccan State PhD Dissertation, University of Casablanca.

142. Golub, M. A. (1974) Type II thermal cyclization in 1,2-polybutadiene and 3,4-polyisoprene. *J. Polym. Sci., Polym. Lett. Ed.*, **12**, 615–622.

143. Chiantore, O., Luda di Cortemiglia, M. P., Guaita, M. and Rendina, G. (1989) Thermal degradation of polybutadiene. 1. Reactions at temperatures lower than 250 °C. *Makromol. Chem.*, **190**, 3143–3152.

144. Kotora, M., Hajek, M., Kvicala, J., Ameduri, B. and Boutevin, B. (1993) Rearrangement of 2-iodo-3-perfluoroalkyl-1-propyl acetates to 1-iodo-3-perfluoroalkyl-2-propyl acetates. *J. Fluorine Chem.*, **64**, 259–267.

145. Hilton, A. (2003) Erosion selective et greffage de copolymères à blocs. PhD dissertation, Université Pierre et Marie Curie, Paris.

146. Oestreich, S. and Antonietti, M. (1999) Novel fluorinated block copolymers. Synthesis and applications. In *Fluoropolymers: Synthesis*, Vol. 10 (Eds, Hougham, G., Cassidy, P. E., Johns, K. and Davidson, T.) Plenum Publishers, New York, pp. 151–166, Chap. 10.

147. Floudas, G., Antonietti, M. and Forster, S. (2000) Dielectric relaxation in poly(styrene-*b*-butadiene) copolymers with perfluorinated side chains. *J. Chem. Phys.*, **113**, 3447–3451.

148. Antonietti, M., Forster, S., Micha, M. A. and Oestreich, S. (1997) Novel fluorinated block copolymers for the construction of ultra-low energy surfaces and as dispersion stabilizers in solvents with low cohesion energy. *Acta Polym.*, **48**, 262–268.

149. Iyengar, D. R., Perutz, S. M., Dai, C. A., Ober, C. K. and Kramer, E. J. (1996) Surface segregation studies of fluorine-containing diblock copolymers. *Macromolecules*, **29**, 1229–1234.

150. Boker, A., Reihs, K., Wang, J., Stadler, R. and Ober, C. K. (2000) Selectively thermally cleavable fluorinated side chain block copolymers: surface chemistry and surface properties. *Macromolecules*, **33**, 1310–1320.

151. Hayakawa, T., Wang, J., Sundararajan, N., Xiang, M., Li, X., Glüsen, B., Leung, G. C., Ueda, M. and Ober, C. K. (2000) Switching surface polarity: synthesis and characterization of a fluorinated block copolymer with surface-active *tert*-butoxycarbonyl groups. *J. Phys. Org. Chem.*, **13**, 787–795.

152. Hwang, S. S., Ober, C. K., Perutz, S. M., Iyengar, D. R., Scheneggen-Burger, L. A. and Kramer, E. J. (1995) Block copolymers with low surface energy segments: siloxane- and perfluoroalkane-modified blocks. *Polymer*, **36**, 1321–1325.

153. Ren, Y., Lodge, T. P. and Hillmyer, M. A. (1998) A new class of fluorinated polymers by a mild, selective, and quantitative fluorination. *J. Am. Chem. Soc.*, **120**, 6830–6831.

154. Ren, Y., Lodge, T. P. and Hillmyer, M. A. (2000) Synthesis, characterization and interaction strengths of difluorocarbene-modified polystyrene–polyisoprene block copolymers. *Macromolecules*, **33**, 866–876.

155. Siddiqui, S. and Cais, R. E. (1986) Addition of difluorocarbene to 1,4-polybutadienes. Synthesis and characterization of novel copolymers. *Macromolecules*, **19**, 595–603.

156. Reisinger, J. J. and Hillmayer, M. A. (2002) Synthesis of fluorinated polymers by chemical modification. *Prog. Polym. Sci.*, **27**, 971–1005.

157. Xiang, M., Li, X., Ober, C. K., Char, K., Genzer, J., Sivaniah, E., Kramer, E. J. and Fischer, D. A. (2000) Surface stability in liquid-crystalline block copolymers with semifluorinated monodendron side group. *Macromolecules*, **33**, 6106–6119.

158. Samuel, J. D. S., Dhamodharan, R. and Ober, C. K. (2000) A solvent-free method for the synthesis of block copolymers with fluorinated pendant groups by a hydrosilylation reaction. *J. Polym. Sci., Part A: Polym. Chem.*, **38**, 1179–1183.

159. Boker, A., Herweg, T. and Reihs, K. (2002) Selective alteration of polymer surfaces by thermal cleavage of fluorinated side chains. *Macromolecules*, **35**, 4929–4937.

160. Genzer, J., Sivaniah, E., Kramer, E. J., Wang, J., Korner, H., Xiang, M., Char, K., Ober, C. K., DeKoven, B. M., Bubeck, B. A., Chaudhury, M. K., Sambasivan, S. and Fischer, D. A. (2000) The orientation of semifluorinated alkanes attached to polymers at the surface of polymer films. *Macromolecules*, **33**, 1882–1887.

161. Li, X., Andruzzi, L., Chiellini, E., Galli, G., Ober, C. K., Hexemer, A., Kramer, E. J. and Fischer, D. A. (2002) Semifluorinated aromatic side-group polystyrene-based block copolymers: bulk structure and surface orientation studies. *Macromolecules*, **35**, 8078–8087.

162. Hackman, M., Repo, T., Jany, G. and Rieger, B. (1998) Zirconocene/MAO-catalyzed homo- and copolymerization of linear asymmetrically substituted dienes with propene. A novel strategy to functional (*co*)poly(α-olefins). *Macromol. Chem. Phys.*, **199**, 1511–1517.

163. Maxson, M. T., Norris, A. W. and Owen, M. (1997) Fluorosilicones. In *Modern Fluoropolymers* (Ed, Scheirs, J.) Wiley, New York, pp. 359–372, Chap. 20.

164. Boutevin, B. and Pietrasanta, Y. (1985) The synthesis and applications of fluorinated silicones, notably in high-performance coatings. *Prog. Org. Coat.*, **13**, 297–331.

165. (a) Pierce, O. R., Holbrook, G. W., Johannson, O. K., Saylor, J. C., Brown, E. D. and Kim, Y. K. (1960). *Ind. Eng. Chem.*, **52**, 783–788; (b) Kim, Y. K. (1971) Original fluorosilicones. *Rubber Chem. Technol.*, **44**, 1350–1359; (c) Kim, Y. K., Loree, L. A. and Pierce, O. R. (1972) Fluorinated silicones. US Patent 3,647,740 (assigned to Dow Corning).

166. Boutevin, B., Guida-Pietrasanta, F. and Ratsimihety, A. (2000) In *Silicon-Containing Polymers* (Eds, Jones, R. G., Ando, W. and Chojnowski, J.) Kluwer Academic Publishers, Dordrecht, pp. 79–112.

167. Marciniec, B. (1992) *Comprehensive Handbook on Hydrosilylation*, Pergamon Press, Oxford.

168. Wilczek, L. (1993) Hydrosilylation of α-fluorinated α-olefins with cobalt catalysts. US Patent 5233071 A (assigned to E.I. DuPont de Nemours Co.).

169. Atherton, J. H. (1973) Fluorine containing silanes and silicones. German Patent DE 2311879 (assigned to Imperial Chemical Industries).

170. Doeff, M. and Lindner, E. (1989) Structure and surface energy characteristics of a series of pseudo-perfluoroalkyl polysiloxanes. *Macromolecules*, **22**, 2951–2957.

171. Chujo, Y. and McGrath, J. E. (1995) Synthesis of α,ω-bifunctional fluorine-containing polysiloxanes by hydrosilation reaction. *J. Macromol. Sci., Pure Appl. Chem.*, **A32**(1), 29–40.

172. Furukawa, Y. and Kotera, M. (2002) Synthesis of fluorosilicone having highly fluorinated alkyl side chains based on the hydrosilylation of fluorinated olefins with polyhydromethylsiloxane. *J. Polym. Sci., Part A: Polym. Chem.*, **40**, 3120–3128.

173. Mera, E. and Morris, R. E. (2001) Hydrosilation in supercritical CO_2: synthesis of fluorinated polysiloxanes. *Macromol. Rapid Commun.*, **22**, 513–518.

174. Kobayashi, H. and Owen, M. J. (1990) Surface tension of poly[(3,3,4,4,5,5,6,6,6-nonafluorohexyl)methylsiloxane]. *Macromolecules*, **23**(23), 4929–4933.

175. (a) Boutevin, B., Guida-Pietrasanta, F. and Ratsimihety, A. (2000) Synthesis of photocrosslinkable fluorinated polydimethylsiloxanes: direct introduction of acrylic pendant groups via hydrosilylation. *J. Polym. Sci., Part A: Polym. Chem.*, **38**, 3722–3728; (b) Guizard, C., Boutevin, B., Guida, F., Ratsimihety, A., Amblard, P., Lasserre, J. C. and Naiglin, S. (2001) VOC vapour transport properties of new membranes based on cross-linked fluorinated elastomers. *Sep. Purif. Technol.*, **22/23**, 23–30; (c) Guizard, C., Boutevin, B., Guida, F., Ratsimihety, A., Amblard, P., Lasserre, J. C. and Naiglin, S. Composition decopolymère et son utilisation pour l'élaboration d'une membrane permselective, French patent 09,669 (assigned to CNRS), 1999.

176. Furukawa, Y. and Yoneda, T. (2003) Synthesis and properties of fluorosilicone with perfluorooctylundecyl side chains. *J. Polym. Sci., Part A: Polym. Chem.*, **41**, 2704–2714.

177. Olander, B., Woisen, A. and Albertsson, A. C. (2002) Argon microwave plasma treatment and subsequent hydrosilylation grafting as a way to obtain silicone biomaterial with well-defined structure. *Biomacromolecules*, **3**, 505–510.

178. (a) Colomines, G., André, S., Andrieu, X., Rousseau, A., Boutevin, B. and Ameduri, B. (2003) Synthesis and characterization of UV-curable fluorinated poly(dimethyl)siloxanes as UV-transparent coatings for optical fiber gratings and for membranes used in VOC elimination. *Fluorine in Coatings V Conference*, Orlando, January 21, 22, 2003; (b) Colomines, G., André, S., Andrieu, X., Rousseau, A. and Boutevin, B. (2003) Synthesis and characterization of ultraviolet-curable fluorinated polydimethylsiloxanes as ultraviolet-transparent coatings for optical fiber grating. *J. Appl. Polym. Sci.* **90**, 2021–2026.

179. Tsibouklis, J., Stone, M., Thorpe, A. A. and Graham, P. (2002) Fluoropolymer coating with inherent resistance to biofouling. *Surf. Coat. Int. Part B, Coat. Trans.*, **85**, 301–308.

180. Pham-Van-Cang, C., Winnik, M., Dorigo, R. and Boileau, S. (1995) Energy transfer studies with fluorosiloxanes. *Eur. Polym. J.*, **31**, 227–231.

181. Abdellah, L., Boutevin, B., Guida-Pietrasanta, F. and Smahi, M. (2003) Evaluation of photocrosslinkable fluorinated poly(dimethyl)siloxanes as gas permeation membranes. *J. Membr. Sci.*, **217**, 295–298.

182. Furukawa, Y. and Kotera, M. (2003) Water and oil repellency of polysiloxanes with highly fluorinated alkyl side chains. *J. Appl. Polym. Sci.*, **87**, 1085–1091.

183. (a) Beyou, E., Bennetau, B., Dunogues, J., Babin, P., Teyssié, D., Boileau, S. and Corpart, J. M. (1995) New fluorinated polysiloxanes containing an ester function in the spacer. II. Surface tension studies. *Polym. Int.*, **38**, 237–244; (b) Boileau, S., Beyou, E., Babin, P., Bennetau, B., Dunoguès, J. and Teyssié, D. (1996) Fluorinated polysiloxanes. A basis for low surface energy materials. *Eur. Coat. J.*, **7–8**, 508–513.

184. (a) Beyou, E., Babin, P., Bennetau, B., Dunogues, J., Teyssié, D. and Boileau, S. (1994) New fluorinated polysiloxanes containing an ester function in the spacer. I. Synthesis and characterization. *J. Polym. Sci., Part A: Polym. Chem.*, **32**, 1673–1681; (b) Beyou, E., Babin, P., Bennetau, B., Dunoguès, J., Teyssié, D. and Boileau, S. (1996) New low surface energy materials based on fluorinated polysiloxanes. *Silicones and Coatings Conference*, Brussels, March 1996, paper 23.

185. Höpken, J., Möller, M. and Boileau, S. (1990) New monomers and polymers with perfluorinated segments. Association and interfacial activity. *Polym. Prepr. (Am. Chem. Soc., Polym. Div.)*, **31**(1), 324–325.

186. (a) Bracon, F., Guittard, F., Taffin De Givenchy, E. and Cambon, A. (1999) Highly fluorinated monomers precursors of side-chain liquid crystalline polysiloxanes. *J. Polym. Sci., Part A: Polym. Chem.*, **37**, 4487–4496; (b) Bracon, F., Guittard, F., Taffin De Givenchy, E. and Geribaldi, S. (2000) New fluorinated monomers containing an ester function in the spacer, precursors of side chain liquid crystalline polysiloxanes. *Polymer*, **41**, 7905–7913.

187. Boutevin, B., Gaboyard, M., Guida-Pietrasanta, F., Hairault, L., Lebret, B. and Pasquinet, E. (2003) Synthesis of 1,3-bis(2-trimethylsilyloxyhexafluoro-2-propyl)-5-allylbenzene and its introduction into polysiloxane chains. *J. Polym. Sci., Part A: Polym. Chem.*, **41**, 1400–1410.

188. Boutevin, B. and Abdellah, L. (1995) Synthesis and applications of photocrosslinkable polydimethyl siloxanes. Part III. Synthesis of polysiloxanes with perfluorinated and acrylated urethane linked pendant groups. *Eur. Polym. J.*, **31**, 1127–1135.

189. Boutevin, B., Youssef, B., Boileau, S. and Garnault, A. M. (1987) Synthèse d'éthers et de thioéthers allyliques fluorés par catalyse par transfert de phase. *J. Fluorine Chem.*, **35**, 399–406.

190. Furukawa, Y., Shin-Ya, S., Saito, M., Narui, S.-I. and Miyake, H. (2002) Fluorosilicone elastomers based on the poly[(3,3,3-trifluoropropyl)methylsiloxane-*co*-(3,3,4,4,5,5,6,6,6-nonafluorohexyl)methylsiloxane]. *Polym. Adv. Technol.*, **13**, 60–65.

191. Giles, M. R., O'Connors, S. J., Hay, J. N., Winder, R. J. and Howdle, S. M. (2000) Novel graft stabilizers for the free radical polymerization of methyl methacrylate in supercritical carbon dioxide. *Macromolecules*, **33**, 1996–1999.

192. Giles, M. R., Griffiths, R. M. T., Aguiar-Ricardo, A., Silva, M. M. C. G. and Howdle, S. M. (2001) Fluorinated graft stabilizers for polymerization in supercritical carbon dioxide: the effect of stabilizer architecture. *Macromolecules*, **34**, 20–25.

193. Lepilleur, C. and Beckman, E. J. (1997) Dispersion polymerization of methyl methacrylate in supercritical CO_2. *Macromolecules*, **30**, 745–756.

194. Jannasch, P. (1998) Surface structure and dynamics of block and graft copolymers having fluorinated poly(ethylene oxide) chain end. *Macromolecules*, **31**, 1341–1347.

195. Pospiech, D., Komber, H., Voigt, D., Jehnichen, D., Häußler, L., Gottwald, A., Kollig, W., Eckstein, K., Baier, A. and Grundke, K. (2003) Synthesis and characterization of semifluorinated polymers and block copolymers. *Macromol. Symp.*, **199**, 173–186.

196. LeNinivin, C. (2003) Elaboration et validation de dérivés polyparaphenylène substitués sulfonés comme électrolyte solide pour piles à combustible à membrane échangeuse de protons, PhD Dissertation, Université de Poitiers, France.

197. Balland Longeau, A., Pereira, F., Capron, P. and Merceir, R. (2002). Polymeres de type polyphényléne, leur procédé de préparation, membrane et dispositif de pile à combustible comprenant ces membranes, French Patent 02 10,008 (assigned to Commissariat à l'Energie Atomique).

198. (a) Kyker, G. S. and Antkowiak, T. A. (1974) Phosphonitrilic fluoroelastomers. New class of solvent resistant polymers with exceptional flexibility at low temperature. *Rubber Chem. Technol.*, **47**, 32–39; (b) Tate, D. P. and Antkowiak, T. A. (1980) Elastomers. *Encyl. Polym. Sci. Tech.*, **10**, 935–954; (c) Allcock, H. R., Rutt, J. S. and Fitzpatrick, R. J. (1991) Surface reaction of poly[bis(trifluoroethoxy)phosphazene] films by basic hydrolysis. *Chem. Mater.*, **3**, 442–453.

199. Allcock, H. R. (2002) *Chemistry and Applications of Polyphosphazenes*, Wiley, New York.

200. Feiring, A. E. and Wonchoba, E. R. (1998) Synthesis of soluble fluorinated vinyl ether copolymers by ionic addition of fluorolefins to poly(vinyl alcohol). *Macromolecules*, **31**, 7103–7104.

201. (a) Yamaguchi, K., Kato, T., Hirao, A. and Nakahama, S. (1987) Protection and polymerization of functional monomers. 9. Synthesis of poly[1-(4-mercaptomethylphenyl)ethylene] by hydrolysis of poly{1[4-(*tert*-butyldimethyl silylthiomethyl)-phenyl]ethylene} produced with a radical initiator. *Makromol. Chem. Rapid Commun.*, **8**, 203–207; (b) Kihara, N., Kanno, C. and Fukutomi, T. (1997) Synthesis and properties of microgel bearing a mercapto group. *J. Polym. Sci., Part A: Polym. Chem.*, **35**, 1443–1451.

202. Trollas, M., Hawker, C. J., Hedrick, J. L. and Carrot, G. (1998) A mild and versatile synthesis for the preparation of thiol-functionalized polymers. *Macromolecules*, **31**, 5960–5963 (and references therein).

203. Ameduri, B., Boutevin, B., Kostov, G. and Petrova, P. (1999) Synthesis and polymerization of fluorinated monomers bearing a reactive lateral group. Part 10. Copolymerisation of vinylidene

fluoride (VDF) with 5-thioacetoxy-1,1,2-trifluoropentene for the obtaining of novel PVDF containing mercaptans side groups. *Des. Monomers Polym.*, **2**, 267–285.

204. (a) Oshibe, Y., Ishigaki, H., Ohmura, H. and Yamamoto, T. (1989) Preparation of block and graft copolymers and their applications. VIII. Preparation of fluorine-containing block copolymers and their solubility characteristics. *Kobunshi Ronbunshu*, **46**, 81–87 (*Chem. Abstr.*, **111**, 8131); (b) Oshibe, Y., Yoshihiro, I., Ishigaki, H., Ohmura, H. and Yamamoto, T. (1989) Preparation of block and graft copolymers and their applications. IX. Surface properties of fluorine-containing block copolymers. *Kobunshi Ronbunshu*, **46**, 89–94 (*Chem. Abstr.*, **111**, 39960).

205. (a) Tobolski, A. V. (1960) Dead end polymerization. *J. Am. Chem. Soc.*, **80**, 5927–5929; (b) Tobolski, A. V., Rogers, C. E. and Brickman, R. D. (1960) Dead end polymerisation. Part II. *J. Am. Chem. Soc.*, **82**, 1277–1280.

206. Kawashima, S. and Yosamura, T. (1984) Poly(vinylidene fluoride) compositions. Jpn. Kokai Tokkyo Koho JP 5,930,847 (assigned to Central Glass Co., Ltd), *Chem. Abstr.*, **101**, 73689v.

207. Guiot, J., Ameduri, B. and Boutevin, B. (2002) Radical homopolymerization of vinylidene fluoride initiated by *tert*-butylperoxypivalate. Investigation of the microstructure by ^{19}F and ^{1}H NMR spectroscopies and mechanisms. *Macromolecules*, **35**, 8694–8707.

208. Kawashima, C. and Yasumuru, T. (1984) Elastic fluorohydrocarbon resin and method of producing the same. US Patent 4,472,557 (assigned to Central Glass Co., Ltd).

209. Suzuki, N., Sugiura, M., Ujikawa, N. and Yamamoto, T. (1988) Preparation of block and graft copolymers and their applications. VII. Preparation of graft copolymers with peroxide monomers. *Kobunshi Ronbunshu*, **45**, 867–871.

210. Chapiro, A. (1962) *Radiation Chemistry of Polymeric Systems*, Wiley, New York.

211. (a) Geymer, D. O. (1972) In *The Radiation Chemistry of Macromolecules*, Vol. 1 (Ed, Dole, M.) Academic Press, New York, Chap. 1; (b) Mandelkern, L. (1972) In *The Radiation Chemistry of Macromolecules*, Vol. 1 (Ed, Dole, M.) Academic Press, New York, Chap. 13.

212. Florin, R. E. (1972) Radiation Chemistry of Fluorocarbon Polymers. In *Fluoropolymers* (Ed, Wall, L. A.) Wiley, New York, pp. 317–331, Chap. 11.

213. (a) Okamoto, J. (1987) Radiation synthesis of functional polymer. *Radiat. Phys. Chem.*, **29**, 469–475; (b) Ivanov, V. S. (1992) *Radiation Chemistry of Polymers, New Concepts in Polymer Science*, Wiley, New York; (c) Singh, A. and Silverman, J. (1992) *Radiation Processing of Polymers*, Hansen, Munich.

214. (a) Lyons, B. J. (1984) The crosslinking of fluoropolymers with ionising radition: a review. In *Second International Conference on Radiation Processing for Plastics and Rubbers*, Chantembury, UK, pp. 1–8; (b) Lyons, B. J. (1997) The radiation crosslinking of fluoropolymers. In *Modern Fluoropolymers* (Ed, Scheirs, J.) Wiley, New York, pp. 335–347, Chap. 18.

215. Uyama, Y., Kato, K. and Ikada, Y. (1998) Surface modification of polymers by grafting. *Adv. Polym. Sci.*, **137**, 1–39.

216. Gupta, B. and Scherer, G. G. (1994) Proton-exchange membranes by radiation-induced graft copolymerisation of monomers into Teflon-FEP films. *Chimia*, **48**, 127–137.

217. (a) Forsythe, J. S. and Hill, D. J. T. (2000) Radiation chemistry of fluoropolymers. *Prog. Polym. Sci.*, **25**, 101–136; (b) Dargaville, T. R., George, G. A., Hill, D. J. T. and Whittaker, A. K. (2003) High energy radiation grafting of fluoropolymers. *Prog. Poly. Sci.*, **28**, 1355–1376.

218. Drysdale, N. E., Wang, L. and Yang, Z. Yu. (1998) Free-radical grafting of polymers with fluorocarbon compounds and products therefrom. Int. Demand WO 9,831,716 (assigned to E.I. DuPont de Nemours and Co.).

219. Kotani, N. and Nakase, Y. (1997) Charge accumulation of grafted open-cell type polyethylene foam by electron beam. *JAERI Rev.*, **97–104**, 39–42.

220. Ciba Geigy (1972) Polymers of a backbone polymer having a perfluoroalkyl group containing monomer derived from fumaric and related acids grafted thereto. British Patent 1, 379,926.

221. Rolando, R. J. and Sax, J. E. (1994) Graft copolymers containing fluoroaliphatic groups. US Patent 5,314,959 (assigned to 3M).

222. Tan, R. A., Wang, W., Hu, G. H. and de Souza Gomes, A. (1999) Functionalisation of polypropylene with fluorinated acrylic monomers in the molten state. *Eur. Polym. J.*, **35**, 1979–1984.

223. Michel, A. and Monnet, C. (1981) Functionalization of isotactic polypropylene by the oxygen–ozone mixture. *Eur. Polym. J.*, **17**, 1145–1152.

224. Niyogi, S. G. and Denicola, A. J. (2001) Polyolefin graft copolymers made with fluorinated monomers, Int. Demand WO 0,109,209 (assigned to Montell Technology Co, BV, The Netherlands).

225. Hiraga, Y., Komatsu, S., Noda, T. and Namimatsu, M. (1998) Method for modifying fluoropolymer. Eur. Patent 1,043,353 A1 (assigned to Daikin, Ind. Ltd).

226. Robin, J. J. (2004) Overview of the use of ozone in the synthesis of new polymers and the modification of polymers. *Adv. Polym. Sci.*, **167**, 35–74.

227. Yoshiba, Y., Kajiyama, M., Tomita, B. and Hosoya, S. (1990) Synthesis of ozonized lignin aminomaleimide resins using the Diels–Alder reaction and their viscoelastic properties. *Mokuzai Gakkaishi*, **36**, 440–447 (*Chem. Abstr.*, **113**, 154548).

228. Platz, G. A. (1993) Depolymerization method for resource recovery polymeric wastes. US Patent 5,369,215 (assigned to S.P. Reclamation Inc.).

229. Kefely, A. A., Rakovski, S. K., Shopov, D. M., Razumovskii, S. D. and Zaikov, G. E. (1981) Kinetic relationships and mechanism of ozone reaction with polystyrene in carbon tetrachloride solution. *J. Polym. Sci.*, **19**, 2175–2182.

230. Razumovski, S. D., Kefeli, A. A. and Zaikov, G. E. (1971) Degradation of polymers in reactive gases. *Eur. Polym. J.*, **7**, 275–285.

231. Michel, A., Cataneda, E. and Guyot, A. (1979) Synthesis of α,ω-difunctional poly(vinyl chloride) sequences and their extension to multisequence copolymers. *Eur. Polym. J.*, **15**, 935–942.

232. Boutevin, B., Mouanda, J., Pietrasanta, Y. and Taha, M. (1985) Synthèse de polymères du polyéthylène et du polypropylène à greffons fluorés. *Eur. Polym. J.*, **21**, 181–188.

233. Brondino, C., Boutevin, B., Parisi, J. P. and Schrynemackers, J. (1999) Adhesive properties onto galvanized steel plates of grafted poly(vinylidene fluoride) powders with phosphonated acrylates. *J. Appl. Polym. Sci.*, **72**, 611–620.

234. Boutevin, B., Pietrasanta, Y. and Robin, J. J. (1991) Synthèse et application de copolymères greffés à base de poly(fluorure de vinylidène). Partie I. *Eur. Polym. J.*, **27**, 815–822.

235. Zeppenfield, G. (1966) Application of the ferrous thiocyanate method to the determination of the peroxide content in radiation-oxidized poly(vinyl chloride). *Makromol. Chem.*, **90**, 169–176.

236. Boutevin, B., Robin, J. J. and Serdani, A. (1992) Synthesis and applications of graft copolymers from ozonized poly(vinylidene fluoride). II. *Eur. Polym. J.*, **28**, 1507–1516.

237. Ying, L., Wang, P., Kang, E. T. and Neoh, K. G. (2002) Synthesis and characterization of poly(acrylic acid)-*graft*-poly(vinylidene fluoride) copolymers and pH sensitive membranes. *Macromolecules*, **35**, 673–679.

238. Ying, L., Kang, E. T. and Neoh, K. G. (2002) Covalent immobilization of glucose oxidase on microporous membranes prepared from poly(vinylidene fluoride) with grafted poly(acrylic acid) side chains. *J. Membr. Sci.*, **208**, 361–374.

239. Riches, J., Knutsen, L., Morrey, E. and Grant, K. (2002) A screening tool for Halon alternatives based on the flame ionisation detector. *Fire Safety J.*, **37**, 287–301.

240. Bressy-Brondino, C., Boutevin, B., Hervaud, Y. and Gaboyard, E. (2003) Adhesive and anticorrosive properties of poly(vinylidene fluoride) powders blended with phosphonated copolymers on galvanized steel plates. *J. Appl. Polym. Sci.*, **83**, 2277–2287.

241. Liu, Y., Lee, J. Y., Kang, E. T., Wang, P. and Tan, K. L. (2001) Synthesis, characterization and electrochemical transport properties of the poly(ethylene glycol)-*grafted*-poly(vinylidene fluoride) nanoporous membranes. *React. Funct. Polym.*, **47**, 201–213.

242. Tarascon, J. M., Gozdz, A. S., Schmutz, C., Shokoohi, F. and Warren, P. C. (1996) Performance Bellcore's plastic rechargeable Li-ion batteries. *Solid State Ionics*, **86**, 49–54.

243. Capiglia, C. (1999) Effects of nanoscale SiO_2 on the thermal and transport properties of solvent-free, poly(ethylene oxide) (PEO)-based polymer electrolyte. *Solid State Ionics*, **118**, 73–79.

244. Akthakul, A. and Hester, J. F. (2002) Preparation of hydrophilic PVDF membranes for oil/water separation via surface segregation of microphase-separated PVDF-*g*-POEM during immersion precipitation. *Polym. Mater. Sci. Technol.*, **87**, 101–102.

245. Iwata, V. H., Oodate, M., Uyama, Y., Amemiya, H. and Ikada, Y. (1991) Preparation of temperature-sensitive membranes by graft polymerization onto a porous membrane. *J. Membr. Sci.*, **55**, 119–130.

246. Ying, L., Kang, E. T. and Neoh, K. G. (2002) Synthesis and characterization of poly(*N*-isopropylacrylamide)-*graft*-poly(vinylidene fluoride) copolymers and temperature-sensitive membranes. *Langmuir*, **18**, 6416–6423.

247. Zhai, G., Kang, E. T. and Neoh, K. G. (2003) Poly(2-vinylpyridine)- and poly(4-vinylpyridine)-*graft*-poly(vinylidene fluoride) copolymers and their pH-sensitive microfiltration membranes. *J. Membr. Sci.*, **217**, 243–259.

248. Dapoz, S. (1996) Perspective d'obtention de biomatériaux par fonctionnalisation de fluoropolymères radiogreffés avec du styrène après irradiation par des ions lourds rapides. PhD Dissertation, University of Caen, France.

249. Betz, N., Le Moel, A., Balanzat, E., Ramillon, J. M., Lamotte, J., Gallas, J. P. and Jaskierowicz, G. (1994) A FTIR study of PVDF irradiated by means of swift heavy ions. *J. Polym. Sci., Part B: Polym. Phys.*, **32**, 1493–1502.

250. (a) Betz, N., Le Moel, A., Duraud, J. P., Balanzat, E. and Darnez, C. (1992) Grafting of polystyrene in poly(vinylidene fluoride) films by means of energetic heavy ions. *Macromolecules*, **25**, 213–219; (b) Balanzat, E., Bouffart, S., Lamotte, J., Gallas, J. P., Le Moel, A. and Betz, N. (1993). *Les Nouvelles du Ganil*, **48**, 25–35.

251. Le Bouedec, A., Betz, N., Esnouf, S. and Le Moel, A. (1999) Swift heavy ion irradiation effects in α-poly(vinylidene fluoride): spatial distribution of defects within the latent track. *Nucl. Instrum. Methods Phys. Res., Sect. B: Beam Int. Mater. Atoms*, **151**, 89–96.

252. Guilmeau, I., Esnouf, S., Betz, N. and Le Moel, A. (1997) Kinetics and characterization of radiation-induced grafting of styrene on fluoropolymers. *Nucl. Instrum. Methods Phys. Res., Sect. B: Beam Int. Mater. Atoms*, **131**, 270–275.

253. Aymes-Chodur, C., Yagoubi, N., Betz, N., Le Moel, A. and Ferrier, D. (2000) SEC with UV and ELSD detection for analysis of gamma-irradiated PS-radiation-grafted P(VDF/HFP) films. *Chromatographia*, **51**, 269–276.

254. Aymes-Chodur, C., Betz, N., Porte-Durrieu, M. C., Baquey, C. and Le Moel, A. (1999) A FTIR and SEM study of PS radiation grafted fluoropolymers: influence of the nature of the ionizing radiation on the film structure. *Nucl. Instrum. Methods Phys. Res., Sect. B: Beam Int. Mater. Atoms*, **151**, 377–385.

255. Betz, N. (1995) Ion track grafting. *Nucl. Instrum. Methods Phys. Res., Sect. B: Beam Int. Mater. Atoms*, **105**, 55–62.

256. Gebel, G., Ottomani, E., Allegraud, J. J., Betz, N. and Le Moel, A. (1995) Structural study of polystyrene grafted in irradiated polyvinylidene fluoride thin films. *Nucl. Instrum. Methods Phys. Res., Sect. B: Beam Int. Mater. Atoms*, **105**, 145–149.

257. Betz, N., Petersohn, E. and Le Moel, A. (1996) Swift heavy ions effects in fluoropolymers: radicals and crosslinking. *Nucl. Instrum. Methods Phys. Res., Sect. B: Beam Int. Mater. Atoms*, **116**, 207–211.

258. (a) Betz, N. (1998) Chemical modifications induced in swift heavy ion irradiated polymers *Proceedings of SPIE—The International Society for Optical Engineering*, 3413 (Materials Modification by Ion Irradiation), pp. 16–26; (b) Ducouret, C., Betz, N. and Le Moel, A. (1996) Study of the molecular weight distribution of polystyrene grafted by means of swift heavy ions. *J. Chem. Phys. Phys.-Chem. Biol.*, **93**, pp. 70–77.

259. Balanzat, E., Betz, N. and Bouffard, S. (1995) Swift heavy ion modification of polymers. *Nucl. Instrum. Methods Phys. Res., Sect. B: Beam Int. Mater. Atoms*, **105**, 46–54.

260. (a) Petersohn, E., Betz, N. and Le Moel, A. (1995) Modification of phase transitions in swift heavy ion irradiated and MMA-grafted ferroelectric fluoropolymers. *Nucl. Instrum. Methods Phys. Res., Sect. B: Beam Int. Mater. Atoms*, **105**, 267–274; (b) Petersohn, E., Betz, N. and Le Moel, A. (1996) A study on radiation grafting of methyl methacrylate induced by swift heavy ions in poly(vinylidene fluoride). *J. Chem. Phys. Phys.-Chem. Biol.*, **93**, 188–193.

261. Betz, N., Begue, J., Goncalves, M., Gionnet, K., Deleris, G. and Le Moel, A. (2003) Functionalisation of PAA radiation grafted PVDF. *Nucl. Instrum. Methods Phys. Res., Sect. B: Beam Int. Mater. Atoms*, **208**, 434–441.

262. (a) Dapoz, S., Betz, N. and Le Moel, A. (1996) A FTIR characterization of a hemocompatible material obtained by swift heavy ion radiation grafting. *J. Chem. Phys. Phys.-Chem. Biol.*, **93**, 58–63; (b) Dapoz-Maglietta, S. (1998) Application of radiation grafting to fluorobiomaterials. *Ann. des Falsifications et de l'Expertise Chim. Toxicol.*, **91**, 215–221.

263. Porte-Durrieu, M. C., Aymes-Chodur, C., Vergne, C., Betz, N. and Baquey, C. (1999) Surface treatment of biomaterials by gamma and swift heavy ions grafting. *Nucl. Instrum. Methods Phys. Res., Sect. B: Beam Int. Mater. Atoms*, **151**, 404–415.

264. Porte-Durrieu, M. C., Aymes-Chodur, C., Betz, N. and Baquey, C. (2000) Development of "heparin-like" polymers using swift heavy ion and gamma radiation. I. Preparation and characterization of the materials. *J. Biomed. Mater. Res.*, **52**, 119–127.

265. Ducouret, C., Petersohn, E., Betz, N. and Le Moel, A. (1995) Fourier transform infrared analysis of heavy ion grafting of poly(vinylidene fluoride). *Spectrosc. Acta, Part A: Mol. Biomol. Spectrosc.*, **51A**, 567–572.

266. Porte-Durrieu, M. C., Aymes-Chodur, C., Betz, N., Brouillaud, B., Rouais, F., Le Moel, A. and Baquey, C. (1997) Synthesis of biomaterials by swift heavy ion grafting: preliminary results of hemocompatibility. *Nucl. Instrum. Methods Phys. Res., Sect. B: Beam Int. Mater. Atoms*, **131**, 364–375.

267. (a) Dapoz, S., Betz, N., Guittet, M. J. and Le Moel, A. (1995) ESCA characterization of heparin-like fluoropolymers obtained by functionalization after grafting induced by swift heavy ion irradiation. *Nucl. Instrum. Methods Phys. Res., Sect. B: Beam Int. Mater. Atoms*, **105**, 120–125; (b) Betz, N., Dapoz, S. and Guittet, M. J. (1997) Heterogeneity of swift heavy ion grafting: an XPS study. *Nucl. Instrum. Methods Phys. Res., Sect. B: Beam Int. Mater. Atoms*, **131**, 252–259.

268. Petersohn, E., Betz, N. and Le Moel, A. (1993) Characterization of radiation grafted PVDF and P(VDF/TrFE) films. In *J. Thin Films, Proc. Jt. Fourth Int. Symp. Trends New Appl. Thin Films 11th Conf. High Vac., Interf. Thin Films* (Eds, Hecht, G., Richter, F. and Hahn, R.) DGM Informationsges, Oberursel, Germany, pp. 650–653.

269. Xu, Y. (1991) *Ferroelectric Materials and their Applications*, Elsevier, Amsterdam, pp. 329–346.

270. Petersohn, E., Betz, N. and Le Moel, A. (1996) Modification of dielectric relaxations and phase transitions in ferroelectric polymers induced by swift heavy ion grafting. *Nucl. Instrum. Methods Phys. Res., Sect. B: Beam Int. Mater. Atoms*, **107**, 368–373.

271. Yasuda, H. (1985) *Plasma Polymerization*, Academic Press, Orlando, FL.

272. d'Agostino, R. (1990) Plasma deposition. *Treatment and Etching of Polymers*, Academic Press, Boston, MA.

273. Ishikawa, Y. and Sasakawa, S. (1992) Development of blood compatible materials by glow discharge-treatment. *Radiat. Phys. Chem.*, **39**(6), 561–567.

274. Gleich, H., Criens, R. M., Mosle, H. G. and Leute, U. (1989) The influence of plasma treatment on the surface properties of high-performance thermoplastics. *Int. J. Adhes.*, **9**, 88–94.

275. Inagaki, N., Tasaka, S., Ohkubo, J. and Kawai, H. (1990) Hydrophilic surface modification of polyethylene by plasma of nitrogen oxides. *J. Appl. Polym. Sci., Appl. Polym. Symp.*, **46**, 399–413.

276. Triolo, P. M. and Andrade, J. D. (1983) Hydrophilic surface modification of polyethylene by plasma of nitrogen oxides. *J. Biomed. Mater. Res.*, **17**, 149–157.

277. Fourimer, G. (1983) *Cinétique de l'oxygène en milieu plasma. Reactivité dans les plasmas. Applications aux lasers et aux agents de surface.* Les Editeurs de Physique, Ecole d'Eté d'Aussois, France, 297.

278. Zhao, Y. and Urban, M. W. (1999) Spectroscopic studies of microwave plasma reactions of maleic anhydride on poly(vinylidene fluoride) surfaces: crystallinity and surface reactions. *Langmuir*, **15**, 3538–3544.

279. Müller, M. and Oehr, C. (1999) Plasma amino functionalisation of PVDF microfiltration membranes: comparison of the plasma conditions and reactants in plasma modifications with a grafting method using ESCA and an amino-selective fluorescent probe. *Surf. Coat. Technol.*, **116–119**, 802–807.

280. Baquey, Ch., Palumbo, F., Porte-Durrieu, M. C., Legeay, G., Tressaud, A. and d'Agostino, R. (1999) Plasma treatment of expanded PTFE offers a way to a biofunctionalization of its surface. *Nucl. Instrum. Methods Phys. Res. B*, **151**, 255–262.

281. Yu, Z. J., Kang, E. T., Neoh, K. G., Cui, C. Q. and Lim, T. B. (2000) Grafting of epoxy resin on surface-modified poly(tetrafluororethylene) films. *J. Adhes.*, **73**, 417–439.

282. Yang, G. H., Zhang, Y., Lee, K. S., Kang, E. T. and Neoh, K. G. (2003) Thermal imidization of poly(pyromellitic dianhydride-4, 4(-oxydianiline) precursors on fluoropolymers modified by surface graft-copolymerization with glycidyl methacrylate. *J. Fluorine Chem.*, **119**, 151–160.

283. Zhang, Y., Tan, K. L. and Liaw, B. Y. (2001) Thermal imidization of fluorinated poly(amic acid) precursors on a glycidyl methacrylate graft-polymerized Si(100) surface. *J. Vac. Sci. Technol.*, **19**, 547–556.

284. Lee, Y. M. and Shim, J. M. (1996) Plasma surface graft of acrylic acid onto a porous poly(vinylidene fluoride) membrane and its riboflavin permeation. *J. Appl. Polym. Sci.*, **61**, 1245–1250.

285. Rafik, M., Mas, A., Elharfi, A. and Schué, F. (2000) Modification de membranes PVDF par plasma d'acide acrylique et de nonafluorobutyléthylène. *Eur. Polym. J.*, **36**, 1911–1919.

286. Makuuchi, K., Asano, M. and Abe, T. (1976) Effect of evolved hydrogen fluoride on radiation-induced crosslinking and dehydrofluorination of poly(vinylidene fluoride). *J. Polym. Sci., Polym. Chem. Ed.*, **14**, 617–625.

287. (a) Brefeld, W. and Gurtler, P. (1991) In *Handbook on Synchrotron Radiation*, Vol. 4 (Eds, Ebashi, S., Koch, M. and Rubenstein, E.) J. Wiley, Amsterdam; (b) Liug, H., Keller, C., Wolf, W., Shani, J., Mizka, H., Zyball, A., Gerve, A., Balaban, A. T., Kellerer, A. M. and Griebel, J. (1993) *Ullmann's Encyclopedia of Industrial Chemistry*, Vol. A22, VCH Verlagsgesellschaft, Weinheim, pp. 499–526.

288. Zhen, Z. X. (1990) Long-lived fluoropolymeric radicals in irradiated fluoropolymer powders at room temperature. *Radiat. Phys. Chem.*, **35**, 194–198.

289. Nasef, M. M., Saidi, H. and Dahlan, K. Z. M. (2001) Investigation of electron irradiation induced-changes in poly(vinylidene fluoride) films. *Polym. Degrad. Stab.*, **75**, 85–92.

290. Dapor, M. (2003) *Electron Beam Interaction with Solids*, Springer, Heidelberg.

291. Machi, S., Sugo, T. and Sugishita, A. (1978) Ion-exchange membrane. Jpn. Patent 7,808,692 (assigned to Japan Atomic Energy Research Institute) (*Chem. Abstr.*, **89**, 25600).

292. Blaise, J. (1981) Vinylidene fluoride graft copolymers useful as membranes. French Patent 2, 476,657 (assigned to Produits Chimiques Ugine Kuhlman).

293. Blaise, J. (1980) Poly(vinylidene fluoride) treated to increase its adherence to metal. French Patent 2,478,649 (assigned to Produits Chimiques Ugine Kuhlman).

294. Mascia, L. and Hashim, K. (1997) Compatibilization of poly(vinylidene fluoride)/nylon 6 blends by carboxylic acid functionalization and metal salts formation. I. Grafting reactions and morphology. *J. Appl. Polym. Sci.*, **66**, 1911–1923.

295. Valenza, A., Carianni, G. and Mascia, L. (1998) Radiation grafting functionalization of poly(vinylidene fluoride) to compatibilize its blends with polyolefin ionomers. *Polym. Eng. Sci.*, **38**(3), 452–460.

296. Jan, T. D., Jeon, J. S. and Raghavan, S. (1994) Surface modification of PVDF membranes by grafting of a vinylphosphonium compound. *J. Adhes. Sci. Technol.*, **8**, 1157–1168.

297. (a) Munari, S., Vigo, F., Nicchia, M. and Canepa, P. (1976) Hyperfiltration membranes prepared by radiochemical grafting of styrene onto poly(tetrafluoroethylene). Influence of the size and shape of emulsion particles used to obtain the PTFE film. *J. Appl. Polym. Sci.*, **20**, 243–253; (b) Vigo, F., Capannelli, G., Uliana, C. and Murani, S. (1981) Asymmetric polyvinylidene fluoride (PVDF) radiation grafted membranes: preparation and performance in reverse osmosis application. *Desalination*, **36**, 63–73.

298. Holmberg, S., Lehtinen, T., Näsman, J., Ostrovoskii, D., Paronen, M., Serimaa, R., Sundholm, F., Sundholm, G., Torell, L. and Torkkeli, M. (1996) Structure and properties of sulfonated poly[(vinylidene fluoride)-g-styrene] porous membranes. *J. Mater. Chem.*, **6**, 1309–1317.

299. Hietala, S., Holmberg, S., Karjalainen, M., Näsman, J., Paronen, M., Serimaa, R., Sundholm, F. and Vahvaselka, S. (1997) Structural investigation of radiation grafted and sulfonated poly(vinylidene fluoride), PVDF, membranes. *J. Mater. Chem.*, **7**, 721–726.

300. Hietala, S., Koel, M., Skou, E., Elomaa, M. and Sundholm, F. (1998) Thermal stability of styrene grafted and sulfonated proton conducting membranes based on poly(vinylidene fluoride). *J. Mater. Chem.*, **8**, 1127–1132.

301. Ennari, J., Hietala, S., Paronen, M., Sundholm, F., Walsby, N., Karjalainen, M., Serimaa, R., Lehtinen, T. and Sundholm, G. (1999) New polymer electrolyte membranes for low temperature fuel cells. *Macromol. Symp.*, **146**, 41–45.

302. Hietala, S., Paronen, M., Holmberg, S., Näsman, J., Juhanoja, J., Karjalainen, M., Serimaa, R., Toivola, M., Lehtinen, T., Parovuori, K., Sundholm, G., Ericson, H., Mattsson, B., Torell, L. and Sundholm, F. (1999) Phase separation and crystallinity in proton conducting membranes of styrene grafted and sulfonated poly(vinylidene fluoride). *J. Polym. Sci., Part A: Polym. Chem.*, **37**, 1741–1753.

303. Hietala, S., Maunu, S. L., Sundholm, F., Lehtinen, T. and Sundholm, G. (1999) Water sorption and diffusion coefficients of protons and water in PVDF-g-PSSA polymer electrolyte membranes. *J. Polym. Sci., Part B: Polym. Phys.*, **37**, 2893–2900.

304. Jokela, K., Serimaa, R., Torkkeli, M., Sundholm, F., Kallio, T. and Sundholm, G. (2002) Effect of the initial matrix material on the structure of radiation-grafted ion-exchange membranes: wide-angle and small-angle x-ray scattering studies. *J. Polym. Sci., Part B: Polym. Phys.*, **40**, 1539–1555.

305. Holmberg, S., Nasman, J. and Sundholm, F. (1998) Synthesis and properties of sulfonated and crosslinked poly[(vinylidene fluoride)-*graft*-styrene] membranes. *Polym. Adv. Technol.*, **9**, 121–130.

306. Elomaa, M., Hietala, S., Paronen, M., Walsby, N., Jokela, K., Serimaa, R., Torkkeli, M., Lehtinen, T., Sundholm, G. and Sundholm, F. (2000) The state of water and the nature of ion clusters in crosslinked proton conducting membranes of styrene grafted and sulfonated poly(vinylidene fluoride). *J. Mater. Chem.*, **10**, 2678–2684.

307. (a) Kallio, T., Jokela, K., Ericson, H., Serimaa, R., Sundholm, G., Jacobsson, P. and Sundholm, F. (2003) Effects of a fuel cell test on the structure of irradiation grafted ion exchange

membranes based on different fluoropolymers. *J. Appl. Electrochem.*, **33**, 505–514; (b) Gode, P., Ihonen, J., Ericson, A., Lindbergh, G., Paronen, M., Sundholm, G., Sundholm, F. and Walsky, N. (2003) Membrane durability in a proton exchange membrane fuel cell studied using PVDF based radiation grafted membranes. *Fuel Cells*, **3**, 21–27; (c) Paronen, M., Sundholm, F., Ostrovskii, D., Jacobsson, P., Jeschker, G., Raeihala, E. and Tikkanen, P. (2003) Preparation of proton conducting membranes by direct sulfonation. 1. Effect of radical and radical decay on the sulfonation of polyvinylidene fluoride film. *Chem. Mater.*, **15**, 4447–4455.

308. Holmberg, S., Holmlund, P., Wilen, C. E., Kallio, T., Sundholm, G. and Sundholm, F. (2002) Synthesis of proton-conducting membranes by the utilization of preirradiation grafting and atom transfer radical polymerization techniques. *J. Polym. Sci., Part A: Polym. Phys.*, **40**, 591–600.

309. Liu, B., Yuan, C. and Hu, C. (2000) Well-defined graft copolymer of styrene and fluorinated acrylate synthesized by living radical polymerization. *Hecheng Xiangjiao Gongye*, **23**, 377 (*Chem. Abstr.*, **134**, 147974).

310. (a) Mayes, A. M., Hester, J. F., Banerjee, P. and Akthakul, A. (2002) Graft copolymers, methods for grafting hydrophilic chains onto hydrophobic polymers, and articles thereof. International demand PCT Int. Appl. WO 0222712 (assigned to Massachusetts Institute of Technology); (b) Hester, J. F., Banerjee, P., Won, Y.-Y., Akthakul, A., Acar, M. H. and Mayer, A. M. (2002) ATRP of amphiphilic graft copolymers based on PVDF and their use as membrane additives. *Macromolecules* **35**, 7652–7661.

311. Brack, H.-P. and Scherer, G. G. (1998) Modification and characterization of thin polymer films for electrochemical applications. *Macromol. Symp.*, **126**, 25–49.

312. Brack, H.-P. and Scherer, G. G. (2000) Grafting of preirradiated poly(ethylene-*alt*-tetrafluoroethylene) films with styrene: influence of base polymer film properties and processing parameters. *J. Mater. Chem.*, **10**, 1795–1803.

313. Williams, D. F. (Ed.) (1987) Definitions in biomaterials. Proceedings of a Consensus Conference of the European Society for Biomaterials, Chester, England (March 3–5, 1986), Progress in Biomedical Engineering 4, pp. 1–72.

314. Marmey, P., Porté-Durrieu, M. C. and Baquey, C. (2003) PVDF multifilament yarns grafted with polystyrene induced by gamma-irradiation: influence of the grafting parameters on the chemical properties. *Nucl. Instrum. Methods Phys. Res. B*, **208**, 429–433.

315. Fougnot, C., Jozefonwicz, J., Samana, M. and Bara, L. (1979) New heparin-like insoluble materials. Part I. *Ann. Biomed. Eng.*, **7**, 429–439.

316. Migonney, V., Fougnot, C. and Jozefowicz, M. (1988) Heparin-like tubings. I. Preparation, characterization and biological in vitro activity assessment. *Biomaterials*, **9**, 145–149.

317. Caix, J., Migonney, V., Baquey, A., Fougnot, C., Perrot Minnot, A., Beziade, A., Vuillemin, L., Baquey, C. and Ducassou, D. (1988) Control and isotopic quantification of affinity of antithrombin III for heparin-like surfaces. *Biomaterials*, **9**, 62–64.

318. Fougnot, C., Jozefowicz, M. and Rosenberg, D. (1984) Catalysis of the generation of thrombin–antithrombin complex by insoluble anticoagulant polystyrene derivatives. *Biomaterials*, **5**, 94–99.

319. Fougnot, C., Dupillier, M. P. and Jozefowicz, M. (1983) Anticoagulant activity of amino acid modified polystyrene resins: influence of carboxylic acid function. *Biomaterials*, **4**, 101–104.

320. Klimova, M., Szöcs, F., Bartos, J., Vacek, K. and Pallanova, M. (1989) ESR and DSC study of the radiation crosslinking effect on macroradical decay in poly(vinylidene fluoride). *J. Appl. Polym. Sci.*, **37**, 3449–3458.

321. Golgerg, E. P., Yahiaoui, A., Mentak, K., Erickson, T. R. and Seeger, J. (2002) Polymer grafting for enhancement of biofunctional properties of medical and prosthetic surface. US Patent 6,387, 379 (assigned to the University of Florida).

322. (a) Jarvis, C. R., Macklin, W. J., Macklin, A. J., Mattingley, N. J. and Kronfli, E. (2001) Use of grafted PVDF-based polymers in lithium batteries. *J. Power Source*, **97–98**, 664–666;

(b) Coowar, F. and Kronfli, E. (2001) Microporous membranes for solid electrolytes for lithium cells manufactured by casting vinylidene fluoride polymer solutions containing nonsolvents for polymers and slowly evaporating the liquid phase. International Demand WO 2001 048063.

323. Flint, S. D. and Slade, R. C. T. (1997) Investigation of the radiation-grafted PVDF-*g*-polystyrene sulfonic acid ion exchange membranes for use in hydrogen oxygen fuel cells. *Solid State Ionics*, **97**, 299–307.

324. Danks, T. N., Slade, R. C. T. and Varcoe, J. R. (2002) Comparison of PVDF- and FEP-based radiation-grafted alkaline anion-exchange membranes for use in low temperature portable DMFCs. *J. Mater. Chem.*, **12**, 3371–3373.

325. Danks, T. N., Slade, R. C. T. and Varcoe, J. R. (2003) Alkaline anion-exchange radiation-grafted membranes for possible electrochemical application in fuel cells. *J. Mater. Chem.*, **13**, 712–721.

326. Watanabe, J. (1990) Preparation of vinylidene fluoride graft copolymers. Jpn. Patent JP 02,034, 608 (assigned to Shin-Etsu Chemical Industry Co., Ltd., Japan).

327. (a) Mislavsky, B. V. and Melnikov, V. P. (1993) Synthesis, properties, and applications of composite materials based on grafted copolymers of perfluoropolymers and perfluorinated monomers with functional groups. *Polym. Prepr. (Am. Chem. Soc., Polym. Div.)*, **34**(1), 377–378; (b) Mislavsky, B. V. (1997) Fluorinated polymers with functional groups. Synthesis and applications In *Fluoropolymers*, Vol. 1 (Eds, Hougham, G., Cassidy, P., Johns, K. and Davidson, T.) Kluvert, New York, pp. 91–110, Chap. 7.

328. Ebnesajjad, S. (2003) *Fluoroplastics, Melt Processable Fluoropolymers*, Plastic Design Library, Norwich, NY, Vol. 2.

329. Hintzer, K. and Lohr, G. (1997) Modified polytetrafluoroethylene—the second generation. In *Modern Fluoropolymers* (Ed, Scheirs, J.) Wiley, New York, pp. 239–256, Chap. 12.

330. Lunkwitz, K., Buerger, W., Lappan, U., Brink, H. J. and Ferse, A. (1995) Surface modification of fluoropolymers. *J. Adhes. Sci. Technol.*, **9**, 297–310.

331. Fuchs, B., Lappan, U., Lunkwitz, K. and Scheler, U. (2002) Radiochemical yields for cross-links and branches in radiation-modified poly(tetrafluoroethylene). *Macromolecules*, **35**, 9079–9082.

332. Brack, H.-P., Buchi, F. N., Huslage, J., Rota, M. and Scherer, G. G. (2000) Development of radiation-grafted membranes for fuel cell applications based on poly(ethylene-*alt*-tetrafluoro-ethylene). *ACS Symp. Ser.*, **744**, 174–188.

333. (a) Bozzi, A. and Chapiro, A. (1987) The nature of the initiating centers for grafting in air-irradiated perfluoro polymers. *Eur. Polym. J.*, **23**, 255–257; (b) Bozzi, A. and Chapiro, A. (1988) Synthesis of permselective membranes by grafting acrylic acid into air-irradiated Teflon FEP films. *Radiat. Phys. Chem.*, **32**, 193–196.

334. Turmanova, S., Trifonov, A., Kalaijiev, O. and Kostov, G. (1997) Radiation grafting of acrylic acid onto polytetrafluoroethylene films for glucose oxidase immobilization and its application in membrane biosensors. *J. Membr. Sci.*, **127**, 1–7.

335. Ruskov, T., Turmanova, S. and Kostov, G. (1997) Study of IR and Mössbauer spectroscopy of grafted metal complexes of poly(acrylic acid) onto low density poly(ethylene) and onto poly(tetrafluoroethylene). *Eur. Polym. J.*, **33**, 1285–1288.

336. Hegazy, El. S., Ishigaki, I. and Okamoto, J. (1981) Radiation grafting of acrylic acid onto fluorine-containing polymers. I. Kinetic study of preirradiation grafting onto poly(tetrafluoro-ethylene). *J. Appl. Polym. Sci.*, **26**, 3117–3124.

337. (a) Omichi, H. and Okamoto, J. (1982) Synthesis of ion-exchange membranes by radiation-induced multiple grafting of methyl α,β,β-trifluoroacrylate. *J. Polym. Sci., Polym. Chem. Ed.*, **20**, 521–528; (b) Omichi, H. and Okamoto, J. (1984) Synthesis of ion-exchange by cografting methyl α,β,β-trifluoroacrylate and propylene. *J. Polym. Sci., Polym. Chem. Ed.*, **22**, 1775–1782; (c) Omichi, H. and Okamoto, J. (1985) Effect of mixed monomers on the synthesis of ion-exchange membranes by radiation-induced grafting. *J. Appl. Polym. Sci.*, **30**, 1277–1284.

338. Bex, G., Chapiro, A., Huglin, M. B., Jendrychowska-Bonamour, A. M. and O'Neill, T. (1968) Preparation of semipermeable membranes with remarkable properties by the radiochemical grafting of poly(tetrafluoroethylene) films. *J. Polym. Sci., Polym. Symp.*, **22**, 493–499.

339. Chapiro, A., Gordon, E. and Jendrychowska-Bonamour, A. M. (1973) Solvent effects on radiochemical grafting of 4-vinylpyridine onto poly(tetrafluoroethylene) films. *Eur. Polym. J.*, **9**, 975–984.

340. Kostov, G. K. and Turmanova, S. C. (1997) Radiation-initiated graft copolymerization of 4-vinylpyridine onto polyethylene and polytetrafluoroethylene films and anion-exchange membranes therefrom. *J. Appl. Polym. Sci.*, **64**, 1469–1475.

341. Turmanova, S., Dimitrova, A., Kostov, G. and Nedkov, E. (1996) Study of the structure of polyethylene and poly(tetrafluoroethylene) radiation grafted ion-exchange copolymers. *Macromol. Chem. Phys.*, **197**, 2973–2981.

342. Burillo, G. and Chapiro, A. (1986) Radiochemical grafting of *N*-vinylpyrrolidone into poly(tetrafluoroethylene) films. *Eur. Polym. J.*, **22**, 653–656.

343. Guzman-Garcia, A. G., Pintauro, P. N., Verbrugge, M. W. and Schneider, E. W. (1992) Analysis of radiation grafted membranes for fuel cell electrolytes. *J. Appl. Electrochem.*, **22**, 204–214.

344. Nasef, M. M. (2000) Thermal stability of radiation grafted PTFE-*g*-polystyrene sulfonic acid membranes. *Polym. Degrad. Stab.*, **68**, 231–238.

345. Nasef, M. M., Saidi, H., Nor, H. M. and Foo, O. M. (2000) Radiation-induced grafting of styrene onto poly(tetrafluoroethylene) films. Part. II. Properties of the grafted and sulfonated membranes. *Polym. Int.*, **49**, 1572–1579.

346. (a) Nasef, M. M., Saidi, H., Dessouki, A. M. and El-Nesr, E. M. (2000) Radiation-induced grafting of styrene onto poly(tetrafluoroethylene) (PTFE) films. I. Effect of grafting conditions and properties of the grafted films. *Polym. Int.*, **49**, 399–406; (b) Nasef, M. M., Saidi, H., Nor, H. M. and Yarmo, M. A. (2000) XPS studies of radiation grafted PTFE-*g*-polystyrene sulfonic acid membranes. *J. Appl. Polym. Sci.*, **76**, 336–349.

347. Liang, G.-Z., Lu, T.-L, Ma, X.-Y., Yan, H.-X. and Gong, Z.-H. (2003) Synthesis and characteristics of radiation-grafted membranes for fuel cell electrolytes. *Polym. Int.*, **52**, 1300–1308.

348. Denamganai, J., Schué, F., Sledz, J. and Molenat, J. (1995) Desorption followed by radioisotopes and membranes potentials in perfluorinated 4-vinylpyridine and quaternary ammonium membranes. *Eur. Polym. J.*, **31**, 1067–1074.

349. Zhao, X. (1982) Decay of free radicals in fluoropolymers irradiated with fast electrons. *He Huaxue Yu Fangshe Huaxue*, **4**, 111–115 (*Chem. Abstr.*, **97**, 39500).

350. Mascia, L., Pak, S. H. and Caporiccio, G. (1994) Radiation assisted compatibilisation of blends of vinylidene fluoride polymers with fluorosilicone elastomers by means of crosslinking coagents. *Polym. Int.*, **35**, 75–82.

351. Mascia, L., Pak, S. H. and Caporiccio, G. (1995) Properties enhancement of fluorosilicone elastomers with compatibilised crystalline vinylidene fluoride polymers. *Eur. Polym. J*, **31**, 459–465.

352. Okoshi, H., Chen, H., Soldani, G., Galletti, P. M. and Goddard, M. (1992) Microporous small diameter PVDF-TrFE vascular grafts fabricated by a spray phase inversion technique. Slide Forum 5, vascular prosthesis/new designs and materials. *Am. Soc. Artif. Internal Organs*, **38**, 201–206.

353. Kerbow, D. L. (1997) Ethylene–tetrafluoroethylene copolymer resins. In *Modern Fluoropolymers* (Ed, Scheirs, J.) Wiley, New York, pp. 301–310, Chap. 15.

354. (a) Kostov, G. K. and Atanassov, A. N. (1991) Synthesis of hydrophilic ion-exchange fluoropolymers containing sulpho- and both sulpho- and carboxyl groups. *Eur. Polym. J.*, **27**, 1331–1333; (b) Kostov, G. K. and Atanassov, A. N. (1993) Radiation initiated grafting of acrylic acid onto tetrafluoroethylene–ethylene copolymers. *J. Appl. Polym. Sci.*, **47**, 361–366; (c) Kostov, G. K. and Atanassov, A. N. (1993) Properties of cation-exchange membranes

prepared by radiation grafting of acrylic acid onto tetrafluoroethylene–ethylene copolymers. *J. Appl. Polym. Sci.*, **47**, 1269–1276.

355. El-Sawy, N. M. and Al Sagheer, F. A. (2002) Physicochemical investigation of radiation-grafted poly(acrylic acid)-*graft*-poly(tetrafluoroethylene–ethylene)copolymer membranes and their use in metal recovery from aqueous solution. *J. Appl. Polym. Sci.*, **85**, 2692–2698.

356. (a) Horsfall, J. A. and Lovell, K. V. (2001) Fuel cell performance of radiation-grafted sulfonic acid membranes. *Fuel Cells*, **1**, 186–195; (b) Horsfall, J. A. and Lovell, K. V. (2002) Comparison of fuel cell performance of selected fluoropolymer and hydrocarbon based grafted copolymers incorporating acrylic acid and styrene sulfonic acid. *Polym. Adv. Technol.*, **13**, 381–390.

357. Horsfall, J. A. and Lovell, K. V. (2003) Synthesis and characterization of acrylic acid-grafted hydrocarbon and fluorocarbon polymers with simultaneous or mutual grafting technique. *J. Appl. Polym. Sci.*, **87**, 230–243.

358. Kaur, I., Misra, B. N., Kumar, R. and Singh, B. (2002) Radiation induced graft copolymerization of methyl acrylate and ethyl acrylate onto poly(tetrafluoroethylene-*co*-ethylene) film. *Polym. Polym. Compos.*, **10**, 391–404.

359. (a) Ellinghorst, G., Fuehrer, J. and Vierkotten, D. (1981) Radiation initiated grafting on fluoro polymers for membrane preparation. *Radiat. Phys. Chem.*, **18**, 889–897; (b) Ellinghorst, G., Niemoeller, A. and Vierkotten, D. (1983) Radiation-initiated grafting of polymer films—an alternative technique to prepare membranes for various separation problems. *Radiat. Phys. Chem.*, **22**, 635–642.

360. Becker, W. and Schmidt-Naake, G. (2002) Proton exchange membranes by irradiation-induced grafting of styrene onto FEP and ETFE: influences of the crosslinker *N,N*-methylene-bis-acrylamide. *Chem. Eng. Technol.*, **25**, 373–377.

361. Elmidaoui, A., Cherif, A., Brunea, J., Duclert, F., Cohen, T. and Gavach, C. (1992) Preparation of perfluorinated ion-exchange membranes and their application in acid recovery. *J. Membr. Sci.*, **67**, 263–271.

362. Arico, A. S., Baglio, V., Creti, P., Di Blasi, A., Antonucci, V., Brunea, J., Chapotot, A., Bozzi, A. and Schoemans, J. (2003) Investigation of grafted ETFE-based polymer membranes as alternative electrolyte for direct methanol fuel cells. *J. Power Sources*, **123**, 107–115.

363. Horsfall, J. A. and Lovell, K. V. (2002) Synthesis and characterisation of sulfonic acid-containing ion exchange membranes based on hydrocarbon and fluorocarbon polymers. *Eur. Polym. J.*, **38**, 1671–1682.

364. Chuy, C., Basura, V. I., Simon, E., Holdcroft, S., Horsfall, J. A. and Lovell, K. V. (2000) Electrochemical characterisation of ethylenetetrafluoroethylene-*g*-polystyrenesulfonic acid solid polymer electrolytes. *J. Electrochem. Soc.*, **147**, 4453–4458.

365. (a) Brack, H.-P., Bonorand, L., Buhrer, H. G. and Scherer, G. G. (1998) Radiation grafting of ETFE and FEP films: base polymer film effects. *Polym. Prepr. (Am. Chem. Soc., Div. Polym. Chem.)*, **39**, 976–977; (b) Brack, H.-P., Bonorand, L., Buhrer, H. G. and Scherer, G. G. (1998) Radiation grafting of ETFE and FEP films: base polymer film effect. Book of Abstracts, 216th *ACS National Meeting Boston*, August 23–27, 1998, Poly-517.

366. (a) Brack, H.-P., Buchi, F. N., Rota, M. and Scherer, G. G. (1997) Development of radiation-grafted membranes for fuel cell applications based on poly(ethylene-*alt*-tetrafluoroethylene). *Polym. Mater. Sci. Eng.*, **77**, 368–369; (b) Brack, H.-P., Bonorand, L., Buhrer, H. G. and Scherer, G. G. (1998) Radiation processing of fluoropolymer films. *Polym. Prepr. (Am. Chem. Soc., Div. Polym. Chem.)*, **39**, 897–898; (c) Brack, H.-P., Bonorand, L., Buhrer, H. G. and Scherer, G. G. (1998) Radiation processing of fluoropolymer films. Book of Abstracts, *216th ACS National Meeting*, Boston, August 23–27, 1998, Poly-487; (d) Brack, H.-P., Wyler, M., Peter, G. and Scherer, G. G. (2003) A contact angle investigation of the surface properties of selected proton-conducting radiation-grafted membranes. *J. Membr. Sci.*, **214**, 1–19.

367. Scott, K., Taama, W. M. and Argyropoulos, P. (2000) Performance of the direct methanol fuel cell with radiation-grafted polymer membranes. *J. Membr. Sci.*, **171**, 119–130.

368. Nezu, S., Akakabe, M., Yamada, C. and Kato, M. (1999) Fuel cell with ion-exchange solid polymer electrolytes having hydrophobic surfaces. Jap. Patent 11,273,696 (assigned to Aisin Seiki Co. Ltd.), *Chem. Abstr.*, **139**, 120001.

369. Stone, C., Steck, A. E. and Choudhury, B. (2002) Graft fluoropolymer membranes and ion-exchange membranes formed therefrom. US Patent 2002-0,137,806 (assigned to Ballard Power Systems, Inc.).

370. Pozzoli, M., Vita, G. and Arcella, V. (1997) Melt processable perfluoropolymers. In *Modern Fluoropolymers* (Ed, Scheirs, J.) Wiley, New York, pp. 373–396, Chap. 21.

371. Florin, R. E. and Wall, L. (1961) Gamma irradiation of fluorocarbon polymers. *J. Res. Natl. Bur. Stand. A*, **65**, 375–387.

372. Lovejoy, E. R., Bro, M. I. and Bowers, G. H. (1965) Chemistry of radiation cross-linking of branched fluorocarbon resins. *J. Appl. Polym. Sci.*, **9**, 401–410.

373. Rosenberg, Y., Siegmann, A., Narkis, M. and Shkolnik, S. (1992). Low dose γ-irradiation of some fluoropolymers: effect of polymer chemical structure. *J. Appl. Polym. Sci.*, **45**, 783–795.

374. Bhattacharya, A., De, V. and Bhattacharya, S. N. (1994) Preparation of polypyrrole composite with acrylic acid-grafted tetrafluoroethylene–hexafluoropropylene (Teflon-FEP) copolymer. *Synth. Met.*, **65**, 35–38.

375. Hegazy, E. S., Ishigaki, I., Dessouki, A. M., Rabie, A. and Okamoto, J. (1982) The study on radiation grafting of acrylic acid onto fluorine-containing polymers. III. Kinetic study of preirradiation grafting onto poly(tetrafluoroethylene–hexafluoropropylene). *J. Appl. Polym. Sci.*, **27**, 535–543.

376. Hegazy, A., Ishigaki, I., Rabie, A., Dessouki, A. M. and Okamoto, J. (1983) The study on radiation grafting of acrylic acid onto fluorine-containing polymers. IV. Properties of membrane obtained by preirradiation grafting onto poly(tetrafluoroethylene–hexafluoropropylene). *J. Appl. Polym. Sci.*, **28**, 1465–1479.

377. Gen, Y. (1987) Preparation of charged micro-mosaic membrane via radiation-induced grafting of styrene and methyl methacrylate onto Teflon FEP films. *Desalination*, **62**, 265–273.

378. (a) Nasef, M. M., Saidi, H., Nor, H. M. and Foo, O. M. (2000) Proton exchange membranes prepared by simultaneous radiation grafting of styrene onto poly(tetrafluoroethylene-*co*-hexafluoropropylene) films. II. Properties of sulfonated membranes. *J. Appl. Polym. Sci.*, **78**, 2443–2453; (b) Nasef, M., Saidi, H. and Nor, H. M. (2000) Proton exchange membranes prepared by simultaneous radiation grafting of styrene onto poly(tetrafluoroethylene-*co*-hexafluoropropylene). *J. Appl. Polym. Sci.*, **76**, 220–227.

379. (a) Nasef, M. M. and Saidi, H. (2000) Thermal degradation behaviour of radiation grafted FEP-*g*-polystyrene sulfonic acid membranes. *Polym. Degrad. Stab.*, **70**, 497–504; (b) Nasef, M. M., Saidi, H. and Yarmo, M. A. (2000) Surface investigations of radiation grafted FEP-*g*-polystyrene sulfonic acid membranes using XPS. *J. New Mater. Electrochem. Syst.*, **3**, 309–317.

380. (a) Rouilly, M. V., Krotz, E. R., Haas, O., Scherer, G. G. and Chapiro, A. (1993) Proton exchange membranes prepared by simultaneous radiation grafting of styrene onto Teflon-FEP films. Synthesis and characterization. *J. Membr. Sci.*, **81**, 89–96; (b) Gupta, B., Buchi, F. N. and Scherer, G. G. (1994) Cation exchange membranes by pre-irradiation grafting of styrene into FEP films. I. Influence of synthesis conditions. *J. Polym. Sci., Part A: Polym. Chem.*, **32**, 1931–1938; (c) Gupta, B., Scherer, G. G. and Highfield, J. G. (1994) Proton-exchange membranes by radiation grafting of styrene onto FEP films. II. Mechanism of thermal degradation in copolymer membranes. *J. Appl. Polym. Sci.*, **51**, 1659–1666.

381. (a) Gupta, B., Buchi, F. N., Staub, M., Grman, D. and Scherer, G. G. (1996) Cation exchange membranes by pre-irradiation grafting of styrene into FEP films. II. Properties of copolymer membranes. *J. Polym. Sci., Part A: Polym. Chem.*, **34**, 1873–1880; (b) Gupta, B., Scherer, G. G. and Highfield, J. G. (1998) Thermal stability of proton exchange membranes prepared by

grafting of styrene into pre-irradiated FEP films and the effect of crosslinking. *Ang. Makromol. Chem.*, **256**, 81–84.

382. (a) Gupta, B., Buchi, F. N., Scherer, G. G. and Chapiro, A. (1993) Materials research aspects of organic solid proton conductors. *Solid State Ionics*, **61**, 213–218; (b) Gupta, B., Buchi, F. N., Scherer, G. G. and Chapiro, A. (1994) Property–structure correlations. *Polym. Adv. Technol.*, **5**, 193–198; (c) Buchi, F. N., Gupta, B., Haas, O. and Scherer, G. G. (1995) Study of radiation-grafted FEP-*g*-polystyrene membranes as polymer electrolytes in fuel cells. *Electrochim. Acta*, **40**, 145–153.

383. Du, C., Xu, M., Zhow, I., Wo, W. and Ma, Z. (1983) Studies on radiation curing of acrylate coatings. II. Preparation of acrylate coatings and electron beam radiation. *Radiat. Phys. Chem.*, **22**, 1007–1010.

384. Herman, H., Slade, R. C. T. and Varcoe, J. R. (2003) The radiation-grafting of vinylbenzyl chloride onto poly(hexafluoropropylene-*co*-tetrafluoroethylene) films with subsequent conversion to alkaline anion-exchange membranes: optimization of the experimental conditions and characterization. *J. Membr. Sci.*, **218**, 147–163.

385. Hintzer, K. and Lohr, G. (1997) Melt processable tetrafluoroethylene–perfluoropropylvinyl ether copolymers (PFA). In *Modern Fluoropolymers* (Ed, Scheirs, J.) Wiley, New York, pp. 223–238, Chap. 11.

386. Dargaville, T. R., George, G. A., Hill, D. J. T., Sheler, U. and Whittaker, A. K. (2003) Cross-linking of PFA by electron beam, an investigation of the thermal and tensile properties of PFA following γ-radiolysis. *Macromolecules*, **36**, 7132–7137.

387. Okada, A., Ichinose, N. and Kawanishi, S. (1996) A novel KrF laser-induced graft reaction of poly(acrylic acid) onto tetrafluoroethylene–perfluoroalkyl vinyl ether copolymer film. *Polymer*, **37**, 2281–2283.

388. Nasef, M. M., Saidi, H., Nor, M. H., Dahlan, K. Z. M. and Hashim, K. (1999) Cation exchange membranes by radiation-induced graft copolymerization of styrene onto PFA copolymer films. I. Preparation and characterization of the graft copolymer. *J. Appl. Polym. Sci.*, **73**, 2095–2102.

389. Nasef, M. M., Saidi, H., Nor, H. M. and Foo, O. M. (2000) Cation exchange membranes by radiation-induced graft copolymerization of styrene onto PFA copolymer films. II. Characterization of sulfonated graft copolymer membranes. *J. Appl. Polym. Sci.*, **76**, 1–11.

390. Nasef, M. M., Saidi, H. and Nor, H. M. (2000) Cation exchange membranes by radiation-induced graft copolymerization of styrene onto PFA copolymer films. III. Thermal stability of the membranes. *J. Appl. Polym. Sci.*, **77**, 1877–1885.

391. Nasef, M. M., Saidi, H. and Yarmo, M. A. (2000) Cation exchange membranes by radiation-induced graft copolymerization of styrene onto PFA copolymer films. IV. Morphological investigations using x-ray photoelectron spectroscopy. *J. Appl. Polym. Sci.*, **77**, 2455–2463.

392. Nasef, M. M. and Saidi, H. (2002) Post-mortem analysis of radiation grafted fuel cell membrane using x-ray photoelectron spectroscopy. *J. New Mater. Electrochem. Syst.*, **5**, 183–189.

393. Cardona, F., Hill, D. J. T., George, G., Maeji, J., Firas, R. and Perera, S. (2001) Thermal characterization of copolymers obtained by radiation grafting of styrene onto poly(tetrafluoroethylene–perfluoropropylvinylether) substrates: thermal decomposition, melting behavior and low-temperature transitions. *Polym. Degrad. Stab.*, **74**, 219–227.

394. Cardona, F., George, G. A., Hill, D. J. T., Rasoul, F. and Maeji, J. (2002) Copolymers obtained by the radiation-induced grafting of styrene onto poly(tetrafluoroethylene-*co*-perfluoropropylvinyl ether) substrates. 1. Preparation and structural investigation. *Macromolecules*, **35**, 355–364.

395. Cardona, F., George, G. A., Hill, D. J. T. and Perera, S. (2002) Spectroscopic study of the penetration depth of grafted polystyrene onto poly(tetrafluoroethylene-*co*-perfluoropropylvinylether) substrates. 1. Effect of grafting conditions. *J. Polym. Sci., Part A: Polym. Chem.*, **40**, 3191–3199.

396. Cardona, F., George, G. A., Hill, D. J. T. and Perera, S. (2003) Comparative study of the radiation-induced grafting of styrene onto poly(tetrafluoroethylene-*co*-perfluoropropylvinyl ether) and polypropylene substrates. I. Kinetics and structural investigations. *Polym. Int.*, **52**, 827–837.

397. Vasilets, V. N., Kovalchuk, A. V., Yuranova, T. I., Ponomarev, A. N., Talroze, R. V., Zubarev, E. R. and Plate, N. A. (1996) Sandwich structure containing liquid crystal polymer grafted on polymer support. *Polym. Adv. Technol.*, **7**, 173–176.

398. Kihara, N., Xu, Y. and Fukutomi, T. (1997) Utilization of chain transfer to graft polymerization. I. Graft polymerization onto the polystyrene bearing allylthio group. *J. Polym. Sci., Part A: Polym. Chem.*, **35**, 817–822.

399. Corner, T. (1984) Free radical polymerisation. The synthesis of graft copolymers. *Adv. Polym. Sci.*, **62**, 95–142.

400. Kostov, G. and Petrov, P. Ch. (1992) Study of synthesis and properties of tetrafluoroethylene–propylene copolymers. *J. Polym. Sci., Part A: Polym. Chem.*, **30**, 1083–1088.

401. Inoue, H., Matsumoto, A., Matsukawa, K. and Ueda, A. (1990) Surface characteristics of fluoroalkylsilane–poly(methylmethacrylate) block copolymers and their PMMA blends. *J. Appl. Polym. Sci.*, **40**, 1917–1938.

402. Perrier, S., Jackson, S. G., Haddleton, D. M., Ameduri, B. and Boutevin, B. (2002) Preparation of fluorinated methacrylic copolymers by copper mediated living radical polymerization. *Tetrahedron*, **58**, 4053–4059.

403. Perrier, S., Jackson, S. G., Haddleton, D. M., Ameduri, B. and Boutevin, B. (2003) Preparation of fluorinated copolymers by copper mediated living radical polymerization. *Macromolecules*, **36**, 9042–9049.

404. Kotzev, A., Laschewsky, A. and Rakotoaly, R. H. (2001) Polymerizable surfactants and micellar polymers bearing fluorocarbon hydrophobic chains based on styrene. *Macromol. Chem. Phys.*, **202**, 3257–3267.

405. Kotzev, A., Laschewsky, A., Adriaensens, P. and Gelan, J. (2002) Micellar polymers with hydrocarbon and fluorocarbon hydrophobic chains. A strategy to multicompartment micelles. *Macromolecules*, **35**, 1091–1101.

406. Fromyr, T., Hansen, F. K., Kotzev, A. and Laschewsky, A. (2001) Adsorption and surface elastic properties of corresponding fluorinated and nonfluorinated cationic polymer films measured by drop shape analysis. *Langmuir*, **17**, 5256–5264.

407. Jehnichen, D., Pospiech, D., Janke, A., Friedel, P., Häußler, L., Gottwald, A., Kummer, S., Kollig, W. and Grundke, K. (2001). *Mater. Sci. Forum*, **378**, 378–381.

408. Pospiech, D. U., Jehnichen, D. E., Gottwald, A., Häußler, L., Sceler, U., Friedel, P., Kollig, W., Ober, C. K., Li, X., Hexemer, A., Kramer, E. J. and Fischer, D. A. (2001). *Polym. Mater. Sci. Eng.*, **84**, 314–315.

409. (a) Pospiech, D., Häußler, L., Eckstein, K., Komber, H., Voigt, D., Jehnichen, D., Friedel, P., Gottwald, A., Kollig, W. and Kricheldorf, H. R. (2001). *High Perform. Polym.*, **13**(2), 275–288; (b) Pospiech, D., Häußler, L., Eckstein, K., Voigt, D., Jehnichen, D. E., Gottwald, A., Kollig, W., Janke, A., Grundke, K., Werner, C. and Kricheldorf, H. R. (2001). *Macromol. Symp.*, **163**, 113–121.

410. Pospiech, D., Häußler, L., Jehnichen, D., Kollig, W., Eckstein, K. and Grundke, K. (2003) Bulk and surface properties of blends with semifluorinated polymers and block copolymers. *Macromol. Symp.*, **198**, 421–434.

411. Chen, Y., Ying, L., Yu, W., Kang, E. T. and Neoh, K. G. (2003) Poly(vinylidene fluoride) with grafted poly(ethylene glycol) side chains via the RAFT-mediated process and pore size control of the copolymer membranes. *Macromolecules*, **36**, 9451–9457.

412. Ameduri, B. and Boutevin, B. New fluorinated grafted copolymers synthesised by free radical controlled polymerisation. French Patent under deposit (assigned to Specific Polymers).

Index

W

X

Z